CLEAN ELECTRICITY FROM PHOTOVOLTAICS

2nd Edition

SERIES ON PHOTOCONVERSION OF SOLAR ENERGY

ISSN: 2044-7701

Series Editor: Mary D. Archer *(Cambridge, UK)*

Vol. 1: Clean Electricity from Photovoltaics
eds. Mary D. Archer & Robert Hill

Vol. 2: Molecular to Global Photosynthesis
eds. Mary D. Archer & Jim Barber

Vol. 3: Nanostructured and Photoelectrochemical Systems for Solar Photon Conversion
eds. Mary D. Archer & Arthur J. Nozik

Vol. 4: Clean Electricity from Photovoltaics, 2nd Edition
eds. Mary D. Archer & Martin A. Green

Series on Photoconversion of Solar Energy — Vol. 4

CLEAN ELECTRICITY FROM PHOTOVOLTAICS

2nd Edition

Editors

Mary D Archer

Imperial College, UK

Martin A Green

University of New South Wales, Australia

ICP Imperial College Press

Published by

Imperial College Press
57 Shelton Street
Covent Garden
London WC2H 9HE

Distributed by

World Scientific Publishing Co. Pte. Ltd.
5 Toh Tuck Link, Singapore 596224
USA office: 27 Warren Street, Suite 401-402, Hackensack, NJ 07601
UK office: 57 Shelton Street, Covent Garden, London WC2H 9HE

Library of Congress Control Number: 2014950883

British Library Cataloguing-in-Publication Data
A catalogue record for this book is available from the British Library.

Series on Photoconversion of Solar Energy — Vol. 4
CLEAN ELECTRICITY FROM PHOTOVOLTAICS
Second Edition

ISBN 978-1-84816-767-4

Typeset by Stallion Press
Email: enquiries@stallionpress.com

Printed in Singapore

This volume is dedicated

to

Peter Theodore Landsberg

8 August 1922–14 February 2010

who saw so far beyond the detailed-balance limit

CONTENTS

ABOUT THE AUTHORS

Mary Archer read chemistry at Oxford University and took her PhD from Imperial College London, in 1968. From 1968 to 1972, she did post-doctoral work in electrochemistry with Dr John Albery at Oxford, and she then spent four years at The Royal Institution in London, working with Lord Porter (then Sir George Porter) on photoelectrochemical methods of solar energy conversion. She taught chemistry at Cambridge University from 1976 to 1986. From 1991 to 1999, she was a Visiting Professor in the Department of Biochemistry at Imperial College London, and from 1999 to 2002, she held a Visiting Professorship at ICCEPT (Imperial College Centre for Energy Policy and Technology). She is President of the UK Solar Energy Society and the National Energy Foundation and a Companion of the Energy Institute. She was awarded the Melchett Medal of the Energy Institute in 2002 and the Eva Philbin Award of the Institute of Chemistry of Ireland in 2007. In 2012, she was appointed a Dame Commander of the British Empire for services to the UK National Health Service.

Christophe Ballif received his MSc and PhD degrees in physics from the Federal Polytechnic School of Lusanne, Switzerland, in 1994 and 1998, respectively, focusing on novel photovoltaic materials. Following post-doctoral research at the National Renewable Energy Laboratory, Golden, USA, on CIGS and CdTe solar cells, he moved to the Fraunhofer Institute for Solar Energy Systems, Freiburg, Germany, where he focused on crystalline silicon photovoltaics until 2003. He then joined the Swiss Federal Laboratories for Materials Testing and Research, Thun, Switzerland, before becoming a full professor at the Institute of Microengineering, University of Neuchâtel, Switzerland, in 2004. In 2009, the Institute of Microengineering was transferred to EPFL. He is the Director of the Photovoltaics and Thin-Film Electronics Laboratory in the Institute, and since 2013 he has also been the Director of the PV-Centre of the Swiss Centre for Electronics and Microtechnology (CSEM), Neuchâtel. His research interests include thin-film silicon, high-efficiency heterojunction crystalline cells, module technology, contributing to technology transfer, and industrialisation of novel devices.

Dieter Bonnet was born in Stuttgart, Germany, in 1937 and obtained his PhD on the photoelectric properties of organic materials at Frankfurt University in 1963. In 1965, he joined Battelle Institute in Frankfurt, where in 1968 he started work on thin-film solar cells based on II–VI compounds, including CdTe. In 1970, he made the world's first CdTe/CdS thin-film solar cell in the presently known configuration. After a period working on other solar cell materials, he resumed work on CdTe technology in 1990, and initiated the successful EU (European Union) projects EUROCAD in 1992 and the CdTe interest group SOLARPACT in 2005. In 1993, he co-founded the pioneer company ANTEC GmbH to manufacture CdTe modules. In May 2001, the road in which ANTEC is located in Arnstadt, Germany, was named Dr. Bonnet Weg in honour of his accomplishments, and he was awarded the Becquerel Prize of the European Commission for outstanding achievements in photovoltaics in 2006. He retired in 2001 and is now an independent consultant.

Dan Credgington received his MSci in Natural Sciences from the University of Cambridge in 2004, and was awarded the Herchel Smith scholarship to study at Harvard University. In 2010, he obtained his PhD for work on the nanoscale microscopy and lithography of conjugated molecules from University College, London, under the supervision of Professor Franco Cacialli, and went on to conduct post-doctoral research on organic solar cells with Professor James Durrant at Imperial College London. He is currently a post-doctoral researcher in the group of Professor Sir Richard Friend at the Cavendish Laboratory of the University of Cambridge, where his interests lie in the study of recombination processes in organic light emitting diodes and solar cells, and hybrid organic/inorganic technologies.

Matthieu Despeisse received his degree in electrical engineering from INSA, Lyon, France, in 2002. He then joined CERN in Geneva, Switzerland, to work on low-noise, low-power and radiation-hard microelectronic circuits and silicon detectors. He obtained his PhD in 2006 for his work on amorphous silicon radiation sensors vertically integrated onto integrated circuits. He then pursued a fellowship from 2006 to 2008 at CERN, working on the design of a beam tracker and the transfer of CERN technologies to medical imaging and biotechnologies. He worked on fast Si 3D sensors, SiPM for PET-scan imaging, MCP-based systems for time-resolved spectroscopy, and high-speed readout electronics. He then joined the PVLAB group of Professor Christophe Ballif, which was part of the University of Neuchâtel, Switzerland, until 2009 and of the Federal Polytechnic School of Lusanne since then. He worked first on the development of amorphous and microcrystalline silicon and silicon alloys, thin-film silicon (TF-Si) single, tandem and triple junction solar cells and metrology equipment. He then became R&D manager of the PVLAB TF-Si activities, leading a research team of 15 people providing innovation to enhance performance and lower the costs of TF-Si PV. Since January 2013, he has headed research activities on crystalline silicon solar cells developments at the CSEM PV-Centre, in Neuchâtel, with special focuses on silicon heterojunction technology, passivating contacts and metallisation. He has authored or co-authored more than 70 papers in peer-reviewed journals, and he is inventor/co-inventor on 5 patents/patent applications.

James Durrant is Professor of Photochemistry in the Department of Chemistry at Imperial College London. After completing his undergraduate studies in physics at the University of Cambridge, he obtained a PhD in biochemistry at Imperial College London, in 1991, studying the primary reactions of plant photosynthesis. After post-doctoral positions and a BBSRC Advanced Fellowship, he joined the Chemistry Department at Imperial College in 1999, taking up his current post in 2005. His additional responsibilities include being Deputy Director of Imperial's Energy Futures Lab 2009–2014 and, from 2013, Sêr Cymru Solar Professor, University of Swansea. His interests are in photochemical approaches to solar energy conversion, including both excitonic solar cells and solar-driven fuel synthesis (artificial photosynthesis). He has published over 280 research papers and was awarded the 2012 Tilden Prize by the Royal Society of Chemistry.

Ned Ekins-Daukes is a Senior Lecturer in the Department of Physics of Imperial College London. He is also director of the IC Energy Futures Centre for Doctoral Training. He researches high efficiency, 'third generation' approaches to photovoltaic energy conversion, in particular III–V multijunction solar cells, intermediate-band and hot-carrier solar cells. He also works on modelling the energy yield from III–V solar concentrator systems. He previously worked in Australia as a lecturer at the School of Physics at the University of Sydney, and later as a visiting research fellow at the ARC Photovoltaics Centre of Excellence, UNSW. Prior to that, he was a JSPS research fellow at the Toyota Technological Institute, Japan.

Timothy Gessert is a Principal Scientist in the National Center for Photovoltaics at the National Renewable Energy Laboratory (NREL) in Golden, Colorado. He holds degrees in physics from the University of Wisconsin-River Falls (BSc), Colorado School of Mines (MSc), and University of Wales — College of Cardiff (PhD). He joined NREL (then known as the Solar Energy Research Institute — SERI) in 1983 where he has worked on various aspects of photovoltaic and thermophotovoltaic devices related to GaAs, InP and CdTe material systems. Activities have also included the development of vacuum and photolithographic processes, transparent-conducting oxides and related electrical contacts. His current research is directed at understanding how choices in processes and device design affect the dominant defects and ultimate performance/stability of thin-film photovoltaic devices. Apart from these research activities at NREL, he is active in the American Vacuum Society, the Materials Research Society and the IEEE. He also teaches specialised courses on photovoltaics and vacuum technology for professional societies and academic institutions.

Michael Grätzel is a Professor at the Federal Polytechnic School of Lusanne, where he directs the Laboratory of Photonics and Interfaces. He pioneered studies of mesoscopic materials and their use in energy conversion systems, in particular photovoltaic cells and photoelectrochemical devices for the solar generation of chemical fuels, as well as lithium ion batteries. He discovered a new type of solar cell based on sensitised nanocrystalline oxide films. His most recent awards include the Leonardo Da Vinci Medal of the European Academy of Science, the Marcel Benoist Prize, the Albert Einstein World Award of Science, the Paul Karrer Gold Medal, the Balzan Prize and the 2010 Millennium Technology Grand Prize. He received a doctoral degree in Natural Science from the Technical University Berlin and holds honorary doctorates from ten European and Asian Universities. He is a member of the Swiss Chemical Society and a Fellow of the European Academy of Science as well as the Royal Society of Chemistry (UK) and the Max Planck Society. He is also an elected honorary member of the Société Vaudoise de Sciences Naturelles and the Bulgarian Academy of Science. As the author of over 1000 publications and inventor of 50 patents, he is with some 120,000 citations and an h–index of 162 one of the three most highly cited chemists in the world.

Martin Green is a Scientia Professor at the University of New South Wales, Sydney, and the Director of the Australian Centre for Advanced Photovoltaics. He was born in Brisbane and educated at the University of Queensland and then McMaster University, Canada. His contributions to photovoltaics include heading the team that has improved silicon solar cell performance by over 50% since the 1980s, and the commercialisation of several different solar cell technologies. Major international awards include the IEEE William R. Cherry Award in 1990, the 1999 Australia Prize, the 2002 Right Livelihood Award, also known as the Alternative Nobel Prize, the 2007 SolarWorld Einstein Award and the 2010 Eureka Prize for Leadership. In 2012, he was appointed as a Member of the Order of Australia in recognition of his contributions to photovoltaics and photovoltaics education. He has been elected to Fellowship of the Australian Academy of Science, the Australian Academy of Technological Sciences and Engineering, the Institute of Electrical and Electronic Engineers and the Royal Society of London. He is the author of four books on solar cells, several book chapters and numerous reports and papers in the area of semiconductor properties, microelectronics and solar cells.

Franz-Josef Haug studied physics at the Universities of Ulm, Germany, and Waikato, New Zealand. For his diploma he investigated nucleation and coalescence phenomena during the initial stages of chemical vapour epitaxy of silicon. He undertook his PhD program at ETH Zürich, Switzerland, working on $Cu(In,Ga)Se_2$ solar cells in the superstrate configuration. After graduation, he worked on plasma processing of super-hard nano-composite coatings of TiN/SiN_x before joining the Jülich Research Centre in Germany, where he worked on surface modifications of the transparent front electrode of thin-film silicon solar cells. In 2005, he joined the Institute of Microengineering, initially affiliated to the University of Neuchâtel, Switzerland, and since 2009 part of the EPFL. There he leads a research group devoted to thin-film silicon solar cells on glass and flexible plastic substrates, investigating novel texturing techniques for light scattering and absorption enhancement. He is author or co-author of some 70 peer-reviewed papers and 4 patents. His research interests include solar cells, semiconductor physics, and optoelectronics.

Arnulf Jäger-Waldau received his Dr. rer. nat. from the Physics Department of the University of Konstanz, Germany, in 1993. He has worked in the field of material research for solar cells since 1987 and holds patents on semiconductor material deposition for thin-film solar cells and solar module design. In 1994 and 1995 he worked as a post-doctoral JSPS fellow at Shinshu University, Nagano, Japan, before joining the Hahn–Meitner Institute Berlin in 1996. Since 2001 he has been a Scientific Officer and Senior Scientist at the Renewable Energy Unit, Institute for Energy and Transport of the European Commission's Joint Research Centre, where he works on the assessment of renewable energy technologies, the effectiveness of their implementation, their integration into energy infrastructures and the role of renewable energy for climate change mitigation. Among other roles, he has been the Technical Chairman of the European Photovoltaic Solar Energy Conference (EUPVSEC) since 2011, and he was a Lead Author for Solar Energy of the Special Report of the IPCC on Renewable Energy and Climate Change Mitigation. He also serves as a member of the Executive Committee of the European Materials Research Society, the Academic Advisory Board of the Chinese Trina State Key Laboratory for Photovoltaics, the International Advisory Board of the Warsaw University Photovoltaic Centre and the Scientific Advisory Board of the Solar Research Centre of the Bulgarian Academy of Science, and he is Vice-Chairman of the Academic Committee of the Asian Photovoltaic Industry Association (APVIA).

Antonio Luque obtained his Doctor of Engineering degree from the Polytechnic University of Madrid in 1967. Today he is also Doctor Honoris Causa of three other Spanish universities. In 1969, he joined the university staff and founded its Semiconductor Laboratory. In 1979, this centre became the Institute of Solar Energy that he leads at present. In 1981, he founded the company Isofotón to manufacture the bifacial cells he had invented, and he chaired its board until 1990. He has written some 300 papers and registered around 20 patents, of which some are in exploitation. He has won several scientific awards, among which are the Spanish National Prize for Technology in 1987 and 2003, the King Jaime I Prize for environmental protection, the Becquerel Prize awarded by the European Commission for PV in 1992, the IEEE William Cherry Award in 1996 and the SolarWorld Einstein Award in 2008. Among other distinctions, he is a member of the Royal Academy of Engineering, and of the Russian Academy of Sciences. He has also been a member of the Advisory Council for Science and Technology, which advises the Spanish Prime Minister.

Ignacio Luque-Heredia is CEO of BSQ Solar, a high concentration photovoltaics (HCPV) manufacturing company that he co-founded in 2009. He received his MSc and PhD degrees in electrical engineering from Polytechnic University of Madrid in 1995 and 2010 respectively. He was co-founder in 1995 of the company Inspira, also operating in the CPV field, which was acquired by the Silicon Valley leading CPV manufacturer Solfocus in 2007. From 2007 to 2009 he was CTO for Solfocus Europe. Leading Inspira's and BSQ Solar's engineering, he has participated in 23 collaborative projects in the field of CPV, 11 of them funded by the European Commission. These range from the EUCLIDES project in 1995, which resulted in the biggest CPV plant of its time, to the most recent NGCPV project, on 3rd generation photovoltaic devices and their integration in CPV technologies. He holds four patents and has led the deployment of CPV pilot systems and large-scale plants, as well as several technology transfer programs in the USA, China, India, Japan, Australia, MENA, Brazil, Mexico and Europe. He is a member of the Scientific Committee of the International Conference on CPV Systems, and of Work Group 7 for the development of CPV standards in Technical Committee 82 for Solar Photovoltaic Energy Systems of the International Electrotechnical Commission. He is also a Senior Member of the IEEE.

Jenny Nelson is a Professor of Physics at Imperial College London, where she has researched novel types of solar cell since 1989. Her current research focuses on photovoltaic energy conversion using molecular materials, characterisation of the charge transport, charge separation and morphological properties of molecular semiconductors, the theory of charge transport in organic semiconductors and modelling of photovoltaic device behaviour. She has published over 200 papers on photovoltaic materials and devices, and a book on the physics of solar cells.

Nicola Pearsall is a Professor at Northumbria University, where she leads their photovoltaic research activities. She holds a degree in physics from the University of Manchester Institute of Science and Technology and obtained her PhD from Cranfield Institute of Technology for her research on indium phosphide solar cells for satellite applications. She has been involved in research in photovoltaics for 30 years, and her current interests relate to the performance and implementation of photovoltaic systems. She is a member of the Steering Committee of the European Photovoltaic Technology Platform and has contributed to the development of their Strategic Research Agenda, as well as working with the Solar Europe Industry Initiative.

Uwe Rau received his PhD in physics in 1991 from the University of Tübingen, Germany, for his work on temporal and spatial structure formation in the low-temperature electronic transport of bulk semiconductors. From 1991 to 1994, he worked at the Max Planck Institute for Solid State Research, Stuttgart, on Schottky contacts, semiconductor heterojunctions and silicon solar cells. From 1994 to 1997, he worked at the University of Bayreuth, Germany, on electrical characterisation and simulation of Si and $CuInSe_2$ solar cells. In 1997, he joined the Institute for Physical Electronics at the University of Stuttgart, where he became leader of the Device Analysis Group. His research interests centre on transport phenomena, especially electrical transport in solar cell heterojunction devices and interface and bulk defects in semiconductors. He has authored or co-authored more than 100 scientific publications.

Hans-Werner Schock received his diploma in electrical engineering in 1974, and doctoral degree in electrical engineering in 1986, from the University of Stuttgart's Faculty of Electrical Engineering. Since the early 1970s, he has worked on the development of polycrystalline II–VI and I–III–VI_2 compound semiconductor thin-film solar cells, taking the development of chalcogenide solar cells from research to pilot fabrication. A series of successful research projects on thin-film solar cells under his guidance resulted in several production lines for thin film solar cells in Europe. From 2004 to 2012, he was director of the Institute of Technology at the Helmholtz Zentrum Berlin for Materials and Energy and he is an Honorary Professor at the Technical University Berlin. He received the Becquerel Prize of the European Commission in 2010 for his achievements in the development of thin-film solar cells. At present he is a consultant in the field of photovoltaics. He is the author or co-author of more than 300 contributions in books, scientific journals and conference proceedings.

Masafumi Yamaguchi is a Professor of the Toyota Technological Institute. He received his BS and PhD degrees from Hokkaido University, Japan, in 1968 and 1978, respectively. In 1968, he joined the NTT Electrical Communications Laboratories, working on radiation damage to Si and III–V compounds, ZnSe blue-light-emitting diodes and III–V compound solar cells. In 1983, he discovered the superior radiation-resistance of InP materials and solar cells, thereby showing the great potential of InP cells for space applications. His group also developed high-efficiency InP, GaAs-on-Si, and AlGaAs/GaAs tandem cells by proposing a double-hetero structure tunnel junction for realising a high performance and stable multijunction cell interconnection in 1987. As Japanese team leader of the EU–Japan Collaborative Research on Concentrator Photovoltaics, he contributed to the attainment of InGaP/GaAs/InGaAs 3-junction cells with efficiencies of 44.4% at 302 Suns of AM1.5D and 37.9% at AM1.5G. He is also the project leader of the Next Generation High Performance Photovoltaics Research and Development Project of the Japanese New Energy Development and Industrial Technology Development Organisation, and the Research Supervisor in the Research Area of the Creative Clean Energy Generation using Solar Energy programme of the Japan Science and Technology Agency. He has published more than 300 original papers and received numerous awards, such as the Becquerel Prize from the European Commission in 2004 and the William Cherry Award from the IEEE in 2008 for outstanding contributions to science and technology development of high-efficiency photovoltaics.

PREFACE TO THE FIRST EDITION

And there the unregulated sun
Slopes down to rest when day is done
And wakes a vague, unpunctual star...

Rupert Brooke, *The Old Vicarage, Grantchester*, May 1912

Since the dawn of history, man has been fascinated by the Sun, the provider of the light and warmth that sustains life on Earth. In pre-industrial times, our major sources of energy — wood, wind and water power — derived from solar energy. The subsequent discovery and massive exploitation of fossil fuels laid down in the Earth's crust by early aeons of photosynthetic activity have conditioned the developed world to be dependent on convenient, readily available energy. But we are living on our energy capital. The Earth's reserves of coal, oil and gas are finite and likely to become resource-depleted in the course of this century. A sense of living on borrowed time was therefore appropriate even before concerns about global climate change, sustainability and energy security combined to raise interest in renewable energy to its current encouraging level.

This book is the first in a series of four multi-authorial works on the photoconversion of solar energy. It was created from my long-held conviction that, despite slow starts and setbacks, solar energy — broadly defined to encompass other renewable energy forms that derive from solar — will become the Earth's major energy source within this century. The Sun is a source of both radiant heat and light, and techniques for using solar energy correspondingly divide into thermal methods (solar power towers, water heaters and so on) and photoconversion (sometimes called direct) methods. Photoconversion is the subject of this book series. A photoconverter is a device that converts sunlight (or any other source of light) into a useful form of energy, usually electrical power or a chemical fuel, in a process that relies, not on a raised temperature, but on the selective excitation of molecules or electrons in a light-absorbing material and their subsequent de-excitation in a way that produces energy in a useful form. Volume I covers the most developed of the man photoconversion devices, photovoltaic (PV) cells, which are solid-state semiconductor devices that produce electrical power on illumination. Volume II will cover the natural photoconversion system of photosynthesis, the potential of biomass as an energy source and the global carbon budget. Volume

III will explore the less developed but exciting possibilities of synthesising artificial 'molecule-based' photoelectrochemical or photochemical photoconverters. Finally, Volume IV will draw together the common themes of photoconversion and provide some background material.

The series is intended mainly for senior undergraduates, graduate students and scientists and technologists working on solar photoconversion. Chapters 1–12 of this book deal with PV cell design, device physics and the main cell types — crystalline and amorphous silicon, cadmium telluride and copper indium diselenide — as well as more advanced or less developed options such as quantum-well and thermophotovoltaic cells. These chapters are mainly technical, requiring sound knowledge of physics, chemistry or materials science for ready understanding. Chapters 13–18 deal with PV systems, manufacturers, markets and economics and are accessible without specialist knowledge.

A multi-authorial work owes its very existence to its authors, and my wholehearted thanks must go to the twenty-five distinguished individuals, all recognised authorities in their own fields, who have contributed to this book and patiently answered my queries during the editing stage. I have also been helped by discussions about PV with many friends and colleagues, and visits to installations throughout the world: I have been up Swiss mountains, onto Japanese rooftops and into the Arizona desert, and thoroughly enjoyed every minute. I am most grateful to those who have read and commented on various parts of this book or provided specialist information in advance of publication: Dennis Anderson, Jeffrey and William Archer, Stephen Feldberg, Martin Green, Eric Lysen, Larry Kazmerski, Bernard McNelis and Nicola Pearsall. I also warmly thank Alexandra Anghel, Barrie Clark, Stuart Honan and my PA Jane Williams for editorial assistance, and Ellen Haigh and John Navas of IC Press and Alan Pui of World Scientific Press for guiding the book to publication.

For me the sad part of writing this preface is that I must do so in the first person, for my co-editor Professor Robert Hill died suddenly on 26 November 1999. Bob was the most knowledgeable champion of photovoltaics in the UK, and his premature death has deprived the British PV community of its cornerstone. He had drafted his chapter with Nicky Pearsall some months before he died, and the flow of emails delivering his astute editorial comments on other chapters continued until the day before his death.

Bob believed unshakeably in the future of PV. Although he knew that system costs will have to fall by another factor of 2–3 if PV is to become cost-competitive in major new grid-accessible markets, there are good grounds for believing this is possible. PV technology is still young, and significant further economies of scale from larger manufacturing facilities, as well as further advances in the fundamental science, can confidently be expected. The world's first-generation televisions and mobile telephones were at least as uncommon and expensive as PV is now.

The Old Vicarage, Grantchester Mary Archer
December 2000

PREFACE TO THE SECOND EDITION

O Sonne, Königin der Welt,
Die unser dunkles Rund erhellt
In lichter Majestät;
Erhab'nes Wunder einer Hand,
Die jene Himmel ausgespannt
Und Sterne hingesät!

Johann Peter Uz (1720–1796), *An die Sonne*

The world of photovoltaics has advanced at a phenomenal pace since the first edition of this book was published in 2001. Then we were able to report with modest pride that 200 MWp of PV had been installed worldwide during 1999, taking global cumulative installed capacity to just over 1 GWp. By the end of 2012, global installed capacity had edged past the iconic figure of 100 GWp. The final figure for global installations in 2013 is yet to be determined, but it will add around another 35 GWp, with plausible estimates for cumulative installed capacity by 2020 topping 300 GWp. PV is now, after hydro and wind, the third most important renewable energy source in terms of installed global capacity.

This prodigious growth has impacted on all sectors of the PV industry. In 1999, much of the small quantity of pure silicon needed by the cell manufacturing industry came from the moderate resistivity *p*-type waste material discarded by the electronics industry. As the PV market grew, a tipping point was reached in 2006 when, for the first time, over half of the world's supply of polysilicon was used for PV production. Silicon feedstock for the PV industry is now made by dedicated plant, and manufacturing capacity has swung sharply away from the USA and Europe to the lower-cost economies of the Far East.

As the PV market has grown, so the price of cells and modules has dropped. In their early years, photovoltaic cells were an expensive and exotic novelty for use in space. In 1955, market leader Hoffman Electronics were offering 2% efficient silicon cells at a price of $1500/Wp (about €9520/Wp in today's money). At the end of 2013, spot prices on the European market were nearly 20,000 times lower at €0.5–0.7/Wp for silicon modules of efficiencies in the range 15–18%.

Between the two editions of this book lie nearly 15 years of scientific discovery, technological advance and market maturation, punctuated by global economic and financial turmoil, international bickering about the need to mitigate climate change, national anxieties about energy security, stability and independence, and political and regulatory instability as governments reconsider their commitment to renewable energy sources and carbon offset.

The architecture of this edition follows the lines of the first edition, with a set of chapters about the main PV technologies following an introductory chapter. Omitted from this edition are thermophotovoltaics, PV for space applications and electricity storage, while 3-junction III–V concentrator cells of nearly 45% efficiency, CZTS (copper–zinc–tin–sulphide/selenide), dye-sensitised and perovskite solar cells are notable newcomers to the stage. A further addition is an overview of the limits to photovoltaic conversion efficiency and how far they can be pushed beyond single-junction detailed-balance constraints. The book concludes with a look at PV systems, applications and markets.

We are deeply indebted to all our authors, without whose generosity with their knowledge and patience with the editorial process this book could not have been produced. We are particularly grateful to those who have stayed with us from the first edition and undertaken the painstaking task of radically updating their chapters: Dieter Bonnet, Antonio Luque, Jenny Nelson (writing about QW solar cells in the first edition and OPV in the second), Nicola Pearsall, Hans-Werner Schock and Masafumi Yamaguchi. We warmly welcome newcomer authors Christophe Ballif, Dan Credgington, Matthieu Despeisse, James Durrant and Michael Grätzel (who jointly contributed a chapter to Volume III of this series), Ned Ekins-Daukes, Timothy Gessert, Franz-Josef Haug, Arnulf Jäger–Waldau, Ignacio Luque-Heredia and Uwe Rau.

The structure of this book series has altered since the preface to the first edition written: Volume III did not deal with solar fuels as originally planned, but with photoelectrochemical and nanostructured devices for solar photon conversion; solar fuels are now planned for the fourth and final volume.

Editing this book series has become a way of life for one of us (MA). She thanks her family for their tolerance of this enduring preoccupation, her PA Carol Burling for her skill at the electronic keyboard, James Archer, Richard Friend, Larry Kazmerski, Bernard McNelis and colleagues at the UK's annual PVSAT conferences for much helpful advice and input. MG wishes to thank his partner, Judy Green, for her ongoing support and Joyce Ho for her assistance on several fronts during book preparation, as well as the many colleagues and students who have helped make photovoltaics such a stimulating field of interest.

We jointly acknowledge with gratitude the editorial assistance of Jacqueline Downs, Leah James and Tom Stottor of Imperial College Press and Catherine Yeo of World Scientific Publishing, who have seen the book through to publication.

January 2014

Mary D. Archer
Martin A. Green

CHAPTER 1

THE PAST AND PRESENT

MARY D. ARCHER

The Old Vicarage, Grantchester, Cambridge CB3 9ND, UK

mda12@cam.ac.uk

Time present and time past
Are both perhaps present in time future.

T. S. Eliot, *Burnt Norton*, Four Quartets, 1935–1942.

1.1 Introduction

Photovoltaic (PV) cells generate electric power when illuminated by sunlight or artificial light. They are by far the most highly developed of the man-made photoconversion devices. Born of the space age in the 1950s, their earliest terrestrial applications emerged in the 1970s and they have enjoyed rapid market expansion since the turn of the millennium. PV technology has a number of advantages over conventional methods of electricity generation. First and foremost, solar energy is the world's major renewable energy resource. PV power can be generated from the Sun anywhere — in temperate or tropical locations, in urban or rural environments, in distributed or grid-feeding mode — where there is adequate light. As a fuel-free distributed resource, PV could in the long run make a major contribution to national energy security and carbon dioxide abatement. PV is uniquely scalable, the only energy source that can supply power on a scale of milliwatts to megawatts from an easily replicated modular technology with excellent economies of scale in manufacture. A standard 156×156 mm crystalline silicon PV cell generates about 4 peak watts[1] (Wp) of DC power and typical PV 60-cell and 72-cell modules about 240 Wp and 300 Wp respectively. The world's largest PV generating facility is currently the Agua Caliente Solar Project in Arizona, which had 251.3 MWp installed as at November 2013.

[1]The power output of a PV cell or module is rated in peak watts (Wp), meaning the power output at 25°C under standard AM1.5 solar radiation of global irradiance $1\,\mathrm{kW\,m^{-2}}$. PV ratings in 'watts' invariably mean peak watts. To convert from peak watt rating to 24-hour average power output in a sunny location, divide by a factor of ~5.

PV cells are made of thin semiconductor wafers or films. They contain small amounts only of (usually non-toxic) materials and, when manufactured in volume, have modest embedded energy, with energy payback times for silicon PV modules in Southern Europe of about one year (Fraunhofer, 2012). They possess no moving parts, generate no emissions, generally require no cooling except when used in concentrator systems and are silent in operation. PV systems are reliable, easy to use and long-lived if properly maintained (most commercial modules have lifetime warranties of 25–30 years, though some balance-of-system components are less reliable and long-lived than this). Carefully designed, PV arrays are not visually intrusive, and can indeed add architectural merit to the aesthetic of a built structure.

The PV industry has advanced dramatically in the decade since the first edition of this book was published. Over the period 2000–2012, solar PV was the fastest-growing renewable power technology worldwide, and its market growth has outstripped even the most bullish of forecasts: cumulative global installed PV capacity in 2000 was only 1.4 GWp, while by the end of 2012 it had just edged past the milestone of 100 GWp installed (EPIA, 2013). Despite the difficult economic situation, global PV module production in 2012 was 35.9 GWp (GTM Research, 2013). Future growth scenarios differ widely, but the average forecast is for some 260 GWp of cumulative installed global PV capacity by 2020, with demand shifting from Europe to countries such as China, India and the USA. These projections show there are huge opportunities for PV in the future.

A number of factors have driven this growth. Many countries, including most in the OECD and several developing countries, have introduced tax or regulatory policies that favour the renewables. Modest renewable set-asides (requirements on major utilities to source some power from renewables) are in place in many countries with liberalised electricity supply industries, and the right to supply power to the grid has been extended to independent power producers (IPPs), sometimes with incentives to source electricity from renewable sources. IPPs can site their plant close to the consumer and avoid the costs of distribution, often as significant as the costs of generation.

The 100,000 roofs programme in Germany, which ran from 1999 to 2004, and other national roofs programmes stimulated demand for distributed PV systems. Feed-in tariffs (FiTs), also pioneered by the German government, have stimulated PV installations by guaranteeing an above-market price for a fixed term for PV-generated power to be fed into the grid. These tariffs, which are now offered in many countries, give assurance of return to investors. Later FiT rounds generally provide less generous rates of return, which has helped to drive system prices down, although abrupt tariff changes can produce boom/bust cycles and create investor uncertainty.

In the USA, residential and commercial PV developments were stimulated by a Treasury Program that ran from 2009 to the end of 2011, which provided an upfront federal grant in lieu of recompense through the less certain mechanism of an Investment Tax Credit. US utilities seeking to fulfil their RPS (Renewable Portfolio Standard) obligations have also developed a number of grid-connected power arrays, and the domestic market for PV in the USA doubled in 2011, and again in 2012. The US Department of Energy's SunShot initiative currently supports Research & Development in most PV technologies with the aim of making solar electricity cost-competitive with other sources of energy by 2020. SUNPATH, which stands for Scaling Up Nascent PV AT Home, is one of the initiatives under SunShot, aimed at supporting innovative, high-tech companies to accelerate cost reductions and commercialisation of their PV technologies.

The PV manufacturing industry has also undergone profound changes over the last decade. Since 2000, global PV cell/module production has increased by almost two orders of magnitude, with annual growth rates in the range 40–90, stalling only in 2012 as a result of massive inventory overhang. Until 1998, USA PV manufacturers accounted for the greatest share of global shipments, but the manufacturing base has switched to Asia, particularly to China, which accounted for 64% of worldwide production in 2012, while Europe's share dropped to 11% and Japan's and the USA's to 5% and 3% respectively (GTM Research, 2013). Several years of explosive growth in China's production capacity led to significant overcapacity (Yang, 2012), causing industry profits to slump, with major manufacturer Suntech finding itself in severe difficulties (PV Magazine, 2013). According to PV Tech (2012), 90% of China's polysilicon raw material producers have stopped production. The Chinese government has called for industry consolidation and sought to stimulate the domestic market for PV in its latest 5-year solar energy development plan by setting a target of 20 GWp installed by 2015 (China Briefing, 2012).

The oil-rich nations of the MENA (Middle East/North Africa) region have enormous solar resources, and are beginning to prepare themselves for the post-hydrocarbon era and increase their oil export revenues by developing their solar resources for domestic use. In May 2012, Qatar Solar Technologies secured an unprecedented investment of US$1 bn from a Qatari bank to build a polysilicon production facility in Ras Laffan Industrial City. Saudi Arabia aims to meet a third of its electricity with solar power by 2032, while Abu Dhabi has targeted 7% of its power from renewable sources by 2020 and Dubai 5% by 2030 (Mahdi, 2012).

Wafer-based crystalline Si remains the dominant commercial technology, with about 89% market share of module production in 2012, although its hegemony was challenged in the period 2005–2009 by the temporary global shortage of Si feedstock. This led to massive expansion of investment in non-silicon thin-film

PV capacity, particularly in CdTe. Thin-film's market share of production grew rapidly to 21% in 2009, but fell back to 11% in 2012 as new production facilities for high-purity polysilicon were brought on line.

Supply of polysilicon outstripped demand in 2011 and 2012, and as a consequence Si module prices dropped rapidly throughout 2011 from an average of about \$1.7 Wp^{-1} to a level approaching \$1 Wp^{-1} later in the year. 2012 saw further price reductions, to an average of \$0.87 Wp^{-1} for Tier 1 (major manufacturers) of crystalline silicon (c-Si) modules (GTM Research, 2012). The c-Si production industry is currently having to consolidate, with idled plant, sector exits and bankruptcies producing a painful shake-out of unprofitable business. Even the venerable BP Solar shut down in 2011 after 40 years in the business, saying it 'simply can't make any money from solar' (MIT, 2011).

The dominant thin-film materials are cadmium telluride (46% of total thin-film production), amorphous Si or amorphous/microcrystalline Si (35%), and CIGS (copper indium gallium selenide, 19%). Thin-film modules save on PV materials, are highly suited for low-cost mass production and open up new applications such as the provision of power for consumer electronics. The active layers of thin-film cells are typically only a few microns thick, about 50 times thinner than standard c-Si cells. Despite this, thin-film modules are not lighter per Wp than Si modules due to their lower efficiency and the need for double glass encapsulation (because of their greater moisture sensitivity). Nevertheless, the generally lower cost per Wp of thin-film modules and their insensitivity to partial shading make them attractive where area is not limited, though their lower efficiency puts them at a disadvantage where it is. Another advantage over c-Si technology, in which wafers, cells and modules must be produced in different steps, is that thin-film cells can be fabricated as series-connected modules in an integrated manufacturing process. Continued efficiency improvements remain important because higher efficiency modules incur lower balance-of-system costs per unit output.

Figure 1.1 shows the full range of PV cell types with the world record efficiencies achieved by each over the past twenty years, as catalogued by Green and his colleagues. These are research cell, not production cell, efficiencies, but they set a benchmark for their commercial cousins, the efficiency of which also improves year on year.

Grid parity — where the cost of PV power has dropped sufficiently that the installed cost per watt is equal to the local cost of fossil fuel power — marks the point that PV should be able to expand without subsidy. This has already been reached or passed in sunny, remote regions and islands where the alternative is to generate electricity using imported oil. Grid parity has also been achieved in parts of Spain, and was expected to be reached by parts of Italy, Brazil, Chile and

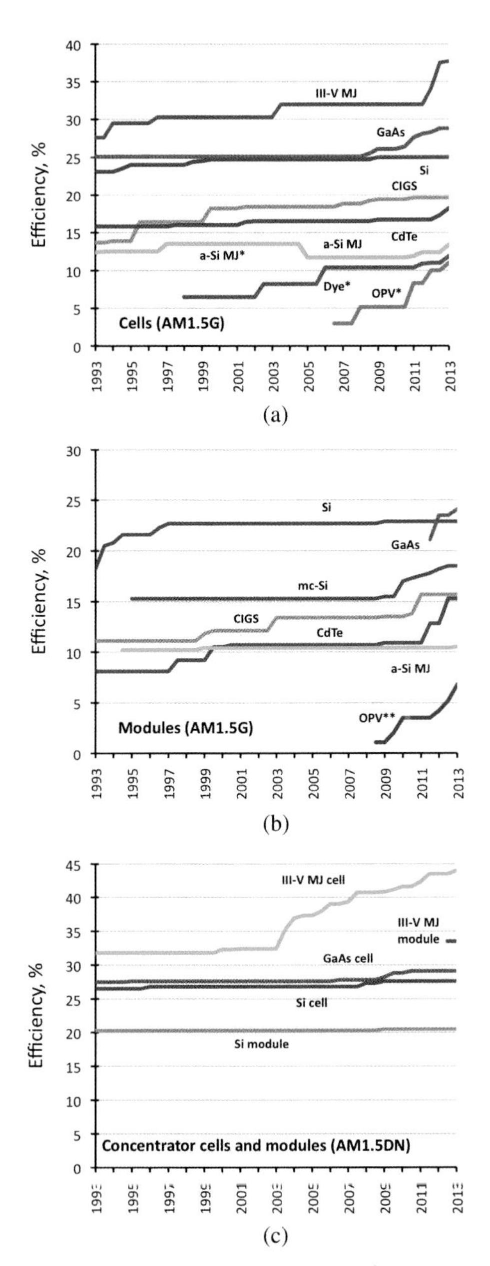

Figure 1.1 Highest confirmed efficiencies for (a) $\geq 1\,\text{cm}^2$ area cells (b) $\geq 800\,\text{cm}^2$ area modules (200–400 cm^2 for OPV); (c) concentrator cells and modules. Source: Green *et al.* (2013).

Australia during 2013 (Brittlebank, 2012). PV can also be cost-effective in peak shaving before it has reached full grid parity.

International policy to mitigate climate change has also provided some stimulus to PV. The Kyoto Protocol to the UN Framework Convention on Climate Change, which came into force in 2005, set binding obligations on industrialised countries to reduce greenhouse gas (GHG) emissions. This led to a number of regional bloc and national programmes to stimulate the use of renewable energy forms, PV among them, as well as a number of 'flexible mechanisms' by means of which signatories could meet their GHG reduction targets by investing in emissions reduction programmes overseas.

Enthusiasm for the Protocol has been at best patchy. The UK, law-abiding as ever, passed the world's first long-term legally binding framework to mitigate climate change in its 2008 Climate Change Act. However, little PV was installed until attractive FiTs were introduced in 2010. By the end of Q2 2013, cumulative installed capacity had risen to 2.4 GWp, and the government forecasts 10 GWp installed by 2020 (DECC, 2013).

The EU–15[2] embraced the Kyoto Protocol wholeheartedly and now has the greatest installed PV capacity of any regional bloc, with a cumulative installed capacity of over 66 GWp at the end of 2012. PV accounted for 56% (21.5 GWp) of new net generating capacity in the EU in 2012, beating wind with 25% (9.4 GW) of market share, into second place. PV now provides 2.6% of EU electricity demand on average, and 5.2% of peak demand (EPIA, 2013).

The US signed, but never ratified, the Kyoto Protocol, and Canada withdrew from it in 2011, arguing that a treaty that does not include the world's two largest GHG emitters — the US and China — would prove ineffectual at reducing emissions. At the 2012 Doha Climate Change Conference of Kyoto signatories, little progress was made on emissions targets: 37 countries agreed to a second round of reductions, but a number of others, notably Japan and Russia, did not take on new obligations. It is becoming clear that the Kyoto Protocol is not achieving its original aim, and it was agreed in principle in Doha that it should be replaced by a new treaty binding on all UN members whereby rich nations will compensate poor nations for the damage caused by climate change; the 2013 UN Climate Change Conference in Warsaw achieved little in this or other respects.

1.2 Milestones in the development of photovoltaic technology

The discovery of photovoltaism is commonly, if inaccurately, ascribed to Becquerel (1839), who observed that photocurrents were produced on illuminating platinum

[2] The 15 states that made up the European Community at the time of the Kyoto negotiations.

electrodes coated with silver chloride or silver bromide and immersed in aqueous solution.[3] The observation by Smith (1873) of photoconductivity in solid selenium led to the discovery of the photovoltaic effect in a purely solid-state device by Adams and Day (1877), who observed photovoltages in a selenium rod to which platinum contacts had been sealed, which they (incorrectly) ascribed to light-induced recrystallisation of the selenium. The first practical photovoltaic device — a light meter consisting of a thin layer of selenium sandwiched between an iron base plate and a semi-transparent gold top layer made by Fritts (1883) — was promoted by the German industrialist Werner von Siemens as demonstrating 'for the first time, the direct conversion of the energy of light into electrical energy' (Siemens, 1885). Photometers based on selenium photocells were commercialised in Germany in the 1930s, and amorphous selenium photodetectors remain of interest today for medical imaging applications.

The selenium photocell is an example of a *barrier layer cell*, so called because it contains an electrical barrier that is highly resistive to current flow in one direction — a rectifying junction, in modern parlance. Two further barrier layer cells, the thallous sulphide cell (Case, 1920) and the copper oxide cell (Kennard and Dieterich, 1917; Grondahl and Geiger, 1927), were developed during the 1920s. Garrison (1923) and later Grondahl (1933) reviewed work on Cu_2O, Garrison being the first to report the logarithmic dependence of open-circuit voltage on irradiance. Fink and Fogle (1934) discovered the efficacy of antireflective coatings accidentally when they found that beeswax, applied to protect cells during chemical etching, enhanced the photocurrent. Schottky (1930) noted and explained the difference in spectral response of 'Vorderwandzelle' (front-wall-illuminated cells) and 'Hinterwandzelle' (back-wall-illuminated cells) while experimenting with semitransparent metal contacts for the Cu_2O/Cu barrier cell.

Lange (1930), like Siemens, foresaw the possible application of the photovoltaic effect to energy conversion, pointing out that 'using a more appropriate semiconductor for the intermediate layer [of a barrier cell], it will be possible to select the cell effectiveness for a specific spectral region ... Use of a more appropriate unipolar [contact] layer is expected to give a further increase in the efficiency of the cell. It is then possible that efficiencies can be reached which allow direct conversion of light into electrical energy.' Sheldon and Geiger (1922), Geiger (1923) and Bergmann and Hänsler (1936) observed and recorded photovoltaic effects with a wide range of other minerals and compounds, Bergman reporting what may be the earliest observation of semiconducting effects in organic dyestuffs. Reviews

[3] Becquerel's observation was strictly speaking a photoelectrochemical effect, but its basis — the rectifying junction formed between two dissimilar electric conductors — is the same as that of the photovoltaic effect in purely solid-state devices.

by Winther (1928) and Crossley *et al.* (1967) and the book by Lange (1938) give an account of these early devices.

The electrical barrier of barrier layer cells was originally thought to lodge in an interfacial foreign layer of high resistivity such as an oxide, but Schottky (1938), and independently Davydov (1939) and Mott (1939), showed that a third phase was not necessarily involved. Rather, metal/semiconductor junctions could in themselves be rectifying by virtue of the space-charge layer created in the semiconductor by electronic equilibration/charge redistribution when contact was made with a metal of different work function. The first diffusion theory of *p-n* junction rectification, which became the basis for Shockley's theory, was published by Davydov (1938).

Metal/semiconductor devices make inefficient solar converters because their dark currents are relatively large and this diminishes the photovoltaic response. Semiconductor/ semiconductor junctions are better in this regard. The first *p-n* junctions to be reported were the germanium homojunctions of Lark-Horovitz's group at Purdue University (Benzer, 1946, 1947) and the lead sulphide quasi-homojunction formed by pressing together two PbS wafers, one enriched with lead and the other with sulphur (Sosnowski *et al.*, 1947).

1.3 The crystalline silicon solar cell

It was Russell Ohl, a metallurgist working at Bell Telephone Laboratories in New Jersey before World War II, who ushered in the modern era of photovoltaics by his observation that crystallisation of a melt of commercial 'high purity' silicon produced a 'well-defined barrier having a high degree of photovoltaic response' (Ohl, 1941). This barrier was in fact a *p-n* homojunction formed from the unequal distribution of impurities as the Si crystal grew from the melt. From this discovery, after a delay occasioned by the war, grew the seminal work of Chapin *et al.* (1954) at Bell on the development of the diffused *p-n* junction single-crystal silicon cell. This technology was commercialised by Hoffman Electronics, and a place on board the 1958 NASA Vanguard I satellite for Hoffman's 9%-efficient c-Si cells pioneered the way for a number of space applications.

Efficiencies improved rapidly. The so-called silicon 'violet cell' with its improved short-wavelength response, developed at COMSAT in the early 1970s (Lindmeyer and Alison, 1973), achieved AM1.5G efficiency of over 15%. Next came the 'black cell' of Haynos *et al.* (1974), in which front-surface reflectance was greatly reduced by a chemical etch of the Si surface, giving the cells a black appearance after the application of antireflection coatings. This took terrestrial cell efficiency above 17% for the first time. The 18% milestone was passed a decade later by the MINP (Metal-Insulator-NP junction) cell of Green *et al.* (1985). From then on, Si cell technology progressed from MINP to PESC (the Passivated Emitter Solar Cell), to rear contact cells, to PERC (Passivated Emitter and Rear Contacts) cells, and ultimately to the PERL (Passivated Emitter, Rear Locally-diffused) cell, which is 24.8% efficient in monocrystalline form (Zhao *et al.*, 1998). Green (1999) describes the development of silicon solar cell technology in more detail.

A small terrestrial market for c-Si modules started to appear for remote, small-power applications and demonstration projects in the 1970s, stimulated by the oil price hikes of that time (Green, 1993; Treble, 1998). Successive markets opened up for silicon PV as costs reduced. In 1970, c-Si cells for use in space cost several hundred dollars per watt. By the mid-1970s, the efforts of Elliot Berman and his Solar Power Corporation (backed by Exxon) had reduced the cost of cells made specifically for terrestrial applications to $\$20\,\mathrm{Wp}^{-1}$. In round terms, the RAPS (remote area power supplies) market opened up in the 1980s at module costs of $\$10\,\mathrm{Wp}^{-1}$. Solar lighting in grid-remote locations opened up in the early 1990s at $\$5\,\mathrm{Wp}^{-1}$. Multicrystalline silicon (mc-Si) cells entered the PV market in 1981 and captured an increasing market share as module efficiencies improved from an early 10–12% to ~16–18% today.

Until the mid-2000s, sufficient high-quality material for c-Si cell manufacture was available from the electronics industry as cut-price off-spec and waste *polysilicon* (small-grained, polycrystalline silicon of 9N purity).[4] This was mostly made from metallurgical-grade silicon (MG-Si) by the energy-intensive Siemens process, which dates back to the 1940s. The main method for producing single-crystal Si from polysilicon is the even more venerable Czochralski (CZ) method, in which a rod of monocrystalline Si is grown by slowly withdrawing a seed crystal from a polysilicon melt. The CZ method dates from the 1916 work of the Polish scientist Jan Czochralski, co-founder of the German Metallurgical Society, and is also an energy-intensive process. Cutting the CZ rod into thin wafers to make c-Si cells incurs further losses and wastage of material, and fabricating cells from wafers is a complex and multi-step process. Multicrystalline silicon (mc-Si) is commonly made by the Bridgman method of directional crystallisation: silicon melt is poured into a cast and solidified from the bottom up by extracting heat from the crucible base, yielding blocks with columnar crystal growth that are then cross-sectioned into large-grained wafers.

By 2005, the market for PV had grown to the point where Si PV manufacturers were competing against chip companies for electronic-grade material. This

[4]The purity of silicon is often expressed as the total number N of nines in 99.99 ... %.

produced a bottleneck in the supply of polysilicon in the period 2006–2008, during which its price rose to the point where lower-cost upgraded metallurgical silicon (UMG-Si) was able to enter the PV supply chain. This had ~6N purity, adequate to make multicrystalline Si cells of reasonable performance. However, the global recession of early 2009 caused the price of polysilicon to fall sharply to the point where wafer and cell manufacturers found the higher cost of 9N polysilicon was justified by the higher cell efficiencies they could achieve using it. Thus UMG-Si fell from favour.

As already outlined in Section 1.1, overcapacity in the Si module market, combined with stagnation of demand in Europe, has led to a sharp drop in c-Si prices. At the time of writing (November 2013), business conditions remain very difficult for manufacturers. Some full-service companies that have substantial installation contracts are doing rather better.

1.4 GaAs and III–V multijunction cells

Gallium arsenide (GaAs) and related III–V materials have a fortunate combination of properties which gives them several advantages as solar cell materials. Their effective carrier masses are small and hence their carrier mobilities are high, and their optical absorption is intense with sharp absorption edges because of their direct-gap band structure. III–V materials can be doped both *n*- and *p*-type, and lattice-matchable ternary (e.g. $Al_xGa_{1-x}As$, $In_xGa_{1-x}As$) and quaternary (e.g. $Al_xGa_{1-x}As_yP_{1-y}$) alloys can be made over wide composition ranges. This allows the fabrication of multijunction (MJ) III–V cells of very high performance in which the bandgap of each sub-cell is optimised by adjusting the alloy composition. Most MJ cells are monolithic devices consisting of two or three 'current-matched' sub-cells stacked vertically with a tunnel junction series connection between each.

A photovoltaic effect in a GaAs *p-n* homojunction was first reported by Welker (1954), followed a year later by Gremmelmaier (1955), who obtained ~1% efficiency in a polycrystalline *p-n* homojunction solar cell. Jenny *et al.* (1956) of RCA Laboratories followed up with a >6%-efficient monocrystalline GaAs *p-n* cell. However, GaAs has the disadvantage of a high surface recombination velocity, so most of the photogenerated carriers recombine in the top layer of a GaAs *p-n* homojunction cell before they can reach the junction. Adding a thin top window layer of $Al_xGa_{1-x}As$ avoids this because carrier recombination at the $Al_xGa_{1-x}As$/GaAs interface is slow. The first lattice-matched AlGaAs/GaAs heterostructures to incorporate such a window layer were independently reported by Rupprecht *et al.* (1967)

at IBM's Thomas Watson Research Center and Alferov *et al.* (1967) at the Ioffe Institute of Leningrad (as it then was).[5] This was quickly followed by the *p*-AlGaAs/*p*-GaAs/*n*-GaAs heteroface cell developed by Woodall and Hovel (1972) at the Thomas Watson Research Center, which in 1977 achieved a record single-junction efficiency for its time of 21.9%.

Tandem solar cells were proposed as far back as 1955 by Jackson (1955) and later by Wolf (1960). However, efficient III–V tandem cells were not achieved until the late 1980s because of difficulties in making high-performance stable tunnel junctions and the effects of oxygen-related defects in the AlGaAs window layer (Ando *et al.*, 1987). Yamaguchi *et al.* (1987) proposed the use of double-hetero (DH) structure tunnel diodes as optically and electrically low-loss cell interconnects. Olson *et al.* (1990) achieved an efficiency of 27.3% in a cell with a $Ga_{0.5}In_{0.5}P$ homojunction grown epitaxially on a GaAs homojunction with a GaAs tunnel junction interconnect, and Japan Energy (Takamoto *et al.*, 1997) broke the (1 Sun) 30% barrier with their InGaP/GaAs tandem cell, which had a DH tunnel junction interconnect in which the InGaP layers were surrounded by high-bandgap AlInP barriers.

The main drawback of III–V cells is their high cost. Aluminium is so reactive in the vapour phase that it is not possible to prepare AlGaAs layers by conventional chemical vapour deposition using elemental sources. Rather it is necessary to use the slower and more expensive method of metal-organic chemical vapour epitaxy (MOVPE), using sources such as trimethyl aluminium. Cell stacks are usually grown epitaxially on substrates such as germanium. However, the high efficiency, good radiation hardness of InP and InGaP (Yamaguchi *et al.*, 1997) and low series resistance of III–V MJ cells have allowed them to dominate two niche markets: provision of PV power in space and use in concentrator PV systems (CPV). III–V MJ concentrator cells have the highest conversion efficiencies in their class of any PV technology. Boeing Spectrolab broke the 40% efficiency barrier in 2006 with their 3-junction GaInP/GaInAs/Ge concentrator cell, which was 40.7% efficient under 240 Suns (King *et al.*, 2007). At the time of writing, the world record holder for CPV is the Fraunhofer Institute's 4-junction cell, which is 44.7% efficient under 297 Suns (Fraunhofer, 2013). In theory, efficiencies of 50% or more are possible with MJ concentrator cells with four or more junctions (Yamaguchi *et al.*, 2008).

[5]Zhores Alferov won the 2000 Nobel Prize for Physics for his development of heterostructures and is probably the only solid-state physicist to have an asteroid named after him.

1.4.1 *Quantum well solar cells*

Quantum well solar cells (QWSCs) are relatively recent constructs, proposed by Barnham and Duggan (1990) and discussed by Jenny Nelson and Ned Ekins-Daukes in Chapter 10. They are based on III–V *p-i-n* or *p-n* homostructure cells into the *i* layer of which several extremely thin layers of a second semiconductor of smaller (bulk) bandgap are introduced, forming quantum wells in which 'extra' charge carriers are created by the additional light absorbed. The aim is that, compared with the parent cell (i.e. the host cell without the QWs), the photocurrent will be enhanced while the voltage is not reduced, so the efficiency is boosted over that of the parent cell.

Photocurrent enhancement has been demonstrated in a range of QWSCs and efficiency enhancement has been observed in materials whose bandgap is larger than the optimum for solar energy conversion. In materials of bandgap close to the optimum, experimental tests on QW cells of equivalent quality to homojunction cells have not yet been possible. Achieving these efficiency improvements needs carefully grown, defect-free structures. Successive layers are often grown in alternately compressive and tensile stress to allow more QWs to be inserted. For a short period in 2011, a strain-balanced GaAsP/InGaAs QWSC set a world record for power conversion efficiency in a single-junction cell.

Because QWSCs are as costly to produce as high efficiency III–V homojunction cells, they are likely to find application only where III–Vs are preferred. At the present time this means space, concentrator PV (where the drop in performance with increasing temperature is less marked than for the parent cell) and thermophotovoltaic systems.

1.5 Concentrator photovoltaics

In CPV systems, direct sunlight is concentrated by optical means on small, high-performing PV cells. The cell area required for a given power output is reduced by roughly the solar concentration ratio, so cell costs per output watt are reduced, although the optics and Sun-tracking required add complexity and increase balance-of-system costs. Concentrations of up to $\times 200$ Suns can be achieved with linear Fresnel lenses or parabolic mirrored troughs, and low-resistance Si cells are suitable in this range. Higher concentrations of 250–1000 Suns are achievable with two-axis solar tracking. III–V MJ cells come into their own at these high irradiances, because their very high efficiency justifies their high cost.

CPV was developed at Sandia National Laboratories in the 1970s, but there was virtually no commercial activity until the 1990s, when firms such as Amonix and Entech were founded, and a number of CPV demonstration plants were built.

In the early 2000s, the combination of the relatively high cost of flat-plate c-Si PV systems and the impressive improvements in III–V MJ cell efficiency stimulated a number of other commercial CPV ventures. By mid-2011 the cumulative global installed CPV capacity was 23 MWp, Spain accounting for 70% of this total.

2012 has seen a more mixed picture. Some commentators predict continued market growth: Lux Research (2012) saw the market growing to 697 MWp p.a. by 2017 and IMS Research (2012) predicted that cumulative installations would reach 1.2 GWp by 2016. However, 2012 has proved to be as difficult a market for CPV as it has for conventional PV, and commercial CPV ventures have retrenched. Amonix, which announced a world record CPV module efficiency of 33.5% in October 2012, closed its manufacturing facility in North Las Vegas in July. The development of III–V multijunction cells of even higher performance may tilt the economics back in favour of CPV: King *et al.* (2012) calculate that 50% efficient MJ CPV cells would be the lowest cost option for solar electricity generation in high direct normal irradiance regions. Other commentators are forecasting that CPV will be unable to compete with the continued low price of flat-plate PV.

1.6 Inorganic thin-film cells

1.6.1 *The cadmium sulphide cell: p-Cu_2S/n-CdS*

From the 1970s, when terrestrial applications of c-Si technology began to emerge, there was a parallel effort to develop other semiconductors in order to make thin-film (polycrystalline) devices of lower cost and better light-absorbing properties than c-Si. The original motive for investigating thin-film cells was not, however, lower cost but their better power-to-weight ratio for space applications. The first commercially developed thin-film PV device was the cuprous sulphide/cadmium sulphide (p-Cu_2S/n-CdS) heterojunction cell, made in 6%-efficient single-crystal form by Reynolds *et al.* (1954) at Wright Patterson Air Force Base in the USA, and in thin-film form by Carlson (1956) at the Clevite Research Center in Cleveland, Ohio. This cell excited much interest because of its low intrinsic costs and the simple way in which the junction is formed, by dipping CdS into a solution containing cuprous ions. Clevite Corporation mounted a major development effort on CdS PV in 1964, and several other companies followed suit. However, in spite of promising results, reviewed by Hill and Meakin (1985), that included the first thin-film cell of $>10\%$ efficiency, these cells suffered from poor stability arising from the high diffusivity of copper ions into the CdS layer, and there were also serious problems in making ohmic contacts to Cu_2S. Thus this technology fell from favour. CdS lives on, however, as the window layer of CdTe and CIGS cells.[6]

1.6.2 *Thin-film (amorphous and microcrystalline) silicon*

The Japanese had effectively already delivered the *coup de grâce* to Cu_2S/CdS technology by the early 1980s by commercialising small amorphous hydrogenated silicon (a-Si:H) PV panels of modest but sufficient (3–4%) efficiency to power small consumer goods such as watches and calculators. Amorphous silicon of good quality (with sufficiently few mid-gap states to be dopable either *n*- or *p*-type) had earlier been made by Spear and Le Comber (1975) in Dundee, Scotland. Independently, David Carlson and Chris Wronski, then both at RCA Laboratories in Princeton, New Jersey, made several square-centimetre sized *n-i-p* and *p-i-n* cells of ~2% efficiency (Carlson and Wronski, 1976), and smaller area MIS cells of 5.5% efficiency. The *p-i-n* configuration was to be the forerunner of modern a-Si:H photovoltaic technology. The Staebler–Wronski effect, which is the ~10–30% diminution of efficiency that occurs on the first prolonged exposure of a cell to light, was discovered soon afterwards, in 1977. Unwelcome as this was, ways to reduce its impact by using thin cells (in which the higher built-in field reduces this volume recombination effect) have been developed.

Single-junction amorphous silicon modules now achieve stabilised efficiencies of 6–8%. However, most commercial product now has the so-called *micromorph* configuration, consisting of a two-junction (tandem) cell with an a-Si:H *p-i-n* top cell and a microcrystalline (μc-Si:H) *p-i-n* bottom cell which utilises the red and near-IR light that passes through the a-Si:H top cell. These cells, originally developed by Meier *et al.* (1994) at IMT Neuchatel with manufacturing technology licensed to the Swiss company Oerlikon Solar in 2003, have achieved stabilised module efficiencies of 11.9% (Bailat *et al.*, 2010) at a reported cost of EUR0.5 Wp^{-1}. Triple-junction cells (a-Si:H/a-SiGe:H/μc-Si:H) of 1.5% stabilised efficiency have recently been reported (Söderström *et al.*, 2012).

1.6.3 *The cadmium telluride cell: n-CdS/p-CdTe*

CdTe has long been familiar to the semiconductor industry from its use, in very pure crystalline form, as a photoconductive high-energy particle detector. Although it can be doped both *n*- and *p*-type, it is hard to make efficient *p-n* homojunction CdTe solar cells because of the difficulty of forming a shallow junction with an

[6]Cadmium sulphide also lives on in the paintings of impressionists such as Monet, whose favourite yellow pigment it was.

active top layer in the face of the material's high surface recombination velocities. The way forward proved to be the *n*-CdS/*p*-CdTe heterojunction cell, in which CdTe forms the active, light-absorbing base layer and CdS the front window layer. The *n*-CdS/*p*-CdTe device structure combines good optical transparency with sufficiently close lattice and thermal matching to form a 'good' (spike-free) junction, albeit after a special activation process. Single-crystal *n*-CdS/*p*-CdTe cells of up to 8% efficiency were made in the 1970s (Saraie *et al.*, 1972; Yamaguchi *et al.*, 1977; Mitchell *et al.*, 1977), and this good performance allied to ease of junction formation and tolerance to materials purity that caught the attention of industry. General Electric was an early market leader in CdTe cell manufacture (Cusano, 1962, 1966). BP Solar followed in the 1990s with a research programme at Sunbury-on-Thames near London on electrodeposition of CdTe cells, and later module production in the US, but axed this programme in late 2002. Dieter Bonnet, who co-authored the account of CdTe PV in Chapter 5, was one of the pioneers of this field. With his co-worker Rabenhorst, he was the first to report the all-thin-film CdS/CdTe heterojunction cell (Bonnet and Rabenhorst, 1972) and he founded ANTEC in Kelkheim, Germany, which for some years fabricated 7%-efficient cells deposited by closed space sublimation (CSS).

The CdTe market phenomenon of the late 2000s — First Solar of Tempe, Arizona — was founded by entrepreneur Harold McMaster in 1990 and is now owned by the investment arm of the family that owns Walmart. The company has focussed aggressively on cost reduction, with impressive results. In 2009, First Solar's CdTe modules were the first to break the US\$1 Wp^{-1} cost barrier (one PV equivalent of the 4-minute mile) and the company was the top PV supplier of that year. Since then, the falling price of c-Si modules has eroded the value proposition for CdTe, and First Solar, like other PV manufacturers, has had to retrench.

Best research-cell efficiency plateaued at 16.7% for a decade but has now reached 18.7% in a cell made with First Solar's commercial-scale manufacturing equipment and materials (First Solar, 2013), and module efficiencies of above 11% are routinely achieved (Semiconductor Today, 2011). Given that the CdTe bandgap of 1.44 eV is almost ideally matched to the terrestrial solar spectrum, cell efficiencies of $>20\%$ should be possible.

Cadmium is highly toxic, and tellurium is toxic if ingested. However, fears about the safety of CdTe technology have been largely allayed by tests showing that module encapsulation in glass prevents escape of CdTe to the environment, though the safe disposal of panels will require attention. Te is currently classified as an extremely rare element (1–5 ppb in the Earth's crust) and this could cause supply difficulties if CdTe PV were to become a mainstream power provider. However, it has recently been discovered that some undersea ridges are rich in the element.

1.6.4 *Copper indium gallium diselenide (CIGS) cells*

CIS/CIGS (copper indium selenide/copper indium gallium selenide) is another promising thin-film PV technology. The very high optical absorptivity of CIGS is an advantage, although its quaternary nature led to past difficulties in composition control during manufacture. Like CdTe, CIS/CIGS can exhibit both *n*- and *p*-type conductivity arising from intrinsic defects, but it is better used in the *p*-type form in a heterojunction device with an *n*-CdS top window layer. 12%-efficient single-crystal *n*-CdS/*p*-$CuInSe_2$ cells were made by Wagner *et al.* (1974) and Shay *et al.* (1975), and thin-film cells of 4–5% efficiency quickly followed (Kazmerski, 1976).

By the end of the 1980s, commercialisation efforts by PV pioneer ARCO through its subsidiary ARCO Solar had achieved thin-film CIS modules with areas of up to $1 \times 4\,ft^2$ and ~10% efficiency. Persistent problems with the process yield were later overcome by control of sodium impurities in the CIS film and improved junction fabrication processes. The work of the EuroCIS consortium in the early 1990s resulted in significant efficiency increases to ~16%.

The current best laboratory cell efficiency is an impressive 20.8% (ZSW, 2013), with sub-module efficiency not far behind at 17.8% (Solar Frontier, 2012). Like other PV technologies, commercial CIGS has not been immune to the continued declining prices of c-Si modules. The 2011 bankruptcy of California-based CIGS start-up Solyndra, which had received substantial US federal support, became a 2012 presidential campaign issue. The Japanese company Solar Frontier, a 100% subsidiary of Showa Shell Sekiyu, continues as the only vendor shipping CIGS product in volume.

1.7 Organic and hybrid technologies

1.7.1 *Organic photovoltaics*

The observation of photoeffects in organic materials has a long history: the photoconductivity of crystalline anthracene was studied over a century ago by Pochettino (1906). Photoconductivity in such organic solids arises because the conjugated π-electronic orbitals of neighbouring molecules ovelap sufficiently to allow charge transport by hopping. Light absorption in organic semiconductors creates short-lived excitons, rather than ‘free’ carriers, as in broadband inorganic semiconductors. However, the presence of an electric field can assist exciton dissociation and the drift-assisted separation of electron-hole pairs. In an organic photoconducting device, the field is supplied by the external electric bias on the electrodes. In an organic photovoltaic cell, the field is usually created at the heterojunction between two different materials, one with donor, and the other with acceptor, properties.

Porphyrins, phthalocyanines and chlorophyll, all strongly-coloured π-conjugated compounds, were among the earliest organic semiconductors to be investigated for photovoltaic activity. Kearns and Calvin (1958) reported a photovoltage of 200 mV, but negligible power output, in a bilayer cell with magnesium phthalocyanine as the electron donor and oxidised tetramethyl p-phenylene diamine as the acceptor. Ghosh *et al.* (1974) found a maximum power conversion efficiency of only 0.01% in an Al/MgPh/Au cell under monochromatic illumination at 690 nm. Tang and Albrecht (1975a; 1975b) obtained an efficiency of 0.05% in a Cr/Chl-a/Hg cell under monochromatic illumination at 745 nm.

The very poor efficiency of these early organic photovoltaic (OPV) cells was due to a number of factors, high electrical resistance, poor materials control and low internal quantum efficiency prominent among them. The thin (70 nm) copper phthalocyanine/ perylene tetracarboxylic acid cell of Tang (1986) had a lower resistance and achieved an improved power conversion efficiency of 1%.

Since these pioneer endeavours, steady and substantial improvement in OPV efficiency has been achieved by better junction engineering and better materials purity and control; at the time of writing, the best reported efficiency is 10.7% in Heliatek's tandem OPV cell (Heliatek, 2013). The development of *bulk heterojunctions* (also called dispersed heterojunctions) was a major step forward. In bulk heterojunction (BHJ) cells, the donor and acceptor materials are intimately mixed together while remaining two separate phases, so that excitons generated anywhere in either material are within a diffusion length of an interface between the two at which they are dissociated and mobile charge carriers are formed. The development of high-quality dopable polymers for the OLED industry and other optoelectronic applications has also been of major assistance to OPV. The first small molecule dye/dye BHJ cell was reported by Hiramoto *et al.* (1991) and the first polymer BHJ cell by Yu *et al.* (1994).

OPV has the advantages of cheapness, lightness and bandgap tunability. Their disadvantage is their poor stability and lifetime. OPV companies, notably Konarka, have been among those PV companies to have been overcome by falling c-Si prices and competition from more durable and efficient thin-film products. However, there should be a place even in today's harsh PV market for off-grid small-power OPV applications such as the Indigo 'pay-as-you-go' power offered by Eight19 (2012), a company spun off from the Cavendish Laboratory of the University of Cambridge. Spanggaard and Krebs (2004) give an excellent account of the history of OPV.

1.7.2 *Dye-sensitised solar cells*

The modern era of dye-sensitised solar cells (DSSC) began with the seminal report by O'Regan and Grätzel (1991) of a 7%-efficient cell containing a ruthenium dye-sensitised mesoporous TiO_2 photoanode and an iodide/triiodide redox couple in a liquid organic solvent. The efficiency of this cell greatly exceeded that of any previous dye-sensitised photoelectrochemical cell, and the reason for this remained unclear until kinetic studies revealed the main factor as the exceptionally slow rate of back electron transfer from TiO_2 into the oxidised dye or I_3^-.

Since the discovery of the Grätzel cell, as it is sometimes known, DSSC efficiencies have increased through improvements in all cell components, but particularly the redox couple and the dye. Cobalt$^{2+/3+}$-based redox couples have increase the open-circuit voltage above that possible with I^-/I_3^-, and donor-π-acceptor dyes absorb much more strongly than the original ruthenium-based complexes (Mishra*et al.*, 2009). The current world-record DSSC cell is 12.3% efficient under standard test conditions (STC), and this increases to 13.1% under low light intensity (Yella *et al.*, 2011). This good performance under low light, coupled with the fact that the cell materials do not need to be rigorously purified, has stimulated considerable commercial interest in BIPV (building-integrated photovoltaic) and other applications of DSSC, applied as translucent films on glass. However, liquid electrolytes are volatile and the I^-/I_3^- couple is corrosive, so a solid-state hole conductor generally replaces both in commercial product. Solid-state DSSC efficiencies have improved over the year, currently to a record 7.2% in a cell with an organic D-π-A sensitiser and *p*-doped *spiro*-MeOTAD as hole conductor (Burschka *et al.*, 2011). Hardin *et al.* (2012) describe improvements that could take DSSC efficiencies towards 20%.

1.7.3 *Perovskite solar cells*

Hybrid organic/inorganic perovskite solar cells based on methylammonium lead halides have made a dramatic debut onto the PV scene, demonstrating efficiencies that have risen to over 15% in only four years of development. Back in the 1990s, there was considerable interest in the electronic properties of solution-processable organometallic perovskites in applications such as ferroelectric and superconducting materials. In particular, David Mitzi and colleagues at IBM examined some electronic applications of these materials (Mitzi *et al.*, 1994; Kagan *et al.*, 1999). However, it was the Tokyo-based group of Tsutomu Miyasaka (Kojima *et al.*, 2009) who made the discovery that the molecular dye sensitiser of the 'classical' DSSC could be replaced by a nanocrystalline film of $CH_3NH_3PbBr_3$ or $CH_3NH_3PbI_3$, yielding a cell of 3.8% efficiency in the case of the triiodide. These materials have the perovskite structure and are direct-gap semiconductors, with a bandgap that can be tuned through the visible and into the near IR by adjusting the mole fraction x in $CH_3NH_3Pb(I_{1-x}Br_x)_3$. The charge carriers generated in them by light absorption are quite long-lived and mobile in these materials, with both holes and electrons having diffusion lengths of ~100 nm. Moreover, little energy is lost in separating the charge carriers (as compared with the losses incurred in exciton dissociation in OPV) so high open-circuit voltages are achieved. Finally, the materials can be prepared in tailored compositions and absorption thresholds by low-cost, low-temperature solution methods.

The first cells that were made were not stable because the perovskites were attacked by the liquid electrolyte of the DSSC . However, it was soon discovered that this could be substituted by a solid-state hole conductor, and then that this could be omitted as the materials are ambipolar and carry both holes and electrons (Etgar *et al.*, 2012). Other features of the DSSC also proved expendable: Henry Snaith's group at Oxford showed that by using $CH_3NH_3PbI_2Cl$ the mesostructured TiO_2 scaffold could be replaced by alumina (Lee *et al.*, 2012), possibly because of the order of magnitude increase in the exciton/charge carrier diffusion length in $CH_3NH_3PbI_2Cl$ compared with $CH_3NH_3PbI_3$ (Stranks *et al.*, 2013). It was also found that the perovskite itself did not have to be nanostructured: a simple planar heterojunction cell consisting of a few hundred nanometres of vapour-deposited perovskite sandwiched between two charge-selective electrodes can have efficiencies of over 15% (Liu *et al.*, 2013). Michael Grätzel's group (Burshka *et al.*, 2013), working with a mesostructured TiO_2 scaffold, showed that sequential deposition of the perovskite, first laying down PbI_2 and then converting it to perovskite by exposure to CH_3NH_3I in solution, yields more consistently high-performance cells than co-deposition.

Perovskite cells will certainly be an active area of research in the next few years, with some commentators predicting that efficiencies of over 20% may be possible (Park, 2013). However, many questions about stability, durability and device repeatability remain to be addressed.

1.8 Overview of photovoltaic cell operation

This book aims to present an in-the-round approach to PV technology, touching on aspects from the choice of semiconductor materials through system design to market status. But PV cells themselves and how they work form its main subject matter. By way of introduction to the specialist chapters that follow this one, I therefore conclude this chapter with an account of the main PV cell types and the basic principles of cell operation. Archer *et al.* (1996) give a more detailed account of analytic solutions for *p-n* junction cells than is here presented.

1.8.1 *The p-n homojunction cell*

All PV cells, with the exception of dye-cells, work in essentially the same way. They contain a junction between two different materials across which there is a

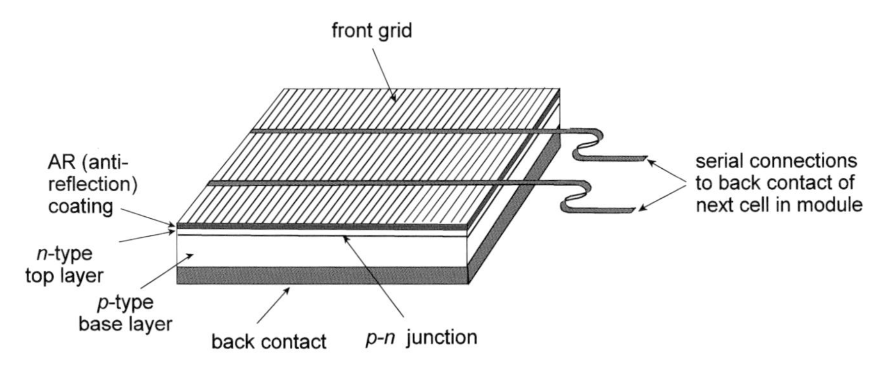

Figure 1.2 The essential features of a *p-n* homojunction Si solar cell.

'built-in' electric field. When the cell absorbs light, excitons (electron-hole pairs) are created. In broadband (most inorganic) semiconductors, these excitons very quickly dissociate into mobile electrons and holes, which flow in opposite directions across the junction. In narrow-band and organic semiconductors, excitons do not dissociate spontaneously but are separated at the junction. In either case the flow of absorbed photons is converted into a flow of DC power from the illuminated cell.

The crystalline silicon (c-Si) cell has a simple junction structure, and provides a good model with which to explore the PV effect. Figure 1.2 shows the essential features of these cells, which are typically square wafers of dimensions $\sim$15 cm $\times$ 15 cm $\times$ 0.2 mm. The top (emitter) region is a 0.2–0.3 μm thick layer of n-type silicon, and the base region is a 300 μm thick layer of p-type silicon.[7] The work function of the p material is greater than that of the n material, so the two layers reach electronic equilibrium (in the dark) by the transfer of some electrons from the n to the p side. The structure as a whole remains electrically neutral, but the junction region contains an electric double layer, consisting of two *space-charge regions* or *depletion regions* (DRs), as shown in Fig. 1.3. These DRs are typically less than a micron thick, and the charges they contain are those of the ionised dopants (P^+ and B^- in the case of c-Si). Beyond the base-layer DR in the c-Si cell (and some other cells) lies a quasineutral region (QNR) — a region that contains no space charge.

Figure 1.4 shows what happens in the illuminated c-Si cell. The absorption of photons of energy greater than the bandgap energy of silicon promotes electrons from the valence band to the conduction band, creating hole-electron pairs

[7] c-Si cells are usually configured n-on-p because this best suits the properties of silicon, but some other p-n cells are configured p-on-n. These cells are also quite thick, because c-Si absorbs light relatively weakly. Most other cells are much thinner.

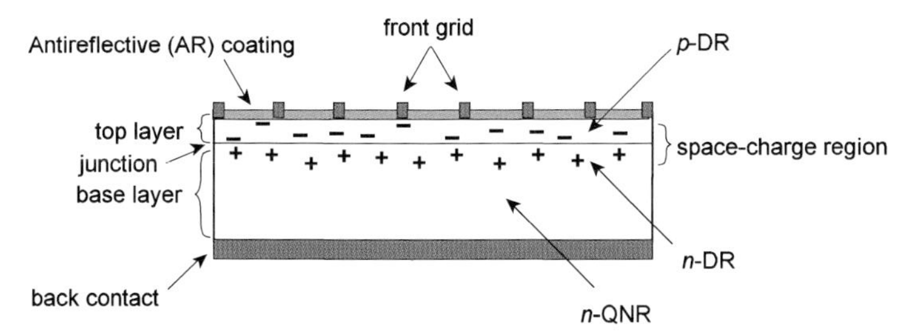

Figure 1.3 Cross section through a p-n homojunction cell, showing the electrical double layer consisting of ionised dopant atoms (denoted + and –) in the junction region, the two depletions regions (DRs) that contain equal and opposite quantities of junction charge, and the base-layer quasineutral region (QNR).

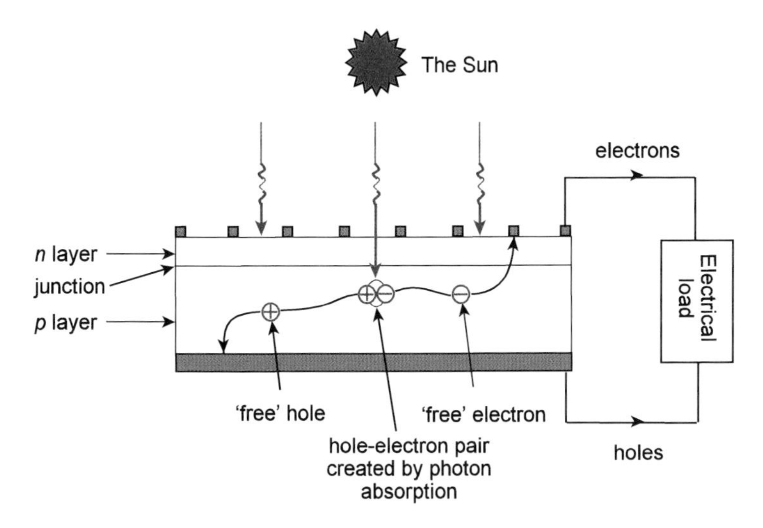

Figure 1.4 Generation and movement of free carriers in a p-n junction cell.

throughout the illuminated part of the cell, which in c-Si cells extends well into the base layer. In c-Si and most other semiconductors, these hole-electron pairs quickly dissociate into 'free' carriers — mobile holes and electrons that move independently of each other. Those free carriers that approach the junction come under the influence of the built-in electric field, which sweeps electrons from the p to the n side, and holes from the n to the p side.

1.8.2 *Junction structure and dark current*

The electric double layer at the p-n junction has an important effect on the semiconductor energy levels, as shown in Fig. 1.5. The separate (uncharged) phases

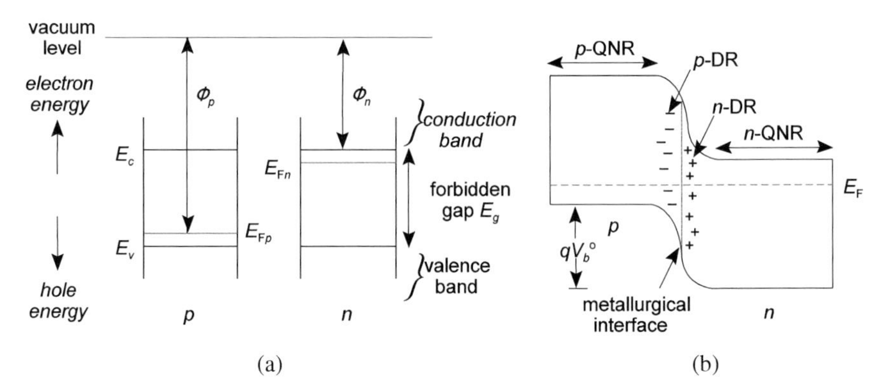

Figure 1.5 Energy band structure of a *p-n* homojunction in the dark: (a) in uncharged blocks of *p*-type and *n*-type semiconductors before contact, showing the conduction and valence-band energies E_c and E_v, the forbidden gap E_g and the Fermi levels E_{Fn} and E_{Fn} (red dashed lines) in the *n* and *p* phases; (b) across the *p-n* homojunction after contact and equilibration of the two phases, showing the electric double layer formed by transient charge transfer, the depletion regions (DRs) and quasi-neutral regions (QNRs) and the common Fermi level E_F throughout the device.

(Fig. 1.5a) have the same conduction and valence band-edge energies E_c and E_v, separated by the forbidden gap E_g, but different work functions Φ_p and Φ_n, and therefore different Fermi levels E_{Fp} and E_{Fn}.[8] In the equilibrated cell (Fig. 1.5b), the Fermi level E_F is the same throughout the device but the band-edge energies E_v and E_c (in common with all the energy levels of the semiconductor) bend across the junction in response to the local electric field. Inspection of Fig. 1.5 shows that the equilibrium band-bending energy is qV_b^o is related to the difference in the work functions of the (separate, uncharged) materials by

$$qV_b^o = \Phi_n - \Phi_p \tag{1.1}$$

Since the Fermi level in a doped semiconductor normally lies within the forbidden gap but near the majority-carrier band edge, qV_b^o is normally slightly smaller than the bandgap energy E_g. Figure 1.6 shows how the band bending is affected and a current is caused to flow when a bias voltage V_j is applied across the cell in the dark. At equilibrium (Fig. 1.6a), no net current[9] flows through any part of

[8] The Fermi level is the energy for which the probability of a state being occupied by an electron is exactly one-half. In an intrinsic (undoped) semiconductor, the Fermi level falls in the middle of the forbidden gap. In a lightly doped semiconductor, the Fermi level remains within the forbidden gap but is near the majority-carrier band edge. In a heavily doped semiconductor, the Fermi level lies within the majority-carrier band.

[9] All the currents given the symbol i in Figs. 1.6–1.8 are strictly speaking current densities.

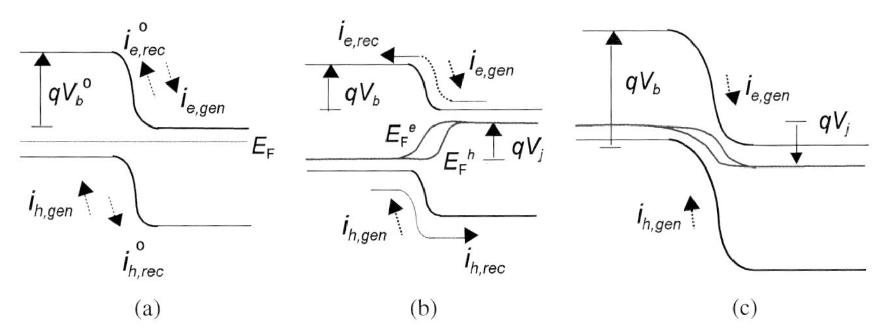

Figure 1.6 Dark *p-n* homojunction cell in the dark (a) at equilibrium; (b) under forward bias V_j; (c) under reverse bias V_j, showing the generation and recombination currents as dotted lines and the Fermi level E_F and hole and electron quasi-Fermi levels E_F^h and E_F^e across the junction region as red dashed lines.

the cell. However, small, balanced fluxes of electrons in the conduction band and holes in the valence band pass each way across the junction. These are referred to as generation and recombination currents. The *(thermal) generation currents* $i_{h,gen}$ and $i_{e,gen}$ shown in Fig. 1.6a come from the minority carriers (electrons in the *p* side and holes in the *n* side) generated throughout the device, albeit at a minuscule rate, by thermal excitation. Those minority carriers that reach the junction without recombining are swept across it in opposite directions by the strong electric field. The *recombination currents* $i^{o}_{h,rec}$ and $i^{o}_{e,rec}$ also shown in Fig. 1.6a come from majority carriers (holes in the *p* side and electrons in the *n* side) that flow 'up' the band-bending barrier (this is energetically unfavourable, but entropically favourable because the carriers move from a region of high to low concentration).

At equilibrium, the generation and recombination currents in each band exactly balance each other. The sum of the hole and electron thermal generation currents is called the *saturation current density* i_o of the junction.

$$i_o = i_{h,gen} + i_{e,gen} = i^{o}_{h,rec} + i^{o}_{e,rec} \tag{1.2}$$

When a forward bias[10] voltage V_j is applied across the junction of the dark cell, the barrier height is reduced to $qV_b = q(V_b^o - V_j)$, as shown in Fig. 1.6b. This does not affect the generation currents, but it strongly increases the recombination currents. The net current across the junction, which is the difference between the recombination current and the generation current, is called the *dark current* or

[10]Forward biasing a junction means applying a voltage across the device that lowers the band-bending barrier. Reverse biasing means applying a voltage in the opposite direction.

junction current i_j.

$$\begin{aligned} i_j(V_j) &= i_{h,rec}(V_j) + i_{e,rec}(V_j) - i_{h,gen} - i_{e,gen} \\ &= i_{h,rec}(V_j) + i_{e,rec}(V_j) - i^{\mathrm{o}}_{h,rec} - i^{\mathrm{o}}_{e,rec} \end{aligned} \quad (1.3)$$

When a reverse bias ($V_j < 0$) is applied, the barrier height is increased as shown in Fig. 1.6c to $qV_b = q(V_b^o + |V_j|)$. The generation currents are unaffected, but the recombination currents are now suppressed. Thus only the very small, bias-independent saturation current passes.

$$i_j(V_j < 0) = -i_{\mathrm{o}} \quad (1.4)$$

The dependence of the recombination currents $i_{h,rec}(V_j)$ and $i_{e,rec}(V_j)$ on V_j is determined by the dominant recombination mechanism of the carriers injected into the junction. In most cells, the dark current–voltage characteristic conforms well to the empirical *diode equation*

$$i_j(V_j) = i_{\mathrm{o}}[\exp(qV_j/\beta kT) - 1] \quad (1.5)$$

where β is called the diode *ideality factor*. For an ideal junction, in which no injected carriers recombine in the junction, $\beta = 1$. For a non-ideal junction, in which some carriers do recombine in the junction, $1 < \beta < 2$. For many cells, Eq. (1.5) is better written as the *two-diode equation*

$$i_j(V_j) = i_{\mathrm{o}1}[\exp(qV_j/kT) - 1] + i_{\mathrm{o}2}[\exp(qV_j/2kT) - 1] \quad (1.6)$$

where the first term corresponds to carriers that move across the junction without recombining, and the second to the carriers that recombine in mid-gap. Regardless of the exact form of the diode equation, all PV cells behave as rectifiers in the dark, showing highly non-linear current–voltage characteristics similar to that labelled ‘dark’ in Fig. 1.8. Junctions must show rectifying properties in the dark if they are to show photovoltaic properties in the light.

1.8.3 *The illuminated cell*

When a PV cell is illuminated, a photocurrent and photovoltage are generated. Figure 1.7 shows how this happens, again using the example of a *p-n* homojunction cell. Absorption of photons of energy greater than the bandgap energy of the semiconductor creates excess minority carriers throughout the illuminated region of the cell (the light intensity in the cell interior falls off exponentially with distance into the cell, but often it penetrates into the base layer). The photogenerated minority carriers in the illuminated cell behave like the much smaller population of thermally generated minority carriers in the dark cell. That is, they diffuse

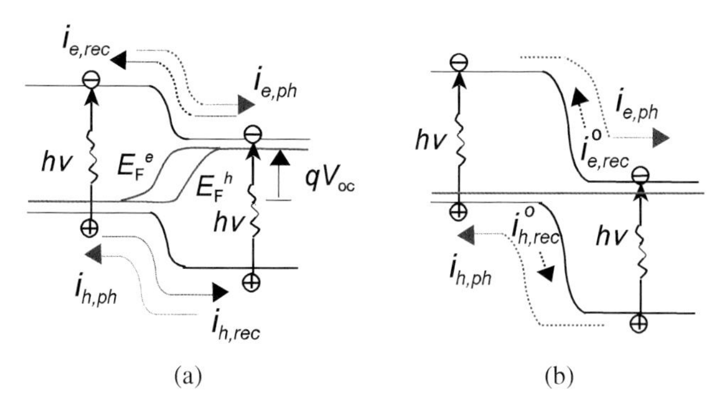

Figure 1.7 Illuminated *p-n* homojunction cell (a) at open circuit, showing the photocurrents as red dotted lines and the hole and electron quasi-Fermi levels E_F^h and E_F^e across the junction region as red dashed lines; (b) at short circuit, assuming that the cell has no internal resistance (i.e. that carrier mobility is infinite).

from the QNRs towards the junction, where they are swept across it by the strong junction field. These fluxes of photogenerated minority carriers give rise to the *photogeneration currents* $i_{e,ph}$ and $i_{h,ph}$ shown in Fig. 1.7a, consisting respectively of photogenerated electrons drifting from the p to the n side of the junction and photogenerated holes drifting the other way. The sum of the two is the overall photocurrent i_{ph}.

$$i_{ph} = i_{h,ph} + i_{e,ph} \tag{1.7}$$

The photocurrent is directly proportional to the absorbed photon flux but independent of bias (provided that the junction field is always high enough to sweep carriers across the junction). At open circuit (Fig. 1.7a), no current is drawn from the cell and the photocurrent must be balanced by the recombination current. The junction self-biases in the forward direction by the open-circuit voltage V_{oc}, at which point the recombination (junction) current exactly opposes the photocurrent, i.e.

$$i_{ph} - i_j(V_{oc}) = 0 \tag{1.8}$$

As shown in Fig. 1.7a, the hole and electron quasi-Fermi levels E_F^h and E_F^n diverge across the junction and converge past the quasi-neutral regions, and qV_{oc} is the difference between the Fermi levels on the two far sides of the junction. Since metal contacts always equilibrate with the local majority carrier Fermi level, V_{oc} is an observable output voltage.

Figure 1.7b shows what happens when the illuminated cell is short-circuited. The cell delivers maximum current but at zero output voltage. Provided internal resistance effects are negligible, the junction bias V_j is also zero, so the band bending is the same as in the dark junction at equilibrium.[11] The short-circuit current is given by

$$i_{sc} = |i_{ph}| - i_o \tag{1.9}$$

Under normal operating conditions, the band bending and junction current are intermediate between the open-circuit and short-circuit cases, and the cell delivers current i at output voltage $V \approx V_j$, where i is given by

$$i = i_{ph} - i_j(V_j) \tag{1.10}$$

Provided the photocurrent i_{ph} is bias-independent, the current–voltage characteristics of the dark and illuminated cells will therefore show *superposition*. That is, they will map onto each other, but are shifted with respect to each other by the constant amount i_{ph}, as shown in Fig. 1.8.

Superposition is an idealisation that is seldom accurately obeyed. It is not to be expected where the photocurrent is bias-dependent, which can happen for a number of reasons. In the amorphous silicon cell, for example, the field in the junction region becomes weaker as the forward bias increases and this makes it more difficult to collect the photogenerated carriers. Cells operating in the high-injection mode, where the concentration of photogenerated minority carriers becomes comparable with

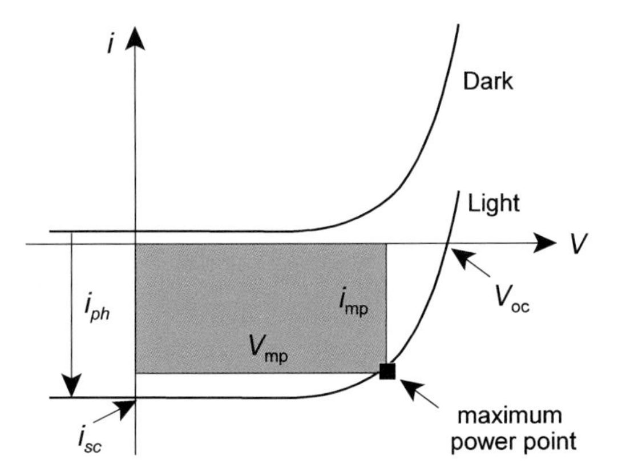

Figure 1.8 Current–voltage curves in the dark and the light for a cell that shows superposition (i.e. one in which the photogenerated current is bias-independent), showing the short-circuit and maximum-power currents i_{sc} and i_{mp}, the open-circuit and maximum-power voltages V_{oc} and V_{mp}, and the maximum power point (■).

[11]If the cell has significant internal resistance, the output voltage V drops below the junction voltage V_j, and a small forward bias remains across the junction when the cell is short-circuited.

that of the majority carriers, do not show superposition because the majority-carrier concentrations and fluxes are not then the same in the light and the dark. Cells with significant internal series resistance or shunt conductance also depart from superposition.

1.8.4 *Cell current–voltage characteristics*

The current–voltage characteristic of the illuminated cell is found by substituting Eq. (1.5) into Eq. (1.10). Assuming superposition and negligible internal resistance effects, the current–voltage characteristic is given by

$$i = i_{ph} - i_o[\exp(qV/\beta kT) - 1] \tag{1.11}$$

The output power is the product iV, which is the area of a rectangle of sides i and V inscribed in the i–V curve. The power is zero for both the open-circuit and short-circuit conditions. The maximum-power condition is reached where the area $i_{mp}V_{mp}$ (shaded in Fig. 1.8) is a maximum. The *fill factor* η_{fill} is a measure of the squareness of the i–V curve and is defined as

$$\eta_{fill} = \frac{i_{mp}V_{mp}}{i_{sc}V_{oc}} \tag{1.12}$$

In efficient cells, the fill factor is around 0.7–0.8. In poor cells, it can be 0.5 or lower.

By setting $i = 0$, $V = V_{oc}$ in Eq. (1.11) and rearranging, the open-circuit voltage of the illuminated cell is found as

$$V_{oc} = \frac{\beta kT}{q}\ln\left(1 + \frac{i_{ph}}{i_o}\right) \approx \frac{\beta kT}{q}\ln\left(\frac{i_{ph}}{i_o}\right) \tag{1.13}$$

For good performance, i_{ph} and V_{oc} must be as large as possible. The maximum value of i_{ph} would be obtained if all photogenerated electron-hole pairs were collected as photocurrent, and i_{ph} can achieve over 90% of this limit if light absorption and minority carrier collection are both highly efficient. If there are no sources of voltage in the cell other than at the junction, the limiting value of V_{oc} is the built-in voltage V_b^o, corresponding to complete flattening of the bands across the junction.[12] This would only happen under extremely intense illumination, and 1 Sun V_{oc} values are usually no more than $\sim$0.75 V_b^o; for the best GaAs cells, $V_{oc} \approx 0.79E_g$. For a high open-circuit voltage, V_b^o should be as large as possible

[12] In lightly doped Si cells, V_{oc} can exceed V_b^o at the junction, with the bulk going into high injection and an additional voltage across the rear back-surface field: SunPower cells effectively work in this way.

given the semiconductor bandgap, so the work function difference between the two sides of the junction should be as large as possible.

Inspection of Eq. (1.13) shows that V_{oc} increases as the saturation current i_o decreases. Interestingly, i_o has no absolute minimum value. In thin cells with well-passivated surfaces, i_o can be driven down toward zero, and V_{oc} towards its upper limit of V_b^o. In thicker cells in which volume recombination occurs, the lower limit on i_o is determined by the rate of radiative recombination of minority carriers. Usually nonradiative recombination also occurs and this raises i_o by several orders of magnitude, lowering V_{oc} accordingly.

1.8.5 *Cell efficiency*

The maximum-power solar conversion efficiency η_{mp} of a solar cell (often called simply the cell efficiency) is defined as

$$\eta_{mp} = \frac{i_{mp}V_{mp}}{I_o{}^S} = \frac{\eta_{fill}\, i_{sc}V_{oc}}{I_o{}^S} \qquad (1.14)$$

where I_o^S (watts per unit area) is the incident solar irradiance. Since i_{ph} normally increases in direct proportion to I_o^S, while V_{oc} increases as ln i_{ph} (Eq. (1.13)), it follows that η_{mp} should increase logarithmically with irradiance, other factors being equal. This is observed for solar concentrations of up to several hundred Suns for some cell types, though ultimately the series resistance of the cell and the increased operating temperature limits the efficiency increase obtainable by using concentrated sunlight.

Most commercial PV cells have (1 Sun) efficiencies in the range ~11–18%; the best commercial cell is now at least 23% efficient. As Fig. 1.1 shows, the best research cells have higher efficiencies, now up to ~28.8% for a single-junction device under AM1.5G Sun. Establishing the theoretical limits of cell efficiency is of considerable practical importance. Since PV cells are direct conversion devices, they are not subject to the Carnot limits that control the efficiency of heat engines. Nevertheless, there are constraints on PV cell efficiency. The major constraint comes from the poor match between the broadband spectral distribution of sunlight and the single bandgap E_g of a given semiconductor. Solar photons of energy $E < E_g$ are not absorbed in the semiconductor (or if they are, they do not create hole-electron pairs). Photons of energy $E \geq E_g$ can be absorbed and create hole-electron pairs, but their initial 'excess' energy $(E - E_g)$ is very quickly lost by thermalisation, that is, dissipation as heat via carrier–phonon collisions.

The bandgap of the photoactive semiconductor determines the upper bound on both the open-circuit voltage V_{oc} and the short-circuit current i_{sc}. A large-bandgap cell has a larger V_{oc} than a small-bandgap one, but it absorbs fewer solar photons so it has a smaller i_{sc}. The 'detailed-balance' AM1.5G limiting efficiency of an 'ideal' isotropic single-junction cell[13] of optimal bandgap $E_g = 1.34\,\text{eV}$ is 33.8%. In real cells, 'non-ideal' loss mechanisms — for example, nonradiative recombination of carriers in the cell interior or at junction defects or cell surfaces — lower the efficiencies below the detailed-balance limit.

The route to high efficiency in a single-junction cell lies in eliminating these non-ideal losses as far as possible. Higher efficiency can be achieved by stacking two or three single-junction devices on top of each other so that each absorbs the portion of the solar spectrum best suited to its bandgap, and the loss of energy from carrier thermalisation is diminished. This is the approach taken in multijunction cells. More ambitiously, if thermalisation losses could be avoided altogether, very high efficiencies of over 80% could be achieved. Green (2000) proposed a number of 'third generation' device designs, such as hot-carrier and thermophotonic cells, that in principle avoid thermalisation, and instigated a programme to bring these to 'proof-of-concept' level.

1.9 Other junction types

The *p*-*n* homojunction examined in the preceding section is not only the simplest type of PV junction, but also the most common, being that found in crystalline silicon cells, but there are others. Figure 1.9 shows how the conduction and valence band-edge energies E_c and E_v, and the equilibrium Fermi level E_F, vary across the main junction types in solar cells.

Homojunctions (Fig. 1.9a) are *p*-*n* junctions formed by the creation of adjacent *p*- and *n*-doped regions in the same semiconductor, of bandgap E_g. Homojunction cells have the advantage that the junction, which is typically about 0.2 m wide, can be formed in an almost defect-free state within a single crystal (or crystallite of a polycrystalline material). Outside the junction region, there is no electric field to assist carrier collection, so the movement of carriers to the junction must rely on diffusion. Good minority lifetimes are therefore necessary, particularly in an indirect-gap material like silicon. The number of semiconductors that can be satisfactorily doped both *n*- and *p*-type is limited, but they include Si, as well as GaAs and InP.

p-*i*-*n junctions* (Fig. 1.9b) are junctions of extended width ($\sim 0.5\,\mu$m) formed by interposing an intrinsic (i) layer of the undoped semiconductor between *p*- and

[13] An ideal isotropic cell is one in which electrons and holes are thermalised to the band edge, the only decay channel for excited states is radiative recombination, and light can enter the cell at all forward angles.

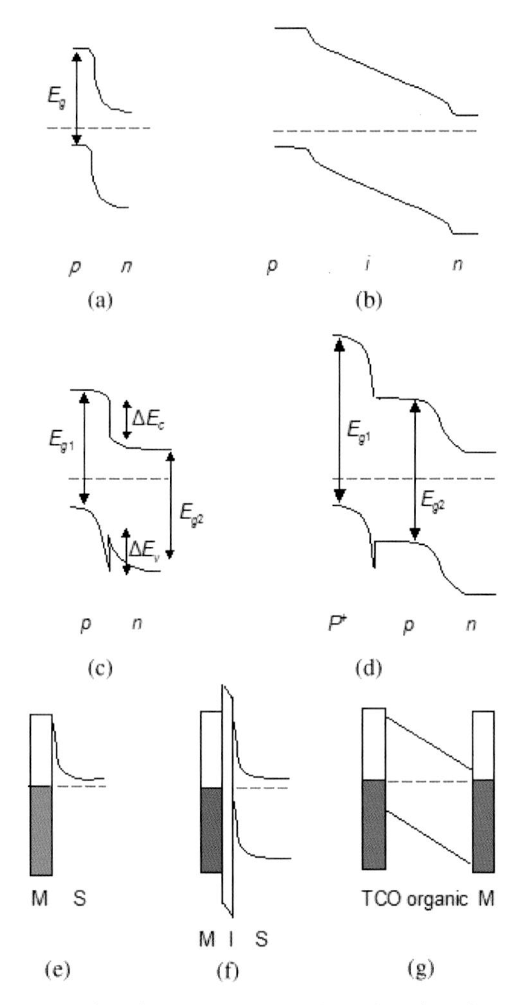

Figure 1.9 Common PV junction types. (a) *p-n* homojunction, formed within a single semiconductor of bandgap E_g; (b) *p-i-n* junction, formed within a single semiconductor of bandgap E_g; (c) anisotype *P-n* heterojunction formed between semiconductors of bandgaps E_{g1} and E_{g2}, showing a valence-band spike ΔE_v and a conduction-band notch ΔE_c; (d) P^+-*p-n* heteroface junction; (e) MS junction between a metal M and an *n*-type semiconductor S; (f) MIS junction with a thin layer of an insulator I interposed between M and S; (g) organic cell containing an organic co-polymer blend between a top transparent conducting oxide (TCO) electrode and a metal electrode M. All these junctions are shown at equilibrium in the dark, so the Fermi levels, shown by the red dashed lines, are the same throughout each junction.

n-layers of the same semiconductor. The i-layer behaves like the dielectric in a capacitor, effectively stretching the electric field of an ordinary p-n junction across itself. The extended electric field throughout the i-layer, which is where most of the light is absorbed, aids in the collection of photogenerated carriers by adding a component of drift (migration in an electric field) to their normal diffusive motion. This is the junction type in amorphous silicon (a-Si:H) cells, and it was adopted because the high density of states and traps in this disordered material causes carrier mobilities to be low.

Heterojunctions (Fig. 1.9c) are junctions formed between two chemically different semiconductors with different bandgaps. The larger bandgap material is often denoted by writing its conductivity type as upper-case N or P, and the smaller bandgap material by a lower-case letter. In the *frontwall* configuration, the top layer is the main light-absorbing layer, and light enters through its front surface. In the *backwall* configuration, the top layer is an optical window material of wide bandgap through which light passes to enter the light-absorbing base layer at the junction. Heterojunctions may be *anisotype*, meaning that the two semiconductors have opposite conductivity types, or *isotype*, meaning that they have the same conductivity type.

The advantage of this junction type is that it allows semiconductors which have good light absorption and carrier lifetime properties, but can only be doped n- or p-type, to be used in solar cells. Its disadvantages are first, that any significant lattice mismatch between the two materials creates numerous junction defects, which diminishes the photovoltage, and second, that the energy band mismatch between the two materials creates notches or spikes in the junction band-edge profile. Figure 1.9c shows a poor junction with a pronounced valence-band spike, which would seriously impede the collection of photogenerated holes from the n-semiconductor. The two most successful heterojunctions for PV applications, n-CdS/p-CdTe and n-CdS/p-CuInSe$_2$, are both N-p anisotype backwall devices with good lattice matches between the two materials, and favourable band alignment with no spike presented to carriers arriving at the junction. *Heteroface junctions* or *buried homojunctions* (Fig. 1.9d) have a p-n homojunction fronted by a highly conducting window layer (*face*) semiconductor of larger bandgap ($E_{g1} > E_{g2}$). This window layer, although not photovoltaically active, is beneficial. As well as acting as current collector from the top layer, it passivates its surface and minimises front-surface recombination. The p^+-Al$_x$Ga$_{1-x}$As/p-GaAs/n-GaAs cell is an important example of this cell type. An unprotected p-n GaAs homojunction would not make an efficient solar cell because the top layer of this direct-gap material must be very thin to allow light to penetrate to the junction region; the dark current would then be undesirably high because of the extent of recombination at the unpassivated front surface.

Surface junctions are formed when a phase of very high carrier density (a metal, degenerate semiconductor or a concentrated electrolyte solution) makes contact with a phase of much lower carrier density, usually a moderately or lightly doped semiconductor. The charge on the highly conducting side of the junction then lies virtually 'in' the interface whereas that on the poorly conducting side is spread out over a space-charge layer as usual. All the band bending and potential drop then occurs on the side of the poor conductor. Photovoltaically active metal–semiconductor (MS) junctions can be made by choosing a metal and semiconductor of such work functions that the semiconductor is in depletion,[14] as shown in Fig. 1.9e. If the semiconductor surface at the MS interface is sufficiently defect-free, a *Schottky barrier* is formed, in which the barrier height is purely determined by the difference in work function of the two materials. If the density of semiconductor surface states is greater than $\sim 10^{17}\,\mathrm{m}^{-2}$, a *Bardeen barrier* is formed in which the surface states contain most of the charge in the semiconductor. Schottky behaviour makes for greater band bending and is preferable.

Surface junctions are encountered in a number of guises. Depositing a thin layer of semiconductor (say, an organic pigment) onto a metal electrode can make an electrode assembly with photovoltaic activity. Semiconductor/electrolyte solution junctions are also surface junctions, in which the electrolyte solution plays the role of the metal in an MS junction. These can perform very well as photoconversion devices, since their dark currents are usually very small and the achievable photovoltages are therefore high. For example, the Grätzel cell (Grätzel, 2000), which contain a dye-sensitised $TiO_2/I_2, I^-$ junction, has an open-circuit voltage of ~ 0.7 V. However, the dark currents of true metal/semiconductor devices (which are determined by thermionic emission from the metal) tend to be large, and the photovoltages correspondingly poor. A better device may be made by interpolating a very thin (<2 nm) layer of an insulator (I) between the metal film and the semiconductor to make an MIS junction (Fig. 1.9f). The I layer impedes thermionic emission and improves the photovoltage without detriment to the photocurrent.

Organic cells (Fig. 1.9g) typically contain one or two thin organic layers sandwiched between a transparent electrode and a metal. Organic materials do not support a space charge. Consequently such cells behave electrically like capacitors, and the band-edge energies drop linearly across the organic layer(s). Photon absorption generates excitons rather than free carriers. These generally dissociate only at an electrode or at the junction between two dissimilar organic layers. Organic

[14]Rather than in accumulation, which would create a photovoltaically inactive ohmic contact.

materials are sufficiently soft that the cell constituents may be co-blended to create a bulk heterojunction cell with interfaces throughout the device. This increases the chance of exciton formation within a diffusion length of the nearest junction. However, even the best OPV cells have low open-circuit voltages for their effective bandgap, arising from their poor radiative efficiencies (i.e. high nonradiative recombination rates), which is due to their blended structure.

1.10 Structure of this book

Martin Green, whose group at the University of New South Wales has been responsible for a string of c-Si efficiency records, discusses the limits to photovoltaic efficiency in Chapter 2 and the current status of crystalline silicon photovoltaics in Chapter 3. Christophe Ballif, Matthieu Despeisse and Franz-Josef Haug provide a full account of amorphous and microcrystalline silicon technology in Chapter 4. Tim Gessert of NREL and thin-film pioneer Dieter Bonnet examine CdTe technology in Chapter 5. Uwe Rau and Hans-Werner Schock cover CIGS technology in Chapter 6. Masafumi Yamaguchi, who has been at the forefront of III–V solar cell technology for over 25 years, discusses GaAs and its use in conjunction with other III–V semiconductors in high-efficiency MJ cells in Chapter 7. Dan Credgington describes the current status of OPV in Chapter 8. Michael Grätzel, the distinguished founding father of DSSC, and James Durrant describe the current status of this technology in Chapter 9. Quantum well solar cells, in which III–V semiconductors are typically used for both the quantum well and the host material, are described by Jenny Nelson and Ned Ekins-Daukes in Chapter 10. The expert father-and-son team of Ignacio Luque-Heredia and Antonio Luque discuss all aspects of CPV in Chapter 11.

The final two chapters of the book are devoted to PV systems and applications: Nicola Pearsall discusses modules, systems and applications in Chapter 12, and Arnulf Jäger-Waldau covers manufacturers and markets in Chapter 13.

References

Adams W. G. and Day R. E. (1877), 'The action of light on selenium', *Proc. Roy. Soc.* **A25**, 113–117.

Ando K., Amano C., Sugiura H., Yamaguchi M. and Salates A. (1987), 'Nonradiative e-h recombination characteristics of mid-gap electron trap in $Al_xGa_{1-x}As$ ($x = 0.4$) grown by molecular beam epitaxy', *Jpn. J. Appl. Phys.* **26**, L266–L269.

Archer M. D., Bolton J. R. and Siklos S. T. C. (1996), A review of analytic solutions for a model *p-n* junction under low-injection conditions,*Solar Energy Mater. Solar Cells* **40**, 133–176.

Bailat J., Fesquet L., Orhan J.-B., Djeridane Y., Wolf B., Madhiger P., Steinhauser J., Benagli S., Borrello D., Castens L., Monteduro G., Marmelo M., Dehbozorgi B., Vallat-Sauvain E., Multone X., Romang D., Boucher J.-F., Meier J., Kroll U., Despeisse M., Bugnon G., Ballif C., Marjanovic S., Kohnke G., Borrelli N., Koch K., Liu J., Modavis R., Thelen D., Vallon S., Zakharian A. and Wiedman D. (2010), 'Recent developments of high-efficiency micromorph tandem solar cells in KAI-M PECVD reactors', *Proc. 5th. World Conf. on Photovoltaic Energy Conversion*, Valencia, Spain, 6–10 September 2010.

Barnham K. W. J. and Duggan G. (1990), 'A new approach to high-efficiency multi-bandgap solar cells', *J. Appl. Phys.* **67**, 3490–3493.

Becquerel A. E. (1839), 'Recherches sur les effets de la radiation chimique de la lumière solaire, au moyen des courants électriques', *Compt. Rend. Acad. Sci.* **9**, 145–149, 561–567.

Benzer S. (1946), The photo-diode and photo-peak characteristics in germanium', *Phys. Rev.* **70**, 105.

Benzer S. (1947), 'Excess-defect germanium contacts', *Phys. Rev.* **72**, 1267–1268.

Bergmann L. and Hansler J. 'Lichtelektrische Untersuchungen an Halbleitern', *Zeits. Physik* **100**, 50– 79.

Bonnet D. and Rabenhorst H. (1972), 'New results on the development of thin film *p*-CdTe/*n*-CdS heterojunction solar cell', *Proc. 9th IEEE PVSC*, Silver Springs, MD (IEEE, 1972), pp. 129–132.

Brittlebank W. (2012), 'Solar PV approaching grid parity', www.climateaction.org, 10 August 2012, retrieved 12 January 2013.

Burschka J., Kessler D. A., Baranoff E., Cevey-Ha N. L., Yi C., Nazeeruddin M. K. and Grätzel M. (2011), 'Tris(2-(1H-pyrazol-1-yl)pyridine)cobalt(III) as *p*-type dopant for organic semiconductors and its application in highly efficient solid-state dye-sensitized solar cells', *J. Amer. Chem. Soc.***133**, 18042–18045.

Burschka J., Pellet N., Moon S.-J., Humphry-Baker R., Gao P., Nazeeruddin M. K. and Grätzel M. (2013), 'Sequential deposition as a route to high performance perovskite-sensitized solar cells', *Nature* **499**, 316–319.

Carlson A. (1956), *Research in Semiconductor Films*, WADC Technical Report, Clevite Corporation.

Carlson D. E. and Wronski C. R. (1976), 'Amorphous silicon solar cells', *Appl. Phys. Lett.* **28**, 671–673.

Case T. W. (1920), '"Thalofide cell" — a new photoelectric substance', *Phys. Rev.* **15**, 289–292.

Chapin D. M., Fuller C. S. and Pearson G. O. (1954), 'A new silicon *p–n* junction photocell for converting solar radiation into electrical power', *J. Appl. Phys.* **25**, 676–677.

China Briefing (2012), 'China releases twelfth five-year plan on solar power development', www.china-briefing.com/news, 19 September 2012, retrieved 7 January 2013.

Crossley P. A., Noel G. T. and Wolf M. (1967), 'Review and evaluation of past solar cell development efforts', Report prepared under contract for NASA. http://ntrs.nasa.gov/archive/nasa/casi.ntrs.nasa.gov/19670022851_1967022851.pdf.

Cusano D. A. (1962), 'Polycrystalline thin-film CdTe solar cells', *IRE Trans. Electron Devices* **ED–9,** 504.

Cusano D. A. (1966), 'The performance of thin film solar cells employing photovoltaic $Cu_{2-x}Te$–CdTe heterojunctions', *Rev. Phys. Appl.* **1**, 195–200.

Davydov B. I. (1938), 'Theory of rectification in the semiconductors', *Bull. Acad. Sci. U.S.S.R.* **5–6**, 625–629.

Davydov B. I. (1939), 'Contact resistance of semiconductors', *Zh. Eksp. Teor. Fiz.* **9**, 451–458.

DECC (2013), *UK Solar PV Strategy Part 1: Roadmap to a Brighter Future*, Department of Energy and Climate Change, October 2013.

Eight19 (2012), 'Eight19 spins out Indigo pay-as-you-go solar', www.eight19.com, 21 August 2012, retrieved 12 January 2013.

EPIA (2013), *Global Market Outlook for Photovoltaics 2013–2017.*

Etgar L., Gao P., Xue Z., Peng Q., Chandiran A. K., Liu B., Nazeeruddin M. K. and Grätzel M. (2012), 'Mesoscopic $CH_3NH_3PbI_3/TiO_2$ heterojunction solar cells', *J. Amer. Chem. Soc.* **134**, 17396–17399.

Fink C. G. and Fogle M. E. (1934),'A study of cuprous oxide solid photoelectric cells', *J. Electrochem. Soc.* **66**, 271–322.

First Solar (2013), 'First Solar sets new world record for CdTe solar cell efficiency', http://investor.firstsolar.com press release 26 February 2013, retrieved 5 April 2013.

Fraunhofer (2013), 'World record solar cell with 44.7% efficiency', www.ise.fraunhofer.de 23 September 2013, retrieved 30 November 2013.

Fraunhofer Institute for Solar Energy Systems ISE (2012), *Photovoltaics Report*, December 11, 2012.

Fritts C. E. (1883), 'On a new format of selenium cell, and some electrical discoveries made by its use', *Am. J. Sci.* **26**, 465–472.

Garrison A. D. (1923), 'The behaviour of cuprous oxide photo-voltaic cells', *J. Phys. Chem.* **27**, 601–622.

Geiger P. H. (1923), 'Spectro-photoelectrical effects in argentite: the production of an electromotive force', *Phys. Rev.* **22**, 461– 469.

Ghosh A. K., Morel D. L., Feng T., Shaw R. F. and Rowe C. A. (1974), 'Photovoltaic and rectification properties of Al/Mg phthalocyanine/Ag Schottky-barrier cells', *J. Appl. Phys.* **45**, 230–236.

Grätzel M. (2000), 'Perspectives for dye-sensitized nanocrystalline solar cells', *Progr. Photovoltaics* **8**, 171–185.

Green M. (1993), 'Silicon solar cells: evolution, high-efficiency design and efficiency enhancements', *Semiconductor Sci. Technol.* **8**, 1–12.

Green M. A. (2000), 'Third generation photovoltaics: advanced structures capable of high efficiency at low cost', *16th. European Photovoltaic Solar Energy Conf.*, Glasgow, 1–5 May 2000.

Green M. A. (2009), 'The path to 25% silicon solar cell efficiency: history of silicon cell evolution', *Progr. Photovoltaics* **17**, 183–189.

Green M. A., Blakers A. W. and Osterwald C. R. (1985), 'Characterization of high-efficiency silicon solar cells', *J. Appl. Phys.* **58**, 4402–4408.

Green M. A., Emery K., Hishikawa Y., Warta W. and Dunlop E. D. (2013), 'Solar cell efficiency tables (version 41)', *Progr. Photovoltaics* **21**, 1–11.

Gremmelmaier R. (1955), 'Gallium-arsenic photoelement', *Z. Naturforsch. A* **19**, 501–502.

Grondahl L. O. (1933), 'The copper-cuprous-oxide rectifier and photoelectric cell', *Rev. Mod. Phys.* **5**, 141–168.
Grondahl L. O. and Geiger P. H. (1927), 'A new electronic rectifier', *J. Am. Inst. Elec. Eng.* **46**, 215–222.
GTM Research (2012), *Thin Film 2012–2016: Technology, Markets and Strategies for Survival.*
GTM Research (2013), 'Yingli gains crown as top producer in a 36 GW global PV market', www.greentechmedia.com 1 May 2013, retrieved 30 November 2013.
Hardin B. E., Snaith H. J. and McGehee M. D. (2012), 'The renaissance of dye-sensitized solar cells', *Nature Photonics* **6**, 162–169.
Haynos J., Allison J., Arndt R. and Meulenberg A. (1974), 'The COMSAT non-reflective silicon solar cell: a second generation improved cell', *Int. Conf. on Photovoltaic Power Generation*, Hamburg, p. 487.
Heliatek (2013), 'Heliatek sets new world record of 10.7% for its organic tandem cell', www.heliatek.com/newscenter 24 July 2012, retrieved 5 April 2013.
Hill R. and Meakin J. D. (1985), 'Cadmium sulphide–copper sulphide solar cells', Ch. 5 in Coutts T. J. and Meakin J. D., eds., *Current Topics in Photovoltaics*, Vol. 1, Academic Press, London.
Hiramoto M., Fujiwarai H. and Yokoyama. M. (1991), 'Three-layered organic solar cell with a photoactive interlayer of co-deposited pigments', *Appl. Phys. Lett.* **58**, 1062–1064.
IMS Research (2012), *The World Market for Concentrated PV*, IMS Research, 26 September 2012.
Jackson W. D. (1955), *Trans. Conf. on the Use of Solar Energy 5*, Tucson: University of Arizona Press, p.122.
Jenny D. A., Loferski J. J. and Rappaport P. O. (1956), 'Photovoltaic effect in GaAs *p-n* junctions and solar energy conversion', *Phys. Rev.* **101**, 1208–1209.
Kagan C. R., Mitzi D. B. and Dimitrakopoulos C. D. (1999), 'Organic-inorganic hybrid materials as semiconducting channels in thin-film field-effect transistors', *Science* **286**, 945–947.
Kazmerski L. L., White F. R. and Morgan G. K. (1976), 'Thin-film $CuInSe_2$/CdS heterojunction solar cells', *Appl. Phys. Lett.* **29**, 268–270.
Kearns D. and Calvin M. (1958), 'Photovoltaic effect and photoconductivity in laminated organic systems', *J. Chem. Phys.* **29**, 950–951.
Kennard E. H. and Dieterich E. O. (1917), 'An effect of light upon the contact potential of selenium and cuprous oxide', *Phys. Rev.* **9**, 58–63.
King R. R., Bhusari D., Larrabee D., Liu X.-Q., Rehder E., Edmondson K., Cotal H., Jones R. K., Ermer J. H., Fetzer C. M., Law D. C. and Karam N. H. (2012), 'Solar cell generations over 40% efficiency', *Progr. Photovoltaics* **20**, 801–815.
King R. R., Law D. C., Edmondson K. M., Fetzer C. M., Kinsey G. S., Yoon H., Sherif R. A. and Karam N. H. (2007), '40% efficient metamorphic GaInP/GaInAs/Ge multijunction solar cells', *Appl. Phys. Lett.* **90**, 183516–183519.
Kojima A., Teshima K., Shirai Y. and Miyasaka T. (2009), 'Organometal halide perovskites as visible-light sensitizers for photovoltaic cells', *J. Amer. Chem. Soc.* **131**, 6050–6051.
Lange B. (1930), 'Uber eine Neue Art von Photozellen', *Phys. Zeits.* **31**, 964–969.
Lange B. (1938), *Photoelements and Their Applications*, Reinhold, New York.

Lee M. M., Teuscher J., Miyasaka T., Murakami T. N. and Snaith H. J. (2012), 'Efficient hybrid solar cells based on meso-superstructured organometal halide perovskites', *Science* **338**, 643–647.
Lindmeyer J. and Alison J. F. (1973), 'The violet cell: an improved silicon solar cell', *COMSAT Technical Review* **3**, 1–22.
Liu M., Johnston M. B. And Snaith H. J. (2013), 'Efficient planar heterojunction perovskite solar cells by vapour deposition', *Nature* **501**, 395–398.
Lux Research (2012), *Putting High-Concentrating Photovoltaics into Focus*, June 2012.
Mahdi W. (2012), 'Qatar to tender 200 megawatt of solar power projects in 2013', www.bloomberg.com, 3 December 2012, retrieved 3 January 2013.
Meier J., Dubail S., Flückiger R., Fischer D., Keppner H. and Shah A. (1994), 'Intrinsic microcrystalline silicon (mc-Si:H) — a promising new thin film solar cell material', *Proc. 1st. World Conf. on Photovoltaic Energy Conversion*, Hawaii, pp. 409–412.
Mishra A., Fischer M. K. R. and Bauerle P. (2009), 'Metal-free organic dyes for dye-sensitized solar cells: from structure–property relationships to design rules', *Angew. Chem. Int. Ed.* **48**, 2474–2499.
MIT (2011), 'Why BP Solar failed', www.technologyreview.com 21 December 2011, retrieved 26 January 2014.
Mitchell K., Fahrenbruch A. L. and Bube R. H. (1977), 'Evaluation of the CdS/CdTe heterojunction solar cell', *J. Appl. Phys.* **48**, 4365–4371.
Mitzi D. B., Feild C. A., Harrison W. T. A. and Guloy A. M. (1994), 'Conducting tin halides with a layered organic-based perovskite structure', *Nature* **369**, 467–469.
Mott N. F. (1939), 'Copper–cuprous oxide photocells', *Proc. Roy. Soc.* **A171**, 281–285.
Ohl R. S. (1941), 'Light-sensitive electric device', U.S. Patent No. 2,402,662; 'Electrical translating device utilizing silicon', U.S. Patent No. 2,402,839; 'Light-sensitive device including silicon', U.S. Patent No. 2,443,542.
Olson J. M., Kurtz S. R., Kibbler A. E. and Faine P. (1990), 'A 27.3% efficient $Ga_{0.5}In_{0.5}P$/GaAs tandem solar cell', *Appl. Phys. Lett.* **56**, 623–625.
O'Regan B. and Grätzel M. (1991), 'A low-cost, high-efficiency solar cell based on dye-sensitized colloidal TiO_2 films', *Nature* **353**, 737–740.
Park N.-G. (2013), 'Organometal perovskite light absorbers toward a 20% efficiency low-cost solid-state mesoscopic solar cell', *J. Phys. Chem. Lett.* **4**, 2423–2429.
Pochettino A. (1906), 'Sul comportamento foto-elettrico dell' antracene', *Acad. Lincei Rendiconti* **15**, 355–363.
PV Magazine (2013), 'Solar stocks: China left for dead; FSLR, SPWR and WFR to rise', www.pv-magazine.com, 2 April 2013, retrieved 6 April 2013.
PV Tech (2012), 'Around 90% of Chinese polysilicon producers stop production', www.pv-tech.org, 13 December 2012, retrieved 8 January 2013.
Reynolds D. C., Leies G., Antes L. L. and Marburger R. E. (1954), 'Photovoltaic effect in cadmium sulfide', *Phys. Rev.* **96**, 533–534.
Saraie J., Akiyama M. and Tanaka T. (1972), 'Epitaxial growth of cadmium telluride by a closed-space technique', *Jpn. J. Appl. Phys.* **11**, 1758–1759.
Schottky W. (1930), 'Über den Entstehungsort der Photoelektronen in Kupfer–Kupferoxydul–Photozellen', *Zeit. Tech. Phys.* **11**, 458–461.
Schottky W. (1938), 'Halbleitertheorie der Sperrschicht', *Naturwiss.* **26**, 843.

Semiconductor Today (2011), 'First Solar raises CdTe PV cell efficiency record from 16.7% to 17.3%', www.semiconductor-today.com, 26 July 2011, retrieved 12 January 2013.
Shay J. L., Wagner S. and Kasper H. M. (1975), 'Efficient $CuInSe_2$/CdS solar cells', *Appl. Phys. Lett.* **27**, 89–90.
Sheldon H. H. and Geiger P. H. (1922), 'The production of an E.M.F. on closed circuit, by a light effect on argentite', *Phys. Rev.* **19**, 389– 390.
Siemens W. (1885), 'On the electromotive action of illuminated selenium discovered by Mr. Fritts, of New York', *Van Nostrand's Engineering Magazine* **32**, 392n.
Smith W. (1873), 'The action of light on selenium', *J. Soc. Telegraph Engineers* **2**, 31–33.
Söderström K., Bugnon G., Biron R., Pahud C., Meillaud F., Haug F.-J. and Ballif C. (2012), 'Thin-film silicon triple-junction solar cell with 12.5% stable efficiency on innovative flat light-scattering substrate', *J. Appl. Phys.* **112**, 114503–114504.
Solar Frontier (2013), 'Solar Frontier achieves record 19.7% CIS efficiency', www.solar-frontier.com, 8 January 2013, retrieved 11 January 2013.
Sosnowski L., Starkiewicz J. and Simpson O. (1947), 'Lead sulfide photoconductive cells', *Nature* **159**, 818–819.
Spanggaard H. and Krebs F. C. (2004), 'A brief history of the development of organic and polymeric photovoltaics', *Solar Energy Mater. Solar Cells* **83**, 125–146.
Spear W. E. and Le Comber P. G. (1975), 'Substitutional doping of amorphous silicon', *Solid State Commun.* **17**, 1193–1196.
Stranks S. D., Peron G. E., Grancini G., Menelaou C., Alcocer M. J. P., Leijtens T., Herz L. M., Petrozza A. and Snaith H. J.(2013), 'Electron-hole diffusion lengths exceeding 1 micrometer in an organometal trihalide perovskite absorber', *Science* **342**, 341–344.
Takamoto T., Ikeda E., Kurita H. and Ohmori M. (1997), 'Over 30% efficient InGaP/GaAs tandem solar cells', *Appl. Phys. Lett.* **70**, 381–383.
Tang C. W. and Albrecht A. C. (1975a), 'Photovoltaic effects of metal–chlorophyll-a–metal sandwich cells', *J. Chem. Phys.* **62**, 2139–2149.
Tang C. W. and Albrecht A. C. (1975b), 'Transient photovoltaic effects in metal–chlorophyll-a–metal sandwich cells', *J. Chem. Phys.* **63**, 953–961.
Treble F. (1998), 'Milestones in the development of crystalline silicon solar cells', *Renewable Energy* **15**, 473–478.
Wagner S., Shay J. L., Migliorato P. and Kasper H. M. (1974), '$CuInSe_2$/CdS heterojunction photovoltaic detectors', *Appl. Phys. Lett.* **25**, 434–435.
Welker H. (1954), 'Semiconducting intermetallic compounds', *Physica* **20**, 893–909.
Winther C. (1928), 'Über den Becquereleffekt. I.', *Zeits. Physik. Chem.* **131**, 205– 213.
Wolf M. (1960), 'Limitations and possibilities for improvement of photovoltaic energy converters. Part I: considerations for Earth's surface operation', *Proc. Inst. Radio Eng.* **48**, 1246–1263.
Woodall J. M. and Hovel H. J. (1977), 'Isothermal etchback-regrowth method for high-efficiency Ga_{1-x}/Al_xAs–GaAs solar cells', *Appl. Phys. Lett.* **30**, 492–493.
Yamaguchi K., Nakayama N., Matsumoto H. and Ikegami S. (1977), 'Cadmium sulphide–cadmium telluride solar cell prepared by vapor phase epitaxy', *Jpn. J. Appl. Phys.* **16**, 1203–1211.

Yamaguchi M., Amano C., Sugiura H. and Yamamoto A. (1987), 'High efficiency AlGaAs/GaAs tandem solar cells', *Proc. 19th IEEE Photovoltaic Specialists Conf.*, IEEE, New York, pp. 1484–1485.

Yamaguchi M., Okuda T., Taylor S. J., Takamoto T., Ikeda E. and Kurita H. (1997), 'Superior radiation-resistant properties of InGaP/GaAs tandem solar cells', *Appl. Phys. Lett.* **70**, 1566–1568.

Yamaguchi M., Takamoto T. and Araki K. (2008), 'Present and future of super high efficiency multi-junction solar cells', *Proc. SPIE* 6889, Physics and Simulation of Optoelectronic Devices XVI, 688906.

Yang G. (2012), 'Should China bail out its solar PV industry?', www.chinadialogue.net, 12 September 2012, retrieved 4 January 2013.

Yella A., Lee H.-W., Tsao H. N., Yi C., Chandiran A. K., Nazeeruddin M. K., Diau E. W.-G., Yeh C.-Y., Zakeeruddin S. M. and Grätzel M., 'Porphyrin-sensitized solar cells with cobalt (II/III)–based redox electrolyte exceed 12% efficiency', *Science* **334**, 629–634.

Yu G., Pakbaz K. and Heeger A. J. (1994), 'Semiconducting polymer diodes: large size, low cost photodetectors with excellent visible-ultraviolet sensitivity', *Appl. Phys. Lett.* **64**, 3422–3424.

Zhao J., Wang A., Green M. A. and Ferrazza F. (1998), 'Novel 19.8% efficient "honeycomb" textured multicrystalline and 24.4% monocrystalline silicon solar cells', *Appl. Phys. Lett.* **73**, 1991–1993.

ZSW (2013), 'ZSW produces world record thin-film solar PV cell, achieves 20.8% efficiency and overtakes multi-crystalline silicon technology', www.solarserver.com 24 October 2013, retrieved 30 November 2013.

CHAPTER 2

LIMITS TO PHOTOVOLTAIC ENERGY CONVERSION EFFICIENCY

MARTIN A. GREEN
Australian Centre for Advanced Photovoltaics
School of Photovoltaic and Renewable Energy Engineering
University of New South Wales
Sydney, N.S.W. Australia, 2052
m.green@unsw.edu.au

There can be economy only where there is efficiency.
Benjamin Disraeli, 1868.

2.1 Introduction

Although the sunlight conversion efficiencies of most photovoltaic systems installed to date have been quite modest, generally in the 10–15% range, there is no fundamental reason why these cannot be very much higher. Historically, each successful photovoltaic technology has evolved to ever-increasing energy conversion efficiency, while simultaneously reducing cost. It seems reasonable to expect this trend to continue as the industry grows, with conversion efficiency increasingly becoming a key differentiator between technologies, as already apparent over the last decade with the thin-film technologies. This makes it relevant to understand fundamental constraints on efficiency and how these may be circumvented.

The publishing of the quantum mechanical theory of semiconductors (Wilson, 1931) coincided with a surge of interest in cuprous oxide solar cells (Grondahl, 1933), stimulating rapid progress in theoretical understanding of the photovoltaic effect. The serendipitous discovery of the silicon *p-n* junction at Bell Laboratories in the early 1940s due to its large photovoltages (Ohl, 1941; Riordan and Hoddeson, 1997) led to the development of *p-n* junction theory (Shockley, 1949) and its extension to illuminated junctions (Cummerow, 1954a). The reporting of greatly improved silicon cells in 1954 (Chapin *et al.*, 1954) was followed by an estimation of the limiting efficiency (Cummerow, 1954b) and investigations of the optimum bandgap (Rittner, 1954; Trivich and Flinn, 1955; Loferski, 1956).

Apart from the Trivich–Flinn analysis, subsequently reviewed, these calculations were empirical in that Shockley's *p-n* junction theory was combined with assessments of ultimately achievable values of the material parameters appearing in the *p-n* junction equation. A significant step forward was made in 1961 (Shockley and Queisser, 1961) when a materials-independent theory of limiting cell performance was developed. This was based on the recognition that recombination in devices of limiting performance would be entirely radiative, the inverse of the optical absorption processes creating the cell photocurrent. Provided other intrinsic effects do not prevent this radiative limit from being reached, all cells of the same bandgap have the same limiting efficiency, with the perfection of the associated cell and materials technology acting as a practical differentiator.

Even prior to this work, it was recognised that cell performance could be improved by using multiple cells of different bandgap (Jackson, 1955). The Shockley–Queisser approach was extended to these multiple cell systems (Henry, 1980; Marti and Araujo, 1996), then successively to thermophotovoltaics (Würfel and Wolfgang, 1980), hot-carrier cells (Ross and Nozik, 1982; Würfel, 1995; Würfel *et al.*, 2005), multiple electron–hole pair (and multiple-exciton) approaches (Werner *et al.*, 1995), intermediate-band (Luque and Marti, 1997) and impurity level (Brown and Green, 2002) devices, and then to up- and down-converters (Trupke *et al.*, 2002a; 2002b). Such 'bottom-up' analyses allow material-independent ranking of these different approaches to circumventing the limits upon conventional cells (Green, 2003) such as shown in Fig. 2.1. The most important of

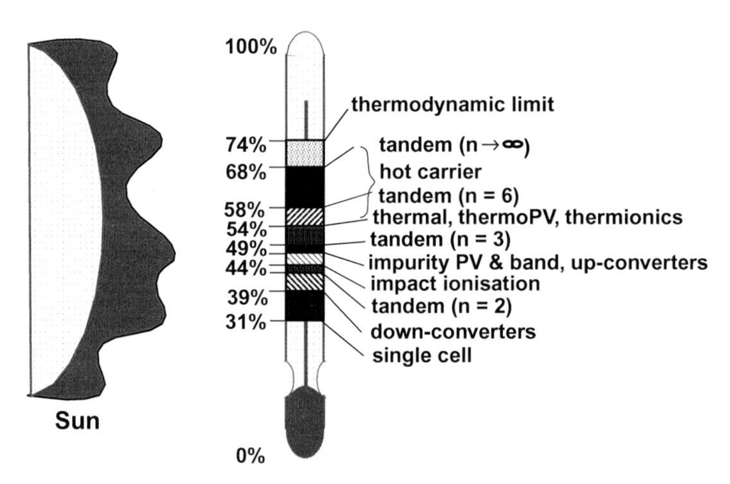

Figure 2.1 Limiting efficiency of various photovoltaic conversion options under global sunlight (6000 K Sun and 300 K ambient assumed; n = number of junctions). Source: Green and Ho (2011).

such approaches remain the single junction and the monolithic stack of multiple-junction tandem devices that provide the main focus of this chapter and of the present volume.

Complementing these 'bottom-up' limiting efficiency calculations have been 'top-down' thermodynamic analyses that are both device- and system-agnostic, based purely on the energy and entropy fluxes associated with the incident sunlight (Landsberg and Tonge, 1980). The corresponding limit for the 'top-down' approach is also shown in Fig. 2.1 and lies well above the highest efficiency from the 'bottom-up' approaches. This was once thought to be because the 'top-down' limit was unattainable, even in principle, since the analysis failed to include unavoidable losses (Pauwels and De Vos, 1981; Marti and Araujo, 1996). However, from the work of Ries (1983), it becomes clear that the difference arises from the implicit assumption of time-symmetry in the 'bottom-up' approaches. Photovoltaic systems that are capable of the thermodynamically limiting performance are possible, in principle, although the only suggested implementation, subsequently discussed, is very complex.

Resonating with the remark by Disraeli opening this chapter, the author has elsewhere emphasised the likely long-term evolution of photovoltaics to highest possible energy conversion efficiencies, as a path to lowest possible costs (Green, 2003). Figure 2.2a effectively demonstrates the enormous leverage of high efficiency in reducing the cost/watt of solar modules. At the system level, when area-related installation costs are included, the leverage increases further, as illustrated in Fig. 2.2b. This highlights the relevance of a clear understanding of both

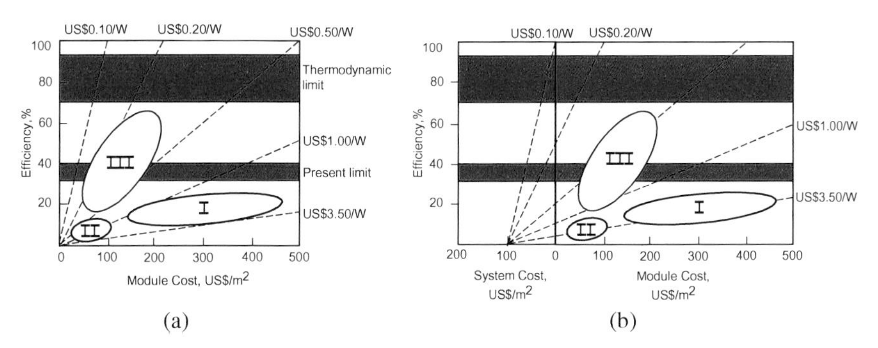

Figure 2.2 (a) Efficiency-cost trade-off for three generations of photovoltaic technology, based on (I) silicon wafers, (II) 'second-generation' thin-films and (III) advanced, high-efficiency, thin-film 'third generation' approaches; (b) Additional leverage provided by high efficiency when additional area-related costs are included. Efficiency limits are shown as bands to accommodate the increased efficiencies possible if sunlight is concentrated (or the acceptance angle of the cell reduced). Source: Green and Ho (2011).

the efficiency limits that apply to the various photovoltaic technologies and also of the improvements required to progress towards them.

2.2 Photovoltaic converters: essential requirements

The development of respectably efficient dye-sensitised and organic solar cells has broadened the conception of what constitutes a photovoltaic converter. No longer are solar cells limited to the heteroface, heterojunction and homojunction *p-n* and *p-i-n* junction devices that previously dominated the field.

Key general features required for a solar photovoltaic converter are a set of ground and excited states, with electrons photoexcited from the ground state able to be replenished from one contact and those in excited states able to be extracted by a second contact. As a practical matter, there needs to be selectivity in this contacting process so that the contact to the ground state contact does not contact excited states to any significant extent and *vice versa* (Fig. 2.3). With such contacts, current can be generated by the device on short-circuit and voltage on open-circuit.

However, for an efficient converter, another feature is required. The voltage able to be generated by a converter depends on its degree of excitation as measured by the different chemical potentials of electrons in the ground and excited states. However, electrostatics additionally dictate that the electric potential across the device must change by a value equal to the voltage between the contacts. Moreover, this potential change is in a direction that opposes the collection of photogenerated

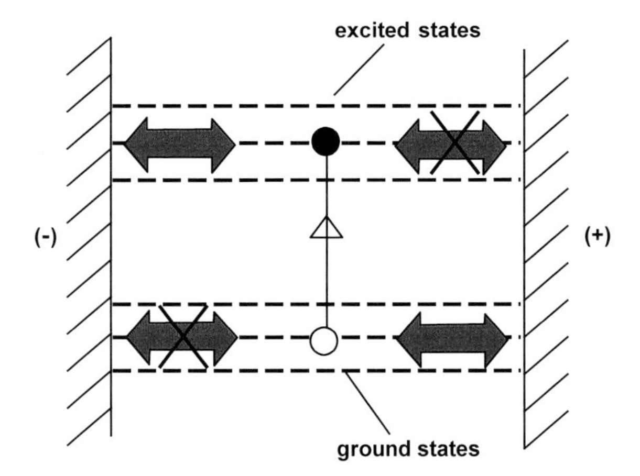

Figure 2.3 Essential requirements for photovoltaics, selectively contacted ground and excited states. Source: Green and Hansen (2002).

carriers. What is also needed in an efficient solar cell is a way of taking up this voltage change without affecting carrier collection.

For *p*-*n* junction devices, this mechanism is provided by the 'built-in' voltage at the junction, which has a value at thermal equilibrium that depends on the product of the doping concentration on either side of the junction. This junction potential drop reduces under light exposure, providing the electrostatic adjustment needed to accommodate the generation of photovoltage. If an attempt is made to exceed the built-in voltage, the consequences depend on the conditions at the contacts. If the doping is higher here than near the junction region, as in the mainstream silicon wafer commercial product, additional adjustment can occur such as at the potential step associated with the rear back surface field (BSF). The doping of a *p*-*n* junction also provides the selective contacting requirement, with contact made selectively to the conduction band on the *n*-type side of the device and to the valence band on the *p*-type side of the device. The selectivity is provided by the different numbers of conduction-band electrons and valence-band holes at the two contacts (Fig. 2.4).

For thin *p*-*i*-*n* structures, such as amorphous silicon devices, a similar built-in voltage is present, corresponding to a nearly uniform field across the *i*-region, with this field aiding carrier collection. This field's reduction on illumination, as photovoltage increases, reduces the effectiveness of carrier collection, resulting in reduced fill factors. Doping again provides the contacting selectivity.

For dye-sensitised cells, a barrier layer at the TiO_2/front contact interface provides a built-in potential reservoir, with an insufficient value of this potential producing low fill factors (Kron *et al.*, 2003; Snaith and Grätzel, 2006). Contact to the excited states is provided by the *n*-type TiO_2, while contact to the ground

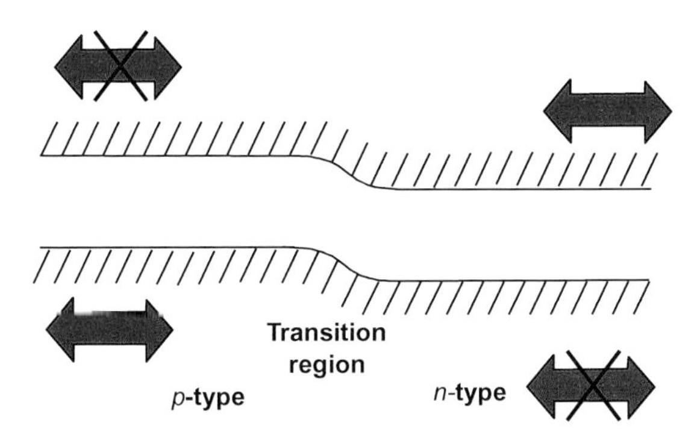

Figure 2.4 Essentials of *p*-*n* junction solar cell operation. Doping provides selective contact to the appropriate band by controlling occupancy in the contact region. Source: Green and Hansen (2002).

states in this case is provided by a completely different process, charge transport across an ionic liquid.

For both bilayer and bulk heterojunction organic solar cells, the work function difference between the contacts provides the voltage reservoir, giving rise to a built-in field at thermal equilibrium (Blom *et al.*, 2007), similar to that in a p-i-n junction device, with similar consequences for device operation. Photogenerated carrier collection reduces as the field reduces under operating conditions, giving a strong dependence of open-circuit voltage, short-circuit current and fill factor on the original built-in field and hence the contact work function difference (Ramsdale *et al.*, 2002; Mihailetchi *et al.*, 2004). Contact selectivity, particularly important for bulk heterojunction devices since both donor and acceptor materials extend to both contacts, is achieved by better alignment of the appropriate contact Fermi level to either the ground or excited states, increasing the occupancy of states in the vicinity for the desired carrier. A 'hole-transporting' PEDOT:PSS layer, and often a LiF or TiO_2 layer at the excited-state contacts (Lee *et al.*, 2008) assist in enforcing this selectivity.

2.3 Thermodynamic properties of sunlight

Black bodies feature predominantly in the theory of the limiting efficiency of solar cells. Not only does emission from the Sun approximate that from a black-body at 6000 K but an ideal solar cell, as a perfect absorber of sunlight at the energies of interest, has properties that are related to those of a black body at the cell temperature (around 300 K).

Figure 2.5a compares the actual spectrum of sunlight as measured outside the Earth's atmosphere (Air Mass Zero or AM0 radiation) with standard terrestrial reference spectra used in solar cell measurements (ASTM G170-03 global and direct normal AM1.5 spectra, hereafter referred to as AM1.5G and AM1.5D).

The spectral energy content of black-body radiation was deduced by Planck (1913) in work that changed the course of twentieth century physics. The energy flux in black-body radiation between photon energies E_1 and E_2 is given by

$$\dot{E}(E_1, E_2) = \frac{2\Lambda}{h^3c^2}\int_{E_1}^{E_2} \frac{E^3 dE}{e^{E/kT} - 1} \tag{2.1}$$

where h is Planck's constant, c is the velocity of light and Λ is the *etendue*, a measure of the areal and angular spread of the radiation, given by $\Lambda = \mathrm{A}\pi \sin^2\theta$ where A is the area of the emitter or receiver and θ is the half-angle describing the divergence of the emitted or received beam. For direct sunlight outside the Earth's atmosphere, $\Lambda_d = \mathrm{A}\pi/46200$ (Green, 2003). For emission into a hemisphere,

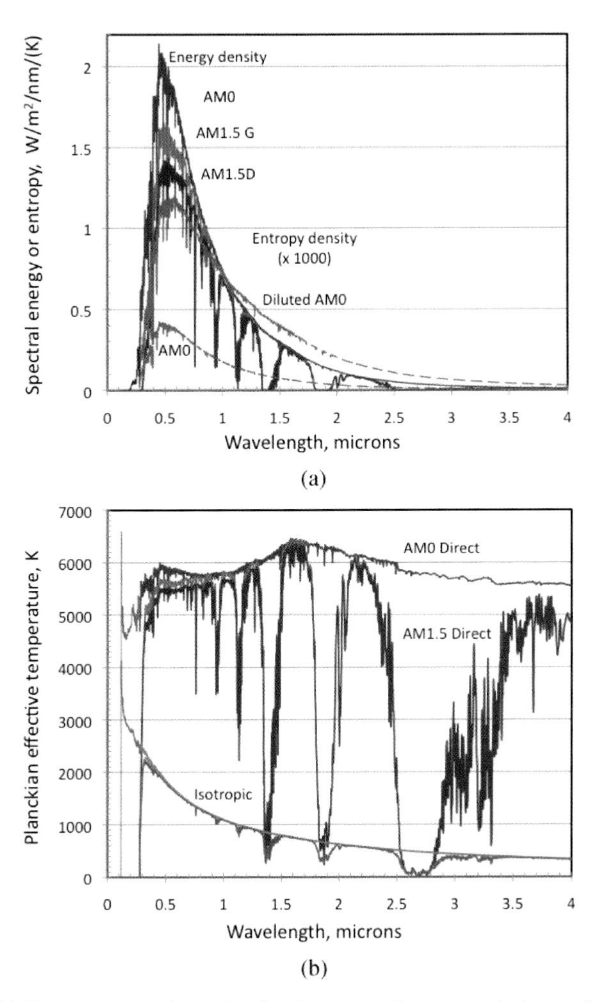

Figure 2.5 (a) Energy spectral density for the presently accepted photovoltaic reference spectra (entropy spectral density is also shown in the lower section of the plot on an enlarged scale as dashed lines for the direct and diluted AM0 spectrum); (b) Planckian equivalent temperature versus wavelength for the three spectra. Source: Green and Ho (2011).

$\Lambda_g = A\pi$. In the latter case, integrating Eq. (2.1) from 0 to ∞ gives the result σT^4 per unit area where σ, the Stefan–Boltzmann constant, equals $2\pi^5 k^4/(15h^3c^3)$. A similar expression to Eq. (2.1) applies to the photon flux $\dot{N}(E_1, E_2)$ but with E^3 in the numerator replaced by E^2.

For radiation that differs from the black-body spectrum, the same expressions can still be used by defining a 'Planckian' temperature $T(E)$, a function of energy, which gives the correct magnitude for the spectral component at each energy. Figure 2.5b (upper curves) shows the calculated Planckian temperatures associated with the AM0 and AM1.5 spectra. The variation in this temperature for the AM0 spectrum is largely due to different absorption strengths in the Sun's photosphere for different wavelengths, resulting in the sampling of temperatures at different depths. For the AM1.5 spectra, there is the additional impact of atmospheric absorption bands, notably the very strong water vapour absorption bands centred at about 0.95, 1.1, 1.4, 1.9 and 2.7 μm, although the latter is at too long a wavelength to be particularly significant for solar energy conversion.

The above Planckian effective temperatures have thermodynamic significance. If the photon occupancy number is defined as

$$n_{\mathrm{ph}} = 1/(\mathrm{e}^{E/kT} - 1) \tag{2.2}$$

where T is the effective Planckian temperature, the entropy flux corresponding to the previous energy flux becomes

$$\dot{S}(E_1, E_2) = \frac{2\Lambda k}{h^3 c^2} \int_{E_1}^{E_2} E^2[(1 + n_{\mathrm{ph}}) \ln(1 + n_{\mathrm{ph}}) - n_{\mathrm{ph}} \ln(n_{\mathrm{ph}})] dE \tag{2.3}$$

The entropy flux calculated in this way for the AM0 spectrum is also included in Fig. 2.5a. If the Planckian temperature is constant for all wavelengths, a property possessed solely by black-body radiation, integration gives the simple result for the entropy flux

$$\dot{S}(0, \infty) = (4/3)\dot{E}(0, \infty)/T = (4/3)\sigma T^3$$

The effective temperatures shown uppermost in Fig. 2.5b are appropriate to systems that respond only to sunlight coming directly from the Sun's disc. The thermodynamic limits calculated using these temperatures apply only to Sun-tracking, direct-light converters. Most photovoltaic devices to date respond to a much wider range of angles and convert not only direct sunlight but light from a range of angles incident on their surface (Fig. 2.6).

In such cases, where the converter takes no advantage of the directionality of sunlight, the thermodynamic limit on efficiency will be the same whether the sunlight is considered directional or spread uniformly over the entire hemisphere. To spread light from the Sun's disc over the whole hemisphere and maintain the same intensity on the cell, the intensity of sunlight would have to be reduced at each photon energy by a factor of 46,200 (equal to the square of the ratio of the Sun–Earth distance to the Sun's radius). Sunlight is effectively diluted by a dilution factor ξ of 1/46,200 at each energy, giving the smaller effective Planckian temperatures shown by the lower curves in Fig. 2.5b (note that $\xi = \Lambda_{\mathrm{d}}/\mathrm{A}$, a relationship used

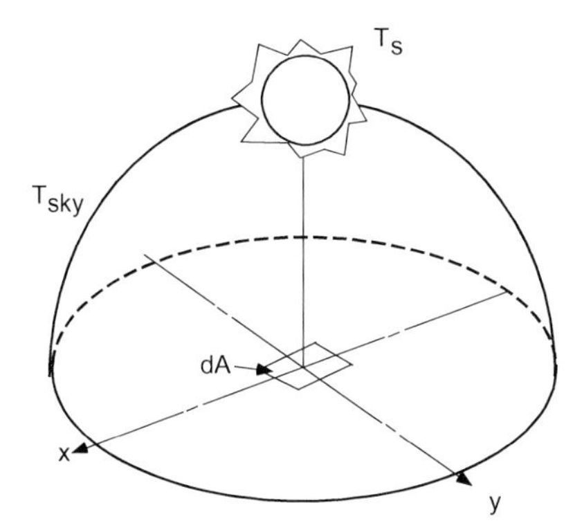

Figure 2.6 Illustration of radiation environment seen by an element on the cell's surface. As well as the direct sunlight at temperature T_S, the cell is also exposed to sky radiation at a temperature T_{sky} assumed to equal the ambient temperature T_A. Source: Green and Hansen (2002).

extensively in later sections). Using these modified Planckian temperatures, the new etendue, Λ_g, would equal $A\pi$.

An important question when dealing with devices responding to light from all directions is how to treat light from directions in the sky other than those occupied by the Sun (Fig. 2.6). The most convenient treatment is to assume a sky temperature equal to the ambient temperature, with dilution factor $(1 - \xi)$. Including the corresponding background thermal fluxes from the sky ensures the Planckian effective temperature of the incident radiation remains above the ambient temperature, even at long wavelengths where this would not otherwise be the case (Fig. 2.5b).

2.4 'Top-down' thermodynamic efficiency limits

If the Sun is regarded as black-body source at 6000 K and the Earth as a sink at 300 K, the Carnot efficiency on converting sunlight into electricity (or other useful work) is 95%, as given by $(1 - T_{\text{sink}}/T_{\text{source}})$. However, this limit applies only when an infinitesimal amount of work is done by the converter, with most of the incident energy recycled back to the Sun. In principle, this Carnot efficiency can be attained by a photovoltaic system (an infinite tandem cell stack under maximal possible sunlight concentration operating close to open circuit), but the infinitesimally small power outputs involved are of no practical interest.

The problem with the Carnot calculation in this context is that, in reality, there is no rebate for energy recycled back to the Sun. This recycled energy should therefore be treated as a loss with attention restricted to the more confined system

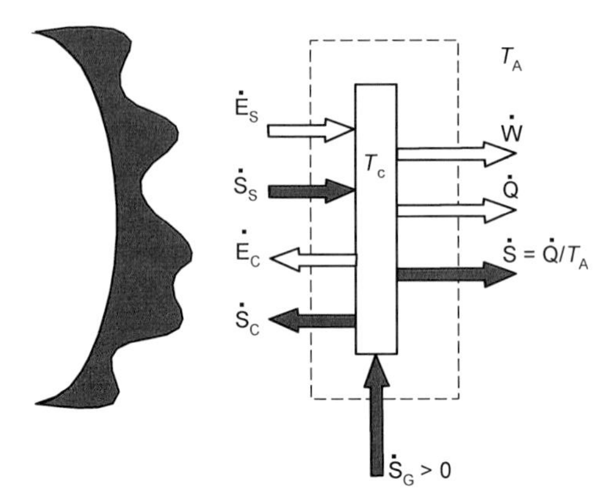

Figure 2.7 System for calculating the Landsberg efficiency limit. Source: Green and Hansen (2002).

of Fig. 2.7. An energy and entropy balance gives (Green, 2003)

$$\eta = \dot{W}/\dot{E}_S = (1 - T_A\dot{S}_S/\dot{E}_S) - (1 - T_A\dot{S}_C/\dot{E}_C)\dot{E}_C/\dot{E}_S T_A\dot{S}_G/\dot{E}_S \quad (2.4)$$

where η is the energy conversion efficiency, $\dot{W}$ is the useful work (electricity) and the $\dot{S}$ and $\dot{E}$ terms correspond to the associated entropy and energy fluxes. Noting $T_C \geq T_A$, and assuming the cell is a perfect absorber over all wavelengths emitted by the Sun (both black bodies), the maximum value of this efficiency occurs when $\dot{S}_G = 0$ and $T_C = T_A$, giving

$$\eta_L = 1 - \frac{4}{3}\frac{T_A}{T_S} + \frac{1}{3}\frac{T_A^4}{T_S^4} \quad (2.5)$$

This limit is known as the Landsberg limit (Landsberg and Tonge, 1980). For a 6000 K Sun and a 300 K ambient, the resulting efficiency limit is 93.3%. While it is generally accepted that this represents an upper bound on the efficiency of solar energy conversion, some have thought this bound cannot be reached in principle, as previously noted, due to unavoidable entropy generation during light absorption (making the $\dot{S}_G = 0$ condition fundamentally unattainable). However, the arguments involved apply only to time-symmetric systems. It has been shown that, in principle, the Landsberg limit can be approached arbitrarily closely by systems that are time-asymmetric (Ries, 1983).

A subtlety is that this limit applies only to systems involving a direct exchange with the Sun. This can be achieved by concentrating optics if sunlight is concentrated to the maximum extent possible (46,200 times). During such concentration, etendue ideally is conserved. The consequence is that, under maximal concentration, the solar intensity at the Sun's surface is reproduced terrestrially, but over an area 46,200 times smaller than the system aperture. Such high concentration levels have been approached experimentally (Cooke *et al.*, 1990). An alternative, but to date more difficult approach giving the same result, in principle, is to restrict the angular absorption, and hence emission, of the cell to the small range of angles required to accept only direct sunlight (same etendue Λ_d for incoming and emitted radiation).

Efficiencies calculated under the above conditions are generally referred to as efficiencies calculated under 'maximal concentration' although, as noted above, light need not be concentrated, at least in principle, to attain them.

As already mentioned, most photovoltaic panels have an approximately isotropic response, responding reasonably well to light regardless of the angle of incidence. This gives the very practical advantages of being able to respond to diffuse sunlight scattered by the atmosphere and also of not having to track the Sun as it moves across the sky.

An appropriate limit in this case, where the converter makes no attempt to exploit the directionality of sunlight, is the 'isotropic limit' that would apply to a converter able to respond equally well to light from any part of the sky. As before, the Sun could be considered as completely filling the sky but with its intensity from each direction reduced to give the same overall flux on the device (the reduction would be the same for all wavelengths, such as obtained by passing through a neutral-density filter). The thermodynamics of such 'diluted black-body radiation' have been examined elsewhere (Landsberg and Tonge, 1979), where it is shown the entropy-to-energy flux ratio increases by a factor χ that depends on the extent of the dilution. Imaging the rest of the sky acting as a black-body source at temperature T_A, the general form of the Landsberg limit, presented in the form below for the first time, becomes

$$\eta_L = 1 - \frac{4}{3}\chi\frac{T_A}{T_S} + \left(\frac{4}{3}\chi - 1\right)\frac{T_A^4}{T_S^4} \tag{2.6}$$

where an accurate approximate expression for χ is (Green, 2011)

$$\chi = 0.9652 - 0.2777\ln(\xi) + 0.0348\,\xi^{0.9} \tag{2.7}$$

For the present case of dilution by 46,200 times, χ has a value of 3.94, giving a limiting Landsberg efficiency for isotropic converters of 73.7% for a 6000 K Sun and a 300 K ambient, the value shown in Fig. 2.1.

Figure 2.8 shows a photovoltaic system capable in principle of reaching such limits (Green, 2003). Each converter in the system consists of a stack of a large

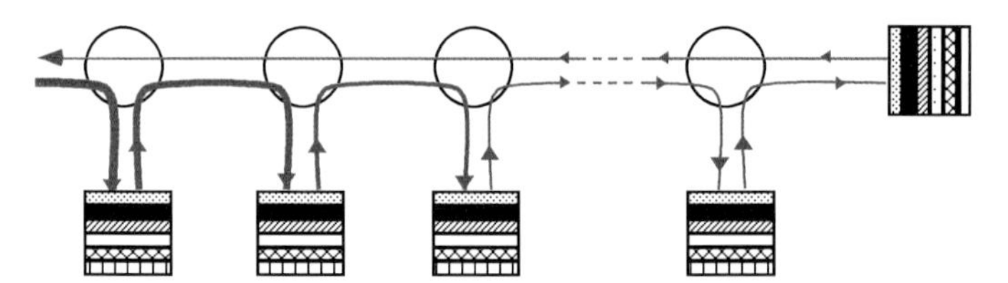

Figure 2.8 Time-asymmetric photovoltaic system using an infinite number of circulators in combination with infinite stacks of tandem cells. Source: Green and Hansen (2002).

number of cells of different bandgap. The first converter accepts the full sunlight intensity from the Sun via a circulator, which allows the light emitted by the cell to be steered to the next cell. In this way, each cell can be operated infinitesimally close to open-circuit, where it converts all wavelengths at the Carnot efficiency, with essentially only ambient temperature black-body radiation eventually sent back to the Sun.

Clearly, a much simplified way of implementing time-asymmetric effects will be required if they are to be used to practical advantage. An encouraging result is that, if only one circulator and two cell stacks are used, the maximal efficiency increases from 86.8% to 90.7%, over half the possible benefit from time asymmetry (Brown, 2003).

2.5 Single-cell efficiency limits

2.5.1 *Trivich–Flinn limit*

The fundamental trade-off associated with photovoltaic energy conversion is shown in Fig. 2.9. Photons of energy $E = h\nu$ less than the bandgap E_g of the photovoltaic converter are not absorbed while photons of energy above the bandgap create an electron-hole pair that quickly loses any energy in excess of the bandgap (h is Planck's constant as before; ν the light frequency).

One of the earliest treatments of the bandgap dependence of photovoltaic conversion efficiency analysed, very literally, the situation of Fig. 2.9 (Trivich and Flinn, 1955). Regarding the incident sunlight as from a black-body source at temperature T_S, using the black-body photon and energy fluxes deduced by Planck and assuming the final outcome of the absorption process was an electron-hole pair separated by the bandgap energy, the Trivich–Flinn (TF) efficiency is given by

$$\eta_{\mathrm{TF}} = \frac{E_g \dot{N}(E_g, \infty)}{\dot{E}(0, \infty)} = \frac{E_g \int_{E_g}^{\infty} E^2/[\exp(E/kT_s) - 1]\,\mathrm{d}E}{\int_0^{\infty} E^3/[\exp(E/kT_s) - 1]\,\mathrm{d}E} \tag{2.8}$$

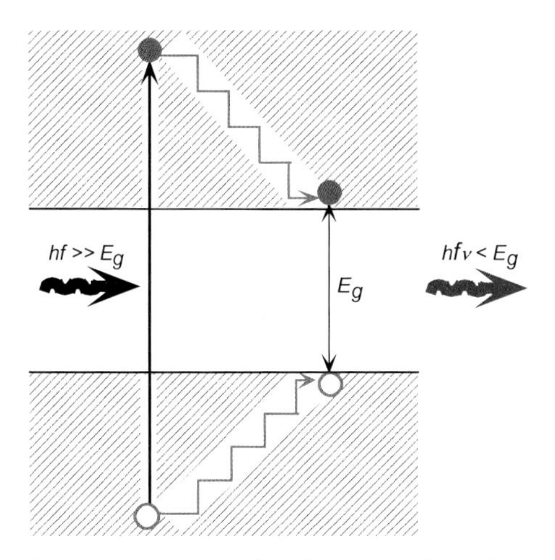

Figure 2.9 Photoexcitation across a semiconductor bandgap showing the fundamental trade-off in photovoltaic conversion. Source: Green and Ho (2011).

The Bose–Einstein integrals involved can be evaluated by noting, with $\varepsilon = E/kT_S$

$$\int_{\varepsilon_g}^{\infty} \frac{\varepsilon^n \mathrm{d}\varepsilon}{\mathrm{e}^{\varepsilon} - 1} = \int_{\varepsilon_g}^{\infty} \varepsilon^n \mathrm{e}^{-\varepsilon}(1 + \mathrm{e}^{-\varepsilon} + \mathrm{e}^{-2\varepsilon} + \cdots)\mathrm{d}\varepsilon = \sum_{m=1}^{\infty} \int_{\varepsilon_g}^{\infty} \varepsilon^n \mathrm{e}^{-m\varepsilon} \mathrm{d}\varepsilon \tag{2.9}$$

This gives, for $n = 2$

$$\begin{aligned} \int_{\varepsilon_g}^{\infty} \frac{\varepsilon^2 d\varepsilon}{\mathrm{e}^{\varepsilon} - 1} &= \sum_{m=1}^{\infty} \mathrm{e}^{-m\varepsilon_g}(m^2\varepsilon_g^2 + 2\varepsilon_g + 2)/m^3 \\ &= \varepsilon_g^2 Li_1(\mathrm{e}^{-\varepsilon_g}) + 2\varepsilon_g Li_2(\mathrm{e}^{-\varepsilon_g}) + 2Li_3(\mathrm{e}^{-\varepsilon_g}) \end{aligned} \tag{2.10}$$

where $Li_s(z)$ is the polylogarithm of order s, defined as (Lewin, 1981)

$$Li_s(z) = \sum_{k=1}^{\infty} \frac{z^k}{k^s} \tag{2.11}$$

Importantly for efficiency calculations (Green, 2012), polylogarithms have the general property

$$\frac{\partial Li_s(z)}{\partial z} = \frac{Li_{s-1}(z)}{z} \quad \text{giving} \quad \frac{\partial Li_s(\mathrm{e}^{\mu})}{\partial \mu} = Li_{s-1}(\mathrm{e}^{\mu}) \tag{2.12}$$

where μ is any variable. For integral $s \leq 1$, polylogarithms can be expressed in terms of elementary functions. Of particular relevance to the present chapter

$$Li_1(z) = -\ln(1-z); \quad Li_0(z) = \frac{z}{1-z}; \quad Li_{-1}(z) = \frac{z}{(1-z)^2} \tag{2.13}$$

For higher orders of s, the power series involved in Eq. (2.11) are quite rapidly convergent for $z < 1$, the case of present interest. Note also for positive arguments in this range

$$z \leq Li_3(z) \leq Li_2(z) \leq Li_1(z) \leq Li_0(z) \leq Li_{-1}(z) = z/(1-z)^2 \tag{2.14}$$

Hence, all orders will have roughly similar values for small z. Repeating the previous integration for $n = 3$ gives

$$\int_{\varepsilon_g}^{\infty} \frac{\varepsilon^3 d\varepsilon}{e^{\varepsilon} - 1} = \varepsilon_g^3 Li_1(\mathrm{e}^{-\varepsilon_g}) + 3\varepsilon_g^2 Li_2(\mathrm{e}^{-\varepsilon_g}) + 6\varepsilon_g Li_3(\mathrm{e}^{-\varepsilon_g}) + 6Li_4(\mathrm{e}^{-\varepsilon_g}) \tag{2.15}$$

Hence, the Trivich–Flinn limit expressed in polylogarithms is

$$\eta_{\mathrm{TF}} = \frac{\varepsilon_g[\varepsilon_g^2 Li_1(\mathrm{e}^{-\varepsilon_g}) + 2\varepsilon_g Li_2(\mathrm{e}^{-\varepsilon_g}) + 2Li_3(\mathrm{e}^{-\varepsilon_g})]}{6Li_4(1)} \tag{2.16}$$

where $Li_4(1)$ is an Euler series with value $\pi^4/90$. The value of ε_g giving the maximum TF efficiency (ε_{g0}) can be found by differentiating the above numerator to give

$$\varepsilon_{g0}^3 Li_0(\mathrm{e}^{-\varepsilon_{g0}}) = \varepsilon_{g0}^2 Li_1(\mathrm{e}^{-\varepsilon_{g0}}) + 2\varepsilon_{g0} Li_2(\mathrm{e}^{-\varepsilon_{g0}}) + 2Li_3(\mathrm{e}^{-\varepsilon_{g0}}) \tag{2.17}$$

This equation has the solution $\varepsilon_{g0} = 2.1657\ldots$ normalised to kT_{S} with the corresponding maximum value of the TF efficiency of 43.88%, regardless of T_{S} (the assumed black-body temperature of the Sun). In terms of ε_{g0}, it can be seen by comparing Eqs (2.16) and (2.17) that this maximum efficiency is given by

$$\eta_{\mathrm{TF}}^m = \frac{15}{\pi^4} \frac{\varepsilon_{g0}^4}{\mathrm{e}^{\varepsilon_{g0}} - 1} \tag{2.18}$$

For a 6000 K black-body source, the optimum bandgap would be 1.120 eV, about the bandgap of silicon.

2.5.2 *Shockley–Queisser limit*

For any actual photovoltaic converter, the voltage output will be less than the 'bandgap voltage' E_g/q (where q is the electronic charge). An elegant approach to finding a fundamental limit on the voltage output of a photovoltaic cell was published by Shockley and Queisser (1961). Shockley and Queisser (SQ) realised that a strongly absorbing cell in thermal equilibrium with its ambient at temperature T_A would be emitting black-body radiation. For photon energies above the cell bandgap, the physical source of the emitted photons would be band-to-band radiative recombination, the inverse of the absorption process of Fig. 2.9.

In a non-equilibrium situation with voltage across the device, SQ were aware that, in a high-quality *p-n* junction device, the electron–hole product, and hence these radiative recombination rates, would be exponentially enhanced throughout the cell volume by the cell voltage, V, normalised to the 'thermal voltage', kT_{A}/q. Hence, a good-quality cell, for photon energies above its bandgap, would emit exponentially enhanced black-body radiation. Assuming 100% quantum efficiency, one electron would flow into the cell terminals for each emitted photon.

Using this insight, but a more general chemical potential model of photon emission (Würfel, 1982), the electrical power density output from a photovoltaic cell in the SQ limit can be expressed as (Green, 2003)

$$\begin{aligned} P_{\mathrm{SQ}} &= VJ \\ &= \frac{2\pi qV}{h^3c^2}\left(\xi\int_{E_g}^{\infty}\frac{E^2\mathrm{d}E}{\exp(E/kT_{\mathrm{S}})-1}-\int_{E_g}^{\infty}\frac{E^2\mathrm{d}E}{\exp[(E-qV)/kT_{\mathrm{A}}]-1}\right) \end{aligned} \tag{2.19}$$

where ξ is the black-body radiation 'dilution' factor. As before, ξ is unity if the whole range of acceptance angles for the cell is filled by direct sunlight, as for a cell operating under the maximal sunlight concentration ratio of 46,200. ξ falls to a value of 1/46,200 if the available sunlight is considered spread over the whole terrestrial hemisphere, an appropriate model for a converter with an isotropic response over this hemisphere.

The first integral can be expressed in terms of polylogarithms as in Section 2.5.1. The second integral can be expanded as in Eqs (2.9)–(2.10) to give

$$\begin{aligned} &\int_{E_g}^{\infty}\frac{E^2\mathrm{d}E}{\exp[(E-qV)/kT_{\mathrm{A}}]-1} \\ &\quad = (kT_{\mathrm{A}})E_g^2 Li_1[\mathrm{e}^{(qV-E_g)/kT_{\mathrm{A}}}] + 2(kT_{\mathrm{A}})^2 E_g Li_2[\mathrm{e}^{(qV-E_g)/kT_{\mathrm{A}}}] \\ &\quad\quad + 2(kT_{\mathrm{A}})^3 Li_3[\mathrm{e}^{(qV-E_g)/kT_{\mathrm{A}}}] \end{aligned} \tag{2.20}$$

This gives the SQ efficiency limit as

$$\begin{aligned} \eta_{\mathrm{SQ}} = {} & \frac{15}{\pi^4}[\varepsilon_g + t\ln(\gamma)]\{\varepsilon_g^2[Li_1(\mathrm{e}^{-\varepsilon_g}) - t\,Li_1(\gamma)/\xi] \\ & + 2\varepsilon_g[Li_2(\mathrm{e}^{-\varepsilon_g}) - t^2 Li_2(\gamma)/\xi] \\ & + 2[Li_3(\mathrm{e}^{-\varepsilon_g}) - t^3 Li_3(\gamma)/\xi]\} \end{aligned} \tag{2.21}$$

where $\gamma = e^{(qV-E_g)/kT_A}$ and $t = T_A/T_S$. Comparison with Eq. (2.16) shows the TF formulation is a limiting case of the SQ formulation for $T_A = 0$. This close relationship will be exploited in the following.

Unlike Eq. (2.16), Eq. (2.21) is a function of not one but three variables ε_g, V and t. Since these variables can be expressed in terms of standard functions supported by common mathematical packages such as Mathematica (2008), even in this form Eq. (2.21) could be simply solved, for example, for its global maximum value at fixed t (built-in Mathematica programs such as FindMaximum require only the entry of the equation, plus essentially a single line of code, to find the solution).

The technique generally used to find the peak SQ efficiency is to fix E_g, find the optimum V for this E_g, and hence steadily map out efficiency as a function of E_g (Shockley and Queisser, 1961). However, partial differentiation of Eq. (2.19) or (2.21) with respect to E_g locates a stationary point when

$$Li_0[e^{(qV-E_g)/kT_A}] = \xi Li_0 e^{(-E_g/kT_S)} \tag{2.22}$$

For the case of undiluted sunlight ($\xi = 1$), this simplifies to the Carnot relationship noted elsewhere (Landsberg, 2000)

$$(E_g - qV_m)/kT_A = E_g/kT_S \quad \text{or} \quad V_m = E_g(1 - T_A/T_S)/q \tag{2.23}$$

where V_m is the optimum voltage for maximum power output.

This Carnot condition must prevail at the global optimum E_g for the SQ efficiency (for undiluted sunlight), but the corresponding value of V will in general be non-optimal for other E_g. Hence, imposing this condition on voltage for all E_g will result in a lower bound on available power output that will nonetheless equal the SQ output at stationary points, including the global maximum.

Inserting Eq. (2.23) into Eq. (2.21) gives the following expression for this lower bound (for $\xi = 1$)

$$\begin{aligned}\eta_{\text{SQLB}}^{\xi=1} = {} & \frac{15}{\pi^4}\varepsilon_g(1-t)^2[\varepsilon_g^2 Li_1(e^{-\varepsilon_g}) \\ & + 2(1+t)\varepsilon_g Li_2(e^{-\varepsilon_g}) + 2(1+t+t^2)Li_3(e^{-\varepsilon_g})]\end{aligned} \tag{2.24}$$

Differentiating Eq. (2.24) gives an analytical expression for the optimum bandgap (alternatively, Eq. (2.21) could be partially differentiated with respect to V with the value given by Eq. (2.23) then inserted, to give the same results). The optimum bandgap is given (for $\xi = 1$) by

$$\varepsilon_{gm}^3 Li_0(\mathrm{e}^{-\varepsilon_{gm}}) = (1-2t)\varepsilon_{gm}^2 Li(\mathrm{e}^{-\varepsilon_{gm}}) + 2(1+t-t^2)\varepsilon_{gm} Li_2(\mathrm{e}^{-\varepsilon_{gm}}) + 2(1+t+t^2) Li_3(\mathrm{e}^{-\varepsilon_{gm}}) \quad (2.25)$$

where since ε_{gm} is in normalised units, the actual bandgap $E_{gm} = \varepsilon_{gm} kT_{\mathrm{S}}$.

Noting the TF formulation is a limiting case of the SQ formulation for $T_{\mathrm{A}} = 0$ and that small $T_{\mathrm{A}}/T_{\mathrm{S}}$ ratios ($\sim$0.05) prevail for solar conversion, the SQ results can be expressed linearly in terms of TF results. The exact value of the linear term in a power series expansion can be found by solving Eq. (2.21) by a single Newton–Raphson iteration, using ε_{g0} as the trial solution, giving

$$\begin{aligned} \varepsilon_{gm}^{\xi=1}(\mathrm{SQ}) &= \varepsilon_{g0} - t[3Li_1(\mathrm{e}^{-\varepsilon_{g0}}) - \varepsilon_{g0} Li_0(\mathrm{e}^{-\varepsilon_{g0}})]/ \\ &\quad [4Li_0(\mathrm{e}^{-\varepsilon_{g0}}) - \varepsilon_{g0} Li_{-1}(\mathrm{e}^{-\varepsilon_{g0}})] \\ &= 2.1657 - 0.422\ T_{\mathrm{A}}/T_{\mathrm{S}} \end{aligned} \quad (2.26)$$

Similarly, expanding Eq. (2.24) to first order in t gives

$$\begin{aligned} \eta_m^{\xi=1}(\mathrm{SQ}) &= \eta_{\mathrm{TF}}\{1 - t[1 + Li_1(\mathrm{e}^{-\varepsilon_{g0}})/Li_0(\mathrm{e}^{-\varepsilon_{g0}})/\varepsilon_{g0}]\} \\ &= 0.4388 - 0.629 T_{\mathrm{A}}/T_{\mathrm{S}} \end{aligned} \quad (2.27)$$

Both expressions are extremely accurate, giving the correct values of bandgap and efficiency for $t = 0.05$ to within one digit in the fifth and fourth significant figure, respectively.

For the case of diluted sunlight, of considerable practical importance since it applies to the general photovoltaic case of non-maximal sunlight concentration and also, as discussed below, to cases where non-radiative recombination occurs in parallel with the modelled radiative recombination, Eq. (2.19) again prevails at the optimum V. At this optimum, γ depends only on ε_g and ξ with a value γ^* equal to or slightly higher than $\xi \mathrm{e}^{-\varepsilon_g}$ as defined below

$$\gamma^* = \frac{\xi}{\mathrm{e}^{\varepsilon_g} - 1 + \xi} \quad (2.28)$$

Proceeding as before, this gives the lower bound on SQ efficiency as

$$\begin{aligned} \eta_{\mathrm{SQLB}} = \frac{15}{\pi^4}[\varepsilon_g + t\ln(\gamma^*)]\{&\varepsilon_g^2[Li_1(\mathrm{e}^{-\varepsilon_g}) - t\, Li_1(\gamma^*)/\xi] \\ &+ 2\varepsilon_g[Li_2(\mathrm{e}^{-\varepsilon_g}) - t^2 Li_2(\gamma^*)/\xi] \\ &+ 2[Li_3(\mathrm{e}^{-\varepsilon_g}) - t^3 Li_3(\gamma^*)/\xi]\} \end{aligned} \quad (2.29)$$

Differentiating Eq. (2.29) again provides the relationship for the optimum bandgap, ε_{gm}(SQ) normalised to kT_S. Maximum efficiency is obtained when the bandgap satisfies the following relationship

$$[\varepsilon_g + t\ln(\gamma^*)][\varepsilon_{gm}^2 Li_0(e^{-\varepsilon_{gm}}) + 2t\varepsilon_{gm} Li_1(\gamma^*)/\xi + 2t^2 Li_2(\gamma^*)/\xi]$$
$$= \varepsilon_{gm}^2[Li_1(e^{-\varepsilon_{gm}}) - t\,Li_1(\gamma^*)/\xi] + 2\varepsilon_{gm}[Li_2(e^{-\varepsilon_{gm}}) - t^2 Li_2(\gamma^*)/\xi]$$
$$+ 2[Li_3(e^{-\varepsilon_{gm}}) - t^3 Li_3(\gamma^*)/\xi] \quad (2.30)$$

Following the same procedure as for $\xi = 1$ and expressing the polylogarithms involved in terms of elementary functions, the optimum bandgap is given by

$$\varepsilon_{gm} = \varepsilon_{g0} + \frac{\ln(1 + A/\xi) - (3A/\xi)\ln(1 + \xi/A)}{4 - \varepsilon_{g0}(1 + 1/A)} t + O(t^2) \quad (2.31)$$

where $A = (e^{\varepsilon_{g0}} - 1) = 7.7221$ with $\varepsilon_{g0} = 2.1657$, as previously evaluated and $O(t^2)$ indicates small terms of the order of t^2. As ξ decreases from unity, the first term in the numerator increases rapidly while the second term increases much more slowly, giving a sign change at $\xi = 0.44$. For the most challenging case of small ξ (=1/46,200), the accuracy of this expression based on only terms linear in t is about 1% at $t = 0.05$. However, since η_{SQ} is relatively insensitive to ε_g near the maximum, inserting this result for ε_{gm} into Eq. (2.29) gives the correct maximum efficiency to close to four significant digits.

Alternatively, expanding Eq. (2.29) to terms linear in t gives

$$\eta_{SQm} = \eta_{TF}\left\{1 - \frac{t}{\varepsilon_{g0}}\left[\ln\left(1 + \frac{A}{\xi}\right) + \frac{A}{\xi}\ln\left(1 + \frac{\xi}{A}\right)\right]\right\} + O(t^2) \quad (2.32)$$

The accuracy obtained using this expression is not as high as with the previous approach, with square and cubic terms in t required for similar accuracy, for the most challenging case of small ξ. However, Eq. (2.32) has the advantage of correctly identifying the important dependencies. Non-ideal device properties in the form of non-radiative recombination occurring in parallel with radiative recombination can be modelled by further dilution of sunlight by a factor equal to the radiative efficiency (the fraction of nett recombination in the device that is radiative).

2.5.3 *Efficiencies under actual AM0 and AM1.5 spectra*

The above treatments in terms of a black-body Sun allow analytical solutions showing the important dependencies. However, limits applying to the tabulated AM0 and AM1.5 reference spectra at the standard cell test temperature of 25°C (298.15 K) are probably of more practical interest. Figure 2.10 summarises efficiencies versus bandgap computed for the different spectra in the extended SQ

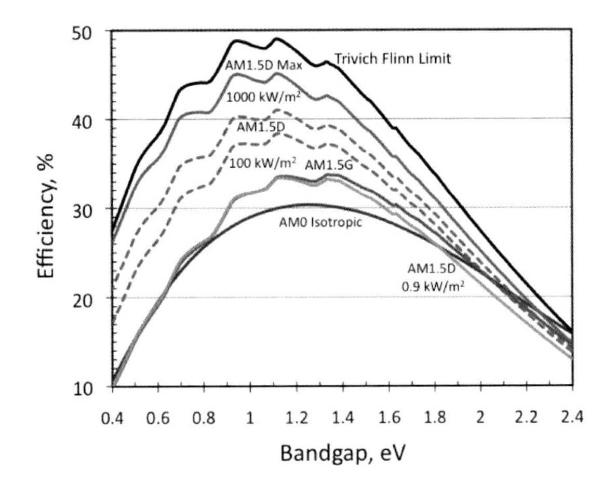

Figure 2.10 Isotropic performance limits for cells under the AM0, AM1.5G and AM1.5D spectral as well as for the AM1.5D spectra, under various concentration levels. The Trivich–Flinn limit for the AM1.5D spectrum is shown for comparison. Source: Green and Ho (2011).

approach. The uppermost curve is the Trivich–Flinn limit for the AM1.5D spectrum, which is approached for small-bandgap cells under high concentration levels (or, more generally, severe angular restriction of the converter response). Note that the limiting efficiency under concentration (or angular restriction) increases more rapidly for small-bandgap cells than for large-bandgap cells (simply because the approximately 60 mV increase in open-circuit voltage for ten times increase in intensity, or decrease in angular acceptance, is proportionately larger for small-bandgap cells).

For the AM0 spectrum for a cell with isotropic response, there is a single peak (30.4% efficiency for $E_g = 1.245\,\mathrm{eV}$). This is similar to the value of 31.0% at a bandgap of 1.31 eV calculated for a 6000 K black-body spectrum at a cell temperature of 300 K. For the terrestrial AM1.5 spectra, there are multiple peaks due to the strong water vapour absorption bands apparent in Fig. 2.5. The most important are the absorption bands in the 920–970 nm (1.28–1.35 eV), 1100–1160 nm (1.07–1.13 eV) and 1300–1500 nm (0.83–0.95 eV) wavelength (energy) ranges.

Loss of radiation in the terrestrial spectra due to the first of these bands eliminates the 'natural' peak apparent in the AM0 spectrum, creating a local maximum on the high-energy side of the absorption band and a local minimum on the low-energy side.

Note that from Fig. 2.10 the AM0 and AM1.5G spectra give relatively higher efficiency for larger bandgaps than does the AM1.5D spectrum. This is due to blue photons being selectively scattered out of the AM1.5D spectrum during passage

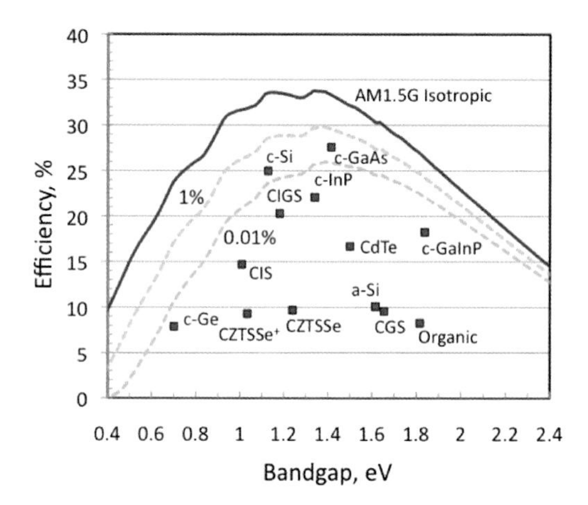

Figure 2.11 Isotropic efficiency limits under the AM1.5G spectrum, and also for devices with 1% and 0.01% radiative efficiency. Also shown are best-certified cell limits efficiencies for various cell technologies. Source: Green and Ho (2011).

through the atmosphere. Combined with the intensity dependence already noted, this shifts the global maximum between three different peaks, each associated with the high-energy side of the three absorption bands already mentioned.

The solid line of Fig. 2.11 shows the limiting efficiency for a conventional cell with isotropic response under the AM1.5G reference spectrum. Compared with an earlier spectrum in use until 2008 (Green *et al.*, 2009), the stronger spectral content over the 650–900 nm range shifts the peak efficiency to higher energy, so that the two peaks around 1.15 eV and 1.35 eV are nearly equal, with peak efficiencies of 33.6% and 33.8% respectively. For the previous reference spectrum (ASTM E892-87, IEC60904-3:1989), the order of the peaks was reversed and both were about 1% lower in efficiency (Tiedje *et al.*, 1984), consistent with expectations discussed elsewhere (Green *et al.*, 2009).

Also shown in Fig. 2.11 as dashed lines are the limiting efficiencies for 0.01% and 1% radiative efficiencies of the cell, with radiative efficiency defined as the fraction of nett recombination in the device that is radiative. Radiative efficiency is a measure of the state of development of the cell material technology. Limiting efficiencies calculated with reduced radiative efficiency are identical to those calculated under a reduced intensity (further dilution) of sunlight. Hence the previous curves under concentration (Fig. 2.10) can also be used to calculate the effects of non-ideal radiative efficiency in this case as well (for example, the curve for 100 kW m^{-2} incident intensity represents the limiting performance for a cell under 1000 kW m^{-2} intensity, if the radiative efficiency is 10% rather than 100%).

Given this correspondence, it is not surprising that the efficiency for low-bandgap cells is most sensitive to low values of the radiative efficiency, since a 60 mV/decade voltage reduction with decreasing radiative efficiency has a proportionately more devastating impact. Peak efficiency is pushed to higher-bandgap cells as the radiative efficiency decreases.

Also shown in Fig. 2.11 are best-certified experimental values for a range of different solar cell technologies (Green *et al.*, 2011). 'Bandgap' in this case corresponds to the energy on the low-energy side of the cell's spectral response curve, where external quantum efficiency (EQE) reaches 50% of its peak value. Silicon (c-Si) and c-GaAs lie above the 0.01% radiative efficiency line. Actual radiative efficiencies are close to or well above 1% for both technologies. The difference arises since experimental devices have additional losses other than the radiative loss assumed including, for example, reflection and resistance losses, but also including additional radiative loss due to non-abrupt absorption thresholds (Kirchartz *et al.*, 2009).

2.5.4 *Radiative efficiency of state-of-the-art photovoltaic cells*

The limiting photovoltaic solar cell performance is obtained when cell recombination is dominated by radiative processes because there are then no non-radiative pathways, a fact used to advantage in the SQ analysis.

The effect of parallel non-radiative processes can be described in terms of what will be known as the *external radiative efficiency* (ERE). This is defined as the fraction of the total recombination in the device that results in radiative emission from the device. Since this fraction may vary in general with the voltage across the device, the open-circuit voltage (V_{oc}) is chosen as the reference voltage. Due to total internal reflection and photon recycling, the ERE will generally be smaller than what will be termed the *internal radiative efficiency* (IRE), the fraction of internal recombination events that are radiative, such as is determined by the ratio of total lifetime to the radiative lifetime. For silicon devices, the IRE can be several times the ERE (Trupke *et al.*, 2003) but obviously the IRE:ERE ratio must decrease towards unity as both approach 100%.

The dashed lines in Fig. 2.11 show the limiting efficiencies for ERE between 1% and 0.01%, with the best cells lying between these limits. As noted above and to be shown below, actual ERE can be appreciably higher, since effects other than radiative inefficiencies also decrease the experimental efficiency below the ideal limits.

Experimental ERE can be deduced by taking advantage of a surprising yet fundamental reciprocal relationship recently identified (Rau, 2007). Actual solar cells do not emit radiation above the cell bandgap exactly as assumed in the Shockley–Queisser analysis (exponentially enhanced by the cell voltage with the spectral distribution of room-temperature black-body radiation). Rather, they emit radiation with this distribution multiplied at each wavelength by the external quantum efficiency (EQE) of the device, as measured for the device operating as a solar cell. This remarkable relationship has been confirmed for a number of experimental devices (Kirchartz and Rau, 2007; Kirchartz *et al.*, 2008a; Kirchartz *et al.*, 2009; Vandewal *et al.*, 2009), as well as being consistent with modelling prior to its recognition (Green *et al.*, 2001).

The cell EQE must be measured for a calibrated measurement of cell efficiency, to adjust for spectral mismatch between the spectrum used for illuminating the cell during measurement and the tabulated reference spectrum (International Standard, 2008). This means that each calibrated cell measurement yields sufficient information to allow calculation of ERE.

On cell open circuit, the photogeneration of carriers within a cell is balanced by recombination within the device, globally if not at each point within the device. The open-circuit voltage measured at the cell terminals is the voltage that increases recombination within the device-active region to the level required to balance the photocurrent able to be collected by the junction. Assuming reasonable device quality, so parasitic resistances are not excessive, and also assuming that the collection probability of carriers is not strongly voltage-dependent, the photocurrent density collected by the junction on open circuit will equal the short-circuit current density. The light emitted by the junction will be the EQE-weighted, exponentially-enhanced black-body radiation described above, allowing the ERE to be calculated as

$$\mathrm{ERE} = \frac{\frac{2\pi q}{h^3c^2}\exp\left(\frac{qV_{oc}}{kT}\right)\int\frac{\overline{EQE}_{\mathrm{abs}}E^2\mathrm{d}E}{\exp(E/kT)-1}}{I_{sc}/A} = \frac{\exp\left(\frac{qV_{oc}}{kT}\right)\int\overline{EQE}_{\mathrm{rel}}N_{\mathrm{BB}}(E)\mathrm{d}E}{\int EQE_{\mathrm{rel}}N_{AM1.5}(E)\mathrm{d}E} \tag{2.33}$$

EQE is the value for the solar cell for near perpendicularly incident light in either absolute or relative terms, with $\overline{\mathrm{EQE}}$ the Lambertian weighted value over all angles of incident light, of absolute value $\overline{\mathrm{EQE}}_{\mathrm{abs}}$ and relative value $\overline{\mathrm{EQE}}_{\mathrm{rel}}$. For a high-quality cell, this will not differ greatly from the near perpendicular value. EQE is typically measured either in absolute or relative terms with a midrange accuracy of about 3% relative. The major contributions to the integration on the numerator come from the long-wavelength region of the spectral response, from the region where this response begins to fall rapidly to the low value near the absorption threshold, while the main contribution to the denominator comes from the region near the peak photon density in the AM1.5 spectrum, which is in the 600–700 nm range.

Some of the sources of inaccuracies in the evaluation of ERE have already been suggested. An isotropic response needs to be assumed for evaluation from standard near-perpendicular EQE data. Collection probabilities may be voltage-dependent so that the short-circuit condition, where EQE is normally evaluated, may not represent conditions at open circuit. The reciprocal relation itself is also only strictly valid under conditions where the quasi-Fermi level separation at the junction is constant (Rau, 2007), which would not be the case for resistive devices even at open circuit, due to circulating currents. However, these effects are expected to be minor for cells of respectable performance, particularly compared with the large differences in ERE noted between different devices and different technologies.

ERE was calculated for a range of representative state-of-the-art cells (Green *et al.*, 2010; Green *et al.*, 2011). The absolute EQEs for these cells are shown in Fig. 2.12a, deduced from the relative EQE by normalising to the experimental short-circuit current density measured for the devices. The calculated spectral luminescence from these devices on open-circuit is shown in Fig. 2.12b. Units in this case are $A/m^2/nm$, representing the current density required on open circuit to support the different spectral emission components shown. Due to large differences between technologies, different scaling factors are applied to the different results, although the same scaling factor is applied to cells of the same type, except for the GaAs cells.

Until recently (Green *et al.*, 2011) the most efficient non-concentrating, single-junction cell was a 26.4% efficient GaAs cell fabricated by the Fraunhofer Institute for Solar Energy (GaAs ISE), with parameters shown in Table 2.1 Integrating the calculated spectral luminescence gives an ERE of 1.26%. Given that this device was fabricated on a GaAs substrate, IRE would have been appreciably higher since n^2 more radiation (where n is the substrate refractive index) would have been emitted into the substrate (with most absorbed there) than into air.

In late 2010, this record was surpassed with 27.6% efficiency measured for a thin-film GaAs device fabricated by Alta Solar (GaAs Alta). As shown in Fig. 2.12a, this has an almost identical spectral response to cell GaAs ISE but, as in Fig. 2.12b, the luminescent intensity on open circuit is almost 20 times higher, corresponding to a greatly improved ERE of 22.5%. This high value suggests photons emitted during radiative recombination events in a direction towards the rear of the device are not wasted as in the GaAs ISE device, but that a reasonable number are reflected back into active device regions.

With radiative recombination forming such a large fraction of total recombination in this device, further improvements in ERE will result in immediate benefits.

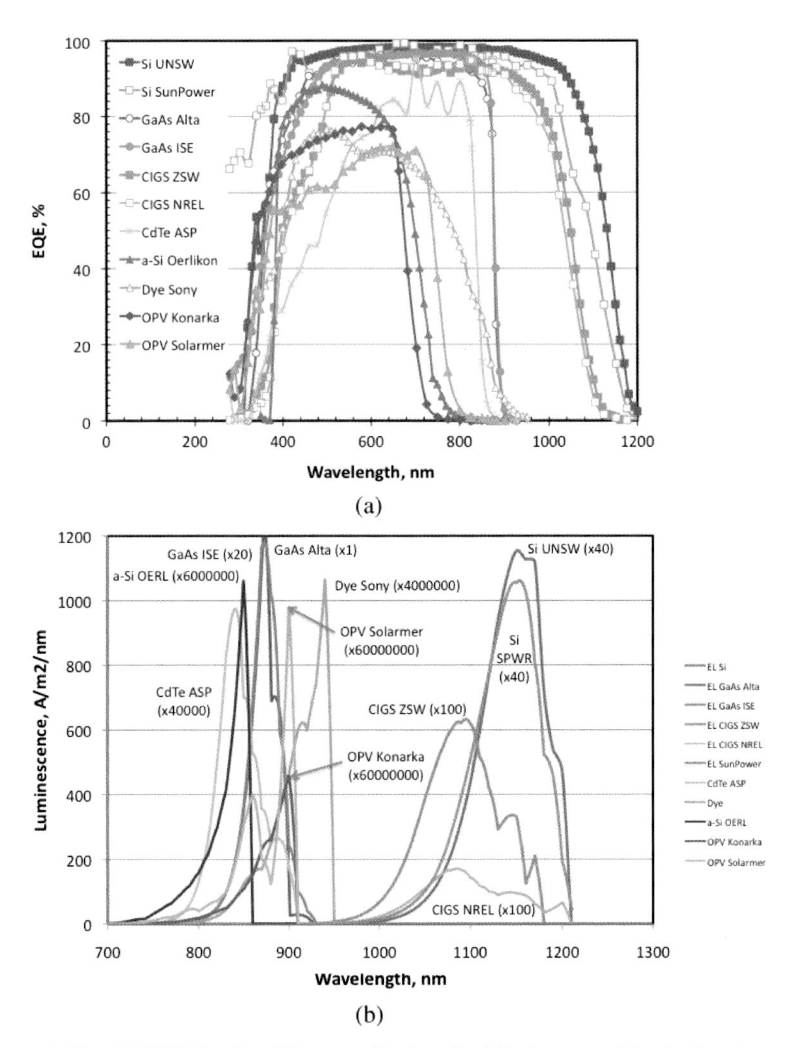

Figure 2.12 (a) EQE for the different cells described in the text; (b) calculated spectral luminescence from the different cells on open circuit, with widely varying multiplication factors. Source: Green and Ho (2011).

As ERE is increased towards 100%, an additional gain in V_{oc} of up 40 mV is predicted with a corresponding fill factor improvement.

The ERE for the two crystalline silicon devices investigated are not appreciably lower than the second-rated GaAs device, with both Si devices displaying an ERE of about 0.6%. This is similar to the electroluminescent quantum efficiency directly measured for similar devices (Green *et al.*, 2001), again confirming the accuracy of

Table 2.1 External radiative efficiency (ERE) and other relevant performance parameters at 25°C for the state-of-the-art devices (Green *et al.*, 2010; 2011) included in the present study.

Device	V_{oc} **mV**	i_{sc} **mA/cm²**	**Effic. %**	**ERE%**
Si UNSW	706	42.7	25.0	0.57
Si SPWR	721	40.5	24.2	0.56
GaAs Alta	1107	29.6	27.6	22.5
GaAs ISE	1030	29.8	26.4	1.26
CIGS ZSW	740	35.4	20.3	0.19
CIGS NREL	713	34.8	19.6	0.057
CdTe* ASP	838	21.2	12.5	1.0E–4
a-Si OERL	886	16.8	10.1	5.3E–6
Dye* Sony	719	19.4	9.9	7.2E–6
OPV Konarka	816	14.5	8.3	2.7E–7
OPV Solarmer	759	15.9	8.1	3.8E–7

*Minimodule: results on a 'per cell' basis.

the present approach. Since the Sunpower device (Si SPWR) with less than 200 μm thickness is appreciably thinner than the University of New South Wales device (Si UNSW) with over 400 μm thickness and has possibly worse rear reflection, the EQE is notably poorer at long wavelengths, with i_{sc} also appreciably lower. This is compensated by a higher V_{oc}, due to the smaller volume of the bulk region, probably combined with better surface passivation. From Fig. 2.12b and Table 2.1, the two devices have almost identical open-circuit voltage luminescence.

Comparison of two recent record CIGS (copper indium gallium selenide) devices also produces interesting results. One is a small area 20.3%-efficient device fabricated by Zentrum für Sonnenenergie-und-Wasserstoff-Forschung (CIGS ZSW), and the second is for a larger (1 cm^2) device of 19.6% efficiency fabricated by the US National Renewable Energy Laboratory (CIGS NREL). From Fig. 2.12a, the two devices can be seen to have similar spectral responses, although the CIGS ZSW device has a much stronger emission on open circuit, with ERE of 0.19% compared to 0.06% for the NREL device. This indicates fundamentally better quality material in the ZSW device, although this may be partly due to its smaller size, given the strong dependence of CIGS cell performance on cell area.

The next three devices discussed include recent record CdTe, a-Si and dye-sensitised devices. The CdTe device is a small 12.5%-efficient submodule with ERE of 1.0×10^{-4}%. Smaller cells could be expected to have higher ERE while commercial CdTe modules, averaging about 11.4% aperture area efficiency in

2011, would be expected to have lower ERE. The best a-Si cell to date has quite a low ERE of $5.3 \times 10^{-6}\%$. However, from Fig. 2.11, its bandgap can be seen to be in an appropriate region for the highest possible conversion efficiency with such low ERE.

The dye-sensitised device is again a small 9.9% efficient submodule, although with energy conversion performance close to the best-performing small-area cell (11.2%). Although the operational principles of a dye-sensitised cell are vastly different from the previous *p-n* junction devices, the reciprocity relationships giving rise to Rau's relationship are expected still to apply (Trupke *et al.*, 1999). ERE is $7.2 \times 10^{-6}\%$, similar to, but slightly higher than, the a-Si device.

The final two devices are two of the first organic photovoltaic devices to exceed 8% energy conversion efficiency, with very different energy absorption thresholds, as apparent from Fig. 2.12a. Both devices, however, display very similar ERE in the $3\text{–}4 \times 10^{-7}\%$ range. A detailed discussion of the electroluminescence from a range of organic cells is presented elsewhere (Vandewal *et al.*, 2009).

Even though organic light-emitting diodes (OLEDs) of quantum efficiency >20% have been reported, the different requirements for photovoltaics (Kirchartz *et al.*, 2008b) greatly reduce the radiative efficiency. In particular, the blending required to produce bulk heterojunction devices with high carrier collection greatly reduces the radiative efficiency.

The above results show that the external radiative efficiency (ERE) can be unambiguously deduced from standard solar cell efficiency measurements and is a useful parameter in comparing the performance of cells of both the same and completely different technologies. As each technology matures, ERE will evolve towards the 100% value required for limiting performance.

2.6 Multiple-junction devices

2.6.1 *The N-cell stack*

Photovoltaic conversion efficiency can be increased by using cells of different bandgaps to convert different parts of the solar spectrum. In the optimal stacked configuration of Fig. 2.13 (Marti and Araujo, 1996), the power output from the *n*th cell in an N-cell stack of radiatively-efficient devices is

$$P_n = \frac{2\pi q V_n}{h^3 c^2} \left\{ \xi \int_{E_{gn}}^{E_{gn}^+} \frac{E^2 dE}{\exp(E/kT_S) - 1} - \int_{E_{gn}}^{E_{gn}^+} \frac{E^2 dE}{\exp[(E - qV_n)/kT_A] - 1} \right\} \tag{2.34}$$

where E_{gn}^+ is the bandgap of cell $(n+1)$ with $E_{gN}^+ = \infty$.

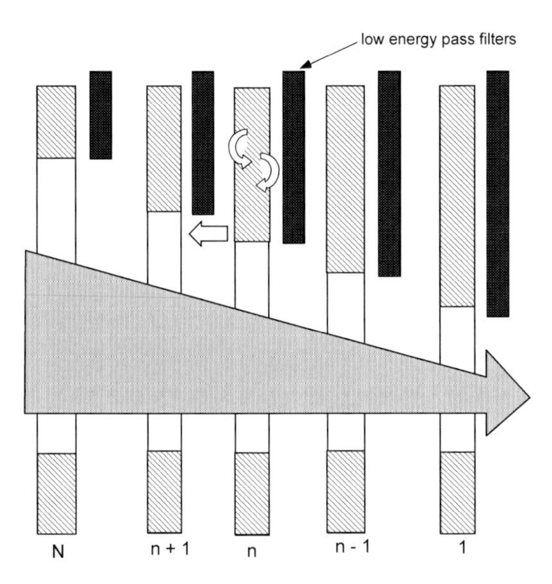

Figure 2.13 Optimal configuration for stacked tandem cells. After Marti and Araujo (1996).

Expressing this in terms of polylogarithms

$$P_n = \frac{2\pi q \xi V_n}{h^3 c^2 (kT_s)^3}\{\varepsilon_{gn}^2[Li_1(e^{-\varepsilon_{gn}}) - t\, Li_1(\gamma_n)] + 2\varepsilon_{gn}[Li_2(e^{-\varepsilon_{gn}}) - t^2 Li_2(\gamma_n)]$$
$$+ 2[Li_3(e^{-\varepsilon_{gn}}) - t^3 Li_3(\gamma_n)] - \varepsilon_{gn}^{+2}[Li_1(e^{-\varepsilon_{gn}^+}) - t Li_1(\gamma_n^+)]$$
$$- 2\varepsilon_{gn}^+[Li_2(e^{-\varepsilon_{gn}^+}) - t^2 Li_2(\gamma_n^+)] - 2[Li_3(e^{-\varepsilon_{gn}^+}) - t^3 Li_3(\gamma_n^+)]\} \quad (2.35)$$

where $\gamma_{\rm n} = e^{(qV_n - E_{gn})/kT_{\rm A}}$ and $\gamma_n^+ = e^{(qV_n - E_{gn}^+)/kT_{\rm A}}$.

With the equations formulated in this way in terms of standard functions, global maxima can be simply found with standard mathematic packages just by entering these equations constrained by current matching if this feature is desired. Figure 2.14 was generated for the unconstrained 3-cell case with a simple contour plot command using Mathematica, involving essentially two steps of code.

2.6.2 *Infinite stack of cells*

The case of an infinite stack of tandem cells is of particular interest since the limit in this case represents the fundamental limit on photovoltaic conversion efficiency for a time-symmetric system (Green, 2003). For an infinite stack of cells, the

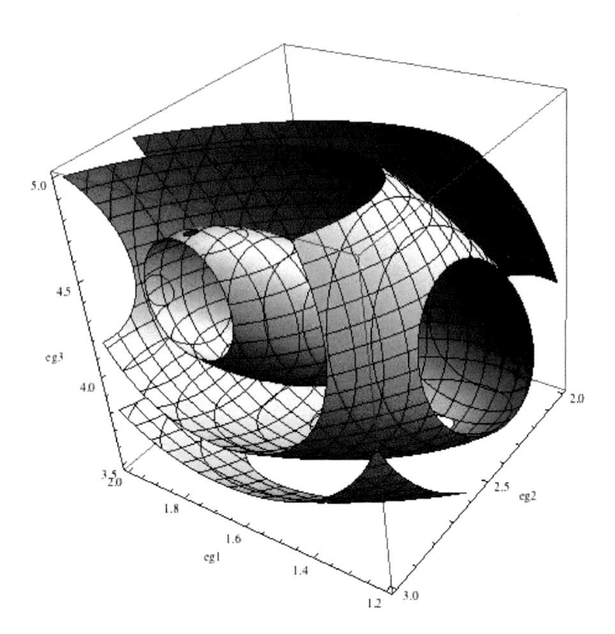

Figure 2.14 Contour plots for 47%, 48% and 49% efficiency versus normalised bandgap ε_{g1}, ε_{g2} and ε_{g3}, for a 3-cell tandem stack (unconstrained) with $t = 0.05$ and $\xi = 1/46{,}200$. Peak efficiency is 49.3% for $\varepsilon_{g1} = 1.58$, $\varepsilon_{g2} = 2.77$ and $\varepsilon_{g3} = 4.36$ (multiply by 0.517 eV to convert to actual bandgap values for $T_A = 300$ K, $T_S = 6000$ K). Source: Green and Ho (2011).

incremental current density at each wavelength is given by (Marti and Araujo, 1996; Green, 2003)

$$\mathrm{d}i = \frac{2\pi q}{h^3c^2}\left[\xi\frac{E_g^2}{\exp(E_g/kT_s)-1} + (1-\xi)\frac{E_g^2}{\exp(E_g/kT_A)-1} - \frac{E_g^2}{\exp[(E_g-qV)/kT_A]-1}\right]\mathrm{d}E_g \tag{2.36}$$

The extra term in $(1-\xi)$ absent in Eq. (2.34) corresponds to thermal background radiation, important if low-bandgap cells ($E_g < kT_A \ln[1/\xi]$) are to be considered in the analysis.

In terms of polylogarithms, the corresponding incremental power density output becomes

$$\mathrm{d}P = \frac{2\pi qVE_g^2}{h^3c^2}[\xi Li_0(\mathrm{e}^{-E_g/kT_s}) + (1-\xi)Li_0(\mathrm{e}^{-E_g/T_A}) - Li_0(\mathrm{e}^{(qV-E_g)/kT_A})]\mathrm{d}E_g \tag{2.37}$$

Partial differentiation with respect to V gives the following two equations for the optimal value V_m

$$
\begin{aligned}
Li_0(\mathrm{e}^{(qV_m-E_g)/kT_\mathrm{A}}) &+ \frac{qV_m}{kT_\mathrm{A}} Li_{-1}(\mathrm{e}^{(qV_m-E_g)/kT_\mathrm{A}}) \\
&= \xi Li_0(\mathrm{e}^{-E_g/kT_\mathrm{s}}) + (1-\xi)Li_0(\mathrm{e}^{-E_g/kT_\mathrm{A}}) \\
&= Li_0(\mathrm{e}^{q(V_\mathrm{oc}-E_g)/kT_\mathrm{A}})
\end{aligned} \tag{2.38}
$$

At open circuit, for $\xi = 1$, Eq. (2.38) shows $V_\mathrm{oc} = E_g(1 - T_\mathrm{A}/T_\mathrm{S})/q$, a result also noted elsewhere (Landsberg, 2000). For other values of ξ, also from Eq. (2.38), V_{oc} for each cell can be expressed explicitly in terms of its specific E_g

$$
qV_\mathrm{oc} = E_g - kT_\mathrm{A} \ln\{1 + 1/[\xi Li_0(\mathrm{e}^{-E_g/kT_\mathrm{s}}) + (1-\xi)Li_0(\mathrm{e}^{-E_g/kT_\mathrm{A}})]\} \tag{2.39}
$$

The above solution has the asymptotic values

$$
qV_\mathrm{oc} = E_g(1 - T_\mathrm{A}/T_\mathrm{s})/[1 + (1/\xi - 1)T_\mathrm{A}/T_\mathrm{s}] \quad \text{for } E_g \ll kT_\mathrm{A} \tag{2.40}
$$

and

$$
qV_\mathrm{oc} = E_g(1 - T_\mathrm{A}/T_\mathrm{s}) + kT_\mathrm{A}\ln(\xi) \quad \text{for } E_g \gg kT_\mathrm{s} \tag{2.41}
$$

with both solutions equivalent for $\xi = 1$.

As for simple ideal-diode cell models, V_m can only be expressed implicitly in terms of either V_oc or E_g

$$
V_m = V_\mathrm{oc} - \frac{kT_\mathrm{A}}{q} \ln\left[\frac{qV_m}{kT_\mathrm{A}}/(1-\gamma_m) + 1\right] + \frac{kT_\mathrm{A}}{q}\ln\left(\frac{1-\gamma_m}{1-\gamma_\mathrm{oc}}\right) \tag{2.42}
$$

where $\gamma_m = \mathrm{e}^{(qV_m-E_g)/kT_\mathrm{A}}$ and $\gamma_{oc} = \mathrm{e}^{(qV_{oc}-E_g)/kT_\mathrm{A}}$.

Since the terms involving V_m on the right-hand side of Eq. (2.42) are relatively small, a good approximation allowing conversion to an explicit form is to replace V_m on this side by V_oc, which is explicitly known from Eq. (2.39). This approximation has been shown to work well for conventional cell analysis (Luque *et al.*, 1978; Green, 1982). It will locate a voltage V_m^* on the output curve close to V_m to give essentially the same, but slightly less, power output where

$$
V_m^* = V_\mathrm{oc} - \frac{kT_\mathrm{A}}{q}\ln\left[\frac{qV_\mathrm{oc}}{kT_\mathrm{A}}/(1-\gamma_\mathrm{oc}) + 1\right] \tag{2.43}
$$

Since the term in the square bracket is already an overestimate, deleting the $(1 - \gamma_m)$ and the $(+1)$ terms will generally improve accuracy, provided V_oc is constrained to values greater than kT_A. Using this modified version of Eq. (2.43) in combination with the asymptotic expansion of Eq. (2.41) gives an explicit relationship for V_m^* which remains useful for $E_g \geq kT_A[1 - \ln(\xi)]/(1 - T_A/T_s)$. For smaller E_g, only a small power contribution would be generated, so setting V_m to zero for such E_g will not cause any significant error in total output. This approximation was found to result in efficiencies which matched exact values to close to three significant digits over the whole range of likely interest.

The resulting expression may also be integrated analytically. Inserting into Eq. (2.37) and expanding to first order in T_A, then integrating, gives the result

$$\eta_\infty = 1.0 - t[A + B\ln(t) + C\ln(\xi)] + O(t^2) \tag{2.44}$$

where

$$B = C = -[Li_3(1)/Li_4(1)]/3 = -0.3702$$
$$A = 1 + [(5 - 2\gamma_E)Li_3(1)/6 + Li_3'(1)/3]/Li_4(1) = 1.6508$$

with $\gamma_E = 0.5772$ (Euler Gamma) and $Li_3'(1) = \partial Li_s(1)/\partial s|_{s=3} = -0.1981$.

This expression gives reasonable accuracy (better than 3% relative) over the range of t and ξ of likely interest. Increasing the dependence on the $\ln(t)$ term by adjusting parameters to $A = 1.1861$, $B = -0.4901$ and $C = -0.3487$ gives the best fit over this range with results accurate to close to three significant digits. This result is to be compared to the similar 'Landsberg' limit on the efficiency of a system where time-symmetry is not imposed. A simple approximate, but accurate, new expression given earlier for this limit based on past work (Press, 1976; Landsberg and Tonge, 1979) is

$$\eta_L = 1 - t[E + D\ln(\xi) + (4/3 - E)\xi^{0.9}] + O(t^4) \tag{2.45}$$

with $E = 11/12 + Li_3(1)/Li_4(1)]/3 = 1.2869$; $D = -[Li_3(1)/Li_4(1)]/3 = -0.3702$.

The key difference is the $\ln(t)$ term present in the time-symmetric case, fundamentally due to the need to move away from open-circuit conditions in this case.

2.7 Other high-efficiency options

Apart from multijunction cells, a range of other options has been suggested to improve on the performance of a single-junction cell, with limiting efficiency from some of the key approaches shown in Fig. 2.1. Each, in its own way, attempts to favourably impact the fundamental trade-off of Fig. 2.9 involved in a single-threshold device. The key options listed in Fig. 2.1 and briefly considered below have been discussed in more detail elsewhere (Green, 2003).

2.7.1 *Hot-carrier cells*

One of the highest efficiency of such options is the 'hot-carrier' cell (Ross and Nozik, 1982), with efficiency potential higher than a 5-tandem cell stack. The key idea, captured in Fig. 2.15, is to avoid, or at least reduce, the loss of the energy in excess of the bandgap imparted to electron-hole pairs generated by photons of energy larger than the bandgap.

In a conventional cell, this extra energy is quickly lost during collisions of the energetic particles with the atoms of the absorber (phonon emission). Within a picosecond or less, the carriers relax back to close to the band edge, as explicitly assumed in the Trivich–Flinn analysis of Section 2.5.1. However, there would appear to be no fundamental reason why this relaxation process could not be considerably slowed, retaining some of this excess energy in a 'hot' carrier population, as in Fig. 2.15b.

To date, the best way of doing this seems to be manipulation of the mechanical properties of the absorber on a microscopic scale. This permits engineering of the properties of allowed mechanical vibrations or phonons in the absorber, particularly the energy ranges allowed for acoustic and optical phonons. Electrons give up heat most effectively to the lattice through the relatively immobile, high-energy optical phonons that then decay to more mobile, lower-energy acoustic phonons. By manipulating the allowed energy ranges for these phonons, it is possible to

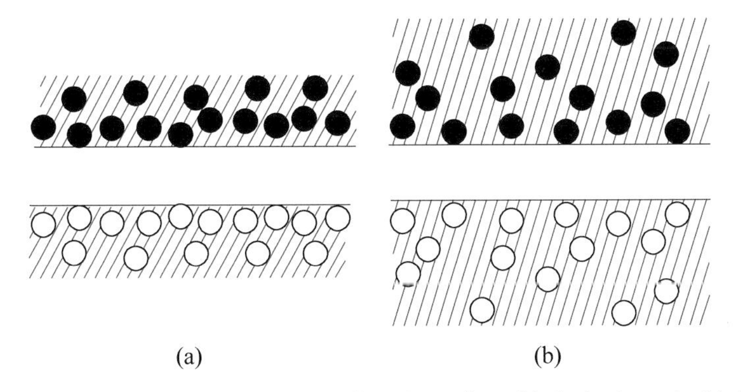

Figure 2.15 Normally photogenerated carriers thermalise with the lattice as in (a). In hot-carrier cells, excess energy is stored in a hot-carrier distribution, as shown in (b). Source: Green and Hansen (2002).

slow the decay of optical into acoustic phonons, and hence the energy relaxation, of carriers by reducing the rate of energy exchange with optical phonons.

Calculations involving the relevant experimental parameters of quantum wells and bulk materials show that hot-carrier effects should be apparent in suitably designed devices operating under highly concentrated sunlight (Guillemoles *et al.*, 2005; Aliberti *et al.*, 2010). Development of new materials will be required to extend this attribute to devices operating under global sunlight.

2.7.2 *Multiple carrier or exciton generation*

It was suggested some time ago (Deb and Saha, 1972) that photovoltaic performance might be improved by creating multiple electron-hole pairs by high-energy photons, as indicated in Fig. 2.16. This has the potential to substantially increase the efficiency of the cell (Werner *et al.*, 1995; Nozik, 2008).

Multiple electron-hole pair generation has been observed in silicon devices (Kolodinski *et al.*, 1994). However, the effect is too small to provide any measurable gain in conventional silicon cell performance.

More recent interest in this effect has been stimulated by evidence for multiple exciton generation (MEG) in Si, PbS, and PbSe quantum dots (Nozik, 2010). Moreover there is an independent interest in replacing the dye molecules that form the absorber in dye-sensitised solar cells by quantum dots, giving at least a conceptual path to possible implementation of devices taking advantage of the MEG effect.

The limiting efficiency of multiple carrier or MEG devices under maximally concentrated direct light is quite close to the time-symmetric limit (85.9% versus

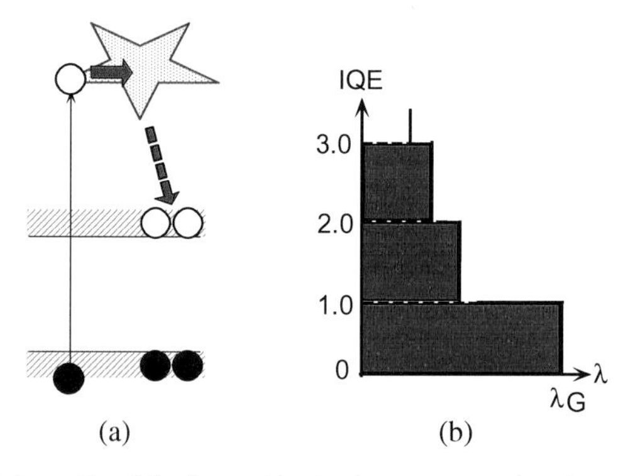

Figure 2.16 (a) Schematic of the impact ionisation process whereby one energetic photon creates multiple electron–hole pairs; (b) energetically feasible internal quantum efficiency, where λ_g is the wavelength corresponding to the threshold energy for electron–hole pair creation. Source: Green and Hansen (2002).

86.8% for a 6000 K black-body source). However, MEG device are less suited to global sunlight, with the limiting efficiency dropping considerably, to just above that of a conventional 2-cell tandem stack (43.6% for a 6000 K black-body source, compared with 42.9% for a 2-cell stack and 68.2% for the time-symmetric limit). This is due to the larger bandgap required for efficient operation at low intensities, reducing the gain from multiple carrier generation (Green, 2003). However, this approach may well provide a way of improving cell performance with specific cell technologies.

2.7.3 *Impurity level and intermediate-band devices*

Many years ago, Wolf (1960) suggested the use of photoexcitation through defect levels in the normally forbidden cell bandgap to boost cell performance as shown in Fig. 2.17. The idea was for some time controversial, since it was thought any gain would be offset by the increased recombination introduced by the defect (Guttler and Queisser, 1970). However, analysis of an extension of the idea, where the defects are present in sufficient quantities to form a mid-gap band (Luque and Marti, 1997), pushed the issue beyond doubt (Brown and Green, 2002). Quite substantial gains in performance are possible if the defect levels/bands have quite specific properties.

The limiting performance of a cell based on this approach is similar to that of a 3-cell tandem, but slightly lower. This is because the effect can be modelled by the specific connection shown in Fig. 2.18 of three different cells of energy bandgap equal to the energy thresholds of the three different absorption processes of Fig. 2.17 (Green, 2003).

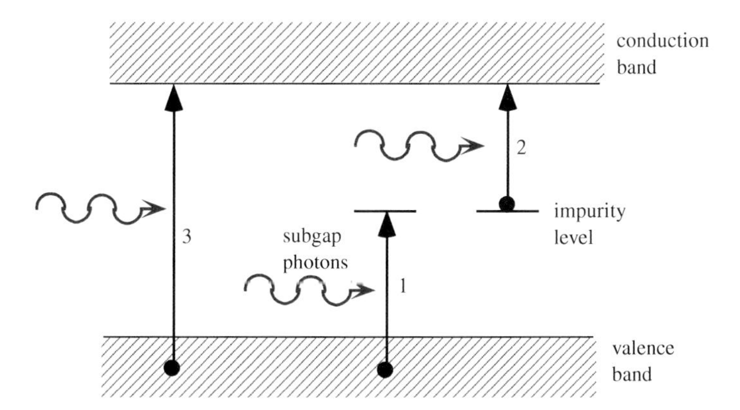

Figure 2.17 Impurity photovoltaic effect where electron-hole pairs are generated by sub-bandgap photons. After Keevers and Green (1994).

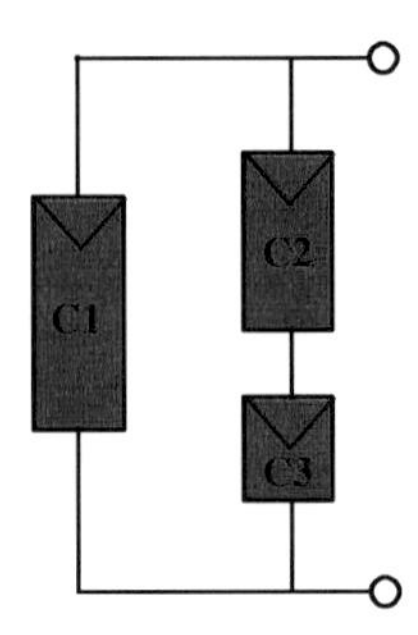

Figure 2.18 Equivalent circuit of the intermediate-band solar cell. The circuit elements shown represent ideal Shockley–Queisser solar cells. Source: Green and Ho (2011).

The multiple quantum well cell (Barnham and Duggan, 1990; also Chapter 10 of this volume) and quantum-dot structures (Luque and Marti, 2010) have become the preferred devices for investigating the implementation of this approach. Although devices with creditable performance have been demonstrated, these have not yet exceeded the performance of controls without the intermediate levels.

2.7.4 *Up- and down-conversion*

A disadvantage of introducing defects into the bulk of the cell is that they are likely to produce additional recombination paths (Guttler and Queisser, 1970). If the associated optical excitations are decoupled from the cell by placing an up-converter at the rear of the cell, as in Fig. 2.19a, this potential disadvantage can be overcome (Trupke *et al.*, 2002a).

Similarly, Fig. 2.20a shows a schematic of how a down-converter might be used on the front of a cell, and Fig. 2.20b shows an equivalent circuit. In this case, the down-converter would strongly absorb light of energy more than twice the main cell bandgap. The intermediate level of the down-converter would lie close to halfway across its effective bandgap so that both the relaxation of electrons to this level and from this level to lower levels produced photons just above the bandgap energy of the main cell. With this additional constraint on energy transition thresholds, the limiting efficiency is close to that of a 2-cell tandem (Trupke *et al.*, 2002b).

2.7.5 *Thermal approaches*

The limiting efficiency of the concentrated solar power (CSP) approach such as used in the PS10 and PS20 solar power towers operating in Spain (Abengoa, 2011)

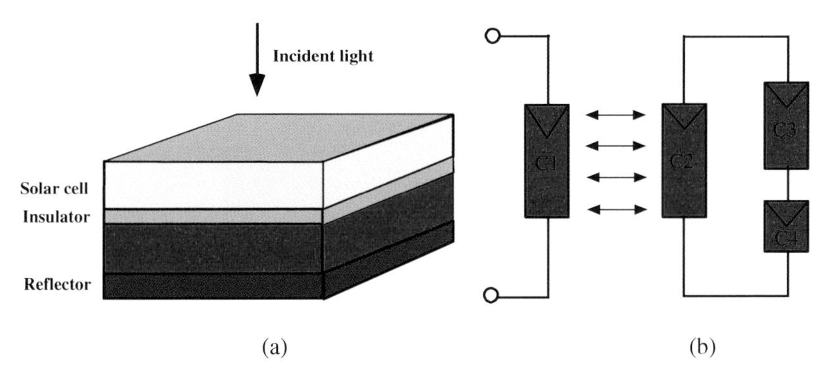

Figure 2.19 (a) Solar cell and an electronically isolated up-converter. Sub-band-gap light transmitted by the solar cell is up-converted into high-energy photons, which are subsequently absorbed in the solar cell; (b) equivalent circuit of the up-conversion system. The solar cell in front of the up-converter is denoted C1. The up-converter is represented by the three cells C2, C3 and C4. The series connected cells C3 and C4 represent the two intermediate transitions while C2 represents the band-to-band transitions. After Trupke *et al.* (2002a).

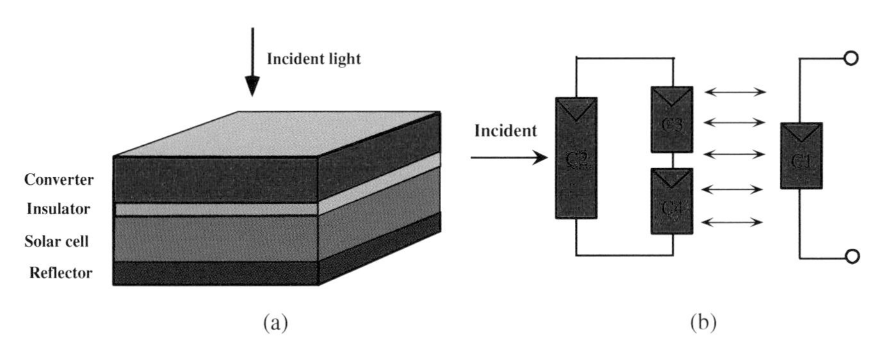

Figure 2.20 (a) Schematic diagram of the down-conversion system with the luminescence converter located on the front surface of a solar cell. High-energy photons with $\hbar\nu > 2E_g$ are absorbed by the converter and down-converted into two lower energy photons with $\hbar\nu > E_g$, which can both be absorbed by the solar cell; (b) equivalent circuit of the system. The luminescence converter is represented by three solar cells C2, C3, and C4 representing the band-to-band transitions and the two types of intermediate transitions, respectively. High-energy photons are absorbed by C2, which then biases the cells C3 and C4 into the forward direction and causes them to emit electroluminescence absorbed by the main cell C1. After Trupke *et al.* (2002b).

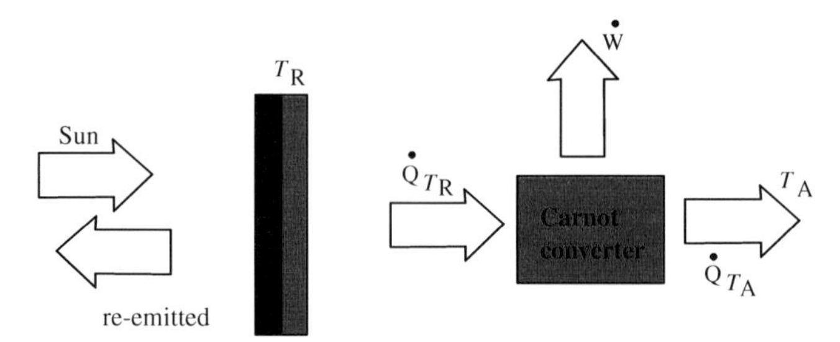

Figure 2.21 Solar thermal system based on the conversion of heat collected by an absorber to electricity at the Carnot efficiency. Source: Green and Hansen (2002).

can be calculated from the schematic of Fig. 2.21. A black absorber receives energy from the Sun, heating it to a high temperature, T_R. Surfaces of the absorber other than those receiving energy from the Sun have low emissivity so that only the receiving surface loses energy back to the Sun. Heat from the absorber drives a Carnot converter.

The limiting efficiency can be written down almost by inspection for a black-body Sun

$$\eta_T = \left(1 - \frac{T_\mathrm{R}^4}{\xi T_\mathrm{S}^4}\right)\left(1 - \frac{T_\mathrm{A}}{T_\mathrm{R}}\right) \tag{2.46}$$

For $\xi = 1$, the optimum T_R is 2544 K for $T_\mathrm{S} = 6000$ K and $T_\mathrm{A} = 300$ K, giving a limiting efficiency of 85.4%, quite close to the time-symmetric limit of 86.8%. However, as the power able to be focussed onto the receiver decreases, the efficiency drops rapidly, as shown by the lower curve of Fig. 2.22, to a limiting value of only 11.7% for undiluted sunlight (including sky radiation, in this case).

The problem is the decreasing Planckian effective temperature of the sunlight at long wavelengths as the intensity reduces (Fig. 2.5b). The solution is to use a receiver with selective absorption properties, so that it absorbs only those energies higher than a threshold value, E_T, determined by the crossover where the Planckian temperature is higher than the receiver temperature.

In this case, with $\varepsilon_R = E_\mathrm{T}/kT_\mathrm{R}$ and $\varepsilon_S = E_\mathrm{T}/kT_\mathrm{S}$, the limiting efficiency becomes

$$\eta_{T_\mathrm{S}} = \left(\frac{\int_{\varepsilon_S}^{\infty} \frac{\varepsilon^3 d\varepsilon}{e^{\varepsilon}-1} - \frac{T_\mathrm{R}^4}{\xi T_\mathrm{S}^4}\int_{\varepsilon_R}^{\infty} \frac{\varepsilon^3 d\varepsilon}{e^{\varepsilon}-1}}{\int_0^{\infty} \frac{\varepsilon^3 d\varepsilon}{e^{\varepsilon}-1}}\right)\left(1 - \frac{T_\mathrm{A}}{T_\mathrm{R}}\right) \tag{2.47}$$

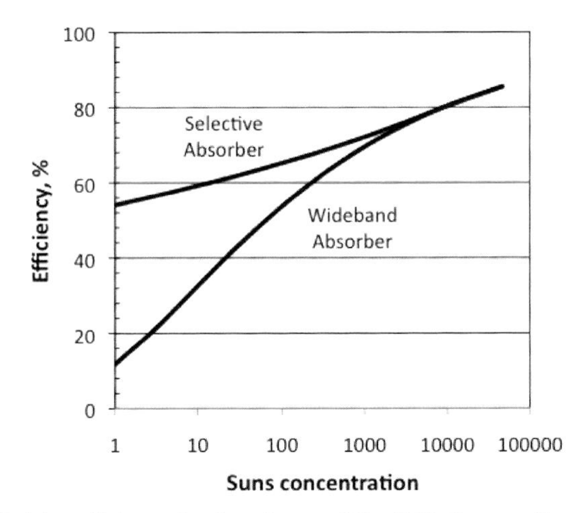

Figure 2.22 Limiting efficiency for the scheme of Fig. 2.21. Source: Green and Ho (2011).

As seen by the upper line in Fig. 2.22, use of such a selective absorber significantly improves the energy conversion efficiency at low incident sunlight intensities.

Similar energy conversion efficiency limits to those shown in Fig. 2.22 apply to photovoltaic devices that follow a thermal route to conversion. In considering these results, it should be borne in mind that these thermal limits are much more difficult to attain in practice than the corresponding limits for conventional photovoltaic devices, since the latter already take into account some of the practical difficulties in obtaining Carnot-like efficiencies.

A solar thermophotovoltaic system uses the radiant energy from the receiver rather than its heat energy. Figure 2.23 shows a schematic illustrating the basic concept. An ideal solar cell with an ideal monochromatic light filter acts as an ideal Carnot converter of radiation emitted by the receiver (Green, 2003). Hence the scheme of Fig. 2.23 fulfils the ideal and is bounded by the same limiting efficiencies as for Fig. 2.22b.

High expectations were held for this scheme in the late 1970s (Swanson, 1979), but results to date have been quite modest. A good explanation of the reasons why one might expect lower than ideal performance is given by Harder and Würfel (2003).

A development of the idea is based on the realisation that a light-emitting diode can convert electricity to light with greater than 100% power conversion efficiency, if heat input to the diode is ignored. This gives rise to the 'thermophotonic' converter of Fig. 2.24, which has advantages in terms of the narrow range of emission energies of the heated device and much stronger emission at low receiver temperatures.

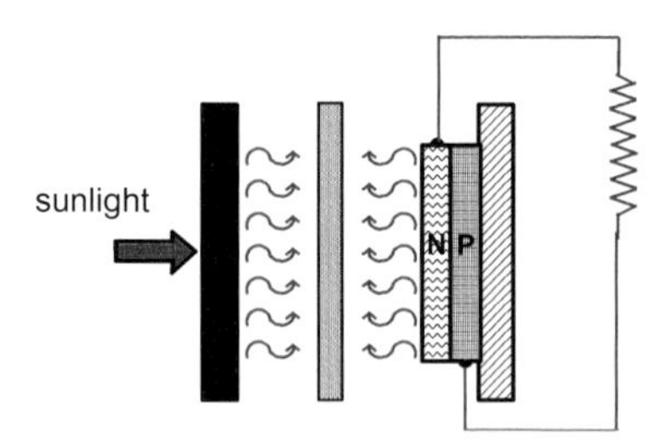

Figure 2.23 Thermophotovoltaic conversion with narrow passband filter. Source: Green and Hansen (2002).

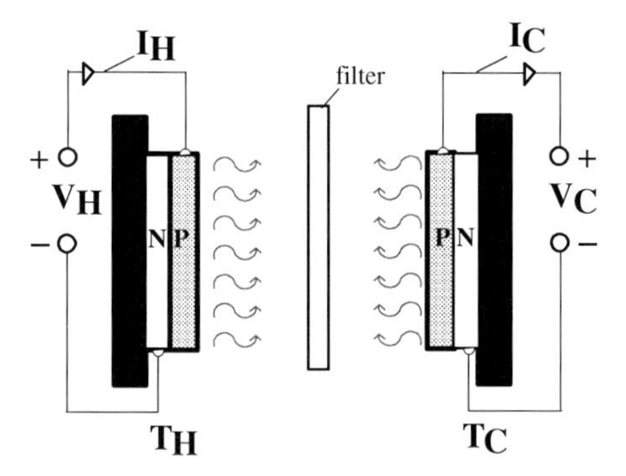

Figure 2.24 Thermophotonic conversion, with narrow-pass filter and independent power supplies. V and I are the voltages and currents at the terminals of the p-n junction devices, while T is the device temperature and the subscripts H and C refer to the hot device heated by sunlight and the cold receiver, respectively. Source: Green and Hansen (2002).

Another way of obtaining Carnot-like efficiencies in converting heat in the receiver to an electric output, at least in principle, is via the thermionic effect of Fig. 2.25 where electrons are 'boiled off' from the receiver (the cathode) and collected by a lower work function anode. In the standard simplified analysis (Schwede *et al.*, 2010), electrons are collected at the anode under 'flat-band' conditions, where the applied voltage is equal to the work function difference between the cathode and anode.

Because the anode work function must be at least greater than the ratio of anode to cathode temperature times that of the cathode work function to ensure a nett electron flux from the cathode to anode, this results in a Carnot efficiency limit for high cathode work functions (and low emitted flux densities). In practice, efficiency is well below the Carnot limit, with 15% often cited as the best to date in converting heat. Solar conversion efficiencies demonstrated to date (Ogino *et al.*, 2004) have

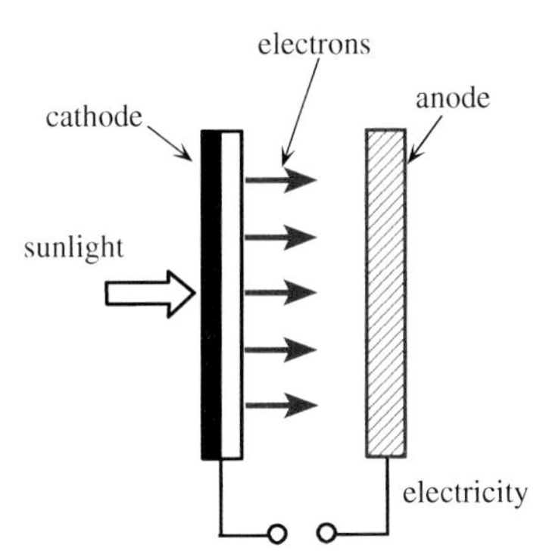

Figure 2.25 Thermionic conversion. Sunlight heats a cathode that emits electrons which are collected by a lower work function anode.

been very modest (well below 1% and even 0.1%). Radiative losses to the cathode (despite attempts to reduce the emissivity of the facing surfaces of the anode and cathode), conduction losses through wiring to the electrodes, resistive losses in the wires and space-charge effects are key contributors to the difference (Hatsopoulos and Gyftopoulos, 1979; Habedank, 1993).

It has been suggested that efficiency can be improved by combining both heat and light excitation of the cathode. In one case, combined photoemission and thermionic emission has been suggested, with the latter possibly augmented by hot-carrier effects (Smestad, 2004). A more recent suggestion has been to use a p-type semiconductor as the heated cathode, with above-bandgap light used to photoexcite the cathode and sub-bandgap light and carrier relaxation losses also used to heat the cathode (Schwede *et al*, 2010).

The advantage of the latter configuration is that the flatband condition is still determined by the relatively high work function of the p-type semiconductor, established by the majority-carrier Fermi level. However, photoexcitation moves the minority-carrier electron quasi-Fermi level to closer to the vacuum level, enhancing photoemission (and also reducing the minimum value of the anode work function). The disadvantage is that to absorb above-bandgap light, the anode must also emit it back to the Sun. For 100% radiative efficiency, this means each absorbed photon will result in a photon of energy just above the bandgap being emitted back to the Sun, a Trivich–Flinn inefficiency. For some bandgaps, this will represent 40% of the incident energy, a huge loss before any serious attempt at conversion begins.

Since the Carnot conversion potential of standard thermionic conversion obviously cannot be exceeded in this way, the advantage lies in offering lower possible operating temperatures for the converter, which may allow some of the previous losses mentioned in traditional converters to be reduced.

The other traditional converter of heat to electricity is a thermoelectric converter. The close relationship to photovoltaics is highlighted by a casual remark by Würfel (2000) that one way of reducing heat loss to the lattice in a hot-carrier cell is to let the lattice heat up to the hot-carrier temperature. A heated absorber with selective energy contacts (Section 2.7.1) acts as a hot-carrier cell in the Würfel limit (Würfel, 1995; Green, 2003). The conversion efficiency limits upon such a cell are identical to the limits plotted in Fig. 2.22. This type of device has been independently suggested as an improved thermoelectric converter (O'Dwyer *et al.*, 2005), demonstrating the close links of thermoelectrics to advanced photovoltaic concepts.

2.8 Summary

Although most commercial single-junction solar cells operate at more than 50% of their theoretical potential, there is scope for improving performance by edging closer to this limit. Even more scope for performance gains arises through advanced 'third generation' options that overcome the basic trade-off in a standard cell: too high a bandgap results in too little current; too low a bandgap gives too little voltage.

Extension of the Shockley–Queisser approach offers a 'bottom-up' method for analysing the ultimate potential of any new scheme, such as the recently suggested photo-enhanced, thermionic conversion (Schwede *et al.*, 2010). Formulation in terms of polylogarithms, as in the present chapter, simplifies the programming of the appropriate analyses. 'Top-down' thermodynamic analyses have already proved useful in identifying the features required for the ultimate photovoltaic converter. Calculations in terms of a black-body Sun have the virtue of timelessness and also of evaluation of results analytically. However, results for the tabulated reference spectra used for experimental cell measurements are also of considerable interest, although these reference spectra are more ephemeral. For example, a recent downscaling of the likely value of the solar constant from 1366.1 $W m^{-2}$ to 1360.8 $W m^{-2}$ (Kopp and Lean, 2011) is likely to have a flow-on effect to first the standard AM0 spectra, then possibly the terrestrial spectra. Some key results for present spectra of interest are compared in Table 2.2.

Apparent from Table 2.2 is the large difference between the limiting performance of standard single-junction cells and the ultimate limits on solar energy conversion. Capitalising on the potential factor of two performance gains would see thin-film module technologies go beyond 25% energy conversion efficiency,

Table 2.2 Summary of photovoltaic efficiency limits under black-body radiation at 300 K and under the three ASTM G173-03 spectra at 25 °C.

	Blackbody (T_S = 6000 K, T_A = 300 K)		**AM0**		**AM1.5G**		**AM1.5D**	
Efficiency	Direct	Isotropic	Direct	Isotropic	Direct*	Isotropic	Direct	Isotropic*
Landsberg (Ultimate)	93.3	73.7	93.1	72.6	92.9	74.0	92.8	73.3
Time-Symmetric (Infinite Tandem)	86.8	68.2	86.3	66.6	86.4	68.0	86.0	67.2
Single Cell (Bandgap eV)	40.9 (1.109)	31.0 (1.306)	40.7 (1.006)	30.4 (1.245)	45.2 (1.119)	33.8 (1.339)	45.1 (1.117)	33.4 (1.135)

*Included for reference.

and possibly beyond 30% in the longer term. For wafer-based technology, the figures would be correspondingly higher.

Several options for obtaining such efficiency improvements have been reviewed. Tandem cell stacks such as those discussed in Chapter 7 of the current volume are clearly the most promising route based on experience to date. Hot-carrier cells have the attraction of offering higher efficiency than a 5-cell tandem stack in a simple 2-terminal configuration, although considerable theoretical and experimental progress will be required before any performance improvement is likely to be demonstrated with this approach. Up- and down-conversion offer the prospects for 'supercharging' the performance of existing cell technologies, although improvements in the performance for both types of converters are required also before any performance gain would be expected. Intermediate-band and multiple exciton generation devices offer improvements for specific material systems.

The 'thermal approaches' — thermophotovoltaics, thermionics, thermoelectric or hot lattice 'hot-carrier' cells — offer tantalisingly high limiting efficiencies. However, the gap between what is likely to be obtained in practice and the theoretical potential is expected to be much larger with such approaches, because of the much higher energy losses that are likely to occur in practice. Such approaches may have potential in concentrating systems, but this is where tandem cell stacks have had the most impact terrestrially and are showing unabated potential for further development.

As is already becoming apparent as the photovoltaic industry matures, efficiency will be a key driver to future cost reduction. Evidence for this is the emphasis given by thin-film companies such as First Solar to quarterly improvements in module efficiency (from under 8% in 2003 to over 13% in 2013) and the increasing number of wafer silicon companies announcing improvements in production cell efficiency to the 19–20% range. Efficiency may be possible without economy, as past experience has demonstrated, but to reach the economies required for photovoltaics to make a large impact on global energy supply, improved efficiency is crucial.

Acknowledgement

This work has been supported by the Australian Government through the Australian Renewable Energy Agency (ARENA). Responsibility for the views, information or advice expressed herein is not accepted by the Australian Government.

References

Abengoa (2011), www.abengoasolar.com.

Aliberti P. Feng Y., Takeda Y., Shrestha S. K., Green M. A. and Conibeer G. (2010), 'Investigation of theoretical efficiency limit of hot carriers solar cells with a bulk indium nitride absorber', *J. Appl. Phys.* **108**, 94507–94517.

ASTM Standard Extraterrestrial Spectrum Reference E-490-00 (tabulated at http://rredc.nrel.gov/solar/spectra/am0/).

Barnham K. and Duggan G. (1990), 'A new approach to high-efficiency multi-band-gap solar cells', *J. Appl. Phys.* **67**, 3490–3493.

Blom W. M., Mihailetchi V. D., Koster L. J. A. and Markov D. E. (2007), 'Device physics of polymer:fullerene bulk heterojunction solar cells', *Adv. Mater.* **19**, 1551–1566.

Brown A. S. (2003), 'Ultimate efficiency limits of multiple energy threshold photovoltaic devices', *Thesis for the Degree of Doctor of Philosophy*, Centre for Third Generation Photovoltaics and the School of Electrical Engineering, University of New South Wales.

Brown A. S. and Green M. A. (2002), 'Impurity photovoltaic effect: fundamental energy conversion efficiency limit', *J. Appl. Phys.* **92**, 1329–1336.

Brown A. S., Green M. A. and Corkish R. (2002), 'Limiting efficiency for multi-band solar cells containing three and four bands', *Physica E* **14**, 121–125.

Chapin D. M., Fuller, C. S. and Pearson G. L. (1954), 'A new silicon *p-n* junction photocell for converting solar radiation into electrical power', *J. Appl. Phys.* **25**, 676–677.

Cooke D., Gleckman P., Krebs H., O'Gallagher J., Sage D. And Winston R. (1990), 'Sunlight brighter than the Sun', *Nature* **346**, 802.

Cummerow R. L. (1954a), 'Photovoltaic effect in *p-n* junctions', *Phys. Rev.* **95**, 16–21.

Cummerow R. L. (1954b), 'Use of silicon *p-n* junctions for converting solar energy to electrical energy', *Phys. Rev.* **95**, 561–562.

Deb S. and Saha H. (1972), 'Secondary ionisation and its possible bearing on the performance of a solar cell', *Solid State Electronics* **15**, 89–1391.

Green M. A. (1982), 'Accuracy of analytical expressions for solar cell fill factors', *Solar Cells* **7**, 337–340.

Green M. A. (2011), 'Limiting photovoltaic efficiency under new ASTM G173-based reference spectra', *Progr. Photovoltaics* **20**, 954–959.

Green M. A. (2012), 'Analytical treatment of Trivich–Flinn and Shockley–Queisser photovoltaic efficiency limits using polylogarithms', *Progr. Photovoltaics* **20**, 127–134.

Green M. A. and Hansen J. (1998), 'Catalogue of photovoltaic drawings', 1st edition, Photovoltaics Special Research Centre, University of New South Wales, Sydney.

Green M. A., Zhao J., Wang A., Reece P. J. and Gal M. (2001), 'Efficient silicon light emitting diodes', *Nature* **412**, 805–808.

Green M. A. and Hansen J. (2002), 'Catalogue of photovoltaic drawings', 2nd edition, Centre for Photovoltaic Engineering, University of New South Wales, Sydney.

Green M. A. (2003), *Third Generation Photovoltaics: Advanced Solar Energy Conversion*, Springer-Verlag, Berlin.

Green M. A., Emery K., Hishikawa Y. and Warta W. (2009), 'Solar cell efficiency tables (Version 33)', *Progr. Photovoltaics* **17**, 85–94.

Green M. A., Emery K., Hishikawa Y. and Warta W. (2010), 'Solar cell efficiency tables (Version 36)', *Progr. Photovoltaics* **18**, 346–352.

Green M. A., Emery K., Hishikawa Y. and Warta W. (2011), 'Solar cell efficiency tables (Version 37)', *Progr. Photovoltaics*, **19**, 84–92.

Green M. A. and Ho J. (2011), 'Catalogue of photovoltaic drawings', 3rd edition, School of Photovoltaic and Renewable Energy Engineering, University of New South Wales, Sydney.

Grondahl, L. O. (1933), 'The copper cuprous oxide rectifier and photoelectric cell', *Rev. Modern Phys.* **5**, 141.

Guillemoles J.-F., Conibeer G. and Green M. A. (2005), 'Phononic engineering with nanostructures for hot carrier solar cells', *PVSEC-15*, Shanghai, pp. 375–376.

Guttler G. and Queisser H. J. (1970), 'Impurity photovoltaic effect in silicon', *Energy Conversion* **10**, 51–55.

Habedank O. D. (1993), 'Analysis of Topaz II and space-R space nuclear power plants using a modified thermionic model', Thesis presented to the Faculty of the School of Engineering of the Air Force Institute of Technology, Air University, USA.

Harder N. P. and Green M. A. (2003), 'Thermophotonics', *Semiconductor. Sci. Tech.* **18**, S270–S278.

Harder N. P. and Würfel P. (2003), 'Theoretical limits of thermophotovoltaic generation of electricity', *Semiconductor. Sci. Tech.* **18**, S151–S157.

Hatsopoulos G. N. and Gyftopoulos E. P. (1979), *Thermionic Energy Conversion*, Volume 2, MIT Press, Cambridge, MA.

Henry C. H. (1980), 'Limiting efficiencies of ideal single and multiple gap terrestrial solar cells', *J. Appl. Phys.* **51**, 4494–4500.

International Standard, IEC 60904–3, Edition 2 (2008), 'Photovoltaic devices — part 3: Measurement principles for terrestrial photovoltaic (PV) solar devices with reference spectral irradiance data', ISBN 2–8318–9705-X, *Int. Electrotechnical Commission*.

Jackson E. D. (1955), *Trans. Conf. on Use of Solar Energy*, Tuscon, AZ, 31 Oct–1 Nov, 122 (also U.S. Patent No. 2,949,498, 16 Aug., 1960).

Keevers M. J. and Green M. A. (1994), 'Efficiency improvements of silicon solar cells by the impurity photovoltaic effect', *J. Appl. Phys.* **75**, 4022–4033.

Kirchartz T., Helbig A. and Rau U. (2008a), 'Note on the interpretation of electroluminescence images using their spectral information', *Solar Energy Mater. Solar Cells* **92**, 1621–1627.

Kirchartz T., Helbig A., Reetz W., Reuter M., Werner J. H. and Rau U. (2009), 'Reciprocity between electroluminescence and quantum efficiency used for the characterization of silicon solar cells', *Progr. Photovoltaics* **17**, 394–402.

Kirchartz T., Mattheis J. and Rau U. (2008b), 'Detailed balance theory of excitonic and bulk heterojunction solar cells', *Phys. Rev. B* **78**, 235320.

Kirchartz T. and Rau U. (2007), 'Electroluminescence analysis of high efficiency Cu(In, Ga)Se_2 solar cells', *J. Appl. Phys.* **102**, 104510.

Kolodinski S., Werner J. H., Wittfchen J. H. and Queisser H. J. (1994), 'Quantum efficiencies exceeding unity in silicon leading to novel selection principles for solar cell materials', *Solar Energy Mater. Solar Cells* **33**, 275–286.

Kopp G. and Lean J. L. (2011), 'A new, lower value of total solar irradiance: Evidence and climate significance', *Geophys. Research Lett.* **38**, L01706.

Kron G., Rau U. and Werner J. H. (2003), 'Influence of built-in voltage on the fill factor of dye-sensitised solar cells', *J. Phys. Chem. B* **107**, 13258–13261.

Landsberg P. (2000), 'Efficiencies of solar cells: where is Carnot hiding?', Conf. Record, *16th European Photovoltaic Solar Energy Conf.*, Glasgow.

Landsberg P. T. and Tonge G. (1979), 'Thermodynamics of the conversion of diluted radiation', *J. Phys. A* **12**, 551–562.

Landsberg P. T. and Tong G. (1980), 'Thermodynamic energy conversion efficiencies', *J. Appl. Phys.* **51**, R1–R20.

Lee K., Kim J. Y. and Heeger A. J. (2008), 'Titanium oxide films as multifunctional components in bulk heterojunction 'plastic' solar cells', in Brabecc, Dyakonovv, Scherfu (eds), *Organic Photovoltaics*, Wiley, Weinheim.

Lewin L. (1981), 'Polylogarithms and associated functions', New York: North-Holland (also see http://en.wikipedia.org/wiki/Polylogarithm).

Loferski J. J. (1956), 'Theoretical considerations governing the choice of the optimum semiconductor for photovoltaic energy conversion', *J. Appl. Phys.* **27**, 777–784.

Luque A. and Marti A. (1997), 'Increasing the efficiency of ideal solar cells by photon induced transitions at intermediate levels', *Phys. Rev. Lett.* **78**, 369.

Luque A. and Marti A. (2010), 'The intermediate band solar cell: progress toward the realization of an attractive concept', *Adv. Mater.* **22**, 160–174.

Luque A., Ruiz J. M., Cuevas A., Eguren J. and Agost J. G. (1978), *Proc. 1st Commission of the European Communities Conf. on Photovoltaic Solar Energy*, Luxembourg, pp. 269–277.

Marti A. and Araujo G. (1996), 'Limiting efficiencies for photovoltaic energy conversion in multigap systems', *Solar Energy Mater. Solar Cells* **43**, 203–222.

Mathematica (2008), Version 7.0 Publisher: Wolfram Research, Inc. Champaign, Illinois.

Mihailetchi V. D., Koster L. J. A. and Blom W. M. (2004), 'Effect of metal electrodes on the performance of polymer:fullerene bulk heterojunction solar cells', *Appl. Phys. Lett.* **85**, 970–972.

Nozik A. J. (2008), 'Fundamentals and applications of quantum-confined structures', in Archer M. D. and Nozik A. J. (eds), *Nanostructured and Photoelectrochemical Systems for Solar Photon Conversion*, Imperial College Press, London.

Nozik A. J. (2010), 'Nanoscience and nanostructures for photovoltaics and solar fuels', *Nano Lett.* **10**, 2735–2741.

O'Dwyer, M. F., Lewis, R. A. and Zhang C. (2005), 'Electronic efficiency in nanostructured thermionic and thermoelectric devices', *Phys. Rev. B* **72**, 205330, 1–10.
Ogino A., Muramatsu1 T. and Kando M. (2004), 'Output characteristics of solar-power-driven thermionic energy converter', *Jpn. J. Appl. Phys.* **43**, 309–314.
Ohl R. S. (1941), 'Light sensitive electric device', U.S. Patent No. 240252, filed 27 March 1941; 'Light sensitive electric device including silicon', U.S. Patent No. 2443542, filed 27 May 1941.
Parrott J. E. (1986), 'Self-consistent detailed balance treatment of the solar cell', *IEEE Proc.* **133**, pp. 314–318.
Pauwels H. and de Vos A. (1981), 'Determination of the maximum efficiency solar cell structure', *Solid State Electronics* **24**, 835–843.
Planck M. (1913), *Vorlesungen über die Theorie der Warmestrahlung*. Leipzig: Barth [translated into English as Planck M, 'The theory of heat radiation', *Dover, New York*, 1959].
Press W. H. (1976), 'Theoretical maximum for energy from direct and diffuse sunlight', *Nature* **264**, 734–735.
Ramsdale C. M., Barker J. A., Arias A. C., MacKenzie J. D., Friend R. H. and Greenham N. C. (2002), 'The origin of the open-circuit voltage in polyfluorence-based photovoltaic devices', *J. Appl. Phys.* **92**,. 4266–4270.
Rau U. (2007), 'Reciprocity relation between photovoltaic quantum efficiency and electroluminescent emission of solar cells', *Phys. Rev. B* **76**, 085303.
Ries H. (1983), 'Complete and reversible absorption of radiation', *Appl. Phys. B* **32**, 153.
Riordan M. and Hoddeson L. (1997), *Crystal Fire: The Birth of the Information Age*, Norton, New York.
Rittner E.S. (1954), 'Use of *p-n* junctions for solar energy conversion', *Phys. Rev.* **96**, 1708–1709.
Ross R. T. and Nozik A. J. (1982), 'Efficiency of hot-carrier solar energy converters' *J. Appl. Phys.* **53**, 3813–3818.
Schwede J. W., Bargatin, I. Riley, D. C., Hardin, B. E., Rosenthal, S. J., Sun Y., Schmitt F., Pianetta, P., Howe R. T. and Shen Z. X. (2010), 'Photon-enhanced thermionic emission for solar concentrator systems', *Nature Mater.* **9**, 762–767.
Shockley W. (1949), 'The theory of *p-n* junctions in semiconductors and *p-n* junction transistors', *Bell System Tech. J.* **28**, 435–89.
Shockley W. and Queisser H. J. (1961), 'Detailed balance limit of efficiency of *p-n* junction solar cells', *J. Appl. Phys.* **32**, 510–519.
Smestad G. P. (2004), 'Conversion of heat and light simultaneously using a vacuum photodiode and the thermionic and photoelectric effects', *Solar Energy Mater. Solar Cells* **82**, 227–240.
Snaith H. J. and Grätzel M. (2006), 'The role of a 'Schottky barrier' at an electron-collection electrode in solid-state dye-sensitised solar cells', *Adv. Mater.* **18**, 1910–1914,
Swanson R. M. (1979), 'A proposed thermophotovoltaic solar energy conversion system', *Proc. IEEE* **67**, pp. 446–447.
Tiedje T., Yablonowitch E., Cody G. D. and Brooks B. G. (1984), 'Limiting efficiency of silicon solar cells', *IEEE Trans. Electron Devices* **ED-31**, 711–716.
Trivich D. and Flinn P. A. (1955), 'Maximum efficiency of solar energy conversion by quantum processes' in Daniels F. and Duffie J. (eds), *Solar Energy Research*, Thames and Hudson, London.

Trupke T., Green M. A. and Würfel P. (2002a), 'Improving solar cell efficiencies by the up-conversion of sub-band-gap light', *J. Appl. Phys.* **92**, 4117–4122.

Trupke T., Green M. A. and Würfel P. (2002b), 'Improving solar cell efficiencies by down-conversion of high-energy photons', *J. Appl. Phys.* **92**, 1668–1674.

Trupke T., Würfel P., Uhlendorf I. and Lauermann I. (1999), 'Electroluminescence of the dye-sensitized solar cell', *J. Phys. Chem. B* **103**, 1905–1910.

Trupke T., Zhao J., Wang A., Corkish R. and Green M. A. (2003), 'Very efficient light emission from bulk crystalline silicon', *Appl. Phys. Lett.* **82**, 2996–2998.

Vandewal K., Tvingstedt K., Gadisa A., Inganäs O. and Manca J. V. (2009), 'On the origin of the open-circuit voltage of polymer–fullerene solar cells', *Nature Mater.* **8**, 904–909.

Werner J. H., Brendel R. and Queisser H. J. (1995), 'Radiative efficiency limit of terrestrial solar cells with internal carrier multiplication', *Appl. Phys. Lett.* **67**, 1028–1030.

Whitehead A. N. (1911), *An Introduction to Mathematics*, Henry Holt & Co., New York, Chapter V.

Wilson A. H. (1931), 'The theory of electronic semiconductors', *Proc. Royal Soc.* London **A133**, 458–491.

Wolf M. (1960), 'Limitations and possibilities for improvement of photovoltaic solar energy converters', *Proc. IRE* **48**, 1246–1263.

Würfel P. (1982), 'The chemical potential of radiation', *J. Phys. C* **15**, 3967–3985.

Würfel P. (1995), 'Is an illuminated semiconductor far from thermodynamic equilibrium', *Solar Energy Materials and Solar Cells* **38**, 23–28.

Würfel P. (2000), *Olympic Conf. on Third Generation Photovoltaics*, Sydney (unpublished work).

Würfel P., Brown A. S., Humphrey T. E. and Green M. A. (2005), 'Particle conservation in hot-carrier solar cell', *Progr. Photovoltaics* **13**, 277–285.

Würfel P. and Wolfgang R. (1980), 'Upper limit of thermophotovoltaic solar-energy conversion', *IEEE Trans. Electron Devices* **ED-27**, 745–750.

CHAPTER 3

CRYSTALLINE SILICON SOLAR CELLS

MARTIN A. GREEN
Australian Centre for Advanced Photovoltaics
School of Photovoltaic and Renewable Energy Engineering
University of New South Wales
Sydney, N.S.W. Australia, 2052
m.green@unsw.edu.au

Vast Power of the Sun Is Tapped by Battery Using Sand Ingredient
Front page headline, *New York Times*, 26 April, 1954.

3.1 Overview

Front page headlines in the New York Times and the Wall Street Journal in 1954 heralded to the world the demonstration of the first reasonably efficient solar cells, an announcement made possible by the rapid development of crystalline silicon technology for miniaturised electronics. The majority of solar cells made since then have been based on silicon in monocrystalline or large-grained polycrystalline form. There are two main reasons for this. One is that silicon is an elemental semiconductor with good stability and a well-balanced set of electronic, physical and chemical properties, the same set of strengths that have made it the preferred material for microelectronics. The second reason why silicon cells have been so dominant is that the success of silicon in microelectronics created an enormous industry where the economies of scale directly benefited the initially much smaller photovoltaics industry.

In the early days of the industry, most silicon cells were made using thin wafers cut from large cylindrical monocrystalline ingots prepared by the exacting Czochralski (CZ) crystal growth process and doped to about one part per million with boron during the ingot growth. A larger proportion now use what are referred to as 'multicrystalline' wafers sliced from ingots prepared by a simpler directional solidification technique, that produces large-grained polycrystalline ingots. To produce a cell, these boron-doped starting wafers generally have phosphorus diffused at high temperatures a fraction of a micron into the surface to form the *p*-*n* junction required for photovoltaic action. Metal contacts to both the *n*- and the *p*-type side of the junction are formed by screen printing metal pastes which are then densified by firing at high temperature. Each cell is typically 12–20 cm either in diameter or along either side if square or rectangular.

Cells generally are sold interconnected and encapsulated into a weatherproof, glass-faced package known as a module, as shown in an exploded view in Fig. 3.1 Each module typically contains 72 cells soldered together in series. Each individual cell gives a maximum output of about 0.63 V in sunlight. The output current depends on cell size and the sunlight intensity (solar irradiance) but generally lies in the 4–12 A range in bright sunshine. The packaging consists of a glass/polymer laminate with the positive and negative leads from the series-connected cells brought out in a junction box attached to the module rear. Such modules have proved extremely reliable in the field, with most manufacturers offering a 25-year warranty on the module power output, one of the longest warranties provided for any commercial product (saucepans have been suggested as one of the few manufactured products

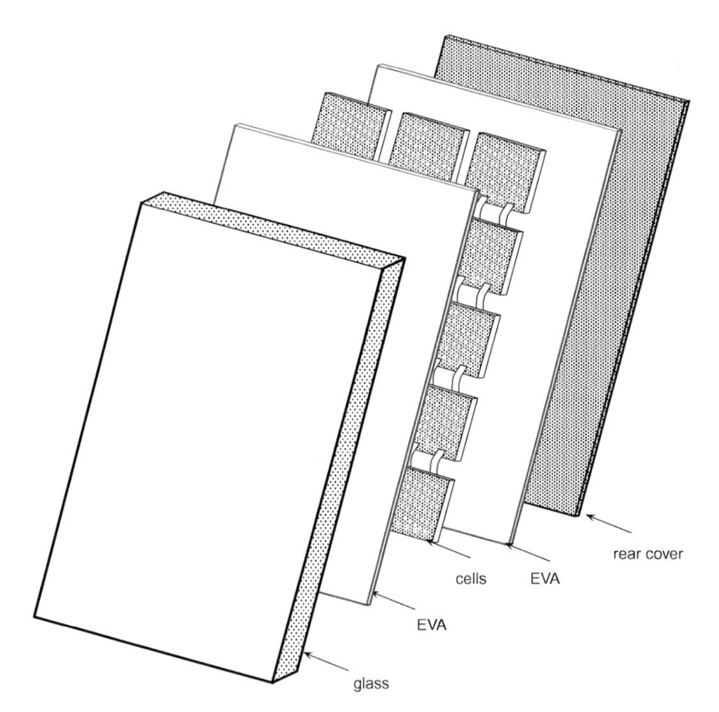

Figure 3.1 Exploded view of a standard silicon photovoltaic module. The different layers shown are laminated together under pressure at a temperature around 140–150 C where the transparent EVA (ethylene vinyl acetate) softens and binds the different layers together on cooling. Source: Green and Hansen (1998).

with a comparable warranty period — my most recently purchased pan only having a 10-year warranty!).

The efficiency of the cells in the module would typically lie in the 15–19% range, appreciably lower than the fundamental 'detailed balance' or Shockley–Queisser limit of 33.5% for silicon (see Chapter 2). Module efficiency is slightly lower than that of the constituent cells due to the area lost by frames and gaps between cells, with module efficiency generally lying in the 12–15% range (Haase and Podewils, 2011). Over the last few years, commercial cells and modules of significantly higher performance have been available in multi-megawatt quantities using more advanced cell processing technology, discussed in more detail later. Cells of 20–22% efficiency and module efficiency in the 18–20% range are now commercially available. (Unless otherwise noted, all efficiencies quoted in this chapter are at standard test conditions, namely with a cell temperature of 25C under 1000 W m^{-2} sunlight intensity with the standard global air mass 1.5 spectral distribution.)

This chapter discusses the historical but steadily weakening links between silicon solar cells and the broader microelectronics industry. Also discussed are standard and improved methods for preparing silicon cell substrates and for processing cells to extract as much performance as possible from these at the lowest possible overall cost. The chapter also describes progress with supported silicon films, which have the advantage of significantly reduced silicon material requirements.

3.2 Silicon cell development

3.2.1 *The early years*

The initial development of silicon photovoltaics was inextricably intertwined with the development of the general silicon electronics field and the subsequent microelectronics industry. The rapid increase in interest in the properties of silicon in the 1940s was triggered by the surprising photovoltaic properties demonstrated by serendipitously formed *p-n* junctions, as described below. This increased interest led directly to the development of point-contact and junction transistors and ultimately to integrated circuits.

The earliest commercial silicon electronic devices were silicon point-contact or 'cat's whisker' diodes dating from the early 1900s (Riordan and Hoddeson, 1997). These devices rectified electrical signals at a junction formed by pressing a thin metal wire against polycrystalline silicon (other semiconductors, such as silicon carbide, were also used). 'Cat's whisker' diodes were key components in early radios. By the 1930s, they were largely replaced by thermionic valves.

However, the evolving field of microwaves created a renewed interest in cat's whisker diodes in the mid-1930s. At Bell Laboratories in the USA, Russell Ohl guessed that impurities caused the erratic behaviour often observed whereby the cat's whisker only operated correctly if located on a 'hot spot' in the silicon. Ohl therefore encouraged colleagues to grow purer silicon, by melting the purest material available in a quartz capsule, and then cooling.

In one specific slow-cooled ingot, the eighteenth in the series (Riordan and Hoddeson, 1997), Ohl and colleagues observed a surprisingly large photovoltage of about half a volt when the ingot was illuminated by a flashlight. This ingot showed two distinct types of properties, dubbed 'positive' (p-type) and 'negative' (n-type), depending on the polarity required for easy current flow between the material and a metal wire placed on the silicon surface, and also the voltage polarity observed under illumination. It was quickly realised that the junction between the p-type and n-type regions, the p-n junction, caused the unusual properties. The first silicon solar cells were formed by cutting the ingot to include sections with both a p- and n-type region and then applying metal contacts (Ohl, 1941). These earliest silicon solar cells, shown in Fig. 3.2a, were only a fraction of percent efficient, but were still very much better in performance than earlier photovoltaic devices based on selenium or cuprous oxide.

The 'grown-in' junctions in the earliest cells arose from the serendipitous distribution of p-type and n-type impurities in the silicon (boron and phosphorus, respectively) resulting from the slow solidification process (Scaff *et al.*, 1949). Ohl realised that more controllable ways of forming the junction would give better performance. In the early 1950s, he was involved in experiments where surface junctions were formed by implanting helium at high energy into the surfaces of p-type polycrystalline silicon slices (Kingsbury and Ohl, 1952). Although this approach produced improved cells of efficiency estimated to be up to 1% (Fig. 3.2b), this work was soon overtaken by independent improvements in silicon technology also made at Bell Laboratories, particularly in two areas.

The first was the development of techniques for preparing single crystals of silicon using the Czochralski growth method. The second was the formation of junctions by high-temperature diffusion of dopant impurities into the silicon surface. Combining these two techniques, researchers at Bell Laboratories were able to announce the first modern silicon solar cell in 1954, in what was one of the early successes of the diffused junction approach (Chapin *et al.*, 1954). Figure 3.2c shows the resulting cell structure.

The impressive performance of these cells by previous standards, up to 6% energy conversion efficiency, created enormous interest at the time, as the newspaper headline at the beginning of this chapter suggests, and also generated unbounded

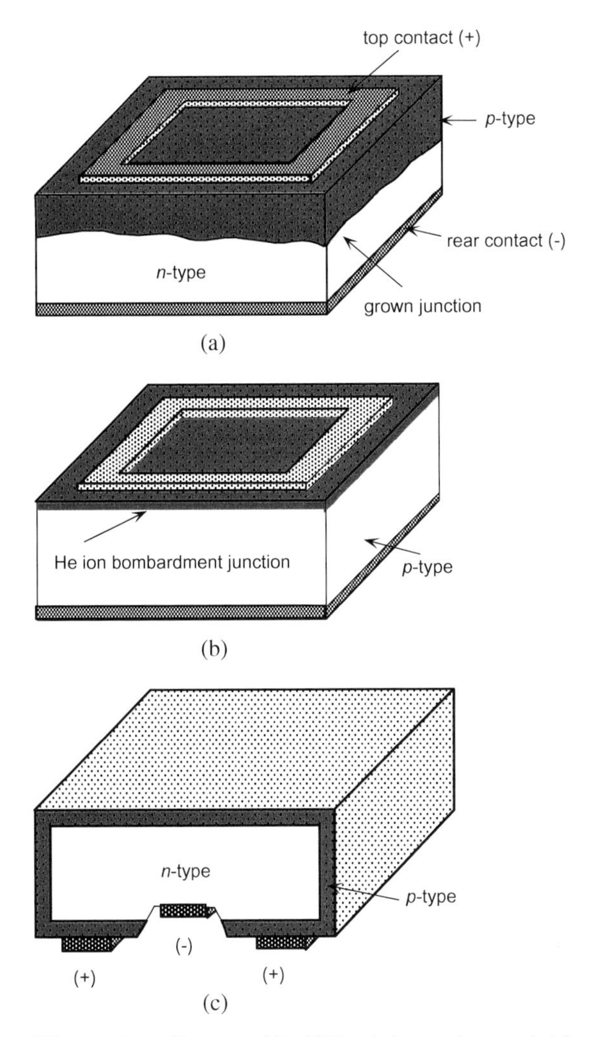

Figure 3.2 (a) Silicon solar cell reported in 1941 relying on 'grown-in' junctions formed by impurity segregation in recrystallised silicon melts; (b) helium-ion bombarded junction device of 1952; (c) first modern silicon cell, reported in 1954, fabricated on single-crystalline silicon wafers with the *p-n* junction formed by dopant diffusion. Source: Green (1995).

enthusiasm for future applications. This enthusiasm proved to be premature, although the cells did find an almost immediate use in spacecraft. Space applications drove the rapid improvement in cell technology such that, by the early 1960s, cell energy conversion efficiency of about 15% under terrestrial sunlight had been demonstrated, with the cells finding a secure market niche powering

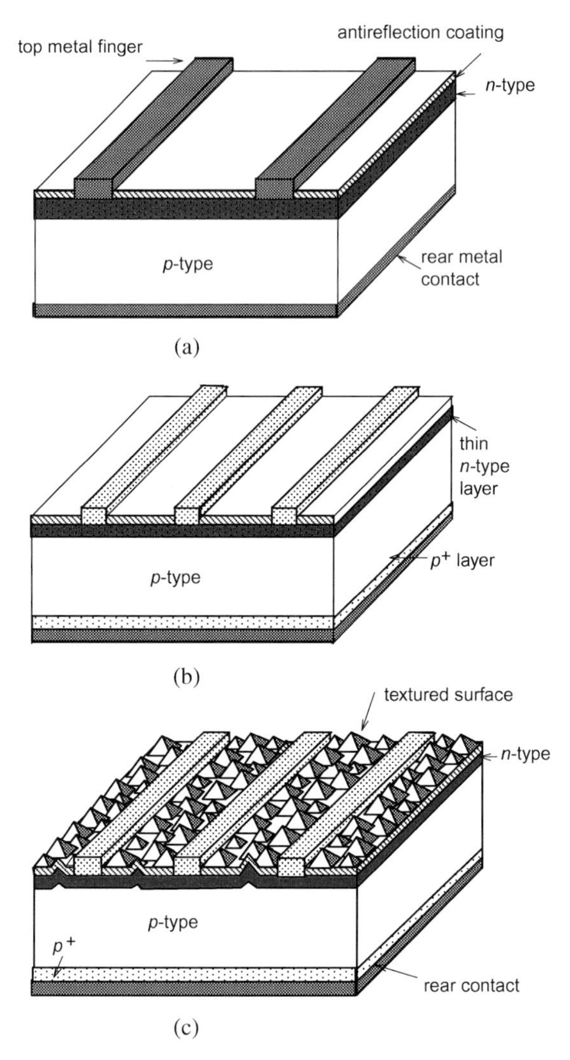

Figure 3.3 (a) Space silicon cell design developed in the early 1960s which became a standard design for over a decade; (b) shallow junction 'violet' cell; (c) chemically textured non-reflecting 'black' cell. Source: Green (1995).

a rapidly increasing number of satellites (Wolf, 1976). The basic cell design that evolved (Fig. 3.3a), remained unchanged from the early 1960s for almost a decade.

In the early 1970s, a reassessment of cell design at COMSAT Laboratories showed that a shallower diffusion combined with more closely spaced metal fingers gave a substantial improvement in the cell performance by improving the response to blue wavelengths (Lindmayer and Allison, 1973). The resulting cells, shown in Fig. 3.3b, known as 'violet' cells due to their characteristic colour arising from the shorter wavelengths reflected, produced the first improvement in cell performance for over a decade.

This improvement was augmented by the realisation that incorporating a thin heavily doped layer under the back contact, creating a so-called 'back surface field', gave unexpected benefits (Godlewski *et al.*, 1973). This approach worked best if the rear doped layer was formed by alloying the underlying silicon with aluminium deposited over the cell rear. Not long afterwards, the idea of using anisotropic chemical etches to form geometrical features on the silicon surface was successfully demonstrated, also at COMSAT Laboratories (Haynos *et al.*, 1974), and resulted in a further boost in performance, taking terrestrial cell efficiency to above 17% (Fig. 3.3c). The surface features consisted of square-based pyramids defined by slowly etching {111} crystallographic planes. These greatly reduced reflection from the cell surface, giving these 'black cells' the appearance of black velvet after antireflection coating.

The improvements of the early 1970s came about primarily by enhancing the ability of the cell to collect carriers generated by the incoming photons. Since cells now appeared to be performing to close to their full potential in this regard, it seemed that any further improvement in cell performance would come from increased open-circuit voltage. Gains in this area became the focus of work directed at improving cell efficiency throughout the second part of the 1970s, largely as a result of a program directed by NASA-Lewis aimed at targeting better space cell performance (Brandhorst and Bernatowicz, 1980).

On the commercial front, the oil embargoes of the early 1970s generated widespread interest in alternative sources of terrestrial energy. A small terrestrial photovoltaic industry came into existence largely as a result of the US Government's photovoltaic program. One component of this program (Christensen, 1985), arguably the most successful in terms of developing the industry and its products, involved a staged series of purchases of photovoltaic modules meeting increasingly stringent specifications.

The first such purchase in 1975/76, known as 'Block I', was remarkable for the diversity of both cell fabrication and module encapsulation approaches used in the product supplied by four different manufacturers. One manufacturer, Spectrolab of Sylmar, California, supplied cells where the contacts had been applied using screen printing (Ralph, 1975), the forerunner of the billions of cells of this type which were to follow. In the 'Block II' purchases under this program (1976/77), the same company combined screen-printed cells with a laminated module design (Fig. 3.1), a combined approach that had been adopted by almost all commercial

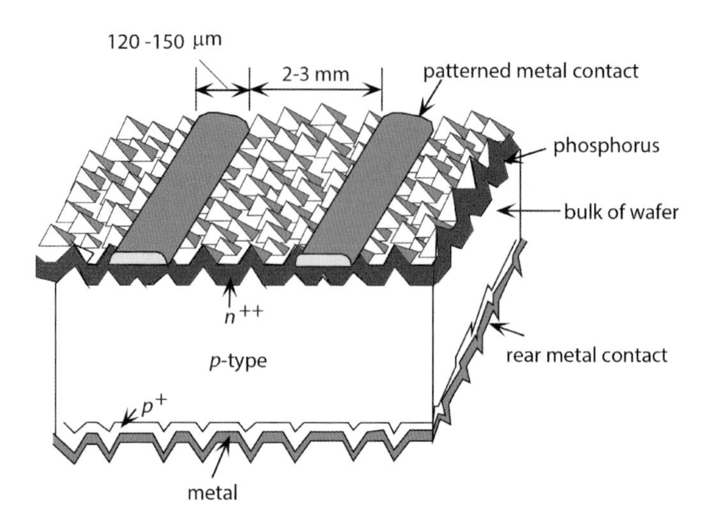

Figure 3.4 Screen-printed crystalline silicon solar cell (not to scale). Source: Green (1995).

manufacturers by the early 1980s and, with relatively minor modification, remains the present commercial standard.

The main features of a commercial screen-printed cell are shown in Fig. 3.4. The basic cell design is similar to that of a standard space cell of the 1960s (Fig. 3.3a), but incorporates the surface texturing of the 'black' cell of Fig. 3.3c as well as the screen-printing approach to applying the front and rear contacts. Present efficiencies are quite similar to those demonstrated by 'black' cells in 1974.

Since the Block II purchases of 1976/77, no major changes were made in either the basic screen-printed approach to cell fabrication or to the cell encapsulation approach until quite recently. Considerable attention, however, has been directed towards reducing the cost of the silicon wafer, initially grown by the Czochralski technique, since this accounts for a large fraction of the cost of a standard silicon module. The most successful approach has been the simplification of the ingot growth processes by using cruder directional solidification approaches to produce multicrystalline ingots (Ferrazza, 1995; Rodriguez *et al.*, 2011).

The first multicrystalline silicon cells developed specifically for the terrestrial market were reported in 1976 (Lindmayer, 1976; Fischer and Pschunder, 1976) and commercial multicrystalline cells have been available since the late 1970s. These multicrystalline approaches involve basically a reversion to the earlier ingot-forming approaches for crystal rectifiers, techniques pre-dating the microelectronics explosion. In recent years, multicrystalline silicon cells accounted for the majority of the total market for photovoltaic product (Hering, 2011). Another major area of emphasis has been to reduce the thickness of the silicon wafer by slicing it more thinly. This resulted in the replacement of inner diameter sawing methods traditionally favoured by the microelectronics industry by wire-cutting approaches, described in more detail in Section 3.3.1.

3.2.2 *The path to 25% cell efficiency*

On the research front, it had become apparent by the late 1970s that oxide passivation of the cell surfaces was the key to improving open-circuit voltage. The early 1980s saw a series of successively improved oxide-passivated cells fabricated by the author's group at the University of New South Wales (UNSW) taking silicon cell efficiency past 18%, then past 19% and finally 20%, the 'four minute mile' of the photovoltaics area. The UNSW group has held the 'world record' for silicon cell performance, almost without interruption, since this time, with this record until recently standing at 25% (Green, 2009).

In 1985, the UNSW-developed microgrooved PESC cell (passivated emitter solar cell) of Fig. 3.5a became the first silicon cell to exceed 20% energy conversion efficiency. The same basic approach was then used by several other groups to produce cells of similar efficiency, with commercial quantities produced for solar car racing and for space. The approach is characterised by the use of a thin thermally grown oxide to 'passivate' (reduce the electronic activity of) the top surface of the junction diffusion layer (the emitter of the cell), combined with the use of a shallow, high sheet resistivity phosphorus diffusion for this emitter. Another feature is the use of photolithography to produce a relatively small contact area to this emitter region by defining openings in the 'passivating oxide'. Photolithography is also used to pattern the top contact fingers and to align these fingers to the oxide openings.

The rear of the cells borrows the 'alloyed-aluminium back-surface-field' approach from earlier space cells. In this approach, a layer of aluminium is deposited on the rear of the cell and alloyed into the cell at temperatures above the silicon–aluminium eutectic. After cooling, this produces a layer of *p*-type silicon heavily doped with aluminium at the rear of the silicon substrate. This reduces rear contact recombination. Some improvement of substrate quality also occurs during alloying through defect and impurity 'gettering'.

Cells of a similar quality to the first 20%-efficient PESC cell also found their way into volume terrestrial cell manufacture through the laser-grooved buried contact cell of Fig. 3.5b. This cell retains an alloyed aluminium rear and also incorporates the improvements to front surface passivation first demonstrated by the PESC cell. To make the approach suitable for low-cost production, the photolithographic metallisation of the PESC cell was replaced by a combination of laser grooving, to define the areas to be metallised, followed by electroless metal plating. The oxide

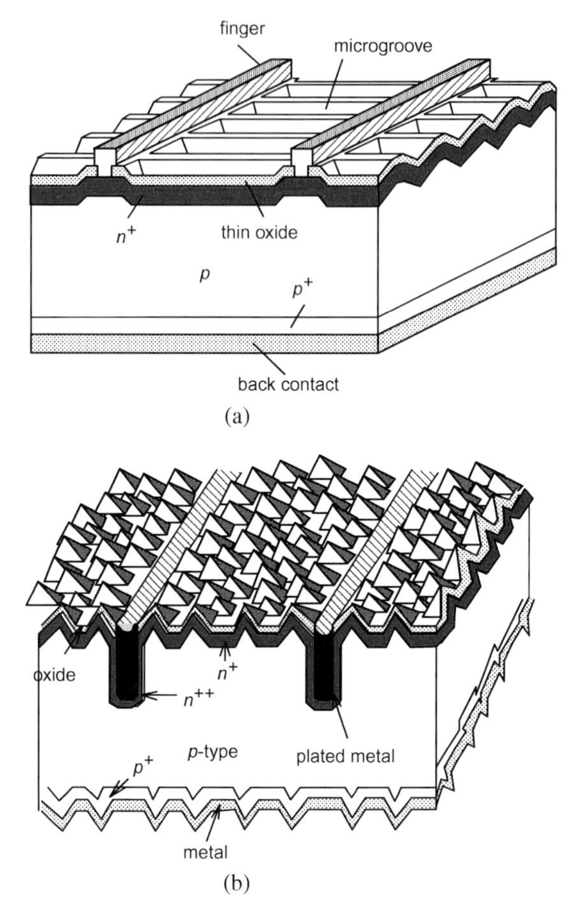

Figure 3.5 (a) The microgrooved passivated emitter solar cell (PESC cell) of 1985, the first silicon cell to exceed 20% efficiency; (b) buried contact solar cell. Source: Green (1995).

or other dielectric coating the top surface in this case not only serves as surface passivation, but also as a diffusion mask to confine the heavy diffusion to the laser-grooved areas and as a plating mask for the subsequent plating of metal into these grooved areas. During the early 1990s, the buried contact approach produced the highest performance terrestrial cells made in any appreciable volume, with efficiency in the 17–18% range obtained using standard commercial silicon wafers (Jordan and Nagle, 1994).

The next major laboratory improvement in silicon cell design came in the use of oxide passivation along both the front and rear surfaces, first demonstrated in the rear point-contact solar cell developed by Stanford University. As shown in

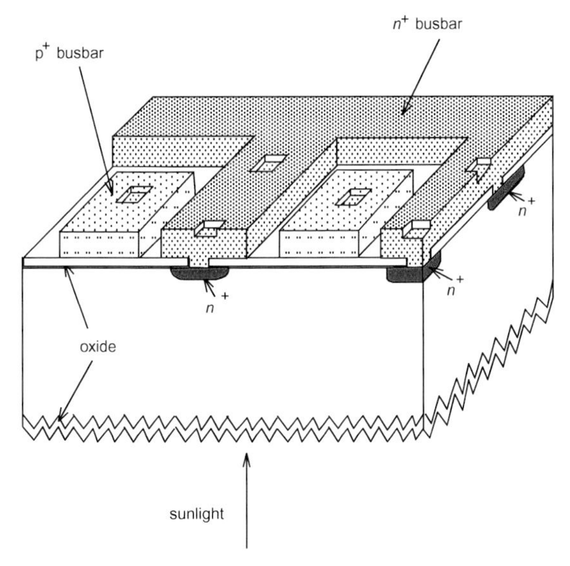

Figure 3.6 Rear point-contact solar cell, which demonstrated 22% efficiency in 1988 (cell rear shown uppermost). Source: Green (1995).

Fig. 3.6, this cell has an unusual design in that both positive and negative contacts are on the rear surface of the cell. Although this might, at first sight, appear to be a regression to a similar design to that used in the first modern silicon cell of Fig. 3.2c, there is a substantial difference in the way the two types of cells operate. The modern rear junction cells take advantage of the excellent quality of silicon now available. Carrier diffusion lengths are several times the cell thickness, allowing carriers photogenerated near the top surface of the cell to diffuse to the rear contacts. In the earlier device, the junctions at top and rear surfaces are electrically connected around the cell edge (Fig. 3.2c). Most carriers in this earlier cell are collected by the top junction and flow around the cell edges to the rear contact. The rear point-contact cell demonstrated 22% efficiency in 1988 and has since been successfully commercialised, as subsequently discussed.

The next improvement in silicon cell efficiency came, again at UNSW, by combining the earlier developments in the PESC cell sequence with the front and rear oxide passivation first demonstrated in the rear point-contact cell. This is possible in a number of ways as shown in Fig. 3.7. In the PERC cell (passivated emitter and rear cell) of Fig. 3.7a, the first successfully demonstrated, rear contact is made directly to the silicon substrate through holes in the rear passivating oxide. This approach works reasonably well provided the substrate is sufficiently heavily doped to ensure low contact resistance between the metal and substrate (below

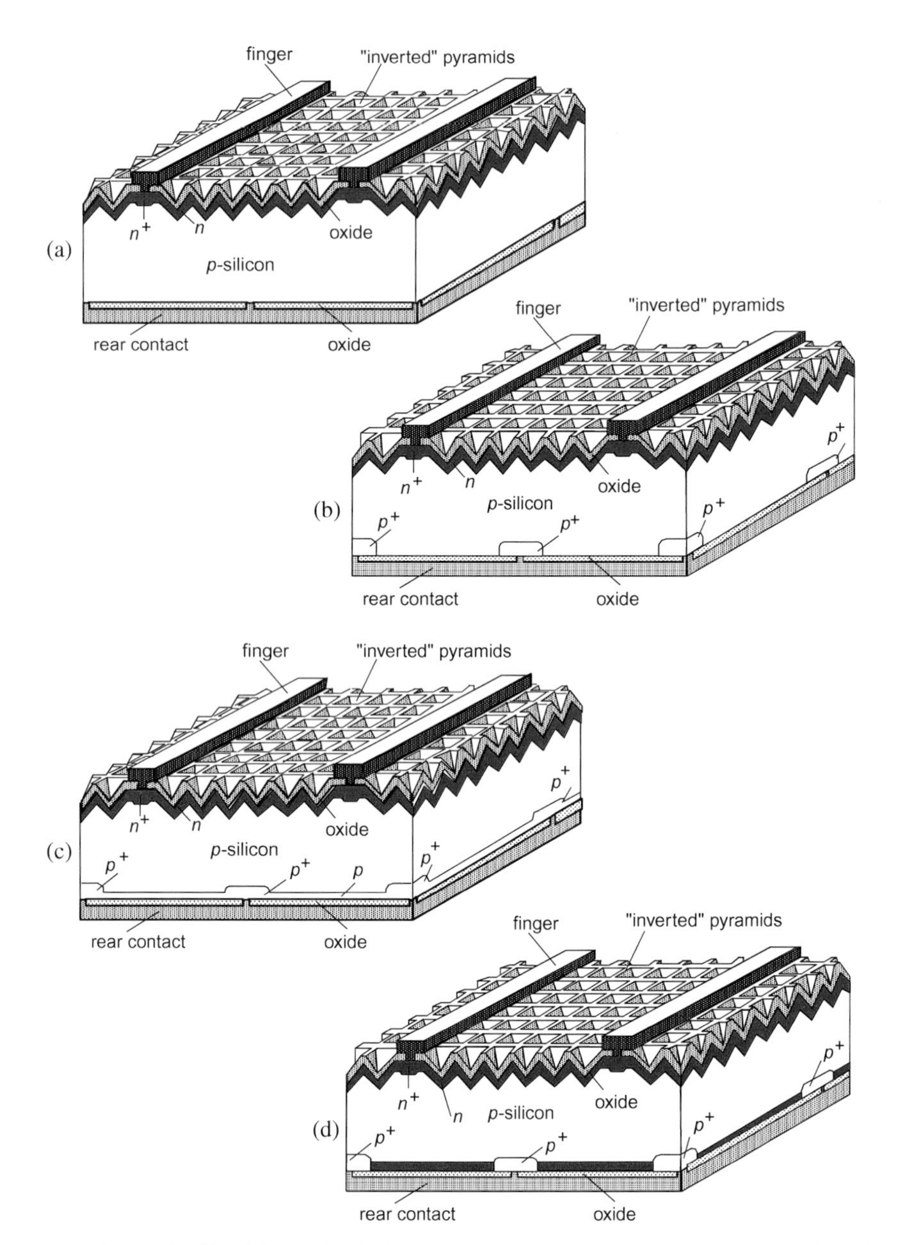

Figure 3.7 A family of four related high-efficiency solar cell structures: (a) the passivated emitter and rear cell (PERC cell); (b) the passivated emitter, rear locally diffused cell (PERL cell) which took efficiency above 24% in the early 1990s and subsequently to 25%; (c) the passivated emitter, rear totally diffused cell (PERT cell); and (d) the passivated emitter, rear floating junction cell (PERF cell). Source: Green and Hansen (1998).

about 0.5 Ω cm resistivity for *p*-type substrates). The PERC cell is often suggested as a relatively low cost way for making silicon cells above 20% efficiency, since it is the simplest of the approaches of Fig. 3.7.

Historically, the next improvement was demonstrated by the PERL cell (passivated emitter, rear locally diffused) of Fig. 3.7b. In this case, local diffusion is used in the area of the rear point contact to provide a minority-carrier-reflecting region between this contact and the substrate and to reduce contact resistance. This approach produced the first 24% efficient silicon cell in 1994 (Zhao *et al.*, 1995) and the world best result for silicon of 25% (Green, 2009) under the standard AM1.5 global spectrum. The PERL cell has been used in reasonably large quantities in solar car racing and in space cells.

The third cell of Fig. 3.7c is the PERT cell (passivated emitter, rear totally diffused). This has given performance almost equivalent to the PERL cell and also offers some fabrication simplifications. The PERT cell has also been used in space cell production. The final structure shown in Fig. 3.7d, the PERF cell (passivated emitter rear floating junction) offers perhaps the best long-term potential for high performance. This structure has produced the highest open-circuit voltage of the cells of Fig. 3.7, with open-circuit voltage up to 720 mV demonstrated under standard test conditions (Wenham *et al.*, 1994), together with efficiencies above 23%.

One feature of these cell designs is the very effective trapping of light within the cell. By depositing metal over the entire rear surface of the cell but ensuring it is displaced from the silicon substrate by an intervening layer of oxide, very high rear reflectance is obtained for light striking this rear reflector from within the cell. When combined with appropriate geometrical structure on the front surface of the cell, weakly absorbed light that is reflected from this rear surface can be trapped quite effectively within the cell, taking advantage of total internal reflection from the front surface. This greatly extends the response of the cell to infrared wavelengths. Cells that convert infrared wavelengths with an efficiency approaching 50% have been demonstrated (Green *et al.*, 1992).

3.2.3 *Subsequent silicon cell developments*

Although the 25% efficiency figure established by the PERL cell remains close to the highest efficiency demonstrated for silicon under standard AM1.5 global test conditions, there have been recent developments that demonstrate paths to future progress.

The most efficient silicon cells to date have been made on *p*-type substrates. This is largely because phosphorus diffusion is, in several ways, easier than boron diffusion, with a wider parameter space giving good results in terms of uniformity of diffusion and minimal degradation of the starting quality of the substrate. Oxide and nitride passivation of phosphorus-diffused surfaces also give good results. This makes phosphorus the preferred dopant for the more critical top doped layer of the cell. For the rear surface, there are fewer constraints on the required properties of the diffusion, allowing boron diffusions to work well there.

The Stanford University rear point-contact cell of Fig. 3.6 is shown as originally conceived, without any diffusion on the surface exposed to sunlight. Adding a non-contacted or 'floating' phosphorus diffusion to the top (sunlight-exposed) surface improved performance, with 22.7% efficiency soon thereafter demonstrated, long the best for any cell on an n-type silicon substrate.

In attempting to adapt their p-type wafer work to n-type substrates, the UNSW group were able to equal this result, but again using a phosphorus diffusion along the top surface. A 'front surface field' structure was used, as in the PERT cell of Fig. 3.7c, but with the substrate n-type rather than p-type. The standard PERL structure of Fig. 3.7b on n-type substrates with all polarities reversed gave slightly poorer results due to the more difficult challenge involved in passivating the B-doped top surface.

The challenge of improving cell performance on n-type substrates was subsequently met by the combined efforts of the Fraunhofer Institute in Freiburg, Germany, and ECN, Netherlands (Benick *et al.*, 2008; 2009). Dielectrics such as silicon oxide and nitride work well on n-type surfaces since they tend to have nett positive electrical charge within them, like many other dielectrics, with this charge tending to attract electrons to underlying silicon surfaces (Godfrey and Green, 1980). This also depletes holes in the surface region, reducing surface recombination rates by squeezing off the supply of these minority carriers. It was found that Al_2O_3 could be prepared with nett negative charge, repelling electrons from the interface to the interior, similarly restricting the supply for recombination on p-type surfaces, such as for the top surface of an n-type cell if B-doped. A significant improvement in n-type cell performance to 23.4% was demonstrated by using the PERL cell structure of Fig. 3.7b with all polarities reversed but with the top surface coating (normally silicon oxide under a ZnS/MgF double layer antireflection coating) replaced by 30 nm Al_2O_3 overlaid by 40 nm Si_3N_4 (Benick *et al.*, 2009). Although below the best PERL cell on p-type substrates on each of the major cell parameters (V_{oc}, i_{sc}, FF), this result is important in demonstrating the ability to passivate B-doped surfaces, increasing future flexibility in cell design.

Another significant development has been the HIT (heterojunction with intrinsic thin layer) cell, developed by Sanyo (now Panasonic) and shown in Fig. 3.8 (Tanaka *et al.*, 1993; Sawada *et al.*, 1994; Maruyama *et al.*, 2006). Here the problems raised by diffusions and their passivation are avoided totally by using deposited layers of hydrogenated amorphous silicon (a-Si:H) for both positive

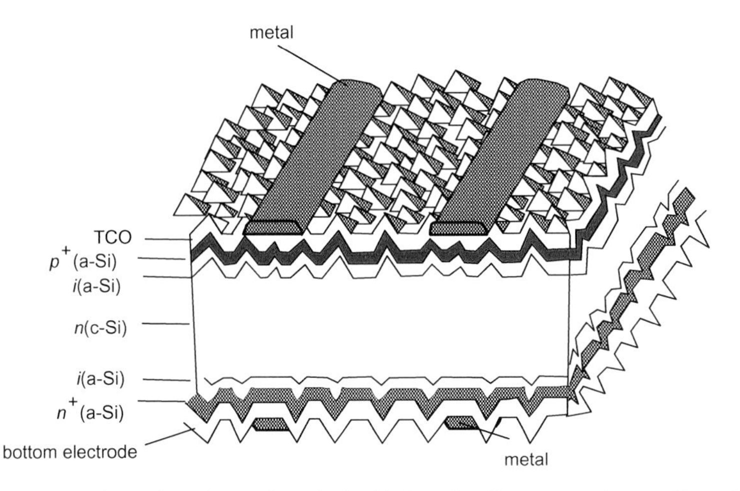

Figure 3.8 HIT (heterojunction with intrinsic thin layer) cell on textured crystalline silicon substrate (after Green and Hansen, 1998).

and negative contacts to top and rear surfaces. The hydrogen, incorporated in the 5–10% range by atomic volume, modifies the electronic structure of the a-Si:H alloy, producing much larger bandgap (about 1.7 eV compared to 1.1 eV for crystalline silicon). The a-Si:H can be doped n- or p-type, although the electronic quality of doped material tends to be poor. This different form of silicon is widely used in its own right for thin-film solar cells (Chapter 4), making the HIT cell structure a marriage between two successful cell technologies. Features such as transparent conducting oxides (TCO) are also used, with these widely used in thin-film technologies but not in standard silicon technology. The HIT also has some similarities with III–V cell technology (Chapter 7), where heterojunctions between semiconductors with different bandgaps are a common feature.

The advantage of using wider bandgap material is that minority-carrier concentrations are greatly suppressed by the wider bandgap, correspondingly reducing recombination rates. In the HIT cell structure, all significant recombination occurs in the wafer substrate or at its surfaces. An important additional feature of the HIT cell is a very thin layer of intrinsic a-Si:H, only 10–20 nm wide, interposed between the wafer and the thin doped a-Si:H regions. This thin intrinsic layer helps reduce interfacial recombination, possibly by reducing defect states in the bandgap of the a-Si:H in these regions that would be created by doping.

The HIT cell has a number of attractive features. One is that all processing of the starting wafer can be effected at low temperature, allowing the initial wafer quality to be retained during processing. The resulting cells have a bifacial structure, so can respond to light from either direction and, in principle, wafers of either polarity could be used, although n-type wafers are the preferred choice.

Standard HIT cells have also demonstrated outstanding efficiencies, up to 24.7% for large area cells (Maruyama *et al.*, 2006), until recently the best ever result for n-type substrates. The outstanding performance parameter has been the open-circuit voltage, with values up to 745 mV demonstrated for 23% efficient cells (compared to 708 mV for 25% efficient PERL cells). This high voltage results from inherently lower surface recombination due to the passivation effects of the overlying a-Si:H. Since there is a very direct and general relationship between the open-circuit voltage and the temperature coefficient of performance of the cell (Green, 2003), this gives a significant advantage at high temperature compared with standard commercial silicon cells (where V_{oc} might be only in the 610–630 mV range).

Technically, the main disadvantages with the standard HIT cell approach are optical. The top doped a-Si:H layer absorbs ultraviolet and blue light and, because of its poor electronic quality, this absorption does not contribute to the cell current. Similarly, the conducting oxide layer will absorb incoming photons, particularly at the infrared end of the spectrum. This will additionally degrade the effectiveness of any light-trapping scheme that attempts to boost the cell's response to poorly absorbed infrared light. Consequently, current output from the cell is 5–10% lower than from other similarly high-efficiency cells. The cost of processing wafers to cells is also reported to be moderately high (Song *et al.*, 2010). The HIT approach has been important in demonstrating that almost perfect surface passivation is possible in silicon cells, showing what ultimately might be possible in the future.

The final recent improvement to be discussed relates to the Stanford rear point-contact cell (Fig. 3.6), an important contributor to the evolution of silicon cell efficiency, as already noted. This cell was commercialised by SunPower (De Ceuster *et al.*, 2007) and now produces the highest efficiency commercial panels. A recent re-evaluation of cell design produced impressive results (Cousins *et al.*, 2010), also for n-type substrates. Figure 3.9 shows the cell structure of this 'Generation III' product.

Improved diffusions for both contacted and non-contacted regions of the rear of the cell have contributed to the improved performance (Cousins *et al.*, 2010). A combination of oxide/nitride passivation is used for non-contacted areas where diffusions are quite light ($\sim 10^{18}/\text{cm}^3$). For the contact regions, a heavier doping is used as a compromise between the optimum for those regions directly contacted by metal and those, also in the contact region, but not so directly contacted. An impressive efficiency of 24.2% was demonstrated on a large-area, commercially-sized cell (156 cm^2) using commercial production equipment, with an impressive V_{oc} of 721 mV (Cousins *et al.*, 2010).

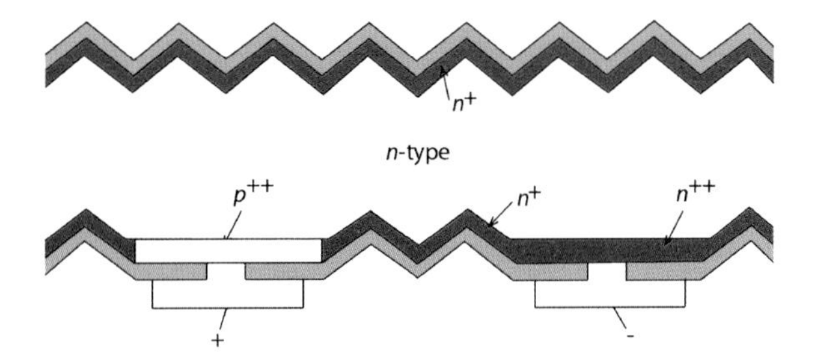

Figure 3.9 SunPower 'Generation III' cell. Source: Green and Ho (2011).

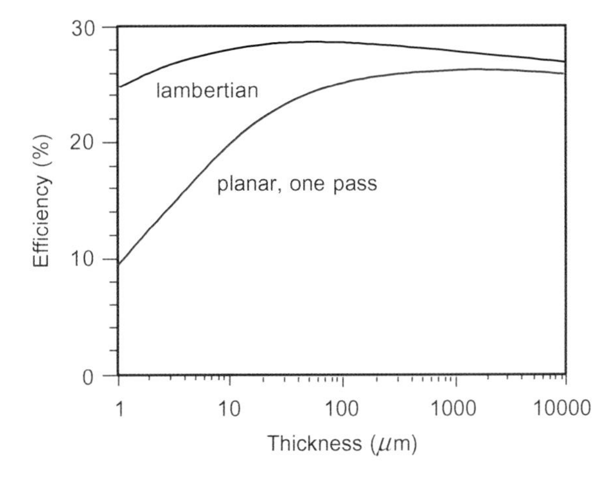

Figure 3.10 Limiting efficiency of a silicon solar cell as a function of cell thickness with and without Lambertian light trapping (global AM1.5 spectrum, 100 mW cm^{-2}, 25°C). Source: Green (1995).

3.2.4 *Opportunities for future improvements*

How will silicon cell design evolve in the future? Some insight is provided by Fig. 3.10, which shows the calculated intrinsic energy conversion efficiency bounds on single-junction silicon solar cells, with and without 'Lambertian' light trapping. In 'Lambertian' light-trapping schemes, the light direction within the cell is randomised (Green, 1995) allowing path length enhancements to be quite readily calculated (these reach about 50 in idealised situations). The limiting efficiency of a silicon cell is somewhat less than the Shockley–Queisser limit of Chapter 2. This is because, as well as intrinsic radiative recombination processes, a competing intrinsic Auger recombination process is also present (Green, 1984; Tiedje *et al.*,

1984). This limits efficiency to about 29%, compared to the Shockley–Queisser limit of 33.5%.

The best laboratory cells have demonstrated around 85% of the achievable efficiency. In the best experimental PERL cells, performance losses of the order of 5% arise from less than ideal values of each of the short-circuit current, open-circuit voltage and fill factor. The short-circuit current losses are most easily identified and reduced. These come from metal finger coverage of the top surface, top surface reflection loss, and less than perfect light trapping in the experimental cells. The voltage loss arises from finite surface and bulk recombination in excess of the lower limit imposed by intrinsic Auger recombination processes (Green, 1984). The fill factor loss comes not only from ohmic series resistance loss within the cell, but also from the same factors producing the open-circuit voltage loss. Parasitic recombination must be sufficiently reduced so that the dominant recombination component at the cell's maximum power point is Auger recombination. This is a more challenging requirement than the corresponding criterion at open-circuit voltage (Green, 1984).

As opposed to the 25% cell efficiency demonstrated by laboratory devices, most present manufacturers of mainstream commercial product would be very pleased to be producing cells of 19% efficiency. Some of the difference between laboratory and commercial cell performance is due to poorer quality of silicon substrate material. A large component, however, is due to limits imposed by the present screen-printing process predominantly used for commercial cell fabrication. The penalty for the processing simplicity offered by this approach is a partly compromised cell design, since a heavily doped emitter layer is required and rear-contact recombination and optical properties are non-ideal. Improved designs subsequently discussed offer the apparently contradictory advantages of both higher cell performance and lower overall manufacturing costs.

It seems that, eventually, efficiencies well above 20% will be the norm for mainstream commercial cells, achieved by paying attention to the passivation of both front and rear surfaces, by thinning the cells to reduce bulk recombination and by modifying the crystal growth processes to produce low-cost silicon customised for photovoltaics, particularly in its ability to withstand high-temperature processing without loss of electronic quality.

An interesting result highlighted by Fig. 3.10 is that light trapping allows high performance, in principle, from silicon cells that are only 1 μm thick. This provides a justification for expecting very high performance, eventually, from the thin, supported silicon cells discussed in Section 3.7. To approach the limiting performance, the demands on bulk quality become less severe as the cell becomes thinner (Green, 1995). However, the demands on light trapping and surface passivation become more severe. Interestingly, Fig. 3.10 shows that the commercial imperative to decrease wafer thickness to decrease silicon costs (SEMI, 2011) may actually lead to higher energy conversion efficiency, with appropriate device structures.

Various approaches have been suggested that have the potential, in principle, for exceeding even the challenging efficiency limits of Fig. 3.10. These include the use of tandem cells, the use of high-energy photons to create more than one electron-hole pair (Werner *et al.*, 1994), or the use of sub-bandgap photons in schemes such as incorporation of regions of lower bandgap (Healy and Green, 1992), multiple quantum wells (Barnham and Duggan, 1990), mid-gap impurity levels (Wolf, 1960) or up- and/or down-conversion (Trupke *et al.*, 2002a; 2002b). The tandem cell approach appears the most likely to have impact in the long term, once the problems are overcome with lattice-matching a top cell with a suitable bandgap to silicon.

3.3 Substrate production

3.3.1 *High-purity silicon*

Silicon solar cell technology initially benefited directly from the economies of scale of the silicon microelectronics industry. Silicon cells were also once able to make use of scrap material from this industry, since the material quality requirements in photovoltaics are less demanding than in the microelectronics field. However, since 2006, the photovoltaic industry has used a larger volume of silicon than microelectronics, changing forever the dynamics of pure silicon supply.

For microelectronics, the starting point for producing the required high-quality 'semiconductor grade' silicon is a lower grade of silicon known as 'metallurgical grade', produced by the reduction in an arc furnace of quartzite by carbon, the latter generally in the form of wood chips. This metallurgical grade silicon is about 98% pure and is produced in large quantities for the steel and aluminium industries. A relatively small quantity is refined for microelectronics and photovoltaics by conversion to a volatile intermediary that can be purified by fractional distillation. The purified intermediate compound is then decomposed to re-extract the silicon in a similarly highly purified form.

Generally the metallurgical grade silicon is converted by hydrochloric acid to trichlorosilane, which is then purified to 99.9999999% (nine 'nines') purity by fractional distillation. Silicon is then extracted from the trichlorosilane by reducing the latter by hydrogen at high temperature. In this process electrically heated silicon rods are exposed to a trichlorosilane/hydrogen mixture that reacts on the surface of the rods, depositing silicon onto them and building up their cross section. These rods grow with a fine grain polycrystalline silicon microstructure. After the rod diameter has increased to the required size, the process is stopped and the rods mechanically broken into smaller chunks, that maintain 'nine-nines' purity. These chunks then become the starting point for the growth of ingots of good crystalline quality.

In the 1970s and early 1980s, several other options for preparing silicon feedstock were investigated as part of a large US government photovoltaic program supported by the Carter administration (Christensen, 1985). Approaches investigated ranged from those involving radically different techniques to those exploring only minor changes from the sequence outlined above, such as the use of different compounds of silicon as the intermediate during the purification process. One such process, based on the use of silane as the intermediate (Christensen, 1985), is now used commercially. Decomposition of the intermediary gas in fluidised bed reactor was also studied in this program and is now also used commercially, greatly reducing process energy requirements. Other projects were based on the use of upgraded metallurgical grade (UMG) silicon, a line of investigation still attracting great attention given its appeal in circumventing the complexity of the traditional purification approach.

3.3.2 *Czochralski growth*

As previously mentioned, crystalline ingots have traditionally been prepared by the Czochralski (CZ) crystal growth process. In this process, the purified silicon chunks are melted in a quartz crucible along with small pieces of silicon heavily doped with boron. This produces a boron-doped melt into which a seed crystal is dipped and slowly withdrawn (Fig. 3.11a). For high-quality crystal growth, good temperature uniformity and slow and steady growth are required. Typically ingots are grown to about 12.5–20 cm in diameter and 1–2 metres in length, weighing 100–200 kg. The crystallographic orientation of the seed is transferred to the grown crystal. Generally, for photovoltaics, the crystal is grown with a preferred orientation so that the wafers sliced from the crystal perpendicular to the growth axis have surfaces parallel to {100} crystallographic planes.

Prior to slicing these ingots into wafers, they were once subject to a centreless grinding operation to remove the slight fluctuations in diameter along the length of the ingot that occur during crystal growth. More commonly, the ingots are now 'squared-off' by sawing off large sections parallel to the growth axis (Fig. 3.11b), giving 'quasi-square' wafers after wafering. The large pieces of silicon sawn off in this approach are generally recycled by re-melting as feedstock for the CZ growth.

The technique traditionally used in microelectronics for sawing wafers from ingots was based on the use of inner diameter saws. In this technique (Fig. 3.12a), thin metal sheet blades are given dimensional solidity by being held in tension

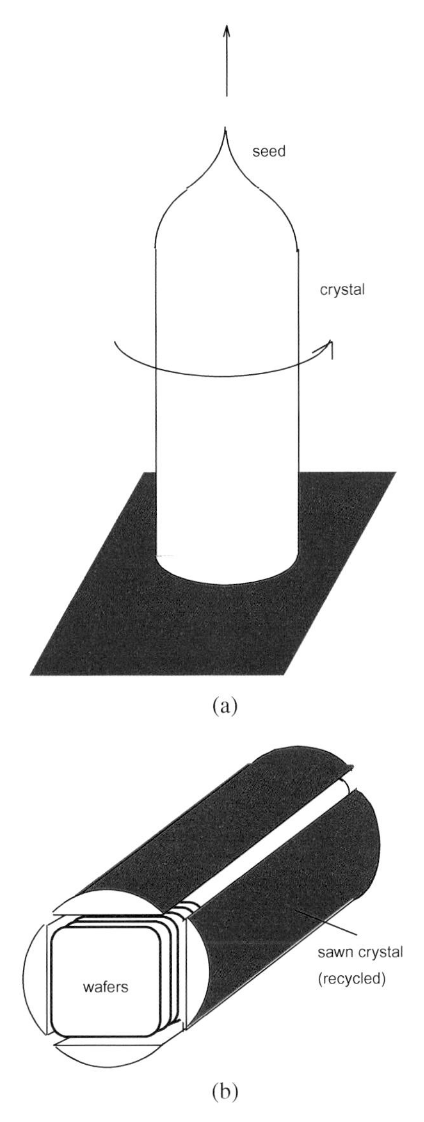

Figure 3.11 (a) Czochralski (CZ) growth; (b) squared-off CZ ingot. Source: Green and Hansen (1998).

around their outer perimeter. The cutting surface is a diamond impregnated edge surrounding a hole within the tensioned metal sheet. This technique gives excellent dimensional tolerance, although there are limitations arising from the thickness of the silicon wafers that is possible to produce while still maintaining high yield.

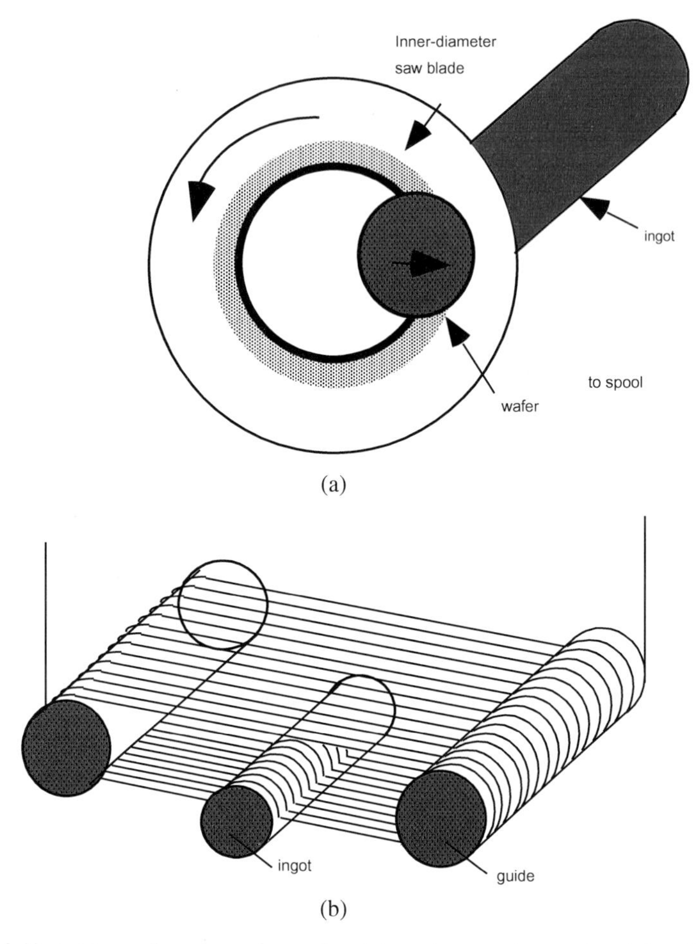

Figure 3.12 (a) Inner diameter wafer sawing; (b) continuous wire sawing. After Dietl *et al.*, 1981.

Other limitations arise from the wastage of silicon as 'kerf' loss during cutting. Generally, about 10–15 wafers per centimetre of ingot length were achieved by this process.

An alternative technique now almost universally used in photovoltaics is based on wire sawing (Fig. 3.12b). In this case, tensioned wire is used to guide an abrasive slurry through the ingot (Schumann *et al.*, 2009). Advantages are thinner wafers and less surface damage for these wafers as well as lower kerf or cutting loss, initially allowing the sawing of over 20 wafers per centimetre. With subsequent reduction in wafer thickness to 160–180 microns in 2011 and of the diameter of the wire to the 100–120 micron range, this figure has increased to over 30 wafers per centimetre, with up to 50 wafers per centimetre of ingot length expected over the coming decade.

There has been a recent resurgence of interest in an alternative wire sawing approach where the wire is similarly thin, but impregnated with diamond particles (Schmidt, 2011). This approach has the advantage of much higher cutting speed as well as reduced kerf loss. The process also offers more potential for the recycling of the silicon that would otherwise be wasted as kerf loss during the sawing process.

An alternative to the standard Czochralski process for producing crystalline ingots is the floatzone (FZ) process. Although some studies have predicted superior economics when compared to the Czochralski process for cell production due to the elimination of consumables such as quartz crucibles, the FZ process, as commercially implemented, is capable of accepting feedstocks only in the form of high-quality cylindrical rods. This makes it unsuitable for using low-cost source material.

Early interest in producing monocrystalline silicon by a directional solidification process (Schmidt, 2011) has been recently revived and will be discussed in the next section. Unlike the FZ process, this approach is very tolerant of low-grade source material (Khattak *et al.*, 1981), similar to the directional solidification approaches to producing multicrystalline silicon, also to be discussed in the next section.

3.3.3 *Multicrystalline silicon ingots*

Over recent years, most of the world's photovoltaic production has been based on multicrystalline silicon wafers. Several companies have developed commercial processes for producing the precursor multicrystalline silicon ingots (Ferrazza, 1995; Rodriguez *et al.*, 2011). Advantages over the Czochralski process are lower capital costs, higher throughput and a higher tolerance to poor feedstock quality.

Basically, the technique involves controllably solidifying molten silicon in a suitable container (crucible) to give silicon ingots with large columnar grains, which generally grow from the bottom of the crucible upwards (Fig. 3.13a). Pioneers of this approach for modern photovoltaics in the mid-1970s were Wacker Chemitronic of Germany (Authier, 1978) and Solarex of the USA (Lindmayer, 1976). In the 1980s, several other manufacturers developed similar processes capable of producing good-quality multicrystalline material. These manufacturers differed in their choices of crucible material, the method of loading the crucible with silicon and the method for controlling the cooling of the melt. A good summary can be found elsewhere (Ferrazza, 1995). The size of nominally rectilinear ingots can be very large, presently up to 100 cm × 100 cm × 30 cm, with each ingot weighing over half

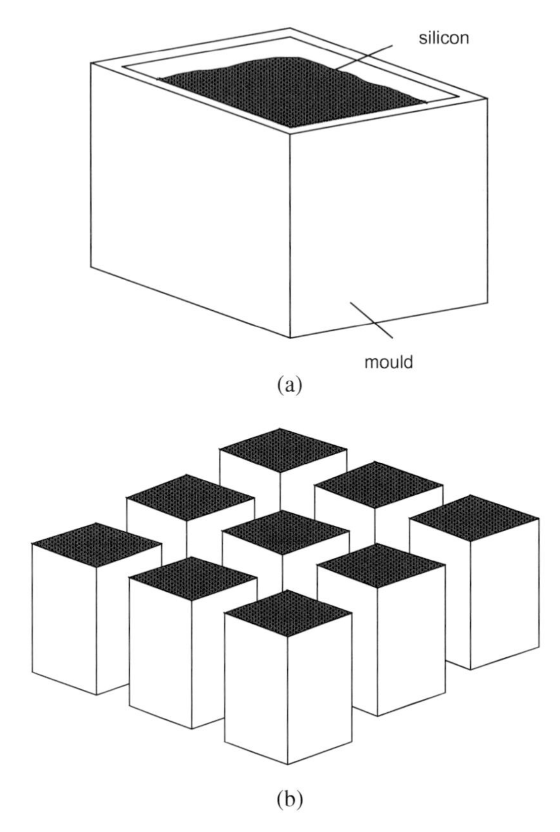

Figure 3.13 (a) Directional solidification of silicon within a mould; (b) sawing of large ingot into smaller sub-sections. Source: Green and Hansen (1998).

a tonne (Rodriguez *et al.*, 2011). This weight is expected to double over the coming decade (SEMI, 2011).

The large ingots are sawn into smaller sections as shown in Fig. 3.13b, which eventually give wafers generally 15.6–20 cm along the sides. These smaller sections can be sawn by the standard continuous wire sawing processes. The resulting multicrystalline wafers are capable of producing cells of over 85% of the performance of a monocrystalline cell fabricated on a CZ wafer. However, because of the higher packing density possible due to their square or rectangular geometry, this performance difference is largely masked at the module level, with multicrystalline module performance lying in the range demonstrated by modules made from monocrystalline cells (Haase and Podewils, 2011).

Interest in producing monocrystalline rather than multicrystalline silicon ingots by such directional solidification approaches dates back to the 1970s

(Schmidt, 2011; Fischer and Pschunder, 1976). However, developmental emphasis was diverted to the multicrystalline approach. More recently, BP Solar have revived interest in the monocrystalline approach (Stoddard *et al.*, 2008). Furnace deliveries based on the monocrystalline approach began in 2011 (AMG, 2010). The monocrystalline directional solidification approach involves seeding the bottom of the melt of Fig. 3.13a with a crystalline template. Close control of temperature is required to allow melt to contact this seed without consuming it.

Eventually, the whole industry may convert to this and related approaches, given the ability to bridge the gap in cell performance between directionally solidified and CZ silicon. Eventually it may be possible to produce material similar to FZ quality, given the better control over oxygen in a directional solidification process compared with silicon CZ growth. The process is also likely to be very tolerant of low-grade silicon (Khattak *et al.*, 1981) and to provide greater purification during directional solidification than is possible with multicrystalline ingots, possibly decreasing the purity requirements for 'solar-grade' feedstock. The renewed interest in this approach has stimulated a range of other 'controlled crystallisation' approaches that have led to steady improvements in the quality of multicrystalline wafers produced commercially.

3.3.4 *Sheet and ribbon silicon*

Although there is the potential for substantial cost reductions in both the cost of preparing the silicon feedstock and in forming crystalline or multicrystalline ingots from it, one unavoidable cost with the silicon wafer approach is the cost of sawing the ingot into wafers. Several studies have suggested that the lower bound on this cost may be something of the order of US$0.20/Watt (Christensen, 1985; Bruton *et al.*, 1997; Song *et al.*, 2010). This has provided the rationale for investigating approaches that produce silicon directly in the form of self-supporting sheets without the need for sawing (Bergin, 1980; Schulz and Sirtl, 1984; Rodriguez *et al.*, 2011).

The first ribbon approach to be used commercially was based on the edge-defined film-fed growth (EFG) method of Fig. 3.14. As originally investigated in the early 1970s, this technique involved the pulling of a thin sheet of silicon ribbon from a strip of molten silicon formed by capillary action at the top of a graphite dye (Fig. 3.14a). Substantially higher throughput was obtained with the more symmetrical configuration shown in Fig. 3.14b, where the ribbon is pulled in the form of a hollow nonagon. Individual wafers are then cut from the sides of the nonagon, normally by laser scribing wafers from each of the sides. The material produced is multicrystalline with elongated grains and is of a similar quality to the standard directionally solidified multicrystalline material. Commercial cells made

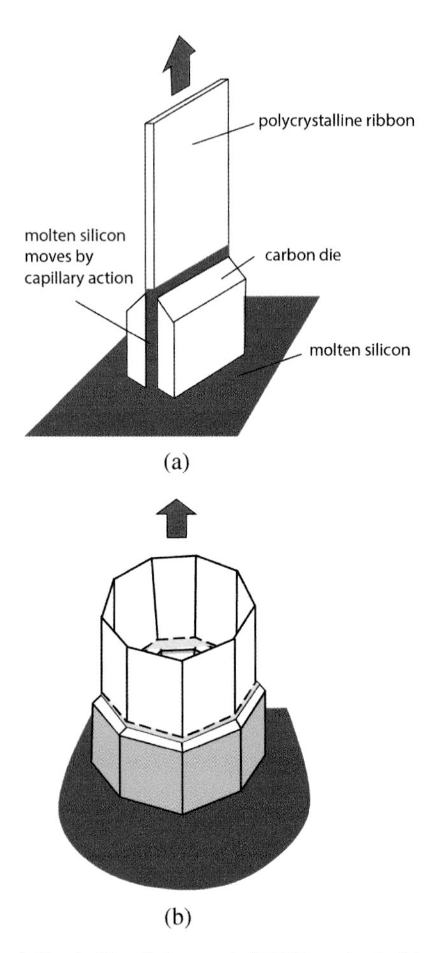

Figure 3.14 (a) Edge-defined, film-fed growth (EFG) method; (b) growth of a nonagonal ribbon of silicon using the EFG method. Source: Green and Hansen (1998).

from this material have been available sporadically since the early 1980s. However, after building up to a manufacturing capacity of about 200 MW/year, a decision was made to cease production in 2009 since 'ingot technology has now proven to be a more economical process' (Schott, 2009).

An even older ribbon growth process is the dendritic web approach of Fig. 3.15, first described by Westinghouse in the 1960s. In this approach, close thermal control is used to cause two dendrites spaced several centimetres from each other to solidify first during the growth step. When these are drawn from the melt, a thin sheet of molten silicon is trapped between them. This quickly solidifies to form a ribbon.

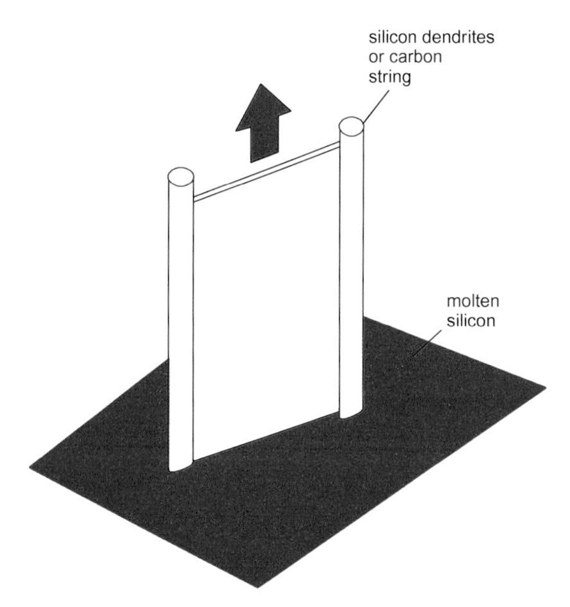

Figure 3.15 Schematic illustrating the dendritic web growth process and the string ribbon approach. Source: Green and Ho (2011).

A somewhat related approach is the string ribbon approach. In this case, the molten silicon is trapped between two graphite strings that are drawn from the melt. This relaxes the requirement on thermal control, compared with the previous dendritic web approach. The string ribbon approach has been commercialised by Evergreen Solar (Janoch *et al.*, 1997; Wallace *et al.*, 1997; Rodriguez *et al.*, 2011). Evergreen produced an estimated 157 MW of modules using this technology during 2010 while a formerly related company, Sovello, produced an additional 145 MW (Hering, 2011).

Despite the obvious attractions of the ribbon approaches, market share steadily declined, representing only 1.2% of production in 2010 after peaking at 5.6% in 2001 (Hering, 2011), with production now phased out. This highlights the difficulty of establishing unique technology in an industry where mainstream manufacturers have access to a skill and resource base far larger than that of any individual manufacturer.

3.4 Cell processing

3.4.1 *Standard process*

In the previous section, standard and non-standard ways of forming the silicon substrate were described. Most commercial substrates are formed by the wafering of monocrystalline and multicrystalline ingots, with a much smaller quantity of ribbon substrates produced commercially.

Some 'vertically integrated' photovoltaic manufacturers prepare their own polysilicon source material, although many bypass this and the crystal growth step by purchasing silicon wafers. Processing begins by chemically cleaning the starting wafers and etching their surfaces, to remove saw damage and to create surface texture (Neuhaus and Münzer, 2007).

For monocrystalline wafers, alkaline sodium or potassium hydroxide solutions are generally used for both steps, with a more dilute solution used for texturing. The composition and temperature of this solution determines the texturing quality (King and Buck, 1991), including the size of the pyramidal features resulting from the texturing and the percentage of wafer surface area successfully covered by such features. For multicrystalline wafers, an acidic etchant produces an effective surface texture (Einhaus *et al.*, 1997; Neuhaus and Münzer, 2007).

The next major stage of processing is the diffusion of the cell junction. This is generally achieved by heating at high temperature in a tube in a furnace down which a gas containing $POCl_3$ is passed (Fig. 3.16). A phosphorus glass is formed on the cell surface, allowing phosphorus dopant atoms to seep into the cell surface by thermal diffusion. Typically, the depth of diffusion is less than 1 μm. The same thermal diffusion approach is widely used in microelectronics. Processing for photovoltaics can involve cruder equipment and techniques, with 'in-line' processing on metal belt furnaces re-emerging as a high-throughput option, with spraying or prior spinning of a compound containing phosphorus onto the wafer surface.

After diffusion, the diffusion oxide is removed by etching. Although the diffusion is required over only one surface of the wafer and processing techniques are generally chosen to encourage such a result, phosphorus invariably seeps into both wafer surfaces to some extent.

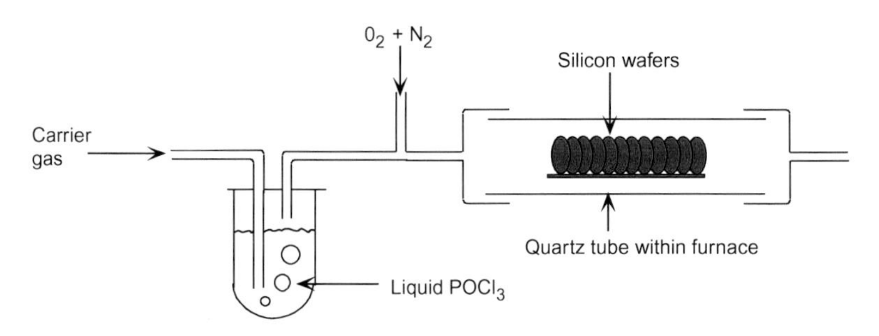

Figure 3.16 Phosphorus diffusions process. Source: Green and Hansen (1998).

To break the connection between the phosphorus diffused into front and rear surfaces, an 'edge junction isolation' step is needed to remove the thin phosphorus layer around the edge of the wafer. This isolation was once achieved by 'coin stacking' the wafers so that only their edges were exposed. The stack was then placed in a plasma etcher to remove a small section of silicon from the wafer edge, hence breaking the conductive link between front and rear surfaces. Single-sided etching is now the preferred approach in which the wafer sits on the etching solution so that only the desired surface is etched.

A silicon nitride quarter-wave antireflection coating is applied to the cell at this stage. Not only does silicon nitride have a reasonably good refractive index for this application, it is deposited using silane (SiH_4) and consequently has a high hydrogen content. This hydrogen can be released into the silicon substrate, giving a beneficial defect passivating effect, particularly for multicrystalline cells.

Cell processing is completed by the screen printing of metal contacts onto the front and rear surface. Silver and aluminium paste (consisting of a suspension of fine metal particles and glass frit in an organic medium together with appropriate binders) is squeezed through a patterned screening mesh onto the cell surface. After application, the paste is dried at low temperature. The Ag-based paste for the top surface is printed in a characteristic finger pattern shown in Fig. 3.17 to minimise the resistive losses in the cell while allowing as much light as possible into it. For the rear, two different pastes are generally applied as shown. An Ag-rich paste is applied to a small fraction of the rear to allow soldering of the cell interconnections, while most of the rear is covered by an Al-based paste. After these three layers

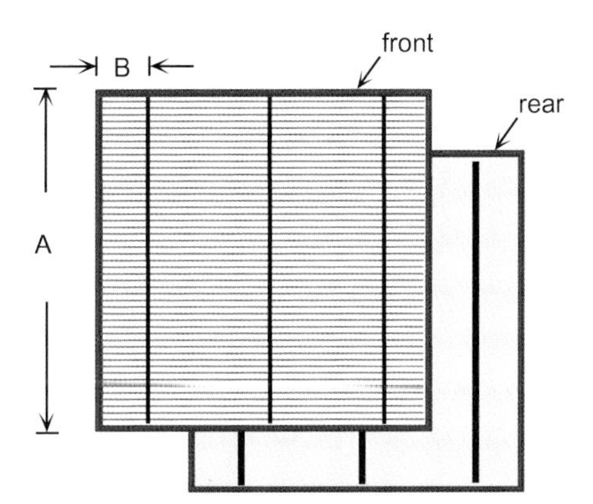

Figure 3.17 Standard 156 mm × 156 mm solar cell 'H' metallisation pattern. Screen-printed silver regions are shown coloured black. The pattern consists of six unit cells of size A × B. Both front and rear views are shown. Source: Green and Ho (2011).

have been individually screened and dried, the pastes are then fired at a higher temperature to drive off the remaining organics and to allow the metal particles in the paste to coalesce. The glass frit is important in promoting adhesion to the silicon substrate. Often pastes are doped with dopants such as phosphorus, in this case to help prevent the screened contact from penetrating the thin phosphorus skin that it is intended to contact.

This screen-printing method for applying the metal contact was borrowed in the early 1970s from the hybrid microelectronics industry (Ralph, 1975). This ensured the ready availability of both screen-printing equipment and the paste firing furnaces suited to this application. Labour and equipment costs associated with this step tend to be very low. However, the pastes themselves can be expensive (Green, 2011) and an even larger cost penalty is paid for the simplicity of this approach by the forfeiture of the inherently available power output from the silicon wafer, as discussed later.

After the paste firing, the cells are then ready for testing under a solar simulator. Cells are usually graded based on their short-circuit current or current at a nominal operating voltage, e.g. 500 mV. Generally, cells are sorted into 5% performance bins. This sorting is required to reduce the amount of mismatch between cells within the completed module. To a large extent, the output current of the module is determined by that of the worst cell in the module, resulting in large power losses within mismatched modules. Even worse, low-output cells can become reverse-biased under some modes of module operation and destroy the module by localised over-heating.

Silicon ribbon substrates often require modifications of the above standard sequence. For example, the rippled surface which is a natural consequence of the EFG ribbon growth process poses continuity hazards for screen-printed metallisation. To accommodate this rough surface, a novel technique was developed whereby the metal paste was squeezed through an orifice and then dropped to the cell surface, much the same as squeezing toothpaste from its tube onto a toothbrush.

3.4.2 *Limitations of the screen-printing approach*

There are four main limitations arising from the screen-printing approach to applying the front contact, causing the simplicity in processing to be at the expense of cell performance. One limitation is that the top-surface phosphorus diffusion has to be heavier than desirable purely from the point of view of cell performance. This is to allow reliable low-resistance contact between the screen-printed metal and the diffusion. Typically, sheet resistivities of this diffusion of less than 60 ohms/square have been required (Green, 1995; De Clercq *et al.*, 1997). Such diffusions generally reduce the quality of the silicon in the region near the cell surface, where blue wavelengths in sunlight are strongly absorbed. A screen-printed cell therefore does not respond well to blue wavelengths in sunlight, wasting up to 10% of the possible current output through this deficiency.

The remaining three limitations relate to the geometry and conductivity of the metal lines it is possible to produce by the standard screen-printing process. Since the paste thickness shrinks to about one-third of its original thickness during firing, it is very difficult to achieve metal lines with high aspect ratio (height/width), which are the key to forming metal grids which result in low overall losses (Serreze, 1978). The nature of the screening meshes that have sufficient ruggedness for use in commercial production means it is very difficult to achieve fine lines using screen printing in production. Typically, 70 μm width is the minimum that can presently be achieved cost-effectively. This limitation means that there will generally be high shading losses in screen-printed cells owing to the large percentage (~10%) coverage of the front surface by the metal (Fig. 3.17). Additionally, the relatively poor conductivity of the fired silver paste (about two times lower than that of pure silver for large features such as busbars but up to six times lower for finer features such as fingers (de Moor *et al.*, 1997)) fundamentally limits the ability to optimise metal contact design in much the same way as does the low aspect ratio previously discussed.

Paste manufacturers are working hard to develop improved pastes that reduce the impact of these deficiencies. Figure 3.18 shows the targeted product introduction of one major paste manufacturer. This manufacturer anticipates ongoing contributions to improving cell efficiency by developing pastes able to contact increasingly lightly diffused top surface diffusions. Practical diffusion sheet resistivity is shown as increasing from 70 ohms/square in 2010 to above 100 ohms/square by 2015. Pb-free pastes are also eventually expected to be used across the industry.

Screen-printing equipment manufacturers have also not been idle. Modern machines are now able to align each screen-printing step very accurately to features on the wafer, including previously printed patterns. This makes it possible to incorporate new design features into cells.

Figure 3.19 shows two such features of present interest. Using the alignment capability previously mentioned, the area under the metal contact can be selectively doped very heavily to allow good contact (Fig. 3.19a), allowing other regions of the surface to be lightly doped, improving the cell blue response (Chunduri, 2010). About 1% absolute gain in performance is possible (e.g. from 18% to 19%) using this 'selective emitter' approach. Figure 3.19b shows another possible improvement by 'double printing'. By printing the metallisation twice, high aspect ratios are possible, reducing metallisation loss. Additionally, the metal layer contacting the

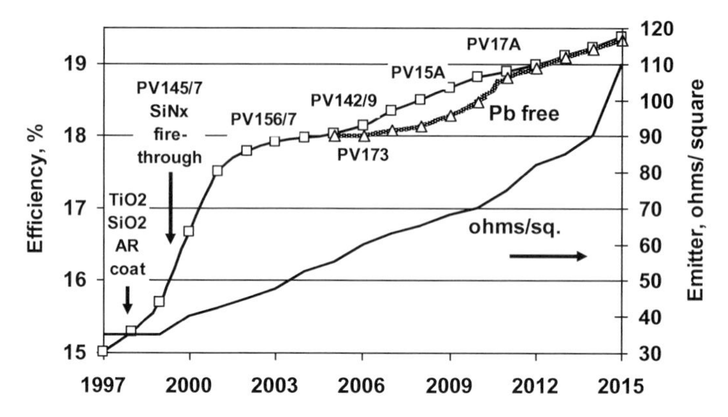

Figure 3.18 Past and planned product introduction schedule for one major paste supplier. Shown are the different generations of paste and the emitter sheet resistivity the pastes are designed to accommodate. The author's estimate of likely production efficiency is based on unscaled estimates provided by the paste manufacturer. Adapted from Laudisio *et al.* (2010).

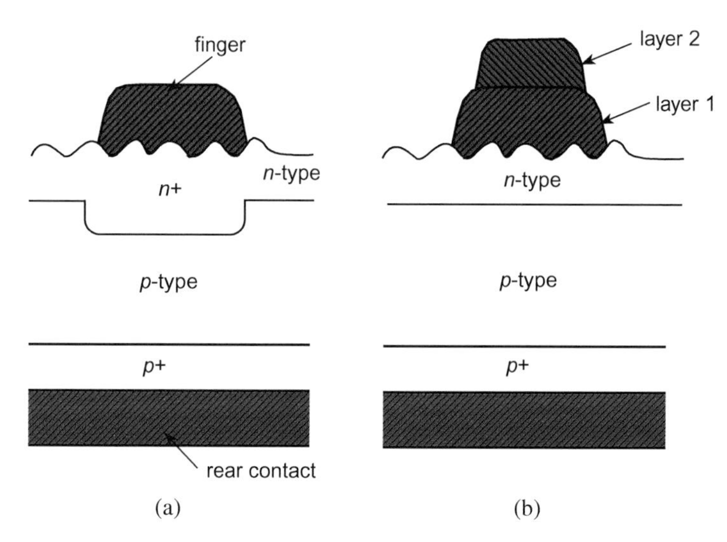

Figure 3.19 (a) 'Selective emitter' approach; (b) 'double printing' to increase aspect ratio of metal fingers. Source: Green and Ho (2011).

silicon can be optimised for this role while the second layer can be optimised for good conductivity.

Recent rapid increases in silver prices cast doubt on the long-term role of this metal in photovoltaics (Green, 2011). The industry seems likely to transition to the use of Cu as the main conductive metal in the cell. Plating Cu to thin screen-printed Ag lines is one possible approach, or it may be that the need for change will stimulate the adoption of more innovative technology.

3.4.3 *Laser-processed solar cells*

As mentioned in Section 3.2, buried contact cells were developed as a way of incorporating some of the efficiency improvements demonstrated in the mid-1980s into low-cost commercial cell production sequences. This was the first use of lasers to define fine-linewidth features in silicon cell processing.

Since that time, two other laser-based cell designs have been used either in production or in large-scale pilot production. The first is the 'semiconductor finger' cell design of Fig. 3.20 which is, to some extent, a hybrid between a buried contact and a conventional screen-printed cell. A heavily doped finger layer is formed by laser doping in the top surface, similar to the grooves of a buried contact solar cell or to the selective emitter regions of the laser-doped selective emitter (LDSE) cell to be discussed below. A screen-printed Ag contact layer is then applied perpendicularly to this semiconductor finger layer as shown. The advantages are that a light diffusion can be used over most of the cell surface and the shading by the Ag fingers is significantly reduced. On large area CZ cells, efficiencies in the 18–19% range have been reported.

A second laser-based cell design is the LDSE cell, with the key processing step shown in Fig. 3.21 (Wenham and Green, 2002). A laser is used to define the areas

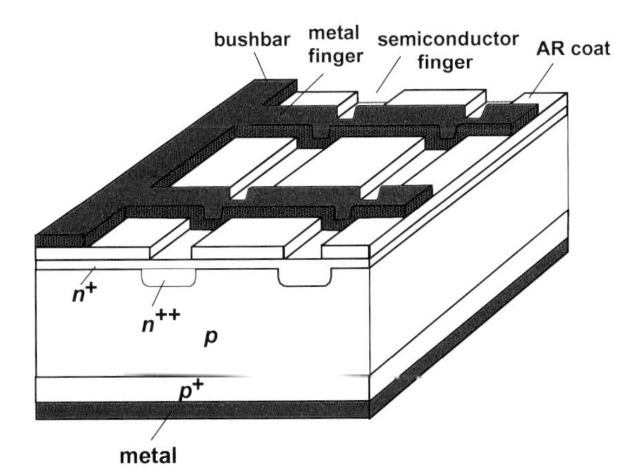

Figure 3.20 Semiconductor finger solar cell. Heavily doped semiconductor regions provide an additional finger layer that is contacted by the traditional metal finger layer. Source: Green and Ho (2011).

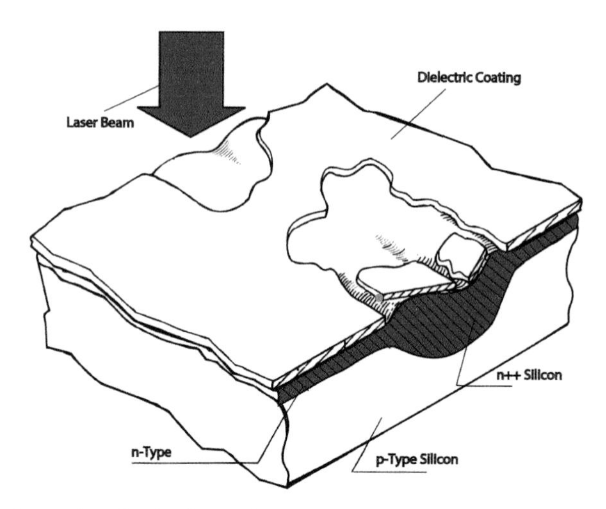

Figure 3.21 Laser-doped selective emitter (LDSE) solar cell. Metal is selectively plated to the heavily doped regions exposed during the laser doping step. Source: Green and Ho (2011).

for top finger contact by removing the dielectric coating by melting the underlying silicon, while simultaneously introducing dopants into the molten region. The front contacts can then be plated as a Ni/Cu/Ag tri-layer. The cell has the advantages of a lightly doped top surface region and of fine, highly conductive metal fingers. Efficiency in the 19–20% range has been confirmed for cells on large area CZ wafers fabricated using largely commercial equipment. This is expected to improve to the 21–22% range once a similar contacting approach is applied to the rear of the cell.

3.4.4 *HIT solar cell*

The HIT cell has already been discussed in Section 3.2.3 (Fig. 3.8). Sanyo introduced this cell into commercial production in 1997, and both efficiency and production volume have increased steadily since then. In 2010, an estimated 300 MW of HIT cell modules were produced (Hering, 2011) with total area module efficiency in the 16.1–19.0% range, well above the market average of about 14% (Haase and Podewils, 2011). The distinctive feature of the technology is the use of amorphous silicon (a-Si) to form heterojunctions with the *n*-type silicon substrate at both front and rear contacts, as previously discussed.

With the expiry of the most important of Sanyo's initial patents in 2010, equipment suppliers began offering turn-key manufacturing facilities for these cells (Strahm *et al.*, 2010).

3.4.5 *Rear junction cells*

In earlier sections, three different rear-junction devices were described: the rear-junction cells of Figs. 3.6 and 3.9 and the early Bell Laboratories cell of 1954 (Fig. 3.2c). The latter could be described as an 'emitter wraparound' cell, a member of a more general group of 'emitter wrap through' (EWT) cells that, like the rear-junction cells, require no metal on the front surface. A related rear-contact device is the 'metal wrap through' (MWT) device that has metal on both front and rear surfaces. Figure 3.22 shows the key features of EWT and MWT devices.

SunPower introduced their rear-junction cell (Fig. 3.9) into production in 2003. This was based on the high-efficiency devices earlier developed at Stanford University (Fig. 3.6) but with simplifications that made the device suitable for commercial

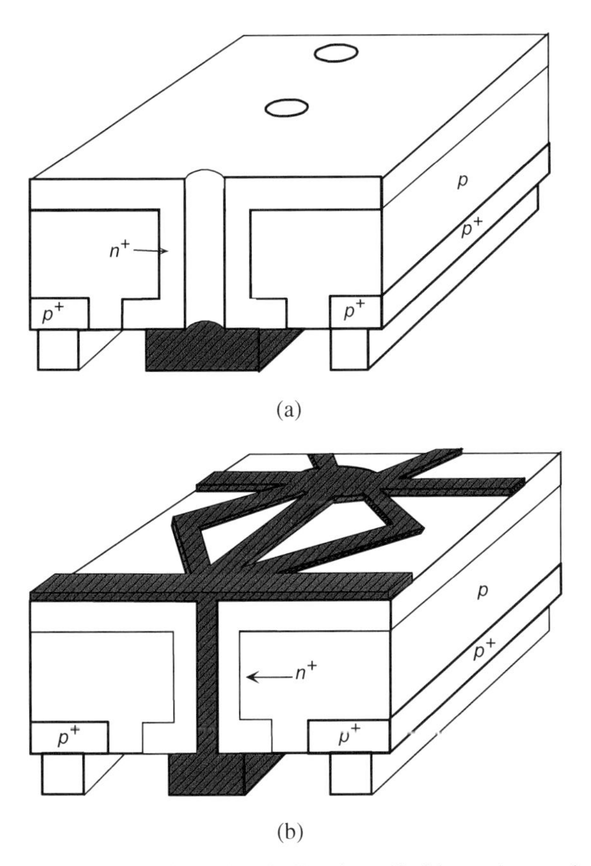

Figure 3.22 (a) Emitter wrap-through (EWT) solar cell; (b) metal wrap-through (MWT) cell. Source: Green and Ho (2011).

production. This was the first commercial cell of efficiency above 20%. In 2008, SunPower introduced an improved 'Generation II' product which took efficiency to 22% (De Ceuster *et al.*, 2007). This has produced commercial modules in large volume with 18.4–19.7% total area efficiency, the highest on the market (Haase and Podewils, 2011). More recently, a Generation III product has been reported (Cousins, 2010) with 24% cell efficiency that is likely to take commercial module efficiency beyond 20% for the first time. Cell processing costs are reported to be higher than for the standard screen-printing approach (Song *et al.*, 2010).

Commercialisation of EWT cells is not as far developed as the rear-junction cell or even the MWT device. A key disadvantage of both compared to the rear-junction device is the need to drill multiple holes through the wafer using a laser, although fewer are required for MWT devices. However, both EWT and MWT cells are more tolerant of low-quality wafers than rear-junction devices. In fact, the first solar modules of over 17% efficiency employing multicrystalline wafers used MWT cells (Lamers *et al.*, 2011). In 2010, at least two manufacturers offered MWT devices commercially (Solland, 2009; Photovoltech, 2010), although manufacturing volume was not large.

3.5 Cell costs

There have been many studies of the costs of the different stages of silicon cell production using different basic assumptions, particularly in relation to the production volume assumed in the study and the cost of polysilicon source material. One of the most authoritative early studies was one conducted under the auspices of the European Union Photovoltaic Programme (Bruton *et al.*, 1997; Bruton, 2002). This study involved representatives of major European photovoltaic manufacturers and research laboratories, and was valuable owing to the breadth of representation and the diversity of approaches explored.

The key assumptions of the study were a manufacturing volume of 500 MWp of solar cells per annum and the availability of silicon source material at US$25/kg. The former seemed like an enormous production volume at the time when large manufacturers were producing only 10–20 MW/year of modules. However, a manufacturer producing 500 MWp/year of modules in 2010 would barely qualify for the list of 'Top 20 Manufacturers' (Hering, 2011). A number of different technologies were compared in the study. Important comparisons were between EFG ribbon, multicrystalline and crystalline wafer technologies, between screen-printed, buried contact and PERL cell processing sequences, in various combinations of wafers and processing, and between two different module encapsulation approaches. However, the results from all possible combinations were not studied (or, at

Table 3.1 Summary of published results of a European Commission study of manufacturing costs for 500 MWp/annum factory (Bruton *et al.*, 1997).

ID	Wafer	Process	Cell efficiency	Estimated cost (€/Wp)	Key variable
#1	DS	SP	15%	0.91	wafer
#2	CZ	SP	16%	1.25	wafer/process
#3	CZ	LGBC	18%	1.15	process
#6	CZ	PERL	20%	1.78	process
#7	EFG	SP	14.4%	0.71	wafer

1€=US$1.2136 at time of study; 1€=US$1.44 in August 2011.
DS: Directional solidification
CZ: Czochralski growth
EFG: Edge-defined film-fed growth
SP: Screen-printed
LGBC: Laser-grooved, buried contact
PERL: Passivated emitter, rear locally diffused (acronym LBSF used in study)

least, not published), but only seven selected combinations, of which Table 3.1 shows five.

From this table, several key results can be deduced. When comparing screen-printed cells on ribbon (EFG), multicrystalline (DS) and monocrystalline (CZ) wafers, the ribbon produced the lowest cost of €0.71/Wp followed by the multicrystalline wafers at €0.91/Wp and the monocrystalline wafers at €1.25/Wp. The advantage of the ribbon stemmed almost entirely from the fact that it does not need to be sawn, as previously mentioned. As noted earlier, EFG production ceased in 2009 due to unfavourable economics. This is probably as a result of the relatively small effort that could be devoted to the development of this technology by a single company compared with the huge effort devoted by almost the entire industry to the directional solidification approach.

Comparing the different processing approaches on single-crystal wafers, the cheapest is the buried contact at €1.15/Wp, followed by the screen-printed at €1.25/Wp, followed by the PERL at €1.78/Wp. The buried contact achieved its cost advantage over the screen-printing approach by virtue of the increased efficiency giving more power per unit processing area. In terms of module costs, the standard laminated module approach was calculated to be slightly cheaper than an alternative resin-fill approach.

At the time these results were published, many regarded the costs deduced as impossibly ambitious. However, actual developments have shown them to be conservative. In 2011, average module manufacturing cost across the industry was estimated as US$1.56/Wp (Bolman *et al.*, 2011), or about €1.08/Wp, with the lowest cost silicon manufacturers producing modules for around US$1.20/Wp (€0.83/Wp). Further significant cost reductions are expected over the coming decade (SEMI, 2011). A rough breakdown of the 2011 costs was that the cost of the pure silicon accounted for 20–25% of the total module cost, while the cost of converting this to a wafer accounted for a similar percentage of total costs, as did the cost of converting to a cell. The cost of encapsulating the cell into the module accounted for a slightly larger 25–30% of total costs (Bolman *et al.*, 2011).

3.6 Opportunities for improvement

3.6.1 *Commercial cells*

The large differential between the efficiencies of a typical screen-printed commercial cell (17–18%) and the best laboratory silicon cell (25%) shows the large potential for further efficiency improvement in commercial devices.

One reason for this differential is the difference between the CZ wafers used in commercial production and the FZ wafers until recently used for the best laboratory cells. CZ-grown wafers are invariably contaminated with oxygen and carbon during growth to a much greater extent than FZ wafers, due to the use of quartz crucibles and graphite heaters in the CZ process. These impurities give rise to a much more subtle dependence on processing conditions of the important silicon material property for producing high performance cells, the minority-carrier diffusion length. For example, applying the high-temperature processing associated with PERL-type sequences to CZ silicon gives a large spread in results depending upon the supplier of the CZ material, and hence most likely on the oxygen and carbon content (Knobloch *et al.*, 1996). Additionally, the quality of B-doped CZ material as compared to FZ falls off quite rapidly as the boron content is increased, which is attributed to boron/oxygen complexes forming within the material (Schmidt and Bothe, 2004). This reduces flexibility in cell design since it eliminates the possibility of using low-resistivity CZ substrates.

Two recent developments may change this situation. One is the progress in improving the laboratory performance of cells on *n*-type substrates (Section 3.2.2). Some manufacturers have now switched to the use of *n*-type wafers for monocrystalline cells, with this expected by some to be an increasing trend over the coming decade (SEMI, 2011). Phosphorus is the dopant in these *n*-type substrates and does not form the same performance-reducing defects with oxygen as does boron. A second change is the growing interest in producing monocrystalline silicon by direction solidification (Section 3.3.2). Better control of oxygen content may be possible in crystalline material prepared this way.

The trend towards thinner cells, which arises mainly from efforts to reduce the costs of the silicon wafer, may actually help to improve the cell efficiency, as already noted. Thin wafers give the opportunity for back-surface fields or other rear-surface passivation approaches to be used to improve cell performance, primarily through increased voltage output. Given better feedstock material or cells below 150 μm thickness, improved rear-surface passivation approaches such as demonstrated by the PERC and PERL cells of Fig. 3.7, as well as rear-junction and HIT cells, will become increasingly relevant.

3.6.2 *Laboratory cells*

For laboratory cells, as previously mentioned, an appropriate reference point for performance is the AM1.5 detailed-balance or Shockley–Queisser efficiency limit of 33.5% for material of the bandgap of silicon. However, it has been shown that another intrinsic process, Auger recombination, provides a more severe fundamental limit for silicon than the radiative recombination processes assumed in the detailed-balance limit (Green, 1984; Tiedje *et al.*, 1984). Unlike the detailed-balance limit, the Auger limit for a silicon cell is dependent on the cell thickness, as shown in Fig. 3.10 (Green, 1995). This difference arises because the detailed-balance calculation includes photon recycling, which makes nett recombination rates independent of cell volume. With Lambertian light trapping, the optimum cell performance in the Auger limit is 29% for a cell of about 80 μm thickness. Such a cell would have an open-circuit voltage of above 760 mV, higher than the highest value ever demonstrated for silicon of 750 mV for a HIT cell, in which surfaces are passivated by a-Si layers.

Figure 3.23 shows the results of efficiency calculations with various amounts of surface recombination added, characterised in terms of the open-circuit voltage limit that this recombination would impose if it were the only recombination process in the cell (Green, 1999). Increasing surface recombination reduces the value of the obtainable efficiency as well as pushing the optimum cell thickness to larger values. Figure 3.23 makes it clear that, to improve silicon cell efficiency much beyond 26%, improved surface passivation (of both surfaces) is essential beyond the 720 mV capability that presently limits all but HIT cells. If this improved quality cannot be achieved, another possibility is to maintain the same quality of surface passivation as presently achieved and reduce the effective threshold energy of the photovoltaic process within the bulk regions of the cell. Techniques such as having sections of the cell alloyed with germanium to reduce its bandgap (Healy and Green, 1992) or doped with a photoactive impurity to give impurity photovoltaic effects in the bulk region (Keevers and Green, 1994) have been suggested and

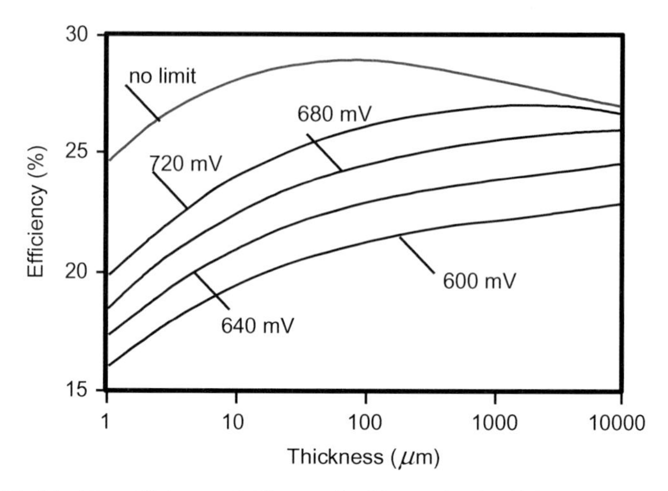

Figure 3.23 Limiting efficiency of silicon cell with Lambertian light trapping as a function of surface recombination velocity, characterised in terms of the voltage limit imposed by this recombination. Source: Green (1999).

shown, in some cases, to have theoretical advantages. However, neither technique has to date demonstrated an experimental performance advantage.

A well-proven approach for improving solar cell efficiency is the use of the tandem cell structure. Efforts to produce tandem cells based on silicon have not yet given good results owing to the inability to find a suitable wide-bandgap partner that is lattice-matched to silicon (Corkish, 1991). For low-quality cells, amorphous silicon/polycrystalline silicon tandems have given improved results over either cell type alone (Yamamoto *et al.*, 1997; Shah *et al.*, 1997). However, the poorer blue response of the a-Si cell and the inability to match the current output of a good-quality c-Si substrate makes this an unlikely path to taking efficiency beyond 26%.

Fuller use of the available photon energy by incorporating efficient impact ionisation process has been suggested as a way of boosting cell performance by generating more than one electron–hole pair from one high-energy photon (Kolodinski *et al.*, 1993; Werner *et al.*, 1994). However, such processes are quite weak in silicon with increases in current density limited to less than 0.1 mA cm^{-2} (Green, 1987). Manipulating the details of silicon's bandgap, for example by alloying with germanium, may improve prospects. However, since the high-energy photons of most interest for this process are absorbed very close to the surface of silicon, such approaches may interfere with the ability to obtain well-passivated surfaces. Limited experimental work with shallow germanium implants has not given any nett performance benefit (Keevers *et al.*, 1995).

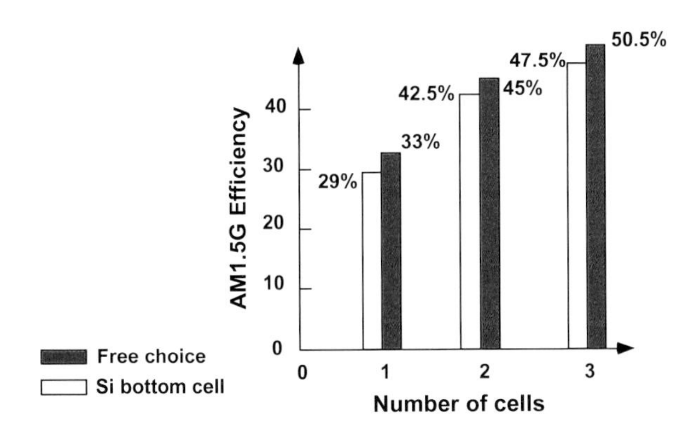

Figure 3.24 Efficiency limits for cells and cell stacks with silicon as the lowermost cell compared to the unconstrained limit for series-connected cells. While silicon is a good choice for a single-cell material with an efficiency limit of 29% compared to the unconstrained limit of 33%, it is an even better choice for the lowermost cell in 2- and 3-cell stacks. Source: Green and Ho (2011).

The tandem cell approach perhaps remains the most likely option, pending technical improvements that allow III–V or related cell technologies to be grown on silicon. Silicon has an ideal bandgap to be the lowermost cell in a 2–4 cell tandem (Fig. 3.24) and also can now be prepared very inexpensively in wafer form. It would provide a clean, low-cost substrate for the subsequent growth of high-performance cells. Low-cost deposition of the subsequent cell layers would be essential. Progress has been reported in depositing lattice-matched III–V compounds on silicon (Grassman *et al.*, 2009), as well as fully-relaxed Ge layers on silicon that might serve as templates for the subsequent growth of a wider range of III–V materials (Wistey *et al.*, 2007; Tsao, 2011).

3.7 Silicon-supported thin films

There has long been an interest in transferring the strengths demonstrated by crystalline silicon wafer technology to cells based on silicon thin films. Historically, work can be divided into two phases: (i) that before the 1980s, when the benefits of light trapping were not fully appreciated; and (ii) that after the mid-1980s, when light trapping has been regarded as an essential feature of any silicon thin-film cell design. The early work laboured under what is now known to be a misconception that quite thick layers ($>20\,\mu$m) of silicon would be required to give reasonable performance due to the poor absorption characteristics of silicon arising from its indirect bandgap (see Fig. 3.8). However, since light trapping can increase the effective optical thickness of a silicon cell by 10–50 times, this means that layers of only

1 μm or so thickness are still inherently capable of producing similar performance to much thicker layers (Fig. 3.10).

Approaches to producing supported silicon films can be divided into high-temperature and low-temperature strategies depending on whether or not the substrate is heated to high temperature during the silicon deposition or subsequent processing.

3.7.1 *High-temperature supported films*

One of the earliest silicon-supported film approaches was the 'silicon-on-ceramic' approach (Christensen, 1985), whereby a ribbon of ceramic material was dipped into a molten silicon bath or pulled across the surface of a silicon melt so that one side was coated with silicon. This produced silicon of modest quality and the approach suffered from difficulties in making rear contact to the cells, since the ceramics used were insulating. This approach was discontinued in the early 1980s. Early work by Ting and Shirley Chu involved the deposition of silicon onto a range of foreign substrates by high-temperature chemical vapour deposition (Chu, 1977). Operational cells were obtained using a number of substrate materials. The best results were obtained by depositing the silicon layers on multicrystalline silicon substrates prepared from metallurgical grade silicon. Given the previous studies which have shown that sawing of wafers represents one of the major costs in any wafer-type approach, the overall economics of such an approach using a wafer substrate are questionable, regardless of the quality of this substrate. Other early work involved the deposition of silicon onto ceramic substrates by high temperature CVD and the subsequent increase in crystal size by melting and directional solidification (Minagawa *et al.*, 1976).

In the post-1980 era, efforts in silicon-supported film were revitalised by the US company AstroPower (Barnett *et al.*, 1985). Initial work was directed at depositing the films on ceramic-coated steel substrates. The use of the steel substrate was soon dispensed with, followed by the use of the ceramic giving a columnar, free-standing silicon sheet (Hall *et al.*, 1994). A somewhat related RGS (ribbon growth on substrate) approach (Lange and Schwirtlich, 1990; Hahn and Schönecker, 2004), whereby the silicon layer is solidified on a re-usable substrate, is still under active development.

3.7.2 *Low-temperature approaches*

One of the first papers addressing silicon photovoltaic thin films described the deposition of silicon by low-temperature chemical vapour deposition onto an aluminium substrate (Fang *et al.*, 1974). A surprisingly large grain size was obtained, attributed to eutectic reaction with the aluminium. In more recent times, laser crystallisation has been used in the active matrix liquid crystal display industry to produce relatively small-grain polycrystalline silicon films from amorphous silicon precursors, generally deposited by low-pressure chemical vapour deposition. Grain sizes are typically less than a micron or so, so that these films would probably not be suitable for photovoltaics. Also, thicknesses for the active matrix display industry tend to be only about 100 nm, which would be too thin for photovoltaic application.

From 1989, a group at Sanyo explored the use of low-temperature solid phase crystallisation of amorphous silicon as a technique for producing thin-film polycrystalline silicon cells. Good results were obtained, with 9.2% (unconfirmed) efficiency reported in 1995 (Baba *et al.*, 1995). These cells were approximately 1 cm^2 in area deposited onto a textured metallic substrate and heated at approximately 600°C for many hours to enable the crystallisation of the originally amorphous films. After crystallisation, the HIT structure developed by Sanyo (Fig. 3.8) was used to complete the cell processing at low temperature.

A similar approach has been developed for depositing thin polycrystalline films of silicon on glass (CSG), as described in more detail elsewhere (Green *et al.*, 2004). This produced small area modules of 10–11% efficiency (Keevers *et al.*, 2007). About 10 MW of 1.4 m^2 modules have been produced commercially with this approach.

Two groups pioneered the development of silicon thin films deposited directly in microcrystalline form (μc-Si) onto glass substrates. In 1997, the University of Neuchatel reported efficiencies of about 7% for 3 μm thick microcrystalline cells deposited at 500°C (Shah *et al.*, 1997). The cell had a *p-i-n* structure with the intrinsic region comprising most of the device thickness. The cell was designed for the intrinsic region to be depleted during normal device operation to create a high electric field to aid carrier collection, as with a standard amorphous silicon cell. Kaneka Corporation (Yamamoto *et al.*, 1997) reported efficiencies of over 10% with a similar device structure (Fig. 3.25). Nearly the same efficiency has been obtained with the total device thickness varied over the 1.5–3.5 μm range. Both the above groups reported even higher efficiencies when amorphous silicon cells were used in a tandem configuration on top of the microcrystalline device (Fischer *et al.*, 1997). Since this early work, the latter tandem device has been commercialised by many groups with tandem a-Si:μc-Si cell modules of typically 8–9% efficiency available from several sources (Haase and Podewils, 2011).

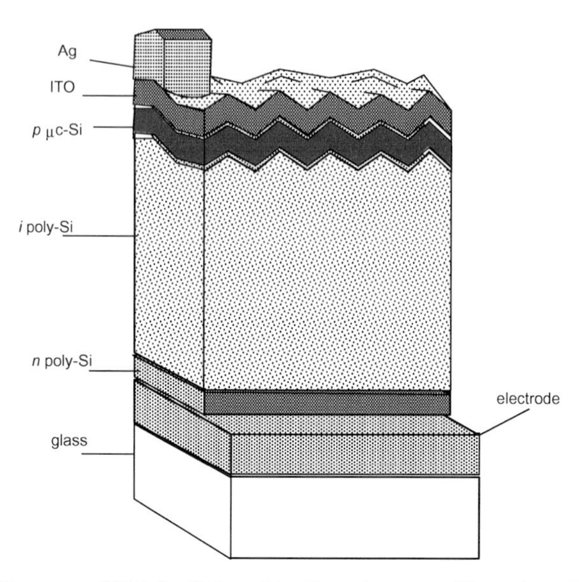

Figure 3.25 Structure of 10.1% efficient thin-film microcrystalline solar cell developed by Kaneka. Cell thickness is typically 1–3 μm. After Yamamoto *et al.* (1997).

3.8 Summary

Although crystalline silicon devices have dominated the commercial marketplace for photovoltaics for more than three decades, there still remains scope for considerable improvement in both the performance and cost of these cells. Current expectations are that manufacturing costs below US$0.50/Wp will soon be achieved, without major changes in present processing sequences. The trend towards thinner silicon wafers to decrease wafer cost is compatible with ongoing increases in cell efficiency provided new cell structures are adopted and improved methods are demonstrated in production for passivating the rear surface of the cell. The large potential for both performance and cost reduction will make these bulk silicon approaches an increasingly challenging target for thin-film approaches. In this context, good progress continues to be made with supported silicon film.

Ultimately, some way of boosting the performance of silicon cells would seem to be required for these to continue to maintain their dominance over the coming decades. The most promising at present would seem to be a tandem cell stack. Such structures would take advantage of the enormous recent cost reductions in producing silicon wafers, which could be used as clean substrates for the deposition of thin, high-quality layers of strongly absorbing, wider-bandgap material. Such an approach would combine the robust commercial infrastructure of silicon cell manufacturing with the outstanding laboratory progress with tandem cell stacks.

Acknowledgement

This work has been supported by the Australian Government through the Australian Renewable Energy Agency (ARENA). Responsibility for the views, information or advice expressed herein is not accepted by the Australian Government.

References

AMG (2010), 'AMG acquires solar silicon casting technology from BP Solar International Inc.', Press Release, 22 December.

Authier B. (1978), 'Poly-crystalline silicon with columnar structure', *Festkörper-probleme XVIII*, Freidr. Vieweg & Son, Wiesbaden.

Baba T., Shima M., Matsuyama T., Tsuge S., Wakisaka K. and Tsuda S. (1995), '9.2% efficiency thin-film polycrystalline solar cell by a novel solid phase crystallisation method', *Conf. Record, 13th European Photovoltaic Solar Energy Conf.*, Nice, October, pp. 1708–1711.

Barnett A. M., Ford D. H., Checchi J. C., Culik S., Hall R. B., Jackson E. L., Kendall C. L. and Rand J. A. (1997), 'Very-large-area silicon-film^{TM} solar cells', *Conf. Proc., 14th European Photovoltaic Solar Energy Conf.*, Barcelona, pp. 999–1002.

Barnett A. M., Hall R. B., Fardig D. A. and Culik J. S. (1985), 'Silicon-film solar cells on steel substrates', *Conf. Record, 18th IEEE Photovoltaic Specialists Conf.*, Las Vegas, October, pp. 1094–1099.

Barnham K. W. J. and Duggan G. (1990), 'A new approach to high-efficiency multi-band-gap solar cells', *J. Appl. Phys.* **67**, 3490–3493.

Benick J., Hoex B., Dingemans G., Richter A., Hermle M. and Glunz S. (2009), 'High efficiency *n*-type silicon solar cells with front side boron emitter', *Proc. 24th European PVSEC*, September, Hamburg, 863–870.

Benick J., Hoex B., van de Sanden M. C. M., Kessels W. M. M., Schultz O. and Glunz S. W. (2008), 'High efficiency *n*-type Si solar cells on Al_2O_3-passivated boron emitters', *Appl. Phys. Lett.* **92**, 253504.

Bergin D. O. (1980), 'Shaped crystal growth — a selected bibliography', *J. Crystal Growth* **50**, 381–396.

Bolman C., Coffey V., Fu B., Song J., Trangucci R. and Zuboff G. (2011), 'The true cost of solar power: the pressure's on', *Photon Consulting*, www.photonconsulting.com.

BP Solar (1991), Data Sheet, BP Saturn Solar Cells.

Brandhorst H. W. and Bernatowicz D. T. (1980), 'Space solar cells: high efficiency and radiation damage',*Conf. Record, 14th IEEE Photovoltaic Specialists Conf.*, San Diego, pp. 667–671.

Bruton T. M. (2002), 'MUSIC FM five years on: fantasy or reality?', Paper OA6.1, Proc. *PV in Europe*, October, Rome.

Bruton T. M., Luthardt G., Rasch K.-D., Roy K., Dorrity I. A., Garrard B., Teale L., Alonso J., Ugalde U., Declerq K., Nijs J., Szlufcik J., Rfauber A., Wettling W. and Vallera A. (1997), 'A study of the manufacture at 500 MWp p.a. of crystalline silicon photovoltaic modules', *Conf. Record, 14th European Photovoltaic Solar Energy Conf.*, Barcelona, pp. 11–16.

Chapin D. M., Fuller C. S. and Pearson G. O. (1954), 'A new silicon *p-n* junction photocell for converting solar radiation into electrical power', *J. Appl. Phys.* **8**, 676–677.

Christensen E. (1985, ed.), *Final Report, Flat Plate Solar Array Project: 10 Years of Progress*, Jet Propulsion Laboratory, Pasadena, CA, JPL Publ. 400–279.

Chu T. L. (1977), 'Silicon films on foreign substrates for solar cells', *J. Crystal Growth* **39**, 45–60.

Chunduri S. K. (2010), 'New choices for 'selective' shoppers', *Photon International*, December, pp. 158–172.

Corkish R. (1991), 'Some candidate materials for lattice-matched liquid phase epitaxial growth on silicon', *Solar Cells* **31**, 537–548.

Cousins P. J., Smith D. D., Luan H. C., Manning J., Dennis T. D., Waldhauer A., Wilson K. E., Harley G. and Mulligan G. P. (2010), 'Gen III: improved performance at lower cost', *35th IEEE PVSC*, June, Honolulu.

De Ceuster D., Cousins P., Rose D., Vicente D., Tipones P. and Mulligan W. (2007), 'Low cost, high volume production of >22% efficiency solar cells', *22nd European Photovoltaic Solar Energy Conf.*, Milan, September, pp. 816–819.

De Clercq K., Frisson L., Szlufcik J., Nijs J. and Mertens R. (1997), 'Design for manufacturability of a silicon solar cell process based on screen printing by means of manufacturing science techniques', *Conf. Proc., 14th European Photovoltaic Solar Energy Conf.*, Barcelona, pp. 135–138.

De Moor H. H. C., Hoornstra J., Weeber A. W., Burgers A. R. and Sinke W. C. (1997), 'Printing high and fine metal lines using stencils', *Conf. Proc. 14th European Photovoltaic Solar Energy Conf.*, Barcelona, pp. 404–407.

Dietl J., Helmreich D. and Sirtl E. (1981), "'Solar" silicon', in *Crystals, Growth, Properties and Applications,* Vol. 5: *Silicon*, Springer-Verlag, pp. 43–107.

Einhaus R., Vazsonyi E., Szlufcik J., Nijs J. and Mertens R. (1997), 'Isotropic texturing of multicrystalline silicon wafers with acidic texturing solutions', *Conf. Proc., 26th IEEE Photovoltaic Specialists Conf.*, pp. 167–170.

Fang P. H., Ephrath L. and Nowak W. B. (1974), 'Polycrystalline silicon films on aluminium sheets for solar cell application', *Appl. Phys. Lett.* **25**, 583.

Ferrazza F. (1995), 'Growth and post-growth processes of multicrystalline silicon for photovoltaic use', in Pizzini S., Strunk H. P. and Werner J. H. (eds), *Polycrystalline Semiconductors IV — Physics, Chemistry and Technology*, Solid State Phenomena" Trans. Tech. Publ., Zug, Switzerland.

Fischer D., Keppner H., Kroll U., Torres P., Meier J., Platz R., Dubail S., Selvan J. A. A., Vaucher N. P., Ziegler Y., Tscharner R., Hof C., Beck N., Goetz M., Pernet P., Goerlitzer M., Wyrsch N., Vuille J., Cuperus J. and Shah A. (1997), 'Recent progress of the 'Micromorph' tandem solar cells', *Conf. Proc., 14th European Photovoltaic Solar Energy Conf.*, Barcelona, pp. 2347–2350.

Fischer H. and Pschunder W. (1976), 'Low cost solar cells based on large area unconventional silicon', *Conf. Record, 12th IEEE Photovoltaic Specialists Conf.*, Baton Rouge, November, pp. 86–82.

Godfrey R. B. and Green M. A. (1980), 'High efficiency silicon miniMIS solar cells — design and experimental results', *IEEE Trans. Elec. Dev.* **ED-27**, 737–745.

Godlewski M. P., Baraona C. R. and Brandhorst H. W. (1973), 'Low-high junction theory applied to solar cells', *10th IEEE Photovoltaic Specialists Conf.*, Palo Alto, pp. 40–49.

Grassman T. J., Brenner M. R., Rajagopalan S., Unocic R., Dehoff R., Mills M., Fraser H. and Ringel S. A. (2009), 'Control and elimination of nucleation-related defects in GaP/Si(001) heteroepitaxy', *App. Phys. Lett.* **94**, 232106–232106-3.

Green M. A. (1984), 'Limits on the open circuit voltage and efficiency of silicon solar cells imposed by intrinsic Auger processes', *IEEE Trans. Electron Devices* **ED-31**, 671–678.

Green M. A. (1987), *High Efficiency Silicon Solar Cells*, Trans Tech Publications, Aedermannsdorf.

Green M. A. (1995), *Silicon Solar Cells: Advanced Principles and Practice*, Bridge Printery, Sydney.

Green M. A. (1999), 'Limiting efficiency of bulk and thin film silicon solar cells in the presence of surface recombination', *Progr. Photovoltaics* **7**, 327–330.

Green M. A. (2003), 'General temperature dependence of solar cell performance and implications for device modelling', *Progr. Photovoltaics* **11**, 333–340.

Green M. A. (2009), 'The path to 25% silicon solar cell efficiency: history of silicon cell evolution', *Progr. Photovoltaics* **17**, 183–189.

Green M. A. (2011), 'Ag requirements for silicon wafer-based solar cells', *Progr. Photovoltaics* **19**, 911–916.

Green M. A., Basore P. A., Chang N., Clugston D., Egan R., Evans R., Ho J., Hogg D., Jarnason S., Keevers M., Lasswell P., O'Sullivan J., Schubert U., Turner A., Wenham S. R. and Young T. (2004), 'Crystalline silicon on glass (CSG) thin-film solar cell modules', *Solar Energy, Special Issue on Thin Film Photovoltaics* **77**, 857–863.

Green M. A. and Hansen J. (1998), *Catalogue of Photovoltaic Drawings*, Photovoltaics Special Research Centre, University of New South Wales, Sydney.

Green M. A. and Ho J. (2011), *Catalogue of Photovoltaic Drawings*, 3rd edition, School of Photovoltaic and Renewable Energy Engineering, University of New South Wales, Sydney.

Green M. A., Zhao J., Wang A. and Wenham S. R. (1992), '45% efficient silicon photovoltaic cell under monochromatic light', *IEEE Electron Device Lett.* **31**, 317–318.

Haase C. and Podewils C. (2011), 'More of everything: market survey on solar modules 2011', *Photon International*, February, pp. 174–221.

Hahn G. and Schönecker A. (2004), 'New crystalline silicon ribbon materials for photovoltaics', *J. Phys.: Condens. Matter* **16**, R1615–R1648.

Hall R. B., Barnett A. M., Brown J. E., Checchi J. C., Ford D. H., Kendall C. L., Mulligan W. P., Rand J. A. and Ruffins T. R. (1994), 'Columnar-grained polycrystalline solar cell and process of manufacture', U.S. Patent No. 5336335, August 9, 1994.

Haynos J., Allison J., Arndt R. and Meulenberg A. (1974), 'The COMSAT non-reflective silicon solar cell: a second generation improved cell', *Int. Conf. on Photovoltaic Power Generation*, Hamburg, p. 487.

Healy S. A. and Green M. A. (1992), 'Efficiency enhancements in c Si solar cells by the incorporation of a region alloyed with germanium', *Solar Energy Mater. Solar Cells* **28**, 273–284.

Hering G. (2011), 'Year of the Tiger: PV cell output roared in 2010 to over 27 GW — beating 2006 through 2009 combined — but can the Year of the Rabbit bring more multiples?', *Photon International*, March, pp. 186–218.

Janoch R., Wallace R. and Hanoka J. I. (1997), 'Commercialization of silicon sheet via the string ribbon crystal growth technique', *Conf. Record, 26th IEEE Photovoltaic Specialists Conf.*, Anaheim, September/October, pp. 95–98.

Jordan D. and Nagle J. P. (1994), 'New generation of high-efficiency solar cells: development, processing and marketing', *Progr. Photovoltaics* **2**, 171–176.

Keevers M. J. and Green M. A. (1994), 'Efficiency improvements of silicon solar cells by the impurity photovoltaic effect', *J. Appl. Phys.* **75**, 4022–4033.

Keevers M. J., Young T. L., Schubert U. and Green M. A. (2007), '10% efficient CSG minimodules', *22nd European Photovoltaic Solar Energy Conf.*, Milan, 3–7 September, pp. 1783–1790.

Keevers M. J., Zhang G. C., Saris F. W., Zhao J. and Green M. A. (1995), 'Screening of optical dopants in silicon solar cells for improved infrared response', *Conf. Proc., 13th European Photovoltaic Solar Energy Conf.*, Nice, pp. 1215–1218.

Khattak C. P., Basaran M., Schmid F., D'Aiello R. V., Robinson P. H. and Firester A. H. (1981), 'Metallurgical-silicon substrates produced by HEM for epitaxial thin film solar cells', *Conf. Record, 15th IEEE Photovoltaic Specialists Conf.*, Orlando, May 12–15, pp. 1432–1437.

King D. L. and Buck E. M. (1991), 'Experimental optimization of an anisotropic etching process for random texturization of silicon solar cells', *Conf. Record, 22nd IEEE Photovoltaic Specialists Conf.*, Las Vegas, October, pp. 303–308.

Kingsbury E. F. and Ohl R. S. (1952), 'Photoelectric properties of ionically bombarded silicon', *Bell Syst. Tech. J.* **31**, 8092.

Knobloch J., Glunz S. W., Biro D., Warta W., Schaffer E. and Wettling W. (1996), 'Solar cells with efficiencies above 21% processed from Czochralski grown silicon', *Conf. Record, 25th IEEE Photovoltaic Specialists Conf.*, Washington, Washington D.C., May, pp. 405–408.

Kolodinski S., Werner J. H., Wittchen T. and Queisser H. J. (1993), 'Quantum efficiencies exceeding unity due to impact ionization in silicon solar cells', *Appl. Phys. Lett.* **63**, 2405–2407.

Lamers M. W. P. E., Tjengdrawira C., Koppes M., Bennett I. J., Bende E. E., Visser T. P., Kossen E., Brockholz B., Mewe A. A., Romijn I. G., Sauar E., Carnel L., Julsrud S., Naas T., de Jong P. C. and Weeber A. W. (2011), '17.9% metal-wrap-through mc-Si cells resulting in module efficiency of 17.0%', *Progr. Photovoltaics*, published online: 23 Mar 2011, DOI: 10.1002/pip.1110.

Lange H. and Schwirtlich I. A. (1990), 'Ribbon growth on substrate (RGS) — a new approach to high speed growth of silicon ribbons for photovoltaics', *J. Cryst. Growth* **104**, p. 108.

Laudisio G., Young R., Li Z. and Getty R. (2010), 'A view of the design challenges involved in the development of advanced n-type contacts using lead-free chemistries', *Proc. 2nd Workshop on Metallization for Crystalline Silicon Solar Cells — Status, Trends and New Directions*, 14–15 April, Constance, Germany.

Lindmayer J. (1976), 'Semi-crystalline silicon solar cells', *Conf. Record, 12th IEEE Photovoltaic Specialists Conf.*, Baton Rouge, November, pp. 82–85.

Lindmayer J. and Allison J. (1973), 'The violet cell: an improved silicon solar cell', *COMSAT Tech. Rev.* **3**, 1–22.

Maruyama E., Terakawa A., Taguchi M., Yoshimine Y., Ide D., Baba T., Shima M., Sakata H. and Tanaka M. (2006), 'Sanyo's challenges to the development of high-efficiency HIT solar cells and the expansion of HIT business', *4th World Conf. on Photovoltaic Energy Conversion (WCEP-4)*, May, Hawaii.

Minagawa S., Saitoh T., Warbisako T., Nakamura N., Itoh H. and Tokuyama T. (1976), 'Fabrication and characterization of solar cells using dendritic silicon thin films grown on alumina ceramic', *Conf. Record, 12th IEEE Photovoltaic Specialists Conf.*, Baton Rouge, November, pp. 77–81.

Neuhaus D.-H. and Münzer A. (2007), 'Industrial silicon wafer solar cells', *Adv. OptoElectronics* **2007**, Article ID 24521, 1–15.

Ohl R. S. (1941), 'Light sensitive electric device', U.S. Patent No. 2,402,622, (27 March), 'Light-sensitive electric device including silicon', U.S. Patent No. 2,443,542 (27 May).

Photovoltech (2010), Doc. D-S&M-008, Rev. A, 22 Oct, www.photovoltech.com.

Ralph E. L. (1975), 'Recent advancements in low cost solar cell processing', *Conf. Proc., 11th IEEE Photovoltaic Specialists Conf.*, Scottsdale, AZ, pp. 315–316.

Riordan M. and Hoddeson L. (1997*), Crystal Fire: The Birth of the Information Age*, Norton, New York.

Rodriguez H., Guerrero I., Koch W., Endrös A. L., Franke D., Häßler C., Kalejs J. P. and Möller H. J. (2011), 'Bulk crystal growth and wafering for PV', in Luque A. and Hegedus S. (eds) *Handbook of Photovoltaic Science and Engineering*, 2nd Edition, Wiley, New York.

Sawada T., Terada N., Tsuge S., Baba T., Takahama T., Wakisaka K., Tsuda S. and Nakano S. (1994), 'High-efficiency a-Si/c-Si heterojunction solar cell', *Conf. Record, 1st World Conf. Photovoltaic Energy Conversion*, Hawaii, pp. 1219–1225

Scaff J. H., Theuerer H. C. and Schumachor E. E. (1949), '*P*-type and *n*-type silicon and the formation of the photovoltaic barrier in silicon ingots', *Trans. Amer. Inst. Mining Met. Engineering*, June, pp. 383–388.

Schmidt F. (2011), 'History of technologies development for silicon cost reduction', in Palz W., *Power for the World*, Pan Stanford, Singapore.

Schmidt J. and Bothe K. (2004), 'Structure and transformation of the metastable boron- and oxygen-related defect center in crystalline silicon', *Phys. Rev. B* **69**, 241071–241078.

Schott (2009), 'Realignment of wafer business at WACKER SCHOTT Solar', Press Release, 30 September.

Schulz M. and Sirtl E. (1984), 'Silicon sheet' in *Landolt-Börnstein: Numerical Data and Functional Relationships in Science and Technology*, Springer-Verlag, Berlin, **17c**, Section 6.1.2.5.3, 52–54 and 442–444.

Schumann M., Orellana Perez T. and Riepe S. (2009), 'The solar cell wafering process', *Photovoltaics International*, August, 53–59.

SEMI (2011), Crystalline silicon PV technology and manufacturing group, *International Technology Roadmap for Photovoltaics: Results 2010*, 2nd Edition, SEMI PV Group Europe, Berlin, March.

Serreze H. B. (1978), 'Optimizing solar cell performance by simultaneous consideration of grid pattern design and interconnect configurations', *Conf. Record, 13th IEEE Photovoltaic Specialists Conf.*, Washington, D.C., pp. 609–614.

Shah A., Meier J., Torres P., Kroll U., Fischer D., Beck N., Wyrsch N. and Keppner H. (1997), 'Recent progress on microcrystalline solar cells', *Conf. Record, 26th IEEE Photovoltaic Specialists Conf.*, Anaheim, September/October, pp. 569–574.

Solland (2009), Preliminary Data Sheet, Sunweb, 14 May, www.sollandsolar.com.

Song J., Boas R., Bolman C., Meyers M., Rogol M. and Trangucci R. (2010), 'The true cost of solar power: how low can you go?', *Photon Consulting*, April.

Stoddard N. G., Wu B., Witting I., Wagener, M. Park Y. Rozgonyi G. and Clark R. (2008), 'Casting single crystal silicon: novel defect profiles from BP Solar's Mono2^{TM} wafers', *Solid State Phenomena* **131–133**, 1–8.

Strahm B., Andrault Y., Baetzner D., Guérin C., Holmes N., Kobas M., Lachenal D., Mendes B., Tesfai M., Wahli G., Wuensch F., Buechel A., Mai J., Schulze T. and Vogt M. (2010), 'Progress in silicon heterojunction solar cell development and scaling for large scale mass production use', *25th European Photovoltaic Solar Energy Conf.*, September, Valencia, Spain.

Tanaka M., Taguchi M., Takahama T., Sawada T., Kuroda S., Matsuyama T., Tsuda S., Takeoka A., Nakano S., Hanfusa H. and Kuwano Y. (1993), 'Development of a new heterojunction structure (ACJ-HIT) and its application to polycrystalline silicon solar cells', *Progr. Photovoltaics* **1**, 85–92.

Tiedje T., Yablonovitch E., Cody G. D. and Brooks B. G. (1984), 'Limiting efficiency of silicon solar cells', *IEEE Trans. Electron Devices* **ED-31**, 711–716.

Trupke T., Green M. A. and Würfel P. (2002a), 'Improving solar cell efficiencies by the up-conversion of sub-band-gap light', *J. Appl. Phys.* **92**, 4117–4122.

Trupke T., Green M. A. and Würfel P. (2002b), 'Improving solar cell efficiencies by down-conversion of high-energy photons', *J. Appl. Phys.* **92**, 1668–1674.

Tsao C.-Y. (2011), 'Fabrication and characterization of Ge thin films and Ge-rich SiGe alloys for photovoltaic applications', *PhD Thesis*, University of New South Wales, submitted on 31 March.

Wallace R. L., Hanoka J. I., Narasimha S., Kamra S. and Rohatgi A. (1997), 'Thin silicon string ribbon for high efficiency polycrystalline solar cells', *Conf. Record, 26th IEEE Photovoltaic Specialists Conf.*, Anaheim, September/October, pp. 99–102.

Wenham S., Robinson R., Dai X., Zhao J., Wang A., Tang, Y. H., Ebong A., Honsberg C. B. and Green M. A. (1994), 'Rear surface effects in high efficiency silicon solar cells', *Conf. Record, 1st World Conf. on Photovoltaic Energy Conversion*, December, pp. 1278–1282.

Wenham S. R. and Green M. A. (2002), 'Self aligning method for forming a selective emitter and metallization in a solar cell', U.S. Patent No. US6,429,037 B1, 6 August.

Werner J. H., Brendel R. and Queisser H. J. (1994), 'New upper efficiency limits for semiconductor solar cells', *Conf. Record, 1st World Conf. Photovoltaic Energy Conversion*, Waikoloa, December, pp. 1742–1745.

Wistey M. A., Fang Y.-Y., Tolle J., Chizmeshya A. V. G. and Kouvetakis J. (2007), 'Chemical routes to Ge/Si(100) structures for low temperature Si-based semiconductor applications', *Appl. Phys. Lett.* **90**, 82108.

Wolf M. (1960), 'Limitations and possibilities for improvement of photovoltaic solar energy converters', *Proc. IRE* **48**, 1246–1263.

Wolf M. (1976), 'Historical development of solar cells', in *Solar Cells*, Backus C. E., ed., New York, IEEE Press.

Yamamoto K., Yoshimi M., Suzuki T., Okamoto Y., Tawada Y. and Nakajima A. (1997), 'Thin film poly-Si solar cell with 'Star Structure' on glass substrate fabricated at low temperature', *Conf. Record, 26th IEEE Photovoltaic Specialists Conf.*, Anaheim, September/October, pp. 575–580.

Zhao J., Wang A., Altermatt P. and Green M. A. (1995), '24% efficient silicon solar cells with double layer antireflection coatings and reduced resistance loss', *Appl. Phys. Lett.* **66**, 3636–3638.

CHAPTER 4

THIN-FILM SOLAR CELLS BASED ON AMORPHOUS AND MICROCRYSTALLINE SILICON

CHRISTOPHE BALLIF*, MATTHIEU DESPEISSE† and FRANZ-JOSEF HAUG‡
Photovoltaics and Thin Film Electronics Laboratory
Institute of Microengineering Ecole Polytechnique
Fédérale de Lausanne Neuchâtel, Switzerland
**christophe.ballif@epfl.ch*
†matthieu.despeisse@csem.ch
‡franz-josef.haug@epfl.ch

One advantage of being disorderly is that one is constantly making exciting discoveries.
A. A. Milne.

4.1 Introduction

Since the first deposition of hydrogenated amorphous silicon out of a silane discharge (Chittik *et al.*, 1969), plasma coating of thin silicon layers has played an increasingly important role for a large number of applications in 'macroelectronics', including flat panel displays (FPD), solar modules and X-ray imagers. Indeed, thin silicon layers acting as active semiconductors were likely coated over more than 140 million square metres in 2010, on glass plates, metal foils and plastic webs. Hydrogenated amorphous silicon used as an active semiconductor n-type channel in thin-film transistors (TFT) in active-matrix liquid crystal displays made the bulk of this surface, whereas around 6% of the surface was dedicated to photovoltaic (PV) applications. The recent synergy between the FPD sector and thin-film silicon PV allowed module producers for the first time in thin-film PV history to purchase thin-film turnkey lines for large area (> 1 m^2) coating. Manufacturing costs below 0.5 €/Wp (Kratzla *et al.*, 2010; Ringbeck and Stutterlueti, 2012) with 10% module efficiency or higher are now demonstrated by several companies that provide equipment, making thin-film silicon a low-cost technology, even at relatively low production volume. Short-term perspective cost in the range of 0.35€/Wp are even indicated. In a growing photovoltaic market, which could reach several hundreds of GW annually, thin-film silicon should be able to remain a player because of many inherent advantages that are the subject of this chapter.

The key feature of 'thin-film silicon' layers for photovoltaic application is the possibility to manufacture a wide variety of materials with similar processes, typically by plasma-enhanced chemical vapour deposition (PECVD):

- Hydrogenated amorphous silicon layers (a-Si:H) with a bandgap (E_g) of 1.7–1.8 eV. Alloying with germanium yields a-Si_xGe_{1-x}:H absorber layers with decreased bandgap; addition of oxygen or carbon allows higher bandgaps. The latter are used essentially as doped or buffer layers, as they tend to suffer from too strong light-induced degradation for use as absorbers.
- Hydrogenated microcrystalline silicon layers (μc-Si:H): This material typically consists of interconnected silicon nanocrystallites, forming micrometre-sized agglomerates. Surface defects of the nanocrystallites are passivated electronically by a surrounding amorphous phase. The μc-Si:H bandgap is similar to crystalline silicon (c-Si), i.e. 1.12 eV. As in the amorphous case, μc-Si_xGe_{1-x}:H is used for absorbers with further reduced bandgap. Incorporation of oxygen or carbon decreases absorption and can be used to fabricate more transparent doped layers.
- Recrystallised or polycrystalline layers, with typical grain size in the micrometre range and with electrical properties (in particular the mobility) closer to those of c-Si. These layers typically exhibit high defect densities at grain boundaries because no hydrogenated amorphous silicon phase is present to passivate dangling bonds.
- Polycrystalline to ideally monocrystalline layers realised by epitaxial growth on monocrystalline wafer substrates.

Table 4.1 summarises the various forms of silicon films as they are used today in PV and FPD on industrial levels. This chapter will focus on the class of *amorphous* and *microcrystalline* materials used as absorber and doped layers in thin-film solar cells, as well as the related technological aspects. Notably, such layers can be made with a variety of electronic properties and nanostructures, which depend on the deposition process and on the substrate (i.e. its surface chemistry and its geometry).

The chapter is organised as follows. The first section gives an overview of device configuration, technology and history. The layers' basic properties are detailed in Section 4.2. Section 4.3 presents the elementary properties of amorphous and multijunction thin-film silicon solar cells. Section 4.4 focuses on device-grade absorber deposition of a-Si:H and μc-Si:H materials. In Section 4.5, aspects related

Table 4.1 Silicon-based materials used in thin-film Si solar cells and transistors.

Material	Application	Typical properties	In production?
intrinsic a-Si:H	channel in thin-film transistor		yes
	absorber layer		yes
intrinsic a-$Si_{1-x}Ge_x$:H	absorber layer	E_g down to 1.4 eV	yes, in *n-i-p* devices
intrinsic a-SiC:H	buffer layer in a-Si:H devices		yes
	high bandgap top cell (Yunaz *et al.*, 2009)	$E_g > 1.7$ eV, suffers from strong degradation and high intrinsic defect density	no
intrinsic μc-Si:H	channel in thin-film transistor	high-mobility *n* and *p* channel	pilot stage
	low-bandgap bottom cell (Meier *et al.*, 1996)	E_g at 1.12 eV, passivation of grains by a-Si:H	yes
intrinsic μc-$Si_{1-x}Ge_x$:H	low-bandgap bottom cell (Matsui *et al.*, 2010)	lower E_g than μc-Si:H	
doped a-Si:H and μc-Si:H	doped contact layers	ultra-low lifetime	yes
p-SiC_x:H	*p*-type window layer (Tawada *et al.*, 1981)	higher transparency	yes
p-SiO_x:H	*p*-type window layer (Sichanugrist *et al.*, 1983)	higher transparency	not disclosed
n- and *p*-μc-SiO_x:H	intermediate reflectors, optical layer (Buehlman *et al.*, 2007; Lambertz *et al.*, 2007; Cuony *et al.*, 2010)	*in situ* material*, containing nanocrystallites, shunt quenching effect (Despeisse *et al.*, 2010)	not disclosed
	reflector buffer layer (Delli Veneri *et al.*, 2010)	*in situ** alternative to ZnO	
n-μc-SiC	transparent contact (Chen *et al.*, 2010)	E_{04} at 3–3.2 eV, C incorporated in crystalline phase	

(*Continued*)

Table 4.1 (Continued)

Material	Application	Typical properties	In production?
(re)-crystallised layers	channel in high mobility transistor	E_g and crystalline structure close to c-Si	yes
	absorber layer (Basore, 2002; van Gestel *et al.*, 2010)	like c-Si, sensitive to grain boundaries and dislocations	

**In situ* means the material can also be deposited by PECVD.

to light trapping and rough front (transparent conducting oxide) and back electrodes (e.g. metallic back reflectors) are presented. Section 4.6 reviews some of the newest device architecture based on novel transparent conducting oxide and doped layers. Finally, in Section 4.7 up-scaling, production costs and environmental impacts of the technology are discussed.

4.1.1 *Basic device configuration and definition*

Figure 4.1 illustrates the variety of thin-film silicon PV devices. All cells require a reflecting back contact and a transparent conducting oxide (TCO) at the front. They include intrinsic layers (*i*-layers) with good electronic properties sandwiched between doped *p*-type and *n*-type layers. Doped layers are required to create the internal electric field but they exhibit poor electronic properties (carriers created in these layers recombine almost immediately); consequently, devices make use of *p-i-n* junctions rather than *p-n* junctions. Because the hole mobility is lower than the electron mobility in a-Si:H, light should enter from the *p*-side of the a-Si:H junction (not necessarily the case for a microcrystalline junction). Hence when depositing on transparent substrates, the deposition sequence usually starts with the *p*-layer and this cell type called *p-i-n* or *superstrate* configuration. When depositing on an opaque substrate (e.g. steel foils or plastic sheets), the *n* layer is deposited first; these cells are called *n-i-p* or *substrate* configuration. Figure 4.1 shows schematic designs.

By combining two or more cells with different bandgaps, tandem or triple junctions can be realised. The micromorph configuration shown in panels a) and b) of Figure 4.1 combines a high-gap amorphous cell with a low-gap microcrystalline cell. Figure 4.2 shows a typical process sequence for the manufacture of such a cell. Single-junction and multijunction thin-film silicon solar cell structures are discussed in detail in Section 4.3.

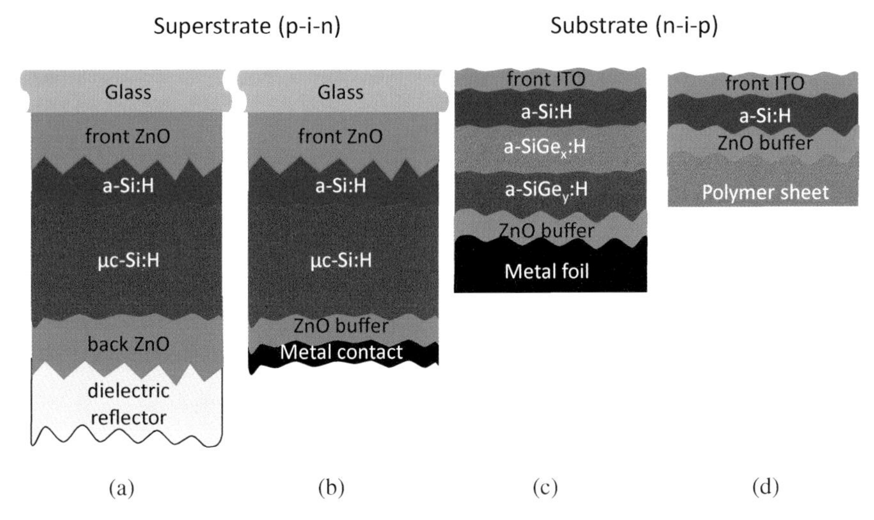

Figure 4.1 Schematic of various device configurations. a) *p-i-n* micromorph cell with thick TCO at the front and back; b) *p-i-n* micromorph with thin TCO covered by reflecting metal at the back; c) triple-junction *n-i-p* cell on metal foil covered with textured metal and thin TCO; d) single-junction a-Si:H cell on textured polymer foil with a metal back reflector and thin TCO.

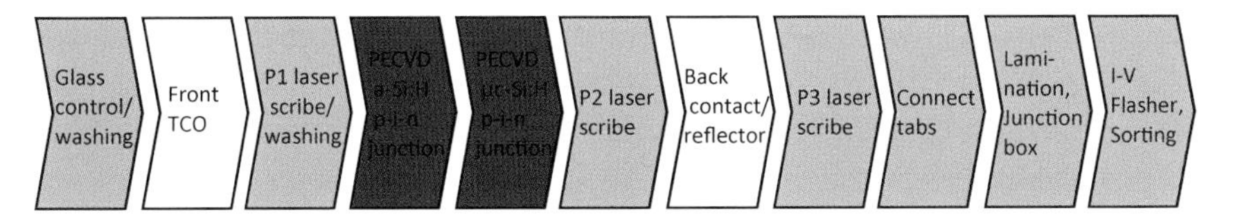

Figure 4.2 Schematic of the process steps in micromorph module production.

All schemes in Figure 4.1 show rough interfaces within the devices. These rough interfaces are essential to ensure efficient light scattering inside the silicon layers and thus to increase the effective light path. The light management schemes that are used in thin-film silicon technology are discussed in Section 4.5, while advanced concepts and substrates are presented in Section 4.6. The properties of the front TCO and back reflector are essential for achieving high device performance; their optimisation is as important as the PECVD deposition process of the silicon films. Two important aspects emerge for the choice of substrate texture:

- In superstrate devices, textured TCOs control the maximum current density via their transmission and light scattering potential.
- The texture of the growth template can lead to defects in the devices, such as collision of growth fronts, which can result in reduced electrical performance.

Intense research is therefore ongoing to find ideal superstrates/substrates which allow for high current density while maintaining optimal electric performance; we discuss this in more detail in Sections 4.3 and 4.5.

4.1.2 *Uniqueness of thin-film silicon for PV: strength and weakness*

The strengths of thin-film silicon PV can be summarised as follows:

- In their most simple form, modules can be made out of glass, a transparent front contact made from zinc oxide or tin oxide, thin layers of silicon and a back electrode. All the materials are abundant (Wadia *et al.*, 2009) and non-toxic. Micromorph modules can be made in compliance with norms on restriction of hazardous substances (RoHS) and can be handled like standard window glass at their end of life. This is presently the only technology fully adapted for transitioning to a terawatt-scale, PV-based society.
- Compared with c-Si, the energy payback time of such modules can be reduced significantly, especially with the use of thin μc-Si absorber layers (Fthenakis *et al.*, 2008a; Fthenakis *et al.*, 2008b; Steimen, 2011). On very large scale, it makes more sense globally to develop technologies with as low as possible energy payback times.
- The temperature coefficient is favourable (typically around –0.2%/°C for a-Si:H modules and –0.25%/°C for micromorph modules), leading to higher energy yield in practice, particularly in hot environments (Shah, 2010; Lechner *et al.*, 2010).
- Proven manufacturing technology has been developed: in recent years a huge industrial effort has been made to industrialise thin-film silicon, tackling most production issues. Production tools are now available to module manufacturers (Kratzla *et al.*, 2010; Klein *et al.*, 2010).
- Eventually the possibility to create materials with various bandgaps should open the route to higher efficiency multiple-junction devices.

The weakness of thin-film silicon technology lies in its relatively modest efficiency, with currently confirmed best stabilised efficiencies of 13.4% and 12.5% for *p-i-n* and *n-i-p* triple-junction cells, respectively (Kim *et al.* 2012; Yue *et al.*, 2010), and 12.3% for *p-i-n* tandem cells (Kroll *et al.*, 2011). However, the latest results from several module and equipment makers indicate that it is possible to manufacture large area modules with 10% stabilised efficiencies in a cost-effective way (Kratzla *et al.*, 2010; Klein *et al.*, 2010), and champion modules of 11% have been reported. In contrast to other technologies, there is only a very small gap between laboratory and real-world efficiencies. Thin-film silicon thus represents a technology that could secure a substantial share of the PV market, in particular for large solar parks in high irradiance areas or for integration of aesthetically pleasing and safe PV products into buildings, but it certainly needs further increased efficiency in the medium term to stay competitive with other mainstream technologies.

4.1.3 *A short historical perspective*

In 1976, Carlson and Wronski demonstrated the first (2.4% initial) 'efficient' a-Si-based PV device (Carlson and Wronski, 1976). One year later, Staebler and Wronski reported that the photoconductivity of a-Si:H suffers on illumination (Staebler and Wronski, 1977). This coined the term 'Staebler–Wronski effect' (SWE) which is often used as a synonym for any kind of light-induced degradation (LID) in thin-film silicon devices. In the 1980s, several companies started the commercialisation of amorphous silicon modules (Solarex based on initial work at RCA, AFG/Chronar, Arco Solar, etc.). In 1986, Arco Solar introduced the first large-area commercial a-Si:H modules (Genesis G-400). In the 1980s and 1990s the PV market was small (a few MWs) and the first a-Si:H modules suffered from reliability issues, including poorly controlled LID, which prevented a successful start to mass production. However, thanks to the simple manufacturing of a-Si:H cells and their good low-light behaviour (the high bandgap results in a low saturation current density i_0), they were successfully integrated in consumer applications such as the pocket calculators commercialised by Sharp and Casio.

At the end of the 1980s and in the 1990s, a-Si:H material stability improved by modifying the plasma deposition processes and, for instance, by adding additional hydrogen to the glow discharge or by introducing very high frequency (VHF) plasma deposition, which resulted in higher deposition rates (Curtins *et al.*, 1987) while maintaining a similar material quality. United Solar Ovonic was the first to introduce triple-junction technology based on a-Si:H/a-Si_xGe_{1-x}:H/a-Si_yGe_{1-y}:H (Yang *et al.*, 1994). They commercialised a flexible large-area product on steel foil. In Japan, Fuji developed flexible modules in a-Si:H/a-Si_xGe_{1-x}:H configuration on polyimide.

During the mid-1990s, the first efficient solar cells based on μc-Si:H were realised by careful control of extrinsic impurities (see Section 4.2.6). Their integration with a-Si:H into high-efficiency micromorph tandem devices was demonstrated (Meier *et al.*, 1994); tandem devices increase stabilised efficiencies by 20% to 30% compared with single-junction amorphous silicon cells. This finding resulted in intense research activities in various laboratories (Shah *et al.*, 2004; Rech *et al.*, 2006; Schropp *et al.*, 2007; Yan *et al.*, 2007). An inherent advantage of μc-Si:H is the fact that it is fabricated from the same hydrogen (H_2) and silane (SiH_4) input gases as a-Si:H. Moreover, the bandgaps of a-Si:H and μc-Si:H coincide with the theoretical optimum of tandem devices for the solar irradiation spectrum (AM1.5G) (DeVos, 1980). Microcrystalline silicon thus offers an alternative to silicon-germanium alloys that were previously used for tandem and triple-junction cells by several research groups and companies.

Japanese companies were the first to put massive R&D efforts into large-scale industrialisation of thin-film silicon technology with a focus on micromorph devices on glass. Sharp, Kaneka and MHI started production of amorphous silicon modules on glass, relying mostly on in-house development of the production equipment for module manufacturing, e.g. the high-rate ladder electrode design of MHI (Takeuchi *et al.*, 2001). Kaneka were the first to commercialise micromorph modules in 2001 at a pilot stage (1∼2 MW). In this time also other flexible activities started up, e.g. Powerfilm (USA), Flexcell (Switzerland) and Xunlight (USA).

Thin-film silicon technology gained momentum in 2005 when equipment makers with experience from FPD entered the scene. The development was further boosted by the silicon raw material shortage in 2007–2008 that prevented newcomers from entering into c-Si cell manufacturing. This strong industrialisation phase required a lot of research resources to solve up-scaling issues (*cf.* Section 4.7) and start up dozens of manufacturing plants, and slowed the quest for improved efficiencies. Recently, however, further improvements in cell performance were achieved; most of the record cell efficiencies listed in Table 4.2 were reported by industrial development labs, for example a new initial record of 16.3% efficiency has been reached for a triple-junction cell (Guha *et al.*, 2011). Close collaboration with research institutes should offer the industry the possibility to develop and use novel materials (e.g. new doped layers) and concepts (e.g. advanced light trapping), which promise further gains in efficiency (Konagai *et al.*, 2011).

4.2 Basic properties of amorphous and microcrystalline silicon alloys

This section gives a brief overview of the electronic and optical properties of thin-film silicon. Following the discovery of a-Si:H and its first use in devices, it soon became necessary to change some of its properties for a variety of reasons.

- Figure 4.3 shows that the bandgap of amorphous silicon is between 1.7 to 1.8 eV; this is generally considered to be too high for efficient absorption of the solar spectrum. Attempts were therefore made to lower the bandgap by alloying with germanium or tin, or to find new materials with lower gaps for the construction of advanced tandem and multijunction devices.
- While the demonstration of doping was surprising because an amorphous network should easily incorporate elements with lower or higher valence, it was soon noted that the achievable doping efficiency is limited; films with better

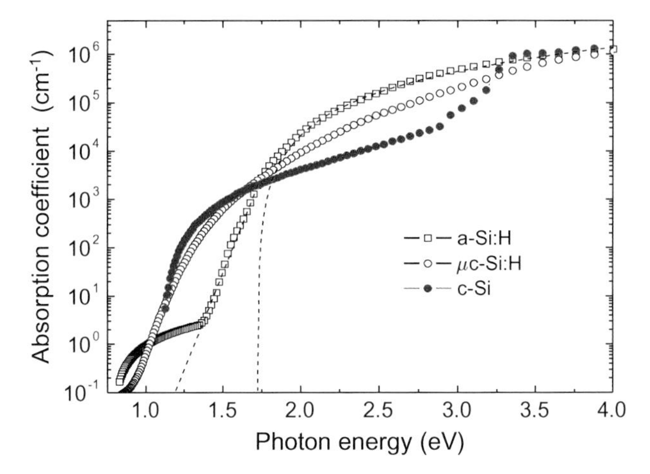

Figure 4.3 Spectral absorption of the thin-film materials a-Si:H and μc-Si:H (open squares and circles, respectively), and c-Si (filled circles). Dashed lines illustrate the Urbach tail between 1.25 and 1.75 eV and the indirect bandgap for energies beyond 1.75 eV.

conductivity became eventually available, but they were found to contain small crystallites of silicon.

- Such crystallite-containing silicon is actually a material in its own right, called microcrystalline or nanocrystalline silicon, useful not only for doped layers but also as an absorber with a low bandgap. Figure 4.3 shows that its absorption is similar to that of crystalline silicon, i.e. it extends into the near IR region, but it is also relatively weak (10^1–10^3 cm^{-1} between 1.5 and 1.12 eV).
- Absorption in the doped layers of solar cells does not contribute to the photocurrent. Recombination losses in doped amorphous layers are unavoidable because defects are inherently created along with doping. In solar cell applications, losses in the doped layers were eventually reduced by using materials with higher bandgap, e.g. by incorporation of carbon, oxygen or nitrogen.

In the remainder of this section these and other aspects will be addressed in more detail.

4.2.1 *Amorphous silicon*

Amorphous silicon can easily be grown by evaporation or sputtering, but the obtained material is generally too defective for use in semiconductor devices. Higher quality amorphous silicon was obtained by deposition from SiH_4 glow discharges (Chittick *et al.*, 1969), essentially the process called PECVD which is illustrated in Figure 4.13. This material showed a thermally activated dark conductivity indicative of transport by charge carriers that are activated over a bandgap, and an acceptable level of photoconductivity. PECVD-grown material differs from evaporated films by the incorporation of hydrogen. Because the hydrogen content can easily reach 10–15 at.%, PECVD material is normally called hydrogenated amorphous silicon.

In contrast to crystalline silicon where the four tetrahedral valence electrons are perfectly accommodated in the diamond lattice, there is no long-range order in amorphous silicon. Nevertheless, a tetrahedral bonding environment reaching as far as the first- and second-order neighbours is maintained (Laaziri *et al.*, 1999). The resulting defects in the amorphous silicon network are associated with dangling bonds which contain an unpaired electron, similar to the dangling bonds at the surface of a freshly cleaved silicon crystal. Their density can thus be measured by the spin resonance of the unpaired electron that resides in the neutral dangling bond (Haneman, 1968; Brodsky and Title, 1969).

Figure 4.3 illustrates that the spectral absorption of amorphous silicon shows the behaviour of an indirect bandgap with $E_g = 1.7\,\text{eV}$, according to $(E - E_g)^2$. Below the bandgap, the absorption becomes exponential and is known as Urbach absorption. The exponential part of the curve can be described by a characteristic energy $E_0 = 50\,\text{meV}$, often called the disorder parameter. In material of inferior quality, a higher degree of disorder yields higher Urbach absorption, and correspondingly higher values of the characteristic energy E_0. For amorphous silicon deposited under different conditions, the transition between the parabolic high-energy part and the exponential Urbach region can vary significantly. Thus, the absorption of amorphous silicon is sometimes characterised by E_{04} or E_{03}, the energy where the absorption coefficient assumes a value of $10^4\,\text{cm}^{-1}$ (or correspondingly $10^3\,\text{cm}^{-1}$). The absorption at energies below 1.4 eV is related to defect levels located around mid-gap; their density can be related linearly to $\alpha_{1.25}$, the value of the absorption coefficient at 1.25 eV (Jackson and Amer, 1982).

Combining optical absorption measurements (Redfield, 1982) with results of drift experiments (Tiedje *et al.*, 1981) and photoemission studies (Jackson *et al.*, 1985) yields the schematic band diagram shown in Figure 4.4. As in ideal semiconductors, valence and conduction bands are associated with freely moving carriers. The band edges in amorphous silicon are not well defined but exhibit exponentially decreasing densities of states which extend from the respective band edges into the gap. The Urbach absorption is dominated by the characteristic energy of the valence band tail whereas the conduction band tail is steeper with a characteristic energy of about 30 meV (Redfield, 1982). As far as conductivity is concerned, carriers residing in tail states are assumed to be immobile and move only after excitation above the band edge, giving rise to the term ‘mobility gap’.

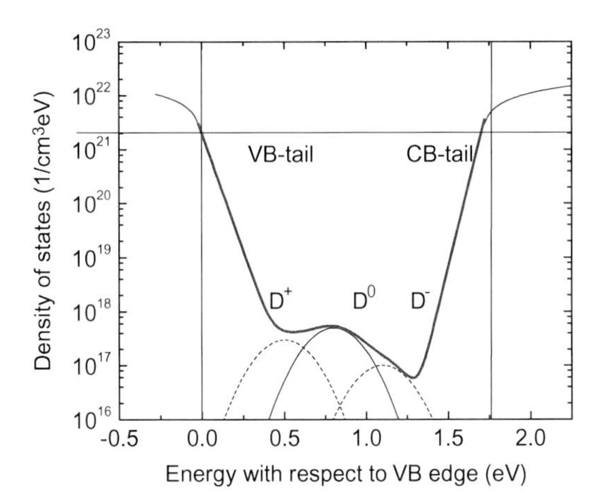

Figure 4.4 Schematic density of states in device-grade a-Si:H. For recombination statistics, often two transition states D^+/D^0 and D^0/D^- are used rather than actual charge states.

Neutral defect states are expected at approximately mid-gap position, but owing to the disordered character of a-Si:H, their density is normally described by a Gaussian distribution. In its neutral state, the defect is occupied by an unpaired electron; therefore it is usually denoted by D^0, but the defect is amphoteric and can also acquire positive or negative charge. For the formation of the negatively-charged defect, D^-, two electrons must be accommodated in the defect state; the amount of energy to overcome electrostatic repulsion is called the *correlation energy U*. Alternatively, the defects can be described by two transition states, D^+/D^0 and D^0/D^-, separated by U.

4.2.2 *Light-induced degradation (LID)*

In addition to defects that are inherently present directly after the growth of amorphous silicon, exposure to light creates additional defects that degrade the photoconductivity as well as the dark conductivity. Subsequent thermal annealing can completely restore the initial state of the samples. This observation, as previously noted, is called the Staebler–Wronski effect (SWE) (Staebler and Wronski, 1977). The salient features are the following (Staebler and Wronski, 1981):

- The creation of light-induced defects does not depend much on the energy of the incident light.
- Fewer defects are created in samples where a bias sweeps out the charge carriers before recombination.
- Samples with rectifying junctions also degrade in the dark when a forward bias is applied.
- The dependence on light intensity is sub-linear.

Based on these observations, it was suggested that the creation of defects in amorphous silicon is actually due to recombination of charge carriers, regardless of their origin. In this context, three more aspects of LID are noteworthy:

- Degradation of the photoconductivity is reduced in films for which hydrogen is added to the discharge (Guha *et al.*, 1981). To date, hydrogen dilution remains an efficient way to obtain more stable absorber material for solar cells. Nevertheless, degradation cannot be completely avoided (*cf.* Figure 4.9).
- The nature of the recombination process matters only little in terms of creating defects. While non-radiative recombination via defects prevails around room temperature, recombination is dominantly radiative at low temperature; however, the degradation of the photoconductivity is virtually unchanged (Stradins, 2003).
- The exact nature of the SWE defects and the role of hydrogen in annealing are not clear. Defect creation by breaking of weak bonds during carrier recombination is a common model (Stutzmann *et al.*, 1985), whereas models involving hydrogen have also been proposed (Winer, 1990; Branz, 1999).

Recovery of an initial state by annealing, and creation of defects by recombination, inspired the so-called equilibrium model; based on a large reservoir (or pool) of weak or strained bonds (Stutzmann *et al.*, 1985), the model assumes that at high temperature breaking and formation of bonds takes place in thermodynamic equilibrium with a given defect formation energy (Smith and Wagner, 1985). The pool of weak bonds is associated with states in the valence-band tail. Their density, in turn, varies exponentially with the disorder parameter E_0, the Urbach energy. Annealing at high temperature establishes equilibrium between bonds that form during annealing and bonds that break by the recombination of thermally activated charge carriers. At a given temperature during cooling, the kinetics of the equilibrium process become too slow to follow the cooling ramp. Thus, the defect density N_D (i.e. the density of broken bonds) gets frozen in at this temperature. If the defect is associated with an energy E_D above the valence-band edge, the equilibrium value of N_D can be calculated statistically by minimising the free energy F at this freeze-in temperature. The idea of freeze-in was very well supported by experiments where samples were cycled repeatedly to high temperature and rapidly quenched in cold water in order to freeze their defect densities at different stages of the cooling ramp (Street and Winer, 1989). Later additions to the model took into account entropy effects by bond breaking (Winer, 1990) and detailed interactions with hydrogen (Powell and Deane, 1996). Common to these models is the existence of a minimum defect density which emerges as inherent consequence of disorder. Being thus an unavoidable property of amorphous silicon, the impact of defects must be taken into account by the device design, e.g. by charge transport via drift rather than diffusion.

4.2.3 *Doping*

Doping of a-Si:H with phosphine (PH_3) to make n-type material was already reported with the first demonstration of a-Si:H (Chittick *et al.*, 1969). Later, p-type doping was also achieved by adding diborane (B_2H_6) to a glow discharge (Spear and Le Comber, 1976), but the doping efficiency in amorphous silicon always remained low. The assumption of amphoteric defects in the middle of the bandgap can explain this behaviour: if n-type doping moves the Fermi level towards the conduction band, the defect eventually lies below the Fermi level. It can be charged with a second electron after overcoming the correlation energy U. Likewise, if p-type doping moves the Fermi level towards the valence band, the unpaired electron residing in the neutral defect D^0 is above the Fermi level and it becomes energetically more favourable to form a charged defect D^+. Free carriers of either type are thus absorbed by the defect state, putting a stop to any further doping.

Figure 4.5 illustrates the energy gain by the two processes, assuming a bandgap of 1.75 eV, a defect state at $E_D = 0.85$ eV above the valence band, and a correlation energy of $U = 0.2$ eV. The predicted defect densities according to the equilibrium model and some experimental data are also illustrated in Figure 4.5; whenever the Fermi level is moved too far from mid-gap, the formation energy of one type

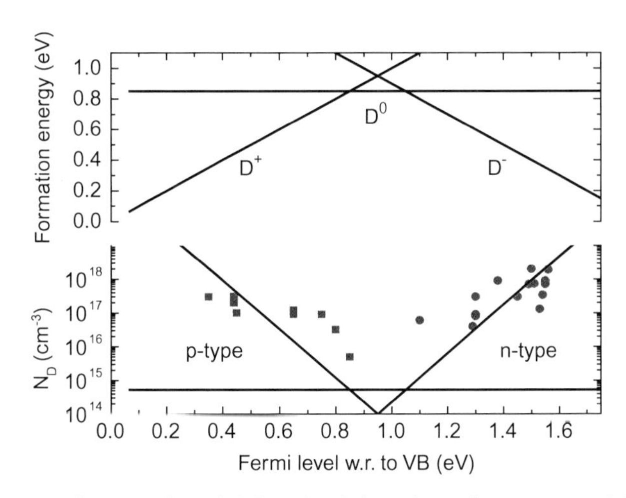

Figure 4.5 Formation energies of defects in their various charge states with respect to the Fermi level (upper panel). The lower panel compares the predicted equilibrium defect densities, squares and circles refer to reported data of p- and n-doped samples, respectively. Data after Winer (1990).

of charged defect starts dropping below the level of that of the neutral defect. Consequently, the formation of this defect becomes more likely and works against any further doping. The lowest possible defect concentrations are found in a narrow region where the neutral defect dominates.

Note that shifts of the Fermi level are not restricted to doped materials, but also occur throughout the intrinsic layer of a *p-i-n* junction. The equilibrium models thus predict increased densities of positively and negatively charged defects close to the *p-i* and *i-n* interfaces, respectively. Charging of the defects close to the *p-i* interface gives rise to the band bending observed in Figure 4.7. The understanding of these effects led to the introduction of buffer layers and bandgap grading close to the interfaces where the defect densities are projected to be the highest (Guha *et al.*, 1989).

While the equilibrium idea is quite powerful, there is still controversy whether such a comparatively simple model with a single defect state and three different charge states is sufficient. Indeed, to date no model can explain consistently how bond breaking and formation takes place on a microscopic level. Also, evidence from cell degradation suggested early on that there is a difference between fast and slow defects (Yang and Chen, 1993). In addition, defect densities determined by two independent methods, such as lifetime and defect absorption measurements, have repeatedly shown very different kinetics during degradation-annealing cycles (Stradins, 2003), suggesting the existence of various defect states rather than isolated dangling bonds.

4.2.4 *The role of hydrogen in amorphous silicon*

The importance of hydrogen in a-Si:H and the differences between a-Si:H and hydrogen-free material grown by sputtering or evaporation were noticed only gradually, and the exact details of the various effects of hydrogen in amorphous silicon are still elusive despite four decades of research. The most prominent effect of hydrogen is a reduction of the defect density; compared with about $10^{20}\,\mathrm{cm}^{-3}$ in sputtered films, the density of dangling bond defects in PECVD material is reduced to values between 10^{16} and $10^{17}\,\mathrm{cm}^{-3}$ (Brodsky and Title, 1969; Brodsky and Kaplan, 1979). The passivation is thought to proceed similarly to the case of crystalline silicon where hydrogen has also been reported to provide excellent passivation of broken bonds on the surfaces (Yablonovitch *et al.*, 1986; Higashi *et al.*, 1990).

The hydrogen content and its bonding environment can be determined by effusion (also called evolution) experiments; on heating a-Si:H above 300°C in vacuum, the effusion of hydrogen from higher hydrides becomes measurable by an increase in the hydrogen partial pressure (Brodsky *et al.*, 1977a); effusion from more stable monohydrides starts only beyond 450°C. In the previous section it was shown that bond breaking is related to the position of the Fermi level. This explains the observation that hydrogen effusion depends on doping (Beyer, 2003). Silicon-hydrogen bonds also have distinct signatures in Fourier transform infrared (FTIR) spectroscopy; stretching modes that absorb at 2010 cm^{-1} are generally attributed to monohydride bonding of dense material whereas a shift of this signature towards 2090 cm^{-1} is attributed to di- or tri-hydrides which are often found in porous material.

While the equilibrium models discussed in Sections 4.2.2 and 4.2.3 were developed around the idea of a continuous random silicon network with isolated defects, experimental evidence suggests that hydrogen can be bonded also at internal surfaces of nano-sized voids that exist even in device quality material (Mahan *et al.*, 1989). This is corroborated by measurements of mass density and hydrogen content that are compatible with the mass deficit of hydrogenated di-vacancies over a wide range of deposition conditions (Smets *et al.*, 2007a). In this context, the above-mentioned IR absorption signatures are attributed to shifted resonance frequencies of hydrogen atoms that are confined to anisotropic volume defects such as hydrogenated di- and poly-vacancies (1980–2010 cm^{-1}, dense material), or to quasi-free oscillations of hydrogen located on the surface of microscopic voids (2090 cm^{-1}, porous material). The integrated intensities of the two contributions are used to define the microstructure factor $R = I_{2090}/(I_{2090} + I_{2010})$ for distinguishing dense and porous material, but more importantly, a dominant presence of monohydride bonding is also associated with reduced defect density (Kroll *et al.*, 1998).

4.2.5 *Amorphous silicon alloys*

Alloying amorphous silicon with carbon widens the bandgap while additions of germanium or tin reduce it. The maximum tolerable germanium content for absorber layers yields a gap of about 1.6 eV (Nakamura *et al.*, 1980); strong hydrogen dilution can extend this range to about 1.5 eV (Matsuda *et al.*, 1986). One of the underlying reasons for the poor performance of a-$Si_{1-x}Ge_x$:H is preferential bonding of hydrogen to silicon compared with germanium, resulting in poor passivation of broken germanium bonds (Lucovsky, 1992). A second complication is the easier dissociation of germane (GeH_4) compared with silane, which can for example, be compensated for by using mixtures of disilane and germane (Matsuda and Ganguly, 1995). Because of their high defect densities, absorber layers made from a-$Si_{1-x}Ge_x$:H are generally graded so that material with higher bandgap but lower defect density is close to the junctions (Guha *et al.*, 1989; Lundszien *et al.*, 2002).

Window layers of p-type a-$Si_{1-x}C_x$:H were introduced by the Hamakawa group at Osaka University (Tawada *et al.*, 1981). Because doped layers are already highly defective, additional defects caused by poor hydrogen passivation are somewhat less destructive; however, increasing the carbon content degrades the mobility in p-type a-$Si_{1-x}C_x$:H, thus putting an upper limit of about 25% on the carbon content. Defect creation does become an issue when a-$Si_{1-x}C_x$:H is used for high-gap absorber layers; problems due to the different reactivity of methane compared with silane can, for example, be avoided by the use of mono-methyl-silane (CH_3-SiH_3, MMS) (Yunaz *et al.*, 2009). For doped layers, amorphous nitrides and oxides have been investigated as high-gap alternatives to carbides (Li *et al.*, 1994). Recently, phosphorus- and boron-doped silicon oxides received renewed attention because they develop a microcrystalline phase when deposited with high amounts of hydrogen; we discuss this further in Section 4.6.

4.2.6 *Microcrystalline films*

PECVD-grown films with a crystalline component were first observed in studies of doped films when high discharge powers or fluorine-containing precursor gases were used (Usui and Kikuchi, 1979; Matsuda *et al.*, 1980); in X-ray diffraction, such films showed the presence of silicon crystallites with sizes between 5 and 10 nm. Because ion bombardment in high-power plasmas is often a concern, microcrystalline films are normally grown by an alternative method using precursor gas mixtures of silane and large quantities of hydrogen, often referred to as hydrogen diluted silane (Matsuda, 1983). The preferential growth of microcrystalline films in fluorine- or hydrogen-containing plasma indicates that the deposition of μc-Si is initiated by etching of unstable nucleation sites.

Figure 4.6 shows schematic phase diagrams of PECVD-grown silicon as the hydrogen content changes. Without additional hydrogen, film deposition is normally completely amorphous. With the addition of hydrogen, isolated crystallites

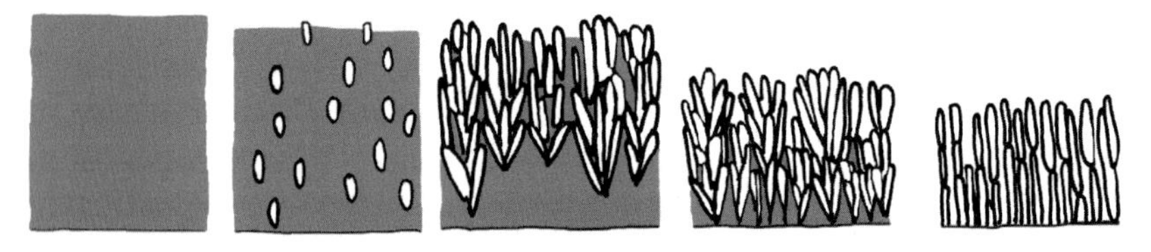

Figure 4.6 Schematic transition from fully amorphous silicon (left) over mixed-phase material to fully microcrystalline silicon (right) with increasing hydrogen content in the precursor gas mix.

nucleate within a matrix of amorphous material; the conductivity of the material changes significantly when the crystallites reach the percolation threshold. Adding yet more hydrogen yields films with a distinct transition between an amorphous nucleation region and conical crystalline clusters which themselves consist of smaller grains. Finally, when using very high hydrogen content the deposition rate is decreased drastically and the resulting films consist of a porous array of elongated grains. Deposition of uniform films requires an interface treatment that establishes a template for nucleation, and a hydrogen gradient during the deposition can ensure uniform crystallinity across the thickness of the film (Koh *et al.*, 1999a; Collins *et al.*, 2003).

Undoped material, i.e. silicon deposited without the addition of doping gases, often exhibits unintentional n-type conductivity, presumably due to oxygen impurities. Truly intrinsic material for use as absorbers is obtained by compensation — also called micro-doping — with $[B_2H_6]/[SiH_4] \approx 10^{-5}$ (Wang and Lucovsky, 1990), by suppressing oxygen impurities with specially purified source gases (Torres *et al.*, 1996), or by carrying out the deposition at low substrate temperature where the oxygen impurities are sufficiently passivated by the incorporation of hydrogen (Nasuno *et al.*, 2001b).

Figure 4.3 compares the spectral absorption of microcrystalline silicon with that of amorphous and crystalline silicon. Because μc-Si:H consists of silicon crystallites embedded in an amorphous matrix, its spectral absorption resembles an interpolation of these extremes. Device-grade material is usually considered to have a crystalline fraction between 50% and 70%. The unique properties of μc-Si:H are attributed to the excellent passivation of defects at the surface of the nanocrystallites which is provided by the surrounding amorphous matrix. Hence, transport of excited carriers proceeds via percolation among the crystallites; collisions at the grain boundaries may reduce their mobility with respect to c-Si, but their lifetime stays at acceptable levels because recombination is suppressed.

Doping of microcrystalline silicon with phosphine (PH_3) is compatible with most processing conditions, but the growth of p-type material requires special attention; diborane (B_2H_6) has been reported to react spontaneously and its shelf life is limited, tri-methyl-boron ($B(CH_3)_3$, TMB) tends to delay the nucleation of microcrystalline films, and boron trifluoride (BF_3) requires very high plasma power due to its chemical stability (Koh *et al.*, 1999b). For a given doping gas ratio, the conductivity in microcrystalline layers is generally about two orders of magnitude higher than in amorphous films (Wang and Lucovsky, 1990).

Other elements (oxygen, carbon, nitrogen, germanium) can be added to modify the properties of intrinsic or doped μc-Si layers, altering the optical and electrical properties, as will be discussed in Section 4.6.

4.3 Thin-film silicon solar cells

In amorphous as well as in microcrystalline silicon, the charge-carrier diffusion lengths are much shorter than the absorption length of near-bandgap light. Moreover, in Section 4.2.1 we showed that every shift of the Fermi level out of the mid-gap position creates additional defects in amorphous silicon. Thin-film silicon solar cells are therefore fabricated as drift devices using *p-i-n* junctions rather than diffusion-controlled *p-n* junctions. The drift length, given by $L_{\text{drift}} = \mu \mathcal{E} \tau$, where $\mathcal{E}$ is the strength of the electrical field and μ and τ are the mobility and lifetime of the carrier, should be larger than the device thickness. At the maximum power point and open-circuit voltage (V_{oc}), L_{drift} is decreased as $\mathcal{E}$ is decreased. Thin devices benefit doubly from the reduced thickness as the transit length is reduced and the electric field is increased. Hence they are much less prone to suffer from localised defects or SWE.

4.3.1 *Single-junction solar cells*

In ideal *p-i-n* devices, the potential variation should be linear across the intrinsic region; however, the formation of charged defects deforms the band structure. In most situations, the band diagram must therefore be determined numerically, for example with software packages like ASA (Pieters *et al.*, 2007) or AMPS (Zhu *et al.*, 1998). Figure 4.7 shows a band diagram for an amorphous *p-i-n* junction and the front TCO, using the defect pool model in ASA. Compared with the ideal case, the region close to the front interface shows a complicated band structure that incorporates a Schottky barrier between the front contact and the *p*-layer, as well as band distortion due to positively charged defects in the *i-p* region; the latter results in a significant reduction of the electric field in the middle of the *p-i-n* junction.

When additional defects are created during illumination, the field in the middle of the junction becomes further weakened. Losses in charge carrier collection are illustrated in Figure 4.8, which shows the external quantum efficiency (EQE) of an *n-i-p* cell before and after light soaking. The EQE represents the spectrally resolved probability of creating and collecting a charge carrier from an incident photon. Figure 4.8 shows that collection losses occur predominantly for wavelengths above 400 nm. Shorter wavelengths are strongly absorbed close to the *i-p* interface; according to Figure 4.7, the electric field in this region is actually enhanced by LID; thus collection here is less affected than in the bulk of the absorber layer.

The EQE is typically measured at zero bias; the corresponding current density represents short-circuit conditions and can be obtained by multiplying the EQE

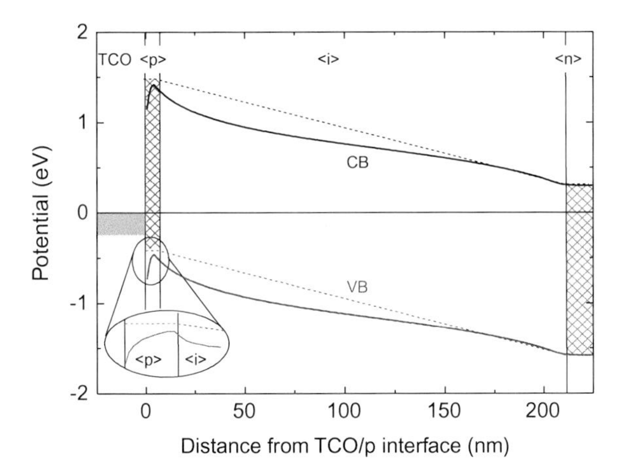

Figure 4.7 Equilibrium band diagram of an amorphous silicon solar cell with a 200 nm thick absorber layer and *p*- and *n*-layers with activation energies of 450 and 250 mV, respectively. Potentials are shown with respect to the Fermi level. Dashed and solid lines represent ideal and realistic band alignment, respectively. The inset illustrates the Schottky junction between the front TCO and the *p*-layer.

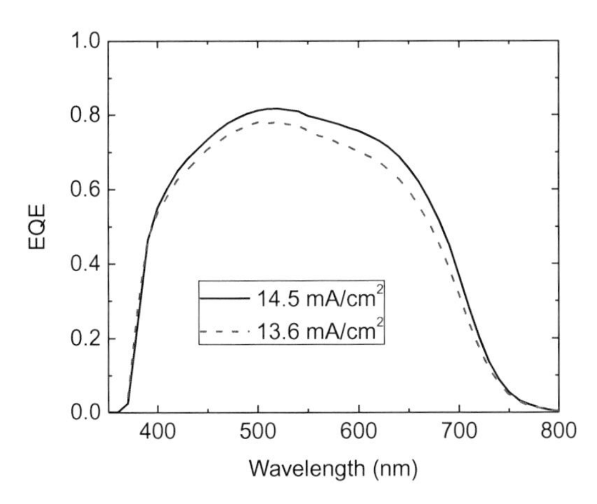

Figure 4.8 EQE of an *n-i-p* a-Si:H single-junction cell without antireflection coating. Solid and dashed lines illustrate the charge-carrier collection in the initial and stabilised states, respectively.

by the illumination spectrum (in units of photons per wavelength interval and per surface area), and integrating over all wavelengths. This determination of the short-circuit current density from the EQE is normally more precise than the value obtained from a current–voltage measurement because the EQE does not depend on the cell area, whereas a reliable determination of the cell area can become difficult for small cells, particularly if they include light-scattering interfaces.

After LID, the drift field is weakened and the collection of photogenerated charge carriers becomes even less effective under the forward bias of operating conditions. As a result, the solar cell efficiency is further reduced via losses in fill factor (FF). Intuitively, V_{OC} should be related to the potential created by the doped layers and thus be minimally influenced by LID. However, this simplistic view does not take into account that V_{OC} can suffer because of increased recombination once the layers become more defective.

Figure 4.9 illustrates the kinetics of cell efficiency degradation. The degradation scales with the absorber layer thickness: thinner layers degrade less because the remaining field in the centre of the junction remains stronger in thinner cells, even when weakened by additional defects. Moreover, the figure shows that cells in which the *i*-layer is deposited with additional hydrogen stabilise faster and at a higher efficiency for a given thickness (Wronski, 1996). In general, there is no simple dependence on illumination duration, intensity or spectrum, and even at moderate temperatures of 60 or 70°C, the degradation is partly reversed by annealing. Therefore, standard degradation experiments are commonly carried out for 1000 hours at 50°C under exposure to light with the intensity and the spectrum of sunlight. Reduced cell thickness is desirable, not only in terms of stability against LID, but also because of throughput considerations in industrial production. A sufficient level of light absorption must nevertheless be maintained. Absorption

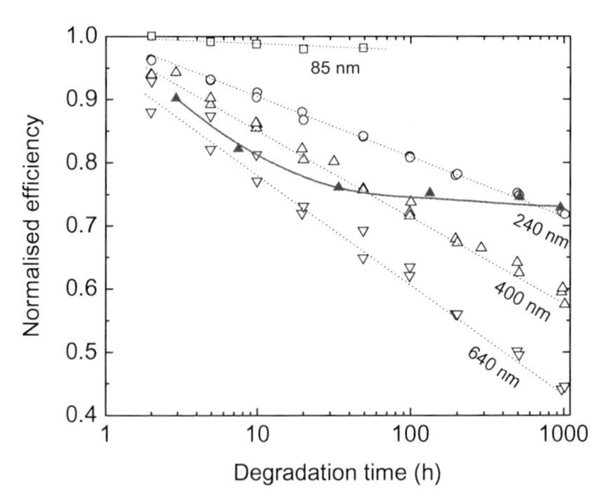

Figure 4.9 Degradation kinetics of solar cells with various absorber layer thicknesses; data from Bennett *et al.* (1986) and Bennett and Rajan (1988). The solid line represents a cell with a 400 nm *i*-layer deposited with hydrogen dilution. Data from Wronski (1996).

enhancement in thin-film silicon solar cells is generally summarised by the term 'light trapping'; we discuss this in more detail in Section 4.5.

4.3.2 *Tandem and multijunction solar cells*

The combination of two bandgaps in tandem solar cells makes a better use of the solar spectrum and therefore yields higher efficiencies. Here, thin-film silicon and its alloys have an inherent advantage because cells with different bandgaps can be stacked monolithically, one on top of another, since no lattice matching is required. Theoretical considerations suggest an optimum for bandgaps of 1.1 and 1.7 eV (DeVos, 1980). The upper value corresponds to a-Si:H and the lower value is in principle accessible with a-$Si_{1-x}Ge_x$:H, but Section 4.2.5 showed the resulting alloy is too defective for working devices. An alternative choice is therefore μc-Si:H. Figure 4.10 shows cross-sections through *p-i-n/p-i-n* and *n-i-p/n-i-p* devices.

Electrically, such cells are connected in series and therefore their current densities must be matched while the output voltage of the tandem cell is obtained by adding the voltages of the individual cells.

The design rules for multijunction cells start by looking into the current that is generated by the bottom cell alone, and then trying to define useful distributions of that current into two or more sub-cells. Technological throughput considerations limit microcrystalline silicon cell thickness to between 1 and 3 μm, resulting in i_{sc} between 25 and 30 mA cm^{-2} with adequate light trapping. The usable composition range of a-$Si_{1-x}Ge_x$:H extends the bandgap range to about 1.4 eV, but its high absorption can yield a similar range of current densities. Amorphous silicon cells with typical thicknesses of 200 nm can yield about 15 mA cm^{-2} (*cf.* Figure ,4.8) and up to 17 mA cm^{-2} if good light trapping schemes and antireflection coatings are used. Without light trapping and back reflectors, a single passage of light through the amorphous cell delivers about 10 to 11 mA cm^{-2}. Considering these values, the following design options are used:

- Historically, the first tandem cells combined two purely amorphous silicon absorbers; while they do not extend the spectral response compared with a single-junction cell, they are nevertheless advantageous because of reduced degradation with respect to a single-junction cell of comparable thickness (Hanak and Korsun, 1982; Bennett and Rajan, 1988; Lechner *et al.*, 2008). Figure 4.11 shows the repartition of 15 mA cm^{-2} between a thin a-Si:H top and a relatively thick a-Si:H bottom cell.
- The combination of amorphous and microcrystalline cells in a tandem with high total current appears unfavourable because the top cell sees only a single

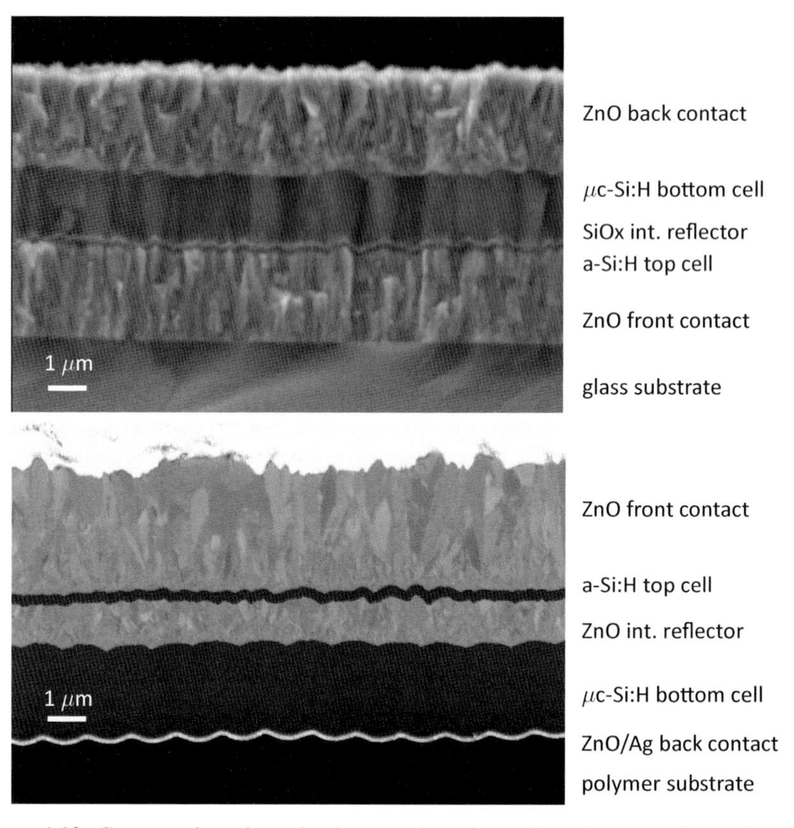

Figure 4.10 Cross-sections through micromorph tandem cells with intermediate reflectors. The upper panel shows a *p-i-n/p-i-n* structure on glass covered with a 4 μm thick ZnO front contact. The intermediate reflector layer appears as a light band between the thin amorphous and the thick microcrystalline films. The lower panel shows an *n-i-p/n-i-p* tandem cell on a periodically textured plastic substrate. In this cell, the intermediate reflector is about 2 μm thick. The growth sequence of both cells is from bottom to top.

pass of light and the resulting tandem would thus be heavily mismatched. This limitation was successfully overcome by the introduction of a ZnO intermediate reflector layer between the component cells, resulting in much better top cell current (Fischer *et al.*, 1996; Yamamoto *et al.*, 2002; Yamamoto *et al.*, 2006). An alternative material for the intermediate reflector is silicon oxide doped with phosphorous (SiO_x:P) (Buehlmann *et al.*, 2007). An advantage of SiO_x:P is the possibility of depositing it between silicon layers without breaking vacuum; this is discussed in more detail in Section 4.6.1.

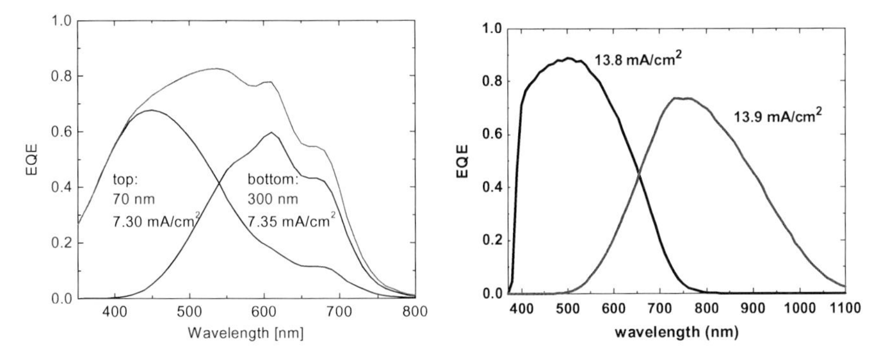

Figure 4.11 EQEs of an a-Si:H/a-Si:H tandem cell (left) and of a micromorph cell with an intermediate reflector layer (right). The micromorph cell has a 240 nm a-Si:H top cell, an 80 nm SiO_x intermediate reflector, and a 2.4 μm thick bottom cell. The currents represent the integrated i_{sc} of the sub-cells. More information on these results can be found in Cubero *et al.* (2008) and Boccard *et al.* (2011), respectively.

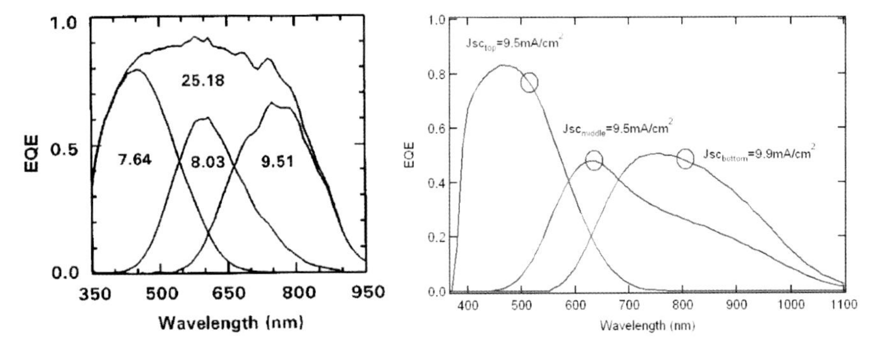

Figure 4.12 EQEs of triple-junction cells. The left panel represents a fully amorphous a-Si/a-SiGe/a-SiGe cell (Yang *et al.*, 1997), the right panel represents an a-Si:H/μc-Si:H/μc-Si:H cell with AR-coating (Söderström, 2013). The currents represent the integrated current densities of the sub-cells.

- Thanks to the continuous variation of bandgaps in the a-Si_xGe_{1-x}:H alloy system, a 10 mA cm^{-2} a-Si top cell can easily be combined with an a-Si_xGe_{1-x}:H bottom cell. Cell design must maximise stabilised efficiency by trading off the germanium content vs. bottom cell thickness trade-off. Going to triple-junction cells, somewhat higher germanium contents are admissible because of the further reduced thicknesses of the component cells. Figure 4.12 shows that triple-junction a-Si:H/a-Si_xGe_{1-x}/a-Si_yGe_{1-y} cells can be matched at about

8.3 mA cm^{-2} (Guha *et al.*, 2000). In the last two approaches no intermediate reflector is required below the a-Si:H top cell.

- An alternative approach distributes the high current of a microcrystalline silicon cell into a triple-junction cell consisting of an amorphous top cell and a microcrystalline tandem bottom (a-Si:H/μc-Si:H/μc-Si:H). Using a slightly thinner and therefore more stable top cell yields current matching between 9 and 10 mA cm^{-2} (Saito *et al.*, 2005; Yue *et al.*, 2010; Söderström, 2013). The potential of going to efficiencies beyond 16% with this approach is discussed in Section 4.6.6.

4.4 Fabrication of device-grade amorphous and microcrystalline silicon

The deposition of thin films is based on physical vapour deposition (PVD) of the bulk material (e.g. by evaporation or sputtering) or on chemical processes using liquid or gaseous precursors. For the production of device-grade thin-film silicon materials, chemical vapour deposition (CVD) is the most viable solution today. It involves the decomposition of a mixture of gases containing silicon and hydrogen, either in a glow-discharge for plasma-enhanced CVD (PECVD) or by heating a filament for thermo catalytic CVD (cat-CVD), also called hot-wire CVD (HWCVD). To date, only PECVD processes have reached industrial implementation. Consequently, their main features are reviewed in detail in this section, from the principles of PECVD to the impact of the plasma properties on layer quality. The transition regions from amorphous to microcrystalline silicon are then detailed, and deposition rate and associated material quality are discussed. The principles of HWCVD are presented in Section 4.4.4.

4.4.1 *PECVD of thin-film silicon*

A PECVD reactor (Figure 4.13) typically consists of a vacuum chamber equipped with an inlet for reaction gases and an outlet to pump out unreacted and reacted gases. The plasma excitation takes place between two parallel electrodes where the grounded one serves as support for the substrate. This electrode also acts as a substrate heater. The other electrode is connected to an electrical power supply and often includes a distribution system for the reaction gases. This so-called shower head concept yields a more uniform flux than feeding from the sides. Before deposition, the desired gas mixture flows into the chamber and a fixed pressure is obtained by controlling the pumping aperture. Typical gases are SiH_4 and H_2, while boron and phosphorus precursors (B_2H_6, $B(CH_3)_3$, BF_3 or PH_3) are used for p-type and

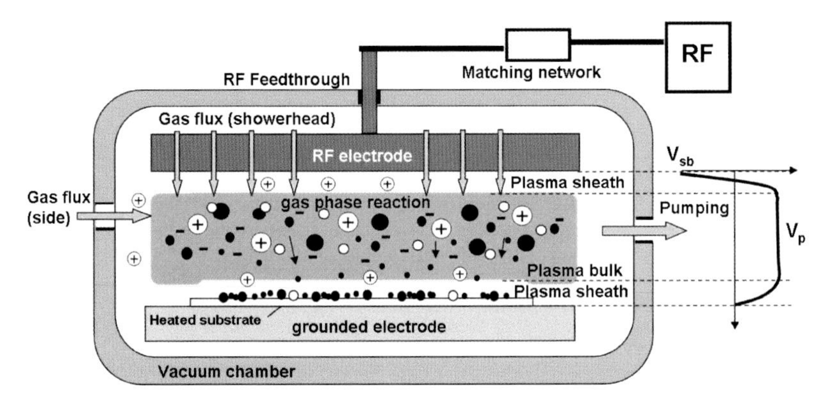

Figure 4.13 Schematic diagram of a capacitive parallel electrode PECVD reactor.

n-type doping, respectively. A voltage is then applied to the parallel electrodes; free electrons present in the reactor are accelerated and gas molecules are dissociated by electron impact ionisation, generating more free electrons which trigger further impact ionisations until the discharge reaches a stable state. Radio frequency (RF) excitation predominantly increases the energy of the light electrons while the heavy ions cannot follow the rapid switching of the alternating field. Thus, RF-PECVD has the advantage of accessing high-temperature chemistry while remaining at typical substrate temperatures of 150 to 250°C (cold plasma).

Electrons have much higher thermal velocities than ions and can thus reach the electrodes faster. As a consequence, electric fields develop near the electrodes which retard electrons and accelerate ions. These regions in front of the electrodes are called sheaths. They extend over small distances (in the mm range at normal deposition pressures) and drive the ion bombardment of the surfaces. The standard RF frequency is 13.56 MHz (as allotted for industrial processes by international authorities). Alternatively, very high frequencies (VHF, 20–150 MHz) proved useful for achieving higher deposition rates (Curtins *et al.*, 1987; Kroll *et al.*, 1997) and are today widely used in the thin-film silicon PV industry. Finally, microwave excitation (2.54 GHz) is also being used in electron-cyclotron resonance PECVD (ECR-PECVD) (Bae *et al.*, 1998) and for inline microwave PECVD remote plasma sources (Soppe *et al.*, 2005).

In the plasma bulk, electron impact dissociation of SiH_4 and H_2 molecules takes place, producing silicon radicals, SiH_x (x = 1–3) and atomic hydrogen, H. Radicals, molecules and ions can further interact in secondary reactions, such as higher silane formation from the iterative insertion reaction $SiH_2 + Si_nH_m \rightarrow Si_{n+1}H_{m+2}$, which triggers powder formation (Bano *et al.*, 2005). Radicals diffuse from the plasma zone where they are generated towards the reactor walls, the electrodes and the substrate. Growth takes place under a complicated non-equilibrium process that is governed by the temperature-dependent sticking coefficients, surface diffusivity of silicon radicals, etching processes by atomic hydrogen, and surface and sub-surface hydrogen recombination. Under adequate conditions, a silicon film with a low hydrogen content of around 10 at.% is deposited.

4.4.2 *Growth and layer quality: from amorphous to microcrystalline silicon*

In RF discharges, SiH_3 radicals are generally considered to be the dominant species contributing to film growth. Coverage of the growing surface with atomic hydrogen allows radicals to diffuse to minimum energy sites, forming a dense random network.

Dangling bonds available for radical fixation can be created by the removal of hydrogen from the surface by thermal excitation or by abstraction with a SiH_3 radical. Radicals such as SiH_2 with short lifetimes can trigger the formation of higher-order silane-related species which yield a higher density of SiH_2 bonds in the growing a-Si:H films and nano-voids, giving rise to reduced stability against LID. The density of such nano-voids in an amorphous silicon film can be assessed via the microstructure factor, as defined in Section 4.2.4. Since the dissociation energy of SiH_4 into SiH_3 radicals is lower than the energy required for SiH_2 radical formation, reduced electron temperature is preferred for stable amorphous silicon growth (Nishimoto *et al.*, 2002). High-quality amorphous silicon materials can be realised using VHF (Benagli *et al.*, 2009), the triode reactor configuration (Matsui *et al.*, 2006) and high substrate temperature for enhancing adatom mobility at the growing surface.

While the formation of polysilanes and powder in the plasma is usually considered to degrade material quality, these conditions were also reported to yield high-quality materials owing to the inclusion of nanoparticles in the amorphous network (Roca i Cabarrocas, 2000). Finally, ion bombardment of the growing surface is a crucial parameter in amorphous and microcrystalline silicon deposition; ions accelerated in the plasma sheaths strongly affect the growth mechanisms and film properties. Their importance depends on their energy and on the ion-to-silicon-atom arrival ratio. For silicon, surface or bulk displacement can happen for high ion energies ($\sim>$ 50 eV) (Smets *et al.*, 2007b; Wittmaack *et al.*, 2003).

The SiH_4/H_2 discharge can then also yield microcrystalline silicon growth, as described by three possible phenomenological models. In the surface diffusion model, high H flux on the surface enhances the mobility of the deposition precursors, strongly influencing the crystalline fraction of the deposited film (Matsuda *et al.*, 1999). In the selective etching model, amorphous and crystalline phases are deposited simultaneously with H preferentially etching the amorphous material, leaving behind the crystallites (Terasa *et al.*, 2000). Finally, the chemical annealing model postulates that H atoms induce crystallisation of amorphous films (Sriraman *et al.*, 2002). These three models imply that a high flux of atomic hydrogen Γ_H with respect to the flux of silicon radicals Γ_{Si} favours microcrystalline silicon growth. A well-known way to reach the transition region between amorphous and microcrystalline silicon is therefore to use silane that is highly diluted in hydrogen, but a high Γ_H/Γ_{Si} ratio can also be attained from pure silane plasma if sufficient H is provided by high silane dissociation efficiency. The Γ_H/Γ_{Si} ratio required to reach the transition zone was experimentally assessed (Dingemans *et al.*, 2008) and also related to the plasma silane concentration c_p (Strahm *et al.*, 2007); this latter parameter is defined by the input silane concentration c and by the depletion fraction of silane in the plasma D as $c_p = c(1-D)$. The transition zone was shown to occur at $0.5\% < c_p < 1.2\%$; higher (lower) c_p systematically results in amorphous (microcrystalline) layers.

Figure 4.14 illustrates two optimum deposition windows for the production of high-quality materials at the transition. In the low silane concentration regime

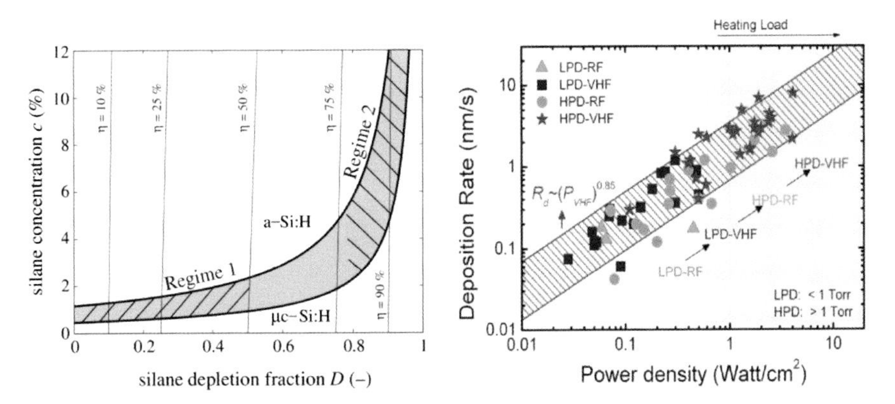

Figure 4.14 Left: Amorphous to microcrystalline transition zone (shaded region) for varying silane concentration and depletion fraction, with low silane concentration labelled as Regime 1 and high silane depletion as Regime 2 (Strahm *et al.*, 2007); right: deposition rate of device-grade microcrystalline material versus the employed power density (data collected from different institutes) from low pressure (<1 Torr) RF and VHF regimes (triangles and squares) to high pressure (>1 Torr) RF and VHF regimes (circles and stars) (Smets *et al.*, 2008).

($c \leq 2.5\%$), transition is achieved for low depletion (D $<$ 0.5): the film microstructure can be controlled with the silane-to-hydrogen gas flow ratio and high deposition rates can be achieved only with the use of high gas flows and high power densities. In the high silane depletion regime (D $>$ 80%), growth in the transition region can be achieved only for higher silane concentration ($c > 5\%$). This regime provides a higher efficiency of silane utilisation and lower hydrogen flows, while the film structure is determined by the depletion which is, for example, accessible by controlling the plasma power. It must be noted that high-quality materials, as assessed by FTPS and FTIR measurements, can be produced in the different process windows presented in this section, and that device-grade materials typically have a Raman crystalline fraction between 50% and 70%. However, it was recently demonstrated that similar high-quality bulk materials, as determined by FTPS and FTIR measurements, can lead to varying device efficiencies (Bugnon *et al.*, 2011). To realise high-efficiency devices, the plasma process conditions must also provide a dense material when the deposition is carried out on a textured substrate in order to limit the creation of localised defects (typically 2D nano-porous material regions) in the sharp valleys of a highly textured substrate. Very importantly, the most efficient process conditions for high-quality devices must therefore provide not only high-quality bulk material but also a minimum density of localised defects or cracks when deposition is carried out on a textured substrate (Bugnon *et al.*, 2011; Despeisse *et al.*, 2011).

4.4.3 *Deposition rate*

Highest material quality and highest resilience to the substrate roughness are generally realised for both a-Si:H and μc-Si:H at low deposition rate (1–3 Ås^{-1}). However, device-grade material must be deposited at higher rates to be economically viable. Higher deposition rates mean increasing precursor and atomic hydrogen fluxes to the growing surface. This can be done by simply increasing precursor gas flows and plasma power density; Figure 4.14 shows a plot of deposition rates versus power density for device-grade μc-Si:H grown in different deposition regimes. However, increasing the power density enhances ion bombardment, leading to amorphisation, defect creation and insufficient relaxation of growth precursors. The use of VHF frequencies (40–110 MHz) was shown to yield high-quality materials at increased deposition rates because it provides higher electron density for lower electron temperature, enhancing the efficiency of silane dissociation in the plasma and pushing the onset of powder formation towards higher deposition rates. Moreover, the energy of the ions that bombard the grounded deposition surface is reduced in VHF, due to decreased sheaths and potentials, giving rise to increased surface mobility of the deposited species on the growing film.

In addition to VHF, or alternatively to VHF for standard RF conditions at 13.56 MHz, the so-called high-pressure depletion (HPD) regime can be used; high-pressure conditions result in long residence time and enhanced silane dissociation, as well as a reduction of ion bombardment energy due to many ion-neutral collisions during the acceleration of ions in the plasma sheath (Smets *et al.*, 2008). The HPD approach led to the production of μc-Si:H single-junction cells with efficiencies of 9.8% and 9.13% at depositions rates of 1 nm s^{-1} and 2 nm s^{-1}, respectively (Mai *et al.*, 2005; Matsui *et al.*, 2006). Secondary gas phase reactions and therefore powder production are enhanced at high pressure. Decreasing the inter-electrode distance in the reactor promotes surface reactions at the electrodes instead of volume reactions in the plasma, so that by means of this simple geometric adjustment higher processing pressure can be used without powder formation (Strahm and Hollenstein, 2010). In addition, a multi-hollow cathode discharge (MHC), which contains filled cavities on the surface of the electrode, has also been proposed as a discharge method to produce higher electron density via intense localised plasmas (Tabuchi *et al.*, 2003). Recently, localised plasma confinement CVD (LPC-CVD), which makes use of a special cathode with regularly arranged nozzles and pumping holes, was demonstrated to be highly effective in the production of high-quality material at deposition rates of 2 nm s^{-1}, thanks to high plasma densities (Aya *et al.*, 2011). The use of a narrow inter-electrode gap, VHF, MHC, LPC-CVD or HPD therefore results in higher material quality at increased deposition rate.

As we already discussed in Section 4.4.2, high-efficiency devices rely on the electric performance of good material and on the optics of textured substrates. This remains a challenge for thin-film silicon: some PECVD processes can lead to films with low bulk defect density, but their growth kinetics might yield strongly defective areas near V-shaped regions of the substrate. Optimum plasma conditions will therefore depend on the exact morphology on which the film is grown.

4.4.4 *Hot-wire chemical vapour deposition (HWCVD)*

In the HWCVD technique, source gases (e.g. SiH_4, H_2, C_2H_2) are introduced into a vacuum chamber with controlled flow ratio and operating pressure. Dissociation is achieved catalytically in the vicinity of a metallic filament (typically tungsten) heated to temperatures of 1700–1900°C. The generated reactive species then induce film growth on a temperature-controlled substrate facing the filament. Amorphous silicon materials deposited using HWCVD exhibit low H content of about 3 at.% and were reported to provide high stability against light soaking (Mahan *et al.*, 2002). In comparison to PECVD, HWCVD is thermal and catalytic in nature:

there are no ions and electric fields present and consequently there is no ion bombardment of the growing surface. This is particularly important when depositing the first part of the intrinsic layer in *p-i-n* microcrystalline junctions, in order to avoid damaging the critical *p-i* interface. The use of intrinsic μc-Si:H buffer layers deposited by HWCVD at the *p-i* interface of μc-Si:H junctions resulted in very high V_{oc} (600 mV) and efficiency (Mai *et al.*, 2005). While this deposition technique is used on an experimental scale for depositing high-quality materials, high deposition rates ($> 10\,\mathrm{nm\,s^{-1}}$) are also achievable (Mahan *et al.*, 2002). However, the performance of HWCVD-deposited solar cells has not yet surpassed that of PECVD-deposited solar cells. Moreover, this deposition technique has not yet been successfully incorporated into large-scale manufacturing facilities, because of problems with deposition uniformity and filament aging. To date, HWCVD has thus been employed only at the laboratory scale for thin-film silicon.

4.5 Light management

A common issue for silicon-based solar cells is that the device thickness must be kept smaller than the optical absorption length. Surprisingly, this is true for all types of silicon, including wafer-based multicrystalline or single-crystal devices where surface corrugations in the range of tens of micrometres are used (Redfield, 1974; Campbell and Green, 1987). Immediately after the demonstration of amorphous thin-film solar cells, surface textures with accordingly reduced feature size were introduced for absorption enhancement (Deckman *et al.*, 1983). Absorption enhancement is equally required for nano- or microcrystalline silicon (Keppner *et al.*, 1999; Vetterl *et al.*, 2000), and for polycrystalline thin-film silicon (Green *et al.*, 2004).

Figure 4.15 presents a naïve but intuitive illustration of the light-trapping process; compared with a cell with flat interfaces, a textured TCO layer scatters light into the silicon absorber layer. At the back reflector, weakly absorbed light gets scattered a second time. Very weakly absorbed light can reach the front contact, where a certain part undergoes total internal reflection because its incident angle

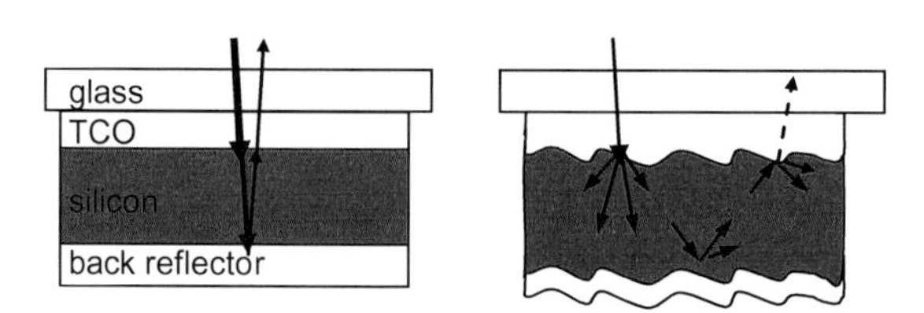

Figure 4.15 Schematic representation of light trapping by interface textures.

with respect to the interface is beyond the critical angle of total internal reflection. Textured interfaces thus multiply the effective absorption length of absorber layers beyond the maximum of two passes through a flat device.

Considering a slab of weakly absorbing material with completely random scattering interfaces, Yablonovitch proposed an upper theoretical limit of $4n_r^2$ for the maximum achievable absorption enhancement, n_r being the refractive index of the absorber material (Yablonovitch and Cody, 1982). It was proposed that deterministic grating structures can exceed the $4n_r^2$ limit in certain wavelength ranges (Sheng *et al.*, 1983; Gee, 2002; Yu *et al.*, 2010). However, in solar cells, the experimentally found path enhancement is normally between 15 and 20, much below the value of $4n_r^2$ of nearly 50 when a refractive index of 3.5 is assumed for silicon. A possible explanation for this discrepancy is the neglect of the supporting layers in most of the theoretical derivations (Haug *et al.*, 2011).

For scattering at real surfaces, a detailed explanation of the light trapping process is far from straightforward. The interaction of electromagnetic radiation with matter can be treated with numerous software packages, but the random nature of typical textures such as those shown in Figure 4.16 is difficult to describe rigorously. Most approaches revert to super-cells, i.e. box-shaped regions that contain the multi-layer stack of the solar cell above a base area of several μm^2 (Rockstuhl

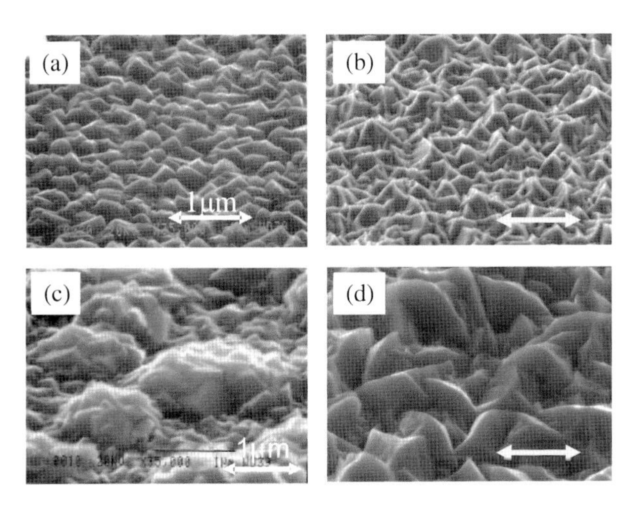

Figure 4.16 Surface morphologies of TCOs with small feature size for a-Si:H solar cells (top) and larger features for microcrystalline cells (bottom). (a) and (c) show AP-CVD-grown SnO_2:F of the Asahi-U and Asahi-W type, respectively (Taneda *et al.*, 2007); (b) and (d) show LP-CVD grown ZnO:B films. The growth processes are described in Faÿ *et al.* (2005) and Faÿ *et al.* (2007). Scale bar: 1 μm.

et al., 2011). Care must be taken to avoid artefacts because most calculation routines assume periodic repetition of the cells. The required computing power has so far limited the use of rigorous models as predictive tools for optimisation; at the time of writing the calculation of an a-Si:H cell can take hours and a full tandem cell may take days.

Approximate methods are typically based on scalar scattering theory which neglects the vectorial nature of electromagnetic fields. Initially developed for remote sensing of ocean surfaces with radar waves, the scaling to visible light and nano-textures proved very successful for solar cells (Zeman *et al.*, 2000; Poruba *et al.*, 2000; Krč *et al.*, 2003). Simulation resembles a ray-tracing approach where scattering at each interface is described by distribution into a specular and diffuse part for both reflection and transmission. The diffuse part is normally referred to as haze; besides its magnitude, a description of the angular intensity distribution is required (ARS, angle resolved scattering) which is used to describe the average path enhancement. In scalar scattering theory, the haze depends only on a single parameter, the root mean square roughness, which is accessible by atomic force microscopy or optical measurements (Carniglia, 1979; Elson and Bennett, 1979). Recently, more detailed scattering models based on Fourier theory have been proposed (Dominé *et al.*, 2010).

Adequate surface textures for light management are commonly incorporated into one of the supporting layers, e.g. the front TCO for *p-i-n* devices on rigid glass substrates. Typical examples of fluorinated tin oxide (SnO_2:F) grown by atmospheric pressure chemical vapour deposition (APCVD) and boron-doped ZnO (ZnO:B) deposited by low pressure CVD (LPCVD) are shown in the scanning electron microscope (SEM) images in Figure 4.16. When deposited under conditions that favour the preferential growth of certain low-energy surfaces, these TCO materials develop facets that resemble small triangular pyramids. The feature size of the pyramids can easily be controlled by the film thickness.

Surfaces that are dominated by pronounced V-shaped valleys between the facets have been found to be detrimental to silicon film growth. Cross-sectional analysis of amorphous solar cells revealed the growth of defective material above such valleys because of reduced adatom mobility (Sakai *et al.*, 1990). Figure 4.18 shows that this effect is pronounced in microcrystalline cells, resulting in severe V_{oc} and FF losses (Nasuno *et al.*, 2001a; Li *et al.*, 2008; Python *et al.*, 2008). The defective zones have also been identified to provide diffusion paths for oxygen and other impurities (Python *et al.*, 2010).

Figure 4.17a shows a surface where the initially pyramidal texture of the LP-CVD ZnO shown in Figure 4.16d has been modified by a plasma treatment that converts the V-shaped valleys between the facets into rounded valleys with smooth,

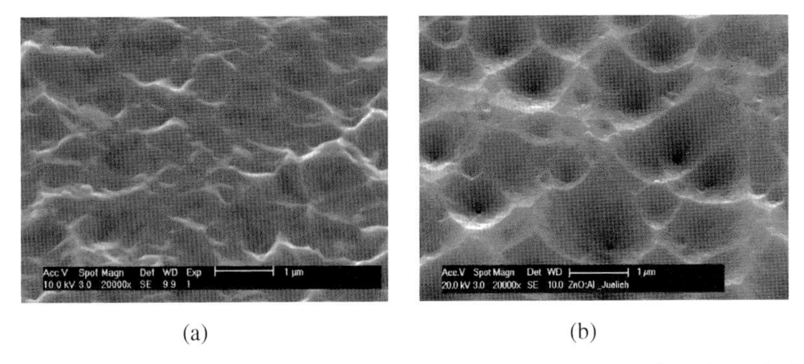

Figure 4.17 Surfaces with rounded features. (a) corresponds to the surface shown in Figure 4.16d after plasma treatment (Bailat *et al.*, 2006); (b) illustrates the crater lime etch pits of sputtered ZnO etched for 30 s in diluted HCl (Kluth *et al.*, 1999). Scale bar: 1 μm.

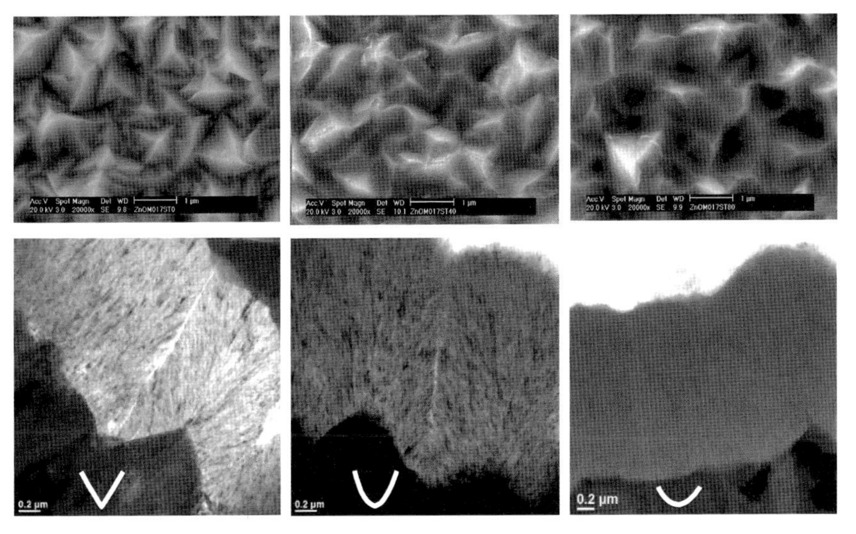

Figure 4.18 Top: top-view SEM images of front electrodes (three TCOs with shapes modified from V to U by plasma treatment); bottom: TEM cross-sections of microcrystalline silicon layers deposited on the TCOs. There is a clear correlation between the ZnO surface morphology and the formation of defective zones within microcrystalline absorber films. Python *et al.* (2008) give more details.

U-shaped bottoms. Unfortunately, U-shaped substrates yield reduced light trapping compared with V-shaped ones; optimisation of cell efficiency must therefore trade off high V_{oc} and FF against reduced i_{sc} (Bailat *et al.*, 2006). Alternatively, valleys with rounded bottoms can be obtained by wet chemical etching; Figure 4.17b shows the crater-like texture of etch pits obtained on sputtered ZnO films after etching in diluted hydrochloric acid (HCl) (Kluth *et al.*, 1999).

An approach that completely separates the texture for light-confinement from the growth process has been suggested for the *n*-*i*-*p* structure (Sai *et al.*, 2011). Starting from a flat silver reflector covered with ZnO, a relief grating is etched into the ZnO and filled with *n*-doped amorphous silicon. Subsequently, the stack is polished in order to provide a flat growth template for the microcrystalline cell. Owing to the complete absence of valleys that induce the growth of defective zones, the resulting device showed the same high voltage as a flat reference cell. However, the *n*-doped filling-layer resulted in parasitic absorption between 650 and 750 nm. An alternative design with an undoped filling-layer in combination with a triple-junction device successfully overcame this limitation (Söderström *et al.*, 2012a; Söderström, 2013).

Traditionally, *n*-*i*-*p* solar cells relied on the natural textures that develop when metals like aluminium or silver are deposited on heated substrates (Hirasaka *et al.*, 1990; Banerjee and Guha, 1991). Because of its relative inertness, the noble metal silver offers an additional handle on texturing if small quantities of aluminium are added to the silver sputtering target: by adding oxygen to the gas mixture during sputtering from such an alloy target, the interface texture is modified by the inclusion of alumina (Al_2O_3) crystallites in the growing film (Franken *et al.*, 2007). The formation of the texture typically requires temperatures of around 300 to 400°C; flexible substrate materials are therefore limited to stainless steel (Banerjee and Guha, 1991) or polyimides (Takano and Kamoshita, 2004). However, in order to fully exploit the promise of reduced production cost with roll-to-roll processing, truly low-cost substrates are desirable. Substrates like polyethylene thus require a different strategy where the substrate itself is textured; this can be achieved by hot embossing (Fonrodona *et al.*, 2005) or by imprinting into a UV-curable polymer (Bailat *et al.*, 2005; Söderström *et al.*, 2010). The interface texture is then transported into the device by conformal growth. An example of a microcrystalline *n*-*i*-*p* cell on a periodic substrate texture obtained by roll-to-roll embossing is shown in the transmission electron microscope (TEM) image in Figure 4.10; Figure 4.19 shows in more detail how the moderate texture of the grating substrate is perfectly reproduced by the sputtered silver and ZnO back contact. However, throughout the microcrystalline absorber layer, the initial sinusoidal texture develops into a structure with a much larger radius of curvature and pinched valleys.

This change of texture is an issue for tandem cells because in the *n*-*i*-*p* configuration the amorphous sub-cell is deposited after the microcrystalline cell. While the large features of the grating substrate can deliver acceptable isc of about 23 to

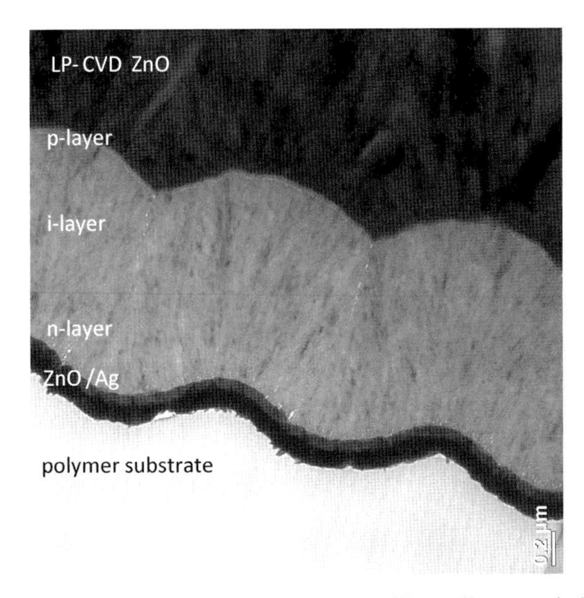

Figure 4.19 TEM cross-section through a microcrystalline cell on a periodically textured back reflector obtained by embossing into a UV-curable polymer (Haug *et al.*, 2009).

25 mA cm^{-2} in bottom cells with thickness varying between 1.2 and 2 μm, the flattening of the interface texture is clearly inadequate for the top cell. The introduction of a thick, textured intermediate reflector can ameliorate this situation: thanks to the texture that ZnO develops when deposited by LP-CVD, Figure 4.10 shows that the top cell can be grown on an adapted texture that is completely independent from the bottom cell (Söderström *et al.*, 2009). Tandem cells in *n-i-p* configuration with stabilised efficiency up to 11.6% have been demonstrated with this approach (Biron *et al.*, 2013).

4.6 Advanced device architecture and record devices

In thin-film silicon PV devices, continuous developments of the constituent layers, interfaces and of their processing conditions are essential to further improve performance. Thanks to the intense industrialisation efforts of the past decade, conversion efficiencies have been significantly increased with the development of efficient light management schemes and with the introduction of low-bandgap partners to a-Si:H to realise multiple-junction devices. At the module level, the successful incorporation of microcrystalline silicon in a-Si:H/μc-Si:H tandem modules triggered direct efficiency improvements. In 2010, progress in the control of the multiple thin-films that constitute such tandem devices over large deposition areas led to the production of the first thin-film silicon modules on glass with stabilised efficiencies surpassing the 10% barrier (Klein *et al.*, 2010; Kratzla *et al.*, 2010). Modules with up to 13.1% initial efficiency and 10.8% stabilised efficiency were demonstrated in 2011 (Kluth *et al.*, 2011).

On metal foil, 12% initial efficiency and 11.2% stabilised efficiency were reported (Banerjee *et al.*, 2011a) for a 400 cm^2 triple-junction cell (a-Si:H/a-SiGe:H/μc-Si:H). These results demonstrate that major bricks for efficient mass production over large areas are in place. To further increase the efficiency, and thus the impact, of thin-film silicon technology, advanced light management schemes, novel nanomaterials and innovative device architectures are currently being developed. Some of these recent advances are reviewed in this section, together with the best efficiencies obtained so far at the cell level.

4.6.1 *Advanced light management with intermediate reflectors*

In monolithically interconnected multiple junctions, the device current is limited by the junction with the smallest current generation. While tailoring the thicknesses of the different absorbers adjusts the photocurrent in each junction, additional layers can also be inserted to selectively reflect photons, thus altering the light distribution in the cell. In micromorph solar cells, the a-Si:H top cell must be kept thin (<300 nm) to minimise its degradation. To realise larger device currents, an intermediate reflecting layer (IRL) can be integrated between the top and bottom cells, to reflect photons not fully absorbed in the first pass through the top cell (wavelengths of 550–700 nm), while ensuring transmittance of red and IR light to the bottom cell (Yamamoto *et al.*, 2004).

The IRL must fulfil three main roles:

- It must have sufficient transverse (out-of-plane) conductivity to avoid electrical losses caused by its series resistance.
- In order to reflect, it must be fabricated from a low refractive index material. Fresnel reflection occurs due to the index contrast between films; the refractive index of silicon n_{rSi} is approximately 3.8 at 600 nm, and n_{rIRL} should therefore ideally be below 2 at this wavelength.
- In addition to the refractive index, the thickness of the IRL controls the intensity and spectral selectivity of back reflection and forward transmission into the bottom cell via interference effects (see Figure 4.20).

The first efficient IRLs were realised using TCO layers such as ZnO (Fischer *et al.*, 1996). More recently, phosphorus-doped silicon oxide materials (n-SiO_x:P) with $n_{rIRL} = 1.6-2$ at 600 nm and with sufficient transverse conductivity were synthesised by PECVD, giving an *in situ* IRL alternative (Buehlmann *et al.*, 2007;

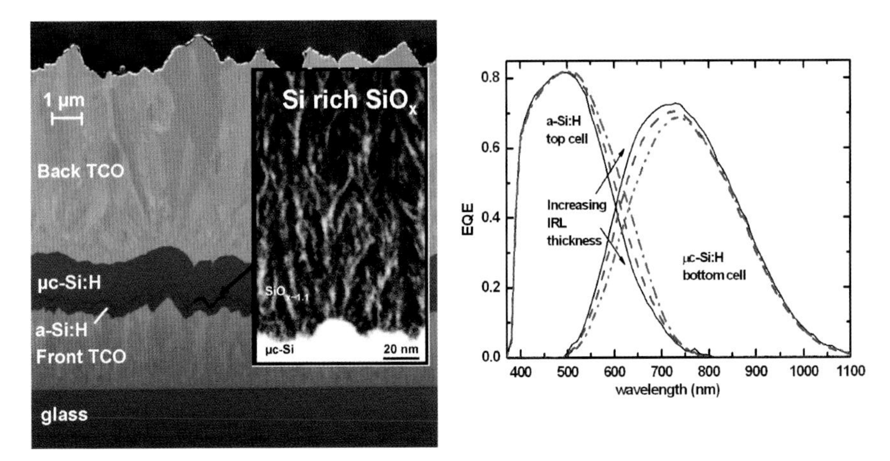

Figure 4.20 Left: SEM image of a focussed ion beam cross-section of a micromorph tandem cell with LP-CVD ZnO front and back contacts, a 250 nm thick a-Si:H top cell, a 40 nm thick silicon-rich silicon oxide IRL and a 1.1 μm thick bottom cell. The nanometre-wide filamentary structure of the silicon phase in the SiOx material is illustrated in the inset EFTEM image. Courtesy of Peter Cuony and Duncan Alexander, EPFL; details can be found in Cuony *et al.*, (2011). Right: spectral response of micromorph tandem cells with no IRL, a 40 nm thick IRL and a 100 nm thick IRL: increasing the IRL thickness results in enhanced (reduced) response in the 500–800 nm range in the top (bottom) cell. Courtesy of Peter Cuony, IMT PV Lab, Neuchâtel.

Das *et al.*, 2008). The best materials were shown to have a mixed-phase structure with nanometre-wide silicon filaments embedded in an amorphous silicon oxide matrix. This advanced nanostructure can be imaged by energy-filtered transmission electron microscopy (EFTEM), as shown in Figure 4.20 (Cuony *et al.*, 2012). Light regions correspond to the doped silicon-rich phase that conducts, whereas dark regions represent the silicon oxide phase that produces the material's low refractive index.

For optimum light harvesting in the top cell, the IRL must be combined with efficient light scattering at the cell entrance, or it must be structured to provide a textured interface to the incident photons, as shown in Figure 4.10 for a tandem device in the substrate configuration with an asymmetric reflector (Söderström *et al.*, 2010). In the superstrate configuration, i_{sc} in 250 nm thick top cells are typically about 11 mA cm^{-2}, while the integration of an IRL can increase i_{sc} up to 14 mA cm^{-2}, demonstrating the impact that such a thin layer can have on efficiency. While n-type silicon oxide nanomaterials are straightforward IRL solutions adaptable to mass production, consecutive approaches such as distributed Bragg reflectors and more sophisticated structures such as 3D photonic crystals are also under development at the cell level to further control the spectral selectivity and the light distribution (Bielawny *et al.*, 2008). Nano-patterned low refractive index dielectric materials (n_r ~1.3–1.5 at 600 nm) were also demonstrated to have a high potential as efficient IRL (Boccard *et al.*, 2013), with in that latter case even a potential for altering the roughness of the surface on which grows the μc-Si:H bottom cell, so as to provide in addition a smoothening of the surface.

4.6.2 *Efficient light in-coupling*

To maximise current generation, reflection and absorption losses at the entrance of the solar cell must be minimised to ensure efficient light in-coupling to the absorber layers. In the superstrate configuration, the refractive indices of glass (n_r ~1.5), TCO (n_r ~2) and silicon (n_r ~3.4–4) lead to optical reflections and to typical transmission losses of 4%, 2% and up to 11% respectively at the air/glass, glass/TCO and flat TCO/silicon interfaces. Textured interfaces and transparent antireflective (AR) coatings have been developed to minimise these reflections.

Textured TCO not only efficiently scatters light but also provides a refractive index grading and multiple reflections of the incoming light, reducing the reflection losses at the TCO/silicon interface. AR coatings with a refractive index intermediate between TCO and silicon can also be used. While titanium dioxide (TiO_2) layers were shown to be efficient (Fujibayashi *et al.*, 2006; Das *et al.*, 2008), high-gap silicon layers adjoining the TCO could also provide an AR effect by index grading. Similarly, texturing of the glass or of a lacquer imprinted on the glass surface (Escarre *et al.*, 2012; Ulbrich *et al.*, 2011) and application of AR coatings with intermediate refractive indices can be successfully used at the air/glass and glass/TCO interfaces.

Absorption in the non-photoactive front TCO and front p-type silicon layers must also be minimised. Doped layers are highly defective and hardly contribute to the photocurrent, so they are designed to be very thin (about 10 nm). Advanced transparent doped materials are under development for use as efficient window layers. Mixed-phase p-type silicon-rich silicon oxide materials (Cuony *et al.*, 2010) and n-type μc-SiC:H layers (Chen *et al.*, 2010) were recently introduced as efficient window layers, leading to EQEs above 70% at 400 nm in μc-Si:H solar cells and $i_{sc} = 29.6\,\mathrm{mA\ cm^{-2}}$ for a 2 μm thick device (Chen *et al.*, 2011). Similarly, minimum absorption must be ensured in the front TCO layer. In devices implementing a μc-Si:H junction, the TCO has to be highly transparent in the 400–1100 nm wavelength range, i.e. up to the near infrared region (NIR); this imposes restrictions on the carrier density as increased free carrier absorption (FCA) leads to a reduction of red and NIR light transmission. High carrier mobility is essential to ensure low sheet resistances (R_{sheet}) for a TCO with reduced free carrier density. Recent advances in the synthesis of doped ZnO have yielded mobilities of 40–60 $\mathrm{cm^2/Vs}$ and absorption levels below 5% over the full spectral range for a layer with $R_{\mathrm{sheet}} = 10\,\Omega/\mathrm{sq}$, and around 2% for $R_{\mathrm{sheet}} = 20\,\Omega/\mathrm{sq}$ (Ruske *et al.*, 2010; Boccard *et al.*, 2010).

Alternatively, hydrogen-doped indium oxide (In_2O_3:H) was recently reported as a highly transparent TCO; even NIR absorption losses are low ($<1\%$) thanks to a high mobility of up to 120 $\mathrm{cm^2/Vs}$ (Koida *et al.*, 2010). These layers do not provide a native or post-process texture, but can be used in next-generation superstrates as thin and highly transparent TCO layers.

4.6.3 *Advanced superstrates with enhanced transparency and multi-scale texturing*

Advanced superstrates with multi-scale texturing are being developed to combine high transparency and conductivity for optimum light-in coupling, efficient light trapping in the full spectral range and the growth of high-quality silicon. Conduction and texture can be decoupled via a bilayer approach: a part of the TCO layer is optimised for conduction with no restriction on its texture. Then, adequate light scattering is achieved with a transparent film whose texture is optimised. Such bilayer TCOs demonstrated increased cell performance thanks to better control of transparency (Selvan *et al.*, 2006; Ding *et al.*, 2011). In a different approach, textured glass can be used in combination with a thin In_2O_3:H TCO layer. As an alternative to glass texturing, Figure 4.21 shows that the texture can equally be applied to a transparent lacquer by nanoimprint lithography (Battaglia *et al.*, 2011a). The efficient light-scattering properties of the textured transparent lacquer combined with the high transparency and conductivity of In_2O_3:H, results in a high summed current of 26 $\mathrm{mA\,cm^{-2}}$ in a micromorph cell using a thin μc-Si.H bottom cell (1.1 μm) and no AR coating.

While the textures typically implemented in thin-film silicon solar cells are dictated by the underlying growth and etching kinetics of the textured TCO, nanoimprint lithography on UV lacquer allows one to replicate an arbitrary structure made from a transparent or opaque arbitrary master material. A large variety of light-trapping structures can thus be designed and combined with a highly transparent TCO. Recently, arbitrary textures from master materials were also demonstrated to be transferable directly to ZnO front electrodes; this nanomoulding of ZnO electrodes enables a similarly wide range of photonic structures to be directly implemented in front electrodes in thin-film silicon solar cells (Battaglia *et al.*,

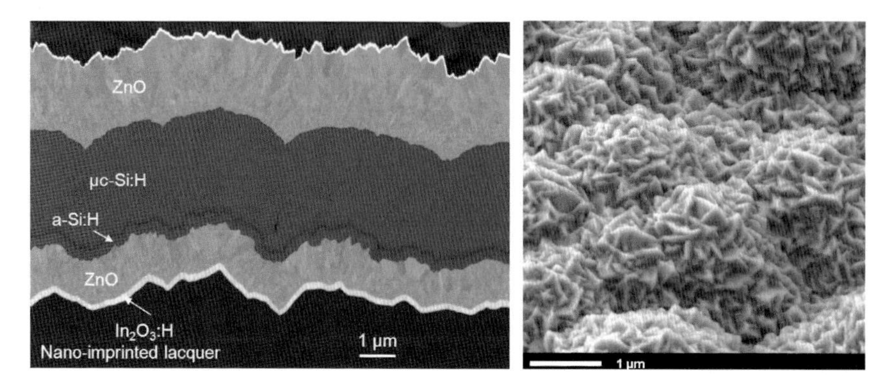

Figure 4.21 Left: SEM cross-section of a-Si:H/μc-Si:H tandem solar cell, developed on a glass superstrate implementing a nanoimprinted UV lacquer that provides large features, a thin In_2O_3:H layer allowing for conduction and high transparency, and a low-doped LP-CVD ZnO layer adding small pyramidal features. Courtesy of Mathieu Boccard; details in Boccard *et al.* (2012). Right: tilted SEM top view image of this multi-scale textured superstrate. Courtesy of Mathieu Boccard, IMT PV Lab, Neuchâtel.

2011b). These recent developments are important tools to determine and implement the most promising light-trapping schemes on the glass or in the electrode itself. Some examples of periodic to random structures transferred to the electrode or replicated on the glass substrate are shown in Figure 4.22.

Multi-scale textures are particularly interesting approaches for tandem cells since they can combine efficient light scattering by small and large textures for the amorphous and microcrystalline cells, respectively. Care must be taken to keep morphologies that are adapted for the growth of the μc-Si:H cell. Two-step etching procedures on sputtered ZnO result in a surface morphology combining large craters with smaller features (Hüpkes *et al.*, 2010). Also, multi-scale texturing of CVD-grown TCO (ZnO and SnO_2) was demonstrated (Taneda *et al.*, 2007). A complementary approach is the combination of textured TCO with textured glass (Bailat *et al.*, 2010; Kroll *et al.*, 2011) or with nanoimprinted lacquer on glass (Boccard *et al.*, 2012; Bessonov, 2011).

Recently, micromorph cells with record stabilised efficiencies of 11.9% (Bailat *et al.*, 2010) and 12.3% (Kroll *et al.*, 2011) were demonstrated by combining textured glass with textured ZnO. Additionally, superstrates combining both multi-scale textures and bilayer TCOs are under development, and Figure 4.21 shows an a-Si:H/μc-Si:H tandem cell realised on such a substrate. Superstrates with further combinations of multi-layer TCOs and multiple textures produced in either TCO or glass by direct texturing or nanoimprinting are still under study to further enhance the efficiency of thin-film silicon solar cells.

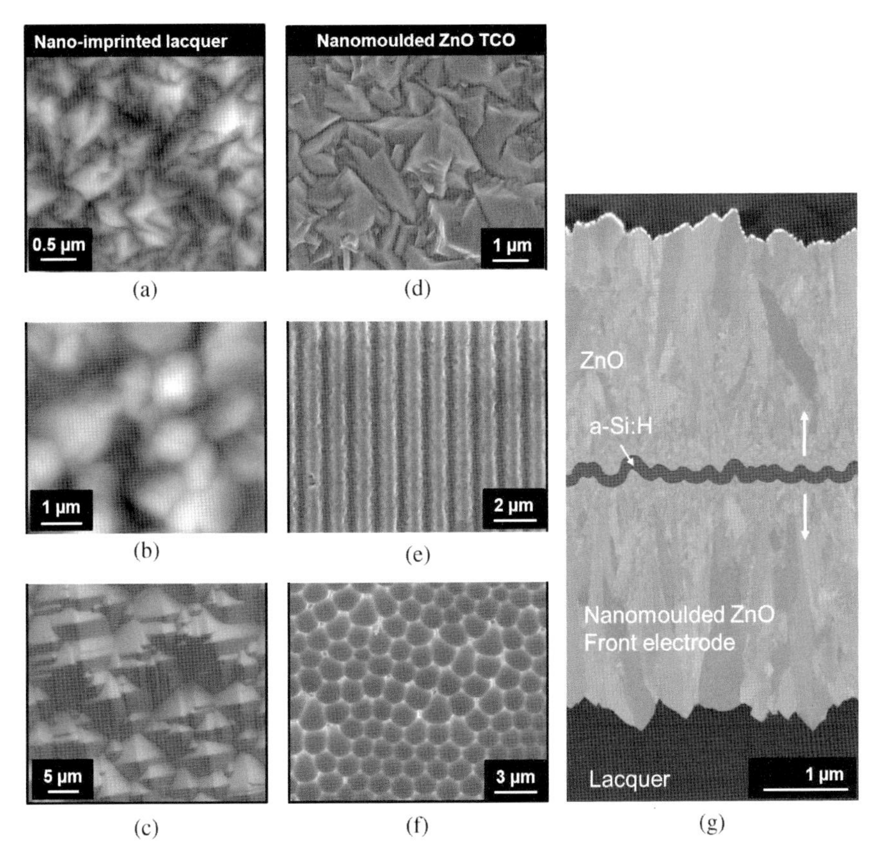

Figure 4.22 SEM images of (a) as-grown random pyramidal LPCVD ZnO; (b) thermally roughened silver; and (c) alkaline-etched crystalline silicon, all realised by nanoimprinting in a transparent lacquer. Courtesy of Jordi Escarre and Karin Söderström, IMT PV Lab, Neuchâtel; details in Söderström *et al.* (2010). SEM images of (d) as-grown thick LP-CVD ZnO; (e) a one-dimensional periodic grating; and (f) quasi-periodic hexagonal dimple pattern of anodic oxidation of aluminium, all transferred to the surface of ZnO front electrodes via nanomoulding. Courtesy of Corsin Battaglia; details in Battaglia *et al.* (2011b). (g): a-Si:H solar cell on a nanomoulded ZnO front electrode with a surface texture corresponding to (f).

4.6.4 *Cell architecture, alternative absorber layers and advanced designs*

The main driver in thin-film silicon module production is currently the a-Si:H/μc-Si:H tandem architecture. Recently, enhanced cell and module resilience to substrate roughness was achieved with the implementation of novel doped layers and with the optimisation of μc-Si:H growth conditions (Bailat *et al.*, 2010; Despeisse *et al.*, 2011; Bugnon *et al.*, 2011). For instance, the implementation of silicon-rich silicon oxide-doped layers was demonstrated to enhance the electrical performance of thin-film silicon solar cells grown on highly textured substrates. These doped layers were shown to mitigate the impact of localised defects on cell electrical performance (Despeisse *et al.*, 2011). Section 4.5 discussed how such defective areas arise when silicon is deposited on highly textured surfaces exhibiting sharp valleys (Python *et al.*, 2008).

Moreover, plasma process conditions during microcrystalline silicon growth were also shown to affect the quality and density of these localised defect regions that occur over sharp underlying surface features (Bugnon *et al.*, 2011), demonstrating that globally dense μc-Si:H material has to be achieved on textured substrates to maximise efficiency. Thin devices were shown to have high potential (Schicho *et al.*, 2010), and the enhanced resilience of the solar cells to the substrate texture allowed for the development of high-efficiency thin superstrate devices, with 11.3% and 10.9% stable cells realised using 1.1 μm and 0.8 μm thick bottom cells, respectively (Despeisse *et al.*, 2011; Bailat *et al.*, 2010). Such thin cells give stabilised efficiencies higher than 10% in modules (Kratzla *et al.*, 2010), and they trigger cost reductions because of reduced deposition and reactor cleaning times, resulting in a promising trade-off between efficiency and costs for the micromorph approach.

Alternatively, high-efficiency concepts with folded 3D cell designs (Vanecek *et al.*, 2010; Naughton *et al.*, 2010) are under consideration for a-Si:H single junctions and micromorph tandem devices. Novel nanostructures such as nano-columns or nano-/micro-holes on superstrates or substrates could open possibilities; the cell layers would be optically thick but electrically thin (*cf.* Figure 4.23). Accordingly, the resulting folded cell would yield thin layers with very low LID (Vanecek *et al.*, 2010).

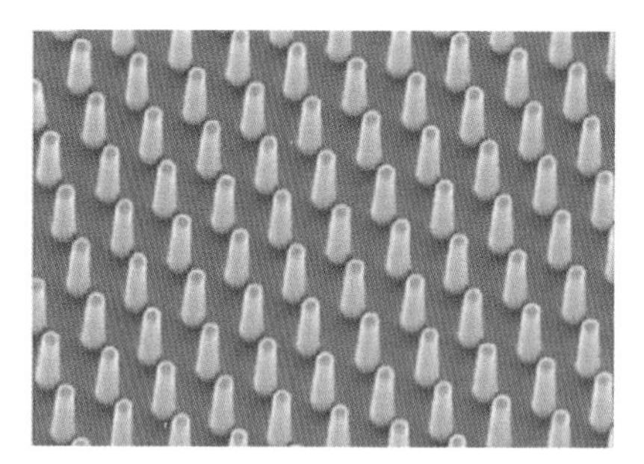

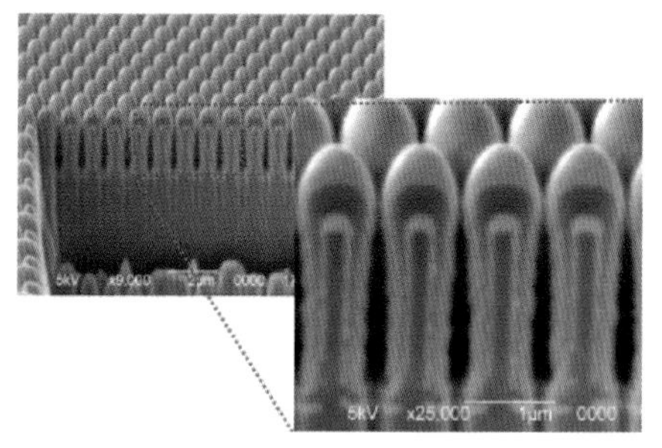

Figure 4.23 SEM image of aligned nano-pillars (left panel (Rizal *et al.*, 2013)) and FIB cross-section through *n-i-p* solar cells deposited on these pillars. The inset illustrates non-conformal growth with preferential deposition of reflector, silicon and ITO layer at the tips (Naughton *et al.*, 2010).

An alternative approach to higher stabilised efficiencies is the use of triple-junction cells; however, the additional junction comes at the price of enhanced spectral sensitivity and further increased cell thickness, i.e. higher costs or harsher requirements on absorber deposition rates. The most efficient combinations demonstrated so far consist of a-Si:H/a-SiGe:H/μc-Si:H or a-Si:H/μc-Si:H/μc-Si:H devices. Both have been successfully developed in substrate *n-i-p* configuration (Xu *et al.*, 2010) by the company Unisolar. They showed stabilised cell efficiencies beyond 12.5% (Yue *et al.*, 2010) which translate to prototype module efficiencies above 11% (Banerjee *et al.*, 2011a). In the *p-i-n* configuration, a-Si:H/μc-Si:H/μc-Si:H triple junction cells with stabilised efficiency of 13.4% have been presented (Kim *et al.*, 2013).

As outlined in Section 4.3.2, the matched cell current is lower than in tandem junctions. This can be exploited to achieve higher stability by further thinning the a-Si:H top cell, or by depositing it yet closer to the transition between the amorphous and microcrystalline phases, which generally increases the bandgap and enhances its stability against light-induced degradation at the same time (Kamei *et al.*, 1999; Koval *et al.*, 1999). Research on absorber layers alloyed with carbon or oxygen for alternative top cells is also being conducted (Yunaz *et al.*, 2009; Inthisang *et al.*, 2011). Complementary research addresses materials with lower bandgap such as μc-$Si_{1-x}Ge_x$:H (Matsui *et al.*, 2010) which successfully enhances the conversion of the IR part of the spectrum. A wide range of thin-film silicon absorber materials is therefore in development (Konagai *et al.*, 2011), and they may lead to novel combinations in triple- or quadruple-junction thin-film silicon solar cells to reach beyond 16% stabilised efficiency.

4.6.5 *Record devices*

Table 4.2 lists the highest certified efficiencies at the time of writing (mid-2013). Single-junction cells based on a-Si:H and μc-Si:H absorber layers reached efficiencies of 10.1% and 10.7% on device areas of 1 cm^2, respectively. The best tandem and triple-junction cells have reached stabilised efficiencies of 12.3% and 13.4%, respectively. Table 4.3 gives an (incomplete) list of notable cells. These results are quoted as reported by the respective authors, generally without certification. Most results refer to cells in their initial states, except where the authors give their stabilised values.

4.6.6 *Roadmap for higher efficiencies*

To survive the competition of crystalline silicon, thin-film silicon technology needs to go faster towards higher efficiencies. Based on the findings of the last decade and

Table 4.2 Overview of certified record efficiencies in cells and modules.

Source	Structure	Eff. (%)	V_{oc}(V)	FF (%)	i_{sc} (mA cm^{-2})	Comments/ Reference
Certified record cells (stabilised)						
Oerlikon Solar	a-Si	10.1	0.886	67	16.75	*p-i-n* on LP-CVD ZnO (Benagli *et al.*, 2009)
AIST	a-Si	10.1	0.906	70	16.05	*p-i-n* on Asahi-VU (Matsui *et al.*, 2013)
AIST	μc-Si	11.0	0.542	73.8	27.44	(Sai *et al.*, 2014)
IMT	a-Si/μc-Si	12.6	1.382	71.3	12.83	(Boccard *et al.*, 2014)
United Solar	a-Si/ a-SiGe/a-SiGe	12.1	2.297	69.7	7.56	*n-i-p* on steel, 0.27 cm^2 (Yang *et al.*, 1997)
United Solar	a-Si/ μc-Si/μc-Si	12.5	2.01	68.4	9.1	*n-i-p* on steel, 0.27 cm^2 (Yue *et al.*, 2010)
LG electronics	a-Si/ μc-Si/μc-Si	13.44	1.96	71.9	9.52	*p-i-n* (Kim *et al.*, 2013)
Certified record (mini-)modules (stabilised)						
TEL	a-Si/μc-Si	12.2	205.2	66.9	1.22 A	Gen5 module, 1.43 m^2 (TEL, 2014)
United Solar	a-Si/ a-SiGe/a-SiGe	11.0	2.34	70.3	6.65	*n-i-p* on steel, 462.3 cm^2, large area cell (Banerjee *et al.*, 2011a)
United Solar	a-Si/ μc-Si/μc-Si	12.0	1.95	69.2	8.93	*n-i-p*, enc. 399.8 cm^2 cell (Banerjee *et al.*, 2011b)

on the requirement to 'decouple' the electrical from the optical transport paths, a simple calculation based on the best top and bottom cell realised so far shows that tandem solar cells are probably limited to stabilised efficiencies between 14% and 15%. For two reasons, triple-junction solar cells such as a-Si:H/μc-Si:H/μc-Si:H

Table 4.3 Overview of noticeable efficiencies (reported by the authors, not certified).

Source	Structure	Eff. (%)	V_{oc}(V)	FF (%)	i_{sc} (mA cm^{-2})	Comments/ Reference
United Solar	a-Si/ a-SiGe/ μc-Si	16.3 (initial)	2.242	77	9.43	*n-i-p*cell, 0.27 cm^2 (Yan *et al.* 2011)
Kaneka	Triple-junction	15 (initial)	2.28	73.5	8.93	(Yamamoto *et al.*, 2006)
Kaneka	a-Si/ μc-Si	14.7 (initial)	1.41	72.8	14.4	(Yamamoto *et al.*, 2004)
Oerlikon Solar	a-Si/ μc-Si	14.4 (initial)	1.419	70.7	14.35	(Kroll *et al.*, 2011)
IMT Neuchâtel	a-Si/ μc-Si	14.1 (initial)	1.41	71.5	14	on IO/LP-CVD, 1.2 cm^2 (Ballif *et al.*, 2011)
IMT Neuchâtel	a-Si/ μc-Si/ μc-Si	13.0 (stable)	1.94	70.3	9.5	$n-i-p$ on polished substate (Söderström, 2013)
Sanyo	a-Si/ μc-Si	13.5 (ini) 12.2 (st)	1.41 –	75.4 –	12.7 –	(Terakawa *et al.*, 2011)

are an attractive alternative to go further; as discussed in Section 4.3.2, matching at a lower current relieves several constraints of cell processing. The intermediate reflector is no longer needed, and the thin top cells can be made with a very high $V_{oc} > 1$ V (Yan *et al.*, 2003). The μc-Si:H middle cell is also thin enough to use a recently demonstrated design for thin cells with high $V_{oc} > 600$ mV (Hänni *et al.*, 2013). With slightly improved bottom cells and the latest advances in light management, a stable device efficiency of 16.5% could thus be targeted with V_{oc} of 2.15 V, FF of 72% and i_{sc} of 10.7 mA cm^{-2}. Alternatively, the use of Ge alloys for the middle cell could lead to similar results. Moving to significantly higher efficiency will probably require the development of novel absorber materials with higher quality, and the use of innovative passivating contacts that extract the full voltage potential of the materials.

4.7 Industrialisation and large-area production technology

4.7.1 *From small-area to large-area devices*

The FPD sector showed impressive development through 2004. Indeed, at that time, more than 10 million square metres were coated annually in the manufacturing of liquid crystal displays (LCDs); the TFTs in these displays are based on amorphous silicon as well as silicon nitride. Moreover, the FPD industry demonstrated a continuous decrease of display production costs (20% per m^2 annually). It was hence not a surprise that equipment makers, who had made a significant investment in the development of coating tools, identified thin-film silicon solar technology as an emerging growth market for their products, with a focus on PECVD equipment that could be used for layer deposition. This led companies such as Oerlikon, Applied Films (later purchased by Applied Material), Ulvac and Leybold Optics to enter the PV market, trying to transfer several of the processes developed in research laboratories to mass production. Most manufacturers work on Gen5 scale (1.4 m^2) but Applied Materials has also delivered tools for Gen8.5 (5.5 m^2). The transition from a mobility-controlled device (as in a TFT) to a lifetime-controlled device (as in PV) was, however, not trivial for three related reasons:

- In a-Si:H and μc-Si:H, the material quality depends very much on the deposition conditions and rate. Reasonable rates at acceptable quality often required equipment adaptation (e.g. narrower inter-electrode gaps for higher pressure or measures to avoid standing-wave patterns for the case of VHF-excitation).
- By definition, μc-Si:H is a material deposited close to a phase transition; consequently its homogeneous deposition on > 1 m^2 required significant hardware adaptation in order to compensate for gas flow and electromagnetic inhomogeneities, as well as boundary effects.
- The two first steps had to be taken in conjunction with the development of the TCO front contact (not fully developed at that time), the back contact and laser scribing.

While solving these issues, several equipment makers were able to sign contracts to sell 'turnkey' production lines, making thin-film silicon the first technology offering an easy entrance into the PV market, while covering the full chain (from glass to module). Indeed from 2006 to 2008, concerned about silicon feedstock problems for c-Si cells manufacturers, over 50 companies announced their entrance into thin-film silicon module manufacturing. Even though solving the three issues noted above was more demanding than initially expected, in 2011 several companies were able to manufacture modules with total-area stabilised efficiencies of 6.5–7.3% and 9–10% for a-Si:H and a-Si:H/μc-Si:H modules on glass, respectively. Note that 'total-area' refers to a widely used module size of 1.4 m^2 which includes a non-active edge area of 6–8%. Commercial a-Si:H/a-$Si_{1-x}Ge_x$:H/a-$Si_{1-y}Ge_y$:H triple-junction modules on flexible substrates reach 8.2% stabilised efficiencies (designated aperture area); a comparison with the record devices in Table 4.2 shows that there is a relatively little gap between small-scale cell results and m^2 sized modules, at least for *p-i-n* cells.

Probably no other technology has made as big an effort at transferring laboratory results to production-size modules, and a large part of this was facilitated by FPD experience. The last half-decade has seen the resolution of all major production issues and a huge amount of knowledge was gleaned on up-scaling issues. The basic manufacturing process sequence on glass was presented in Figure 4.2. We review some of its key aspects here and reserve a specific section for roll-to-roll processes.

4.7.2 *TCO deposition on glass*

The front contact made from transparent conducting oxides (TCOs) is a key element for modules prepared on glass. Its sheet resistance R_{sheet} (5–20 Ω/sq) and the segment width W (typically 5–15 mm) control the relative TCO resistive power losses according to the formula $\Delta P/P = R_{sheet} \times W^2 \times I_{max}/3V_{max}$, where I_{max} and V_{max} are the current and voltage at the maximum power point of the $I-V$ curve. Additional losses linked to the interconnection areas (from segment to segment) can occur as well. The TCO must scatter light efficiently to achieve high current in devices (30–50% current increase compared with flat substrates), but if the free carrier concentration required for lowering the sheet resistance is too high, strong parasitic FCA will take place.

Usually, reducing the doping and thickening the layer help reduce FCA, because it scales super-linearly to the TCO carrier concentration (Faÿ *et al.*, 2010). At large thicknesses, the shape of the TCO grains can lead to issues with the growth of the subsequent devices, as detailed in Section 4.5. Advanced device processing (e.g. using specially tailored doped layers, as discussed in Section 4.6) can be used to mitigate such effects. A drawback of reducing the free-carrier concentration too much is a stronger sensitivity to moisture ingress; humidity creates trap states at the grain boundaries, leading to a conductivity decrease and thus adequate encapsulation schemes must be considered.

There are currently four major routes used for the fabrication of TCOs on large area substrates ($>1\,m^2$). Some of the advantages and weaknesses of the various approaches are mentioned in Table 4.4. Figure 4.16 shows images of the various TCOs. The four routes are:

- Inline deposition of $SnO_2{:}F$ by APCVD; the TCO deposition is directly integrated into the cooling stage of the float glass line (McCurdy *et al.*, 1999). Deposition takes place between 640–590°C (depending on the location on the line, T > 600°C means the glass is still liquid). The deposition rate is around 100 nm/s. Commercially used precursors are tin-tetrachloride, dimethyl-tin-dichloride and butyl-tin-dichloride. Usually a barrier layer against sodium migration is included in the stack as well as a thin AR layer between the glass

Table 4.4 Overview of advantages and challenges linked to the use of various front TCOs in amorphous and a-Si:H/μc-Si:H production.

TCO	Advantages	Challenges
SnO_2:F inline	Minimal hardware costs, medium to high current in top cell	Process control, high transparency
SnO_2:F off-line	Well-controlled process, medium to high current in top cell	High transparency and scattering, high temperature process
Sputter-etched ZnO	Substrate suited to the growth of microcrystalline cells, good light scattering for total current	Uniformity, cost of hardware and targets, lower current in top cell, even with IRL
LP-CVD-ZnO	High current in top cell, very high with IRL, doping and scattering easily tuneable	Control of subsequent PECVD growth, encapsulation

and the TCO. Typical TCO thickness is 700 nm. After SnO_2:F deposition, the glass is cut and the edges are ground.

- Off-line deposition of SnO_2:F by APCVD: deposition on glass plates that are reheated to typical temperatures of 540–590°C. Deposition rates are 50 nm/s and the same precursors can be used as for the inline process. Typical TCO thickness is 700 nm.
- Off-line sputter-etched ZnO: aluminium-doped ZnO is sputtered and subsequently wet-etched to create crater-like nanostructures (Kluth *et al.*, 2003). ZnO rotary ceramic targets containing 1–2 wt.% of Al_2O_3 are used for DC sputtering (Zhu *et al.*, 2009) at typical temperatures of 300–370°C. After deposition, the cooled substrates are etched in a 0.3–0.5% HCl solution. Typical TCO thickness is 700 nm.
- Off-line deposition of ZnO by LPCVD (Vogler *et al.*, 2008): deposition typically takes place at 170–200°C, at pressures of a few mbar, using di-ethyl-zinc (DEZ, $(C_2H_5)_2Zn$), water and di-borane for doping. Typical deposition rates are 2–4 nm per second, and typical layer thickness is 1.4–2 μm.

Notably, the processes run by research laboratories might differ from those employed by industry. At the laboratory level, all of these TCOs have demonstrated good results. Between 2009 and 2011, certified world record efficiencies were reported for amorphous cells deposited on LP-CVD ZnO (Benagli *et al.*, 2009) and for micromorph cells on LP-CVD ZnO grown on textured glass (Bailat *et al.*, 2010). Micromorph modules with efficiencies of over 10% have been reported for most of the above TCOs (Klein *et al.*, 2010; Kratzla *et al.*, 2010; Aya *et al.*, 2011).

4.7.3 *Deposition of thin-film silicon by PECVD*

This section concentrates on deposition of the silicon-based layers by PECVD because this is the only technique implemented in production. The discussion is moreover focussed on microcrystalline layers, which are thicker and more challenging to deposit with high quality and low cost. The challenges involved in upscaling PECVD processes from the laboratory to production can be summarised as follows:

- A major issue is plasma uniformity, which is particularly critical for materials close to the a-Si:H to μc-Si:H phase transition. On large area, the gas composition (e.g. ratio of Si radical H) can be non-uniform because of educts that are not incorporated in the layers. Uniformity can be influenced by varying the gas distribution (e.g. different gas flow at the border compared with the centre).
- RF and VHF waves couple along the lateral dimensions of reactors and may create standing wave effects whose suppression requires special hardware corrections. As a general rule, if the size of the parallel plate electrodes are smaller than a tenth of the excitation wavelength (i.e. 2.21 m and 0.74 m respectively for 13.56 MHz and 40.68 MHz), electromagnetic non-uniformities start to have a direct impact on uniformity (Sansonnens *et al.*, 2006).
- Edge effects, including the telegraph effect, parasitic plasma at the chamber corners and edges, and lateral duct pumping. All will influence the homogeneity of the coating. In some cases, they can be employed to compensate for inhomogeneity of other origins.
- Powder formation tends to occur in μc-Si:H processes and can lead to local non-uniformities and plasma instabilities. Local pumping might help cope with such effects (Aya *et al.*, 2011).
- The same chamber should be suitable for running different processes, but this is not trivial since the effects described above tend to depend on the applied power, pressure and gas composition. Also, contaminations from doping and chamber cleaning have to be considered carefully (Ballutaud *et al.*, 2004).
- A sufficient deposition rate (0.5 to, ideally, several nm/s), or very low-cost equipment at a lower rate (0.3 to 0.5 nm/s), is required for economical production.

Surmounting these difficulties was the main objective of much system development during the last decade; meeting new challenges on the way was the reason it took longer than expected to bring the technology to maturity. During this time, several companies and equipment providers worked on deposition hardware and processing for μc-Si:H with solutions based on parallel plate RF PECVD. In most approaches, the inter-electrode gap is low (8–20 mm) and the pressure is kept high (5–15 mbars); under these conditions, collisions slow down ions that are accelerated in the high voltage drop at the sheath. Most deposition regimes use high dilution (at most a few percent of silane in hydrogen). In very large coating systems (such as Gen8.5 tools), a combination of multiple small hollow cathodes with variable geometry may be integrated in the RF electrode and combined with a gas diffuser consisting of holes in the showerhead plate. Additionally varying the spacing or the gas conductance is varied, such designs can compensate standing wave and edge effects.

A wide variety of system configurations exists (e.g. inline, vertical, batch, cluster). TEL (formerly Oerlikon) Solar uses the design shown in Figure 4.24 where

Figure 4.24 Top: Example of a compact production system (KAI systems for PECVD, with two process modules with stacks of 10 reactors). Bottom: Flexible amorphous silicon modules for roof-top integration.

a stack of ten reactors is placed inside a single vacuum chamber which reduces the required components (common gas supply, pressure control, and vacuum pump for all reactors). Only the 40.68 MHz VHF excitation is supplied from individual power supplies. The standing wave effect is suppressed using a magnetic lens compensation system (Sansonnens and Schmitt, 2003). Several other equipment manufacturers, some of them not from the FPD industry, also proposed low-cost systems for amorphous silicon, usually based on 13.56 MHz parallel plate reactors and sometimes with lateral ingress of the processing gas (e.g. with one RF generator powering one electrode and with two counter electrodes, allowing for coating of four glass substrates with one generator).

Most of today's equipment is suitable for deposition rates of 0.2–0.4 nm/s for a-Si:H and 0.3–0.6 nm/ for μc-Si. Approaches for much higher deposition on large-area cells use arrays of nozzle electrodes and local pumping (Aya *et al.*, 2011) which probably prevents excessive powder formation and plasma perturbation, or VHF at 60 MHz with ladder-shaped electrodes (Takeuchi *et al.* 2001; Takatsuka *et al.*, 2005; Mashima *et al.*, 2006). This last configuration creates an asymmetry favouring low-energy bombardment of the substrate. A dual frequency mode and duty cycle with suitable spacing of the ladder allows compensating the standing wave issue. These last two approaches have facilitated good initial module efficiencies over 10% at high deposition rate (>2 nm/s). However, for such systems, input power and thermal load management are not trivial.

4.7.4 *Back reflectors and back contacts*

Various back reflectors and back contacts are used industrially. A common back contact for a-Si:H is a doped ZnO/aluminium stack, with typically 50–100 nm ZnO and 200 nm aluminium. Aluminium is a medium quality reflector in the NIR and IR, and amorphous modules can gain a few Wp/m^2 when using silver instead of aluminium. For micromorph modules, aluminium induces excessive optical losses and must be replaced by silver. Aluminium and silver are sometimes covered by nickel/vanadium for easier soldering. All contacts are applied by sputtering. ZnO/silver contacts require careful optimisation, as modules that employ them tend to be more prone to shunting. Indeed, the low sheet resistance of the metallic back contact can allow a gain in FF. If localised shunts exist, the highly conducting metal layer can also be detrimental since it supports larger leakage currents as a TCO-based back reflector. This can be made, for example, with a 1.5–2 μm thick LP-CVD ZnO with $R_{sheet} = 10-20\,\Omega$/sq. In this case a white reflector must be added as back reflector, either in the form of screen-printed white paste, or using a white polymer in the encapsulation process.

4.7.5 *Module technology*

Laser scribing following the ideas described by Nakano and co-workers (Nakano *et al.*, 1986; Haas *et al.*, 2007; Bartlomé *et al.*, 2010) is performed in the three steps shown in Figure 4.2, referred to as P1, P2 and P3. The lasers are typically diode-pumped solid state lasers emitting at 1064 nm or 355 nm (third harmonic) for the P1 step, and at 532 nm (second harmonic) for the P2 and P3 processes. Typical pulse lengths and energies are in the few ns and μJ range, with a typical repetition rate of 30–100 kHz ensuring a scribing speed of over 1 m/s for a spot size of 30–40 μm. The typical dead area is in the range of 200 to 350 μm. This can be reduced with optimised mechanical and optical systems. Narrower scribing allows for more versatility, such as thinner segment widths and the use of more resistive TCO with less FCA. The realisation of low-voltage modules (with parallel connection of series of segments) can also require the laser scribes to be adapted (to go from high to low voltage instead of low to high).

The inorganic materials used in thin-film silicon modules are suitable for long module lifetimes. In the case of a-Si:H/a-Si:H modules, systems monitored for 14 and 16 years showed a minor efficiency decrease of 0.2% p. a. (Lechner *et al.*, 2010). However, care has to be taken to avoid TCO delamination (especially SnO_2), which can result from ion migration at high system voltage (Jansen and Delahoy, 2003) and careful packaging is often required as some layers can be sensitive to moisture. Most thin-film modules rely on glass/glass protection (where none, one or both of the glasses are tempered, depending on the manufacturer) and some on glass/back sheet protection, in which case the back sheet usually contains a moisture-tight aluminium foil barrier. Lateral moisture ingression needs to be controlled as well, by the use of a suitable encapsulant (modern PVB) or by introduction of edge sealants (such as butyl, possibly in conjunction with EVA). Moreover, good control of glass cleanliness and encapsulant adhesion is required to avoid delamination or the ingress of water vapour.

4.7.6 *Roll-to-roll technology*

The advantage of roll-to-roll technology lies in flexible, lightweight unbreakable products, which are attractive for building integration as shown in Figure 4.24. In most cases the *n-i-p* configuration is used. Notably some of the best results for all medium-size thin-film modules have been reported by United Solar working on steel foils (12% initial on 400 cm^2, in line with the aperture area of the best module on glass).

The dynamic roll-to-roll process requires uniformity only in one dimension, facilitating the use of fast plasma sources, e.g. a combination of VHF and a very narrow gap in 1D (Strobel *et al.*, 2009; Zimmermann *et al.*, 2010). When deposited at high temperature, aluminium and particularly silver develop surface textures that are suitable for light trapping (Banerjee and Guha, 1991). This 'hot silver' reflector is also applicable to polyimide substrates, but low-cost plastic such as polyethylene does not withstand the required temperatures. Alternatively, flat silver covered by textured ZnO can act as a high-quality reflector (Guha *et al.*, 2011), or light trapping textures may be embossed into the plastic substrate itself (Bailat *et al.*, 2005; Escarré *et al.*, 2011).

Table 4.5 summarises the status of some flexible thin-film silicon companies. Usually, *n-i-p* modules incorporate EVA/ETFE protection at the front side, which is not vapour tight but is compatible with the front ITO used by most manufacturers. Contrary to the case of glass substrates, there are no official turnkey suppliers for roll-to-roll technology and most developments have been made in-house by the respective companies. Helianthos uses the *p-i-n* structure on a sacrificial aluminium substrate coated with a SnO_2 front contact, a substrate that is also compatible with micromorph technology (van den Donker *et al.*, 2007).

Table 4.5 Overview of some flexible thin-film silicon suppliers and developers.

Company	**Country**	**Configuration**	**Module efficiency**	**Capacity (2011)**	**Notes**
Unisolar (discontinued in 2012)	USA	*n-i-p* triple a-Si/a-SiGe/a-SiGe	up to 8.2% aperture area	~180 MWp	On steel foil, shingle connection
Powerfilm	USA	*n-i-p* single	3.3% total area	n.a.	On polyimide substrate
Flexcell (discontinued in 2012)	CH	*n-i-p* single a-Si	5% aperture area	25 MWp	On PET, light trapping not yet introduced
Sanyo	Japan	*n-i-p* single	4.1–4.5% total area	n.a.	Amorton product line
Fuji	Japan	*n-i-p* tandem a-Si/a-SiGe	6–7%	12 MWp	On polyimide, light trapping by textured metal
Xunlight Corp	USA	Triple a-Si/a-SiGe/nc-Si	6–7% aperture area	25 MWp	
Hyet Solar (ex Helianthos)	NL	*p-i-n* on SnO_2	(4–5)	n.a.	Textured AP-CVD SnO_2, on sacrificial Al film

Table 4.6 Example of cost structure for a 140 MW Oerlikon Solar THINfab II in China in 2013 for a 10–11% stabilised module efficiency based on micromorph tandem technology (courtesy of Dr. R. Benz Oerlikon Solar).

Contribution	**Amount (€/W$_p$)**	**Type of cost**
Capital investment	0.08–0.12	fixed
Labour	0.01	fixed
Materials	0.12–0.15	variable
Gas	0.07–0.08	variable
Utilities	0.03	variable
Maintenance	0.03–0.04	variable
Yield loss	0.01	variable
Total module manufacturing cost	0.435–0.44	

4.7.7 *Costs of thin-film silicon production*

Recent technological improvements, such as the use of thinner microcrystalline absorber layers (Schicho *et al.*, 2010; Kratzla *et al.*, 2010; Despeisse *et al.*, 2011), the increase in efficiency of devices as well as the decrease in the cost of production equipment, promise low production costs. Table 4.6 shows what a modern production line could achieve in terms of module production costs, according to a major equipment supplier. Production costs in the range of 0.35 €/Wp should be possible with a modest production volume of 140 MW. This calculation is given for a line from glass to module (i.e. a line including TCO coating), and the equipment is depreciated over seven years.

Other companies quote similar cost potential for micromorph modules. Note that the process uses cleaning gases for the PECVD reactors. Even though such gases are expensive, their consumption is reduced with the use of thinner microcrystalline absorber layers. In terms of life cycle assessment, micromorph modules perform well. This is the case even if the reactor is cleaned with a gas such as nitrogen trifluoride (NF_3) (which has a high CO_2-equivalent greenhouse effect), provided that thin layers are used and the gas is abated properly. This is also the case for thick layers in combination with on-site fluorine (F_2) generation (with no greenhouse gas effect) or with mechanical cleaning. For the technology quoted in Table 4.6, with a μc-Si:H i-layer thickness below 1 μm, an equivalent CO_2 emission of 34 g/kWh is calculated, and the energy payback time at the module level is below one year in sunny places (>1700 kWh/m^2) (Steimen, 2011). These values are in line with those reported for other thin-film technologies and, logically, lower than those given for thicker devices (Fthenakis *et al.*, 2008a; Fthenakis *et al.*, 2008b; Bravi *et al.*, 2011).

4.7.8 *Conclusion and outlook*

Thin-film silicon modules use abundant, non-toxic materials and have an excellent temperature coefficient, promising ultra-wide-scale implementation. Thanks to huge industrialisation efforts during the last decades, dozens of companies can now produce hundreds of MW annually, and the best production lines are reported to work at close to 98% production yield. Thin-film silicon is also the first thin-film technology where efficient turnkey production lines are available for new investors entering the market. The best micromorph modules with areas $>1\ m^2$ reach 10% stabilised efficiencies on glass (demonstrated at the pilot line level and in some production lines). In parallel, several companies are developing amorphous silicon modules at ultra-low cost. There are still a lot of options to further improve efficiencies, including improved light trapping structures, new materials, interfaces and doped layers as presented in this review. The large choice of materials accessible within a single technology, as well as the possibility to easily integrate multiple junctions, promises that significant increases in cell and module efficiency will be achieved in coming years.

Acknowledgements

We would like to thank Z. Holman for careful reading and corrections, D. Sheel (CVtech) and J. Hüpkes (HZB Jülich) for providing information on TCO preparation, Reinhardt Benz and Irene Steimen (Oerlikon Solar) for providing cost calculation data, the personnel of PV Lab for providing figures and data, as well as D. Alexander and M. Cantoni (CIME EPFL) and M. Leboeuf (CSEM) for creating some of the TEM and SEM images.

References

Aya Y., Shinohara W., Matsumoto M., Murata K., Kunii T., Nakagawa M., Terakawa A. and Tanaka M. (2011), 'Progress of thin-film silicon photovoltaic technologies SANYO', *Progr. Photovoltaics: Res. Applications*, published online.

Bae S., Kaan Kalkan A., Cheng S. and Fonash S. (1998), 'Characteristics of amorphous and polycrystalline silicon films deposited at 120°C by electron cyclotron resonance plasma-enhanced chemical vapor deposition', *J. Vac. Sci. Technol. A* **16**, 3, 1992.

Bailat J., Dominé D., Schlüchter R, Steinhauser J., Faÿ S., Freitas F., Bucher C., Feitknecht L., Niquille X., Tscharner R., Shah A. and Ballif C. (2006), 'High efficiency pin microcrystalline and micromorph thin-film silicon solar cells deposited on LPCVD ZnO coated glass substrates', *Proc. 4th World PVSEC*, Honolulu, HI.

Bailat J., Fesquet L., Orhan J.-B., Djeridane Y., Wolf B., Madliger P., Steinhauser J., Benagli, Borrello D., Castens L., Monteduro G., Dehbozorgi B., Vallat-Sauvain E., Multone X., Romang D., Boucher J.-F., Meier J., Kroll U., Despeisse M., Bugnon G., Ballif C., Marjanovic S., Konhke G., Borrelli N., Koch K., Liu J., Modavis R., Thelen D., Vallon S., Zakharian A. and Weidman, D. (2010), 'Recent developments of high-efficiency micromorph tandem solar cells in KAI-M PECVD reactors',*Proc. 5th World PVSEC*, Valencia, pp. 2720–2724, 3BO.11.5.

Bailat J., Terrazzoni-Daudrix V., Guillet J., Freitas F., Niquille X., Shah A., Ballif C., Scharf T., Morf R., Hansen A., Fischer D., Ziegler Y. and Closset A. (2005), 'Recent development of solar cells on low-cost plastic substrates', *Proc. 20th EU Photovoltaic Solar Energy Conf.*, Barcelona, pp. 1529–1532.

Ballif C., Barraud L., Battaglia C., Boccard M., Bugnon G., Charrière M., Cuony P., Despeisse M., Ding L., Escarré J., Haug F.-J., Hänni S., Löfgren L., Sculatti-Meillaud S., Nicolay S., Parascondolo G., De Wolf S., Söderström K. and Stückelberger M. (2011), 'Novel Materials and Superstrates for High-Efficiency Micromorph Solar Cells', *Proc. 26th EUPVSC*, Hamburg, pp. 2384–2391, 3CO.1.1.

Ballutaud J., Bucher C., Hollenstein Ch., Howling A. A., Kroll U., Benagli S., Shah A. and Buechel A. (2004), 'Reduction of the boron cross-contamination for plasma deposition of *p-i-n* devices in a single-chamber large area radio-frequency reactor', *Thin Solid Films* **468**, 222–225.

Banerjee A., Beglau D., Su T., Pietka G., Yue G., Yan B., Yang J. and Guha S. (2011a), '11.0% stable efficiency on large area, encapsulated a-Si:H and a-SiGe:H based multijunction solar cells using VHF technology', *Proc. MRS Spring Meeting*, San Francisco, CA, A14-02.

Banerjee A. and Guha S. (1991), 'Study of back reflectors for amorphous-silicon alloy solar-cell application', *J. Appl. Phys.* **69**, 1030–1035.

Banerjee A., Liu F., Beglau D., Su T., Pietka G., Yang J. and Guha S. (2011c), '12.0% efficiency on large-area, encapsulated, multijunction nc-Si:H triple cell', *Proc. 37th. IEEE PVSC*, Seattle, WA, pp. 1013–1017.

Banerjee A., Su T., Beglau D., Pietka G., Liu F., Yan B., Yang J. and Guha S. (2011b), 'High efficiency, large area, nanocrystalline silicon based triple junction solar cells', *Proc. MRS Spring Meeting*, San Francisco, CA, A01-03.

Bano G., Horvath P., Rozsa K. and Gallagher A. (2005), 'The role of higher silanes in silane-discharge particle growth', *J. Appl. Phys.* **98**, 013304.

Bartlomé R., Strahm B., Sinquin Y., Feltrin A. and Ballif C. (2010), 'Laser applications in thin-film photovoltaics', *Appl. Phys. B* **100**, 427–436.

Basore P.A. (2002), 'Pilot production of thin-film crystalline silicon on glass modules', *Proc. 29th IEEE PVSC*, New Orleans, LA, pp. 49–52.

Battaglia C., Escarre J., Söderström K., Erni L., Ding L., Bugnon G., Billet A., Boccard M., Barraud L., De Wolf S., Haug F.-J., Despeisse M. and Ballif C. (2011a), 'Nanoimprint lithography for high efficiency thin-film silicon solar cells', *Nano Letters* **11**, 661–665.

Battaglia C., Escarre J., Söderström K., Charrière M., Despeisse M., Haug F.-J. and Ballif C. (2011b), 'Nanomoulding of transparent zinc oxide electrodes for efficient light trapping in solar cells', *Nature Photonics*, DOI 10.10038.

Benagli S., Borrello D., Vallat-Sauvain E., Meier J., Kroll U., Hoetzel J., Bailat J., Steinhauser J., Marmelo M., Monteduro G. and Castens L. (2009), 'High efficiency amorphous silicon devices on LPCVD-ZnO TCO prepared in industrial KAI-M R&D reactor', *Proc. 24th. European PVSEC*, Hamburg, pp. 2293–2398, 3BO.9.3.

Bennett M. S., Newton J. L. and Rajan K. (1986), 'The influence of electric field on stability', *Proc. 7th. European PVSEC*, Sevilla, pp. 544–548.

Bennett M. S. and Rajan K. (1988), 'The stability of multi-junction a-Si solar cells', *Proc. 20th. IEEE PVSC*, Las Vegas, NV.

Bessonov A., Cho Y., Jung S.-J., Park E.-A., Hwang E.-S., Lee J.-W., Shin M. and Lee S. (2011), 'Nanoimprint patterning for tunable light trapping in large-area silicon solar cells', *Solar Energy Mater. Solar Cells* **95**, 2886–2892.

Beyer W. (2003), 'Diffusion and evolution of hydrogen in hydrogenated amorphous and microcrystalline silicon', *Solar Energy Mater. Solar Cells* **78**, 235–267.

Bielawny A., Upping J., Miclea P. T., Wehrspohn R. B., Rockstuhl C., Lederer F., Peters M., Steidl L., Zentel R., Lee S.-M., Knez M., Lambertz A. and Carius R. (2008), '3D photonic crystal intermediate reflector for micromorph thin-film tandem cell', *Physica Status Solidi A* **205**, 2796–2810.

Biron R., Hänni S., Boccard M., Pahud C., Söderström K., Duchamp M., Dunin-Borkowski R., Bugnon G., Ding L., Nicolay S., Parascandolo G., Meillaud F., Despeisse M., Haug F.-J. and Ballif C. (2013), 'New progress in the fabrication of *n-i-p* micromorph solar cells for opaque substrates', *Solar Energy Mater. Solar Cells* **114**, 147–155.

Boccard M., Battaglia C., Blondiaux N., Pugin R., Despeisse M. and Ballif C. (2013), 'Smoothening intermediate reflecting layer for tandem thin-film solar cells', *Solar Energy Mater. Solar Cells*, http://dx.doi.org/10.1016/j.solmat.2013.03.040.

Boccard M., Cuony P., Battaglia C., Despeisse M. and C. Ballif (2010), 'Unlinking absorption and haze in thin-film silicon solar cells', *Physica Status Solidi Rapid Research Letters* **4**, 326–328.

Boccard M., Cuony P., Battaglia C., Hänni S., Nicolay S., Ding L., Benkhaira M., Bugnon G., Billet A., Charrière M., Söderström K., Escarré J., Sculati-Meillaud F., Despeisse M. and Ballif C. (2012), 'Nanometer- and micrometer-scale texturing for high-efficiency micromorph thin-film silicon solar cells', *IEEE J. Photovoltaics* **2**, 83–87.

Boccard M., Despeisse M., Escarre J., Niquille X., Bugnon G., Hänni S., Bonnet-Eymard M., Meillaud F. and Ballif C. (2014), 'Low-refractive-index silicon-oxide interlayers for high-stable-efficiency multi-junction thin-film silicon solar cells', presented at the 40th IEEE PVSC, Denver.

Branz H. M. (1999), 'The hydrogen collision model: quantitative description of metastability in amorphous silicon', *Phys. Rev. B* **59**, 5498–5512.

Bravi M., Parisi M. L., Tiezzi E. and Basosi R. (2011), 'Life cycle assessment of a micromorph photovoltaic system', *Energy* **36**, 4297–4306.

Brodsky M., Cardona M. and Cuomo J. J. (1977b), 'Infrared and Raman spectra of silicon-hydrogen bonds in amorphous silicon prepared by glow discharge and sputtering', *Phys. Rev. B* **16**, 3556–3571.

Brodsky M., Frisch M. A. and Ziegler J. F. (1977a), 'Quantitative analysis of hydrogen in glow discharge amorphous silicon', *Appl. Phys. Lett.* **30**, 561–563.

Brodsky M. and Kaplan D. (1979), 'Hydrogenation and the density of defect states in amorphous silicon', *J. Non-Cryst. Solids* **32**, 431–435.

Brodsky M. and Title R. (1969), 'Electron spin resonance in amorphous silicon, germanium, and silicon carbide', *Physical Review Letters* **23**, 11, 581–585.

Buehlmann P., Bailat J., Domine D., Billet A., Meillaud F., Feltrin A. and Ballif C. (2007), 'In situ silicon oxide based intermediate reflector for thin-film silicon micromorph solar cells', *Appl. Phys. Lett.* **91**, 143505.

Bugnon G., Parascandolo G., Söderström T., Bartlome R., Cuony P., Hänni S., Boccard M., Holovsky J., Despeisse M., Meillaud F. and Ballif C. (2011), 'High rate deposition of microcrystalline silicon with silicon oxide doped layers: highlighting the competing roles of both intrinsic and extrinsic defects on the cells performances', to be published in *Proc. 37th IEEE PVSC*, Seattle, WA.

Campbell P. and Green M. A. (1987), 'Light trapping properties of pyramidally textured surfaces', *J. Appl. Phys.* **62**, 243–249.

Carlson D. E. and Wronski C. R. (1976), 'Amorphous silicon solar cells', *Appl. Phys. Lett.* **28**, 671–673.

Carniglia C. K. (1979), 'Scalar scattering theory for multilayer optical coatings', *Optical Engineering* **18**, 104–115.

Chen T., Huang Y. L., Yang D., Carius R. and Finger F. (2010), 'Microcrystalline silicon thin-film solar cells with microcrystalline silicon carbide window layers and silicon absorber layers both prepared by hot-wire CVD', *Physica Stat. Solidi, Rapid Research Lett.* **4**, 61–63.

Chen T., Huang Y., Yang D., Carius R. and Finger F. (2011), 'Development of microcrystalline silicon carbide window layers by hot-wire CVD and their applications in microcrystalline silicon thin-film solar cells', *Thin Solid Films* **519**, 4523–4526.

Chittick R., Alexander J. and Sterling H. (1969), 'The preparation and properties of amorphous silicon', *J. Electrochem. Soc.* **116**, 77–81.

Collins R., Ferlauto A., Ferreira G., Chen C., Koh J., Koval R., Lee Y., Pearce J. and Wronski C. (2003), 'Evolution of microstructure and phase in amorphous, protocrystalline, and microcrystalline silicon studied by real time spectroscopic ellipsometry', *Solar Energy Mater. Solar Cells* **78**, 143–180.

Cuony P., Alexander D., Perez-Wurfl Y., Despeisse M., Bugnon G., Boccard M., Söderström T., Hessler-Wyser H., Hébert C. and Ballif C. (2012), 'Silicon filaments in silicon oxide for next generation photovoltaics', *Adv. Mater.* **24**, 1182–1186.

Cuony P., Marending M., Alexander D. T. L., Boccard M., Bugnon G., Despeisse M. and Ballif C. (2010), 'Mixed-phase *p*-type silicon oxide containing silicon nanocrystals and its role in thin-film silicon solar cells', *Appl. Phys. Lett.* **97**, 213502-1–21502-3.

Curtins H., Favre M., Wyrsch N., Brechet M., Prasad K. and Shah A. (1987), 'High-rate deposition of hydrogenated amorphous silicon by the VHF-GD method', *Proc. 19th IEEE PVSC*, New Orleans, LA, pp. 695–698.

Das C., Lambertz A., Huepkes J., Reetz W. and Finger F. (2008), 'A constructive combination of antireflection and intermediate-reflector layers for a-Si/μc-Si thin-film solar cells', *Appl. Phys. Lett.* **92**, 053509-1–053509-3.

Deckman H. W., Wronski C. R., Witzke H. and Yablonovitch E. (1983), 'Optically enhanced amorphous-silicon solar-cells', *Appl. Phys. Lett.* **42**, 968–970.

Delli P., Veneri L. Mercaldo V. and Usatii L. (2010), 'Silicon oxide based n-doped layer for improved performance of thin-film silicon solar cells', *Appl. Phys. Lett.* **97**, 023512–023515.

Despeisse M., Battaglia C., Boccard M., Bugnon G., Charrière M., Cuony P., Hänni S., Löfgren L., Meillaud F., Parascandolo G., Söderström T. and Ballif C. (2011),

'Optimization of thin-film silicon solar cells on highly textured substrates', *Physica Stat. Solidi A* **208**, 1863–1868.

Despeisse M., Bugnon G., Feltrin A., Stueckelberger M., Cuony P., Meillaud F., Billet A. and Ballif C. (2010), 'Resistive interlayer for improved performance of thin-film silicon solar cells on highly textured substrate', *Appl. Phys. Lett.* **96**, 073507-1–073507-3.

DeVos A. (1980), 'Detailed balance limit of the efficiency of tandem solar cells', *J. Phys. D: Appl. Phys.* **13**, 839–846.

Ding L., Nicolay S., Bugnon G., Benkhaira M. and Ballif C. (2011), 'New generation transparent LPCVD ZnO electrodes for enhanced photocurrent in micromorph solar cells and modules', *IEEE J. Photovoltaics* **2**, 88–93.

Dingemans G., van den Donker M. N., Hrunski D., Gordijn A., Kessels W. M. M. and van de Sanden M. C. M. (2008), 'The atomic hydrogen flux to silicon growth flux ratio during microcrystalline silicon solar cell deposition', *Appl. Phys. Lett.* **93**, 111914-1–111914-3.

Dominé D., Haug F. J., Battaglia C. and Ballif C. (2010), 'Modeling of light scattering from micro- and nanotextured surfaces', *J. Appl. Phys.* **107**, 044504–8.

Elson J. and Bennett J. (1979), 'Relation between the angular dependence of scattering and the statistical properties of optical surfaces', *J. Opt. Sci. Amer.* **69**, 31–47.

Escarré J., Söderström K., Battaglia C., Bugnon G., Parascandolo G., Nicolay S., Despeisse M. and Ballif C. (2012), 'Geometric light trapping scheme to improve the efficiency of micromorph thin-film silicon solar cells', *Solar Energy Mater. Solar Cells* **98**, 185–190.

Faÿ S., Kroll U., Bucher C., Vallat-Sauvain E. and Shah A. (2005), 'Low pressure chemical vapour deposition of ZnO layers for thin-film solar cells: temperature-induced morphological changes', *Solar Energy Mater. Solar Cells* **86**, 385–397.

Faÿ S., Steinhauser J., Nicolay S. and Ballif C. (2010), 'Polycrystalline ZnO: B grown by LPCVD as TCO for thin-film silicon solar cells', *Thin Solid Films* **518**, 2961–2966.

Faÿ S., Steinhauser J., Oliveira N., Vallat-Sauvain E. and Ballif C. (2007), 'Opto-electronic properties of rough LP-CVD ZnO:B for use as TCO in thin-film silicon solar cells', *Thin Solid Films* **515**, 8558–8561.

Fischer D., Dubail S., Anna Selvan J. D., Pellaton Vaucher N., Platz R., Hof C., Kroll U., Meier J., Torres P., Keppner H., Wyrsch N., Goetz M., A. Shah and Ufert K.-D. (1996), 'The micromorph solar cell: extending a a-Si:H technology towards thin-film crystalline silicon', *Proc. 25th IEEE PVSC*, Washington D. C., pp. 1053–1056.

Fonrodona M., Escarré J., Villar F., Soler D., Asensi J., Bertomeu J. and Andreu J. (2005), 'PEN as substrate for new solar cell technologies', *Solar Energy Mater. Solar Cells* **89**, 37–47.

Franken R., Stolk R., Li H., van der Werf C., Rath J. and Schropp R. (2007), 'Understanding light trapping by light scattering textured back electrodes in thin-film *n-i-p*-type silicon solar cells', *J. Appl. Phys.* **102**, 014503-1–014503-7.

Fthenakis V., Gualtero S., van der Meulen R. and Kim H. C. (2008a), 'Comparative life cycle analysis of photovoltaics based on nano-materials: a proposed framework', *Proc. MRS Spring Meeting* **1041**, 25–32.

Fthenakis V., Kim H. C. and Alsema E. (2008b), 'Emissions from photovoltaic life cycles', *Environmental Sci. Technol.* **42**, 2168–2174.

Fujibayashi T., Matsui T and Kondo M. (2006), 'Improvement in quantum efficiency of thin-film Si solar cells due to the suppression of optical reflectance at transparent conductive oxide/Si interface by TiO_2/ZnO antireflection coating', *Appl. Phys. Lett.* **88**, 183508-1–183508-3.

Gee J. M. (2002), 'Optically enhanced absorption in thin silicon layers using photonic crystals', *Proc. 29th IEEE PVSC*, New Orleans, LA, pp. 150–153.

Green M. A., Basore P., Chang N., Clugston D., Egan R., Evans R., Hogg D., Jarnason S., Keevers M. and Lasswell P. (2004), 'Crystalline silicon on glass (CSG) thin-film solar cell modules', *Solar Energy* **77**, 857–863.

Guha S., Narasimhan K. and Pietruszko S. (1981), 'On light induced effect in amorphous hydrogenated silicon', *J. Appl. Phys.* **52**, 859–860.

Guha S., Yang J. and Banerjee A. (2000), 'Amorphous silicon alloy photovoltaic research — present and future', *Progr. Photovoltaics: Res. Applications* **8**, 141–150.

Guha S., Yang J., Pawlikiewicz A., Glatfelter T., Ross R. and Ovshinsky S. (1989), 'Band gap profiling for improving the efficiency of amorphous silicon alloy solar cells', *Appl. Phys. Lett.* **54**, 2330–2332.

Guha S., Yang J. and Yan B. (2011), 'Amorphous and nanocrystalline silicon solar cells and modules', in Bhattacharya, P., Fornari, R. and Kamimura, H. (eds) *Comprehensive Semiconductor Science and Technology*, Elsevier, London.

Haas S., Gordijn A. and Stiebig H. (2007), 'High speed laser processing for monolithical series connection of silicon thin-film modules', *Progr. Photovoltaics: Res. Applications* **16**, 195–203.

Hanak J. J. and Korsun V. (1982), 'Optical stability studies of a-Si: H solar cells', *Proc. 16th IEEE PVSC*, San Diego, CA, pp. 1381–1383.

Haneman D. (1968), 'Electron paramagnetic resonance from clean single-crystal cleavage surfaces of silicon', *Phys. Rev.* **170**, 705–718.

Hänni S., Battaglia C., Boccard M., Bugnon G., Cuony P., Despeisse M., Nicolay S., Meillaud F. and Ballif C. (2011), 'Towards better understanding of long-term stability in thin-film microcrystalline solar cells', *Proc. 26th PVSEC*, Hamburg, pp. 2699–2703, 3AV.2.42.

Haug F.-J., Söderström K., Naqavi A. and Ballif C. (2011), 'Resonances and absorption enhancement in thin-film silicon solar cells', *J. Appl. Phys.* **109**, 084516-1–084516-8.

Haug, F.-J. Söderström, T., Python, M., Terrazzoni-Daudrix, V., Niquille, X. and Ballif, C. (2009), 'Development of micromorph tandem solar cells on flexible low cost plastic substrates', *Solar Energy Mater. Solar Cells* **93**, 884–8877.

Higashi G., Chabal Y., Trucks G. and Raghavachari K. (1990), 'Ideal hydrogen termination of the Si (111) surface', *Appl. Phys. Lett.* **56**, 656–658.

Hirasaka M., Suzuki K., Nakatani K., Asano M., Yano M. and Okaniwa H. (1990), 'Design of textured Al electrode for a hydrogenated amorphous silicon solar cell', *Solar Energy Mater.* **20**, 99–110.

Hüpkes J., Owen J. I., Bunte E., Zhu H., Pust S. E., Worbs J. and Jost G. (2010), 'New texture etching of zinc oxide: tunable light trapping for silicon thin-film solar cells', *Proc. 5th World PVSEC*, Valencia, pp. 3224–3227, 3AV.2.25.

Inthisang S., Krajangsang T., Yunaz I. A., Yamada A., Konagai M. and Wronski C. (2011), 'Fabrication of high open circuit voltage a-$Si_{1-x}O_x$:H solar cells by using *p*-a-$Si_{1-x}O_x$:H as window layer', *Physica Stat. Solidi* (*c*), 2990–2993.

Jackson W. and Amer N. (1982), 'Direct measurement of gap-state absorption in hydrogenated amorphous silicon by photothermal deflection spectroscopy', *Phys. Rev. B* **25**, 5559–5562.

Jackson W., Kelso S., Tsai C., Allen J. and Oh S. (1985), 'Energy dependence of the optical matrix element in hydrogenated amorphous and crystalline silicon', *Phys. Rev. B* **31**, 5187–5198.

Jansen K. W. and Delahoy A. E. (2003), 'A laboratory technique for the evaluation of electrochemical transparent conductive oxide delamination from glass substrates' *Thin Solid Films* **423**, 153–160.

Kamei T., Stradins P. and Matsuda A. (1999), 'Effects of embedded crystallites in amorphous silicon on light-induced defect creation', *Appl. Phys. Lett.* **74**, 1707–1709.

Keppner H., Meier J., Torres P., Fischer D. and Shah A. (1999), 'Microcrystalline silicon and micromorph tandem solar cells', *Appl. Phys. A: Mater. Sci. Processing* **69**, 169–177.

Kim S., Chung J.-W., Lee H., Park J., Heo Y. and Lee H.-M. (2013), 'Remarkable progress in thin-film silicon solar cells using high-efficiency triple-junction technology', *Solar Energy Mater. Solar Cells* **119**, 26–35.

Klein S., Wieder S., Buschbaum S., Schwanitz K., Stolley T., Severin D., Obermeyer P., Kress M., Sommer E., Marschner T., Martini M., Noll-Baumann S., Haack J., Schmidt U. I., Straub A., Ahmed K. and Schuegraf K. (2010), 'Large area thin-film solar modules with 10% efficiency for mass production', *Proc. 5th. World PVSEC*, Valencia, pp. 2709–2713, 3B.11.1.

Kluth O., Fecociu-Morariu M., Losio P., Hötzel J., Klindworth M., Eisenhammer T. and Bailat J. (2011), 'The way to 11% stabilized efficiency based on 1.4 m^2 micromorph tandem', *Proc. 26th PVSEC*, Hamburg, pp. 2354–2358, 3BO.36.

Kluth O., Rech B., Houben L., Wieder S., Schöpe G., Beneking C., Wagner H., Löffl A. and Schock H. W. (1999), 'Texture-etched ZnO: Al coated glass substrates for silicon based thin-film solar cells', *Thin Solid Films* **351**, 247–253.

Kluth O., Schöpe G., Hüpkes J., Agashe C., Müller J. and Rech B. (2003), 'Modified Thornton model for magnetron sputtered zinc oxide: film structure and etching behaviour', *Thin Solid Films*, **442**, 80–85.

Koh J., Ferlauto A., Rovira P., Wronski C. and Collins R. (1999a), 'Evolutionary phase diagrams for plasma-enhanced chemical vapor deposition of silicon thin-films from hydrogen-diluted silane', *Appl. Phys. Lett.* **75**, 2286–2688.

Koh J., Fujiwara H., Koval R., Wronski C. and Collins R. (1999b), 'Real time spectroscopic ellipsometry studies of the nucleation and growth of *p*-type microcrystalline silicon films on amorphous silicon using B_2H_6, $B(CH_3)_3$ and BF_3 dopant source gases', *J. Appl. Phys.* **85**, 4141–4153.

Koida T., Sai H. and Kondo M. (2010), 'Application of hydrogen-doped In_2O_3 transparent conductive oxide to thin-film micro-crystalline silicon solar cells', *Thin Solid Films* **518**, 2930–2933.

Konagai M. (2011), 'Present status and future prospects of silicon thin-film solar cells', *Japan. J. Appl. Phys.* **50**, 030001-1–030001-12.

Koval R. J., Koh J., Lu Z., Jiao L., Collins R. W. and Wronski C. R. (1999), 'Performance and stability of Si:H *p-i-n* solar cells with *i*-layers prepared at the thickness-dependent amorphous-to-microcrystalline phase boundary', *Appl. Phys. Lett.* **75**, 1553–1555.

Kratzla T., Zindel A. and Benz R. (2010), 'Oerlikon Solar's key performance drivers to grid parity', *Proc. 5th. World PVSEC*, Valencia, pp. 2807–2810, 3CO.14.1.

Krč J., Smole F. and Topič M. (2003), 'Analysis of light scattering in amorphous Si:H solar cells by a one-dimensional semi-coherent optical model', *Progr. Photovoltaics* **11**, 15–26.

Kroll U., Meier J., Torres P., Pohl J. and Shah A. (1998), 'From amorphous to microcrystalline silicon films prepared by hydrogen dilution using the VHF (70 MHz) GD technique', *J. Non-Cryst. Solids* **227–230**, 68–72.

Kroll U., Meier J., Steinhauser J., Fesquet L., Benagli S., Orhan J.-B., Djeridane Y., Wolf B., Borrello D., Madliger P., Vallat-Sauvain E., Boucher J.-F., Castens L., Multone X., Marjanovic S., Konhke G., Borrelli N., Koch K., Liu J., Modavis R., Thelen D., Vallon S., Zakharian A. and Weidman D. (2011), 'Recent developments of high-efficiency micromorph tandem solar cells in KAI-M PECVD reactors', *Proc. 26th. PVSEC*, Hamburg, pp. 2340–2343, 3BO.2.6.

Kroll U., Shah A., Keppner H., Meier J., Torres P. and Fischer D. (1997), 'Potential of VHF plasmas for low-cost production of a-Si:H solar cells', *Solar Energy Mater. Solar Cells* **48**, 343–350.

Laaziri K., Kycia S., Roorda S., Chicoine M., Robertson J., Wang J. and Moss S. (1999), 'High-energy x-ray diffraction study of pure amorphous silicon', *Phys. Rev.* **B60**, 13520–13533.

Lambertz A., Dasgupta A., Reetz W., Gordijn A., Carius R. and Finger F. (2007), 'Microcrystalline silicon oxide as intermediate reflector for thin-film silicon', *Proc. 22nd EU-PVSEC*, Milan, pp. 1838–1842.

Lechner P., Frammelsberger W., Psyk W., Geyer R., Maurus H., Lundszien D., Wagner H. and Eichhorn B. (2008), 'Status of performance of thin-film siliocn solar cells and modules', *Proc. 23rd EU-PVSEC*, Valencia, pp. 2023–2026.

Lechner P., Geyer R., Haslauer A. and Roehrl T. (2010), 'Long-term performance of ASI tandem junction thin-film solar modules', *Proc. 5th. World PVSEC*, Valencia, pp. 3283–3287, 3AV.2.46.

Li H., Franken R. H., Stolk R. L., Rath J. K. and Schropp R. E. I. (2008), 'Mechanism of shunting of nanocrystalline silicon solar cells deposited on rough Ag/ZnO substrates', *Solid State Phenomena* **131/133**, 27–32.

Li Y.-M., Jackson F., Yang L., Fieselmann B. F. and Russel L. (1994), 'An exploratory survey of *p*-layers for a-Si:H solar cells', *Proc. MRS Spring Meeting*, San Francisco, CA, pp. 663–668.

Lucovsky G. (1992), 'Hydrogen in amorphous silicon: local bonding and vibrational properties', *J. Non-Cryst. Solids* **141**, 241–256.

Lundszien D., Finger F. and Wagner H. (2002), 'Band-gap profiling in amorphous silicon-germanium solar cells', *Appl. Phys. Lett.* **80**, 1655.

Mahan A. H., Williamson D. L., Nelson B. P. and Crandall R. S. (1989), 'Small angle X-ray scattering studies of microvoids in a-SiC:H and a-Si:H', *Solar Cells* **27**, 465–476.

Mahan A. H., Xu Y., Iwaniczko E., Williamson D. L., Nelson B. P. and Wang Q. (2002), 'Amorphous silicon films and solar cells deposited by HWCVD at ultra-high deposition rates', *J. Non-Cryst. Solids*, **299–302**, 2–8.

Mai Y., Klein S., Carius R., Stiebig H., Geng X. and Finger F. (2005), 'Open circuit voltage improvement of high-deposition-rate microcrystalline silicon solar cells by hot wire interface layers', *Appl. Phys. Lett.* **87**, 073503-1–073503-12.

Mashima H., Yamakoshi H., Kawamura K., Takeuchi Y., Noda M., Yonekura Y., Takatsuka H., Uchino S. and Kawai Y. (2006), 'Large area VHF plasma production using a ladder-shaped electrode', *Thin Solid Films* **506–507**, 512–516.

Matsuda A. (1983), 'Formation kinetics and control of microcrystallite in μc-Si: H from glow discharge plasma', *J. of Non-Crystalline Solids* **59**, 767–774.

Matsuda A. (1999), 'Growth mechanism of microcrystalline silicon obtained from reactive plasmas', *Thin Solid Films* **337**, 1–6.

Matsuda A. and Ganguly G. (1995), 'Improvement of hydrogenated amorphous silicon germanium alloys using low power disilane–germane discharges without hydrogen dilution', *Appl. Phys. Lett.* **67**, 1274–1276.

Matsuda A., Koyama M., Ikuchi N., Imanishi Y. and Tanaka K. (1986), 'Guiding principle in the preparation of high-photosensitive hydrogenated amorphous Si-Ge alloys from glow-discharge plasma', *Japan. J. Appl. Phys.* **25**, L54–L56.

Matsuda A., Yamasaki S., Nakagawa K., Okushi H., Tanaka K., Iizima S., Matsumura M. and Yamamoto H. (1980), 'Electrical and structural properties of phosphorous-doped glow-discharge Si:F and Si:H films', *Japanese J. Appl. Phys.* **19**, L305–L308.

Matsui T., Jia H. and Kondo M. (2010), 'Thin-film solar cells incorporating microcrystalline $Si_{1-x}Ge_x$ as efficient infrared absorber: an application to double junction tandem solar cell', *Progr. Photovoltaics* **18**, 48–53.

Matsui T., Kondo M., Nasuno Y., Sonobe H. and Shimizu S. (2006), 'Key issues for fabrication of high quality amorphous and microcrystalline solar cells', *Thin Solid Films* **501**, 243–246.

Matsui T., Sai H. and Kondo. (2013), 'Development of highly stable and efficient amorphous silicon based solar cells', *Proc. 28th European PV Conf.*, Paris, p. 3DO.7.2.

McCurdy R. J. (1999), 'Successful implementation methods of atmospheric CVD on a glass manufacturing line', *Thin Solid Films* **351**, 66–72.

Meier J., Dubail S., Flückinger R., Fischer D., Keppner H. and Shah A. (1994), 'Intrinsic microcrystalline silicon — a promising new thin-film solar cell material', *Proc. 1st World PVSEC*, Honolulu, HI, pp. 409–412.

Meier J., Spitznagel J., Kroll U., Bucher C., Faÿ S., Moriarty T. and Shah A. (2003), 'High efficiency amorphous and micromorph silicon solar cells', *Proc. 3rd World PVSEC*, Osaka, pp. 2801–2805.

Meier J., Torres P., Platz R., Dubail S., Kroll U., Selvan J. A., Pellaton Vaucher N., Hof C., Fischer D., Keppner H. and Shah A. (1996), 'On the way towards high-efficiency thin-film silicon solar cells by the 'micromorph' concept', *Proc. MRS Spring Meeting*, **420**, 3–14.

Nakamura G., Sato K., Yukimoto Y. and Shirahata K. (1980), 'Amorphous $Si_{1-x}Ge_x$ for high performance solar cell', *Proc. 3rd EU-PVSEC*, Cannes, pp. 835–839.

Nakano S., Matsuoka T., Kiyama S., Kawata H., Nakamura N., Nakashima Y., Tsuda S., Nishiwak H., Ohnishi M., Nagaoka I. and Kuwano Y. (1986), 'Laser patterning method for integrated type a-Si solar cell submodules', *Japan. J. Appl. Phys.* **25**, 1936–1943.

Nasuno Y., Kondo M. and Matsuda A. (2001a), 'Effects of substrate surface morphology on microcrystalline silicon solar cells', *Japan. J. Appl. Phys.* **40**, L303–L305.

Nasuno Y., Kondo M. and Matsuda A. (2001b), 'Passivation of oxygen-related donors in microcrystalline silicon by low temperature deposition', *Appl. Phys. Lett.* **78**, 230–232.

Naughton M. J., Kempa K., Ren Z. F., Gao Y., Rybczynski J., Argenti N., Gao W., Wang Y., Peng Y., Naughton J. R., Burns M. J., Shepard A., Clary M., Ballif C., Haug F.-J., Söderström T., Cubero O. and Eminian C. (2010), 'Efficient nano-coax based solar cells', *Physica Stat. Solidi Rapid Research Lett.* **4**, 181–183.

Nishimoto T., Takai M., Miyahara H., Kondo M. and Matsuda A. (2002), 'Amorphous silicon solar cells deposited at high growth rate', *J. Non-cryst. Solids* **299–302**, 1116–1122.

Pieters B., Krč J. and Zeman M. (2007), 'Advanced numerical simulation tool for solar cells-ASA5', *Proc. 4th World PVSEC*, Honolulu, HI, pp. 1513–1516.

Poruba A., Fejfar A., Remes Z., Springer J., Vanecek M., Kocka J., Meier J., Torres P. and Shah A. (2000), 'Optical absorption and light scattering in microcrystalline silicon thin-films and solar cells', *J. Appl. Phys.* **88**, 148–160.

Powell M. and Deane S. (1996), 'Defect-pool model and the hydrogen density of states in hydrogenated amorphous silicon', *Phys. Rev. B* **53**, 10121–10132.

Python M., Dominé D., Söderström T., Meillaud F. and Ballif C. (2010), 'Microcrystalline silicon solar cells: effect of substrate temperature on cracks and their role in post oxidation', *Progr. Photovoltaics* **18**, 491–499.

Python M., Vallat-Sauvain E., Bailat J., Dominé D., Fesquet L., Shah A. and Ballif C. (2008), 'Relation between substrate surface morphology and microcrystalline silicon solar cell performance', *J. Non-Cryst. Solids* **354**, 2258–2262.

Rech B., Repmann T., van den Donker M. N., Berginski M., Kilper T., Hupkes J., Calnan S., Stiebig H. and Wieder S. (2006), 'Challenges in microcrystalline silicon based solar cell technology', *Thin Solid Films* **511**, 548–555.

Redfield D. (1974), 'Multiple pass thin-film silicon solar cell', *Appl. Phys. Lett.* **25**, 647–648.

Redfield D. (1982), 'Energy-band tails and the optical absorption edge; the case of a-Si:H', *Solid State Commun.* **44**, 1347–1349.

Ringbeck R. and Stutterlueti J. (2012), 'BOS costs: status and optimization to reach industrial grid parity', *Proc. 27th EU-PVSEC*, Frankfurt, pp. 2961–2975, 4DP.2.2.

Rizal B., Ye F., Dhakal P., Chiles T.C., Shepard S., McMahon G., Burns M.J. and Naughton M.J. (2013), 'Imprint-templated nanocoax array architecture: Fabrication and utilization', *Nano-Optics for Enhancing Light-Matter Interactions on a Molecular Scale*, NATO Science for Peace and Security Series B: Physics and Biophysics, 359–370.

Roca i Cabarrocas P. (2000), 'Plasma enhanced chemical vapour deposition of amorphous, polymorphous and microcrystalline silicon films', *J. Non-Cryst. Solids*, **266–269**, 31–37.

Rockstuhl C., Fahr S., Bittkau K., Beckers T., Carius R., Haug F.-J., Söderström T., Ballif C. and Lederer F. (2011), 'Comparison and optimization of randomly textured surfaces in thin-film solar cells', *Optics Express* **18**, A335–A342.

Ruske F., Roczen M., Lee K., Wimmer M., Gall S., Hüpkes J., Hrunski D. and Rech B. (2010), 'Improved electrical transport in Al-doped zinc oxide by thermal treatment', *J. Appl. Phys.* **107**, 013708–013708-8.

Sai H., Kanamori Y. and Kondo M. (2011), 'Flattened light-scattering substrate in thin film silicon solar cells for improved infrared response', *Appl. Phys. Lett.* **98**, 113502-1–113502-3.

Sai H., Matsui T., Matsubara K., Kondo M. and Yoshida I. (2014), '11.0%-Efficient Thin Film Microcrystalline Silicon Solar Cells with Honeycomb Textured Substrates', presented at the 40th IEEE PVSC, Denver.

Saito K., Sano M., Okabe S., Sugiyama S. and Ogawa K. (2005), 'Microcrystalline silicon solar cells fabricated by VHF plasma CVD method', *Solar Energy Mater. Solar Cells* **86**, 565–575.

Sakai H., Yoshida T., Hama T. and Ichikawa Y. (1990), 'Effects of surface morphology of transparent electrode on the open-circuit voltage in a-Si: H solar cells', *Japan. J. Appl. Phys.* **29**, 630–635.

Sansonnens L., Schmidt H., Howling A. A., Hollenstein C., Ellert C. and Buechel A. (2006), 'Application of the shaped electrode technique to a large area rectangular capacitively coupled plasma reactor to suppress standing wave nonuniformity', *J. Vac. Sci. Technol.* A **24**, 1425–1430.

Sansonnens L. and Schmitt J. (2003), 'Shaped electrode and lens for a uniform radio-frequency capacitive plasma', *Appl. Phys. Lett.* **82**, 182–184.

Schicho S., Hrunski D., van Aubel R. and Gordijn A. (2010), 'High potential of thin ($<1\ \mu$m) a-Si:H/μc-Si:H tandem solar cells', *Progr. Photovoltaics* **18**, 83–89.

Schropp R. E. I., Carius R. and Beaucarne G. (2007), 'Amorphous silicon, microcrystalline silicon, and thin-film polycrystalline silicon solar cells', *Proc. Mater. Research Symposium* **32**, 219–224.

Selvan J. A., Delahoy A. E., Guo S. and Li Y.-L. (2006), 'A new light-trapping TCO for nc-Si:H solar cells', *Solar Energy Mater. Solar Cells* **90**, 3371–3376.

Shah A. (2010), 'Thin-film Silicon solar cells', *EPFL Press*, ISBN 1420066749.

Shah A., Schade H., Vanecek M., Meier J., Vallat-Sauvain E., Wyrsch N., Kroll U., Droz C. and Bailat J. (2004), 'Thin-film silicon solar cell technology, *Progr. Photovoltaics* **12**, 113–142.

Sheng P., Bloch A. N. and Stepleman R. S. (1983), 'Wavelength-selective absorption enhancement in thin-film solar cells', *Appl. Phys. Lett.* **43**, 579–581.

Shockley W. and Queisser H. J. (1961), 'Detailed balance limit of efficiency of *p-n* junction solar cells', *J. Appl. Phys.* **32**, 510–519.

Sichanugrist P., Yoshida T., Ichikawa Y. and Sakai H. (1993), 'Amorphous silicon oxide with microcrystalline phase', *J. Non-Cryst. Solids* **164–166**, 1081–1084.

Smets A. H. M., Kessels W. M. M. and van de Sanden M. C. M. (2007b), 'The effect of ion-surface and ion-bulk interactions during hydrogenated amorphous silicon deposition', *J. Appl. Phys.* **102**, 073523-1–073523-12.

Smets A. H. M., Matsui T. and Kondo M. (2008), 'High-rate deposition of microcrystalline silicon *p-i-n* solar cells in the high pressure depletion regime', *J. Appl. Phys.* **104**, 034508-1–034508-11

Smets A. H. M., and van de Sanden M. C. M. (2007a), 'Relation of the Si-H stretching frequency to the nanostructural Si-H bulk environment', *Phys. Rev. B* **76**, 073202-1–073202-4.

Smith Z. and Wagner S. (1985), 'Intrinsic dangling-bond density in hydrogenated amorphous silicon', *Phys. Rev. B* **32**, 5510–5513.

Söderström K. (2013), 'Coupling light into thin silicon layers for high-efficiency solar cells', *PhD Thesis, EPFL No. 5714.*

Söderström K., Bugnon G., Biron R., Pahud C., Meillaud F., Haug F.-J. and Ballif C. (2012b), 'Thin-film silicon triple-junction solar cell with 12.5% stable efficiency on innovative flat light-scattering substrate', *J. Appl. Phys.* **112**, 114503–114503-4.

Söderström K., Bugnon G., Haug F.-J. and Ballif C. (2012a), 'Electrically flat/optically rough substrates for efficiency above 10% in *n-i-p* thin-film silicon solar cells', *Proc. MRS Spring Meeting*, San Francisco, CA, pp. 39–44.

Söderström K., Escarré J., Cubero O., Haug F.-J., Perregaux S. and Ballif C. (2010), 'UV nano imprint lithography technique for the replication of back reflectors for *n-i-p* thin-film silicon solar cells', *Progr. Photovoltaics* **19**, 202–210.

Söderström T., Haug F. J., Niquille X., Terrazoni-Daudrix V. and Ballif C. (2009), 'Asymmetric intermediate reflector for tandem micromorph thin-film silicon solar cells', *Appl. Phys. Lett.* **94**, 063501–063501–3.

Soppe W. J., Muffler H. J., Biebericher H. C., Devilee C., Burgers A. R., Poruba A., Hodakova L. and Vanecek M. (2005), 'Optical and structural properties of microcrystalline silicon grown by microwave PECVD', *Proc. 20th EU-PVSEC*, Barcelona, pp. 1604–1607, 3DV.3.21.

Spear W. and Le Comber P. (1976), 'Electronic properties of substitutionally doped amorphous Si and Ge', *Phil. Mag.* **33**, 935–949.

Sriraman S., Agarwal S., Aydil E. S. and Maroudas D. (2002), 'Mechanism of hydrogen induced crystallization of amorphous silicon', *Nature* **418**, 62–65.

Staebler D. L. and Wronski C. R. (1977), 'Reversible conductivity changes in discharge-produced amorphous Si', *Appl. Phys. Lett.* **31**, 292–294.

Staebler D. L. and Wronski C. R. (1981), 'Stability of *n-i-p* amorphous silicon solar cells', *Appl. Phys. Lett.* **39**, 733–735.

Steimen I. (2011), Private communication based on Oerlikon Solar process and LCA analyses performed with EMPA CH.

Stradins P. (2003), 'Light-induced degradation in a-Si: H and its relation to defect creation', *Solar Energy Mater. Solar Cells* **78**, 349–367.

Strahm B. and Hollenstein C. (2010), 'Powder formation in SiH_4-H_2 discharge in large-area capacitively coupled reactors: a study of the combined effect of interelectrode distance and pressure', *J. Appl. Phys.* **107**, 023302-1–023301-7.

Strahm B., Howling A., Sansonnens L. and Hollenstein C. (2007), 'Plasma silane concentration as a determining factor for the transition from amorphous to microcrystalline silicon in SiH_4/H_2 discharges', *Plasma Sources Sci. Technol.* **16**, 80–89.

Street R. and Winer K. (1989), 'Defect equilibria in undoped a-Si: H', *Phys. Rev. B* **40**, 6236–6249.

Strobel C., Zimmermann T., Albert M., Bartha J. W. and Kuske J. (2009), 'Productivity potential of an in-line deposition system for amorphous and microcrystalline silicon solar cells', *Solar Energy Mater. Solar Cells* **93**, 1598–1607.

Stutzmann M., Jackson W. B. and Tsai C. C. (1985), 'Light-induced metastable defects in hydrogenated amorphous silicon: a systematic study', *Phys. Rev. B* **32**, 23–47.

Tabuchi T., Takashiri M. and Mizukami H. (2003), 'Hollow electrode enhanced RF glow plasma for the fast deposition of microcrystalline silicon', *Surface Coatings Technol.* **173**, 243–248.

Takagi T., Ueda M., Ito N., Watabe Y. and Kondo M. (2006), 'Microcrystalline silicon solar cells fabricated using array-antenna-type very high frequency plasma-enhanced chemical vapor deposition system', *Japan. J. Appl. Phys.*, **45**, 4003–4005.

Takano A. and Kamoshita T. (2004), 'Light-weight and large-area solar cell production technology', *Japan. J. Appl. Phys.* **34**, 7976–7983.

Takatsuka H., Takeuchi Y., Yamauchi Y., Shioya T. and Kawai Y. (2005), 'Production of large area VHF plasma using ladder-shaped electrode', *Surface Coatings Technol.* **2005**, 972–975.

Takeuchi Y., Nawata Y., Ogawa K., Serizawa A., Yamauchi Y. and Murata M. (2001), 'Preparation of large uniform amorphous silicon films by VHF-PECVD using a ladder-shaped antenna', *Thin Solid Films* **386**, 133–136.

Taneda N., Oyama T. and Sato K. (2007), 'Light scattering effects of highly textured transparent conductive oxide films', *Techical Digest 17th PVSEC*, Fukuoka, pp. 309–312, 5O-B6-01.

Tawada Y., Okamoto H. and Hamakawa Y. (1981), 'a-SiC:H/a-Si:H heterojunction solar cell having more than 7.1% conversion efficiency', *Appl. Phys. Lett.* **39**, 237–239.

TEL (2014), TEL Solar press release, 9 July 2014.

Terakawa A., Hishida M., Aya Y., Shinohara W., Kitahara A., Yoneda H., Iseki M. and Tanaka M. (2011), 'R&D on thin-film silicon solar cells in Sanyo', *Proc. 26th PVSEC*, Hamburg, pp. 2362–2365, 3BO.4.2.

Terasa R., Albert M., Grüger H., Haiduk A. and Kottwitz A. (2000), 'Investigation of growth mechanisms of microcrystalline silicon in the very high frequency range', *J. Non-cryst. Solids* **266–269**, 95–99.

Tiedje T., Cebulka J., Morel D. and Abeles B. (1981), 'Evidence for exponential band tails in amorphous silicon hydride', *Phys. Rev. Lett.* **46**, 1425–1428.

Torres P., Meier J., Flückiger R., Kroll U., Selvan J., Keppner H., Shah A., Littelwood S., Kelly I. and Giannoules P. (1996), 'Device grade microcrystalline silicon owing to reduced oxygen contamination', *Appl. Phys. Lett.* **69**, 1373–1375.

Ulbrich C., Zahren C., Noll J., Lambertz A., Gerber A. and Rau U. (2011), 'Analysis of the short-circuit current gains by anti-reflective texture coating on silicon thin-film solar cells', *Proc. 26th PVSEC*, Hamburg, pp. 302–305, 1CV.3.29.

Usui S. and Kikuchi M. (1979), 'Properties of heavily doped GD-Si with low resistivity', *J. Non-Cryst. Solids* **34**, 1–11.

van den Donker M. N., Gordijn A., Stiebig H., Finger F., Rech B., Stannowski B., Bartlb R., Halmers E. A. G., Schlatmann R. and Jongerden G. J. (2007), 'Flexible amorphous and microcrystalline silicon tandem solar modules in the temporary superstrate concept', *Solar Energy Mater. Solar Cells* **91**, 572–580.

van Gestel D., Chahal M., van der Wilt P. C., Qiu Y., Gordon I., Im J. S. and Portmans J. (2010), 'Thin-film polycrystalline silicon solar cells with low intragrain defect density made via laser crystallization and epitaxial growth', *Proc. 35th IEEE PVSC*, Honolulu, HI, pp. 279–282.

Vanecek M., Neykova N., Babchenko O., Purkrt A., Poruba A., Remes Z., Holovsky J., Hruska K., Meier J. and Kroll U. (2010), 'New 3-dimensional nanostructured thin-film silicon solar cells', *Proc. 5th World PVSEC*, Valencia, pp. 2763–2766, 3CO.12.2.

Vepřek S. and Marečeka V. (1968), 'The preparation of thin layers of Ge and Si by chemical hydrogen plasma transport', *Solid-State Electronics* **11**, 683–684.

Vetterl O., Finger F., Carius R., Hapke P., Houben L., Kluth O., Lambertz A., Mück A., Rech B. and Wagner H. (2000), 'Intrinsic microcrystalline silicon: a new material for photovoltaics', *Solar Energy Mater. Solar Cells* **62**, 97–108.

Vogler B., Kerschbaumer J., Kuhn H., Mark A., Poppeller M., Schneider S., Zimin D. and Zindel A. (2008), 'TCO 1200°C Oerlikon production tool for transparent conductive oxide thin-films', *Proc. 23rd EU-PVSEC*, Valencia, pp. 2492–2493.

Wadia C., Alivisatos A. P. and Kammen D. M. (2009), 'Materials availability expands the opportunity for large-dcale photovoltaics deployment', *Environmental Sci. Technol.* **43**, 2072–2077.

Wang C. and Lucovsky G. (1990), 'Intrinsic microcrystalline silicon deposited by remote PE-CVD: a new thin-film photovoltaic material', *Proc. 21st IEEE PVSC*, Orlando, FL, pp. 1614–1618.

Winer K. (1990), 'Defect formation in a-Si:H', *Phys. Rev. B* **41**, 12150–12161.

Wittmaack K. (2003), 'Analytical description of the sputtering yields of silicon bombarded with normally incident ions', *Phys. Rev. B* **68**, 235211–235221.

Wronski C. (1996), 'Amorphous silicon technology: coming of age', *Solar Energy Mater. Solar Cells* **41**, 427–439.

Xu X., Zhang J., Beglau D., Su T., Ehlert S., Li Y., Pietka G., Worrel C., Lord K., Yue G., Yan B., Banerjee A., Yang J. and Guha S. (2010), 'High efficiency large-area a-SiGe:H and nc-Si:H based multi-junction solar cells: a comparative study', *Proc. 5th World PVSEC*, Valencia, pp. 2783–2787, 3CO.13.1.

Yablonovitch E., Allara D., Chang C., Gmitter T. and Bright T. (1986), 'Unusually low surface-recombination velocity on silicon and germanium surfaces', *Phys. Rev. Lett.* **57**, 249–252.

Yablonovitch E. and Cody G. D. (1982), 'Intensity enhancement in textured optical sheets for solar cells', *IEEE Trans. Electron Devices*, **29**, 300–305.

Yamamoto K., Nakajima A., Yoshimi M., Sawada T., Fukuda S., Suezaki T., Ichikawa M., Goto M., Meguro T., Matsuda T., Sasaki T. and Tawada Y. (2006), 'High efficiency thin-film silicon hybrid cell and module with newly developed innovative interlayer', *Proc. 4th World PVSEC*, Honolulu, HI, pp. 1489–1492.

Yamamoto K., Nakajima A., Yoshimi M., Sawada T., Fukuda S., Suezaki T., Ichikawa M., Koi Y., Goto M., Meguro T., Matsuda T., Kondo M., Sasaki T. and Tawada Y. (2004), 'A high efficiency thin-film silicon solar cell and module', *Solar Energy* **77**, 939–949.

Yamamoto K., Toshimi M., Suzuki T., Tawada Y., Okamoto T. and Nakajima A. (1998), 'Thin-film poly-Si solar cell on glass substrate fabricated at low temperature', *Proc. MRS Spring Meeting*, San Francisco, CA, pp. 131–138.

Yamamoto K., Yoshimi M., Tawada Y., Fukuda S., Sawada T., Meguro T., Takata H., Suezaki T., Koi Y. and Hayashi K. (2002), 'Large area thin-film Si module', *Solar Energy Mater. Solar Cells* **74**, 449–455.

Yan B., Yang J. and Guha S. (2003), 'Effect of hydrogen dilution on the open-circuit voltage of hydrogenated amorphous silicon solar cells', *Appl. Phys. Lett.* **83**, 782–784.

Yan B., Yue G. and Guha S. (2007), 'Status of nc-Si:H solar cells at United Solar and roadmap for manufacturing a-Si:H and nc-Si:H based solar panels', *Proc. MRS Spring Meeting*, San Francisco, CA, A15–01.

Yan B., Yue G., Sivec L., Yang J., Guha S. and Jiang C.-S. (2011), 'Innovative dual function nc-SiO_x:H layer leading to a $>16\%$ efficient multi-junction thin-film silicon solar cell', *Appl. Phys. Lett.* **99**, 113512-1–113512-3.

Yang J., Banerjee A., Glatfelter T., Hoffman K., Xu X. and Guha S. (1994), 'Progress in triple-junction amorphous silicon based alloy solar cells and modules using hydrogen dilution', *Proc. 1st World PVSEC*, Honolulu, HI, pp. 380–385.

Yang J., Banerjee A. and Guha S. (1997), 'Triple-junction amorphous silicon alloy solar cell with 14.6% initial and 13.0% stable conversion efficiencies', *Appl. Phys. Lett.* **70**, 2975–2977.

Yang L. and Chen L. (1993), '"Fast" and "slow" metastable defects in hydrogenated amorphous silicon', *Appl. Phys. Lett.* **63**, 400–402.

Yu Z., Raman A. and Fan S. (2010), 'Fundamental limit of light trapping in grating structures', *Optics Express* **18**, A366–A380.

Yue G., Sivec L., Yan B., Yang J. and Guha S. (2010), 'High efficiency hydrogenated nanocrystalline silicon based solar cells deposited by optimized Ag/ZnO back reflectors', *Proc. 25th EU-PVSEC*, Valencia, pp. 3196–3200, 3AV.2.17.

Yunaz I. A., Hashizume K., Miyajima S., Yamada A. and Konagai M. (2009), 'Fabrication of amorphous silicon carbide films using VHF-PECVD for triple-junction thin-film solar cell applications', *Solar Energy Mater. Solar Cells* **93**, 1056–1061.

Zeman M., van Swaaij R. A. C. M. M., Metselaar J. W. and Schropp R. E. I. (2000), 'Optical modeling of a-Si:H solar cells with rough interfaces: effect of back contact and interface roughness', *J. Appl. Phys.* **88**, 6436–6443.

Zhu H., Bunte E., Hüpkes J., Siekmann H. and Huang S. M. (2009), 'Aluminium doped zinc oxide sputtered from rotatable dual magnetrons for thin-film silicon solar cells', *Thin Solid Films* **517**, 3161–3166.

Zhu H., Kalkan A., Hou J. and Fonash S. (1998), 'Applications of AMPS-1D for solar cell simulation', *Proc. AIP Conf.* **462**, pp. 309–314.

Zimmermann T., Strobel C., Albert M., Beyer W., Gordijn A., Flikweert A. J., Kuske J. and Bartha J. W. (2010), 'Inline deposition of microcrystalline silicon solar cells using a linear plasma source', *Physica Stat. Solidi C* **7**, 1097–1100.

CHAPTER 5

POLYCRYSTALLINE CADMIUM TELLURIDE PHOTOVOLTAIC DEVICES

TIMOTHY A. GESSERT
National Renewable Energy Laboratory
Golden, Colorado, USA 8040
tim.gessert@nrel.gov

and

DIETER BONNET
Private Consultant, formerly Antec Solar

5.1 Introduction

Thin-film photovoltaic (PV) devices based on CdTe absorbers represent one of the fastest growing segments of all PV technologies (Mehata, 2010; Coggeshall and Margolis, 2011). It is even more remarkable that most of this impact has occurred within the past few years. Much of the reason for this rapid development can be traced to two factors: (i) Thin-film PV has been designed specifically to embody production advantages over historic PV products (i.e. wafer-based technologies); (ii) CdTe PV presently enjoys production advantages over other thin-film technologies.

In this chapter, we review the development and present state of CdTe PV technology to describe the advantages that have made this technology one of two dominant forces in the present PV market. Although much is understood regarding the physics and material science of the CdTe thin-film device, we will see that producing PV modules that are cost-effective in the present market(s) remains a complex technological and financial undertaking. Furthermore, although current CdTe commercial technology demonstrates the lowest specific production cost per watt of any PV technology, we will also suggest that additional improvements and further market penetration remain likely.

Finally, we will argue that as the economic viability of CdTe PV products improves further, the impacts of less-obvious considerations will emerge. These include developing reliability extending beyond the warranty period of typically 25 years, the likelihood of large-scale mineral resource availability, principally of Te, and how CdTe PV sustainability compares with other base-load energy sources in terms of energy payback time and toxicity.

5.2 Brief history of CdTe PV devices

One of the first uses of CdTe was for gamma and X-ray detectors, and it was through research related to these applications that much of the early material understanding was established (Zanio, 1978). Additionally, because the photoelectric absorption coefficient of CdTe is similar to that of skin tissue, several uses related to nuclear medicine have been investigated (Meyer *et al.,* 1972; Garcia *et al.*, 1974).

The first notable investigations of CdTe as a solar absorber were published by Vodakov *et al.* (1960) and Naumov and Nikolaeva (1961). Although few details are provided in these papers, they reported devices with efficiencies up to ~6% which were fabricated by diffusing semitransparent surface coatings into crystalline *n*-CdTe.

More detailed work was reported by Cusano (1963), who was seeking to understand the formation and operation of (what would eventually be called) the Cu_xS/CdS cell. In this work, Cusano substituted *n*-CdTe for the more typically used *n*-CdS to form $Cu_{2-x}Te$/CdTe devices. Both thin-film and crystalline *n*-CdTe layers were used, and the process involved dipping the CdTe into a solution containing Cu(I) ions.

These studies provided considerable insight into how the *p*-type $Cu_{2-x}Te$ layer is formed during cell processing, and indicated that the $Cu_{2-x}Te$/CdTe device performs more like a heterojunction than a CdTe homojunction. This study also revealed several problems related to CdTe PV devices that remain today. Noteworthy among these were a close relationship between effects occurring at the CdTe contact electrodes and junction characteristics, and the observation that the best II–VI devices almost always involved the use of Cu.

In the same year, Nicoll (1963) became one of the first to report on the use of close-space vapour transport (CSVT) for the deposition of CdTe. During the next few years, measurements of minority-carrier lifetimes indicated that the maximum lifetime of holes in *n*-CdTe is only about 10 ns, with even shorter electron lifetimes observed in *p*-CdTe (Cusano and Lorenz, 1964; Cusano, 1967; Artobolevskaya *et al.*, 1968). Subsequent modelling suggested that efficiencies > 10% were unlikely for CdTe homojunction devices without drift assistance unless the minority-carrier lifetimes were at least 100 ns (Bell *et al.,* 1975).

This benchmark lifetime was an order of magnitude longer than that observed for the lowest-doped, monocrystalline CdTe. Because the homojunction emitter would have to be heavily doped to reduce series resistance effects, most research efforts during the last 40 years have been directed toward development of CdTe heterojunction devices. Recently, however, reports have emerged suggesting that monocrystalline homojunction CdTe-based devices may demonstrate a significant opportunity for multijunction concentrator PV devices (Carmody *et al.*, 2010).

In the CdTe heterojunction configuration, the choice for an appropriate heteroface material is limited by several considerations. For example, because of the near-optimum bandgap of CdTe, the heterojunction should be designed so that most of the absorption occurs within the CdTe bulk. Thus, the heteroface partner must act as a highly transparent, low-resistance window layer and not be responsible for carrier generation. Because all known wide-gap, low-resistivity window materials are n-type, the CdTe layer must be p-type. The choice of window layer is further constrained in that it is better for it to have a small lattice mismatch with CdTe to avoid excessive interface recombination. Finally, to promote long-term stability, the window material should be composed of elements that are slow to diffuse into CdTe, preferably with the cation from Group II (e.g. Zn, Cd, Hg) to optimise conduction-band alignment.

Although these considerations provide a guide for the choice of the wide-bandgap heteroface window, a wide range of materials have been investigated, including CdS (Bonnet and Rabenhorst, 1972; Yamaguchi *et al.*, 1977; Bube *et al.*, 1977), ZnO (Aranovich *et al.*, 1977) and ZnSe (Bube *et al.*, 1975). It is noteworthy that the 5–6%-efficient CdS/CdTe devices fabricated by Bonnet in 1972 were not only the first CdS/CdTe devices reported but were also the first all-thin-film CdS/CdTe devices to be made, in that gas-phase transport was used to deposit thin films of both layers.

Some of the other window materials may eventually demonstrate advantages over CdS. However, the most efficient cells produced to date have the CdS/CdTe configuration (Bube, 1988; Chu, 1988; Wu, 2004; Green, 2013). Additionally, because the CdS/CdTe configuration represents a 3-element system (rather than 4-element, or more), it may produce fewer potential native defects, and ultimately present fewer unforeseen industrial (production) problems arising from interdiffusion.

5.3 Initial attempts towards commercial modules

With the success of the CdS/CdTe all-thin-film device reported by Bonnet and Rabenhorst in 1972, many entrepreneurs who had previously worked to develop

Table 5.1 Historical and present commercial companies involved in CdTe PV.

Approximate time period	Name	Location	CdTe absorber deposition
1975–1984 1984–1992 1992–1997	Photon Power Inc. Photon Energy Inc. Golden Photon Inc.	El Paso, USA Golden, USA	Chemical spraying
1977–c.1997	Matsushita Corp.	Japan	Screen-printed sublimation
1980–1984 1984–1999 1999–2002	Monosolar Inc. BP Solar BP/Solarex	Santa Monica, USA London, UK Baltimore, USA	Electrodeposition
1980–1989	AMETEK	Harleysville, USA	Electrodeposition
1994–2002 2002–present	ANTEC GmbH Antec Solar Energy	Arnstadt, Germany	Vacuum sublimation
1990–1999 1999–present	Solar Cells Inc. First Solar	Toledo, USA	Vacuum sublimation
2003–2007 2007–present	Solar Fields Calyxo USA	Toledo, USA	Non-vacuum sublimation
2006 2006–2008 2008–present	Ziax Corp. Primestar Solar Primestar GE Solar	Arvada, USA	Vacuum sublimation
2006–present	Solexant Corp.	San José, USA	Non-vacuum printing
2007–2009 2009–2012	AVA Solar Abound Solar	Fort Collins, USA	Vacuum sublimation
2008–present	Xunlight 26 Solar	Toledo, USA	Sputtering
2008–present	SunPrint/Alion Inc	Richmond, USA	Non-vacuum printing
2008–present	Advanced Solar Power	Hangzhou, China	Vacuum sublimation
2007–present	WK Solar Group	Toledo, USA	Vacuum sublimation
2008–c.2011	Clean Cell Int.	Sunnyvale, USA	Vacuum sublimation
2009–present	REEL Solar	Menlo Park, USA	Electrodeposition
2010–present	Encore Solar	Fremont, USA	Electrodeposition

In some cases, several entries are listed, indicating name transitions and/or involvements by different corporate entities. This table has been assembled from various sources, including reports, websites and personal communications, and is intended to be representative rather than exhaustive.

Cu_xS/CdS PV products began to consider if CdTe might demonstrate advantages compared with Cu_xS-based devices. Table 5.1 highlights early attempts at CdTe commercialisation.

As Table 5.1 shows, several significant commercial endeavours were initiated in the ten years between 1975 and 1985 to develop thin-film CdTe PV products. Many of the early attempts were based on a belief that chemical and/or atmospheric pressure processes were a more attractive commercialisation avenue compared with the vacuum-sublimation processes used to produce the laboratory devices of the early 1970s. Although the only companies presently in commercial (or pre-commercial) production are using vacuum sublimation, several companies are now re-exploring the potential advantages of non-vacuum CdTe deposition processes. Braun and Skinner (2007) provide a more detailed discussion of the successes and difficulties that can be encountered during CdTe PV module commercialisation.

5.4 Review of present commercial industry/device designs

Nearly all of the companies listed in Table 5.1 have been involved in developing or producing thin-film CdTe devices configured in the *superstrate* design which is shown in Fig. 5.1. In this design, light enters through a transparent glass superstrate

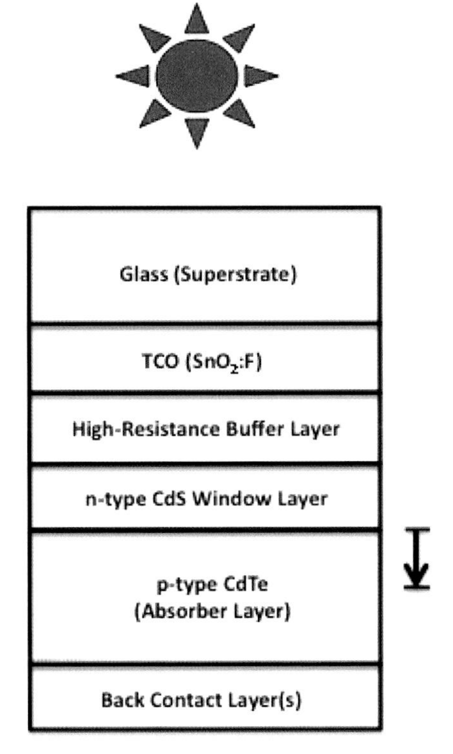

Figure 5.1 Schematic diagram showing main functional components of a typical thin-film CdTe PV solar cell device configured as a superstrate structure. The arrow on the right indicates the direction and approximate location of the junction electric field.

and passes through one or more transparent conducting oxide (TCO) layers and a CdS window layer. It is then absorbed in the CdTe absorber layer.

Because the glass superstrate provides the mechanical surface onto which all subsequent layers are deposited, the sequence of layer deposition is TCO, then CdS, then CdTe, and then back contact. Although the superstrate design is currently the most widely used, and the present world record thin-film CdTe device is a superstrate structure (Rose *et al.*, 1999; Wu, 2004; Green *et al.*, 2013; Gloeckler *et al.*, 2013), the *substrate* design continues to attract both research and commercial interest. Here the device is typically grown on a non-transparent substrate that is often either a Mo or Mo-coated stainless steel foil. This results in a deposition sequence opposite to the superstrate device (i.e. back contact, then CdTe, then CdS, then TCO, then encapsulation). Although the currently reported maximum performance of substrate devices is significantly lower than that of high-performance superstrate devices (~13% efficiency vs. ~20% efficiency: Romeo *et al.*, 1992; Singh *et al.*, 1999; Dhere*et al.*, 2012; Duenow *et al.*, 2012; Gretener *et al.*, 2012; Kranz *et al.*, 2013; Green *et al.*, 2013), the substrate configuration has advantages, in particular in avoiding the use of glass, which could be exploited in certain commercial products. These include applications requiring high specific power (i.e. watts/kg), the ability to produce flexible products for aerospace and/or building-integrated PV markets, and/or envisioned use of roll-to-roll processing during module manufacture.

Figure 5.2 shows a scanning electron microscopy (SEM) cross-section of a superstrate CdTe device produced at NREL. This particular device was produced on barium silicate glass (Corning 7059), and had an SnO_2:F TCO layer and SnO_2 buffer layer produced by MOCVD, a CdS layer produced by chemical bath deposition (CBD), a CdTe layer produced by close-space sublimation (CSS), a dry $CdCl_2$ activation process, a chemical pre-contact etch step and a graphite-paste contact.

5.5 General CdTe material properties

One reason many technologists believe CdTe PV devices embody industrial advantages is that the phase diagram of this material is very simple. As Fig. 5.3 shows, single-phase CdTe can exist only very near the 50/50% Cd/Te ratio. If the composition deviates from this stoichiometric ratio during crystal growth, the CdTe material will phase-separate into CdTe and Cd or Te — whichever is present in the greater abundance. Because the vapour pressure of both Cd and Te are much higher than that of CdTe, any elemental regions will tend to re-evaporate before they can be incorporated into the growing crystal. However, if a sufficient Cd or Te overpressure exists, and/or if the rate of crystal growth exceeds the elemental evaporation rates, elemental inclusions can be incorporated. If the temperature is

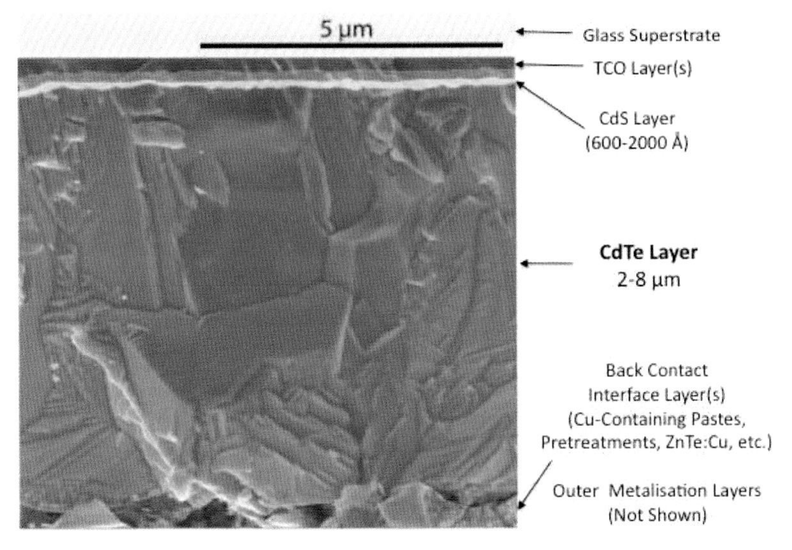

Figure 5.2 SEM micrograph showing cross-section of CdS/CdTe superstrate device produced at NREL. The layers have been colour-enhanced for clarity.

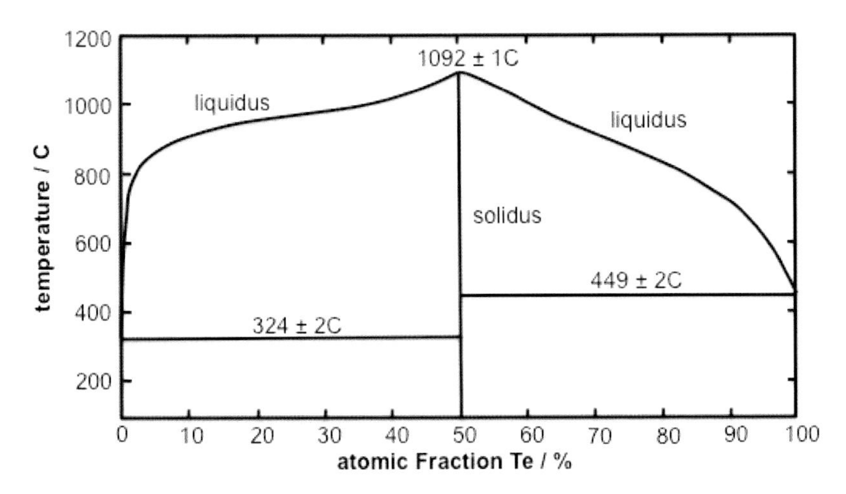

Figure 5.3 Phase diagram of CdTe. Source: Zanio (1978).

sufficiently high (>324°C for Cd and >449°C for Te), liquid elemental regions can reside on the surface of the crystal. The CdTe itself will melt, as shown by the outside of the overall curve, and the melting point is a function of off-stoichiometry. The highest melting point at ~1092°C is for stoichiometric CdTe.

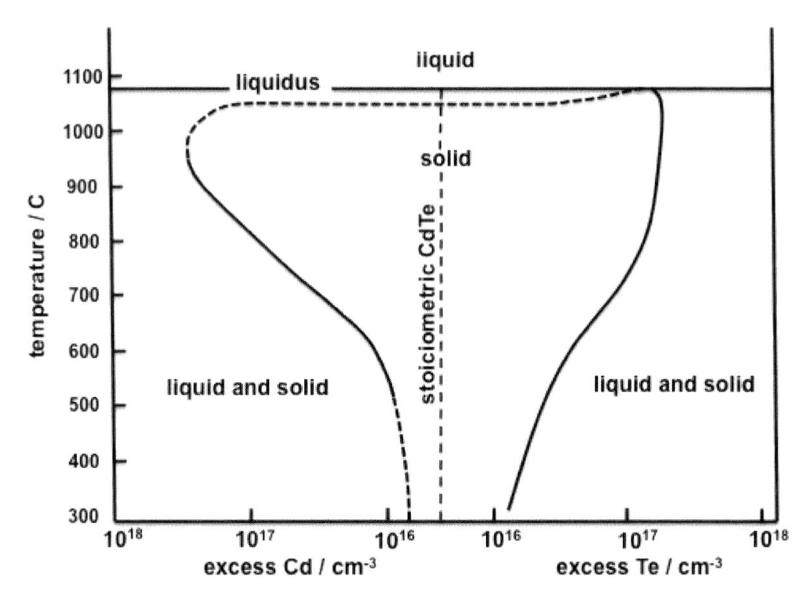

Figure 5.4 Temperature vs. composition diagram for CdTe showing extent of electrically active centres for both Cd- and Te- rich concentration. Source: Zanio (1978).

Although single-phase CdTe exists only very near the 50/50% composition, the material can sustain a Cd or Te deficiency to a small extent. As shown in Fig. 5.4, this off-stoichiometry can extend to a maximum value of about $5 \times 10^{17}\,\text{cm}^{-3}$ before the Te or Cd deficiency will spontaneously produce separate Cd or Te phases. Recalling that most materials have a density of atoms on the order of $1 \times 10^{22}\,\text{cm}^{-3}$, this means that the maximum off-stoichiometry that can be sustained in crystalline CdTe at thermodynamic equilibrium is only about 0.01%. Analysis of off-stoichiometry also suggests that, for equilibrium conditions, temperatures greater than 600°C can yield higher excess Te or Cd than temperatures below 500°C (Greenberg, 1996). Depending on the concentration of other defects, the excess Te or Cd can manifest in polycrystalline films as a material with net acceptors or net donors. Therefore, if a high degree of stoichiometry control can be maintained at high growth temperatures, it may be possible to have these defects assist with various types of junction functionality. However, we shall see that the influence of the $CdCl_2$ treatment and Cu incorporation during contacting also has a significant impact on electrical properties.

In addition to a relatively simple diagram space, another important reason CdTe possesses production advantage is that it demonstrates a property known as congruent sublimation and condensation. When a source of bulk CdTe is heated to a temperature at which the CdTe surface sublimes into Cd and Te_2 gas, the Cd and Te atoms evaporate at roughly the same rate (i.e. congruent sublimation). Although the vapour pressure of Cd is higher than that of Te_2 at any given temperature, Te_2 has two atoms, so the arrival rates of Te and Cd atoms at a surface are very similar. Furthermore, when Cd and Te_2 vapours condense onto a (cooler) surface, they again do so at about the same rate (i.e. congruent condensation). These two congruent processes enable a bulk source of stoichiometric CdTe to evaporate and condense into a film of nearly stoichiometric CdTe. If for some reason Cd and Te arrive at a growing film surface at different rates, then because the vapour pressure of Cd (or Te_2) above a pure Cd (or Te) surface is much higher than that of either above a CdTe surface, any depositing Cd or Te surface atoms that do not quickly coordinate into the CdTe material will re-evaporate from the surface, limiting the formation of Cd or Te secondary phases.

The previous discussion pertains to high-purity, crystalline CdTe formed under equilibrium conditions. However, for thin-film PV devices, the CdTe will be formed using less pure materials and ambients, and relatively non-equilibrium conditions. These differences can result in polycrystalline thin films that can contain numerous structural and impurity defects that can affect the material properties of the film. Luckily, because the kinetics of film growth for CdTe are fairly rapid, even high-rate film growth processes tend to retain much of the structural regularity that is observed for bulk crystalline materials.

Although CdTe is ambipolar (i.e. it can be doped both n-type and p-type), the material tends to self-compensate. This means if one attempts to incorporate acceptors by substituting Group I dopants on the cation [Cd] site, or Group V dopants on the anion [Te] site, a dopant concentration will be reached where the lattice will spontaneously create compensating donors, with the result that the further change in net acceptor concentration is minimal (Zunger, 2003).

Production of epitaxial CdTe films using molecular beam epitaxy (MBE) techniques can achieve n- and p-type doping into the mid-$10^{18}\,cm^{-3}$ range using extrinsic dopants (Taterenko *et al.*, 1993; Dhese *et al.*, 1994). For polycrystalline PV devices, the CdTe is typically deposited without extrinsic dopants added. The as-deposited CdTe is often believed to be p-type as a result of cadmium vacancies (V_{Cd}); however, the electrical resistance of as-deposited CdTe is very high, so the electrical properties of as-deposited polycrystalline thin films are not well established. Most device technologists know only that a working junction results if the CdTe is deposited onto an n-type heteroface layer, such as CdS, and a back contact is fabricated onto the CdTe. The conclusion is that the CdTe must have been, or have become, p-type. As we will see in the following discussion, for some CdTe deposition processes, it is possible that the as-deposited CdTe may be p-type, intrinsic, or even *n*-type, and it is the post-deposition processes ($CdCl_2$ processing and/or contacting with a Cu(I)-containing layer) that establish the necessary *p*-type properties.

5.6 Layer-specific process description for superstrate CdTe devices

5.6.1 *Superstrate materials*

CdTe thin-film PV devices are typically grown on a glass superstrate. At this time, nearly all published research on CdTe PV has been performed using barium silicate glass (e.g. Corning 7059), borosilicate glass or soda-lime float glass. There are presently no definitive studies that compare material and/or performance differences resulting from these different glass choices. This is partly because different glass materials have different strain points, and thus can allow higher-temperature processes, and partly because some transparent conducting oxides (TCOs) are only available on certain types of glass. The following discussion notes some of the parameters that should be considered when choosing a glass superstrate for a CdTe PV device.

Temperature stability

CdTe technologists suggest that higher substrate temperatures minimise the defect density in the CdTe film. Although firm evidence of this has not been reported, it is true that the devices with higher performance tend to have had the CdTe deposited at higher temperature. In the case of commonly used soda-lime float glass substrates, it is essential to keep the processing temperatures below 550–575°C to avoid mechanical distortion due to the low strain and softening points, and to avoid altering the degree of temper of the as-fabricated glass. For some industrial processes, the temperature may be much lower to avoid even small distortions in the glass and/or to limit loss of glass temper. For higher processing temperatures, borosilicate (and historically, barium silicate/Corning 7059) glass substrates have been used for laboratory devices to allow studies of material functionality at higher temperature. Because of the size of the potential market for thin-film CdTe PV, many commercial glass companies are considering the use of other types of glass compositions that may demonstrate higher strain points than soda-lime glass, or may embody other commercial advantages.

Coefficient of thermal expansion

A difference in the coefficient of thermal expansion (CTE) between the substrate and the film can induce stress in the deposited layers following deposition and cooling. This gives rise to structural defects and can also affect the adhesion of the film(s). Although this is another area where studies have not been conducted and/or reported, it remains prudent to choose a glass with a CTE that closely matches that of the material(s) being deposited.

Glass composition and impurity diffusion

Because glass constituents can diffuse out from the glass at elevated processing temperatures and/or during long-term field deployment, the effect on the device layers of mobile impurities from the glass must be considered. For alkali glasses, impurities (e.g. sodium, potassium, calcium etc.) can diffuse into the films during processing. If impurity diffusion is found to impact underlying device layers adversely, additional diffusion-barrier layers and/or glass compositions that mitigate these issues may be required. It may also be possible to produce barrier layers that incorporate designs which reduce optical reflection, especially at non-normal incidence. Alternatively, and as in CIGS PV device technology, improved understanding of the CdTe material may indicate that certain types of alkali diffusion could be beneficial.

5.6.2 *TCO layer*

Following deposition of any alkali diffusion barrier layer(s), the next layer deposited onto the glass is the TCO layer. Because this provides a lateral conduction pathway for electrons from the *n*-type side of the device, this must be highly conductive. However, because incident light must pass through the TCO layer for subsequent absorption in the junction, it must also be highly transparent. Presently, most CdTe modules incorporate a TCO layer that has a sheet resistivity of 10–15 ohms/square, and optical transparency of $\geq$90% within the spectral region ~600–700 nm.

For many production CdTe technologies, fluorine-doped SnO_2 (SnO_2:F) has been the historical choice for the TCO layer. SnO_2:F-coated glass has been widely available for many years because of its use in low-emissivity (low-e) glazing for residential and appliance applications (such as commercial/residential windows, and heat-reflecting glass in oven windows). SnO_2:F is chemically inert, is not greatly affected by environmental moisture or diffusion from other materials in the PV device, and may provide a degree of protection from Na diffusion from soda-lime glass. Finally, SnO_2:F demonstrates very high stability in its electrical and optical properties at CdTe processing temperatures (>600°C).

Detriments of SnO_2:F include that it has lower visible transmission than either ITO (indium tin oxide) or ZnO:Al, and this limits current generation. Although a complete explanation for this difference has not been reported, it partly arises because the maximum electron mobility of present commercial-grade SnO_2:F is lower than in other TCOs (in the range of 20–25 $cm^2\ V^{-1}\ sec^{-1}$). This means that to achieve both low sheet resistance and high optical transmission, higher carrier concentration must be used in the film ($\sim 7 \times 10^{20}\ cm^{-3}$). This higher carrier concentration will produce more free-carrier absorption, limiting the transmitted near infrared (NIR) light entering the PV device, and thus the short-circuit current.

Many research groups are presently investigating avenues to improve the performance of SnO_2:F, and/or develop alternative TCOs for CdTe devices (Dhere *et al.*, 2010). For example, a recent high-performance thin-film CdTe device uses a Cd_2SnO_4 TCO layer (Wu *et al.*, 1996; Coutts *et al.*, 2000; Wu, 2004). An alternative avenue to improve the transmission of TCOs for PV applications involves forming TCO alloys with higher dielectric permittivity (Gessert *et al.*, 2011). Because the dielectric permittivity affects the plasma wavelength, this technique can yield a TCO with much less NIR absorption.

5.6.3 *Buffer layer*

In most high-performance CdTe (and CIGS) PV devices, a high-resistance oxide layer is placed between the TCO and the CdS layers. This layer is often called a *buffer layer* (BL) but also has been called a high-resistance transparent (HRT) layer. Many attributes have been ascribed to this layer, and it is not presently clear if all yield benefits in all CdTe device processes, or if certain BL attributes assist only certain types of device designs and/or fabrication processes. Suggested benefits include:

- The BL allows thinner CdS layers to be used. One proposed reason for this is that the space charge on the *n*-type side of the junction can expand into the BL when the CdS becomes too thin to balance the charge needed to maintain the space-charge width in the *p*-type side of the junction. This enables the space charge in the *p*-CdTe region to remain sufficiently wide as the CdS layer is thinned, thereby avoiding device functionality dominated by voltage-dependent collection (i.e. low fill factor arising from loss of minority-carrier collection at high forward bias). A different explanation is that the BL reduces surface recombination at the CdS/TCO interface. The suggested benefit is that reduced recombination at this interface may improve device performance if thinner CdS layers position this interface closer to the CdS/CdTe junction region.
- The BL provides tolerance to shunt (or short) paths from the back contact to the front contact. These pathways will be present if the CdTe layer contains

cracks or pinholes that may form during CdTe deposition, or during pre-contact treatments such as chemical etching. These shunt paths are particularly problematic if vacuum-deposited metals are used in the contact (as opposed to mechanically applied paste contacts).

- Certain types of BL can allow higher temperatures or longer times to be used during the $CdCl_2$ treatment. The issue here is that during the $CdCl_2$ process (see Section 5.6.7), delamination can occur between the CdS and TCO layers if the $CdCl_2$ process temperature is excessively high. Indeed, the optimum $CdCl_2$ processing conditions are typically defined as the maximum time and temperature that can be tolerated before delamination. It has been suggested this benefit may be more likely for a chemically active BL, such as Zn_2SnO_4 or related alloys that can form mixed alloys with the CdS.
- A chemically active buffer layer, such as Zn_2SnO_4 or related alloys, may consume some of the CdS layer, reducing the amount of (optically absorbing) CdS, and producing in its stead a wider bandgap CdZnS material (Wu, 2004).

The particular choices of glass superstrate, TCO, buffer layer and CdS will significantly affect the amount of light that can enter the junction region of a CdTe thin-film PV device, and thus will impact the device short-circuit current. This is illustrated in Fig. 5.5, which compares the quantum efficiency consistent with the 16.7% CdTe device produced at NREL ($i_{sc} =\sim 26\,\text{mA cm}^{-2}$, dotted quantum efficiency curve) with a historic commercial device ($i_{sc} =\sim 19\,\text{mA cm}^{-2}$, solid curve) (Wu *et al.*, 2002; Green *et al.*, 2012).

The comparison reveals significant loss differences in both the NIR region (600–850 nm) and the UV region (300–600 nm). About one half of the $\sim 3\,\text{mA cm}^{-2}$ NIR loss is believed to be due to absorption in the (high-Fe) commercial glass and the other half is due to the free-carrier absorption of commercial TCO. For the NREL device shown, barium silicate glass (Corning 7059) and a Cd_2SnO_4 TCO layer were used to minimise both these NIR losses. It should also be noted that the commercial TCO modelled for this figure (using Drude Theory approximations) might actually suggest less NIR absorption than typical commercial SnO_2:F. The difference in UV loss between the commercial and NREL devices shown in this figure ($\sim 4\,\text{mA cm}^{-2}$) is primarily due to the use of relatively thick layers (typically $\sim$150–300 nm) of CdS in the commercial device. In the NREL device, a Zn_2SnO_4-alloy BL was used to allow a CdS layer that is typically thinned to $\sim$70 nm thick. From this discussion, it is clear that additional performance benefits remain possible in commercial modules by optimisation of the CdS thickness. However, this will require improved understanding of the function(s) and optimisation of specific BLs.

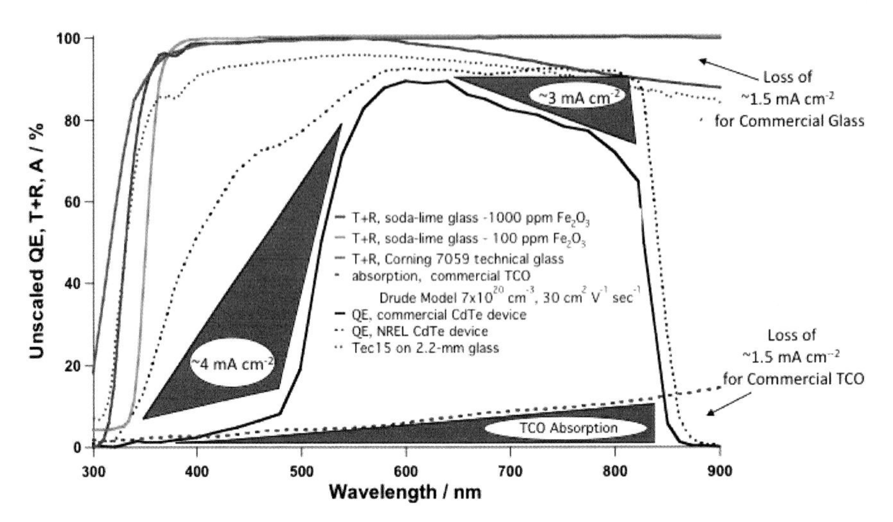

Figure 5.5 Comparison of quantum efficiency curves from the 16.7% efficiency CdS/CdTe device at NREL ($i_{sc} = \sim 26\,\text{mA cm}^{-2}$) and a representative commercial CdS/CdTe device ($i_{sc} = \sim 19\,\text{mA cm}^{-2}$). The red triangles indicate the approximate amount of loss in each region. The figure also shows several different types of glass superstrate and a modelled absorption curve for a typical commercial TCO.

5.6.4 *CdS layer*

Although the CdS layer is formally the *n*-type heteroface partner of the CdS/CdTe device, for most device processes the resulting electrical junction is cited between the CdTe absorber layer and an intermixed CdSTe layer that forms during high-temperature CdTe deposition. Because of this intermixing, the device is not generally considered a heterojunction — but instead a quasi-homojunction (Dhere *et al.*, 2008). Although materials other than CdS have been tested for the heteroface partner layer, CdS has been found to yield the highest device performance. Moreover, and unlike the situation in $CuInSe_2$-alloy devices, because a much thicker Cd-containing layer still exists in the device (the CdTe layer!), there has been little motivation to eliminate the Cd from the heteroface layer based solely on perceived toxicity concerns.

Many different CdS deposition processes can be used to make the CdTe device. The previous NREL world-record cell uses chemical-bath deposition (CBD) (Rose *et al.*, 1999). However, sputter deposition, evaporation and various forms of gas-phase transport (close-space sublimation, vapour-transport deposition etc.) are also used by research and industry. When processes other than CBD are used, CdTe technologists have found that adding oxygen to the CdS deposition process appears to produce beneficial diffusion during the subsequent CdTe deposition process. In the case of sputter deposition of CdS, oxygen addition can also lead to the formation of a nanocrystalline CdS phase (Wu *et al.*, 2002). It is believed that quantum-confinement effects yield a wider bandgap for nanocrystalline CdS, allowing increased device short-circuit current from the CdS:O/CdTe device.

5.6.5 *CdTe layer*

A primary production advantage of CdTe, relative to other known thin-film absorber layers, is that it can be deposited very quickly (at rates $>20\,\mu$m/min). This means that the length of the CdTe deposition zone (and related deposition hardware costs) can be much smaller compared with other thin-film PV technologies that require slower deposition rates.

A variety of deposition techniques have been used to deposit CdTe including: various forms of vacuum sublimation (e.g. evaporation, close-spaced sublimation, vapour transport deposition) (Chu *et al.*, 1992; Sasala *et al.*, 1996; Rose *et al.*, 1999; McCandless *et al.*, 1999), sputtering (Li *et al.*, 1993; Wendt *et al.*, 1998), electrodeposition (Fulop *et al.*, 1982; Basol, 1984), screen printing (Suyama *et al.*, 1990), spray techniques (Jordan, 1993; Kester *et al.*, 1996) and printing of nanoparticles (Gur *et al.*, 2005). Following proper post-deposition processing, nearly all these techniques have yielded devices with efficiency > 10%. The first method used to deposit CdTe films for high-efficiency solar cell applications used vacuum sublimation in which a crystal of CdTe was heated in a partial vacuum to its sublimation point, and the vapour was transported by a gas and re-condensed on a substrate (Bonnet and Rabenhorst, 1972). Figure 5.6 illustrates two laboratory-scale methods that use higher-rate vacuum sublimation for deposition of CdTe (and possibly CdS). Powell *et al.* (1999) provided a description of one high-rate commercial vacuum sublimation process. Because of the dominant importance of various methods of high-rate vacuum sublimation in today's commercial CdTe industry, the follow discussion will focus on parameters relevant to this deposition process only.

One of the primary parameters that affects CdTe material properties is the deposition temperature. Material properties can be categorised broadly into films deposited at low temperature (<~425°C), intermediate temperature (~450–575°C) and high temperature (~600–625°C). Because of void formation and substrate limitations at temperatures greater than ~625°C, this substrate temperature represents the approximate upper limit for present high-rate vacuum sublimation processes. Although devices have been produced on as-deposited CdTe and heat-treated films (without $CdCl_2$ treatment), it is worth noting that the only device to exceed 10% efficiency without $CdCl_2$ was produced from CdTe layers deposited in the high-temperature range (Tyan and Perez-Albuerne, 1982).

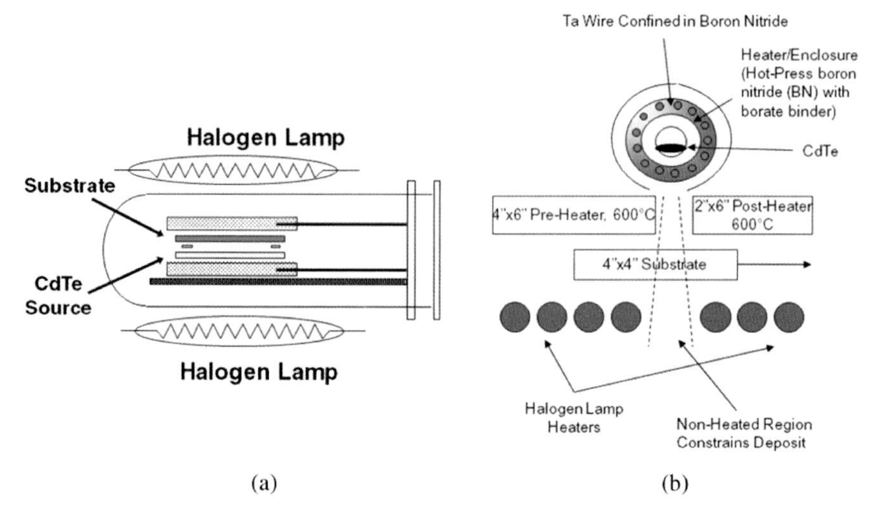

Figure 5.6 (a) Cross-section schematic of research-scale CSS system; (b) cross-section of research-scale gas-phase transport system.

Low-temperature growth yields films with maximum grain diameters $<\sim 0.5\,\mu$m, and which are highly oriented and exhibit significant inbuilt stress (Moutinho *et al.*, 1999; McCandless *et al.*, 1999). Generally, these films must undergo higher-temperature heat treatment (or more typically, the $CdCl_2$ treatment as described in Section 5.6.7) to yield high device performance. In contrast, intermediate- and high-temperature growth yields films with maximum grain diameters $>\sim 0.5\,\mu$m that demonstrate progressively less orientation and lower inbuilt stress as the substrate temperature increases (Moutinho *et al.*, 1999; McCandless *et al.*, 1999). Presently, because all commercial modules are produced using soda-lime glass that has an effective upper temperature limit in the range of ~550–575°C (owing to production issues related to glass softening), most commercial CdTe films are produced in the intermediate temperature range.

Closely linked to substrate temperature, the deposition ambient exerts a significant effect on material properties. Although sublimation occurs primarily in a relatively non-reactive environment (usually He, Ar, or N_2), deposition in a small partial pressure of O_2 (either intentional or unintentional) will lead to higher device performance (Tyan *et al.*, 1984) and nucleation of smaller grains (Rose, 1997). As growth proceeds, these smaller grains yield a denser film that provides resistance to shunting between the front and back contacts, and may reduce recombination at deep defect states (e.g. O_{Te} is thermodynamically favoured over V_{Te}) (Corwine *et al.*, 2005). Following nucleation, grain size increases with film thickness such that the maximum grain size tends to be equal to the film thickness. However, the initially small grains have significant implications for superstrate devices because the grain size nearest to the junction region (where light absorption occurs) is generally small, and therefore issues arising from interface and/or grain-boundary recombination will be more pronounced.

Reported electrical properties of as-deposited polycrystalline CdTe films are based primarily on resistivity and hot-probe studies (Zanio, 1978; McCandless, 2003; Dhere and Li, 2012). These suggest that low deposition temperatures produce n-type films with resistivity at or above $\sim 10^8$ ohm cm. Intermediate-temperature deposition yields as-deposited resistivity in the range of 10^7–10^8 ohm cm, and the films tend to be intrinsic but can demonstrate either n- or p-type characteristics. Research devices that have been produced at high temperatures suggest as-deposited films are slightly p-type (Dhere *et al.*, 2008). The transition of conductivity type from n to p with increasing temperature may be linked to the formation of more cadmium vacancy defects. The $CdCl_2$ treatment reduces the resistivity by 3–4 orders of magnitude and can assist conversion of intrinsic or n-type films to lightly p-type (this is discussed more fully in Section 5.6.7). Although more detailed information on the carrier concentration and type of as-deposited films would be valuable, results of CV and Hall analyses of as-deposited and $CdCl_2$-treated CdTe films are generally inconclusive because the low carrier concentration produces surface and interface depletion regions that are equal to or wider than typical film thicknesses. Therefore, nearly all of the electrical data reported for polycrystalline CdTe are for films that have been treated after deposition, typically by $CdCl_2$ and Cu contacting.

As we note below, the assumed interplay between different process steps is one of the things that make CdTe a difficult material to process reproducibly. However, we will also note that improved understanding of any of these process steps may not only lead to their elimination, but also elimination of related process issues.

5.6.6 *CdSTe layer*

CdTe substrate temperatures higher than $\sim$425°C yield progressively more interdiffusion between the CdTe and the underlying CdS layer, forming what is often called the CdSTe alloy region (or CdSTe layer) (Moutinho *et al.*, 1999, McCandless *et al.*, 1999). For devices formed at high temperature, the CdSTe-alloy region forms primarily during CdTe growth, and is altered further during the $CdCl_2$ process (discussed below). If a low-temperature CdTe deposition process is used, the CdSTe-alloy region is typically established entirely during the $CdCl_2$ treatment (Moutinho *et al.*, 1999). Although most high-performance CdTe devices contain

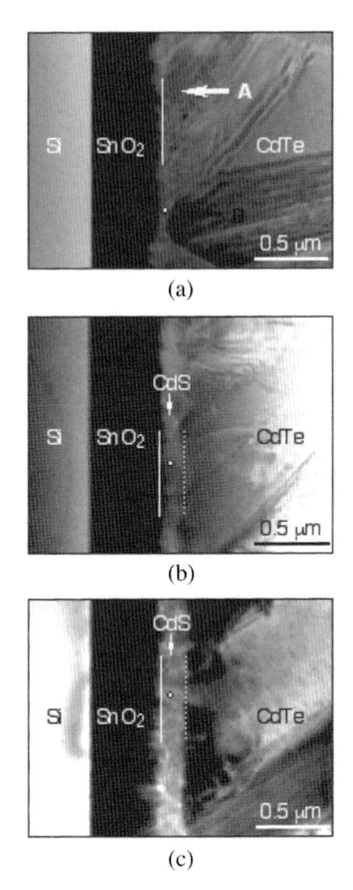

Figure 5.7 SEM cross section of CdS/CdTe device illustrating the effect of oxygen in the CdS on CdS consumption during CdTe deposition: (a) CSS-deposited CdS containing no oxygen. Note significant consumption of CdS; (b) CSS-deposited CdS that has been post-deposition treated to incorporate oxygen. Note much less consumption of CdS; (c) CBD-deposited CdS. Note very little consumption of CdS. Source: Albin *et al.* (2002).

a CdSTe-alloy region, its function and preferred attributes remain areas of active debate.

In addition to the CdTe deposition temperature and $CdCl_2$ treatment (McCandless *et al.*, 1999), another important parameter affecting the CdSTe-alloy region is the amount of oxygen available within this region (Albin *et al.*, 2002). Figure 5.7 shows that the amount of oxygen in the CdS layer alters the interdiffusion between the CdS and CdTe layers during high-temperature CdTe deposition. In general, adding oxygen to the CdS layer reduces the amount of CdS consumed during high-temperature CdTe deposition. The reduced CdS consumption not only limits the extent of the CdSTe alloy that forms on the CdTe side of the device but also appears to limit the amount of Te that diffuses into the CdS layer.

It has been suggested that the resulting moderation of the extent of the CdSTe thickness by oxygen limits the depth of the CdSTe/CdTe interface, which limits the depth of the quasi-homojunction (Dhere *et al.*, 1996; Dhere *et al.*, 2008). One possible explanation for these effects of oxygen is that oxygen acts as an isoelectronic substitutional defect for Te and/or S vacancies. The formation of these defects could limit diffusion, and possibly reduce interface recombination. However, at this time, the effects of this diffusion on junction formation and ultimate cell performance are not well understood.

5.6.7 *$CdCl_2$ activation*

Since about 1985, nearly all research and commercial thin-film CdTe PV devices have incorporated a $CdCl_2$ treatment or activation step. The incorporation of this step is one of the main reasons why so many different groups, using different CdTe deposition processes and temperatures, have been able to achieve conversion efficiencies greater than 10%. Many call the $CdCl_2$ process an equaliser step because it results in CdTe polycrystalline thin-film material that demonstrates minority-carrier lifetimes that are similar and sufficiently high to enable good device performance.

For PVD CdTe (e.g. GPT, evaporated, sputtered, etc.), the $CdCl_2$ process usually involves spraying a $CdCl_2$-saturated methanol or water solution onto the CdTe surface after deposition, or dipping the CdTe into a $CdCl_2$ solution, followed by drying and annealing in an oxygen-containing ambient at ~400°C for ~10 min. The $CdCl_2$ treatment can also be performed by combining the application and thermal steps into a $CdCl_2$ vapour process at ~400°C. For solution-based CdTe deposition processes, compounds that contain chlorine can be included in the CdTe precursors with results similar to a post-deposition $CdCl_2$ treatment.

For small-grained CdTe ($\sim<0.5\,\mu$m diameter) produced from solution or by low-temperature CdTe growth, the $CdCl_2$ process typically leads to complete recrystallisation that produces larger polycrystalline grains (Mountiho *et al.*, 1999; McCandless *et al.*, 1999). Although for larger-grained (higher-temperature CdTe growth) material, the $CdCl_2$ process does not appear to produce significant recrystallisation, both small- and large-grained material demonstrate higher minority-carrier lifetime after $CdCl_2$ treatment. For large-grained material, the improvement is due primarily to passivation at grain boundaries (Gessert *et al.*, 2001).

The particular defect modifications associated with the $CdCl_2$ treatment continue to be debated (Corwine *et al.*, 2005). In general, adding either excess Cd or Cl to CdTe results in the formation of donor defects (Zanio, 1978). However, many technologists associate improved *p*-type CdTe material quality following the $CdCl_2$ treatment with the formation of what has become known as the Cl A-centre, which is believed to be an acceptor level located ~150 meV above the valence band, and ascribed to a Cl-on-Te vacancy defect (Cl_{Te}) paired with a Cd vacancy defect (V_{Cd}-Cl_{Te}) (Zanio, 1978; Halliday *et al.*, 2001; Valdna *et al.*, 2001; Wei and Zhang, 2002).

One artefact of the $CdCl_2$ process that is often not discussed relates to residuals left behind on the CdTe surface. It has been shown that, if the $CdCl_2$ process is wet (application of a saturated methanol or aqueous solution), residuals are either cadmium oxychlorides or oxytellurides (Gessert *et al.*, 2001), as shown in Fig. 5.8. These residuals are often removed during the pre-contact chemical etching step(s), and are therefore not a major concern for wet-contact processing. If, however, a dry contacting process is to be used, sufficient removal of $CdCl_2$ residuals requires the use of a dry removal process (e.g. ion-beam or ion-etch processes) (Gessert *et al.*, 2001). If a vapour process is used for the $CdCl_2$ treatment, the resulting residual is $CdCl_2$, and this can be removed using a thermal pre-contact processing (Waters *et al.*, 1988).

In addition to the effect of $CdCl_2$ treatment on the electrical properties of the CdTe layer, it also alters CdSTe interdiffusion. Figures 5.9a and 5.9b compare secondary ion mass spectrometry (SIMS) of two CdS/CdTe devices that were deposited at two different CSS substrate temperatures. Both figures also show the sulphur (S) profile before and after a $CdCl_2$ process. By comparing these profiles, one notes that the extent of S diffusion from the CdS layer during the $CdCl_2$ treatment is much greater for the device shown in Fig. 5.9a where the CdTe was

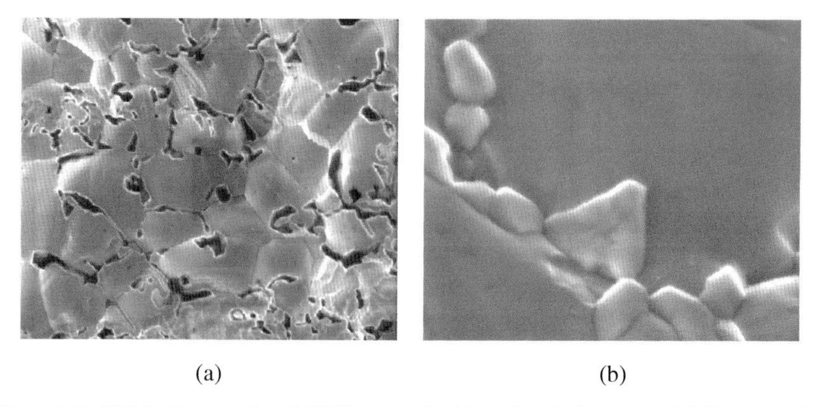

(a) (b)

Figure 5.8 SEM micrographs of $CdCl_2$ oxychloride residuals from wet $CdCl_2$ processing treatment: (a) low magnification; (b) higher magnification. Source: Gessert *et al.* (2001).

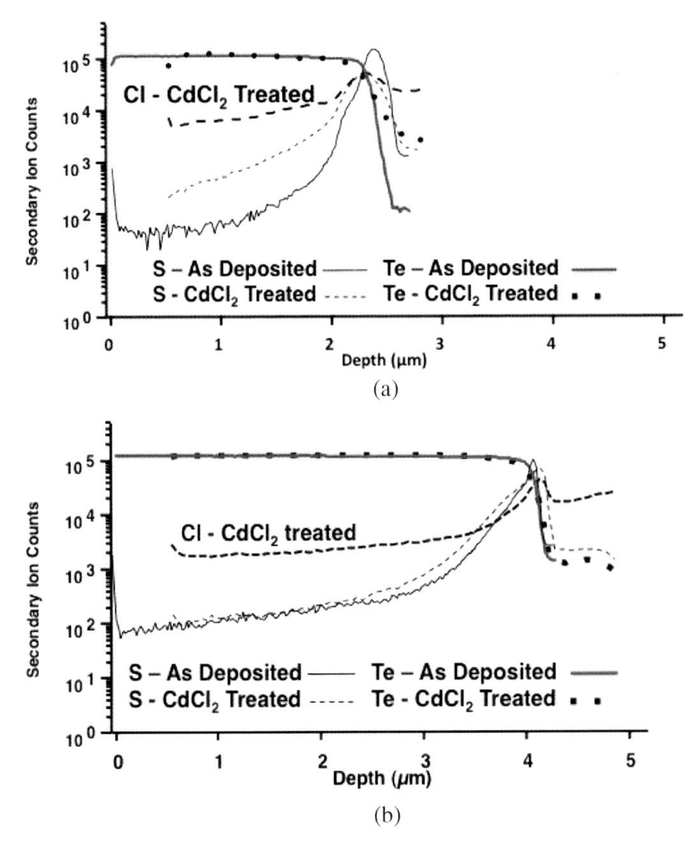

Figure 5.9 SIMS analysis of the effect of $CdCl_2$ treatment on S diffusion as a function of CdTe deposition temperature. (a) 550°C; (b) 600°C. Source: Dhere *et al.* (1996).

deposited at lower temperature (550°C). Recalling that the amount of oxygen available at the CdS/CdTe interface also affects the extent of the CdSTe layer, the result suggests controlling the formation of the CdSTe layer requires (at least) controlling the CdTe temperature, the $CdCl_2$ treatment temperature and time, and the amount of available oxygen. The interrelationship between these parameters begins to explain why developing a reproducible and cost-effective CdTe PV module process requires a deep understanding of the mechanisms at work and the ability to control these mechanisms.

The above description of how the $CdCl_2$ treatment affects CdSTe and CdTe material properties provides clues to how the treatment also affects device performance. Figures 5.10a and 5.10b show the effect of the $CdCl_2$ treatment process on device performance as a function of CdTe CSS substrate temperature (Dhere

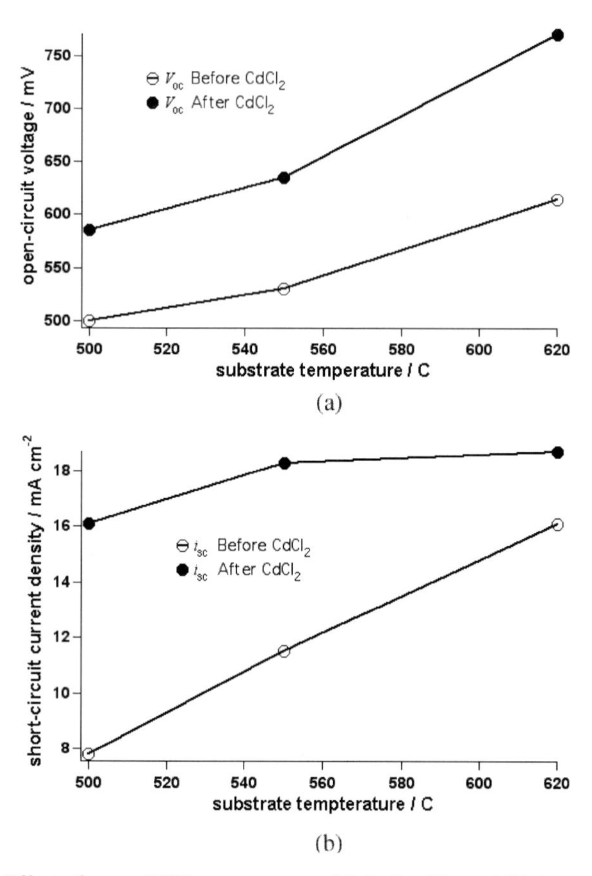

Figure 5.10 Effect of a wet $CdCl_2$ treatment on (a) device V_{oc} and (b) i_{sc} as a function of CSS CdTe substrate temperature. Source: Dhere *et al.* (1996).

et al., 1996). The figure shows that while higher CdTe deposition temperatures lead to improved device performance, the $CdCl_2$ process improves performance at all substrate temperatures. In this data set, a device process parameter (substrate temperature) was altered to yield a device that responds better to a given (wet) $CdCl_2$ treatment. However, depending on process constraints, the $CdCl_2$ treatment could also be varied to produce a higher performance for a given device process. Because these process alterations can be subtle, and are often made without full knowledge of what aspect of cell functionality is being altered, it may be difficult to know precisely how a specific $CdCl_2$ process parameter may be affecting junction performance.

5.6.8 *Back contact*

For superstrate CdTe PV devices, the last step in device formation is fabrication of the back contact. Until about 2000, most CdTe technologists believed the primary function of the back contact was to establish a low-resistance pathway for electrons to enter the CdTe layer during device operation. It was believed that the semiconductor junction properties (n-type and p-type regions, respective space-charge regions, and minority-carrier lifetimes) were established prior to contacting. However, more recent studies have shown that the back contact process not only provides a low-resistance current pathway, but significantly alters the electrical properties of the underlying CdTe layer, thereby significantly affecting junction functionality.

The first consideration in developing a back-contacting process is to determine if/how any residuals from the $CdCl_2$ process should be removed. Depending on how the $CdCl_2$ process was performed, this removal can be easy or difficult: removing $CdCl_2$ residual following a vapour process is easy, while removing Cd oxychloride residuals after a wet process is more difficult (Gessert *et al.*, 2001). Removal process options include thermal, chemical and physical (e.g. ion-beam treatment) methods. Historically, a chemical option has been preferred because it often produces a Te-rich layer that assists certain types of contact formation (Li *et al.*, 1999; Levi *et al.*, 2000). It has also been suggested that gettering of excess Cu by etch-formed Te inclusions can enhance device stability (Albin *et al.*, 2006). In addition, one also needs to consider that the process used to remove $CdCl_2$ residuals must also produce the CdTe surface stoichiometry required for specific contact functionality — a p-ZnTe contact interface requires a stoichiometric CdTe surface, whereas many other contact designs benefit from a Te-rich surface layer. Finally, it has been suggested that certain chemical surface treatments may be effective at reducing the effect of weak diodes and/or microshunts that may exist in some types of CdTe layers (Karpov *et al.*, 2003).

Once the type of surface treatment is determined, one must next realise that the electrical quality of the as-deposited and/or $CdCl_2$-treated bulk CdTe layer is insufficient for effective junction operation. Depending on the CdTe source material used, the specifics of the CdTe deposition, and $CdCl_2$ treatment processes, the net acceptor density will generally be too low ($<10^{-12}\,cm^{-3}$) for optimal device operation, and/or the material may be n-type. Furthermore, even if the acceptor density is sufficiently high, the minority-carrier lifetime is generally too short ($<0.5\,ns$).

Most contacting processes utilise a contact interface layer (CIFL) that can diffuse one or more dopant species into the CdTe layer. Although Cu has historically been used for the active diffusing dopant species, other Group I species (Au or Ag) or Group V species (P, As, Sb or Bi) have also demonstrated potential (Zanio, 1978). While the precise defect formation that occurs during dopant diffusion remains debated, it is generally accepted that a successful diffusion alters the electrical properties of the underlying CdTe layer so that it becomes sufficiently p-type to establish a strong field (i.e. narrow space-charge region) in the ~0.5–1 μm of the device nearest the CdS layer (Gessert *et al.*, 2006). Furthermore, Cu diffusion at appropriate temperatures (~250–350°C) has been found to increase electron lifetime near the CdS/CdTe interface (Gessert *et al.*, 2009; Gessert *et al.*, 2010).

As mentioned in Section 5.6.5, it is possible that the $CdCl_2$-treated CdTe layers may be intrinsic or n-type. This suggestion is supported by contact studies where dopant diffusion can be limited to very low concentrations. These show, for CdTe deposited at intermediate temperatures (450–575°C), devices formed without Cu dopant diffusion from the contact typically demonstrate very little (if any) rectification (Gessert *et al.*, 2002). However, when even a small amount of Cu is allowed to diffuse from the contact, strong junction rectification is observed, suggesting that contact diffusion may be converting the layer from n-type to p-type.

Additional differences in the response of the CdTe layer to contact diffusion are observed between layers grown at intermediate (450–575°C) and high temperature (600–625°C). Capacitance–voltage analysis of devices grown at intermediate temperatures, shown in Fig. 5.11, indicates that initial dopant diffusion from the contact causes the junction space charge to contract systematically. This suggests that even the initial dopant diffusion from the contact leads to an increase in net acceptor concentration (Gessert *et al.*, 2005; Gessert *et al.*, 2006]. In contrast, contact diffusion into material grown at high temperatures (600–625°C) causes the space charge to first *expand*, and then contract (Fig. 5.11). This suggests that the net acceptor density decreases with initial dopant diffusion, and thereafter increases (Chin *et al.*, 2010).

It has recently been reported that this seemingly confusing functionality can be explained by noting that higher deposition temperature will possibly yield CdTe layers in which the effective ionisation energy changes with the extent of Cu diffusion (Ma *et al.*, 2011). Specifically, high-temperature deposition yields CdTe with more cadmium vacancy defects (V_{Cd}) than would be present at intermediate temperatures. Further, the V_{Cd} exists as a double acceptor, with both a lower (~130 meV) and a higher ionisation energy (~210 meV). As Cu diffuses into the CdTe, it preferentially substitutes onto available V_{Cd} sites and becomes an acceptor Cu_{Cd}. However, the ionisation energy of Cu_{Cd} (~220 meV) is greater than either of the V_{Cd} activation energies, and so the observed net acceptor density (N_A) *decreases*. This reduction in N_A continues until the available cadmium vacancies

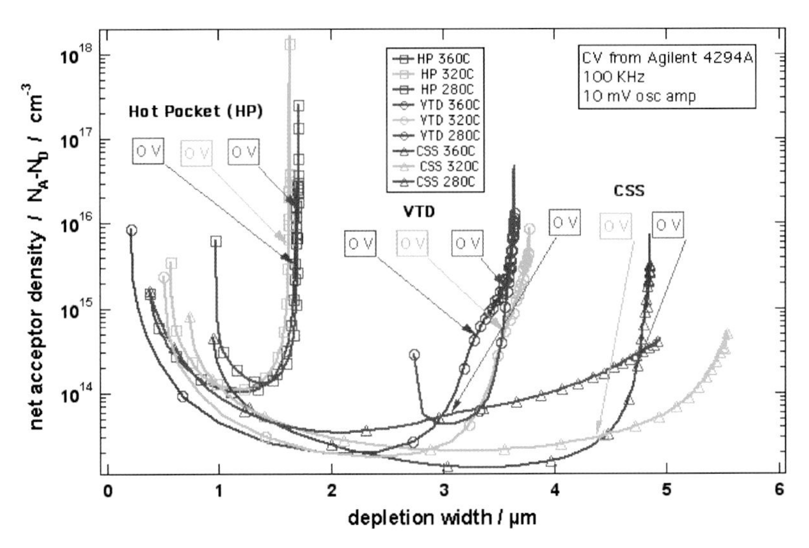

Figure 5.11 Capacitance–voltage analysis of two sets of devices produced at intermediate temperatures (450–575°C, labelled Hot Pocket (HP) and VTD) and one set produced at higher temperature labelled CSS. For the two sets produced at intermediate temperature, the space charge decreases systematically with increasing contacting temperature (i.e. increasing Cu incorporation). However, for the device produced at high temperature (labelled CSS), the space charge initially increases and then decreases, suggesting that the net acceptor density initially decreases and then increases.

are consumed. At this point, remaining Cu can either displace the Cd in the CdTe lattice (the formation energy of Cu_{Cd} in CdTe is ~1.31 eV) to form additional Cu_{Cd} acceptors, or form Cu interstitial donors (Cu_i) (Wei and Zhang, 2002). Recent studies have shown, depending on chemical potential (i.e. whether CdTe is formed with excess Cd or Te concentration), the Cu_{Cd} defect formation is preferred over Cu_i up to a certain defect concentration, at which point formation of the compensating Cu_i will commence and pin the Fermi level (Ma *et al.*, 2011). Further, depending on the particular thermal process conditions following Cu diffusion, the Cu_i can begin to dominate, type-converting the material back to n-type.

In contrast to doping the CdTe layer from the contact and producing the $n-p$ junction functionality described above, an alternative contact design is intended to yield a low-resistance CdTe/CIFL interface while sustaining a CdTe layer that is electrically intrinsic (i.e. not extrinsically doped p-type) while still demonstrating a long minority-carrier lifetime (i.e. $n-i-p$ junction functionality) (Sites and Pan, 2007). For this junction design, if the intrinsic CdTe layer demonstrates sufficiently long lifetime, it may also be advantageous for the CIFL to function as an electron reflector. Modelling suggests wider-bandgap tellurides or their alloys (e.g. ZnTe,

CdZnTe or CdMgTe) could be appropriate CIFL choices in this application (Hsiao and Sites, 2009).

Regardless of the type of junction functionality with which the contacting process must align, achieving high device performance requires that the resistance at the CdTe/CIFL interface be minimised. The strategies to accomplish this fall in one of two broad categories: Category 1 — A CIFL is chosen that will form a relatively low barrier height between the CIFL and CdTe, while the barrier width is reduced as much as possible by increasing p-type doping in the CdTe through diffusion from the CIFL (Sites and Pan, 2007). CIFLs that fall into this category include those where a Te or CuTe layer is formed or deposited, and Cu is diffused from or through it into the CdTe; Category 2 — choosing a CIFL that facilitates valence-band alignment between the CdTe and CIFL, and can also be degenerately doped p-type. For this contact, holes in the valence band will not experience a barrier as they move across the CdTe/CIFL interface (i.e. the interface will be a low-resistance pathway), while transport from the (degenerately doped) CIFL into an outer metallisation will be via low-resistance quantum-mechanical tunnelling (Gessert *et al.*, 2003; Gessert *et al.*, 2007). One CIFL that falls into this category is p-ZnTe:Cu; however CIFLs that include SbTe or alloys of HgTe may also benefit from this functionality.

Once an appropriate CIFL is deposited onto the CdTe, metal layers are generally deposited on top of the CIFL. These layers provide several functions including reducing lateral resistivity, limiting moisture ingress, and/or providing functionality to enhance stability. For commercial modules, these outer layers must also be developed to provide attributes such as allowance for high deposition temperature without diffusion, high adhesion and low cost.

5.7 Where is the junction?

Having established a basic appreciation of what each layer of a CdTe PV device may do in various device designs, a final question remains as to where the semiconductor junction is actually located. We now know that the specific parameters of the CdS and CdTe deposition combine with the parameters of the $CdCl_2$ treatment to affect the formation and extent of the CdSTe layer. Furthermore, we know that diffusion from the CIFL additionally alters the electrical properties of the CdTe, thereby altering the junction. However, we still have not discussed where the junction is.

To understand this better, we must also consider that when sulphur is diffused into the CdTe layer (forming a CdSTe interdiffused layer), the bandgap of the CdSTe alloy decreases relative to CdTe or CdS. This is called band bowing, and is shown schematically in Fig. 5.12a. At a temperature of ~600°C, depending on

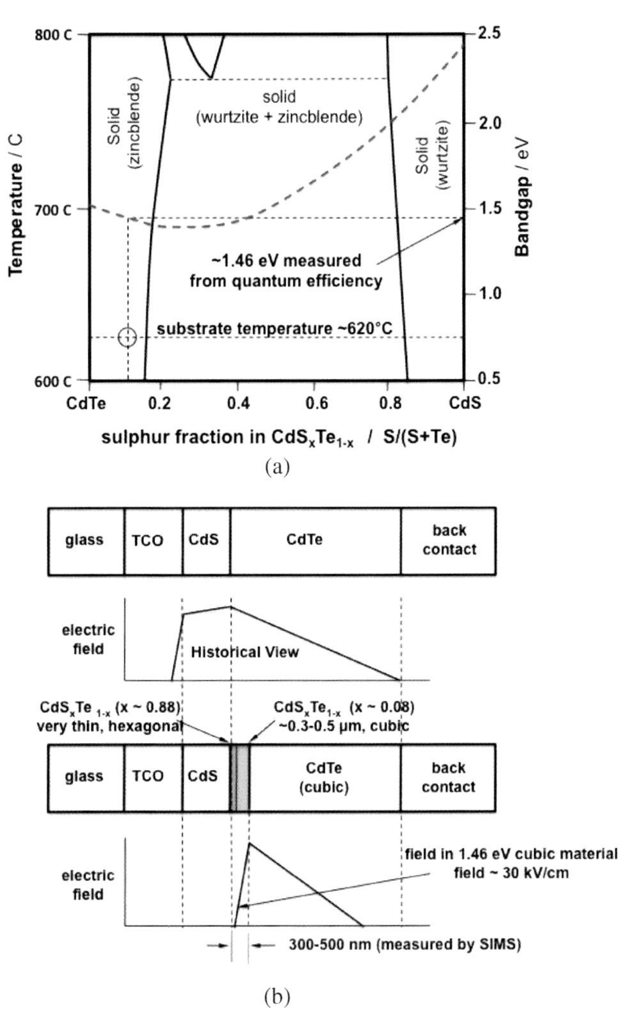

Figure 5.12 (a) Diagram illustrating the effect of band bowing and phase separation in the CdS_xTe_{1-x} layer. The circle at $x = \sim 0.1$ and 620°C indicates the most likely sulphur concentration, based on the effective bandgap and the available single-phase region; (b) diagram illustrating the location of the junction in a CdS/CdTe PV device from a historic (top) and more recent (bottom) perspective.

the stoichiometry of the CdSTe alloy, this can exist as a single-phase, hexagonal material (wurtzite, ~85–100% sulphur), a single-phase cubic material (zincblende, 0–~15% sulphur), or a mixed-phase CdSTe material (between ~15 and ~85% sulphur) (Nunoue *et al.*, 1990). During intermixing of the CdS and CdTe layers at the CdTe deposition temperature, it is believed that all three of these phases will be present in what has so far been described simply as the CdSTe layer. Many technologists believe that it is reasonable to assume the CdSTe layer will be primarily in the wurtzite (i.e. S-rich) phase nearest the CdS layer, primarily in the zincblende (Te-rich) phase nearest the CdTe layer, and in the mixed phase between these two (primarily) single-phase endpoints. However, only a few studies of this region have been reported (Dhere *et al.,* 2008).

For the type of devices produced at NREL, we believe the electrical junction forms between the non-alloyed CdTe, and a diffusion-formed zincblende CdSTe (Dhere *et al.*, 2008). Analysis of the IR absorption characteristics of quantum efficiency data indicates the effective sulphur concentration in the junction region is ~10%. This is estimated by calculating the effective bandgap from device quantum efficiency (~1.46 eV), while accounting for the known band bowing (Compaan *et al.*, 1996). This estimation process is shown graphically in Fig. 5.12a. Two schematic diagrams illustrating both the historic and more recent understanding of the layer structure, as well as the associated electric fields, are shown in Fig. 5.12b.

5.8 Considerations for large-scale deployment

The previous discussion suggests that CdTe thin-film PV modules are a viable option for large-scale, long-term, renewable electricity generation. This is because the specific *cost* of these thin-film modules is (or is quickly becoming) less than the *price* of delivering electricity in many parts of the world. Although economic viability is a necessary element for both continued production expansion and societal acceptance, the promise of thin-film PV as a source of *sustainable* base energy requires additional issues to be considered. Even though a thorough treatment of any of the following topics is beyond the scope of this chapter, the following discussion suggests why many believe these topics are rapidly becoming just as important as specific cost in the widespread establishment of any PV energy infrastructure.

5.8.1 *Reliability*

As any PV technology matures from the laboratory to commercial production, a point is reached when product reliability must be understood well enough to assign a product warranty. Presently, most CdTe thin-film PV module manufacturers like to align module warranties with established PV alternatives (particularly wafer-based crystalline Si). This level of reliability is typically consistent with the panel delivering at least 80% of its initial power after 25 years of field deployment. Assuming a constant exponential degradation, this equates to about 0.9% power loss per year. Because thin-film manufacturers need to ensure their products demonstrate reliability consistent with their warranty, developing techniques that can rapidly, accurately and cost-effectively probe reliability expectations remains one of the more critical challenges for both public- and private-sector PV researchers.

Another evolving value consideration relates to expectations of post-warranty deployment. Like many older cars that are still on the road after their warranty has expired, many PV modules have exceeded their warranty period, but remain usefully deployed. Although degraded modules may not be appropriate for many applications, modules with de-rated performance will continue to have significant value in many other off-grid and/or remote-power applications. It therefore seems reasonable to assume that designing CdTe module reliability for time periods exceeding the warranty could impart (marketing) advantages to new PV products.

5.8.2 *Minerals availability*

As production of all types of PV technologies expand, there is increasing debate as to whether minerals availability will become a significant factor in module cost. This is especially true for present thin-film alternatives because each uses a mineral component that is perceived to be relatively scarce. The minerals most typically mentioned in this context are Ge (for some amorphous Si devices), In (for CIGS devices) and Te (for CdTe devices). A direct answer in this debate is complicated because, like other (secondary) minerals that are considered scarce, most of these minerals are presently co-processed along with other (primary) minerals.

Because of this interdependence, parameters that will influence Te availability include the following: (i) the amount of production of the primary ore (i.e., presently Cu is the primary ore for Te); (ii) concentration of the secondary mineral in the primary ore; (iii) industry incentive to co-process the secondary mineral (i.e. emerging and/or multiple markets for the secondary ore); and (iv) environmental, governmental or practical constraints related to extracting the mineral. The main message is that the business plan of anyone considering the establishment of a large-scale production facility for any PV technology should include thorough advisement from a reputable mineral geologist.

5.8.3 *Environmental considerations*

It is well known that although PV modules do not produce toxic waste or greenhouse gas when operating, toxic products and/or by-products can be produced during the initial production and recycling phases of their life cycle. This waste is the primary component of what is often referred to as the 'total life-cycle eco-toxicity' of any PV technology. Although this eco-toxicity derives from many sources, much of it depends on the particular chemicals used to process subcomponents, as well as the burning of fossil fuels to mine, transport or refine raw materials, or provide the electricity used to produce the modules.

Realising that different module-production methods may use different amounts and types of chemicals, and that different countries use different amounts, types and mixtures of fossil fuels to produce and deliver their electricity, it is easy to see how associating a given level of eco-toxicity with any particular PV technology involves significant assumptions about both production process and location. Furthermore, the energy mix will continue to evolve as more PV-derived electricity is used to fuel the production of PV modules (often called the breeder PV energy-balance scenario). Nevertheless, reports are beginning to suggest some useful guidance for technology-dependent reduction in PV eco-toxicity (Fthenakis, 2004; Raguei *et al.*, 2007; Held and Ilg, 2008). For example, a primary source of eco-toxicity in glass-encapsulated CdTe thin-film modules arises from the production of the glass. This suggests that significant opportunities may exist to reduce both eco-toxicity (as well as to reduce the energy used to make the module) by developing module designs that incorporate either less glass or glass produced in such a way to generate less eco-toxicity.

5.8.4 *Energy payback time*

The energy payback time (EPT) of a PV module is the amount of time a module must produce power to recover the energy it took to produce the module initially. Although assumptions vary, the calculation of the energy to produce the module should be as inclusive as possible, accounting for everything from the energy needed to mine, transport, refine, produce and deliver all module subcomponents to that required to deposit/assemble/package the module, deploy it, and eventually recycle it at the end of its life.

Within this definition, if a particular module is expected to last for 25 years, and it demonstrates an EPT of five years, one can roughly expect to get about five times more energy out of the module than was used to produce it. Although this ratio of energy returned-to-invested sounds great, a five-year EPT may still be too long when one accounts for current production-growth scenarios.

The growth scenario discussed previously might be consistent with PV production and efficiency levels of several years ago. In this case, as long as the PV module produced more energy in its expected lifetime (~15–25 years) than it took to produce the module, it was considered a net-energy producer. In contrast, we presently have PV production technologies and module efficiencies that yield EPTs of two years — or less. An EPT of two years would allow a module with a 25-year warranty to produce more than ten times the original energy investment during its lifetime.

Although this sounds both environmentally and financially attractive, a serious complication arises when the growth in PV production becomes so rapid that the expected doubling time of production rate becomes shorter than the EPT. In this rapid-growth scenario, an energy balance at any given time during production expansion can only be achieved if the EPT is equal to or less than the expected production doubling time. Otherwise, additional amounts of traditional energy resources (e.g. fossil fuels) may have to be supplied to provide the energy to sustain the rapid growth in PV production. Hence, if PV is to be a *net energy producer* during this period of rapidly expanding production, the EPT may need to be *significantly less* than the production doubling time. Specifically, the 46% exponential growth in PV production observed during the past ten years indicates that EPTs need to be less than two years. Therefore, the primary question is this: what is the EPT for different PV technologies, and specifically for thin-film CdTe PV technologies?

Recent reports have indicated that most thin-film technologies may have advantages over wafer Si in this respect, with CdTe demonstrating EPTs of between six months and one year (Raugei *et al.*, 2007; Held and Ilg, 2008).

5.9 Conclusions

This chapter has presented a technological description of components and process alternatives consistent with the present generation of CdTe thin-film PV devices. We have also suggested why this technology may embody considerable opportunity for becoming a significant part of future large-scale electricity production. However, even when economic requirements are met, the establishment of a sustainable energy-production infrastructure from CdTe technologies will require additional consideration of longer-term issues. These include reliability, materials availability, environmental considerations and energy payback time.

Acknowledgements

The authors wish to thank R. G. Dhere, D. S. Albin, T. M. Barnes, J. N. Duenow, S.-H. Wei and T. J. Coutts of NREL for assistance and thoughtful suggestions regarding the preparation of this chapter. Portions of this work were supported under DOE Contract No. DE-AC36-08-GO28308 to NREL.

References

Albin D. S., Demtsu S. H. and McMahon T. J. (2006), 'Film thickness and chemical processing effects on the stability of cadmium telluride solar cells', *Thin Solid Films* **515**, 2659–2668.

Albin D. S., Yan Y. and Al-Jassim M. M. (2002), 'The effect of oxygen on interface microstructure evolution in CdS/CdTe solar cells', *Progr. Photovoltaics* **10**, 309–322.

Aranovich J., Golmayo D., Fahrenbruch A. L. and Bube R. H. (1980), 'Photovoltaic properties of ZnO–CdTe heterojunctions prepared by spray pyrolysis', *J. Appl. Phys.* **51**, 4260–4268.

Artobolevskaya E. S., Afanaseva E. A., Vodopyanov L. K. and Sushkov V. P. (1967), 'Photoelectromagnetic effect in cadmium telluride', *Sov. Phys. Semicond.* **1**, 1531.

Basol B. M. (1984), 'High-efficiency electroplated heterojunction solar cell', *J. Appl. Phys.* **55**, 601–603.

Bell R. O., Serreze H. B. and Wald F.V. (1975), 'A new look at CdTe solar cells', *Proc. 11th IEEE Photovoltaic Specialists Conf.*, Scotsdale, AZ, pp. 497–502.

Bonnet D. and Rabenhorst H. (1972), 'New results on the development of thin film *p*-CdTe/*n*-CdS heterojunction solar cell', *Proc. 9th IEEE Photovoltaic Specialists Conf.*, Silver Springs, MD, pp. 129–132.

Braun G. W. and Skinner D. E. (2007), 'Experience scaling-up manufacturing of emerging photovoltaic technologies', Natl. Renewable Energy Laboratory Subcontract Report, NREL/SR-640-39165, January 2007.

Bube R. H. (1988), 'CdTe junction phenomena', *Solar Cells* **23**, 1–17.

Bube R. H., Buch F., Fahrenbruch A. L., Ma Y. Y. and Mitchell K. W. (1977), 'Photovoltaic energy conversion with *n*-CdS-*p*-CdTe heterojunctions and other II–VI junctions', *IEEE Trans. Electron. Dev.* **ED-24**, 487–492.

Bube R. H., Fahrenbruch A., Aranovich J., Buch F., Chu M. and Mitchell K. (1975), 'Applied research in II–VI compound materials for heterojunction solar cells', NSF Report No. NSF/RANN/SE/AER-75-1679/75/4.

Carmody M., Mallick S., Margetis J., Kodama R., Biegala T., Xu D., Bechmann P., Garland J. W. and Sivananthan S. (2010), 'Single-crystal II–VI on Si single-junction and tandem solar cells', *Appl. Phys. Lett.* **96**, 153502–153502-3.

Chin K. K., Gessert T. A. and Wei S.-H. (2010), 'The roles of Cu impurity states in CdTe thin film solar cells', *Proc. 35th IEEE Photovoltaic Specialists Conf.*, Piscataway, NJ, pp. 1915–1918.

Chu T. L. (1988), 'Cadmium telluride solar cells', in Coutts T. J. and Meakin J. D., (eds), Vol. 3 of *Current Topics in Photovoltaics*, Academic Press, New York.

Chu T. L., Chu S. S., Britt J., Ferekides C., Wang C., Wu C. Q. and Ullal H. S. (1992), '14.6% Efficient thin-film cadmium telluride heterojunction solar cell', *IEEE Electron Dev. Lett.* **13**, 303–304.

Coggeshall C. and Margolis R. (2011), *Solar Vision Study*, U.S. Dept. of Energy Report.

Compaan D., Feng Z., Contreas-Puente G., Narayanswamy C. and Fisher A. (1996), 'Properties of pulsed laser deposited CdS_xTe_{1-x} films on glass', *MRS Proc.* **426**, 367–372.

Corwine C., Sites J. R., Gessert T. A., Metzger W. K., Dippo P., Li J., Duda A. and Teeter G. (2005), 'CdTe photoluminescence: comparison of solar-cell material with surface-modified single crystals', *Appl. Phys Lett.* **86**, 221909–221909-3.

Coutts T. J., Young D. L. and Li X. (2000), 'Characterization of transparent conducting oxides', *MRS Bulletin* **25**, 58–65.

Cusano D. A. (1963), 'CdTe solar cells and photovoltaic heterojunctions in II–IV Compounds', *Solid-State Electron.* **6**, 217–218.

Cusano D. A. (1967), 'Thin film studies and electro-optical effects', in Aven M. and Prener J. S. (eds.), *Physics and Chemistry of II–VI Compounds*, North-Holland Publishing Co., Amsterdam, Ch. 14, pp. 709–766.

Cusano D. A. and Lorenz M. R. (1964), 'CdTe hole lifetime from the photovoltaic effect', *Solid State Commun.* **2**, 125–128.

Dhere R. G., Albin D. S., Rose D. H., Asher S. E., Jones K. M., Al-Jassim M. M., Moutinho H. R. and Sheldon P. (1996), 'Intermixing at the CdS/CdTe interface and its effect on device performance', *MRS Proc.* **426**, 361–366.

Dhere R. G., Bonnet-Eymard M., Charlet E., Peter E., Duenow J. N., Li J. V., Kuciauskas D. and Gessert T. A. (2011), 'CdTe solar cell with industrial Al:ZnO on soda-lime glass', *Thin Solid Films* **519**, 7142–7145.

Dhere R. G., Duenow J. N., DeHart C. M., Li J. V., Kuciauskas D. and Gessert T. A., (2012), 'Development of substrate structure CdTe photovoltaic devices with performance exceeding 10%', *Proc. 38th IEEE Photovoltaic Specialists Conf.*, IEEE, Piscataway, NJ, pp. 3208–3211.

Dhere R. G. and Li X. (2012), NREL, private communication.

Dhere R. G., Zhang Y., Romero M. J., Asher S. E., Young M., To B., Noufi R. and Gessert T. A. (2008), 'Investigation of junction properties of CdS/CdTe solar cells and their correlation to device properties', *Proc. 33rd IEEE Photovoltaic Specialists Conf.*, San Diego, CA, Manuscript No. 279.

Dhese K. A., Devine P., Ashenford D. E., Nicholls J. E., Scott C. G., Sands D. and Lunn B. (1994), 'Photoluminescence and *p*-type conductivity in CdTe:N grown by molecular beam epitaxy', *J. Appl. Phys.* **76**, 5423–5428.

Duenow J. N., Dhere R. G., Kuciauskas D., Li J. V., Pankow J. W., Dippo P. C., DeHart C. M. and Gessert T. A., 'Oxygen incorporation during fabrication of substrate CdTe photovoltaic devices', *Proc. 38th IEEE Photovoltaic Specialists Conf.*, IEEE, Piscataway, NJ, pp. 3225–3229.

Fthenakis V. M. (2004), 'Life cycle impact analysis of cadmium in CdTe PV production', *Renewable and Sustainable Energy Reviews* **8**, 303–334.

Fulop G., Doty M., Meyers P., Betz J. and Liu C. H. (1982), 'High-efficiency electrodeposited cadmium telluride solar cells', *Appl. Phys. Lett.* **40**, 327–328.

Garcia D. A., Entine G. and Tow D. E. (1974), 'Detection of small bone abscesses with a high-resolution cadmium telluride probe', *J. Nucl. Med.* **15**, 892–895.

Gessert T. A., Asher S., Johnston S., Duda A., Young M. R. and Moriarty T. (2006), 'Formation of ZnTe:Cu/Ti contacts at high temperature for CdS/CdTe devices', *Proc. 4th World Conf. Photovoltaic Energy Conversion*, pp. 432–435.

Gessert T. A., Asher S., Johnston S., Young M., Dippo P. and Corwine C. (2007), 'Analysis of CdS/CdTe devices incorporating a ZnTe:Cu/Ti contact', *Thin Solid Films* **515**, 6103–6106.

Gessert T. A., Burst J., Li X., Scott M. and Coutts T. J. (2011), 'Advantages of transparent conducting oxide thin films with controlled permittivity for thin film photovoltaic solar cells', *Thin Solid Films* **519**, 7146–7148.

Gessert T. A., Dhere R. G., Duenow J. N., Kuciauskas D., Danevce A. and Bergeson J. D. (2010), 'Comparison of minority carrier lifetime measurements in superstrate and substrate CdTe PV devices', *Proc. 37th IEEE Photovoltaic Specialists Conf.*, IEEE, Piscataway, NJ, pp. 335–339.

Gessert T. A., Metzger W. K., Dippo P., Asher S. E., Dhere R. G. and Young M. R. (2009), 'Dependence of carrier lifetime on Cu-contacting temperature and ZnTe:Cu thickness in CdS/CdTe thin film solar cells', *Thin Solid Films* **517**, 2370–2371.

Gessert T. A., Perkins C. L., Asher S. E., Duda A. and Young M. R. (2003), 'Study of ZnTe:Cu/metal interfaces in CdS/CdTe photovoltaic solar cells', *Mater. Res. Soc. Symp.*, Vol. 796, pp. 79–84.

Gessert T. A., Romero M. J., Johnston S., Keys B. and Dippo P. (2002), 'Spectroscopic cathodoluminescence studies of the ZnTe:Cu contact process for CdS/CdTe solar cells', *Proc. 29th IEEE Photovoltaic Specialists Conf.,* IEEE, Piscataway, NJ, pp. 535–538.

Gessert T. A., Romero M. J., Perkins C. L., Asher S. E., Matson R., Moutinho H. and Rose D. (2001), 'Microscopic residuals on polycrystalline CdTe following wet $CdCl_2$ treatment', *MRS Proc.* **668**, H1.10.1–H1.10.6.

Gessert T. A., Smith S., Moriarty T., Young M., Asher S., Johnston S., Duda A. and DeHart C. (2005), 'Evolution of CdS/CdTe device performance during Cu diffusion', *Proc. 31st IEEE Photovoltaic Specialists Conf.*, IEEE, Piscataway, NJ, pp. 291–294.

Gloeckler M., Sankin I. and Zhao Z. (2013), 'CdTe solar cells at the threshold to 20% efficiency', *IEEE J. Photovoltaics* **3**, 1389–1398.

Green M. A., Emery K., Hishikawa Y., Warta W. and Dunlop E. D. (2012), 'Solar cell efficiency tables (Version 39)', *Progr. Photovoltaics* **20**, 12–20.

Green M. A., Emery K., Hishikawa Y., Warta W. and Dunlop E. D. (2013), 'Solar cell efficiency tables (Version 42)', *Progr. Photovoltaics* **21**, 827–837.

Greenberg J. H. (1996), 'P–T–X phase equilibrium and vapor pressure scanning of non-stoichiometry in CdTe', *J. Cryst. Growth* **161**, 1–11.

Gretener C., Perrenoud J., Kranz L., Kneer L., Schmitt R., Buecheler S. and Tiwari N. (2012), 'CdTe/CdS thin film solar cells grown in substrate configuration', *Progr. Photovoltaics* **21**, 1580–1586.

Gur I., Fromer N. A., Geier M. L. and Alivisatos A. P. (2005), 'Air-stable all-inorganic nanocrystal solar cells processed from solution', *Science* **310**, 462–465.

Halliday D. P., Potter M. D. G., Boyle D. S. and Durose K. (2001), 'Photoluminescence characterization of ion implanted CdTe', *MRS Proc.***668**, H1.8.1.

Held M. and Ilg R. (2008), 'Life cycle assessment (LCA) of CdTe thin film PV modules and material flow analysis (MFA) of cadmium within EU27', *Proc. 23rd Eur. Photovoltaic Solar Energy Conf.*, Sept. 1–5, Valencia.

Hsiao K. J. and Sites J. R. (2009), 'Electron reflector strategy for CdTe solar cells', *Proc. 34th IEEE Photovoltaic Specialists Conf.*, Fort Collins, CO, pp. 1846–1850.

Jordan J. F. (1993), 'Photovoltaic cell and method', U.S. Patent No. 5261968.

Karpov V. G., Shvydka D., Roussillon Y. and Compaan A. D. (2003), 'The mesoscale physics of large-area photovoltaics', *Proc. 3rd World Conf. Photovoltaic Energy Conversion*, WCPEC–3, Osaka, pp. 495–498.

Kester J. L., Albright S., Kaydanov V., Ribelin R., Woods L. M. and Phillips J. A. (1996), 'CdTe solar cells: electronic and morphological properties', *AIP Conf. Proc. Ser.* **394**, 162–169.

Kranz L., Schmitt R., Gretener C., Perrenoud J., Pianezzi F., Uhl A. R., Keller D., Buecheler S. and Tiwari A.N. (2013), 'Progress towards 14% efficient CdTe solar cells in substrate configuration', *Proc. 39th IEEE Photovoltaic Specialists Conf.*, Tampa, FL, 1644–1648.

Levi D., Albin D. and King D. (2000), 'Influence of surface composition on back-contact performance in CdTe/CdS PV devices', *Progr. Photovoltaics* **8**, 591–602.
Li X., Gessert T. A., Matson R. J., Hall J. F. and Coutts T. J. (1993), 'Microstructural study of sputter-deposited CdTe thin films', *J. Vac. Sci. Technol. A* **12**, 1608–1613.
Li X., Niles D. W., Hasoon F. S., Matson R. J. and Sheldon P. (1999), 'Effect of nitric–phosphoric acid etches on materials properties and back-contact formation in CdTe-based solar cells', *J. Vac. Sci. Technol. A* **17**, 805–809.
Ma J., Wei S.-H., Gessert T. A. and Chin K. K. (2011), 'Carrier density and compensation in semiconductors with multiple dopants and multiple transition energy levels: case of Cu impurities in CdTe', *Phys. Rev. B* **83**, 245207–245214.
McCandless B. (2003), 'Effects of treatments on CdTe film conductivity', NREL Subcontract Report #ADJ-1-30630-12, 23 December.
McCandless B. E., Youm I. and Birkmire R. W. (1999), 'Optimization of vapor post-deposition processing for evaporated CdS/CdTe solar cells', *Progr. Photovoltaics* **7**, 21–30.
Mehata S. (2010, 'PV news annual data collection result: 2010 cell, module production explodes past 20 GW', *PV News* **29**, May 2010.
Meyer E., Martini M. and Sternberg J. (1972), 'Measurement of the disappearance rate of Se-75 sodium selenite in the eye of the rat by a CdTe medical probe', *IEEE Trans. Nucl. Sci.* **19**, 237–243.
Moutinho H. R., Dhere R. G., Al-Jassim M. M., Levi D. H. and Kazmerski L. L. (1999), 'Investigation of induced recrystallization and stress in close-spaced sublimated and radio-frequency magnetron sputtered CdTe thin films', *J. Vac. Sci. Technol. A* **17**, 1793–1798.
Naumov G. P. and Nikolaeva O. V. (1961), 'The efficiency of transformation of direct solar radiation energy into electric energy using a CdTe photocell', *Soviet Phys. Solid State* **3**, 2718.
Nicoll F. H. (1963), 'The use of close spacing in chemical-transport systems for growing epitaxial layers of semiconductors', *J. Electrochem. Soc.* **110**, 1165–1167.
Nunoue S.-Y., Hemmi T. and Kato E. (1990), 'Mass spectrometric study of the phase boundaries of the CdS–CdTe system', *J. Electrochem. Soc.*, **137**, 1248–1251.
Powell R. C., Dorer G. L., Reiter N. A., McMaster H. A., Cox S. M. and Kahle T. D. (1999), 'Apparatus and method for depositing a material on a substrate', U.S. Patent No. 5945163.
Raugei M., Bargigli S. and Ulgiati S. (2007), 'Life cycle assessment and energy pay-back time of advanced photovoltaic modules: CdTe and CIS compared to poly-Si', *Energy* **32**, 1310–1318.
Romeo N., Bosio A. and Canevari V. (1992), 'Large crystalline grain CdTe thin films for photovoltaic application', *Int. J. Solar Energy* **12**, 183–186.
Rose D. H (1997), 'The effects of oxygen on CdTe–absorber solar cells deposited by close-spaced sublimation', PhD Thesis, University of Colorado, Boulder, CO.
Rose D. H., Hasoon F. S., Dhere R. G., Albin D. S., Ribelin R. M., Li X. S., Mahathongdy Y., Gessert T. A. and Sheldon P. (1999), 'Fabrication procedures and process sensitivities for CdS/CdTe solar cells', *Progr. Photovoltaics* **7**, 331–340.
Sasala R. A., Powell R. C., Dorer G. L. and Reiter N. (1996), 'Recent progress in CdTe solar cell research at SCI', *AIP Conf. Proc Ser.* **394**, 171–186.

Singh V. P., McClure J. C., Lush G. B., Wang W., Wang X., Thompson G. W. and Clark E. (1999), 'Thin film CdTe-CdS heterojunction solar cells on lightweight metal substrates', *Solar Energy Mater. Solar Cells* **59**, 145–161.

Sites J. and Pan J. (2007), 'Strategies to increase CdTe solar-cell voltage', *Thin Solid Films* **515**, 6099–6102.

Suyama N., Arita T., Nishiyama Y., Ueno N., Kitamura S. and Murosono M. (1990), 'CdS/CdTe solar cells by the screen-printing–sintering technique', *Proc. 21st IEEE Photovoltaic Specialists Conf.*, Kissimmee, FL, pp. 498–503.

Tatarenko S., Bassani F., Saminadayar K., Cox R. T., Jouneau P. H. and Megnea N. M. (1993), 'Indium doping of (001), (111) and (211) CdTe layers grown by molecular beam epitaxy', *J. Cryst. Growth* **127**, 318–322.

Tyan Y.-S. and Perez-Albuerne E. A (1982), 'Efficient thin-film CdS/CdTe solar cells', *Proc. 16th IEEE Photovoltaic Specialists Conf.*, San Diego, CA, pp. 794–800.

Tyan Y.-S., Vazan F. and Barge T. S. (1984), 'Effect of oxygen on thin-film CdS/CdTe solar cells', *Proc. 17th IEEE Photovoltaic Specialists Conf.*, Kissimmee, FL, pp. 840–844.

Valdna V., Hiie J. and Gavrilov A. (2001), 'Defects in Cl-doped CdTe thin films', *Solid State Phenom.* **80**, 155–162.

Vodakov Y. A., Lomakina G. A., Naumov G. P. and Maslakovets Y. P. (1960), 'Properties of $p - n$ junctions in cadmium telluride photocells', *Soviet Phys. Solid State* **2**, 11.

Waters D. M., Niles D., Gessert T., Albin D., Rose D. and Sheldon P. (1988), 'Surface analysis of CdTe after various pre-contact treatments', *Proc. 2nd World Conf. Photovoltaic Solar Energy Conversion*, Vienna, European Commission, Luxembourg, p. 1031.

Wei S.-H. and Zhang S. B. (2002), 'Chemical trends of defect formation and doping limit in II–VI semiconductors', *Phys. Rev. B* **66**, 155211–155221.

Wendt R., Fischer A., Grecu D. and Compaan A. D. (1998), 'Improvement of CdTe solar cell performance with discharge control during film deposition by magnetron sputtering', *J. Appl. Phys.* **84**, 2920–2925.

Wu X. (2004), 'High efficiency polycrystalline CdTe thin-film solar cells', *Solar Energy* **77**, 803–814.

Wu X., Dhere R. G., Yan Y., Romero M. J., Shang Y., Zhou J., DeHart C., Duda A., Perkins C. and To B. (2002), 'High efficiency polycrystalline CdTe thin-film solar cells with an oxygenated amorphous CdS (a-CdS:O) window layer', *Proc. 29th IEEE Photovoltaic Specialists Conf.*, New Orleans, LA, pp. 531–535.

Wu X., Keane J. C., Dhere R. G., DeHart C., Albin D. A., Duda A., Gessert T. A., Asher A. S., Levi D. H. and Sheldon P. (2002) '16.5% Efficient CdS/CdTe polycrystalline thin-film solar cell,' *Proc. 17th European Photovoltaic Solar Energy Conf.*, October 22–26, 2001,WIP-Renewable Energies, Munich, pp. 995–1000.

Wu X., Mulligan W. P. and Coutts T. J. (1996), 'Recent developments in RF sputtered cadmium stannate films', *Thin Solid Films* **286**, 274–276.

Yamaguchi K., Nakayama N., Matsumoto H. and Ikegami S. (1977), 'CdS–CdTe solar cell prepared by vapor phase epitaxy', *Jpn. J. Appl. Physics* **16**, 1203–1211.

Zanio K. (1978), *Cadmium Telluride*, Vol. 13 in *Semiconductors and Semimetals*, Academic Press, New York.

Zunger A. (2003), 'Practical doping principles', *Appl. Phys. Lett.* **83**, 57–59.

CHAPTER 6

$Cu(In,Ga)Se_2$ AND RELATED SOLAR CELLS

UWE RAU* and HANS W. SCHOCK†
*IEK5-Photovoltaik, Forschungszentrum Jülich GmbH, Germany
†Helmholtz-Zentrum Berlin für Materialien und Energie, Germany
*u.rau@fz-juelich.de
†hans-werner.schock@helmholtz-berlin.de

Mephistopheles: *Wer will Lebendigs erkennen und beschreiben,*
Sucht erst den Geist heraus zu treiben,
Dann hat er die Teile in der Hand,
Fehlt leider ! nur das geistige Band.
Encheiresin naturae nennt's die Chemie,
Spottet ihrer selbst und weiß nicht wie.
Johann Wolfgang von Goethe, *Faust*, 1808.

6.1 Introduction

From the early days of photovoltaics until today, thin-film solar cells have always competed with technologies based on single-crystal materials such as Si and GaAs. Owing to their amorphous or polycrystalline nature, thin-film solar cells always suffered from power conversion efficiencies lower than those of the bulk technologies. This drawback was and still is counterbalanced by several inherent advantages of thin-film technologies. As in the early years of photovoltaics, space applications were the driving force for the development of solar cells, the argument in favour of thin films was their potential lighter weight as compared to bulk materials.

An extended interest in solar cells as a source of renewable energy emerged in the mid-1970s as the limitations of fossil energy resources were widely recognised. For terrestrial power applications the cost arguments and the superior energy balance strongly favoured thin films. However, from the various materials under consideration since the 1950s and 1960s, only three thin-film technologies, namely amorphous (a-)Si and the polycrystalline heterojunction systems CdS/CdTe, and $CdS/CuInSe_2$, have achieved industrial production in any volume.

$CuInSe_2$ was synthesised for the first time by Hahn in 1953 (Hahn *et al.*, 1953). In 1974, this material was proposed as a photovoltaic material (Wagner *et al.*, 1974) with a power conversion efficiency of 12% demonstrated for a single-crystal cell. Thin-film development achieved a historical milestone in the years 1980 to 1982, when Boeing Corp. boosted the efficiencies of thin-film solar cells obtained from a three-source co-evaporation process from 5.7% (Mickelsen and Chen, 1980) to over 10% (Mickelsen and Chen, 1982). (We will discuss the co-evaporation process in Section 6.3.1.) The Boeing result was surpassed in 1987 by ARCO Solar with a long-standing record efficiency for a thin-film cell of 14.1% (Mitchell *et al.*, 1988). This company used a different approach for absorber preparation, namely the selenisation of stacked metal layers by H_2Se (see Section 6.3.1). The lack of reproducibility and resulting low production yield considerably delayed the pilot production envisaged at that time. It took a further ten years before Siemens Solar Industries in the US, after inheriting the ARCO Solar approach, entered the stage of production. In 1998 they produced the first commercially available $Cu(In,Ga)Se_2$ solar modules. During the early 2000s, pilot production lines were set up at various places in the US, Europe and Japan. During the first boom years of photovoltaics, 2003–2007, most of these attempts resulted in industrial-scale production capacities. However, from 2008 up to the time of writing (2013) the overproduction crisis in the photovoltaic industry has led to a considerable shake out of cell, module and equipment manufacturers, which has also affected the producers of $Cu(In,Ga)Se_2$ solar modules.

On the laboratory scale, the successful series of record efficiencies established by the National Renewable Energy Laboratory (Contreras *et al.*, 1999; Contreras *et al.*, 2005; Repins *et al.*, 2008) was eventually topped by the Centre for Solar Energy and Hydrogen Research in 2011 with a confirmed cell efficiency of 20.3% on a 0.5 cm^2 laboratory cell (Jackson *et al.*, 2011). A remarkable improvement in flexible cell efficiency was demonstrated by the Swiss Federal Laboratories for Materials Science and Technology (EMPA) (Čhirila, 2013) with a 20.4%-efficient small cell (0.5 cm^2) on a polymide substrate. The Centre for Solar Energy and Hydrogen improved the laboratory cell efficiency to 20.8%. A record ($4 \times 4\,cm^2$) mini-module of 17.4% efficiency was developed by Solibro Research (Wallin *et al.*, 2012) and champion modules from various industrial production lines exceed the efficiency level of 14%. With the perspective of production costs lower than 0.5\$/W_p combined with high module efficiencies, $Cu(In,Ga)Se_2$ still plays the role of one of the leading thin-film photovoltaic technologies.

This chapter aims to summarise our present knowledge of $Cu(In,Ga)Se_2$-based heterojunction solar cells. We focus on four main areas: (i) The description of basic material properties such as crystal properties, phase diagram and defect physics in Section 6.2; (ii) cell technology starting from the growth of the polycrystalline $Cu(In,Ga)Se_2$ absorber up to device finishing by heterojunction formation and window layer deposition (Section 6.3). This section also discusses options which can be used to design the electronic properties of the absorber material as well as basic technologies for module production; (iii) the electronic properties of the finished heterostructure and some methods of analysing them (Section 6.4); (iv) finally, Section 6.5 discusses the photovoltaic potential of wide-gap chalcopyrites, namely $CuGaSe_2$ and $CuInS_2$, as well as that of the pentenary alloy system $Cu(In,Ga)(S,Se)_2$ and the possibility of building graded-gap structures with these alloys. Regarding In and Ga as rare materials has stimulated the development of the kesterite compounds, in which indium and gallium are replaced by a pair of Group II and Group IV elements such as Zn–Sn.

This chapter can only very briefly cover those scientific issues that are relevant for photovoltaic applications. For other important points and for more detailed information, we refer the reader to the literature. More about the structural properties of $Cu(In,Ga)Se_2$ can be found, for example, in Shay and Wernick (1975), Kazmerski and Wagner (1985), Coutts *et al.* (1986), Rockett and Birkmire (1991), Schock (1996), Bube (1998) and Rau and Schock (1999). Interface properties of $Cu(In,Ga)Se_2$ and related compounds have been reviewed by Scheer (1997). For up-scaling and module technologies see, for example, Dimmler and Schock (1996), and for economic aspects, see Zweibel (1995). A comprehensive overview of $Cu(In,Ga)Se_2$ (and CdTe)-based photovoltaics is further given in the book by Scheer and Schock (2011).

6.2 Material properties

6.2.1 *Basics*

$CuInSe_2$ and $CuGaSe_2$, the materials that form the alloy $Cu(In,Ga)Se_2$, belong to the semiconducting I–III–VI_2 materials family that crystallise in the tetragonal chalcopyrite structure. The chalcopyrite structure of, for example, $CuInSe_2$, is obtained from the cubic zinc-blende structure of II–VI materials like ZnSe by occupying the Zn sites alternately with Cu and In atoms. Figure 6.1 compares the two unit cells of the zinc-blende structure with the chalcopyrite unit cell. Each I (Cu) or III (In) atom has four bonds to the VI atom (Se). In turn each Se atom has two bonds to Cu and two to In. Because the strengths of the I–VI and III–VI bonds are in general different, the ratio of the lattice constants c/a is not exactly 2.1. The quantity $2 - c/a$ (which is -0.01 in $CuInSe_2$, $+0.04$ in $CuGaSe_2$) is a measure of the tetragonal distortion in chalcopyrites.

The bandgap energies of I–III–VI_2 chalcopyrites are considerably smaller than those of their binary analogues (this is the binary material where the I/III elements are replaced by their average II element; thus ZnSe is the binary analogue of

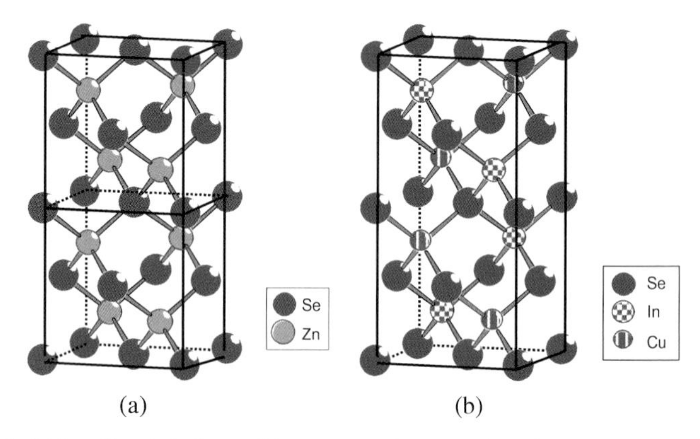

Figure 6.1 Unit cells of chalcogenide compounds. (a) Sphalerite or zinc-blende structure of ZnSe (two unit cells); (b) chalcopyrite structure of $CuInSe_2$. The metal sites in the two unit cells of the sphalerite structure of ZnSe are alternately occupied by Cu and In in the chalcopyrite structure.

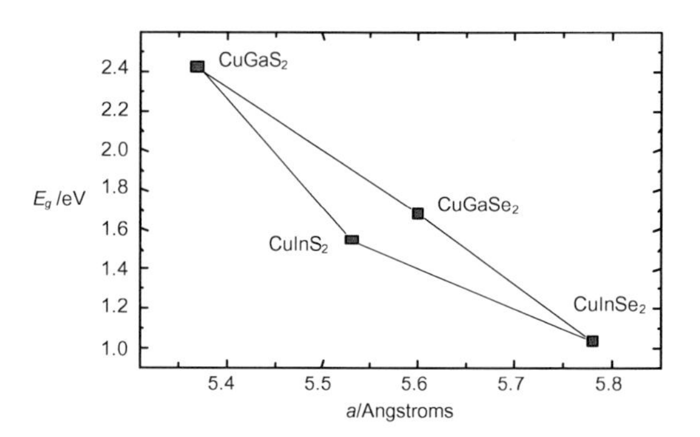

Figure 6.2 Bandgap energies E_g *vs*. lattice constants a of $Cu(In,Ga)(S,Se)_2$ system.

$CuGaSe_2$ and $Zn_{0.5}Cd_{0.5}Se$ that of $CuInSe_2$). This difference is because the Cu $3d$ band, together with the Se $4p$ band, forms the uppermost valence band in the Cu-chalcopyrites, which is not so in II–VI compounds. However, the system of copper chalcopyrites covers a wide range of bandgap energies E_g from 1.04 eV in $CuInSe_2$ up to 2.4 eV in $CuGaS_2$, covering most of the visible spectrum. The kesterite compounds Cu_2ZnSnS_4 and $Cu_2ZnSnSe_4$ have similar bandgaps to the corresponding chalcopyrite compounds $CuInS_2$ and $CuInSe_2$. Figure 6.2 summarises lattice constants a and bandgap energies E_g of this system. Any desired alloys between these compounds can be produced, as there is no miscibility gap in the entire system. We will discuss the status and prospects of this system in more detail in Section 6.5.

6.2.2 *Phase diagram*

Compared with all other materials used for thin-film photovoltaics, Cu(In,Ga)Se_2 has by far the most complicated phase diagram. Figure 6.3 displays the ternary phase diagram, which comprises all ternary Cu–In–Ga compounds. This complex ternary phase diagram can be reduced to a simpler pseudo-binary phase diagram along the tie line between Cu_2Se and In_2Se_3 (solid line in Fig. 6.3). Figure 6.4 shows the phase diagram of $CuInSe_2$ given by Haalboom *et al.* (1997). This investigation had a special focus on temperatures and compositions relevant for the preparation of thin films. The phase diagram in Fig. 6.4 shows the four different phases which have been found to be relevant in this range: the α-phase ($CuInSe_2$), the β-phase ($CuIn_3Se_5$), the δ-phase (the high-temperature sphalerite phase) and $Cu_{2-y}Se$. An interesting point is that all neighbouring phases to the α-phase have a similar structure. The β-phase is actually a defect chalcopyrite phase built by ordered arrays of defect pairs (Cu vacancies V_{Cu} and In–Cu anti-sites In_{Cu}). Similarly, $Cu_{2-y}Se$ can be viewed as constructed from the chalcopyrite by using Cu–In anti-sites Cu_{In} and Cu interstitials Cu_i. The transition to the sphalerite phase arises from disordering the cation (Cu, In) sub-lattice, and leads back to the zinc-blende structure (*cf.* Fig. 6.1a).

The existence range of the α-phase in pure $CuInSe_2$ on the quasi-binary tie line Cu_2Se–In_2Se_3 extends from a Cu content of 24% to 24.5%. Thus, the existence range of single-phase $CuInSe_2$ is astonishingly small and does even not include the

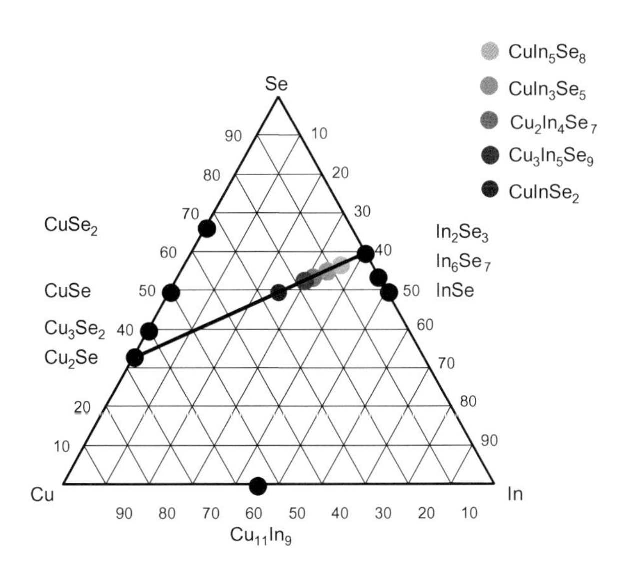

Figure 6.3 Ternary phase diagram of the Cu–In–Se system.

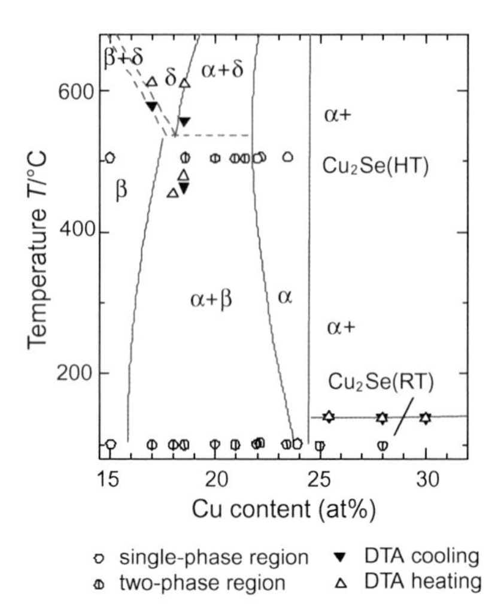

Figure 6.4 Quasi-binary phase diagram of $CuInSe_2$ established by Differential Thermal Analysis (DTA) and microscopic phase analysis. Note that at 25% Cu no single phase exists. After Haalboom *et al.* (1997).

stoichiometric composition of 25% Cu. The Cu content of absorbers for thin-film solar cells typically varies between 22 and 24 at.% Cu. At the growth temperature this region lies within the single-phase region of the α-phase. However, at room temperature it lies in the two-phase $\alpha + \beta$ region of the equilibrium phase diagram in Haalboom *et al.* (1997). Hence one would expect a tendency for phase separation in photovoltaic-grade $CuInSe_2$ after deposition. Fortunately, it turns out that partial replacement of In with Ga, as well as the use of Na-containing substrates, considerably widens the single-phase region in terms of (In + Ga)/(In + Ga + Cu) ratios (Herberholz *et al.*, 1999). Thus, the phase diagram hints at the substantial improvements actually achieved in recent years by the use of Na-containing substrates, as well as by the use of Cu(In,Ga)Se_2 alloys.

6.2.3 *Defect physics of Cu(In,Ga)Se_2*

Basics

The role of defects in the ternary compound $CuInSe_2$, and even more in Cu(In,Ga)Se_2, is of special importance because of the large number of possible intrinsic defects and the role of deep recombination centres in the performance of the solar cells. For insight into the defect physics of Cu(In,Ga)Se_2, see Cahen (1987) and Burgelman *et al.* (1997). The challenge of defect physics in Cu(In,Ga)Se_2, according to Zhang *et al.* (1998), is to explain three unusual effects in this material: (i) the ability to dope $Cu(In,Ga)Se_2$ with native defects; (ii) the structural tolerance to large off-stoichiometries; and (iii) the electrically neutral nature of the structural defects. It is obvious that the explanation of these effects significantly contributes to the explanation of the photovoltaic performance of this material. It is known that the doping of $CuInSe_2$ is controlled by intrinsic defects. Samples with p-type conductivity are grown if the material is Cu-poor and annealed under high Se vapour pressure, whereas Cu-rich material with Se deficiency tends to be n-type (Migliorato *et al.*, 1975; Noufi *et al.*, 1984). Thus, the Se vacancy V_{Se} is considered to be the dominant donor in n-type material (and also the compensating donor in p-type material), and the Cu vacancy V_{Cu} the dominant acceptor in Cu-poor p-type material.

Theoretical considerations

By calculating the metal-related defects in $CuInSe_2$ and $CuGaSe_2$, Zhang *et al.* (1998) found that the defect formation energies for some intrinsic defects are so low that they can be heavily influenced by the chemical potential of the components (i.e. by the composition of the material) as well as by the electrochemical potential of the electrons. For V_{Cu} in Cu-poor and stoichiometric material, the calculated formation energy actually becomes negative. This would imply the spontaneous formation of large numbers of these defects under equilibrium conditions. Low (but positive) formation energies are also found for the Cu-on-In anti-site Cu_{In} in Cu-rich material (this defect is a shallow acceptor which could be responsible for the p-type conductivity of Cu-rich, non-Se-deficient $CuInSe_2$). The dependence of the defect formation energies on the electron Fermi level could explain the strong tendency of $CuInSe_2$ to self-compensation and the difficulties of achieving extrinsic doping. The work of Zhang *et al.* (1998) provides a good theoretical basis for the calculation of defect formation energies and defect transition energies, which exhibit good agreement with experimentally obtained data.

Further important results in Zhang *et al.* (1997) are the formation energies of defect complexes such as $(2V_{cu},In_{Cu})$, (Cu_{In},In_{Cu}) and $(2Cu_i,Cu_{In})$, where Cu_i is an interstitial Cu atom. These formation energies are even lower than those of the corresponding isolated defects. Interestingly, $(2V_{cu},In_{Cu})$ does not exhibit an electronic transition within the forbidden gap, in contrast to the isolated In_{Cu}-anti-site, which is a deep recombination centre. As the $(2V_{cu},In_{Cu})$ complex is most likely to occur in In-rich material, it can accommodate a large amount of excess In (or likewise deficient Cu) and, at same time, maintain the electrical performance of the material. Furthermore, ordered arrays of this complex can be thought as the building blocks of a series of Cu–In–Se compounds such as $CuIn_3Se_5$ and $CuIn_5Se_8$ (Zhang *et al.*, 1997).

Additionally, the microscopic reasons for metastable changes have been intensively investigated in recent years. The (V_{cu},V_{Se}) divacancy complex (Lany and Zunger, 2006) and the In_{Cu} anti-site defect (Lany and Zunger, 2008) have been identified as sources for the metastabilities. Since both defects exist in multiple charge states their influence on the electronic behaviour of CIGS solar cells is rather complex. For example, even in equilibrium the divacancy complex has three different charge configurations, namely $(V_{cu},V_{Se})^{+}$, $(V_{cu},V_{Se})^{-}$, and $(V_{cu},V_{Se})^{3-}$, dependent on the position in the band diagram of the ZnO/CdS/CIGS heterostructure (Urbaniak and Igalson, 2009; Siebentritt *et al.*, 2010).

Device-relevant defects

Let us now concentrate on the defects experimentally detected in photovoltaic grade (and thus In-rich) polycrystalline films. In-rich material is in general highly compensated, with a net acceptor concentration of the order of $10^{16}\,cm^{-3}$. The shallow acceptor level V_{Cu} (which lies about 30 meV above the valence band) is assumed to be the main dopant in this material. As compensating donors, the Se-vacancy V_{Se} as well as the double donor In_{Cu} are considered. The most prominent defect is an acceptor level about 270–300 meV above the valence band, which has been reported by several groups from deep-level transient spectroscopy (Igalson and Schock, 1996) and admittance spectroscopy (Schmitt *et al.*, 1995; Walter *et al.*, 1996b). This defect is also present in single crystals (Igalson *et al.*, 1995).

As an example, Fig. 6.5 displays a defect density spectrum obtained from admittance spectroscopy by the method of Walter *et al.* (1996a). The transition at ~300 meV exhibits a broadened energy distribution with a tail in the defect density towards larger energies. This tail-like distribution is best described by a characteristic energy E^*, as shown in Fig. 6.5. This defect is detected, not only in In-rich, but in equal amounts also in Cu-rich polycrystalline materials (Herberholz, 1998). An assignment of this defect to the Cu_{In} anti-site is in agreement with the theoretical calculations of Zhang *et al.* (1998) as well as with the proposition of several experimentalists. The importance of this transition derives from the fact that its concentration is related to the open-circuit voltage of the device (Herberholz *et al.*, 1997a) and that the defect seems to be involved in the defect metastability (Igalson and Schock, 1996) (*cf.* Section 6.4.5).

The lower-energy transition in Fig. 6.5 is attributed to interfacial defects rather than to a bulk defect (Herberholz *et al.*, 1998) because its activation energy can vary between 50 meV and 250 meV depending on air-annealing prior to the measurement (Rau *et al.*, 1999a). Thus, the activation energy of this transition measures the depth ΔE_{Fn} from the vacuum level of the (electron) Fermi level and the conduction-band

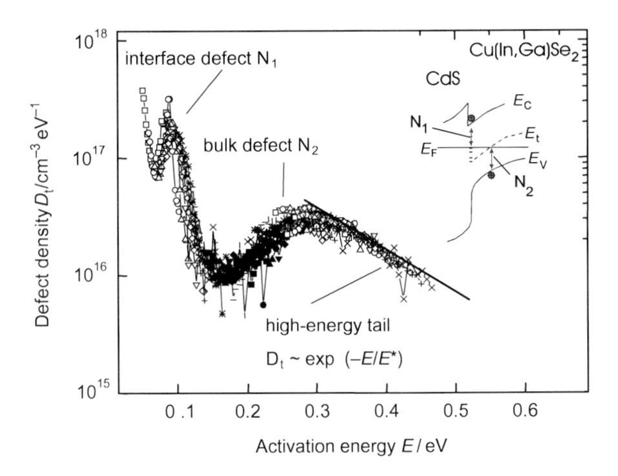

Figure 6.5 Defect density spectrum obtained from admittance spectroscopy of a ZnO–CdS–CuInGaSe$_2$ heterojunction. The peaks N$_1$ and N$_2$ can be related to interface and bulk defects (see inset).

energy at the Cu(In,Ga)Se$_2$ surface (Herberholz *et al.*, 1998), as shown in the top right-hand of Fig. 6.5.

6.3 Cell and module technology

6.3.1 *Structure of the heterojunction solar cell*

The complete layer sequence of a ZnO/CdS/Cu(In,Ga)Se$_2$ heterojunction device is shown in Fig. 6.6. It consists of a typically 1 μm thick Mo layer deposited on a soda-lime glass substrate and serving as the back contact for the solar cell. The Cu(In,Ga)Se$_2$ is deposited on top of the Mo back electrode as the photovoltaic absorber material. This layer has a thickness of 1–2 μm. The heterojunction is then completed by chemical bath deposition (CBD) of CdS (typically about 50 nm) and sputter deposition of a nominally undoped (intrinsic) i-ZnO layer (usually of thickness 50–70 nm) and then a heavily doped ZnO layer. As ZnO has a bandgap energy of 3.2 eV it is transparent for the main part of the solar spectrum and therefore serves as the window layer of the solar cell.

We will first mention four important technological innovations which, during the last decade, have led to a considerable improvement of the efficiencies. These steps are the key elements of the present Cu(In,Ga)Se$_2$ technology.

- The film quality has been substantially improved by the crystallisation mechanism induced by the presence of Cu$_y$Se ($y < 2$). This process is further

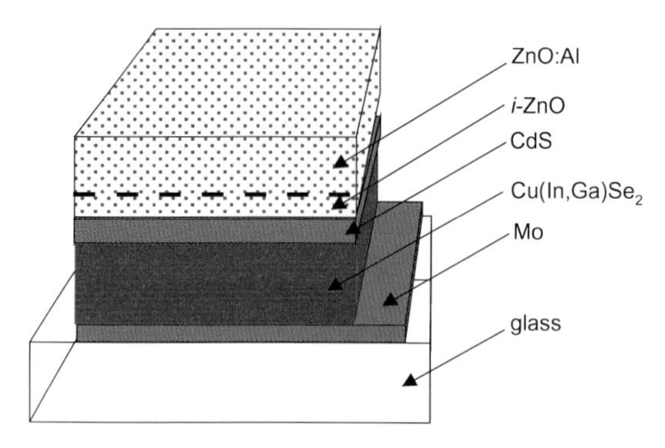

Figure 6.6 Schematic layer sequence of a standard ZnO-CdS-Cu(In,Ga)Se_2 thin-film solar cell.

supported by a substrate temperature close to the softening point of the glass substrate (Stolt *et al.*, 1993).

- The glass substrate has been changed from Na-free glass to Na-containing soda-lime glass (Hedström *et al.*, 1993; Stolt *et al.*, 1993). This has led to an enormous improvement of the efficiency and reliability of the solar cells, as well as to a larger process tolerance. It was first assumed that this improvement was due to better match of thermal expansion coefficients, but the beneficial impact of Na — diffusing from the substrate through the Mo back contact — on the growth of the absorber layer and its structural and electrical properties was soon recognised.
- Initially, the absorbers consisted of pure $CuInSe_2$. The partial replacement of In with Ga (Devaney *et al.*, 1990) made a further noticeable improvement, which increased the bandgap of the absorber from 1.04 eV to 1.1–1.2 eV for high-efficiency devices. The benefit of 20–30% Ga incorporation stems not only from the better bandgap match to the solar spectrum but also from the improved electronic quality of Cu(In,Ga)Se_2 as compared with pure $CuInSe_2$ (Hanna *et al.*, 2000; Herberholz *et al.*, 1999).
- The counter-electrode for the $CuInSe_2$ absorber of the earlier cells was a 2 μm thick CdS layer laid down by Physical Vapour Deposition (PVD). This has been replaced by a combination of a 50 nm thin CdS buffer layer laid down by chemical bath deposition (Potter *et al.*, 1985; Birkmire *et al.*, 1989; Mauch *et al.*, 1991) and a highly conductive ZnO window layer.

The effect of these four items on the electronic properties and performance of Cu(In,Ga)Se_2 solar cells will be considered in detail below, as we discuss the preparation of a Cu(In,Ga)Se_2 solar cell step by step.

6.3.2 *Absorber preparation techniques*

Basics

The preparation of Cu(In,Ga)Se_2-based solar cells starts with the deposition of the absorber material on an Mo-coated glass substrate (preferably soda-lime glass). The properties of the Mo film and the choice of the glass substrate are of primary importance for the final device quality, because of the importance of Na, which diffuses from the glass through the Mo film into the growing absorber material. Some processes use blocking layers such as SiN_x, SiO_2 or Cr between the glass substrate and the Mo film to prevent the out-diffusion of Na. Instead, Na-containing precursors like NaF, Na_2Se or NaS are now deposited prior to absorber growth to provide a controlled, more homogeneous, incorporation of Na into the film. The control of Na incorporation in the film from precursor layers allows the use of other substrates like metal or polymer foils. There seems to be no fundamental efficiency limitation due to the substrate as long as a proper supply of sodium is provided.

During absorber deposition, a $MoSe_2$ film forms at the Mo surface (Wada *et al.*, 1996; Takei *et al.*, 1996). $MoSe_2$ is a layered semiconductor with p-type conduction, a bandgap of 1.3 eV and weak van der Waals bonding along the c-axis. If the layer were oriented parallel to the plane of contact, the $MoSe_2$ would inhibit adhesion of the absorber as well as leading to unfavourable electronic transport. Fortunately, the c-axis is found to be in parallel with, and the van der Waals planes thus perpendicular to, the interface (Wada *et al.*, 1996). Because of the larger bandgap of the $MoSe_2$ compared with that of standard Cu(In,Ga)Se_2 films, the $MoSe_2$ layer provides an electronic mirror for the photogenerated electrons and at the same time provides a low-resistance contact for the holes (see Section 6.4.1).

Photovoltaic-grade Cu(In,Ga)Se_2 films have a slightly In-rich overall composition. The allowed stoichiometry deviations are astonishingly large, yielding a wide process window with respect to composition. Devices with efficiencies above 14% are obtained from absorbers with (In + Ga)/(In + Ga + Cu) ratios between 52% and 64% if the sample contains Na (Ruckh *et al.*, 1994a). Cu-rich Cu(In,Ga)Se_2 shows the segregation of a secondary $Cu_{2-y}Se$ phase preferentially at the surface of the absorber film. The metallic nature of this phase does not allow the formation of efficient heterojunctions. Even after *removal* of the secondary phase from the surface by etching the absorber in KCN, the utility of Cu-rich material for photovoltaic applications is limited, probably due to the high doping density of $10^{18}\,cm^{-3}$ in the bulk and the surface defects. However, the importance of the Cu-rich composition is given by its role *during* film growth. Cu-rich films have grain sizes in excess of 1 μm whereas In-rich films have much smaller grains. A model for film growth under Cu-rich compositions comprises the role of $Cu_{2-y}Se$ as a flux agent

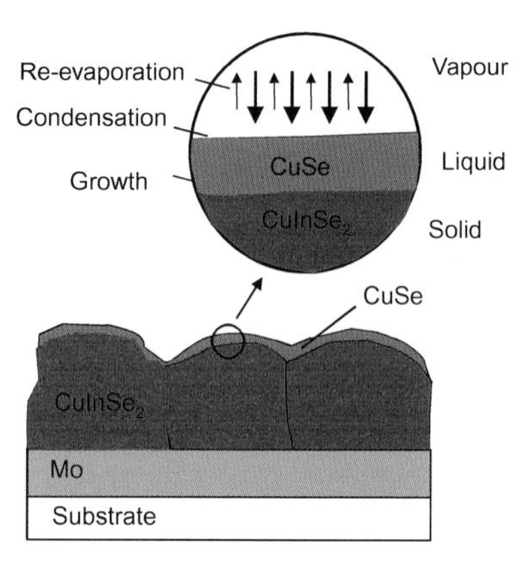

Figure 6.7 Schematic illustration of the growth of a Cu(In,Ga)Se_2 film under Cu-rich conditions. A quasi-liquid Cu–Se phase acts as a flux in a vapour-liquid solid growth mechanism.

during the growth process of co-evaporated films (Klenk *et al.*, 1993). This model for the growth of Cu(In,Ga)Se_2 in the presence of a quasi-liquid surface film of Cu_ySe is highlighted in Fig. 6.7. For Cu(In,Ga)Se_2 prepared by selenisation, the role of Cu_{2-y}Se is similar (Probst *et al.*, 1996); therefore growth processes for high-quality absorber materials have to go through a copper-rich stage and end with an indium-rich composition.

Co-evaporation processes

The absorber material yielding the highest efficiencies is Cu(In,Ga)Se_2 with a Ga/(Ga + In) content of ~20%, prepared by co-evaporation from elemental sources. Figure 6.8 sketches a co-evaporation set-up as used for the preparation of laboratory-scale solar cells and mini-modules. The process requires a maximum substrate temperature of ~550°C for a certain time during film growth, preferably towards the end of growth. One advantage of the evaporation route is that material deposition and film formation are performed during the same processing step. A feed-back loop based on a quadrupole mass spectrometer or an atomic absorption spectrometer controls the rate of each source. The composition of the deposited material with regard to the metals corresponds to their evaporation rates, whereas Se is always evaporated in excess. This precise control over the deposition rates allows for a wide range of variations and optimisations with different sub-steps or stages for film deposition and growth. These sequences are defined by the evaporation rates of the different sources and the substrate temperature during the course of deposition. Figure 6.9 illustrates some of the possibilities, starting with a simple

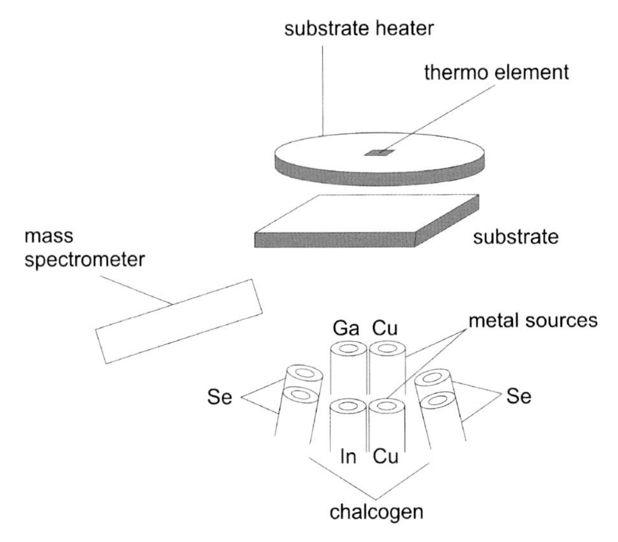

Figure 6.8 Arrangement for the deposition of $Cu(In,Ga)Se_2$ films by co-evaporation on a heated substrate. The rates of the sources are controlled by mass spectrometry.

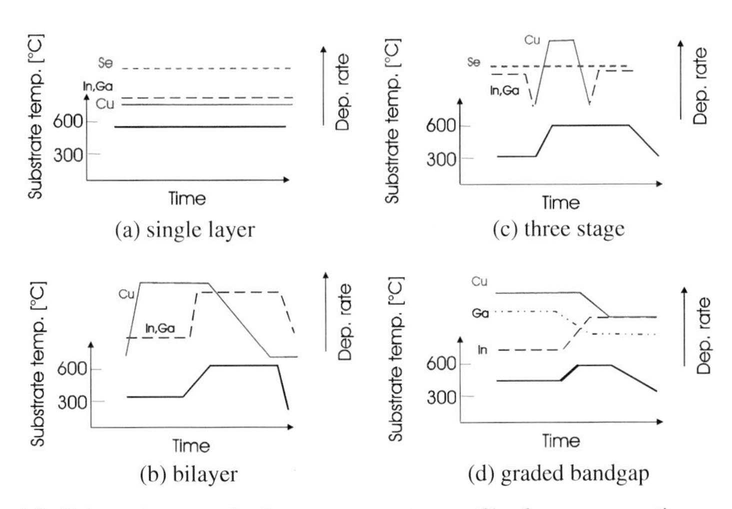

Figure 6.9 Schematic rate and substrate temperature profiles for co-evaporation processes. All processes lead to single-phase films. (a) Single-layer process without Cu-rich growth step; (b) bilayer process ('Boeing recipe') with Cu-rich growth at the start; (c) three-stage inverted process with intermediate Cu-rich growth; (d) growth of graded-gap films under Cu-poor conditions.

single-step process where all rates as well as the substrate temperature are kept constant during the whole process (Fig. 6.9a).

Advanced preparation sequences always include a Cu-rich stage during the growth process and end up with an In-rich overall composition in order to combine the large grains of the Cu-rich stage with the otherwise more favourable electronic properties of the In-rich composition. The first example of this kind of procedure is the so-called Boeing or *bilayer process* (Mickelsen and Chen, 1980), which starts with the deposition of Cu-rich Cu(In,Ga)Se_2 and ends with an excess In rate, as illustrated in Fig. 6.9b. Another possibility is the *inverted process*, where first $(In,Ga)_2Se_3$ (likewise In, Ga, and Se from elemental sources to form that compound) is deposited at a lower temperatures (typically around 300°C). Then Cu and Se are evaporated at an elevated temperature until an overall composition close to stoichiometry is reached (Kessler *et al.,* 1992). This process leads to smoother film morphology than the bilayer process. The most successful version of the inverted process is the so-called *three-stage process* (Gabor *et al.*, 1994) shown in Fig. 6.9c This process puts the deposition of In, Ga, and Se at the end of an inverted process to ensure the overall In-rich composition of the film even if the material is Cu-rich during the second stage. The three-stage process currently leads to the best solar cells. Variations of the Ga/In-ratio during deposition, as shown in Fig. 6.9d, allow the design of graded-bandgap structures (Gabor *et al.*, 1996).

Selenisation processes

The second class of absorber preparation routes is based on the separation of deposition and compound formation into two different processing steps. High efficiencies are obtained from absorber prepared by the selenisation of metal precursors in H_2Se (Binsma and Van der Linden, 1982; Chu *et al.*, 1984; Kapur *et al.*, 1987) and by rapid thermal processing of stacked elemental layers in a Se atmosphere (Probst *et al.*, 1996). These sequential processes have the advantage that large-area deposition techniques such as sputtering can be used for the deposition of the materials. The Cu(In,Ga)Se_2 film formation then requires a second step, the selenisation.

The very first large-area modules were prepared by the selenisation of metal precursors in the presence of H_2Se some 25 years ago (Mitchell *et al.*, 1988). A modification of this process provided the first commercially available Cu(In,Ga)Se_2 solar cells, manufactured by Siemens Solar Industries. This process is schematically drawn in Fig. 6.10. First, a stacked layer of Cu, In and Ga is sputter-deposited on the Mo-coated glass substrate. Then selenisation takes place under H_2Se. To improve device performance, a second thermal process under H_2S is added, resulting in an absorber that is Cu(In,Ga)$(S,Se)_2$ rather than Cu(In,Ga)Se_2.

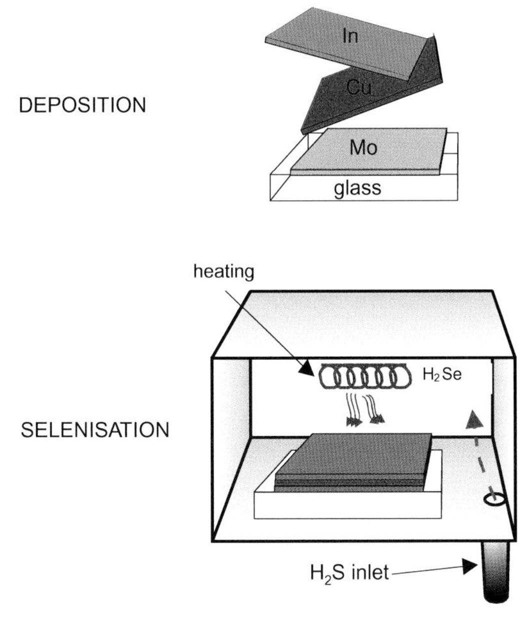

Figure 6.10 Illustration of the sequential process. Stacked metal layers are selenised and converted into $CuInSe_2$ in an H_2Se atmosphere.

A variation of this method that avoids the use of the toxic H_2Se during selenisation is the rapid thermal processing of stacked elemental layers (Probst *et al.*, 1996). Here the precursor includes a layer of evaporated elemental Se. The stack is then selenised by a rapid thermal process (RTP) in either an inert or a Se atmosphere. The highest efficiencies are obtained if the RTP is performed in an S-containing atmosphere (either pure S or H_2S).

On the laboratory scale, the efficiencies of cells made by these preparation routes are smaller by about 3% (absolute) as compared with the record values. However, on the module level, co-evaporated and sequentially prepared absorbers have about the same efficiency. Sequential processes need two or even three stages for absorber completion. These additional processing steps may counterbalance the advantage of easier element deposition by sputtering. Also the detailed and sophisticated control over composition and growth achieved during co-evaporation is not possible for the selenisation process. Fortunately, the distribution of the elements within the film grown during the selenisation process turns out to be close to what one could think to be an optimum, especially if the process includes the sulphurisation stage. Since the formation of $CuInSe_2$ is much faster than that of $CuGaSe_2$, and because film growth starts from the top, Ga is concentrated towards the back surface of the film. An increasing Ga content implies an increase in bandgap energy. This introduces a back-surface field, improving carrier collection at the same time as minimising back-surface recombination. In turn, S from the sulphurisation step is found preferentially towards the front surface of the film, where it reduces recombination losses and also increases the absorber bandgap in the space-charge region (SCR) of the heterojunction.

Other absorber deposition processes

Besides selenisation and co-evaporation, other deposition methods have been studied, either to obtain films with very high quality or to reduce cost of film deposition on large areas. Methods that are used to form epitaxial III–V compound films, such as molecular beam epitaxy (MBE) (Niki *et al.*, 1994) or metal organic chemical vapour deposition (MOCVD) (Gallon *et al.*, 1998) have revealed interesting features for fundamental studies, such as phase segregation and defect formation, but cannot be used to form the base material for high-efficiency solar cells.

Attempts to develop so-called low-cost processes include electrodeposition, (Abken *et al.*, 1998; Lincot *et al.*, 1998), screen printing and particle deposition (Eberspacher *et al.*, 1998). Electrodeposition can be carried out in either one or two steps. The crucial step is final film formation in a high-temperature annealing process. The recrystallisation process competes with the decomposition of the material, so process optimisation is quite difficult. Cells with good efficiencies were obtained by electrodeposition of a Cu-rich $CuInSe_2$ film and subsequent conditioning by a vacuum evaporation step of In(Se) (Ramanathan *et al.*, 1998a). Films prepared by spray pyrolysis did not lead to high-performance devices.

Influence of alkali ions sodium and potassium

The outstanding role played by Na in the growth of $Cu(In,Ga)Se_2$ films was recognised some years ago (Hedström *et al.*, 1993; Stolt *et al.*, 1993; Ruckh *et al.*, 1994a). In most cases, the Na comes from the glass substrate and diffuses into the absorber. But there are also approaches where Na is incorporated by the use of Na-containing precursors such as NaSe (Holz *et al.*, 1994; Nakada *et al.*, 1997), Na_2O_2 (Ruckh *et al.*, 1994a), NaF (Contreras *et al.*, 1997a) or Na_2S (Nakada *et al.*, 1998). Other alkali precursors have been investigated by Contreras *et al.* (1997a), who found that Na-containing precursors yield the best cell efficiencies. The most obvious effects of Na incorporation are better film morphology and higher film conductivity (Ruckh *et al.*, 1994a). Furthermore, the incorporation of Na induces beneficial changes in the defect distribution of the absorber films (Keyes *et al.*, 1997; Rau *et al.*, 1998a).

There are many explanations for the beneficial impact of Na, and it is likely that the incorporation of Na results in a variety of consequences. During film growth, the incorporation of Na leads to the formation of $NaSe_x$ compounds. This slows down the growth of $CuInSe_2$ and could at the same time facilitate the incorporation of Se into the film (Braunger *et al.*, 1998b). Also the widening of the existence range of the α-($CuInSe_2$) phase in the phase diagram, discussed above, as well as the reported larger tolerance to the Cu/(In + Ga) ratio of Na-containing thin films, could be explained in this picture. Furthermore, the higher conductivity of Na-containing films could result from the diminished number of compensating V_{se} donors. Wolf *et al.* (1998) investigated the influence of Na incorporation on $CuInSe_2$ film formation from stacked elemental layers by means of thin-film calorimetry. The addition of Na inhibits the growth of $CuInSe_2$ at temperatures below 380°C. The retarded phase formation is responsible for the better morphology in the case of Na-containing samples.

Another explanation put forward by Kronik *et al.* (1998) is that Na promotes oxygenation and passivation of grain boundaries. This could account for the observed enhancement of the net film doping by Na incorporation, through the diminished positive charge at the grain boundaries. It has in fact been observed that the surfaces of Na-containing films are more prone to oxygenation than are Na-free films (Braunger *et al.*, 1998a).

The above explanations deal with the role of Na during growth. However, the amount of Na in device-quality Cu(In,Ga)Se_2 films is of the order 0.1 at.%, which is a concentration of $10^{20}\,cm^{-3}$ (Niles *et al.*, 1997), and one may ask the question: where are these tremendous quantities of Na in the finished absorber? The electronic effect, i.e. the change of effective doping resulting from Na incorporation, is achieved at concentrations of $\sim 10^{16}\,cm^{-3}$, four orders of magnitude below the absolute Na content. It has long been believed that the main part of the Na is situated at the film surface and the grain boundaries. Final evidence for this hypothesis was obtained by Niles *et al.* (1999) with the help of high spatial resolution Auger electron spectroscopy.

Heske *et al.* (1996, 1997) investigated the behaviour of Na on the surface of polycrystalline Cu(In,Ga)Se_2 films by X-ray photoelectron spectroscopy (XPS). They found two different species of Na: (i) The first, denoted 'reacted', was observed on the air-exposed sample or after storing the sputter-cleaned sample for three days in an ultra-high vacuum (UHV). The second, denoted 'metallic', was found on clean samples either after annealing at 410 K in UHV or after deliberate Na deposition from a metallic source. The latter species is considered to be the active one during crystal growth. In addition, Heske *et al.* found an increase of band bending of $\sim$150 meV induced by the deposition of Na. This finding, as well as the occurrence of two different Na species, is consistent with results obtained from vacuum-cleaved single crystals (Klein and Jaegermann, 1996).

Another interpretation of the beneficial effect of Na is based on the incorporation of Na into the $Cu(In,Ga)Se_2$ lattice (Niles *et al.*, 1997). Niles and co-workers identified Na–Se bonds by means of XPS and concluded that the Na is built into the lattice, replacing In or Ga. The extrinsic defect $Na_{In/Ga}$ should then act as an acceptor and improve the *p*-type conductivity. The incorporation of Na into the $Cu(In,Ga)Se_2$ lattice is supported by X-ray diffraction measurements that indicate an increased volume of the unit cell (Contreras *et al.*, 1997a). Here, the authors assume that Na in a Cu site prevents the formation of the deep double donor In_{Cu}. Schroeder and Rockett (1997) found that Na driven into *epitaxial* $Cu(In,Ga)Se_2$ films at a temperature of 550°C decreases the degree of compensation by up to a factor 10^4. Schroeder and Rockett attributed their findings to an Na-enhanced reorganisation of the defects, which allows them to build electrically passive clusters.

We see from these numerous approaches that, despite the significance of Na incorporation, the benefit is far from being explained in terms of simple models. However, we feel that in view of the amounts of Na (~0.1 at.%) necessary for optimum film preparation, arguments based on its effect on film growth are slightly favoured over those based on the incorporation of Na into the completed film.

A substantial new finding is the role of potassium as key to the improvement of the quality of CIGS films (Würz *et al.*, 2012). A post-deposition treatment of Na containing films with K (Čhirila, 2013) led to the first device on a flexible polymer substrate with efficiency >20%. This treatment appears to passivate the surface of the grains by replacing Cu by K in the Cu-depleted surface (see below).

Influence of oxygen

Air annealing has been an important process step, crucial for good efficiency, especially of the early solar cells based on $CuInSe_2$. Also, though often not mentioned explicitly, an oxygenation step is still used for most of the present-day high-efficiency devices. The beneficial effect of oxygen was explained within the defect chemical model of Cahen and Noufi (1989). In this model, the surface defects at grain boundaries are positively charged Se vacancies V_{Se} (Fig. 6.11a) During air annealing, these sites are passivated by O atoms (Fig. 6.11b). Because of the decreased charge at the grain boundary, the band bending and the recombination probability for photogenerated electrons are reduced. The surface donors and their neutralisation by oxygen are important for the free $Cu(In,Ga)Se_2$ surface as well as for the formation of the $CdS/Cu(In,Ga)Se_2$ interface (Kronik *et al.*, 2000). Electrical analysis of oxidised and unoxidised samples revealed the validity of the Cahen–Noufi model for the earlier $CdS/CuInSe_2$ devices (Sasala and Sites, 1993),

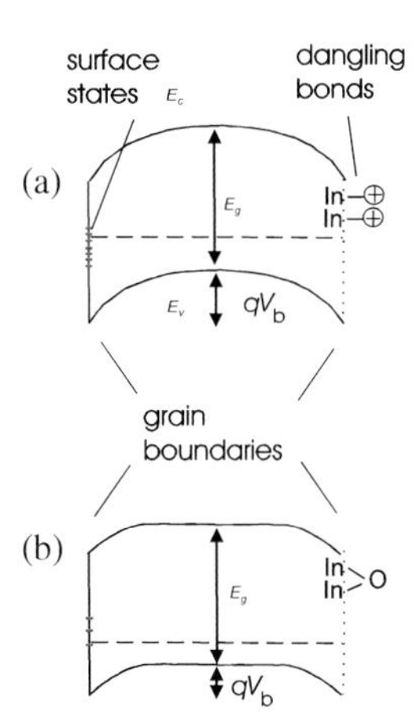

Figure 6.11 Band diagram of the conduction- and valence-band energies across a single grain of Cu(In,Ga)Se_2. (a) The electronic states at the grain boundaries are positively charged. This surface charge is compensated by the negative charges in the depleted grain. This induces the band-bending electronic states at the grain boundaries shown at the left-hand grain boundary, and the defect chemical equivalent, dangling bonds, shown at the right-hand boundary; (b) oxygen passivates these dangling bonds and reduces the band bending.

as well as for the later ZnO/CdS/Cu(In,Ga)Se_2 heterostructures (Rau *et al.*, 1999b; Kronik *et al.*, 1998).

The intriguing interplay between surface oxygenation and the deposition of the CdS buffer layer is visualised in Fig. 6.12. In the initial state (Fig. 6.12a), the film surface as well as the grain boundaries are electrically active due to the positively charged dangling bonds. These charges create a large space-charge region within the grain. Air annealing passivates the dangling bonds at both interfaces. The bands become essentially flat, and space charge essentially vanishes (Fig. 6.12b). Eventually, the chemical bath removes the passivating oxygen and thus re-establishes the beneficial type inversion of the film surface (Fig. 6.12c).

6.3.3 *The free $Cu(In,Ga)Se_2$ surface*

The surface properties of CIGS thin films are especially important, as this surface becomes the active interface of the completed solar cell. However, the band diagram of the ZnO/CdS/Cu(In,Ga)Se_2 heterojunction, especially the detailed structure close to the CdS/Cu(In,Ga)Se_2 interface, is still under debate. Figure 6.13 depicts three different possibilities: (a) the ordered defect compound model of Schmid *et al.* (1993); (b) the surface-state model of Rau *et al.* (1999a); and (c) the

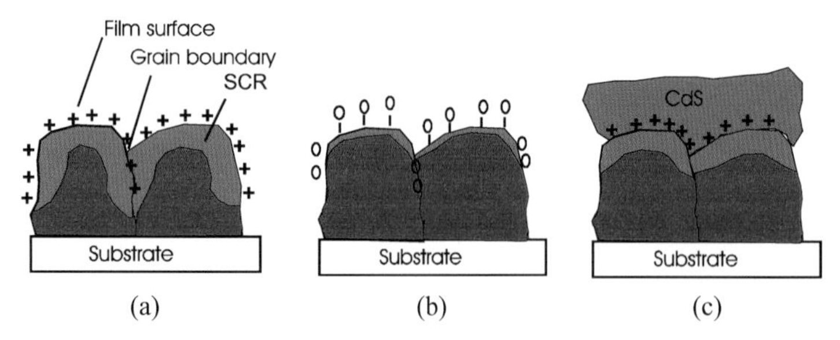

Figure 6.12 Illustration of grain boundary and interface charges during the fabrication of a Cu(In,Ga)Se_2/CdS heterojunction. (a) On as-prepared films, both grain boundaries and the film surface are positively charged because of the presence of the dangling bonds; (b) air exposure or air annealing neutralises these charges; (c) the oxide passivation of the surface is removed by the chemical bath deposition process of CdS. The re-established positive charges give rise to a type inversion of the $CuInSe_2$ surface.

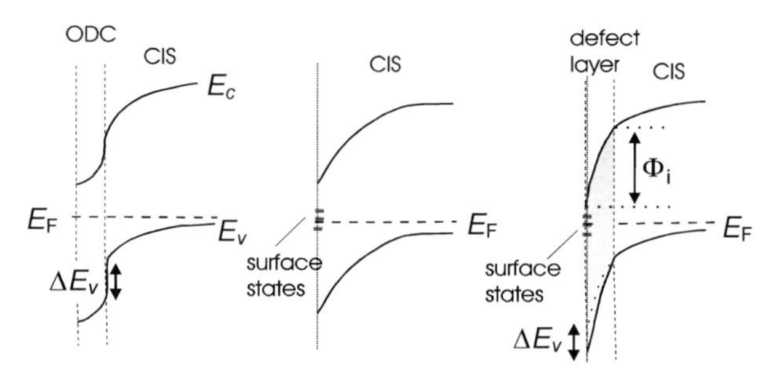

Figure 6.13 Models for the $CuInSe_2$ surface: (a) Segregation of a Cu-poor $CuIn_3Se_5$ ODC on the surface; (b) band bending due to surface charges; (c) band bending induces Cu-depletion at the surface, creating a surface defect layer. The valence-band energy is lowered due to Cu depletion. The defect layer provides an internal barrier Φ_i to the electron transport (Niemegeers *et al.*, 1998).

defect layer model of Niemegeers *et al.* (1998) and Herberholz *et al.* (1999). The free surfaces of as-grown Cu(In,Ga)Se_2 films exhibit two prominent features:

- The valence band-edge energy E_v lies above the surface-Fermi level E_F by about 1.1 eV for $CuInSe_2$ films (Schmid *et al.*, 1993). This energy is larger than the bandgap energy E_g^a of the bulk of the absorber material. This was taken as an indication of a widening of bandgap at the surface of the film. For the surfaces of $Cu(In_{1-x}Ga_x)Se_2$ thin films it was found that $E_F - E_v = 0.8\,eV$ (almost independent of the Ga content if $x > 0$) (Schmid *et al.*, 1996b).
- The surface composition of Cu-poor $CuInSe_2$, as well Cu(In,Ga)Se_2 films, corresponds to a surface composition of (Ga + In)/(Ga + In + Cu) of about 0.75

for a range of bulk compositions of $0.5 < (Ga+In)/(Ga+In+Cu) < 0.75$. Both observations have led to the assumption that a phase segregation of $Cu(In,Ga)_3Se_5$, the so-called *Ordered Defect Compound* (ODC), occurs at the surface of the films. The segregation of this β-phase would be compatible with the phase diagram (see Fig. 6.3) and, as this material displays n-type conductivity, could yield the explanation for the surface type inversion.

Unfortunately, the existence of a separate phase on top of standard $Cu(In,Ga)Se_2$ thin films has, to our knowledge, not yet been confirmed by structural methods such as X-ray diffraction, high resolution transmission electron microscopy or electron diffraction. Furthermore, if the surface phase exhibited the weak n-type conductivity of bulk $Cu(In,Ga)_3Se_5$, simple charge neutrality estimates (Herberholz *et al.*, 1999) show this would not be sufficient to achieve type inversion. The space-charge width in a CIGS absorber of doping density $3 \times 10^{16}\,cm^{-3}$ is approximately 300 nm. This would require a charge density of $2 \times 10^{18}\,cm^{-3}$ in a 15 nm thick ODC layer to warrant charge neutrality. This required n-type doping density is considerably more than that usually found in $Cu(In,Ga)_3Se_5$ compounds.

Based on these arguments, another picture of the surface of $Cu(In,Ga)Se_2$ thin films and of junction formation has emerged. As sketched in Fig. 6.13b, the type inversion can be viewed as resulting from the presence of shallow surface donors. This is the classical Bardeen picture (Bardeen, 1947) of Fermi level pinning by electronic states at semiconductor surfaces. Here, a surface-charge density of a few times $10^{12}\,cm^{-2}\,eV^{-1}$ is sufficient to pin the Fermi level at the neutrality level of free semiconductor surfaces. The positively charged surface donors in Fig. 6.13b are expected to be present in the metal-terminated (112) surface of $CuInSe_2$ because of the dangling bond to the missing Se (Cahen and Noufi, 1989). The type inversion vanishes on air exposure because the surface donors are passivated by the reaction of oxygen with the metal-terminated surface, as discussed in Section 6.3.2 in the context of the Cahen–Noufi model.

Surface states also play an important role in the completed heterostructure, where they become interface states at the absorber/buffer interface. As we noted above, one prominent feature of the defect spectrum of the ZnO/CdS/CIGS heterostructure (transition N_1 in Fig. 6.5) arises from the charging and discharging of these states.

The defect layer model shown in Fig. 6.13c represents in some sense a compromise between the ODC model and the surface defect model, as it takes into account a modification of the band structure due to the Cu deficiency of the surface as well as the presence of positively charged surface states due to the missing surface Se.

However, the defect layer model considers the surface layer not as n-type bulk material (as does the ODC model) but as a p^+-layer (*cf.* Fig. 6.13c). Furthermore, the defect layer is viewed not as the origin, but rather as the consequence of the natural surface type inversion. In contrast to the ODC model, surface states are responsible for the band bending. In turn, this band bending leads to the liberation of Cu from its lattice sites and to Cu migration towards the neutral part of the film (Herberholz *et al.*, 1999). The remaining copper vacancies V_{Cu}^- close to the surface result in a high density of acceptor states, i.e. the p^+-defect layer at the film surface.

6.3.4 *Grain boundaries*

As CIGS is a polycrystalline semiconductor the question of which way grain boundaries (GBs) influence the device performance and the question of how it is possible that a material with a grain size $<1\,\mu$m can have a performance similar to multicrystalline Si (grain size $\gg$1 mm) led to considerable research effort during the last decade. Many aspects have been reviewed by Rau *et al.* (2009). The electronic activity of GBs in CIGS appears controlled almost equally by crystallography and defect chemistry. Most GBs in properly grown films are of low energy, most are even twin boundaries (Abou-Ras *et al.*, 2007). Even GBs with a less favourable crystallography leading to a high concentration of structural defects may possibly not have as many electronically active defects (Yan *et al.*, 2007). Thus, GBs in CIGS do not seem to be very recombination-active from the start, which already provides a good base for a reasonable photovoltaic device. On top of that, the beneficial effect of O and Na appears to result at least partly from GB passivation, as discussed above. Additionally, it has been suggested that an important ingredient for the high performance of polycrystalline CIGS thin-film solar cells stems from an internal valence band offset ΔE_V at the GB, much like that at the CIGS surface (Persson and Zunger, 2003). This fact, sometimes referred to as a neutral barrier, provides a barrier to majority carrier transport and therefore prevents recombination with minority carriers. In fact such neutral barriers have been found experimentally (Siebentritt *et al.*, 2006; Hafemeister *et al.*, 2010). Though being certainly an additional beneficial feature, this valence-band offset cannot be the decisive reason for the low recombination activity of GBs in CIGS as has been shown by extensive numerical electronic simulations (Gloeckler *et al.*, 2005; Taretto and Rau, 2008).

6.3.5 *Heterojunction formation*

The ZnO/CdS/CIGS solar cell structure is usually completed by chemical bath deposition (CBD) of CdS (typically 50 nm) and by the sputter deposition of a nominally undoped (intrinsic) *i*-ZnO layer (usually of thickness 50–70 nm) and then by heavily doped ZnO. This three-step process appears at the moment to be mandatory for high-efficiency devices, but a convincing explanation of the need for such a relatively complicated three-layer structure, especially the role of the *i*-ZnO, is not available at the moment.

Buffer layer deposition

Surface passivation and junction formation is most easily achieved by the CBD deposition of a thin CdS film from a chemical solution containing Cd ions and thiourea (Kessler *et al.*, 1992). The CdS layer has several advantages:

- CBD deposition of CdS provides complete coverage of the rough polycrystalline absorber surface at a film thickness of only 10 nm (Friedlmeier *et al.*, 1996).
- The layer provides protection against damage and chemical reactions resulting from the subsequent ZnO deposition process.
- The chemical bath removes the natural oxide from the film surface (Kessler *et al.*, 1992) and thus re-establishes positively charged surface states and, as a consequence, the natural type inversion at the CdS/Cu(In,Ga)Se_2 interface.
- The Cd ions, reacting first with the absorber surface, remove elemental Se, possibly by the formation of CdSe.
- The Cd ions also diffuse to a certain extent into the Cu-poor surface layer of the absorber material (Ramanathan *et al.*, 1998b; Wada *et al.*, 1998), where they possibly form Cd_{Cu} donors, thus providing positive charges which support the type inversion of the buffer/absorber interface.
- As we discuss below, the open-circuit voltage limitations imposed by interface recombination can be overcome by a low surface recombination velocity in addition to the type inversion of the absorber surface. Thus one might conclude that interface states (except those shallow surface donors responsible for the type inversion) are also passivated by the chemical bath.

Owing to the favourable properties of CdS as a heterojunction partner and the chemistry of the CBD process, it is difficult to find a replacement, although avoiding CdS and the chemical bath step would be advantageous from the point of view of production. First, a toxic material such as CdS requires additional safety regulations and, second, the chemical bath deposition does not fit with the vacuum deposition steps of an in-line module fabrication. Research and development in this area relates to two issues: (i) the search for alternative materials for chemical deposition, and (ii) the development of ways to deposit the front electrode without an intermediate step in a chemical bath. Promising materials to replace CdS are In(OH,S), Zn(OH,S) and ZnSe. However, all these materials require additional precautions to be taken for the preparation of the absorber surface or front electrode deposition. Dry methods like evaporation or sputtering of (In,S) or ZnS are developed as a prerequisite for inline fabrication of cells without breaking the vacuum.

Window layer deposition

The most commonly used material for the preparation of the front electrode is ZnO doped with B or Al. In some cases, doping with Ga or In is claimed to be advantageous. The first large-area modules produced by ARCO Solar (later Siemens Solar Industries) had a ZnO:B window layer deposited by chemical vapour deposition (CVD). Later production facilities at Boeing and EuroCIS use sputtering processes. Present pilot production lines also favour sputtering. As mentioned above, an undoped i-ZnO layer with a thickness of about 100 nm is needed at the heterojunction in order to achieve optimum performance. Stability of the ZnO layer in humid environments is a major concern. Indium tin oxide (ITO) is superior in this respect.

6.3.6 *Module production and commercialisation*

Monolithic interconnects

One inherent advantage of thin-film technology for photovoltaics is the possibility of using monolithic integration for series connection of individual cells within a module. In contrast, bulk Si solar cells must be provided with a front metal grid, and each of these front contacts has to be connected to the back contact of the next cell to make a series connection. The interconnect scheme, shown in Fig. 6.14, has

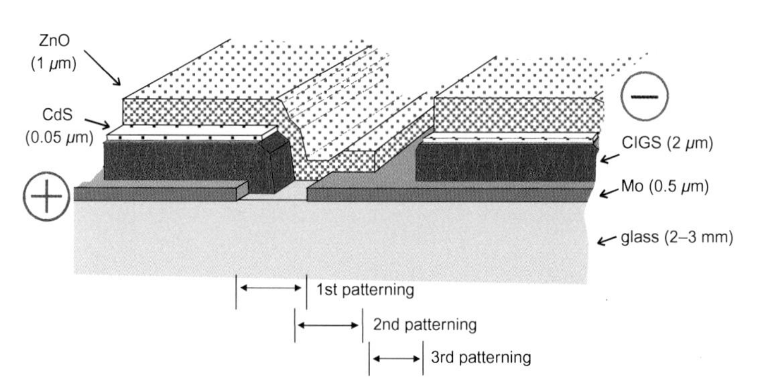

Figure 6.14 Interconnect scheme and patterning of a $Cu(In,Ga)Se_2$ (CIGS) based module.

to ensure that the front ZnO of one cell is connected to the back Mo contact of the next one. Three different patterning steps are necessary to obtain this connection.

The first interrupts the Mo back contact by a series of periodical scribes and thus defines the width of the cells, which is about 0.5–1 cm. A laser is normally used to pattern the Mo in this way. The second patterning step is performed after absorber and buffer deposition, and the final one after window deposition. Scribing of the semiconductor layer is performed by mechanical scribing or laser scribing. The total width of the interconnect depends not only on the scribing tools, but also on the reproducibility of the scribing lines along the entire module. The typical interconnect width is of the order of 300 μm. Thus, about 3–5% of the cell area must be sacrificed to the interconnects.

Module fabrication

Figure 6.15 shows the typical sequence for the production of a Cu(In,Ga)Se_2 module. The technologies for absorber, buffer and window deposition used for module production are the same as those discussed above for the production of small laboratory cells. However, the challenge of modules is to transform the laboratory-scale technologies to much larger areas. Basically, the scheme in Fig. 6.15 applies to both of the two concepts currently used for large-area absorber preparation, selenisation

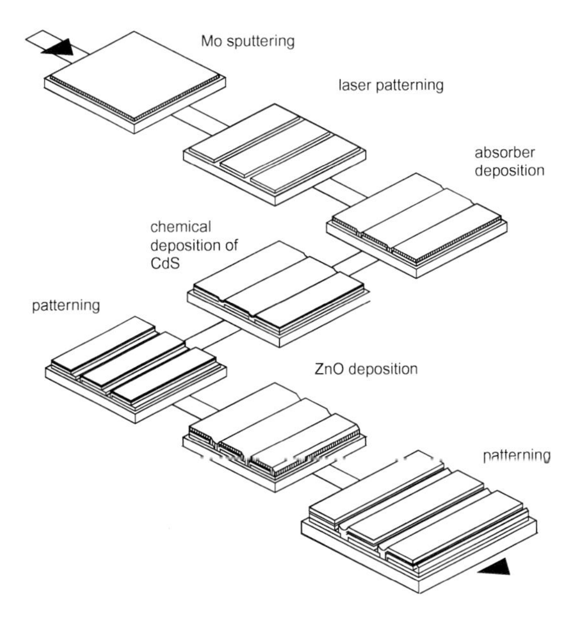

Figure 6.15 Process sequence for the fabrication process of a Cu(In,Ga)Se_2 module.

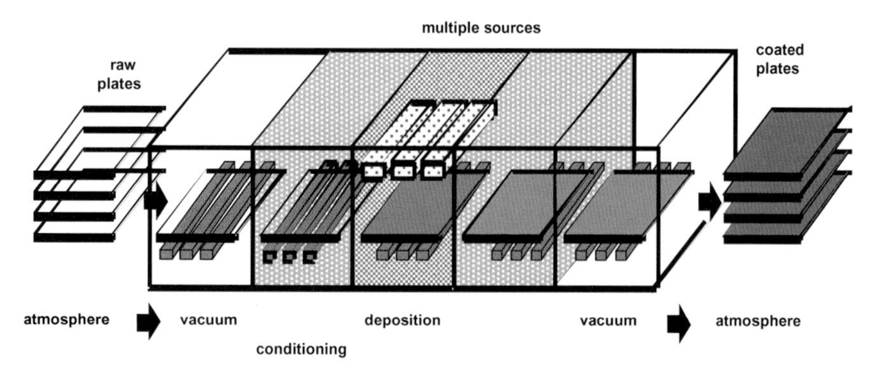

Figure 6.16 In-line evaporation system for the deposition of $CuInSe_2$ thin film. The line source evaporates the material from top to bottom (Dimmler and Schock, 1996).

and co-evaporation. The *selenisation* process uses as much off-the-shelf equipment and processing as possible (e.g. sputtering of the metal precursors) for fabricating $Cu(In,Ga)Se_2$ films. For *co-evaporation* on large areas, the Centre for Solar Energy and Hydrogen Research in Stuttgart (ZSW) has designed its own equipment, shown in Fig. 6.16, for an in-line process. Line-shaped evaporation sources allow continuous deposition of large-area, high-quality $Cu(In,Ga)Se_2$ films. The relatively high substrate temperatures that are necessary for high-quality material impose problems in handling very large-area glass sheets. Future process optimisation therefore implies reduction of the substrate temperature.

A bottleneck for the production is the deposition of the buffer layer in a chemical bath. First, it is not straightforward to integrate this process in a line consisting mainly of dry PVD processes. Second, it would be favourable to replace the CdS layer used at present. Replacement of CdS by a Zn or In compound from solution solves part of the problem (Kushiya, 1999). Complete, Cd-free dry processes have been demonstrated on large areas, but a large enough process window has still to be demonstrated for high-throughput production of large-area modules.

The transparent ZnO front electrode is put down by either CVD or sputtering. Each method has specific advantages as regards process tolerance, throughput, cost and film properties. The cell widths within the module — and therefore the relative losses from the patterning — mainly depend on the sheet resistance of the ZnO.

Module encapsulation is an important issue because module stability depends on proper protection against humidity. Low-iron cover glasses provide good protection, and hermetic sealing of the edges is mandatory to obtain stable modules.

Today's production lines use either co-evaporation of the elements or selenisation of metal layers. Modules with sizes of about 1 m^2 reach efficiencies up to 16% with either absorber deposition method. Typical efficiencies in production are about 13%. For reaching competitive cost the production volume has to exceed 100 MW/year; Solar Frontier has established a capacity of about 1GW/year (Kushiya, 2013).

6.4 Device physics

The device physics underlying electronic transport in thin-film solar cells is the same as for (say) bulk silicon solar cells. However, in view of the fact that thin-film Cu(In,Ga)Se_2 (and also CdTe cells) are heterojunction cells, and because the absorber layer is only a few times thicker than the space-charge region of the heterojunction, we shall develop a description which concentrates more on these specific features.

6.4.1 *Band diagram*

The equilibrium band diagram of the ZnO/CdS/Cu(In,Ga)Se_2/Mo heterostructure in Fig. 6.17 shows the conduction- and valence-band energies $E_{c,v}$ of the Cu(In,Ga)Se_2 absorber, the CdS buffer layer and the ZnO window. The latter consists of the intrinsic and the highly Al-doped layer. Here, we completely

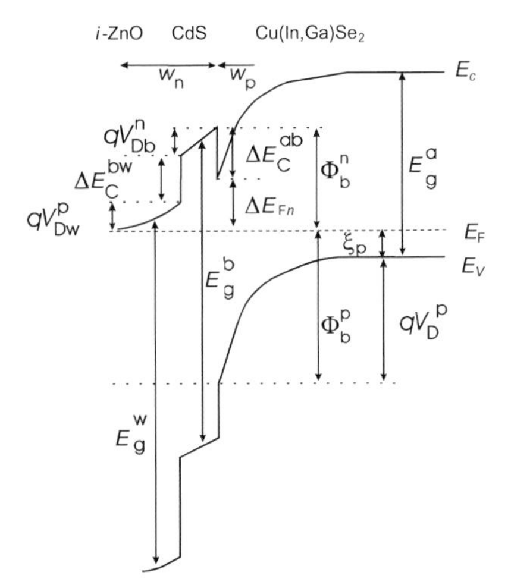

Figure 6.17 Band diagram of the CIGS heterojunction showing the conduction and valence band-edge energies E_c and E_V. The quantities Φ_b^p and Φ_b^n are the barriers for holes and electrons, ΔE_{Fn} is the energy distance between the Fermi level and the conduction-band energy at the CdS/CuIn(Ga)Se_2 heterointerface, ΔE_v^{ab} and ΔE_c^{ab} are the valence-band and conduction-band discontinuities at the buffer/absorber interface, and w_p and w_n are the widths of the space-charge regions in the p-type absorber and the n-type window/buffer respectively.

neglect the polycrystalline nature of the semiconductor materials, which in principle requires a two- or three-dimensional band diagram. We will restrict ourselves in the following to the implication of the one-dimensional diagram of Fig. 6.17.

Even in the one-dimensional model, some details of the band diagram are still not perfectly clear. The diagram in Fig. 6.17 concentrates on the heterojunction and does not show the contact between the Mo and Cu(In,Ga)Se_2 at the back side of the absorber. Another feature under debate but neglected here is a 10–30 nm thick defect layer on top of the Cu(In,Ga)Se_2 absorber, already discussed in Section 6.3.3.

The energetic quantities describing the band diagram in Fig. 6.17 are the bandgap energies E_g^x, where x = a, b, w for the absorber, buffer and window, respectively. The conduction/valence band offsets between the semiconductors are denoted $\Delta E_{c/v}^{xy}$. The built-in or diffusion voltage of the p-type absorber is V_D^p whereas that of the n-type window/buffer is the sum of the contributions V_{Db}^n from the buffer V_{Db}^n and V_{Dw}^n from the window layer. Note that the quantities $V_D^{p/n}$ as drawn in Fig. 6.17 are zero-bias quantities and change when an external voltage is applied. The important barriers Φ_b^p and Φ_b^n can be calculated from $\Phi_b^p = qV_b^p + \varsigma_p$ and $\Phi_b^n = \Delta E_c^{\mathrm{ab}} + \Delta E_{\mathrm{F}n} = \Delta E_c^{w\mathrm{b}} + \Delta E_{b1}^n + \Delta E_{b2}^n$, where $\xi_p = E_{\mathrm{F}} - E_v$ refers to the neutral bulk of the absorber and $\Delta E_{\mathrm{F}n} = E_v - E_{\mathrm{F}}$ to the Cu(In,Ga)Se_2/CdS interface. We discuss the impact of Φ_b^p on interface recombination and that of Φ_b^n on the fill factor in Section 6.4.3. Rau *et al.* (1999a) give a simplified approach to describing the heterojunction and computing Φ_b^p and Φ_b^n.

The most important quantities to be considered in the band diagram are the band discontinuities between the different heterojunction partners. Band discontinuities in terms of valence-band offsets ΔE_v^{ab} between semiconductors a and b are usually determined by photoelectron spectroscopy (for a discussion with respect to Cu-chalcopyrite surfaces and interfaces, see Scheer, 1997). The valence-band offset between a (011)-oriented Cu(In,Ga)Se_2 single crystal and CdS deposited by PVD at room temperature is determined as $\Delta E_v^{\mathrm{ab}} = -0.8\,(\pm 0.2)$ eV (Nelson *et al.*, 1993; Löher *et al.*, 1995), and therefore $\Delta E_c^{\mathrm{ab}} = E_g^{\mathrm{CdS}} - E_g^{\mathrm{CIS}} + \Delta E_v^{\mathrm{ab}} \approx 0.55$ eV, with the bandgaps $E_g^{\mathrm{CdS}} \approx 2.4$ eV and $E_g^{\mathrm{CIS}} \approx 1.05$ eV of CdS and $CuInSe_2$, respectively. Several authors have investigated the valence-band discontinuity between polycrystalline Cu(In,Ga)Se_2 films and CdS, and found values between 0.6 and 1.3 eV with a clear centre of mass around 0.9 eV, corresponding to a conduction-band offset of 0.45 eV (Scheer, 1997). Wei and Zunger (1993) calculated a theoretical value of $\Delta E_v^{\mathrm{ab}} = 1.03$ eV, which would lead to $\Delta E_c^{\mathrm{ab}} \approx 0.3$ eV.

The band alignment of polycrystalline $CuInSe_2$ and Cu(In,Ga)Se_2 alloys was examined by Schmid *et al.* (1993, 1996a), who found that the valence-band offset is almost independent of the Ga content. In turn, the increase of the absorber bandgap leads to a change of ΔE_c^{ab} from positive to negative values. The conduction-band offset between the CdS buffer and the ZnO window layer was determined by Ruckh *et al.* (1994b) to be 0.4 eV.

6.4.2 *Short-circuit current*

Optical losses

The short-current density i_{sc} that can be obtained from the standard 100 mW cm^{-2} solar spectrum (AM1.5) is determined, on the one hand, by *optical losses*, that is, by the fact that photons from a part of the spectrum are either not absorbed in the solar cell or are absorbed without generation of electron-hole pairs. On the other hand, not all photogenerated electron-hole pairs contribute to i_{sc} because they recombine before they are collected. We denote these as *recombination losses.*

Figure 6.18 shows the maximum short-circuit current that can be obtained from the standard AM 1.5G spectrum (Green, 1995, Appendix C) for photon energies $h\nu > E_g$, i.e. the photons that can contribute to the short-circuit current of a semiconductor of bandgap energy E_g. For pure polycrystalline $CuInSe_2$ with $E_g =$ 1.04 eV, this value is 46.8 mA cm^{-2}. For $E_g = 1.11$ eV, the bandgap energy of the best $Cu(In,Ga)Se_2$ solar cell, $i_{sc} = 43.6$ mA cm^{-2}. Now we estimate how much absorber material is needed to achieve this photocurrent. The light absorption in a semiconductor is described by the Lambert–Beer law. The irradiance I decays exponentially with depth x into the semiconductor according to

$$I(x) = I_o \exp(-\alpha x) \tag{6.1}$$

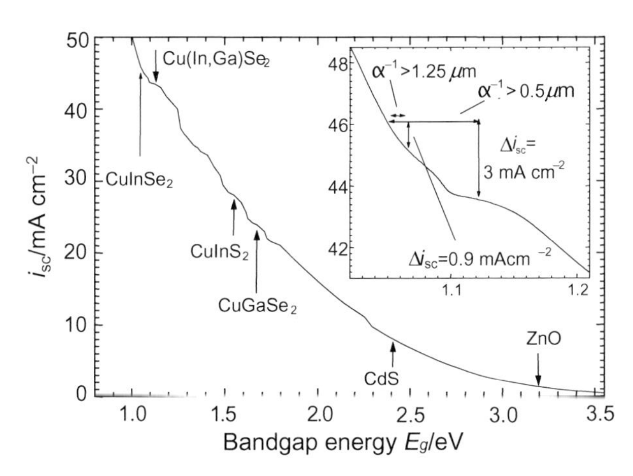

Figure 6.18 Short-circuit current density from the AM 1.5 solar spectrum and correspondence to the bandgap energies of various chalcopyrite compounds, of a typical $Cu(In,Ga)Se_2$ alloy ($E_g = 1.12$ eV), and of the heterojunction partners. The inset shows the losses that occur if less than 1.25/0.5 μm material is available for light absorption.

where I_0 is the incident irradiance and α the absorption coefficient. For direct semiconductors, α depends on the photon energy $h\nu$ according to

$$\alpha(h\nu) = a\frac{\sqrt{h\nu - E_g}}{h\nu} \tag{6.2}$$

The absorption coefficient of $Cu(In,Ga)Se_2$ with a low Ga content is reasonably described by Eq. 6.2 and $a \approx 8 \times 10^4\,\mathrm{eV}^{1/2}\,\mathrm{cm}^{-1}$. By reorganising Eq. 6.2 we can calculate the excess energy

$$\Delta h\nu = h\nu - E_g = \left(\frac{h\nu}{a}\alpha\right)^2 \tag{6.3}$$

of photons that have an absorption coefficient larger than a given α. For instance, with absorption lengths $L_\alpha = \alpha^{-1} = 1.255\,\mu\mathrm{m}$ and $0.5\,\mu\mathrm{m}$, we have $\Delta h\nu = 10\,\mathrm{meV}$ and $62.5\,\mathrm{meV}$. The inset in Fig. 6.18 shows that the losses Δi_{sc} corresponding to the photons that are not absorbed within 1.25 or $0.5\,\mu\mathrm{m}$ of $CuInSe_2$ are $0.9\,\mathrm{mA\,cm^{-2}}$ and $3.0\,\mathrm{mA\,cm^{-2}}$, respectively. The lower-energy photons are either absorbed at the back-metal/absorber interface or reflected out of the cell. Thus a typical absorber of thickness $1.5\,\mu\mathrm{m}$ absorbs all the light from the solar spectrum except for a negligible remnant corresponding to less than $1\,\mathrm{mA\,cm^{-2}}$.

Next, we have to recognise that light absorbed in the ZnO window layer does not contribute to the photocurrent. This loss affects absorption and photogeneration for photons of energy $>3.2\,\mathrm{eV}$ (the bandgap energy of ZnO). As shown in Fig. 6.18, this loss of high-energy photons costs about $1.3\,\mathrm{mA\,cm^{-2}}$. In addition, photons in the energy range $h\nu < 1.4\,\mathrm{eV}$ are absorbed by free carriers within the conduction band of the window material. High-efficiency heterojunction solar cells avoid free-carrier absorption by optimising the conductivity of the front electrode, not by a high concentration, but by the high mobility of the free carriers. In any case, for wide-gap absorbers with $E_g > 1.4\,\mathrm{eV}$ this loss can be neglected.

Another portion of the solar light is absorbed in the buffer layer. If, for instance, a CdS buffer layer caused a sharp cut-off of the spectral response at the bandgap of 2.4 eV, only a total of $38\,\mathrm{mA\,cm^{-2}}$ or $35.5\,\mathrm{mA\,cm^{-2}}$ would be available for the short-circuit current of a $CuInSe_2$ or $Cu(In,Ga)Se_2$ ($E_g = 1.11\,\mathrm{eV}$) solar cell, respectively. However, measurements of the external quantum efficiency (EQE) of a typical $ZnO/CdS/Cu(In,Ga)Se_2$ heterostructure reveal that the EQE typically drops by a factor of only ~0.8 in the wavelength range between the bandgap of CdS and that of the ZnO window layer. About 70–80% of the photons in the wavelength range 440–510 nm contribute to i_{sc} because the thin buffer layer does not absorb all photons and about 50% of the electron-hole pairs created in the buffer layer still contribute to the photocurrent (the hole recombination probability at the buffer/absorber interface is relatively low (Engelhardt *et al.*, 1999)).

Collection loss analysis

The most common characterisation method of solar cells other than current–voltage analysis is the measurement of the quantum efficiency. The EQE at a given wavelength λ is defined as the number of electron-hole pairs contributing to the photocurrent divided by the number of photons incident on the cell. A quantitative evaluation of the EQE can be used to determine the diffusion length L_e if the data are corrected for reflection losses and absorption losses in the window material and if the absorption data of the absorber material are known for the wavelength regime where the absorption length is in the order of L_e (Arora *et al.*, 1980). This analysis has been performed in the past by several authors for different types of devices (Klenk and Schock, 1994; Parisi *et al.*, 1998).

An alternative way to determine the diffusion length in solar cells is provided by Electron Beam Induced Current (EBIC) measurements. Two approaches are possible: planar EBIC, where the electron beam is scanned over the device surface, and junction EBIC, where the device is cleaved and the beam is scanned along the cross section (Jäger-Waldau *et al.*, 1991). For CIGS, the values for L_e extracted from EQE and EBIC measurements are $\sim$0.5–1.5 μm.

6.4.3 *Open-circuit voltage*

Diode characteristics

At open circuit, no current flows across the device and all photogenerated charge carriers have to recombine within the solar cell. The possible recombination paths for the photogenerated charge carriers in the $Cu(In,Ga)Se_2$ absorber are indicated in the band diagram of Fig. 6.19. Here we have considered recombination in the neutral bulk (A) and at the back surface of the absorber (A$'$), recombination in the space-charge region (B), and recombination at the buffer/absorber interface (C). The dotted lines indicate that the latter two mechanisms may be enhanced by tunnelling in the presence of a high built-in electrical field.

At the back contact we have drawn the thin $MoSe_2$ layer which forms during the first minutes of absorber deposition. As drawn here, the $MoSe_2$ has a small conduction-band offset with respect to the $Cu(In,Ga)Se_2$ bulk material and a small Schottky barrier at the Mo back contact. Both features are beneficial for device performance, because the conduction-band offset between the $Cu(In,Ga)Se_2$ absorber and the $MoSe_2$ acts as an electronic mirror (the so-called back-surface field) for the photogenerated electrons and diminishes back-surface recombination, and the narrow Schottky barrier provides no substantial resistance for holes between the absorber and the metallic back contact. We emphasise, however, that the details of this band diagram are still under debate.

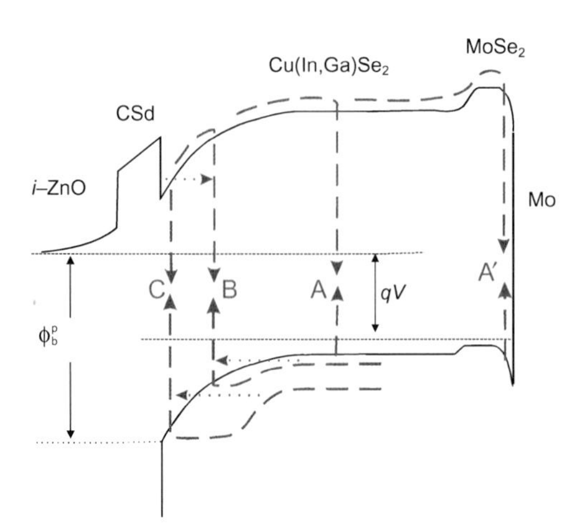

Figure 6.19 Recombination paths in a CdS–Cu(In,Ga)Se_2 junction. The paths A and A′ represent bulk and back-contact recombination. B and C result from space-charge and interface recombination. The dotted arrows indicate tunnelling.

The basic equations for the recombination processes (A–C) can be found in Bube (1992). All recombination current densities i_{rec} for processes A–C can be written in the form of a diode law

$$i_{rec} = i_o \left\{ \exp\left(\frac{qV}{\beta kT}\right) - 1 \right\} \tag{6.4}$$

where V is the applied voltage, β the diode quality factor, and kT/q the thermal voltage. The saturation current density i_o is in general a thermally activated quantity and may be written in the form

$$i_o = i_{oo} \exp\left(\frac{\tilde{E}_a}{kT}\right) \tag{6.5}$$

where $\tilde{E}_a$ is the activation energy and the prefactor i_{oo} is only weakly temperature-dependent. The quantities i_o and β depend on the details of each recombination mechanism. Since mechanisms A–C are connected in parallel, the strongest one will dominate the recombination loss.

At open circuit, the total recombination current density i_{rec} exactly compensates the short-circuit current density i_{sc}. Hence we can write the open-circuit voltage as

$$V_{oc} = \frac{\beta \tilde{E}_a}{q} - \frac{\beta kT}{q} \ln\left(\frac{i_{oo}}{i_{sc}}\right) = \frac{E_a}{q} - \frac{\beta kT}{q} \ln\left(\frac{i_{oo}}{i_{sc}}\right) \tag{6.6}$$

where we have assumed that $V_{\rm oc} > 3\beta kT/q$, which allows us to consider only the exponential term in Eq. (6.4). We have also replaced the activation energy $\tilde{E}_a$ by $E_a = \beta\tilde{E}_a$, which will prove in the following to be the 'true' activation energy of the carrier recombination processes.

We now discuss the recombination processes A–C in more detail.

Recombination in the absorber

In the following we shall assume an $n^+ - p$ junction, i.e. that the doping density on the n-side is much higher than on the p-side. Shockley's diode equation for such a single-sided junction yields the saturation current density for recombination in the neutral region of the (p-type) absorber. Knowing the square of the intrinsic carrier density $n_i^2 = N_c N_v \exp(-E_g/kT)$ we calculate the open-circuit voltage as

$$V_{\rm oc} = \frac{E_g}{q} - \frac{kT}{q}\ln\left(\frac{q D_e N_c N_v}{i_{\rm sc} N_{\rm A} L_e}\right) \tag{6.7}$$

where D_e is the diffusion constant for electrons, and $N_{c/v}$ the effective density of states in the conductance/valence band. For our calculations we have used the values $N_c = 6.7 \times 10^{17}\,{\rm cm}^{-3}$ and $N_v = 1.5 \times 10^{19}\,{\rm cm}^{-3}$ resulting from the density-of-states effective masses $m_e = 0.09\,m_{\rm o}$ and $m_h = 0.71\,m_{\rm o}$ for electrons and holes, respectively, where $m_{\rm o}$ is the free electron mass (Neumann, 1986). The quantity $N_{\rm A}$ is the acceptor density, and L_e is the diffusion length of the electrons. If this becomes comparable with the thickness d of the quasi-neutral region (QNR) of the absorber, the recombination velocity S_b at the back contact has to be taken into account (recombination path A′ in Fig. 6.19), and L_e in Eq. (6.7) has to be replaced by

$$L_{\rm eff} = L_e \frac{\cosh(l^{-1}) + s_b \sinh(l^{-1})}{s_b \cosh(l^{-1}) + \sinh(l^{-1})} \tag{6.8}$$

where $s_b = S_b L_e D_e$ and $l = L_e/d$.

Since the width of the SCR in thin-film solar cells is comparable with the film thickness, recombination in the SCR is important. The $V_{\rm oc}$-limitation due to recombination in the SCR of the absorber may be written in a form comparable to Eq. (6.7), namely

$$V_{\rm oc} = \frac{E_g}{q} - \frac{2kT}{q}\ln\left(\frac{kT D_e \pi/2\sqrt{N_c N_v}}{i_{\rm sc}\mathcal{E}_m L_e^2}\right) \tag{6.9}$$

where $\mathcal{E}_m = (2qN_{\rm A}V_{bm}/\epsilon_0\epsilon_s)^{1/2}$ is the electrical field at the position of maximum recombination. The quantity $\mathcal{E}_m$ depends on the doping density $N_{\rm A}$, the band bending V_{bm}, and the dielectric constant ϵ_s of the absorber. The dependence of

Eqs (6.7) and (6.9) on the doping density N_A is equal in that an increase of N_A by one order of magnitude yields an increase of V_{oc} of $\Delta V_{oc} = (kT \ln 10)/q \approx 60\,\text{mV}$. However, improving the open-circuit voltage by increasing the doping density is limited by the increased Auger recombination in the QNR and the enhancement of tunnelling in the SCR (see Green, 1996a and below).

Note that in Eq. (6.9) the activation energy E_a is given by $E_a = A\tilde{E}_a = 2\tilde{E}_a = 2(E_g/2) = E_g$ whereas in the diode equation for space-charge recombination, the saturation current density is $i_o \propto \exp(E_g/2kT)$ and the activation energy is only $E_g/2$. This demonstrates that we have to correct the activation energies obtained from, for example, Arrhenius plots of the saturation current density for the effect of non-ideal diode behaviour in order to obtain the activation energy relevant to V_{oc}.

In Fig. 6.20 we display the open-circuit voltage limitations given by Eqs (6.7) and (6.9) for a Cu(In,Ga)Se_2 solar cell with an absorber layer thickness of 1.5 μm, a bandgap energy E_g of 1.11 eV and a short-circuit current density i_{sc} of 35.4 mA cm^{-2}. The top and the bottom axes, showing the electron diffusion length L_e and the lifetime τ_e, are connected by $L_e = (D_e\tau_e)^{1/2}$ and a diffusion constant which is here assumed to be $D_e = 2.59\,\text{cm}^2\text{s}^{-1}$. As the open-circuit voltages in Eqs (6.7) and (6.9) can be shifted by the bandgap energy, we have used the right-hand axis of Fig. 6.20 to display the difference $(E_g/q) - V_{oc}$.

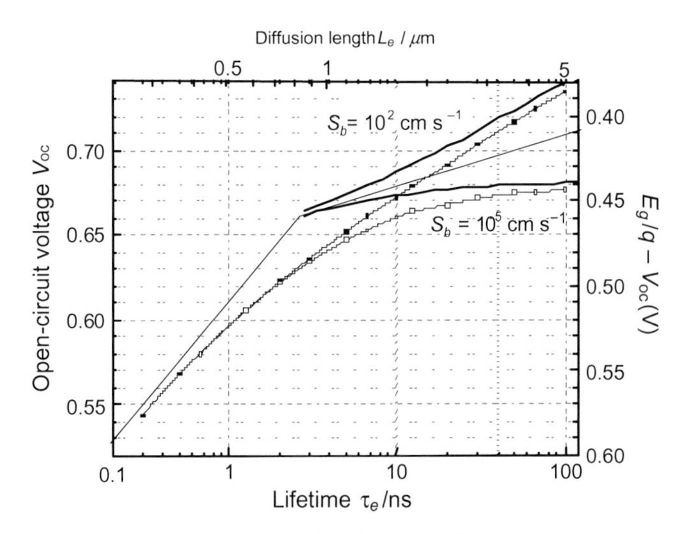

Figure 6.20 Correlation of the open-circuit voltage with lifetimes and diffusion lengths for a device with a bandgap of 1.12 eV. Solid lines are the results of Eq. (6.9) for $L_e < 0.7\,\mu$m. For $L_e > 0.7\,\mu$m, Eq. (6.7) holds if effective diffusion lengths that take back-surface recombination into account are introduced. The lines with symbols are the results of a complete device simulation.

For the high-efficiency $Cu(In,Ga)Se_2$ solar cell of Contreras *et al.* (1999), this difference is only (1.12–0.68) V $= 0.44$ V. The open-circuit voltage of this device requires a lifetime of 30 ns or more, corresponding to a diffusion length of over 2 μm, thus exceeding the absorber thickness. Hence recombination at the back contact also has some influence on V_{oc} — a recombination velocity $S_b > 10^5\,\mathrm{cm\,s^{-1}}$ would hardly allow a V_{oc} of 680 mV.

Since the open-circuit voltage of reasonable $Cu(In,Ga)Se_2$ devices ($E_g/q - V_{oc} \approx 0.5$ V) is just at the threshold between SCR and QNR recombination, we have also conducted some numerical simulations using the software package SCAPS-1D (Niemegeers and Burgelman, 1996). Results for assumed back-surface recombination velocities $S_b = 10^2\,\mathrm{cm\,s^{-1}}$ and $10^5\,\mathrm{cm\,s^{-1}}$ are also displayed in Fig. 6.20. Here we see that recombination can be well described only outside a transition regime of 1 ns $< \tau_e <$ 30 ns (0.5 μm $< L_e <$ 3 μm) by the analytical approaches for SCR or QNR recombination. Within this parameter range, the recombination paths A, A′ and B contribute to recombination. Note that we have suppressed the interface recombination (path C) by setting the recombination velocity for holes at the front contact to $S_p = 10^2\,\mathrm{cm\,s^{-1}}$ and assuming a hole barrier $\Phi_b^p = 1$ eV.

Effective lifetimes for polycrystalline semiconductors

$Cu(In,Ga)Se_2$ solar cells are based on polycrystalline absorbers. Electronic transport in such devices is not completely covered by one-dimensional models. However, quasi-one-dimensional approaches are possible as long as the influence of grain boundaries on the recombination and charge distribution is not too strong. A first-order approximation is the replacement of the minority-carrier lifetime τ_e by an effective lifetime $\tau_{\mathrm{eff}}^{\mathrm{poly}}$, which includes the interface recombination velocity S_g at the grain boundaries. This is given by

$$\frac{1}{\tau_{\mathrm{eff}}^{\mathrm{poly}}} = \frac{1}{\tau_e^b} + \frac{2S_g}{g} \tag{6.10}$$

where τ_e^b is the minority-carrier lifetime within the grain volume and g denotes the grain size. With the help of Eq. (6.10), we can still use Eqs (6.8) and (6.10) if we also use the effective diffusion length $L_{\mathrm{eff}}^{\mathrm{poly}}$ for polycrystalline materials, given by

$$L_{\mathrm{eff}}^{\mathrm{poly}} = [(L_{\mathrm{eff}}^{\mathrm{mono}})^{-2} + 2S_g/(D_n g)]^{-1/2} \tag{6.11}$$

instead of $L_{\mathrm{eff}} = L_{\mathrm{eff}}^{\mathrm{mono}}$. For more details and especially for the limitations of Eqs (6.10) and (6.11), see Green (1996b), Brendel and Rau (1999) and Jensen *et al.* (2000).

Distribution of recombination centres

An approach to describing the temperature dependence of current–voltage curves which is useful for $Cu(In,Ga)Se_2$ devices was introduced in Walter *et al.* (1996b). This approach does not use recombination centres of a single energy within the forbidden gap, but rather a distribution of the form $D_T(E) = D_{T0}\exp(-E/kT^*)$, where the centres are exponentially distributed in energy. The defect density $D_T(E)$ has units of $cm^{-3}eV^{-1}$ and kT^* denotes the characteristic energy of the exponential distribution. The characteristic energy $E^* = kT^*$ is also seen in the defect density spectra obtained from admittance spectroscopy (see Fig. 6.5). Rau *et al.* (2000) give a rigorous mathematical treatment for the recombination current density under this assumption. The recombination current density can be written in the form

$$i_{\text{rec}} = i_{\text{oo}} \exp\left(\frac{qV}{\beta kT}\right) \exp\left(-\frac{E_g}{\beta kT}\right) \tag{6.12}$$

where the pre-exponential term i_{oo} is weakly temperature-dependent, and the diode quality factor is given by

$$\frac{1}{\beta} = \frac{1}{2}\left(1 + \frac{T}{T*}\right) \tag{6.13}$$

The importance of this approach is on the one hand that a defect distribution with a characteristic energy kT^* of the order of 100–150 mV is often observed in $Cu(In,Ga)Se_2$ as well as in $CuGaSe_2$. On the other hand, it has been shown by Walter *et al.* (1996b) and Engelhardt *et al.* (1998) that the temperature dependence of the current–voltage characteristics of high-efficiency $Cu(In,Ga)Se_2$ solar cells in the temperature range $200\,\text{K} < \text{T} < 350\,\text{K}$ is reasonably well described by Eq. (6.16). For lower temperatures and for $CuGaSe_2$ absorbers, the contribution of tunnelling becomes more important (Nadenau *et al.*, 2000).

Interface recombination

Here we consider the simple case of an inverted interface, as shown in Fig. 6.13. If the recombination centres are not too close to the conduction band, the recombination rate is dominated by the concentration $p|_{if}$ of free holes at the interface. From the band diagram and taking the interface recombination velocity for holes as S_p and the diffusion potential at the p-side of the heterojunction as V_D^p, the recombination current density is given by

$$i_{\text{rec}} = qS_p p|_{if} = qS_p N_v \exp\left(-\frac{qV_D^p + \xi}{kT}\right) = qS_p N_v \exp\left(-\frac{\Phi_b^p - qV}{kT}\right) \tag{6.14}$$

with $\xi = E_F - E_v$ in the QNR of the Cu(In,Ga)Se$_2$ absorber. In general, a voltage applied over the heterojunction does not drop only across the p-type part of the junction. Rather, a change ΔV in the externally applied bias is shared between the p-type and n-type part according to $\Delta V = (1-\alpha)\Delta V_D^p + \alpha\,\Delta V_D^n$ where $\Delta V_D^p / \Delta V_D^n$ is the ratio of the diffusion potential in the p-type and n-type components. Note that the calculation of the voltage share between the two heterojunction partners in a complicated heterojunction like that shown in Fig. 6.17 is not straightforward. Here we simply use a linear approach with the coefficient α $(0 \leq \alpha \leq 1)$. The diode law can be written equivalently by the use of a voltage-dependent barrier $\Phi_b^p(V) = \Phi_{b0}^p + \alpha V$, where Φ_{b0}^p is the barrier at zero bias. With this voltage-dependent barrier, we can rewrite the recombination current density as

$$\begin{aligned} i_{\rm rec} &= qS_p N_v \exp\left(-\frac{\Phi_{b0}^p + \alpha q V}{kT}\right)\exp\left(\frac{qV}{kT}\right) \\ &= qS_p N_v \exp\left(-\frac{\Phi_{b0}^p}{kT}\right)\exp\left(\frac{qV(1-\alpha)}{kT}\right) \end{aligned} \tag{6.15}$$

By comparison of the coefficients we find that the coefficient α is linked to the diode quality factor by $\beta = (1-\alpha)^{-1}$. Finally, we write the open-circuit voltage for interface recombination as

$$V_{\rm oc} = \frac{\Phi_{bf}^p}{q} - \frac{\beta kT}{q}\ln\left(\frac{qS_p N_v}{i_{\rm sc}}\right) \tag{6.16}$$

where $\Phi_{bf}^p = \beta\Phi_{b0}^p \geq \Phi_{b0}^p$ is the so-called flat-band barrier.

Early CuInSe$_2$ solar cells consisted of a thick CdS window layer laid down by physical vapour deposition onto the absorber. The doping density in the CdS and the CuInSe$_2$ absorber were approximately equal, so the band bending in the two heterojunction partners was also equal. In terms of Eq. (6.16) and the band diagram, the open-circuit voltage was determined by a relatively low value of Φ_b^p and a high value of S_p. The use of a highly doped ZnO layer and a CdS buffer layer of only several tens of nm in thickness in actual high-efficiency ZnO/CdS/Cu(In,Ga)Se$_2$ devices increases the band bending in the absorber and decreases the recombination velocity for holes at the hetero-interface. For these recent state-of-the-art devices, it is believed that recombination in the bulk of the absorber is the dominant loss mechanism.

Tunnelling

In the presence of high electrical fields, interface recombination as well as space-charge recombination may be enhanced by tunnelling. As shown in Fig. 6.19, the charge carriers do not have to overcome the entire energetic barrier, but only a part of it, in order to recombine. We denote this transport path as tunnelling-enhanced recombination because it can be described in very similar terms to the classical recombination mechanism merely by using modified recombination rates. A useful quantity within the theory of thermally assisted tunnelling is the tunnelling energy

$$E_{00} = \frac{q\hbar}{2}(N_{\mathrm{A},\epsilon_0\epsilon_s}/m*)^{1/2} \tag{6.17}$$

The tunnelling energy E_{00} (Padovani and Stratton, 1966) depends only on the material parameters N_A, the permittivity $\epsilon_0\epsilon_s$ and the effective tunnelling mass m^*.

First we consider the case of *tunnelling-enhanced bulk recombination* in the SCR. Here the modified recombination rate for tunnelling-assisted recombination can be integrated over the width of the space-charge region. For an exponential distribution of trap states, a convenient form of the diode law gives the diode quality factor as (Rau, 1999)

$$\frac{1}{\beta} = \frac{1}{2}\left(1 + \frac{T}{T^*} - \frac{E_{00}^2}{3(kT)^2}\right) \tag{6.18}$$

The recombination current density is again expressed by Eq. (6.12), but with the quality factor given by Eq. (6.18).

The case of tunnelling-enhanced interface recombination can be treated by analogy with the thermionic field emission theory of Schottky contacts (Padovani and Stratton, 1966). The recombination current density is

$$i_{\mathrm{rec}} = qS_pN_v\frac{\sqrt{\pi qV_b(x)E_{00}}}{kT\cosh(E_{00}/kT)}\exp\left(\frac{qV}{\beta kT}\right)\exp\left(\frac{-\Phi_b^p}{\beta kT}\right)\exp\left(\frac{\xi}{kT}\left(\frac{1}{\beta}-1\right)\right) \tag{6.19}$$

and the quality factor

$$\beta = \frac{E_{00}}{kT}\coth\left(\frac{E_{00}}{kT}\right) \tag{6.20}$$

is given in Padovani and Stratton (1966). Tunnelling-enhanced interface recombination is often invoked in interpretations of current–voltage curves of wide-bandgap materials such as $CuInS_2$ (Hengel *et al.*, 2000).

Recombination loss analysis

The analysis of the temperature dependence of current–voltage curves is one way to get closer access to the recombination mechanism relevant to the open-circuit voltage. Usually, current–voltage curves are recorded with different illumination

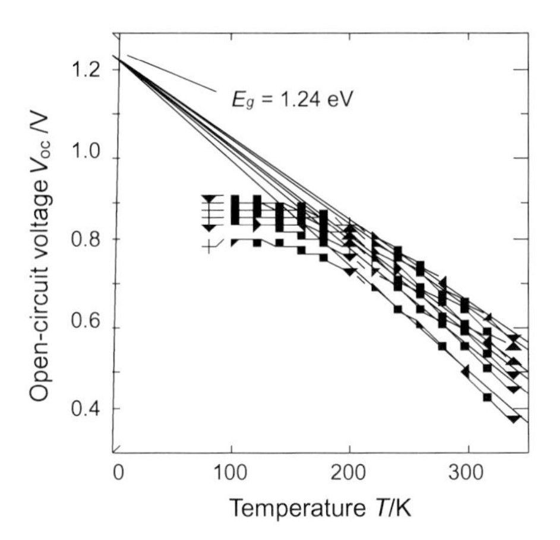

Figure 6.21 The temperature dependence of the open-circuit voltage under various illumination intensities extrapolates at 0 K to the bandgap energy.

intensities and at different temperatures. One evaluation method makes use of the temperature dependence of the open-circuit voltage. Following Eq. (6.7), the open-circuit voltages obtained at different temperatures T but identical illumination intensities (corresponding to roughly the same i_{sc}) display linear temperature dependence. Figure 6.21 shows that the $V_{oc}(T)$ curves in the temperature range 200–350 K extrapolate to an activation energy of 1.24 eV, indicating that the bandgap energy of the $Cu(In,Ga)Se_2$ absorber material is the relevant activation energy for recombination.

Such an evaluation works well as long as the diode quality factor in Eq. (6.6) is not too temperature-dependent. An evaluation scheme that is also able to invoke temperature-dependent quality factors makes use of the fact that the saturation current density of all recombination processes discussed above can be written in the form

$$i_o = i_{oo} \exp\left(\frac{\tilde{E}_a}{kT}\right) = i_{oo} \exp\left(\frac{E_a}{\beta kT}\right) \tag{6.21}$$

Reorganisation of Eq. (6.21) yields

$$\beta \ln\left(\frac{i_o}{i_{oo}}\right) = \frac{E_a}{kT} \tag{6.22}$$

where $E_a = \Phi_b^p$ for interface and $E_a = E_g$ for bulk recombination. Thus a modified Arrhenius plot of $\beta \ln i_o$ *vs.* $1/T$ should yield a straight line with the relevant activation energy E_a as the slope.

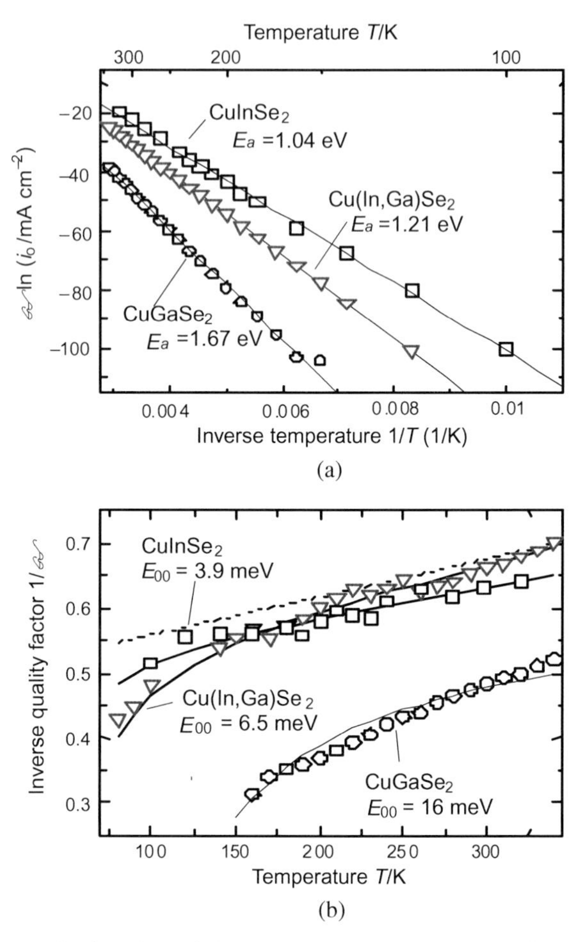

Figure 6.22 (a) Arrhenius plots of the reverse saturation current multiplied by the diode quality factor. Straight lines indicating an activation energy equal to the respective bandgap energy are evidence for bulk recombination; (b) the experimental data of the inverse quality factor vs. temperature can be fitted to the model that includes tunnelling.

Figure 6.22a compares such modified Arrhenius plots of the saturation current densities i_0 derived from three different $Cu(In,Ga)Se_2$ based heterojunctions. The three curves stem from absorbers with different Ga contents $x = 0$, 0.28 and 1. The activation energies $E_a = 1.04$ eV, 1.21 eV and 1.67 eV extracted from the plots correspond to the bandgap energies of the respective absorber materials, indicating that recombination in all these devices occurs in the volume of the absorber.

The corrected Arrhenius plots yield the activation energy of the recombination process whether or not the transport processes are enhanced by tunnelling. More specific information on the transport mechanism is obtained from the temperature dependence of the quality factor β. We evaluate the quality factors with the help of Eq. (6.18), as shown in Fig. 6.22b. From the fits we find for the CuInSe$_2$ ($x = 0$) cell $E_{00} = 3.9\,\text{meV}$ and $kT^* = 95\,\text{meV}$, for the Cu(In,Ga)Se$_2$ cell $E_{00} = 6.5\,\text{meV}$ and $kT^* = 74\,\text{meV}$, and for CuGaSe$_2$ $E_{00} = 16\,\text{meV}$ and $kT^* = 281\,\text{meV}$. Thus an increasing Ga content in Cu(In,Ga)Se$_2$, in general, increases the contribution of tunnelling as expressed by the increase of the characteristic tunnelling energy E_{00} in Fig. 6.22b. However, for CuInSe$_2$ and Cu(In,Ga)Se$_2$ with *moderate* Ga content at room temperature the tunnelling contribution to recombination is insignificant. In contrast, for CuGaSe$_2$ one extracts significant values of E_{00} (Nadenau *et al.*, 2000).

6.4.4 *Fill factor*

The fill factor η_{fill} of a solar cell can be expressed in a simple way as long as the solar cell is well described by a diode law. Green (1986) gives the following phenomenological expression for the fill factor

$$\eta^0_{\text{fill}} = \frac{1 - \ln(\upsilon_{\text{oc}} - 0.7)}{1 + \upsilon_{\text{oc}}} \tag{6.23}$$

where υ_{oc} is the dimensionless quantity $\upsilon_{\text{oc}} = qV_{\text{oc}}/\beta kT$ and the fill factor η^0_{fill} results from the diode law alone. Thus η^0_{fill} depends on temperature as well as on the quality factor of the diode. Effects from series resistance R_s and shunt resistance R_{sh} also contribute to the fill factor losses. A good approximation is given by

$$\eta_{\text{fill}} = \eta^x_{\text{fill}}\left(1 - \frac{\upsilon_{\text{oc}} + 0.7}{\upsilon_{\text{oc}}}\frac{\eta^x_{\text{fill}}}{r_{\text{sh}}}\right) \tag{6.24}$$

where $\eta^x_{\text{fill}} = \eta^0_{\text{fill}}\,(1 - r_s)$, $r_s = R_s i_{\text{sc}}/V_{\text{oc}}$ and $r_{\text{sh}} = R_{\text{sh}} i_{\text{sc}}/V_{\text{oc}}$. The description of Cu(In,Ga)Se$_2$ solar cells in terms of Eqs (6.23) and (6.24) works reasonably well. For example, the typical high-efficiency cell of Contreras *et al.* (1999) has a fill factor of 79% and the values calculated from Eqs (6.23) and (6.24) are 78.0% ($V_{\text{oc}} = 678\,\text{mV}$, $i_{\text{sc}} = 35.2\,\text{mA}\,\text{cm}^{-2}$, $R_s = 0.2\,\Omega\,\text{cm}^2$, $R_{\text{sh}} = 10^4\,\Omega\,\text{cm}^2$, $\beta = 1.5$).

The dependence of the fill factor on the quality factor (bearing in mind that this also determines the open-circuit voltage) highlights the importance of this parameter for the output power of the solar cell. The diode quality factor β_{L} obtained from the illuminated current–voltage curve is often different from β_{D} of the dark curve, with $\beta_{\text{L}} > \beta_{\text{D}}$. One explanation for this important fact could be the finite barrier Φ^n_b for the electron transport across the CdS/Cu(In,Ga)Se$_2$ heterointerface.

Looking at Fig. 6.17, a band offset $\Delta E^{\text{ab}}_c \approx 0.3\,\text{eV}$ between the absorber and the buffer layer seems to represent a substantial barrier hindering the collection of photogenerated electrons. The buffer/absorber interface shown in Fig. 6.6 can be roughly looked at as a Schottky barrier with a height $\Phi_b^n = \Delta E_c^{\mathrm{ab}} + \Delta E_{\mathrm{F}n}$ and the photocurrent as the reverse current over this barrier. A simple calculation shows that the barrier does not provide a large series resistance to the cell as long as $\Phi_b^n < 0.5\,\mathrm{eV}$. The conductance of a Schottky contact with an effective Richardson constant $A^* = 100\,\mathrm{A\,cm^{-2}\,K^{-2}}$ is $G = (qA^*T/k)\exp[-\Phi_b^n/kT] = 1.4\,\mathrm{S\,cm^{-2}}$ for $\Phi_b^n = 0.5\,\mathrm{eV}$ and at a temperature $T = 300\,\mathrm{K}$. A series resistance of $0.7\,\Omega\,\mathrm{cm}^2$ is acceptable for a device with a reasonable open-circuit voltage. It has been shown by numerical calculations that a conduction-band offset ΔE_c^{ab} below 0.4 eV does not affect the device performance (Niemegeers *et al.*, 1995). However, at lower temperatures or for devices where the type inversion at the Cu(In,Ga)Se_2 is less pronounced, i.e. the quantity $\Delta E_{\mathrm{F}n}$ is larger than 200 meV, the barrier Φ_b^n becomes important as shown by experiments (Schmidt *et al.*, 2000) and numerical simulations (Topic *et al.*, 1997).

6.4.5 *Electronic metastabilities*

The slow change (over hours and days) of the open-circuit voltage of Cu(In,Ga)Se_2-based solar cells during illumination is a commonly observed phenomenon (Ruberto and Rothwarf, 1987; Sasala and Sites, 1993). Fortunately, it turns out that in most cases the open-circuit voltage increases with illumination time, a situation which is more favourable than that encountered in a-Si:H (Staebler and Wronski, 1977).

A first model for this slow change in open-circuit voltage was proposed in Ruberto and Rothwarf (1987). This relies on the reduction of interface recombination at the CdS/Cu(In,Ga)Se_2 interface by additional charges introduced into the CdS buffer layer either by illumination under open-circuit conditions or by application of forward bias in the dark. The model is based on the assumption that interface recombination is the dominant recombination mechanism in the solar cells. The increase of positive charges in the buffer layer is assumed to increase the barrier Φ_b^p and thus reduce interface recombination. However, as we noted above, the open-circuit voltages of high-efficiency devices are limited by recombination in the bulk (i.e. in the SCR) rather than at the interface. Since these devices also show light-soaking effects, another mechanism, possibly additional to that proposed in Ruberto and Rothwarf (1987), must be at work.

An important observation is that of persistent photoconductivity in Cu(In,Ga)Se_2 thin films (Rau *et al.*, 1998b) and single crystals (Seifert *et al.*, 1997). Meyer *et al.* (1999) relate the trapping of electrons as the origin of persistent photoconductivity (Fig. 6.23a) to the persistent increase of the charge density in the SCR of the heterojunction, as shown in Fig. 6.23b. This leads to another model for the

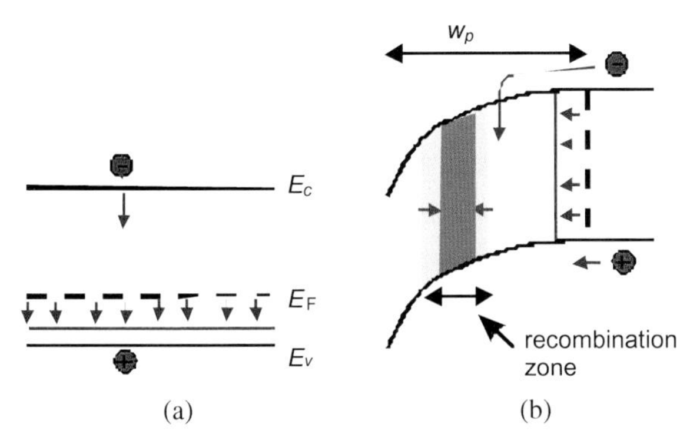

Figure 6.23 Illustration of persistent changes of (a) the density of free charge carriers in the bulk; (b) the charge density in the space-charge region.

open-circuit voltage relaxation in $Cu(In,Ga)Se_2$ solar cells: the gradual decrease of the electrical field in the SCR leads to a decrease of space-charge recombination, and finally to the increase of the open-circuit voltage during illumination.

The (V_{cu},V_{Se}) divacancy complex (Lany and Zunger, 2006) and the In_{Cu} antisite (Lany and Zunger, 2008) have been identified as sources for the metastable behaviour of $Cu(In,Ga)Se_2$. Both defects exist in multiple charge states and, consequently, their influence on the electronic behaviour of CIGS solar cells is rather complex. Figure 6.24a shows that even in the equilibrium band diagram of the heterostructure the divacancy complex has three different charge configurations, namely $(V_{cu},V_{Se})^+$, $(V_{cu},V_{Se})^-$ and $(V_{cu},V_{Se})^{3-}$ (Urbaniak and Igalson, 2009; Siebentritt *et al.*, 2010). Injection of excess minority-charge carriers via illumination or electrical bias, i.e. electrons into the neutral bulk, as well as holes into the inverted CdS/CIGS interface region, will have manifold consequences on the charge distribution and on the electrical properties of the device (Fig. 6.24b).

The reaction $(V_{Cu},V_{Se})^+ + e^- \rightarrow (V_{Cu}, V_{Se})^- + h^+$ will preferably take place in the neutral bulk and at the edge of the SCR, leading to an increase of net doping density and thereby reducing the series resistance, as well as the width of the SCR, as discussed above.

In contrast, the reaction $(V_{Cu},V_{Se})^{3-} + 2h^+ \rightarrow (V_{Cu}, V_{Se})^-$ taking place closer to the interface region will decrease the net doping and widen the SCR. Furthermore, the capture of holes into the $(V_{Cu}, V_{Se})^{3-}$ state will further diminish the negative charge density in the region close to the interface. As a consequence, the collection/injection barrier (Kniese *et al.*, 2008) for electrons at the CdS/CIGS interface will be decreased. Eventually, injection of minority carriers as sketched in Fig. 6.24b will result in at least two quite different effects: (i) the reduction of the

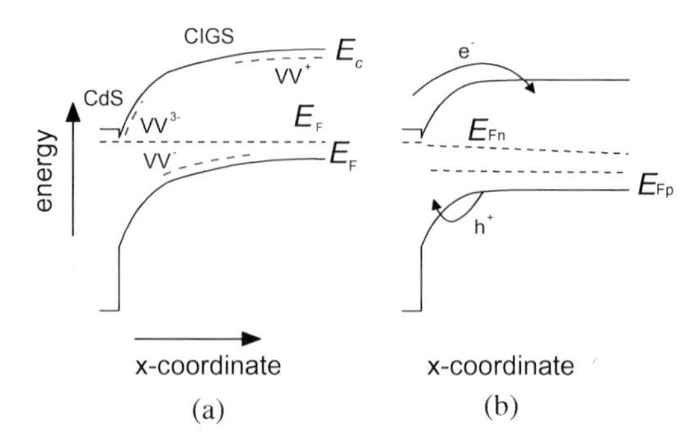

Figure 6.24 (a) Sketch of the equilibrium band diagram of the CdS/CIGS heterojunction involving the three charge states $(V_{Cu},V_{Se})^{+}$, $(V_{Cu},V_{Se})^{-}$, and $(V_{Cu},V_{Se})^{3-}$ of the divacancy; (b) under light or current bias, excess holes diminish the negative charge close to the CdS/CIGS interface and excess electrons increase the effective doping in the bulk of the material.

recombination current, and (ii) the reduction of the series resistance. Notably, these two effects correspond to the commonly observed features during light soaking of CIGS modules or solar cells: the increase of the open-circuit voltage and the increase of the fill factor.

6.5 Wide-gap chalcopyrites

The system $Cu(In,Ga)(S,Se)_2$ provides the possibility of fabricating alloys with a wide range of bandgap energies between the 1.04 eV of pure $CuInSe_2$ up to 2.4 eV for $CuGaS_2$. The achievement of higher voltages in the individual cells by increasing the bandgap of the absorber material is important for thin-film modules. An ideal range for the open-circuit voltage would be between 1.4 and 1.6 eV. This increased voltage would reduce the number of scribes needed for monolithic integration of the cells into a module. The thickness of the front and back electrodes could also be reduced because of the reduced current density. Higher bandgap materials also lower the temperature coefficient dP_{max}/dT of the maximum output power P_{max}, i.e. the loss of conversion efficiency with increasing cell temperature. Higher bandgap materials thus perform better at elevated temperatures. However, the highest efficiency in the chalcopyrite system is achieved with the relatively low-bandgap energy E_g of 1.12 eV, and attempts to maintain the high efficiency level at bandgaps of $E_g > 1.3$ eV have so far failed.

Table 6.1 compares the output parameters of the best chalcopyrite-based solar cells. This compilation clearly shows the superiority of $Cu(In,Ga)Se_2$ with a relatively low Ga content. The fact that the best $CuInSe_2$ devices have an efficiency of

Table 6.1 Operating parameters of the best $Cu(In,Ga)Se_2$, $CuGaSe_2$ and $Cu(In,Ga)S_2$ cells.

Material	E_g/eV	η/%	V_{oc}/mV	i_{sc}/mA cm^{-2}	η_{fill}/%	A/cm^2	Ref.
$Cu(In,Ga)Se_2$	1.12	20.3[a]	730	35.7	77.7	0.5	1
$Cu(In,Ga)Se_2$	1.18[b]	20.4	736	35.1	78.9	0.5	4
$Cu(In,Ga)Se_2$		20.8					5
$CuGaSe_2$	1.68	9.53[a]	905	14.88	70.8	0.471	2
$Cu(In,Ga)S_2$	1.53	12.6[a]	879	20.4	70.0	0.5	3

[a]Confirmed total area values; [b]estimate. References: 1. Jackson *et al.* (2011); 2. Young *et al.* (2003); 3. Merdes *et al.* (2011); 4. Čhirila *et al.* (2013); 5. ZSW press release (2013).

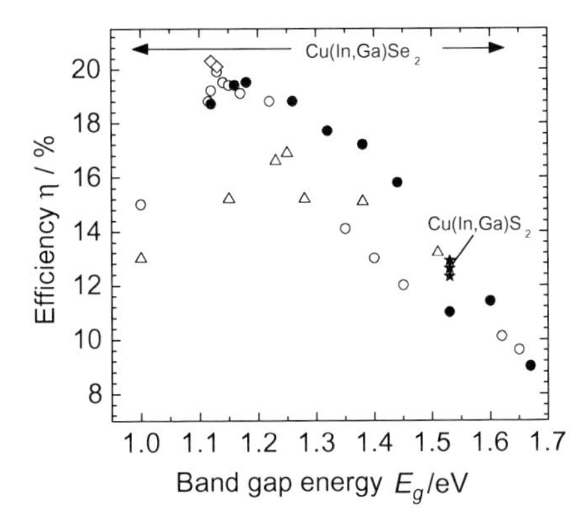

Figure 6.25 Efficiency η *vs.* bandgap energy E_g for several series of $Cu(In_{1-x}Ga_x)Se_2$ devices with varying Ga content x ($0 \leq x \leq 1$). The plot is redrawn after Contreras *et al.* (2012), and the data correspond to original data from this reference (full circles) and earlier data from Contreras *et al.* (2005) (open circles), Eisenbarth *et al.* (2009) (open triangles), as well as the record data from Jackson *et al.* (2011) (open diamonds). Additionally shown are $Cu(In,Ga)S_2$ results (stars) from Merdes *et al.* (2011).

3% below that of the best $Cu(In,Ga)Se_2$ device is due not only to the less favourable bandgap energy but also to the lack of the beneficial effect of small amounts of Ga on film growth, discussed above.

The difficulty of obtaining wide-gap devices with high efficiencies is illustrated by plotting the absorber bandgap of a series of chalcopyrite alloys vs. the attained efficiency. Figure 6.25 shows that efficiencies in the 18–20% range peak sharply at bandgap energies $E_g \sim 1.1-1.2$ eV.

One reason for the low performance of wide-gap devices is the less favourable band offset constellation at the absorber/CdS-buffer interface. Figure 6.26 shows the band diagram of a $CuGaSe_2$-based heterojunction. As the increase of bandgap in

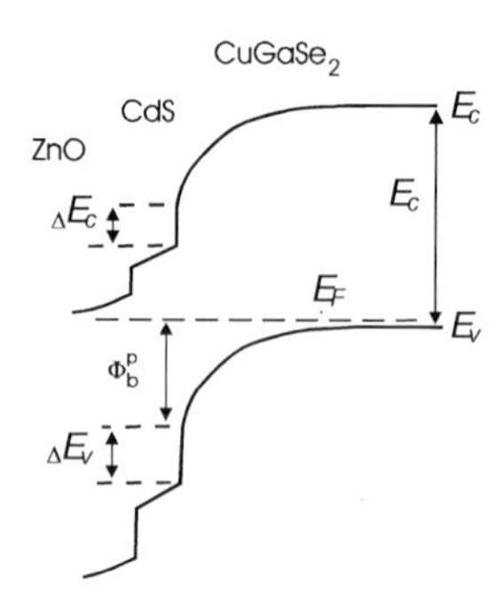

Figure 6.26 Energy band diagram for the ZnO/CdS/$CuGaSe_2$ heterojunction.

going from $CuInSe_2$ to $CuGaSe_2$ takes place almost exclusively by increase of the energy of the conduction band, the positive band offset ΔE_c^{ab} between the absorber and the buffer in Fig. 6.17 turns into a negative one in Fig. 6.26. This implies that the barrier Φ_b^p that hinders the holes from the absorber from recombining with the electrons from the buffer does not increase proportionally with increase in the bandgap energy. Thus the importance of interface recombination (dominated by the barrier Φ_b^p) grows considerably relative to that of bulk recombination (dominated by E_g of the absorber; see Herberholz *et al.*, 1997b).

Using the same arguments with respect to the $MoSe_2$/absorber interface, the back-surface field produced by this type of back contact turns into its contrary because the built-in field acts in the opposite direction when the conduction-band energy of the absorber is increased. At the moment, this drawback of Cu(In,Ga)Se_2 with high Ga contents seems to be of minor importance and does not explain the poorer performance of $CuInSe_2$ devices either. More substantial are the changes of the electronic quality of the bulk material with increasing x. Figure 6.27 (Hanna *et al.*, 2000) compares the defect density spectra of different $CuIn_{1-x}Ga_xSe_2$ alloys with $x = 0$, 0.26, and 0.57.

Two features are obvious:

- The bulk defect at activation energy $E_a = 300$ meV displays a minimum defect density D for the composition $x = 0.26$. Higher bandgap material ($x = 0.56$), as well as lower bandgap material ($x = 0$), has higher defect densities. It might not be incidental that the superior electronic quality of the material is found at that composition which is used to produce the highest-efficiency Cu(In,Ga)Se_2 devices.
- The interface-related peak at lower activation energies for the compositions $x = 0$ and 0.26 is no longer visible at $x = 0.56$. This is just what we would expect from the band diagram shown in Fig. 6.28b: the Fermi level at the

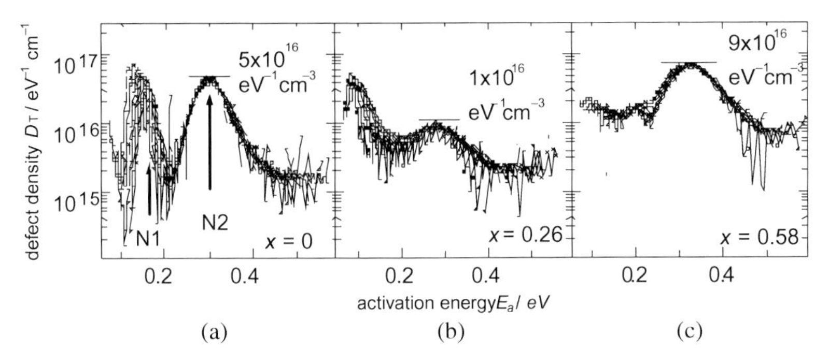

Figure 6.27 Admittance defect density spectra of solar cells with different $CuIn_{1-x}Ga_xSe_2$ alloy compositions. The Ga content is varied from (a) $x = 0$ to (c) $x = 0.58$. The lowest density for the acceptor defect with an activation energy of ~300 meV is found at (b) $x = 0.26$. Source: Hanna *et al.* (2000).

interface moves towards mid-gap and the type inversion is no longer seen in admittance spectroscopy.

Thus we see that bulk as well as interface properties change when going from low Ga content $Cu(In,Ga)Se_2$ towards higher bandgap compositions over a limit which appears to be at $x = 0.3$, corresponding to a bandgap energy of 1.3 eV. A similar limit for S/Se alloying is not yet well established. It appears, however, that making use of the full alloy system $Cu(In,Ga)(S,Se)_2$ enables us to go beyond the limit of 1.3 eV while preserving a high efficiency level.

6.5.1 $CuGaSe_2$

$CuGaSe_2$ has a bandgap of 1.68 eV and therefore would represent an ideal partner for $CuInSe_2$ in an all-chalcopyrite tandem structure. However, a reasonable efficiency for the top cell of any tandem structure is about 15%, far higher than has been reached by the present polycrystalline $CuGaSe_2$ technology. Despite intense research during the last two decades the record efficiency is only 9.5% (Young *et al.*, 2003). In principle, the electronic properties of the material are not so far from those of $CuInSe_2$. However, in detail, all the differences quantitatively point in a less favourable direction. In general, the net doping density N_A in $CuGaSe_2$ appears too high (Jasenek *et al.*, 2000). Together with the charge of deeper defects, the high doping density leads to a strong electrical field in the space-charge region, which enhances recombination by tunnelling.

Furthermore, a structural difference exists between the bulk $CuGaSe_2$ (chalcopyrite) phase and the surface (defect chalcopyrite) phase. Regardless of whether or not this 135-phase is perfectly formed at the film surface or is a defect layer,

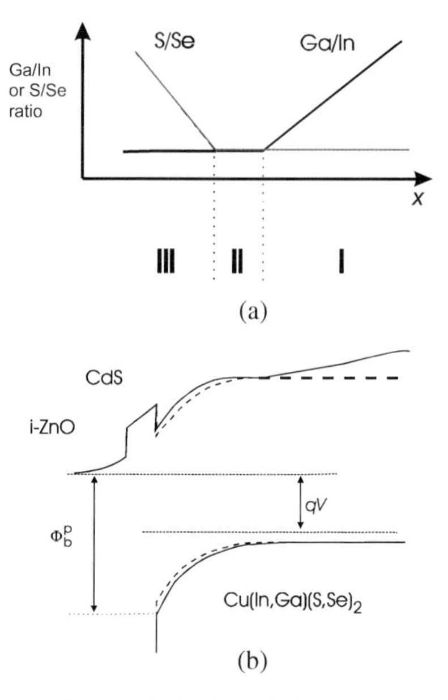

Figure 6.28 Band diagram of an optimised graded-gap device with an increasing Ga/In ratio towards the back surface (region I) and an increasing S/Se-ratio towards the front (region III). Region II has no grading. The dotted lines correspond to the conduction- and valence-band energies of the non-graded device.

a lattice mismatch between the surface layer and the bulk material could account for the increased defect density, which seems operative for $CuGaSe_2$ as well as for $Cu(In,Ga)Se_2$ alloys with high Ga contents (Contreras *et al.*, 1997b).

6.5.2 *Cu(In,Ga)S*$_2$

The major difference between $CuInS_2$ and $Cu(In,Ga)Se_2$ is that the former cannot be prepared with an overall Cu-poor composition. Cu-poor $CuInS_2$ displays an extremely low conductivity, making it almost unusable as a photovoltaic absorber material (Walter *et al.*, 1992). Even at small deviations from stoichiometry on the In-rich side, segregation of the spinel phase is observed (Walter and Schock, 1993). Instead, the material of choice is Cu-rich $CuInS_2$. As in the case of $CuInSe_2$, a Cu-rich preparation route implies the removal of the unavoidable secondary phase (here, $Cu_{2-y}S$) by etching the absorber in KCN (Scheer *et al.*, 1993). Such an etch may involve some damage to the absorber surface as well as the introduction of shunt paths between the front and back electrodes. However, as shown in Table 6.1, the best $Cu(In,Ga)S_2$ device has an efficiency above 12% (Merdes *et al.*, 2011). Remarkably, this is achieved by a sulphurisation process rather than by co-evaporation. The higher quality of the material from sulphurisation might be due to the higher activity of sulphur, and the consequently lower concentration of sulphur vacancies. Although results from co-evaporated samples come close to those of sulphurised cells, $CuInS_2$ is at the moment the only chalcopyrite for which two-step preparation proves itself to be superior to co-evaporation. The main limitation of efficiency results from an open-circuit voltage that is too low as compared with the bandgap. By adding Zn (Walter *et al.*, 1995) or Ga (Neisser *et al.*, 1999) to the absorber layer, voltages above 0.8 V can be achieved. In fact, mastering the difficult task of incorporating small amounts of Ga ($\sim 5\%$) in the active part of the absorber layer was the key for the recent improvement of efficiencies (Merdes *et al.*, 2011).

6.5.3 *Graded-gap devices*

An interesting property of the $CuIn_{1-x}Ga_x(S_{1-y}Se_y)_2$ alloy system is the possibility of designing graded-gap structures that optimise the electronic properties of the final device (Gray and Lee, 1994; Dhingra and Rothwarf, 1996; Gabor *et al.*, 1996; Dullweber *et al.*, 2000). Such bandgap gradings are achieved during co-evaporation by the control of the elemental sources, but selenisation/sulphurisation processes also lead to beneficial compositional gradings.

The art of designing optimum bandgap gradings is to push back charge carriers from critical regions, i.e. regions with high recombination probability within the device. Such critical regions are (i) the interfaces between the back contact and the absorber layer; (ii) the heterojunction, including the absorber/buffer interface. Figure 6.28 shows a band diagram of a grading structure that fulfils the requirements for minimising recombination losses.

To keep the back contact region clear from electrons, one can use a Ga/In grading. The increase of the Ga/(Ga + In) mole fraction x causes a movement of the conduction-band minimum upward with respect to its position in pure $CuInSe_2$. An increase of x towards the back surface leads to a gradual increase of the conduction-band energy, as illustrated in Fig. 6.28a and b. The resulting back-surface field, as in the $Cu(In,Ga)Se_2/MoSe_2$ hetero-contact, drives photogenerated electrons away from the metallic back contact towards the buffer/absorber junction.

The minimisation of junction recombination, both at the point of equal capture rates of holes and electrons and at the metallurgical interface between absorber and buffer, requires a larger bandgap towards the front contact of the absorber. If one had the choice, one would clearly favour a decrease of the valence-band energy, as shown in Fig. 6.28, over an increase of the conduction-band energy. This favours a grading with the help of S/Se alloying, as at least a part of the increasing bandgap energy is supported by a decrease of valence-bandedge energy. The decreased valence-band-edge energy in Fig. 6.28 leads to an increase of the barrier Φ_b^p and thus minimises interface recombination. The favourable combination of In/Ga-grading towards the back contact and S/Se grading towards the front contact appears to be an inherent feature of the combined selenisation/sulphurisation processes, which has led to the highest efficiency devices obtained to date from sequential processing of stacked metallic or stacked elemental precursors.

6.6 Kesterite (CZTS) solar cells

The ongoing discussions about the availability of indium and gallium in a terawatt future have driven research on indium-free materials with properties similar to those of the very efficient chalcopyrite compounds. The kesterites have a similar crystal structure to the chalcopyrites, with the difference that the Group III elements In or Ga are replaced by a pair of Group II and Group VI elements, preferably the abundant elements Zn and Sn. These compounds are the basis of copper–zinc–tin–sulphide/selenide (CZTS) solar cells. A review (Polizzotti *et al.*, 2013) and the analysis of actual devices present a discerning picture of the material. In the quaternary compound $Cu_2ZnSn(S,Se)_4$, disorder caused by secondary phases makes the presence of anti-site defects very likely and the control of composition is an additional challenge. The resulting band tails and potential fluctuations severely affect the open-circuit voltage (Gokmen *et al.*, 2013). In growing thin CZTS films, conditions that reflect the region of the phase diagram where the occurrence of defects and secondary phases is minimised should be chosen. In open systems, considerable loss of Sn is observed (Weber, 2011). Therefore, methods that allow higher chalcogenide pressures in combination with suitable precursors are preferable for making high-quality films. Nevertheless, the maximum efficiency of 12.6% (Wang, 2013) for CZTS devices is as yet much lower than that of chalcopyrite CIGS devices, where an amazing value of 20.8% has been reached.

6.7 Conclusions

The objective of this chapter was not only to describe the achievements of $Cu(In,Ga)Se_2$-based solar cells, but also to give an account of our present understanding of the physical properties of the materials involved and the electronic behaviour of the devices. The fortunate situation of $Cu(In,Ga)Se_2$, which is in a leading position among polycrystalline thin-film solar cell materials, arises from the benign, forgiving nature of the bulk materials and interfaces. Nevertheless, we want to draw the attention of the reader also to the work that has still to be done. Know-how must be transferred into know-why.

Three of the four cornerstones for the recent achievements mentioned in Section 6.3 concern the growth of the films: the optimised deposition conditions, and the incorporation of Na and Ga. However, there is still no detailed model available definitively to describe the growth of $Cu(In,Ga)Se_2$, and especially the impact of Na, which in our opinion is the most important of the different ingredients available to tune the electronic properties of the absorber. A clearer understanding of $Cu(In,Ga)Se_2$ growth would allow us to find optimised conditions in the wide parameter space available, and thus to reduce the number of recombination centres and compensating donors and optimise the number of shallow acceptors.

The deposition of the buffer layer, or more generally speaking, the formation of the heterojunction, is another critical issue. The surface chemistry during heterojunction formation and post-deposition treatments is decisive for the final device performance. Both processes greatly affect not only the surface defects (i.e., recombination and charge), and therefore the charge distribution in the device, but also the defects in the bulk of the absorber. Concentrated effort and major progress in these tasks would not only allow us to push the best efficiencies further above 21%, but would also provide a sound knowledge base for the various attempts at commercialisation of $Cu(In,Ga)Se_2$ solar cells.

References

Abken A., Heinemann F., Kampmann A., Leinkühler G., Rechid J., Sittinger V., Wietler T. and Reineke-Koch R. (1998), 'Large area electrodeposition of $Cu(In,Ga)Se_2$ precursors for the fabrication of thin film solar cells', *Proc. 2nd World Conf. on Photovoltaic Solar Energy Conversion*, Vienna, European Commission, pp. 1133–1136.

Abou-Ras D., Schorr S. and Schock H. W. (2007), 'Grain-size distributions and grain boundaries of chalcopyrite-type thin films,' *J. Appl. Crystallogr.* **40**, 841–848.

Arora N. D., Chamberlain S. G. and Roulston D. J. (1980), 'Diffusion length determination in *p-n* junction diodes and solar cells', *Appl. Phys. Lett.* **37**, 325–327.

Bardeen J. (1947), 'Surface states and rectification at a metal/semiconductor contact', *Phys. Rev.* **71**, 717–727.

Binsma J. J. M. and van der Linden H. A. (1982), 'Preparation of thin $CuInS_2$ films via a two-stage process', *Thin Solid Films* **97**, 237–243.

Birkmire R. W., McCandless B. E., Shafarman W. N. and Varrin R. D. (1989), 'Approaches for high efficiency $CuInSe_2$ solar cells', *Proc. 9th Eur. Photovoltaic Solar Energy Conf.*, Freiburg, Kluwer Academic Publishers, Dordrecht, pp. 134–137.

Braunger D., Hariskos D. and Schock H. W. (1998a), 'Na-related stability issues in highly efficient polycrystalline $Cu(In,Ga)Se_2$ solar cells', *Proc. 2nd World Conf. on Photovoltaic Solar Energy Conversion*, Vienna, European Commission, pp. 511–514.

Braunger D., Zweigart S. and Schock H. W. (1998b), 'The influence of Na and Ga on the incorporation of the chalcogen in polycrystalline $Cu(In,Ga)(S,Se)_2$ thin-films for photovoltaic applications', *Proc. 2nd World Conf. on Photovoltaic Solar Energy Conversion*, Vienna, European Commission, pp. 1113–1116.

Brendel R. and Rau U. (1999), 'Injection and collection diffusion lengths of polycrystalline thin-film solar cells', *Solid State Phenomena* **67–68**, 81–86.

Bube R. H. (1992), *Photoelectronic Properties of Semiconductors*, Cambridge University Press, Cambridge.

Bube R. H. (1998), *Photovoltaic Materials*, Imperial College Press, London.

Burgelman M., Engelhardt F., Guillemoles J.-F., Herberholz R., Igalson M., Klenk R., Lampert M., Meyer T., Nadenau V., Niemegeers A., Parisi J., Rau U., Schock H.-W., Schmitt M., Seifert O., Walter T. and Zott S. (1997), 'Defects in Cu(In,Ga)Se_2 semiconductors and their role in the device performance of thin-film solar cells', *Progr. Photovoltaics* **5**, 121–130.

Cahen D. (1987), 'Some thoughts about defect chemistry in ternaries', *Proc. 7th Int. Conf. on Ternary and Multinary Compounds*, Mat. Res. Soc., Pittsburgh, PA, pp. 433–442.

Cahen D. and Noufi R. (1989), 'Defect chemical explanation for the effect of air anneal on CdS/$CuInSe_2$ solar cell performance', *Appl. Phys. Lett.* **54**, 558–560.

Čhirila A., Reinhard P., Pianezzi F., Bloesch P., Uhl A. R., Fella C., Kranz L., Keller D., Gretener C., Hagendorfer H., Jaeger D., Erni R., Nishiwaki S., Buecheler S. and Tiwari A. N. (2013), 'Potassium-induced surface modification of Cu(In,Ga)Se_2 thin films for high-efficiency solar cells', *Nature Mater.* **12**, 1107–1111.

Chu T. L., Chu S. C., Lin S. C. and Yue J. (1984), 'Large grain copper indium diselenide films', *J. Electrochem. Soc.* **131**, 2182–2184.

Contreras M. A., Egaas B., Dippo P., Webb J., Granata J., Ramanathan K., Asher S., Swartzlander A. and Noufi R. (1997a), 'On the role of Na and modifications to Cu(In,Ga)Se absorber materials using thin MF (M = Na, K, Cs) precursor layers', *Conf. Record 26th IEEE Photovoltaic Specialists Conf.*, Anaheim, CA, IEEE Press, Piscataway, pp. 359–362.

Contreras M. A., Egaas B., Ramanathan K., Hiltner J., Swartzlander A., Hasoon F. and Noufi R. (1999), 'Towards 20% efficiency in Cu(In,Ga)Se_2 polycrystalline solar cells', *Progr. Photovoltaics* **7**, 311–316.

Contreras M. A., Mansfield L. M., Egaas B., Li J., Romero M., Noufi R., Rudiger-Voigt E. and Mannstadt W. (2012), 'Wide bandgap Cu(In,Ga)Se_2 solar cells with improved energy conversion efficiency,' *Progr. Photovoltaics* **20**, 843–850.

Contreras M. A., Ramanathan K., AbuShama J., Hasoon F., Young D. L., Egaas B. and Noufi R. (2005), 'Diode characteristics in ZnO/CdS/Cu($In_{(1-x)}Ga_x$)Se_2 solar cells,' *Progr. Photovoltaics* **13**, 209–216.

Contreras M. A., Wiesner H., Tuttle J., Ramanathan K. and Noufi R. (1997b), 'Issues of the chalcopyrite/defect-chalcopyrite junction model for high-efficiency Cu(In,Ga)Se_2 solar cells', *Solar Energy Mater. Solar Cells* **49**, 239–247.

Coutts T. J., Kazmerski L. L. and Wagner S. (1986, ed.), *Ternary Chalcopyrite Semiconductors: Growth, Electronic Properties, and Applications*, Elsevier, Amsterdam.

Devaney W. E., Chen W. S., Steward J. M. and Mickelson R. A. (1990), 'Structure and properties of high efficiency ZnO/CdZnS/$CuInGaSe_2$ solar cells', *IEEE Trans. Electron Devices* **ED–37**, 428–433.

Dhingra A. and Rothwarf A. (1996), 'Computer simulation and modeling of graded bandgap $CuInSe_2$/CdS based solar cells', *IEEE Trans. Electron Devices* **43**, 613–621.

Dimmler B. and Schock H. W. (1996), 'Scaling-up of CIS technology for thin-film solar modules', *Progr. Photovoltaics* **4**, 425–433.

Dullweber T., Hanna G., Shams-Kolahi W., Schwartzlander A., Contreras M. A., Noufi R. and Schock H. W. (2000), 'Study of the effect of gallium grading in $Cu(In,Ga)Se_2$', *Thin Solid Films* **361–362**, 478–481.

Eberspacher C., Pauls K. L. and Fredric C. V. (1998), 'Improved processes for forming $CuInSe_2$ films', *Proc. 2nd World Conf. on Photovoltaic Solar Energy Conversion*, Vienna, European Commission, pp. 303–306.

Eisenbarth T., Unold T., Caballero R., Kaufmann C. A., Abou-Ras D. and Schock H. W. (2009), 'Origin of defects in $CuIn_{1-x}Ga_xSe_2$ solar cells with varied Ga content,' *Thin Solid Films* **517**, 2244–2247.

Engelhardt F., Bornemann L., Köntges M., Meyer Th., Parisi J., Pschorr-Schoberer E., Hahn B., Gebhardt W., Riedl W. and Rau U. (1999), '$Cu(In,Ga)Se_2$ solar cells with a ZnSe buffer layer — interface characterization by quantum efficiency measurements', *Progr. Photovoltaics* **7**, 423–436.

Engelhardt F., Schmidt M., Meyer Th., Seifert O., Parisi J. and Rau U. (1998), 'Metastable electrical transport in $Cu(In,Ga)Se_2$ thin films and ZnO/CdS/ $Cu(In,Ga)Se_2$ heterostructures', *Phys. Lett. A* **245**, 489–493.

Friedlmeier T. M., Braunger D., Hariskos D., Kaiser M., Wanka H. N. and Schock H. W. (1996), 'Nucleation and growth of the CdS buffer layer on $Cu(In,Ga)Se_2$ thin films', *Conf. Record 25th IEEE Photovoltaic Specialists Conf.*, Washington D.C., IEEE Press, Piscataway, pp. 845–848.

Gabor A. M., Tuttle J. R., Albin D. S., Contreras M. A., Noufi R. and Hermann A. M. (1994), 'High-efficiency $CuIn_xGa_{1-x}Se_2$ solar cells from $(In_xGa_{1-x})_2Se_3$ precursors', *Appl. Phys. Lett.* **65**, 198–200.

Gabor A. M., Tuttle J. R., Bode M. H., Franz A., Tennant A. L., Contreras M. A., Noufi R., Jensen D. G. and Hermann A. M (1996), 'Bandgap engineering in $Cu(In,Ga)Se_2$ thin films grown from $(In,Ga)_2Se_3$ precursors', *Solar Energy Mater. Solar Cells* **41**, 247–260.

Gallon P. N. R., Orsal G., Artaud M. C. and Duchemin S. (1998), 'Studies of $CuInSe_2$ and $CuGaSe_2$ thin films grown by MOCVD from three organometallic sources', *Proc. 2nd World Conf. on Photovoltaic Solar Energy Conversion*, Vienna, European Commission, pp. 515–518.

Gilson M., Bacewicz R. and Schock H. W. (1995), '"Dangling bonds" in $CuInSe_2$ and related compounds', *Proc. 13th Eur. Photovoltaic Solar Energy Conf.*, Nice, H. S. Stephens & Associates, Bedford, pp. 2076–2079.

Gloeckler M., Sites J. R. and Metzger W. K. (2005), 'Grain-boundary recombination in $Cu(In,Ga)Se_2$ solar cells,' *J. Appl. Phys.*, **98**, 113704–113704-10.

Gokmen T., Gunawan O., Todorov T. K. and Mitzi D. B. (2013), 'Band tailing and efficiency limitation in kesterite solar cells', *Appl. Phys. Lett.* **103**, 103506.

Gray J. L. and Lee Y. J. (1994), 'Numerical modeling of graded bandgap CIGS solar cells', *Proc. 1st World Conf. on Photovoltaic Solar Energy Conversion*, Waikoloa, IEEE Press, Piscataway, NJ, pp. 123–126.

Green M. A. (1986), *Solar Cells: Operating Principles, Technology and System Applications*, University of New South Wales, Sydney.

Green M. A. (1995), *Silicon Solar Cells: Advanced Principles and Practice*, University of New South Wales, Sydney.

Green M. A. (1996a), 'Depletion region recombination in silicon thin-film multilayer solar cells', *Progr. Photovoltaics* **4**, 375–380.

Green M. A. (1996b), 'Bounds upon grain boundary effects in minority carrier semiconductor devices: a rigorous 'perturbation' approach with application to silicon solar cells', *J. Appl. Phys.* **80**, 1515–1521.

Haalboom T., Gödecke T., Ernst F., Rühle M., Herberholz R., Schock H.-W., Beilharz C. and Benz K. W. (1997), 'Phase relations and microstructure in bulk materials and thin films of the ternary system Cu–InSe', *Inst. Phys. Conf. Ser.* **152E**, 249–252.

Hafemeister M., Siebentritt S., Albert J., Lux-Steiner M. C. and Sadewasser S. (2010), 'Large neutral barrier at grain boundaries in chalcopyrite thin films,' *Phys. Rev. Lett.* **104**, 196602–196606.

Hahn H., Frank G., Klingler W., Meyer A. and Störger G. (1953), 'Über einige ternäre Chalkogenide mit Chalkopyritstruktur', *Z. anorg. u. allg. Chemie* **271**, 153–170.

Hanna G., Jasenek A., Rau U. and Schock H. W (2000), 'Open circuit voltage limitations in $CuIn_{1-x}Ga_xSe_2$ thin film solar cells — dependence on alloy compositions', *Physica Status Solidi A.* **179**, R7–R8.

Hedström J., Ohlsen, H., Bodegard M., Kylner A., Stolt L., Hariskos D., Ruckh M. and Schock H.-W. (1993), 'ZnO/CdS/Cu(In,Ga)Se_2 thin film solar cells with improved performance', *Conf. Record 23rd IEEE Photovoltaic Specialists Conf.*, Louisville, IEEE Press, Piscataway, NJ, pp. 364–371.

Hengel I., Neisser A., Klenk R. and Lux-Steiner C.-M. (2000), 'Current transport in $CuInS_2$:Ga/CdS/ZnO solar cells', *Thin Solid Films* **361–362**, 458–462.

Herberholz R. (1998), 'Defect characterisation in chalcopyrite-based heterostructures', *Inst. Phys. Conf. Ser.* **152E**, 733–740.

Herberholz R., Braunger D., Schock H. W., Haalboom T. and Ernst F. (1997a), 'Performance and defects in Cu(In,Ga)Se_2 heterojunctions: combining electrical and structural measurements', *Proc. 14th Eur. Photovoltaic Solar Energy Conf.*, Barcelona, H. S. Stephens & Associates, Bedford, pp. 1246–1249.

Herberholz R., Nadenau V., Rühle U., Köble C., Schock H. W. and Dimmler B. (1997b), 'Prospects of wide-gap chalcopyrites for thin film photovoltaic modules', *Solar Energy Mater. Solar Cells* **49**, 227–237.

Herberholz R., Igalson M. and Schock H. W. (1998), 'Distinction between bulk and interface states in $CuInSe_2$/CdS/ZnO by space charge spectroscopy', *J. Appl. Phys.* **83**, 318–325.

Herberholz R., Rau U., Schock H. W., Hallboom T., Gödecke T., Ernst F., Beilharz C., Benz K. W. and Cahen D. (1999), 'Phase segregation, Cu migration and junction formation in Cu(In,Ga)Se_2', *Eur. Phys. J.: Appl. Phys.* **6**, 131–139.

Heske C., Fink R., Umbach E., Riedl W. and Karg F. (1996), 'Na-induced effects on the electronic structure and composition of Cu(In,Ga)Se_2 thin-film surfaces', *Appl. Phys. Lett.* **68**, 3431–3432.

Heske C., Richter G., Chen Z., Fink R., Umbach E., Riedl W. and Karg F. (1997), 'Influence of Na and H_2O on the surface properties of Cu(In,Ga)Se_2 thin films', *J. Appl. Phys.* **82**, 2411–2420.

Holz J., Karg F. and von Phillipsborn H. (1994), 'The effect of substrate impurities on the electronic conductivity in CIGS thin films, *Proc. 12th Eur. Photovoltaic Solar Energy Conf.*, Amsterdam, H. S. Stephens & Associates, Bedford, pp. 1592–1595.

Igalson M. and Schock H. W. (1996), 'The metastable changes of the trap spectra of CuInSe$_2$-based photovoltaic devices', *J. Appl. Phys.* **80**, 5765–5769.

Jackson P., Hariskos D., Lotter E., Paetel S., Wuerz R., Menner R., Wischmann W. and Powalla M. (2011), 'New world record efficiency for Cu(In,Ga)Se$_2$ thin-film solar cells beyond 20%,' *Progr. Photovoltaics* **19**, 894–897.

Jäger-Waldau G., Schmid D. and Jäger-Waldau A. (1991), 'Diffusion length measurement of heterojunction thin films by junction-EBIC', *J. Phys. IV* **1**, C6, 131–132.

Jasenek A., Rau U., Nadenau V. and Schock H. W. (2000), 'Electronic properties of CuGaSe$_2$-based heterojunction solar cells. Part II — Defect spectroscopy', *J. Appl. Phys.* **87**, 594–602.

Jensen N., Rau U., Hausner R. M., Uppal S., Oberbeck L., Bergmann R. B. and Werner J. H. (2000), 'Recombination mechanisms in amorphous silicon/crystalline silicon heterojunction solar cells', *J. Appl. Phys.* **87**, 2639–2645.

Kapur V. K., Basol B. M. and Tseng E. S. (1987), 'Low-cost methods for the production of semiconductor films for CuInSe$_2$/CdS solar cells', *Solar Cells* **21**, 65–72.

Kazmerski L. L. and Wagner S. (1985), 'Cu-ternary chalcopyrite solar cells', in Coutts T. J. and Meakin J. D., *Current Topics in Photovoltaics*, Academic Press, Orlando.

Kessler J., Velthaus K. O., Ruckh M., Laichinger R., Schock H. W., Lincot D., Ortega R. and Vedel J. (1992), 'Chemical bath deposition of CdS on CuInSe$_2$, etching effects and growth kinetics', *Proc. 6th Int. Photovoltaic Solar Energy Conf.*, New Delhi, Oxford IBH Publishing, New Delhi, pp. 1005–1010.

Keyes B. M., Hasoon F., Dippo P., Balcioglu A. and Aboulfotuh F. (1997), 'Influence of Na on the electro-optical properties of Cu(In,Ga)Se$_2$', *Conf. Record 26th IEEE Photovoltaic Specialists Conf.*, Anaheim, IEEE Press, Piscataway, NJ, pp. 479–482.

Klein A. and Jaegermann W. (1996), 'Chemical interaction of Na-cleaved (001) surfaces of CuInSe$_2$', *J. Appl. Phys.* **80**, 5039–5043.

Klenk R. and Schock H. W. (1994), 'Photocurrent collection in thin film solar cells — calculation and characterization for CuGaSe$_2$/(Zn,Cd)S', *Proc. 12th Eur. Photovoltaic Solar Energy Conf.*, Amsterdam, H. S. Stephens & Associates, Bedford, pp.1588–1591.

Klenk R., Walter T., Schock H. W. and Cahen D. (1993), 'A model for the successful growth of polycrystalline films of CuInSe$_2$ by multisource physical vacuum evaporation', *Adv. Mat.* **5**, 114–119.

Kronik L., Cahen D. and Schock H. W. (1998), 'Effects of sodium on polycrystalline Cu(In,Ga)Se$_2$ and its solar cell performance', *Adv. Mater.* **10**, 31–36.

Kronik L., Rau U., Guillemoles J.-F., Braunger D., Schock H. W. and Cahen D. (2000), 'Interface redox engineering of Cu(In,Ga)Se$_2$-based solar cells: oxygen, sodium, and chemical bath effects', *Thin Solid Films* **361–362**, 353–359.

Kushiya K. (2013), 'CIS-based thin-film PV technology in solar frontier K.K.', *Solar Energy Mater. Solar Cells* **122**, 309–313.

Kushiya K., Tachiyuli M., Nagoya Y., Fujimaki A., Sang B., Okumara D., Satoh M. and Yamase O. (1999), 'Progress in large-area Cu(In,Ga)Se$_2$-based thin-film modules with a Zn(O,S,OH)$_x$ buffer layer', *Tech. Digest 11th Int. Sci. Eng. Conf.*, Tokyo University of Agriculture and Technology, Tokyo, Japan, pp. 637–640.

Lany S. and Zunger A. (2006), 'Light- and bias-induced metastabilities in $Cu(In,Ga)Se_2$ based solar cells caused by the (V_{Se}-V_{Cu}) vacancy complex,' *J. Appl. Phys.* **100**, 113725–113740.

Lany S. and Zunger A. (2008), 'Intrinsic DX centers in ternary chalcopyrite semiconductors', *Phys. Rev. Lett.* **100**, 016401–016405.

Lincot D., Guillemoles J.-F., Cowache P., Marlot A., Lepiller C., Canava B., Yousfi F. B. and Vedel J. (1998), 'Solution deposition technologies for thin film solar cells: status and perspectives', *Proc. 2nd World Conf. on Photovoltaic Solar Energy Conversion*, Vienna, European Commission, pp. 440–445.

Löher T., Jaegermann W. and Pettenkofer C. (1995), 'Formation and electronic properties of the $CdS/CuInSe_2$ (001) heterointerface studied by synchrotron-induced photoemission', *J. Appl. Phys.* **77**, 731–737.

Mauch R. H., Ruckh. M., Hedström J., Lincot D., Kessler J., Klinger R., Stolt L. Vedel J. and Schock H. W. (1991), 'High efficiency $Zno/CdS/CuInSe_2$ solar cells — recent results of EUROCIS collaboration', *Proc. 9th Eur. Photovoltaic Solar Energy Conf.*, Lisbon, Portugal; Kluwer Academic Publishers, Dordrecht, pp. 1415–1419.

Merdes S., Mainz R., Klaer J., Meeder, A., Rodriguez-Alvarez H., Schock H. W., Lux-Steiner M. C. and Klenk R. (2011), '12.6% efficient $CdS/Cu(In,Ga)S_2$-based solar cell with an open circuit voltage of 879 mV prepared by a rapid thermal process', *Solar Energy Mater. Solar Cells* **95**, 864–869.

Meyer T., Schmidt M., Engelhardt F., Parisi J. and Rau U. (1999), 'A model for the open-circuit voltage relaxation in $Cu(In,Ga)Se_2$ heterojunction solar cells', *Eur. Phys. J.: Appl. Phys.* **8**, 43–52.

Mickelsen R. A. and Chen W. S. (1980), 'High photocurrent polycrystalline thin-film $CdS/CuInSe_2$ solar cell', *Appl. Phys. Lett.* **36**, 371–373.

Mickelsen R. A. and Chen W. S. (1982), 'Polycrystalline thin-film $CuInSe_2$ solar cells', *Conf. Record 16th IEEE Photovoltaic Specialists Conf.,* San Diego, CA, IEEE Press, Piscataway, NJ, pp. 781–785.

Migliorato P., Shay J. L., Kasper H. M. and Wagner S. (1975), 'Analysis of the electrical and luminescent properties of $CuInSe_2$', *J. Appl. Phys.* **46**, 1777–1782.

Mitchell K. C., Ermer J. and Pier D. (1988), 'Single and tandem junction $CuInSe_2$ cell and module technology', *Conf. Record 20th IEEE Photovoltaic Specialists Conf.*, Las Vegas, IEEE Press, Piscataway, NJ, pp.1384–1389.

Nadenau V., Rau U., Jasenek A. and Schock H. W. (2000), 'Electronic properties of $CuGaSe_2$-based heterojunction solar cells. Part I — Transport analysis', *J. Appl. Phys.* **87**, 584–593.

Nakada T., Iga D., Ohbo H. and Kunioka A. (1997), 'Effects of sodium on $Cu(In,Ga)Se_2$-based thin films and solar cells', *Jpn. J. Appl. Phys.* **36**, 732–737.

Nakada T., Mise T., Kume T. and Kunioka A. (1998), 'Superstrate type $Cu(In,Ga)Se_2$ thin film solar cells with ZnO buffer layers — a novel approach to 10% efficiency', *Proc. 2nd World Conf. on Photovoltaic Solar Energy Conversion*, Vienna, European Commission, pp. 413–417.

Neisser A., Hengel I., Klenk R., Matthes Th. W., Alvarez-Garcia J., Perez-Rodriguez A., Romano-Rodriguez A. and Lux-Steiner M. C. (1999), 'Effect of Ga incorporation in sequentially prepared $CuInS_2$ thin film absorbers', *Tech. Digest 11th Int. Sci. Eng. Conf.*, Tokyo University of Agriculture and Technology, Tokyo, Japan, pp. 955–956.

Nelson A. J., Schwerdtfeger C. R., Wei S.-H., Zunger A., Rioux D., Patel R. and Höchst H. (1993), 'Theoretical and experimental studies of the ZnSe/$CuInSe_2$ heterojunction band offset', *Appl. Phys. Lett.* **62**, 2557–2559.

Neumann H. (1986), 'Optical properties and electronic band structure of $CuInSe_2$', *Solar Cells* **16**, 317–333.

Niemegeers A. and Burgelman M. (1996), 'Numerical modelling of ac-characteristics of CdTe and CIS solar cells', *Proc. 25nd IEEE Photovoltaic Specialists Conf.*, Washington D.C., April 1996, IEEE, New York, NY, pp. 901–904.

Niemegeers A., Burgelman M. and de Vos A. (1995), 'On the CdS/$CuInSe_2$ conduction band discontinuity', *Appl. Phys. Lett.* **67**, 843–845.

Niemegeers A., Burgelman M., Herberholz R., Rau U., Hariskos D. and Schock H. W. (1998), 'Model for electronic transport in Cu(In,Ga)Se_2 solar cells', *Progr. Photovoltaics* **6**, 407–421.

Niki S., Fons P. J., Yamada A., Suzuki R., Ohdaira T., Ishibashi S. and Oyanagai H. (1994), 'High quality $CuInSe_2$ epitaxial films — molecular beam epitaxial growth and intrinsic properties', *Inst. Phys Conf. Ser.* **152E**, 221–227.

Niles D. W., Al-Jassim M. and Ramanathan K. (1999), 'Direct observation of Na and O impurities at grain surfaces of $CuInSe_2$ thin films', *J. Vac. Sci. Technol. A* **17**, 291–296.

Niles D. W., Ramanathan K., Haason F., Noufi R., Tielsh B. J. and Fulghum J. E. (1997), 'Na impurity chemistry in photovoltaic CIGS thin films: investigation with X-ray photoelectron spectroscopy', *J. Vac. Sci. Technol. A* **15**, 3044–3049.

Noufi R., Axton R., Herrington C. and Deb S. K. (1984), 'Electronic properties versus composition of thin films of $CuInSe_2$', *Appl. Phys. Lett.* **45**, 668–670.

Padovani F. A. and Stratton R. (1966), 'Field and thermionic field emission in Schottky barriers', *Solid State Electronics* **9**, 695–707.

Parisi J., Hilburger D., Schmitt M. and Rau U. (1998), 'Quantum efficiency and admittance spectroscopy on Cu(In,Ga)Se_2 solar cells', *Solar Energy Mater. Solar Cells* **50**, 79–85.

Persson C. and Zunger A. (2003), 'Anomalous grain boundary physics in polycrystalline $CuInSe_2$: the existence of a hole barrier,' *Phys. Rev. Lett.*, **91**, 266401–266405.

Polizzotti A., Repins I. L., Noufi R., Wei S.-H. and Mitzi D. B. (2013), 'The state and future prospects of kesterite photovoltaics', *Energy Environ. Sci.* **6**, 3171–3182.

Potter R. R., Eberspacher C. and Fabick L. B. (1985), 'Device analysis of $CuInSe_2$/ (Cd,Zn)S solar cells', *Conf. Record 18th IEEE Photovoltaic Specialists Conf.*, Las Vegas, NV, IEEE Press, Piscataway, pp. 1659–1664.

Probst V., Karg F., Rimmasch J., Riedl W., Stetter W., Harms H. and Eibl O. (1996), 'Advanced stacked elemental layer progress for Cu(InGa)Se_2 thin film photovoltaic devices', *Mat. Res. Soc. Symp. Proc.* **426**, 165–176.

Ramanathan K., Bhattacharya R. N., Granata J., Webb J., Niles D., Contreras M. A., Wiesner H., Haason F. S. and Noufi R. (1998a), 'Advances in CIGS solar cell research at NREL', *Conf. Record 26th IEEE Photovoltaic Specialists Conf.*, Anaheim, CA, IEEE Press, Piscataway, NJ, pp. 319–325.

Ramanathan K., Wiesner H., Asher S., Niles D., Bhattacharya R. N., Keane J., Contreras M. A. and Noufi R. (1998b), 'High efficiency Cu(In,Ga)Se_2 thin film solar cells without intermediate buffer layers', *Proc. 2nd World Conf. on Photovoltaic Solar Energy Conversion*, Vienna, European Commission, pp. 477–481.

Rau U. (1999), 'Tunneling-enhanced recombination in Cu(In,Ga)Se_2 heterojunction solar cells', *Appl. Phys. Lett.* **74**, 111–113.

Rau U., Braunger D., Herberholz R., Schock H. W., Guillemoles J.-F., Kronik L. and Cahen D. (1999a), 'Oxygenation and air-annealing effects on the electronic properties of Cu(In,Ga)Se_2 films and devices', *J. Appl. Phys.* **86**, 497–505.

Rau U., Braunger D. and Schock H. W. (1999b), 'Air-annealing effects on polycrystalline Cu(In,Ga)Se_2 heterojunctions', *Solid State Phenomena* **67–68**, 409–414.

Rau U., Jasenek A., Schock H. W., Engelhardt F. and Meyer T. (2000), 'Electronic loss mechanisms in chalcopyrite based heterojunction solar cells', *Thin Solid Films* **361–362**, 298–302.

Rau U., Schmitt M., Engelhardt F., Seifert O., Parisi J., Riedl W., Rimmasch J. and Karg F. (1998a), 'Impact of Na and S incorporation on the electronic transport mechanisms of Cu(In,Ga)Se_2 solar cells', *Solid State Commun.* **107**, 59–63.

Rau U., Schmitt M., Parisi J., Riedl W. and Karg F. (1998b), 'Persistent photoconductivity in Cu(In,Ga)Se_2 heterojunctions and thin films prepared by sequential deposition', *Appl. Phys. Lett.* **73**, 223–226.

Rau U. and Schock H. W. (1999), 'Electronic properties of Cu(In,Ga)Se_2 heterojunction solar cells — recent achievements, current understanding and future challenges', *Appl. Phys. A* **69**, 131–147.

Rau U., Taretto K. and Siebentritt S. (2009), 'Grain boundaries in Cu(In,Ga)(Se,S)$_2$ thin-film solar cells', *Appl. Phys. A — Mater. Sci. Process.* **96**, 221–234.

Repins I., Contreras M. A., Egaas B., DeHart C., Scharf J., Perkins C. L., To B. and Noufi R. (2008), '19.9%-efficient ZnO/CdS/CuInGaSe_2 solar cell with 81.2% fill factor,' *Progr. Photovoltaics* **16**, 235–239.

Rincon C., Bellabarba C., Gonzalez J. and Sanchez Perez G. (1986), 'Optical properties and characterization of CuInSe_2 ', *Solar Cells* **16**, 335–349.

Rincon C. and Wasim S. M. (1987), 'Defect chemistry of $A^{I}B^{III}C_2^{VI}$ chalcopyrite semiconducting compounds', *Proc. 7th Int. Conf. Ternary and Multinary Compounds*, Mater. Res. Soc., Pittsburgh, PA, pp. 443–452.

Rockett A. and Birkmire R. W. (1991), 'CuInSe_2 for photovoltaic applications', *J. Appl. Phys.* **70**, 81–97.

Ruberto M. N. and Rothwarf A. (1987), 'Time-dependent open-circuit voltage in CuInSe_2/CdS solar cells: theory and experiment', *J. Appl. Phys.* **61**, 4662–4669.

Ruckh M., Schmid D., Kaiser M., Schäffler R., Walter T. and Schock H. W. (1994a), 'Influence of substrates on the electrical properties of Cu(In,Ga)Se_2 thin films', *Proc. 1st World Conf. on Photovoltaic Solar Energy Conversion*, Waikoloa, IEEE Press, Piscataway, NJ, pp. 156–159.

Ruckh M., Schmid D. and Schock H. W. (1994b), 'Photoemission studies of the ZnO/CdS interface', *J. Appl. Phys.* **76**, 5945–5948.

Sasala R. A. and Sites J. R. (1993), 'Time-dependent voltage in CuInSe_2 and CdTe solar cells', *Conf. Record 23rd IEEE Photovoltaic Specialists Conf.*, Louisville, KT, IEEE Press, Piscataway, NJ, pp. 543–548.

Scheer R. (1997), 'Surface and interface properties of Cu-chalcopyrite semiconductors and devices', *Research Trends in Vacuum Sci. Technol.* **2**, 77–112.

Scheer R. and Schock H. W. (2011), *Chalcogenide Photovoltaics*, Wiley-VCh Verlag, Weinheim, Germany.

Scheer R., Walter T., Schock H. W., Fearheiley M. L. and Lewerenz H. J. (1993), '$CuInS_2$ based thin film solar cells with 10.2% efficiency', *Appl. Phys. Lett.* **63**, 3294–3296.

Schmid D., Ruckh M., Grunwald F. and Schock H. W. (1993), 'Chalcopyrite/defect chalcopyrite heterojunctions on basis of $CuInSe_2$', *J. Appl. Phys.* **73**, 2902–2909.

Schmid D., Ruckh M., Grunwald F. and Schock H. W. (1996a), 'Photoemission studies on $Cu(In,Ga)Se_2$ thin films and related binary selenides', *Appl. Surf. Sci.* **103**, 409–429.

Schmid D., Ruckh M. and Schock H. W. (1996b), 'A comprehensive characterization of the interfaces in Mo/CIS/CdS/ZnO solar cell structures', *Solar Energy Mater. Solar Cells* **41**, 281–294.

Schmidt M., Braunger D., Schäffler R., Schock H. W. and Rau U. (2000), 'Influence of damp heat on the electrical properties of $Cu(In,Ga)Se_2$ solar cells', *Thin Solid Films* **361–362**, 283–287.

Schmitt M., Rau U. and Parisi J. (1995), 'Investigation of deep trap levels in $CuInSe_2$ solar cells by temperature dependent admittance measurements', *Proc. 13th Eur. Photovoltaic Solar Energy Conf.*, Nice, H. S. Stephens & Associates, Bedford, pp. 1969–1972.

Schock H. W. (1996), 'Thin film photovoltaics', *Appl. Surf. Sci.* **92**, 606–616.

Schroeder D. J. and Rockett A. A. (1997), 'Electronic effects of sodium in epitaxial $CuIn_{1-x}Ga_xSe_2$', *J. Appl. Phys.* **82**, 5982–5985.

Seifert O., Engelhardt F., Meyer Th., Hirsch M. T., Parisi J., Beilharz C., Schmitt M. and Rau U. (1997), 'Observation of a metastability in the DC and AC electrical transport properties of $Cu(In,Ga)Se_2$ single crystals, thin films and solar cells', *Inst. Phys. Conf. Ser.* **152E**, 253–256.

Shay J. L. and Wernick J. H. (1975), *Ternary Chalcopyrite Semiconductors: Growth, Electronic Properties, and Applications*, Pergamon Press, Oxford.

Siebentritt S., Igalson M., Persson C. and Lany S. (2010), 'The electronic structure of chalcopyrites-bands, point defects and grain boundaries,' *Progr. Photovoltaics* **18**, 390–410.

Siebentritt S., Sadewasser S., Wimmer M., Leendertz C., Eisenbarth T. and Lux-Steiner M. C. (2006), 'Evidence for a neutral grain-boundary barrier in chalcopyrites', *Phys. Rev. Lett.* **97**, 146601–146605.

Staebler D. L. and Wronski C. R. (1977), 'Reversible conductivity changes in discharge-produced amorphous Si', *Appl. Phys. Lett.* **31**, 292–294.

Stolt L., Hedström J., Kessler J., Ruckh M., Velthaus K. O. and Schock H. W. (1993), 'ZnO/CdS/$CuInSe_2$ thin-film solar cells with improved performance', *Appl. Phys. Lett.* **62**, 597–599.

Suryawanshi M. P., Agawane G. L., Bhosale S. M., Shin S. W., Patil P. S., Kim J. H. and Moholkar A. V. (2013), 'CZTS based thin film solar cells: a status review', *Mater. Technol.* **28**, 98–109.

Takei R., Tanino H., Chichibu S. and Nakanishi H. (1996), 'Depth profiles of spatially-resolved Raman spectra of a $CuInSe_2$-based thin-film solar cell', *J. Appl. Phys.* **79**, 2793–2795.

Taretto K. and Rau U. (2008), 'Numerical simulation of carrier collection and recombination at grain boundaries in $Cu(In,Ga)Se_2$ solar cells,' *J. Appl. Phys.* **103**, 094523–094523-11.

Topic M., Smole F. and Furlan J. (1997), 'Examination of blocking current–voltage behavior through defect chalcopyrite layer in ZnO/CdS/Cu(In,Ga)Se_2 solar cell', *Solar Energy Mater. Solar Cells*, 311–317.

Urbaniak A. and Igalson M. (2009), 'Creation and relaxation of light- and bias-induced metastabilities in Cu(In,Ga)Se_2', *J. Appl. Phys.* **106**, 063720.

Wada T., Hayashi S., Hashimoto Y., Nishiwaki S., Sato T., Negami T. and Nishitani M. (1998), 'High efficiency Cu(In,Ga)Se_2 (CIGS) solar cells with improved CIGS surface', *Proc. 2nd World Conf. on Photovoltaic Solar Energy Conversion*, Vienna, European Commission, pp. 403–408.

Wada T., Kohara N., Negami T. and Nishitani M. (1996), 'Chemical and structural characterization of Cu(In,Ga)Se_2/Mo Interface in Cu(In,Ga)Se_2 solar cells', *Jpn. J. Appl. Phys.* **35**, 1253–1256.

Wagner S., Shay J. L., Migliorato P. and Kasper H. M. (1974), 'CuInSe_2/CdS heterojunction photovoltaic detectors', *Appl. Phys. Lett.* **25**, 434–435.

Wallin E., Malm U., Jarmar T., Lundberg O., Edoff M. and Stolt L. (2012), 'World-record Cu(In,Ga)Se_2-based thin-film sub-module with 17.4% efficiency,' *Progr. Photovoltaics* **20**, 851–854.

Walter T., Braunger D., Hariskos D., Köble C. and Schock H. W. (1995), 'CuInS_2: film growth, devices, submodules and perspectives', *Proc. 13th Eur. Photovoltaic Solar Energy Conf.*, Nice, H. S. Stephens & Associates, Bedford, pp. 597–600.

Walter T., Content A., Velthaus K. O. and Schock H. W. (1992), 'Solar cells based on CuIn$(Se,S)_2$', *Solar Energy Mater. Solar Cells* **26**, 357–368.

Walter T., Herberholz R., Müller C. and Schock H. W. (1996a), 'Determination of defect distributions from admittance measurements and application to Cu(In,Ga)Se_2 based heterojunctions', *J. Appl. Phys.* **80**, 4411–4420.

Walter T., Herberholz R. and Schock H. W. (1996b), 'Distribution of defects in polycrystalline thin films', *Solid State Phenomena* **51**, 309–316.

Walter T. and Schock H. W. (1993), 'Structural and electrical investigations of the anion exchange in polycrystalline CuIn$(S,Se)_2$ thin films', *Jpn. J. Appl. Phys.* **32–3**, 116–119.

Wang W., Winkler M. T., Gunawan O., Gokmen T., Todorov T. K., Zhu Y. and Mitzi D. B., 'Device characteristics of CZTSSe thin-film solar cells with 12.6% efficiency', *Advanced Energy Mater.*, published online: 27 Nov. 2013, DOI: 10.1002/aenm.201301465.

Weber A., Mainz R. and Schock H. W. (2010), 'On the Sn loss from thin films of the material system Cu–Zn–Sn–S in high vacuum', *J. Appl. Phys.* **107**, 013516–013516.

Wei S. H. and Zunger A. (1993), 'Band offsets at the CdS/CuInSe_2 heterojunction', *Appl. Phys. Lett.* **63**, 2549–2551.

Wolf D., Müller G., Stetter W. and Karg F. (1998), 'In-situ investigation of Cu–In–Se reactions: impact of Na on CIS formation', *Proc. 2nd World Conf. on Photovoltaic Solar Energy Conversion*, Vienna, European Commission, pp. 2426–2429.

Würz R., Eicke A., Kessler F., Paetel S., Efimenko S. and Schlegel C. (2012), 'CIGS thin-film solar cells and modules on enamelled steel substrates', *Solar Energy Mater. Solar Cells* **100**, 132–137.

Yan Y., Jones K. M., Jiang C. S., Wu X. Z., Noufi R. and Al-Jassim M. M. (2007), 'Understanding the defect physics in polycrystalline photovoltaic materials,' *Physica B* **401**, 25–32.

Young D. L., Keane J., Duda A., AbuShama J. A. M., Perkins C. L., Romero M. and Noufi R. (2003), 'Improved performance in ZnO/CdS/CuGaSe$_2$ thin-film solar cells', *Progr. Photovoltaics* **11**, 535–541.

Zhang S. B., Wei S. H. and Zunger A. (1997), 'Stabilization of ternary compounds via ordered arrays of defect pairs', *Phys. Rev. Lett.* **78**, 4059–4062.

Zhang S. B., Wei S. H., Zunger A. and Katayama-Yoshiba H. (1998), 'Defect physics of the CuInSe$_2$ chalcopyrite semiconductor', *Phys. Rev B.* **57**, 9642–9656.

Zweibel K. (1995), 'Thin films: past, present, future', *Progr. Photovoltaics* **3**, 279–293.

CHAPTER 7

SUPER-HIGH-EFFICIENCY III–V TANDEM AND MULTIJUNCTION CELLS

MASAFUMI YAMAGUCHI

Toyota Technological Institute, Nagoya 468–8511, Japan

masafumi@toyota-ti.ac.jp

The reasonable man adapts himself to the world: the unreasonable one persists in trying to adapt the world to himself. Therefore all progress depends on the unreasonable man.

George Bernard Shaw, *Man and Superman*, 1903.

7.1 Introduction

Although solar electricity, including solar photovoltaics, is expected to make a great contribution as a major energy source, providing a share of about 20% of global electric power in 2050 and about 70% in 2100 (WBGU, 2003), nuclear power is still a major energy source because of its huge power-generation capacity and relatively low electricity cost. In order to realise the vision of a solar electricity future, high-performance solar cells are very attractive.

The development of high-performance solar cells offers a promising pathway toward achieving high power per unit cost for many applications. Substantial increases in conversion efficiency can be realised by using multijunction solar cells rather than single-junction cells. The principles of multijunction cells were suggested as long ago as 1955 (Jackson, 1955), and they have been investigated since 1960 (Wolf, 1960), as shown in the historical timeline of Table 7.1. However, no significant progress was made in multijunction cell conversion efficiency during the period 1960–75 because of poor thin-film fabrication technologies. It is thanks to progress in the technology of liquid-phase epitaxy (LPE) and vapour-phase epitaxy and to optical devices such as semiconductor lasers that more efficient AlGaAs/GaAs tandem cells were developed during the 1980s. These included tunnel junctions (Hutchby *et al.*, 1985) and metal interconnections (Ludowise *et al.*, 1982; Flores, 1983; Chung *et al.*, 1989).

At that time, the predicted efficiency of close to 30% (Fan *et al.*, 1982) was not achieved because of difficulties in making high-performance, stable tunnel

Table 7.1 Progress of III–V tandem solar cell technologies.

Date	Milestone	Location or individual
1955	Principle of multijunction solar cells	Jackson
1960	Proposal of multijunction solar cells	Wolf
1982	Efficiency calculation for tandem cells	MIT
1982	15.1% AlGaAs/GaAs two-junction (2-J) cell (1 Sun)	RTI
1987	Proposal of double-hetero structure tunnel junction for multijunction interconnection	NTT
1987	20.2% AlGaAs/GaAs 2-J cell (1 Sun)	NTT
1988	29.6% GaAs/Si 2-J cell (mechanically stacked, 350 Suns concentration)	Sandia
1989	27.6% GaAlAs/GaAs 2-J cell (1 Sun)	Varian
1989	32.6% GaAs/GaSb concentrator 2-J cell (mechanically stacked, 100 Suns concentration)	Boeing
1990	Proposal of InGaP as top cell material	NREL
1993	29.5% InGaP/GaAs 2-J cell (1 Sun)	NREL
1994	30.2% InGaP/GaAs 2-J cell (180 Suns)	NREL
1996	30.3% InGaP/GaAs 2-J cell (1 Sun)	Japan Energy
1997	Discovery of radiation resistance of InGaP top cell	Toyota T.I.
1997	33.3% InGaP/GaAs//InGaAs 3-J cell (1 Sun) (mechanically stacked)	Japan Energy, Sumitomo, Toyota T.I.
1997	Commercial satellite with 2-J cells	Hughes
2000	31.7% InGaP/InGaAs/Ge 3-J cell (1 Sun)	Japan Energy
2000	32.4% InGaP/GaAs/Ge 3-J cell (42 Suns)	Spectrolab
2001	33.5% InGaP/GaAs/InGaAs 3-J cell (308 Suns) (mechanically stacked)	Fraunhofer ISE
2003	32.0% InGaP/GaAs/Ge 3-J cell (1 Sun)	Spectrolab
2004	38.9% InGaP/InGaAs/Ge 3-J cell (489 Suns)	Sharp, Toyota T.I.
2005	38.9% InGaP/InGaAs/Ge 3-J cell (236 Suns)	Spectrolab
2006	31.5% large-area (5445 cm^2) InGaP/InGaAs/Ge 3-J cell module (outdoor)	Daido Steel, Daido Metal, Sharp, Toyota T.I.
2006	40.7% InGaP/GaAs/Ge 3-J cell (236 Suns)	Spectrolab
2008	33.8% InGaP/GaAs/InGaAs 3-J cell (1 Sun)	NREL
2009	41.1% InGaP/InGaAs/Ge 3-J cell (454 Suns)	Fraunhofer ISE
2009	41.6% InGaP/InGaAs/Ge 3-J cell (364 Suns)	Spectrolab
2009	35.8% InGaP/GaAs/InGaAs 3-J cell (1 Sun)	Sharp
2010	42.3% InGaP/GaAs/InGaAs 3-J cell (406 Suns)	Spire
2011	36.9% InGaP/GaAs/InGaAs 3-J cell (1 Sun)	Sharp
2012	43.5% InGaP/GaAs/GaInNAs 3-J cell (418 Suns)	Solar Junction
2013	37.9% InGaP/GaAs/InGaAs 3-J cell (1 Sun)	Sharp
2013	44.4% InGaP/GaAs/InGaAs 3-J cell (240–300 Suns)	Sharp

junctions (Yamaguchi *et al.*, 1987), and also because of the presence of oxygen-related defects in the AlGaAs (Ando *et al.*, 1987). High-performance, stable tunnel junctions with a double-hetero (DH) structure (in which the GaAs tunnel junction is sandwiched between AlGaAs layers) were developed by Sugiura *et al.* (1988) of NTT Electrical Communications Laboratories (NTT). The use of InGaP for the top cell was introduced by Olson *et al.* (1990) of National Renewable Energy Laboratory (NREL), and as a result, a GaInP/GaAs tandem cell with an efficiency of 29.5%, though a small area of only 0.25 cm^2, was made by Bertness *et al.* (1994). Monolithically grown InGaP/GaAs two-junction solar cells achieved a then-record efficiency of 30.3% in 1997 (Takamoto *et al.*, 1997a) at 1 Sun AM 1.5. As regards concentrator systems, over 30% efficiency was attained by the mechanically stacked GaAs/GaSb cells of Fraas *et al.* (1990).

InGaP/GaAs-based multijunction (MJ) solar cells have drawn increased attention for space applications because of the superior radiation resistance of InGaP top cells and materials, which was discovered by the author and co-workers (Yamaguchi *et al.*, 1997). In addition we also achieved the possibility of high conversion efficiency of over 30%. The commercial satellite HS 601HP with two-junction GaInP/GaAs on Ge solar arrays was launched in 1997 (Brown *et al.*, 1997).

InGaP/GaAs-based MJ solar cells have also drawn increased attention for terrestrial applications because operating MJ cells under concentrated sunlight has great potential for providing high-performance, low-cost solar modules. For concentrator applications, the cell contact grid structure should be designed so as to reduce the energy loss due to series resistance; in this way 38.9% (AM 1.5G, 489 Suns) efficiency was demonstrated by Sharp (Takamoto *et al.*, 2005). The achievement of 41.1% efficiency under 454 Suns with $In_{0.65}Ga_{0.35}P/In_{0.17}GaAs_{0.83}/Ge$ three-junction concentrator cells by Fraunhofer ISE (Bett *et al.*, 2009) and 41.6% efficiency under 364 Suns with lattice-matched InGaP/InGaAs/Ge three-junction concentrator cells by Spectrolab (King *et al.*, 2009) has also been reported. Later, 42.3% efficiency under 406 Suns with bifacial epitaxially grown InGaP/GaAs/InGaAs three-junction concentrator cells was reported by Spire (Wojtczuk *et al.*, 2010); and Solar Junction (Solar Junction, 2011; Sabnis, 2012) achieved 43.5% efficiency under 418 Suns with lattice-matched InGaP/GaAs/GaInNAs three-junction concentrator cells. Most recently, 37.9% efficiency under 1 Sun and 44.4% efficiency under 240 300 Suns have been demonstrated with InGaP/GaAs/InGaAs three-junction cells by Sharp (Sharp, 2013; Sasaki, 2013; Yamaguchi and Luque, 2013; Green *et al.*, 2013). In addition, high-efficiency large-area (5445 cm^2) concentrator InGaP/InGaAs/Ge three-junction modules for 500 Suns used with an outdoor efficiency of 31.5% (Araki *et al.*, 2005) have been developed as a result of incorporating high-efficiency InGaP/InGaAs/Ge three-junction cells, low-optical-loss Fresnel lenses and homogenisers, and designing low-thermal-conductivity modules.

In this chapter, we describe the principles of MJ solar cells, candidate materials, epitaxial technologies, MJ cell configurations and cell interconnection technologies. We also review recent progress with laboratory cells and make some predictions about the future of MJ cells.

7.2 Principles of super-high-efficiency multijunction solar cells

7.2.1 *Conversion efficiency analysis and candidate materials for high-efficiency multijunction solar cells*

While single-junction cells may be capable of attaining AM 1.5 efficiencies of up to 27%, MJ structures were recognised early on as being capable of realising efficiencies of up to 35% (Hutchby *et al.*, 1985). Figure 7.1 shows the principle of wide photoresponse using MJ solar cells for the case of a triple-junction cell. Solar cells with different bandgaps are stacked one on top of the other so that the cell facing the Sun has the largest bandgap (in this example, this is the InGaP top cell). This top cell absorbs all the photons at and above its bandgap energy and transmits

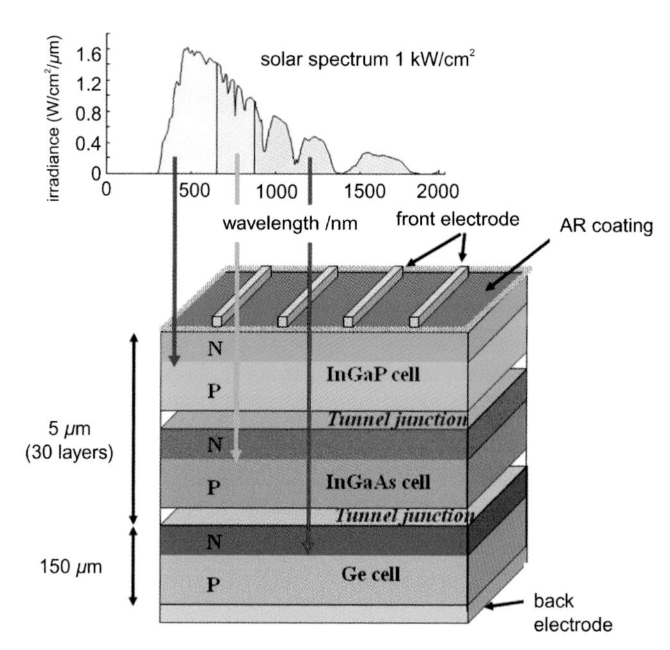

Figure 7.1 Principle of wide photoresponse by using a multijunction solar cell, for the case of an InGaP/InGaAs/Ge triple-junction solar cell.

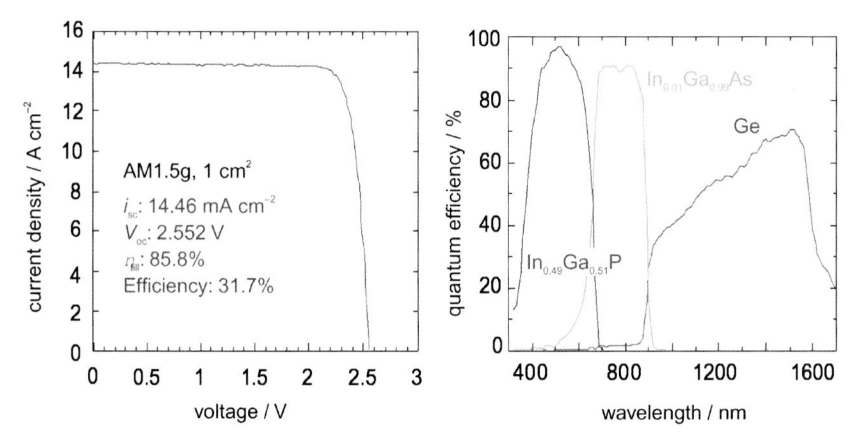

Figure 7.2 Current–voltage curve and spectral response for an InGaP/InGaAs/Ge monolithic, two-terminal three-junction cell (Takamoto *et al.*, 2000).

the less energetic photons to the cells below. The next cell in the stack (here the GaAs middle cell) absorbs all the transmitted photons with energies equal to or greater than its bandgap energy, and transmits the rest downward in the stack (in this example, to the Ge bottom cell). As shown in Fig. 7.2, the current–voltage curve and spectral response for an InGaP/GaAs/Ge monolithic, two-terminal triple-junction cell (Takamoto *et al.*, 2000) shows the wideband photoresponse of MJ cells. In principle, any number of cells can be used in tandem.

Computer analysis of the performance of MJ solar cells has been carried out by several researchers (Loferski, 1976; Lamorte and Abbott, 1980; Mitchell, 1981; Fan *et al.*, 1982; Nell and Barnett, 1987; Amano *et al.*, 1989; Letay and Bett, 2001). The following explanations are based on the findings of Letay and Bett (2001). Figure 7.3 shows their AM 1.5 iso-efficiency plots for a three-cell, two-terminal tandem structure with Ge as the bottom cell at 25°C and 1 Sun. The maximum theoretical efficiency for this system is 38% at AM 1.5. For optimal efficiencies in the two-terminal structure, the allowable range of bandgaps for the top and bottom cells is very narrow. The top cell must have a bandgap of about 1.8 eV, and the middle cell about 1.1 eV. In this case, one of the candidate material combinations is InGaP/InGaAs/Ge. However, because the optimal bandgap combination, which is 1.8 eV/1.1 eV/0.66 eV, is a lattice-mismatched system, 1.85 eV/1.4 eV/0.66 eV and 2 eV/1.4 eV/0.66 eV lattice-matched systems have been realistically developed. The optimal efficiencies of such systems are 31–34.5%. In these cases, InGaP/GaAs/Ge, AlInGaP/GaAs/Ge and AlGaAs/GaAs/Ge are candidate material combinations.

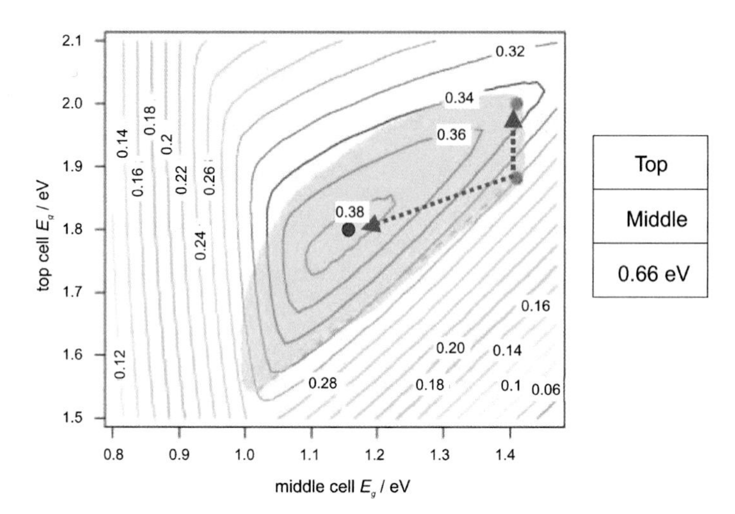

Figure 7.3 Calculated AM 1.5 iso-efficiency plots for a three-cell, two-terminal tandem structure with Ge as the bottom cell at 25°C and 1 Sun. From Letay and Bett (2001).

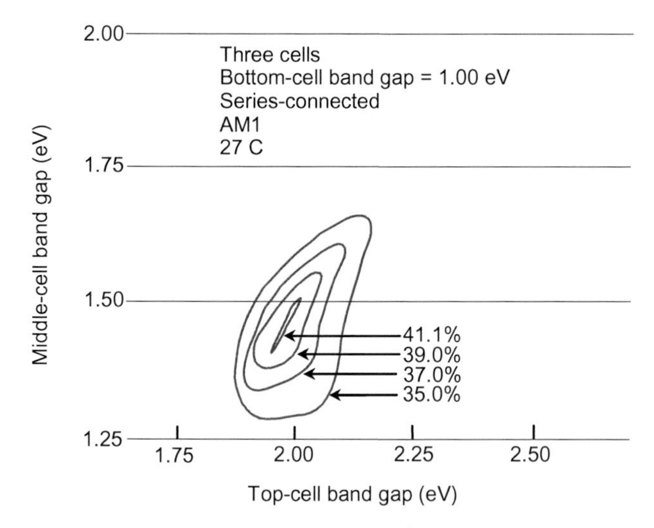

Figure 7.4 Calculated AM 1 iso-efficiency plots at 27°C and 1 Sun for the three-cell tandem structure with the cells connected in series. The bottom cell has a fixed energy gap of 1.0 eV (Fan *et al.*, 1982).

Figure 7.4 shows the AM 1 iso-efficiency plots at 27°C for the three-cell tandem structure with the cells connected in series (Fan *et al.*, 1982). The curves are plotted for a bottom-cell bandgap of 1.0 eV, because the maximum calculated efficiencies are obtained for a range of 0.95–1.0 eV. The optimal top/middle/bottom

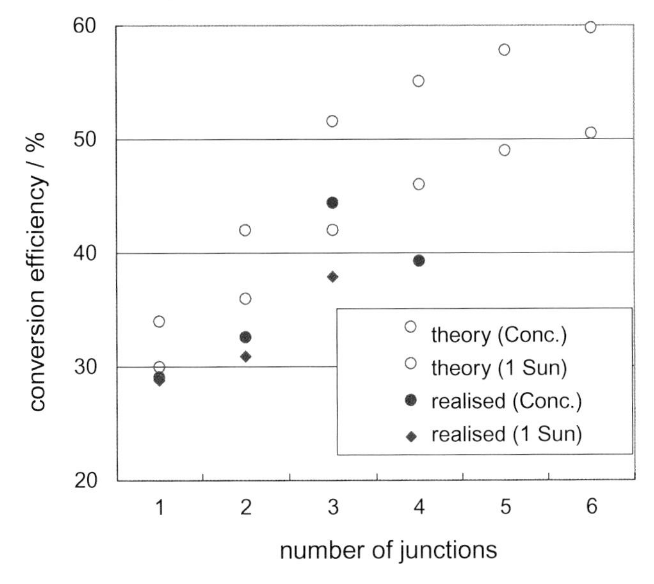

Figure 7.5 Theoretical conversion efficiencies of single-junction and multijunction solar cells in comparison with experimentally realised efficiencies for 1 Sun intensity and under concentration.

cell bandgap combination is 1.95/1.45/1.0 eV for the series-connected structure, and the maximum theoretical efficiency at AM 1 is 41.1%. In this case, (Al)In GaP/GaAs/InGaAs, (Al)InGaP/GaAs/InGaAsP, (Al)InGaP/GaAs/InGaAsN, (Al) InGaP/GaAs/Si and AlGaAs/GaAs/Si are candidate material combinations.

Figure 7.5 compares theoretical and realised energy efficiences for single-junction and MJ solar cells, under unconcentrated and concentrated sunlight. Multijunction solar cells have the potential for achieving ultra-high conversion efficiencies of over 50%, and are promising for both space and terrestrial applications due to their wide photoresponse.

7.2.2 *Cost analysis*

The allowable cost per unit area of solar cell modules depends strongly on module efficiency (Bowler and Wolf, 1980; Yamaguchi *et al.*, 1994). For example, a 30%-efficient cell costing 3.5 times as much as a 15%-efficient cell of the same area will yield equivalent overall photovoltaic (PV) system costs (Bowler and Wolf, 1980). Therefore high-efficiency cells will have a substantial economic advantage over low-efficiency cells as long as the cost of fabricating them is low enough. For space applications, high-efficiency cells also have significant payload advantages.

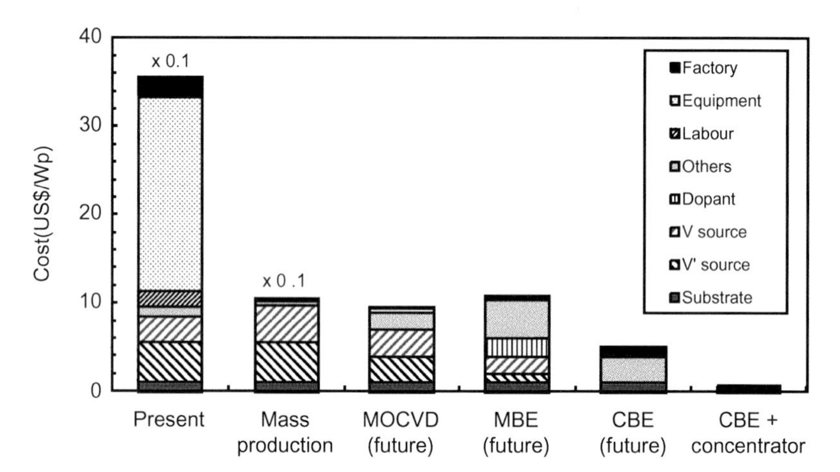

Figure 7.6 Cost estimates for monolithically integrated two-junction solar cells consisting of a III–V compound solar cell combined with a Si cell, fabricated by metal-organic chemical vapour deposition (MOCVD), molecular beam epitaxy (MBE) or chemical beam epitaxy (CBE) (Yamaguchi *et al.*, 1994).

The use of concentrating systems can further enhance the cost advantage of high-efficiency cells. Figure 7.6 shows cost estimates for monolithically integrated two-junction solar cells with a III–V compound solar cell and a Si cell, fabricated by metal–organic chemical vapour deposition (MOCVD), molecular beam epitaxy (MBE) or chemical beam epitaxy (CBE) (Yamaguchi *et al.*, 1994). Through source material cost reduction combined with the use of the tandem structure and a concentrator system, high-efficiency cells costing less than $\$1/W_p$ should be possible.

Concentrator operation is very effective for cost reduction of solar cell modules and thus also of PV systems. Figure 7.7 shows a summary of the estimated cost for the concentrator PV systems vs. concentration ratio (Yamaguchi, 2003). As shown in Fig. 7.7, concentrator PV systems using MJ solar cells have great potential for cost reduction to under $50¢/W_p$ if one could fabricate 35%-efficient modules for 5000 Suns use.

The effectiveness of high-performance, low-cost concentrator PV systems is also discussed in Chapter 11 of this book by Luque-Heredia and Luque, and separately by Yamaguchi and Luque (1999).

7.3 Epitaxial technologies for growing III–V compound cells

Figure 7.8 shows the chronological improvements in the efficiencies of GaAs solar cells fabricated by LPE, MOCVD and MBE. Liquid-phase epitaxy was used to

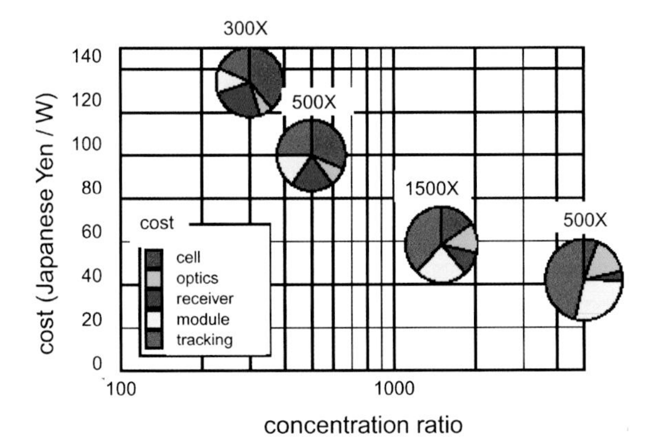

Figure 7.7 Summary of estimated cost for the concentrator PV systems vs. concentration ratio (Yamaguchi, 2003).

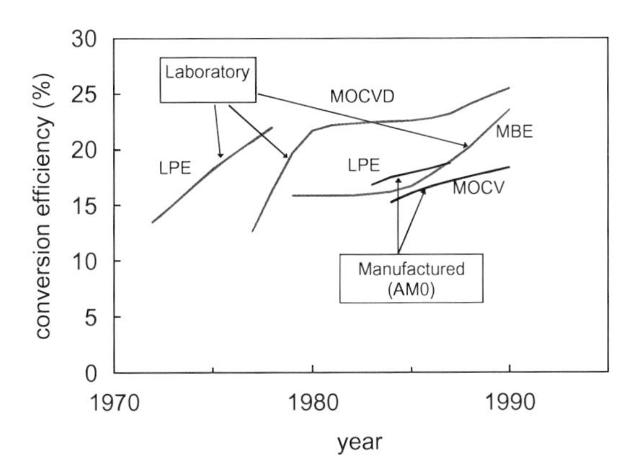

Figure 7.8 Chronological improvements in the efficiencies of GaAs solar cells fabricated by the LPE, MOCVD and MBE methods.

fabricate GaAs solar cells in 1972 because it produces high-quality epitaxial film and has a simple growth system. However, it is not as useful for devices that involve multilayers because of the difficulty of controlling layer thickness, doping, composition and speed of throughput. Since 1977, MOCVD has been used to fabricate large-area GaAs solar cells because it is capable of large-scale, large-area production and has good reproducibility and controllability.

Table 7.2 compares the advantages and disadvantages of the various epitaxial technologies. While LPE can produce high-quality epitaxial films, MOCVD is effective for the large-scale, large-area production of solar cells. Molecular beam

Table 7.2 Advantages and disadvantages of epitaxial technologies.

Characteristics	LPE	MOCVD	MBE	CBE
Quality	****	***	**	***
Multiple quantum well (MQW)	*	***	****	****
Abrupt interface	*	***	****	****
Heavy doping	**	**	***	****
Large area	*	****	**	***
Throughput	***	****	**	***
Efficient use of source materials	***	**	**	****
Equipment cost	***	**	**	*

****Excellent; ***very good; **fairly good; *bad.

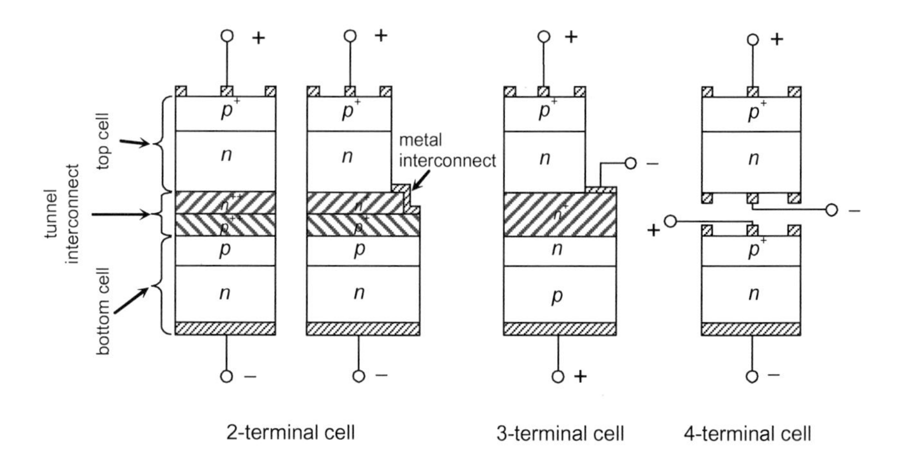

Figure 7.9 Schematic diagrams of various configurations of two-junction cells.

epitaxy and CBE are advantageous for realising novel multilayer structures such as MJ solar cells because they provide excellent controllability of monolayer abruptness and thickness due to the nature of the beam (Yamaguchi *et al.*, 1994). However, so far there have been few reports of CBE-grown solar cells.

7.4 Monolithic vs. multi-terminal connection modes

Figure 7.9 shows the connection options for two-junction cells: the two cells can be connected to form two-terminal, three-terminal or four-terminal devices. In a monolithic, two-terminal device, the cells are connected in series with an optically transparent tunnel junction intercell electrical connection. In a two-terminal structure, only one external circuit load is needed, but the photocurrents in the two cells must be equal for optimal operation. Key issues for maximum-efficiency monolithic cascade cells (two-terminal MJ cells series connected with tunnel junction) are the formation of tunnel junctions of high performance and stability for cell interconnection, and the growth of optimum bandgap top- and bottom-cell structures on lattice-mismatched substrates, without permitting propagation of deleterious misfit and thermal stress-induced dislocations.

In contrast, the photocurrents in three-and four-terminal cells do not have to be equal. However, it is very difficult to connect three-terminal devices in series, so three-terminal tandem cells do not appear to be viable. In the four-terminal case, two separate external circuit loads are used. Since the two individual cells are not coupled, the photocurrents do not have to be the same. Consequently, a much larger range of bandgap energy combinations is possible, and the changes in photocurrents with changing solar spectral distributions do not pose serious limits. This approach avoids the problems of lattice-mismatched epitaxial growth, current matching and the internal electrical connection of the two-terminal device. Important issues for obtaining high-efficiency mechanically stacked cells are the development of MJ cell fabrication techniques such as thinning the top cell, bonding the bottom cell to the top cell, and cell connections.

7.5 Key issues for realising high-efficiency multijunction solar cells

7.5.1 *Selection of cell materials and improvement of quality*

Multijunction cells with different bandgaps are stacked in tandem so that the cells cover a wide wavelength region from 300 nm to 1800 nm. Cell materials are selected by choosing bandgap energies close to the optimal bandgap energy combination found from theoretical efficiency calculations, and by considering lattice matching to substrates and reducing impurity problems. Since, for example, oxygen in AlGaAs acts as recombination centre (Ando *et al.*, 1987), reduction in residual oxygen is necessary for realising higher efficiencies. Figure 7.10 shows the minority-carrier diffusion length dependence of GaAs single-junction solar cell efficiency. It is clear that a high minority-carrier diffusion length L (minority-carrier lifetime $\tau = L^2/D$, where D is the minority-carrier diffusion coefficient) is substantially necessary to realise high-efficiency solar cells.

Figure 7.11 shows the carrier concentration dependence of minority-carrier lifetime in p-type and n-type GaAs (Ahrenkiel *et al.*, 1993). The minority-carrier lifetime τ depends on the carrier concentration N of solar cell layers as expressed by

$$\tau = 1/BN \tag{7.1}$$

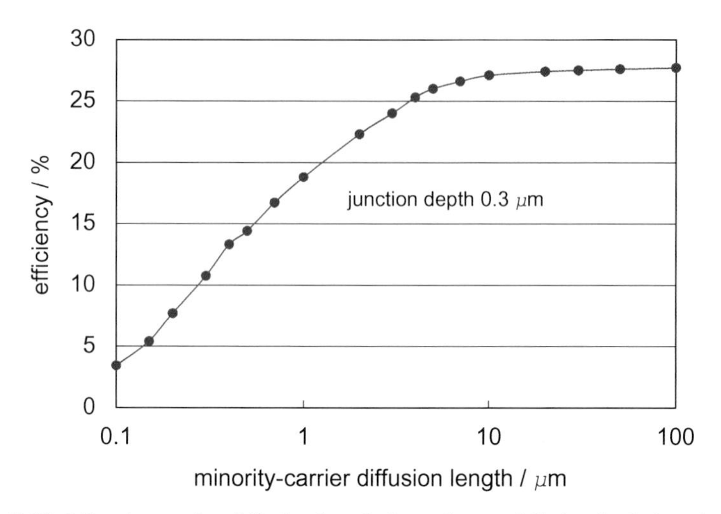

Figure 7.10 Minority-carrier diffusion length dependence of GaAs single-junction solar cell efficiency.

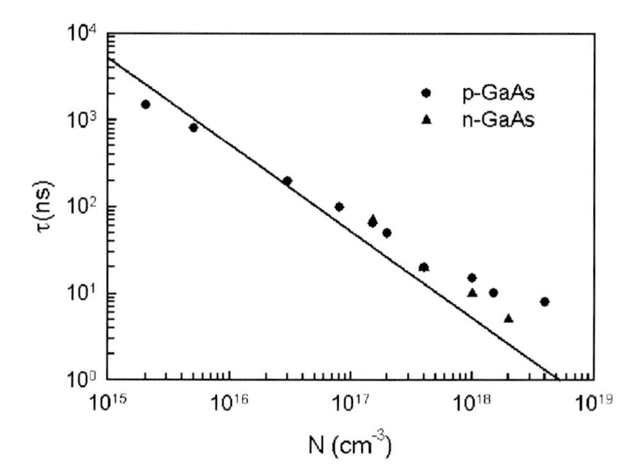

Figure 7.11 Carrier concentration dependence of minority-carrier lifetime in *p*-type and *n*-type GaAs (Ahrenkiel *et al.*, 1993).

where B is the radiative recombination coefficient. B is $2 \times 10^{-10}\,\mathrm{cm^3s^{-1}}$ for GaAs and $1.7 \times 10^{-10}\,\mathrm{cm^3s^{-1}}$ for InGaP. The carrier concentration of cell layers is optimised by considering minority-carrier lifetimes, built-in potential and series resistance of *p-n* junction diodes.

Selection of cell materials, especially the top-cell materials, is also important for achieving high-efficiency tandem cells. It has been found by the authors (Ando

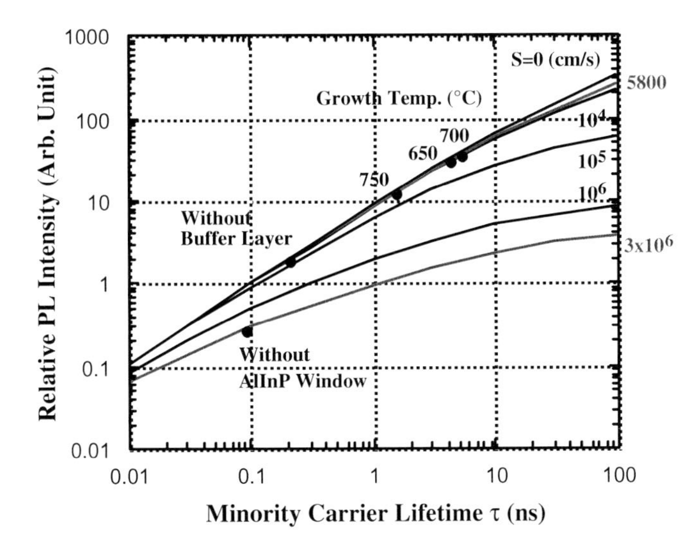

Figure 7.12 Changes in photoluminescence (PL) intensity of the solar cell active layer as a function of the minority-carrier lifetime and surface recombination velocity in InGaP, grown by MOCVD.

et al., 1987) that an oxygen-related defect in the AlGaAs top-cell materials acts as a recombination centre. As a top-cell material lattice-matched to GaAs or Ge substrates, InGaP has some advantages (Olson *et al.*, 1990), such as a lower interface recombination velocity, a lower oxygen-related defect problem and a good window layer material compared with AlGaAs. The top-cell characteristics depend on the minority-carrier lifetime in the top-cell layers.

Figure 7.12 shows changes in photoluminescence (PL) intensity of the solar cell active layer as a function of the minority-carrier lifetime τ of the p-InGaP base layer grown by MOCVD and surface recombination velocity S. The lowest level of S was obtained by introducing an AlInP window layer, and the highest minority-carrier lifetime τ was obtained by introducing a buffer layer and optimising the growth temperature. The best conversion efficiency of the InGaP single-junction cell was 18.5% (Yang *et al.*, 1997).

7.5.2 *Lattice matching between cell materials and substrates*

Lattice mismatching of cell materials to substrates should be minimised because misfit dislocations are generated in the upper cell layers by lattice mismatch, which decreases cell efficiency. Figure 7.13 shows the calculated and experimental dislocation density dependence of the minority-carrier lifetime in GaAs (Yamaguchi and Amano, 1985). The dependence of the dislocation density N_d on minority-carrier

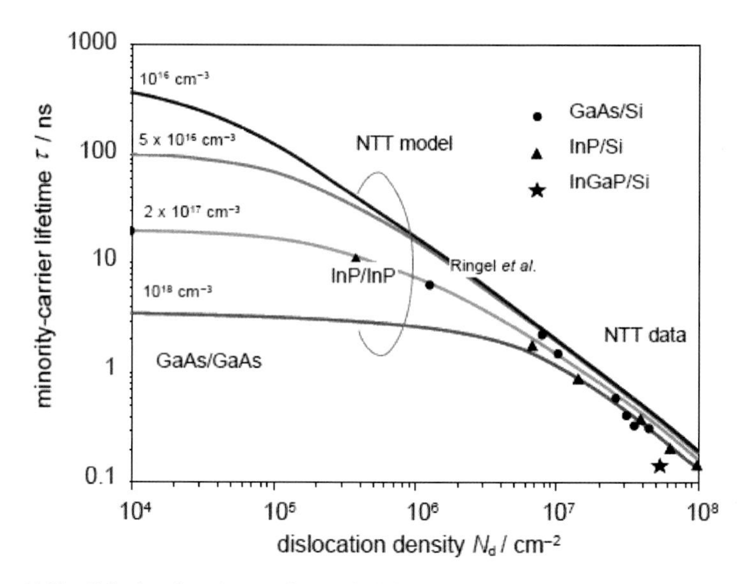

Figure 7.13 Calculated and experimental dislocation density dependence of minority-carrier lifetimes in GaAs, InP and InGaP. NTT model (Eq. (7.2)), NTT data and Ringel's data relate to Yamaguchi and Amano (1985), Yamaguchi *et al.* (2001) and Carlin *et al.* (2000).

lifetime τ was calculated from the following equation (Yamaguchi and Amano, 1985):

$$1/\tau = 1/\tau_r + 1/\tau_0 + \pi^3 D N_d/4 \quad (7.2)$$

where D is the minority-carrier diffusion coefficient, τ_r is the radiative recombination lifetime and τ_0 is the minority-carrier lifetime associated with recombination at other unknown defects.

Application of an InGaAs middle cell (Takamoto *et al.*, 2003) lattice-matched to Ge substrates has been demonstrated to increase the open-circuit voltage (V_{oc}) due to improved lattice matching, and also the short-circuit current density (i_{sc}) due to the decrease in bandgap energy of the middle cell.

7.5.3 *Effectiveness of wide-bandgap windows and back-surface field layers*

Figure 7.14 shows the effect of surface recombination velocity on the short-circuit current density of $In_{0.14}Ga_{0.86}As$ homojunction solar cells as a function of junction depth; because GaAs-related materials have high surface recombination velocity of about 10^7 cm/s^{-1}, formation of a shallow junction with a junction depth of less

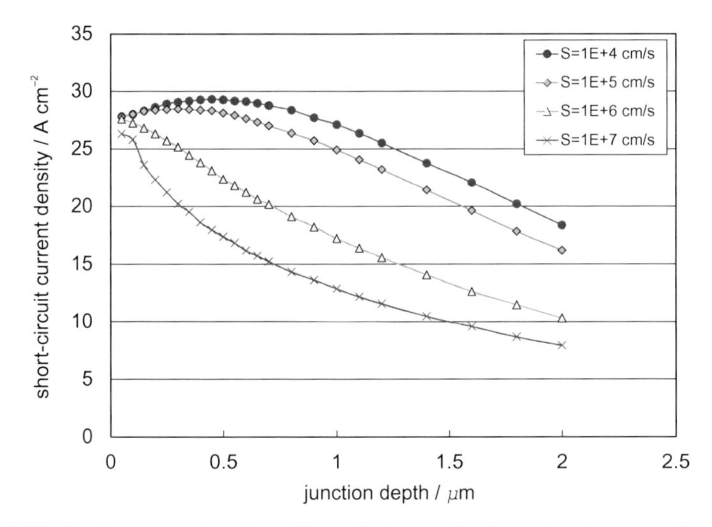

Figure 7.14 Surface recombination effect velocity on the short-circuit current density of a $In_{0.14}Ga_{0.86}As$ homojunction solar cell as a function of junction depth.

than 50 nm is necessary. In order to decrease the efficiency drop due to front and rear surface recombination as shown in Fig. 7.15, formation of a heteroface or double-hetero structure is necessary.

Figure 7.15 shows the changes in V_{oc} and i_{sc} of InGaP single-junction cells as a function of the potential barrier ΔE. A wide-bandgap back-surface field (BSF) layer (Takamoto *et al.*, 2003) is found to be the most effective for confinement of minority carriers.

7.5.4 *Low-loss tunnel junctions for intercell connection and preventing impurity diffusion from the tunnel junction*

Another important issue for realising high-efficiency monolithic-cascade type tandem cells is the achievement of optically and electrically low-loss interconnection of two or more cells. A degenerately doped tunnel junction is attractive because creating it only involves one extra step in the growth process. To minimise optical absorption and achieve higher short-circuit current density for the bottom cell, as shown in Fig. 7.16, the formation of physically thin wide-bandgap tunnel junctions is necessary. In order to form a wide-bandgap tunnel junction, doping of higher concentration impurities into the tunnel junction is necessary because the tunnelling current decreases exponentially with increase in bandgap energy.

In addition, impurity diffusion from a highly doped tunnel junction during overgrowth of the top cell increases the resistivity of the tunnel junction. As shown

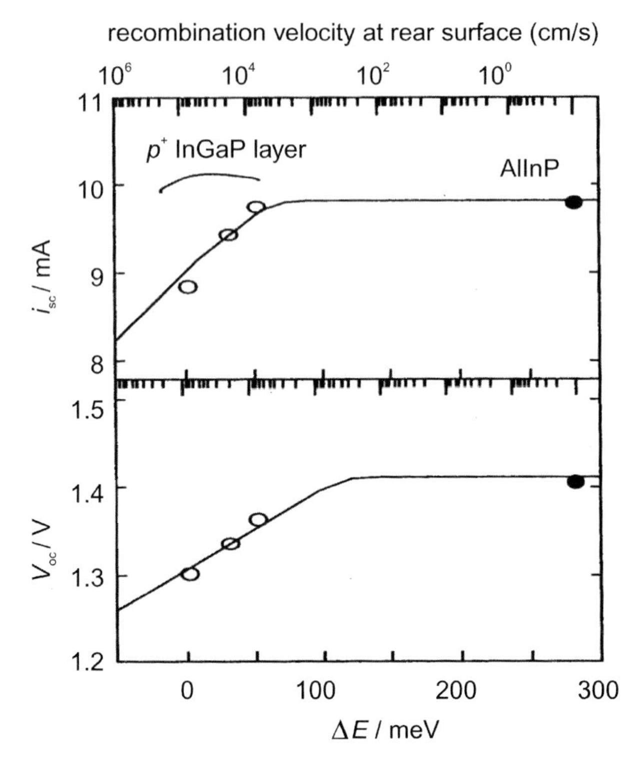

Figure 7.15 Change in V_{OC} and i_{SC} of InGaP single-junction cells as a function of the potential barrier ΔE.

in Fig. 7.17, a double-hetero (DH) structure was found by the authors to be useful for preventing diffusion (Sugiura *et al.*, 1988). An InGaP tunnel junction was tried for the first time for an InGaP/GaAs tandem cell in our work (Takamoto*et al.*, 1997b). Zn and Si were used as p-type and n-type dopants, respectively. The peak tunnelling current of the InGaP tunnel junction increased from 5 mA cm^{-2} up to 2 A cm^{-2} on making a DH structure with AlInP barriers. Effective suppression of the Zn diffusion from tunnel junction by the InGaP tunnel junction with the AlInP-DH structure is thought to be due to the lower diffusion coefficient (Takamoto *et al.*, 1999) for Zn in the wider bandgap energy materials such as the AlInP barrier layer and InGaP tunnel junction layer. In conclusion, the InGaP tunnel junction is very effective for obtaining high tunnelling currents, and the DH structure is useful for preventing diffusion.

Table 7.3 summarises factors of importance in achieving MJ cells of very high efficiency.

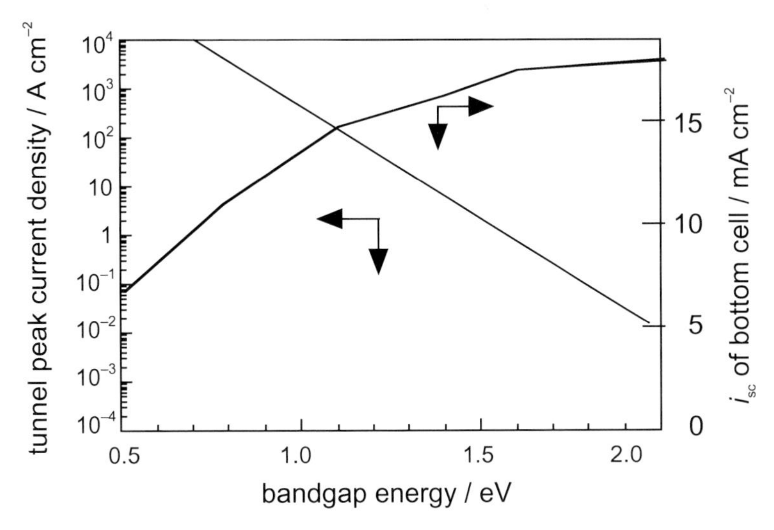

Figure 7.16 Calculated tunnel peak current density and short-circuit current density of a GaAs bottom cell as a function of the bandgap energy of the tunnel junction.

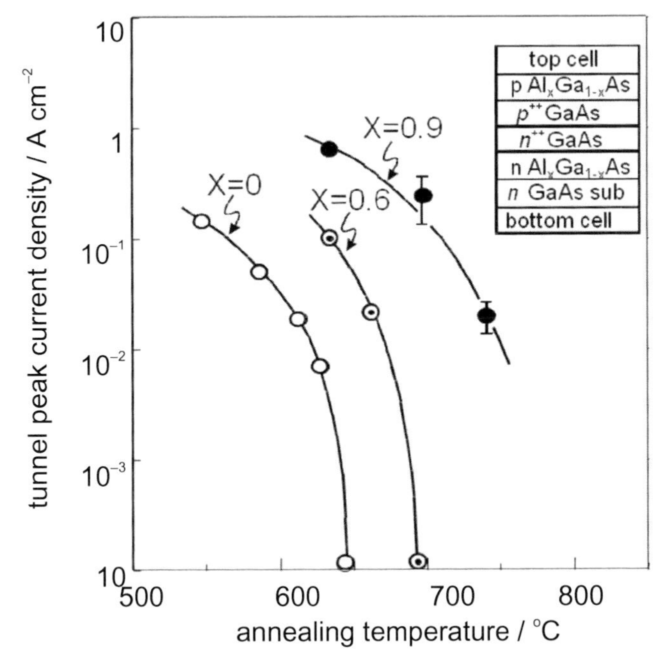

Figure 7.17 Annealing temperature dependence of tunnel peak current densities for double-hetero structure tunnel diodes. X is the Al mole fraction in the $Al_xGa_{1-x}As$ barrier layers.

Table 7.3 Key issues for realising super-high-efficiency multijunction solar cells.

Key issue	Past	Present	Future
Top cell materials	AlGaAs	InGaP	AlInGaP
Third cell materials	None	Ge	InGaAsN etc.
Bottom cell materials	GaAs	Ge	Si, Ge, InGaAs
Substrate	GaAs	Ge	Si, Ge, GaAs, metal
Tunnel junction (TJ)	DH-structure GaAs TJ	DH-structure InGaP TJ	DH-structure InGaP or GaAs TJ
Lattice matching	GaAs middle cell	InGaAs middle cell	(In)GaAs middle cell
Carrier confinement	InGaP-BSF	AlInP-BSF	Wide-gap-BSF Quantum dots (QDs)
Photon confinement	None	None	Bragg reflector, QDs, etc.
Others	None	Inverted epitaxial growth (i.e. epitaxial growth on substrate from top cell to bottom cell)	Inverted epitaxial growth, epitaxial lift off (i.e. thin film peeled off)

7.6 High-efficiency InGaP/GaAs/Ge three-junction solar cells and their space applications

7.6.1 *Development of high-efficiency InGaP/GaAs/Ge three-junction solar cells*

As one of the projects in the Sunshine Programme in Japan, a research and development (R&D) project for super-high-efficiency MJ solar cells was started in 1990. The conversion efficiency of InGaP/GaAs-based multijunction solar cells was improved under this programme by the application of the following technologies:

- Selection and high-quality growth of InGaP as a top-cell material.
- Use of double-hetero structures and wide-bandgap tunnel junctions for cell interconnection.
- Precise lattice matching of the InGaP top cell and the InGaAs middle cell with the Ge substrate.
- Use of AlInP as a back-surface-field layer for the InGaP top cell.
- Use of InGaP-Ge heteroface structure bottom cells.

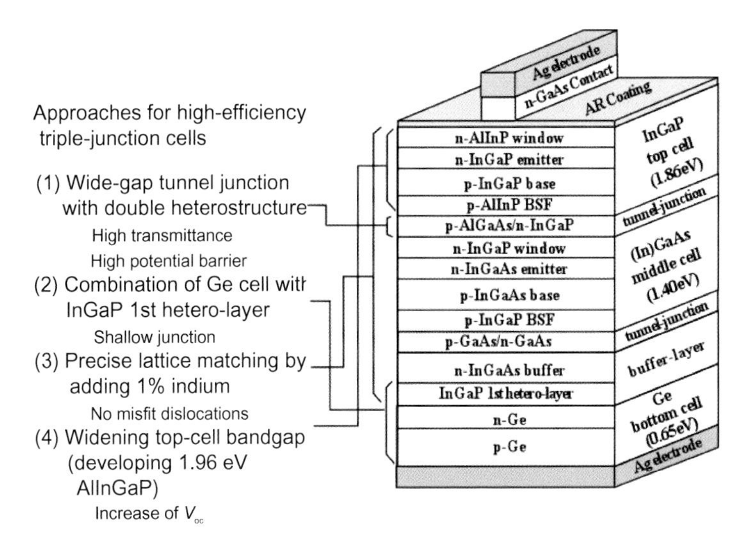

Figure 7.18 Schematic illustration of a triple-junction cell and approaches for improving its efficiency.

As a result of the above innovations and performance improvements, the 1 Sun conversion efficiency of InGaP/(In)GaAs/Ge three-junction solar cells was improved to 31.7% (AM 1.5G) by the turn of the century (Takamoto *et al.*, 2000). Figure 7.18 shows a schematic illustration of the InGaP/(In)GaAs/Ge triple-junction solar cell and the key technologies for improving its conversion efficiency. Although most recently, a 1 Sun efficiency of 37.9% has been achieved with InGaP/GaAs/InGaAs three-junction solar cells by Sharp (Sharp, 2013; Sasaki, 2013; Yamaguchi and Luque, 2013; Green *et al.*, 2013), the key technologies for improving efficiency are based on the following key issues.

Wide-bandgap tunnel junction

Use of a wide-bandgap tunnel junction which consists of the double-hetero structure *p*-Al(Ga)InP/*p*-AlGaAs/*n*-(Al)InGaP/*n*-Al(Ga)InP increases the amount of light incident on the (In)GaAs middle cell and produces effective potential barriers for minority carriers generated in the top and middle cells. Both the open-circuit voltage and the short-circuit current of the cells are improved by such wide-gap tunnel junctions without suffering absorption and recombination losses (Takamoto *et al.*, 1997b). It is difficult to obtain a high tunnelling peak current with wide-gap tunnel junctions, so thinning the depletion layer width by formation of a highly doped junction is necessary. Since impurity diffusion occurs during growth of the top cell (Sugiura *et al.*, 1988), carbon and silicon, which both have low diffusion coefficients, are used as dopants for *p*-type AlGaAs and *n*-type (Al)InGaP, respectively. Furthermore, the double-hetero structure appears to suppress impurity diffusion from the highly doped tunnel junction (Takamoto *et al.*, 1999). The second tunnel junction between the middle and bottom cells consists of *p*-InGaP/*p*-(In)GaAs/*n*-(In)GaAs/*n*-InGaP, which has a wider bandgap than the middle cell materials.

Heteroface-structure Ge bottom cells

InGaP/GaAs cell layers are normally grown on a *p*-type Ge substrate. A *n*-*p* junction is formed automatically during MOCVD growth by diffusion of the V-group atom from the first layer grown on the Ge substrate, so the material of the first heterolayer is important for the performance of the Ge bottom cell. An InGaP layer is thought to be suitable as material for the first heterolayer, because phosphorus has a lower diffusion coefficient in Ge than arsenic, and indium has a lower solubility in Ge than gallium.

Figure 7.19 shows the change in spectral response of the triple-junction cell on changing the first hetero-growth layer on Ge from GaAs to InGaP. The quantum efficiency of the Ge bottom cell was improved by using the InGaP hetero-growth layer. In the case of the GaAs hetero-growth layer, the junction depth was measured to be around 1 μm. On the other hand, the thickness of the *n*-type layer produced by phosphorus from the InGaP layer was 0.1 μm. Takamoto *et al.* (2003, 2005) confirmed that the increase in Ge quantum efficiency was due to this reduction in junction depth.

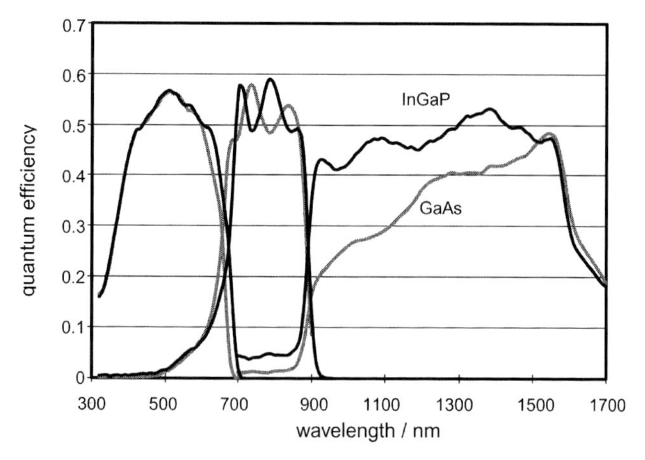

Figure 7.19 Change in the spectral response due to modification of the first heterolayer from GaAs to InGaP (without antireflection coating).

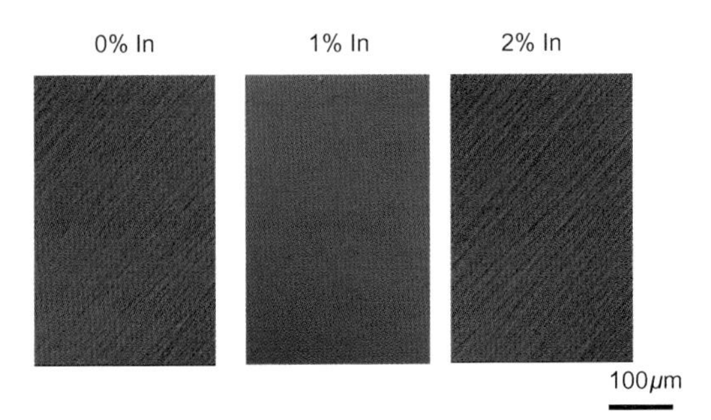

Figure 7.20 Surface morphology of InGaAs with various indium compositions grown on Ge.

Precise lattice-matching to the Ge substrate

Although the 0.08% lattice mismatch between GaAs and Ge is thought to be negligibly small, misfit dislocations are found to be generated in thick GaAs layers, which diminishes cell performance. By adding about 1% indium into the InGaP/GaAs cell layers, all cell layers are lattice-matched precisely to the Ge substrate. As a result, the crosshatch pattern caused by misfit dislocations due to lattice mismatch disappears in the surface morphology of the cells, as shown in Fig. 7.20. Eliminating the misfit dislocations in this way was found to influence the open-circuit voltage of the cell but not the short-circuit current. In addition, the wavelength of the absorption edge increased and the short-circuit current of the both top and middle cells was increased.

7.6.2 *Radiation resistance of InGaP-based multijunction solar cells*

Radiation in space is severe, particularly in the Van Allen radiation belt, and lattice defects are induced in semiconductors by the high-energy electron and proton irradiation of devices sited in this belt. These defects cause a decrease in the output power of solar cells deployed in space. Figure 7.21 shows the effectiveness of radiation resistance and high conversion efficiency of space cells in order to increase power density (W m^{-2}) for space missions. Before 1995, single-crystal Si solar cells were used in space but their BOL (beginning-of-life) and EOL (end-of-life) efficiencies (about 15% and 10%, respectively) were low and insufficient. Developing higher conversion efficiency and highly radiation-resistant space cells was necessary for widespread applications of PV in space missions. InGaP/GaAs-based MJ solar cells have attracted increased attention in this regard because of the possibility of high conversion efficiencies of over 40% and good radiation resistance.

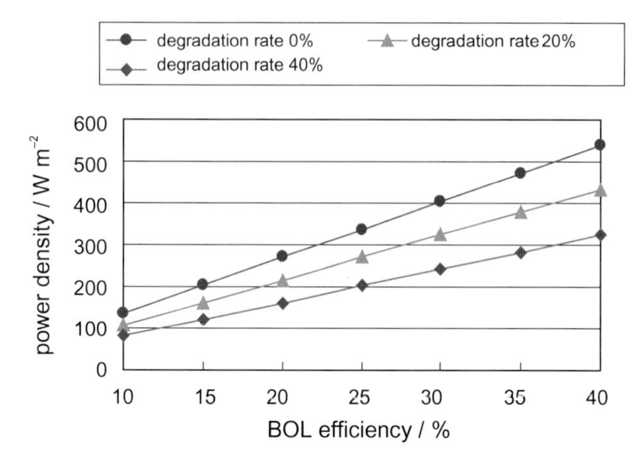

Figure 7.21 Effectiveness of radiation resistance and high conversion efficiency of space cells in increasing the power density of space missions. Beginning-of-life efficiency, BOL, is the efficiency of the space solar cells before space satellite launching.

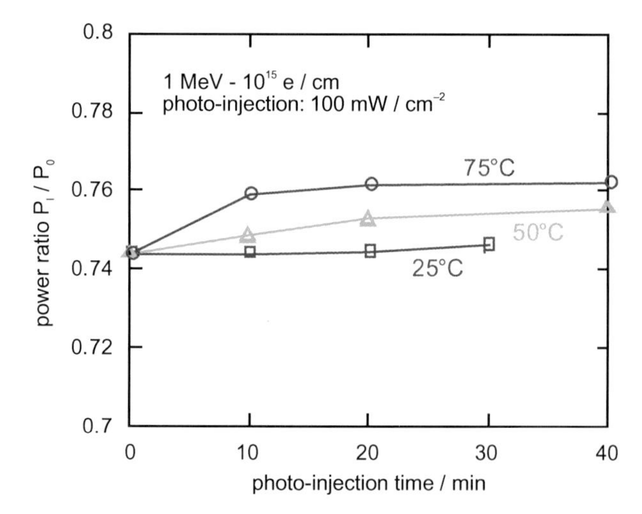

Figure 7.22 Maximum power recovery of the InGaP/GaAs tandem cell due to light illumination at various temperatures.

Figure 7.22 shows the recovery of maximum power produced by light of $100\,\mathrm{mW\,cm^{-2}}$ intensity at various temperatures for InGaP/GaAs tandem cells that have been irradiated with a 1-MeV $1 \times 10^{15}\,\mathrm{cm^{-2}}$ electron flux (Yamaguchi *et al.* 1997). The ratios of maximum power after injection, P_I, to maximum power before irradiation, P_0, are shown as a function of injection time. Even at room temperature,

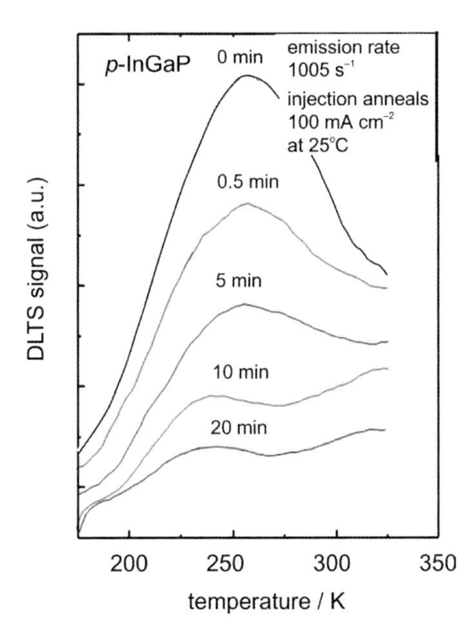

Figure 7.23 Deep-level transient spectroscopy (DLTS) spectrum of trap H2 (Ev + 0.55 eV) for various injection times at 25°C with an injection density of 100 mA cm^{-2}.

photoinjection-enhanced annealing of radiation damage to InGaP/GaAs tandem cells was observed. The recovery ratio increases with increase in ambient temperature within the operating range for space use. Such a recovery is attributed to damage recovery in the InGaP top cell layer (Yamaguchi *et al.*, 1997). These results indicate InGaP/GaAs tandem cells have superior radiation-resistance under device operation conditions.

Figure 7.23 shows DLTS (deep-level transient spectroscopy) spectra of the hole trap H2 (Ev+0.55 eV) for various injection times at 25°C with a forward bias injection density of 100 mA cm^{-2}. Khan *et al.* (2000) found from DLTS measurements that this major defect level recovers under forward bias or light illumination, i.e. the defect signal decreases with prolonged light exposure. Moreover, the H2 centre was confirmed to act as a recombination centre by using the double carrier pulse DLTS method. The enhancement of defect annealing in the InGaP top-cell layer under minority-carrier injection conditions is thought to occur as a result of the non-radiative electron-hole recombination process (Lang *et al.*, 1976), whose energy E_R enhances the defect motion. The thermal activation energy E_A (1.1 eV) of the defect is reduced to E_I (0.48 $\sim$ 0.54 eV) by an amount E_R (0.56 $\sim$ 0.62 eV). Thus electronic energy from a recombination event

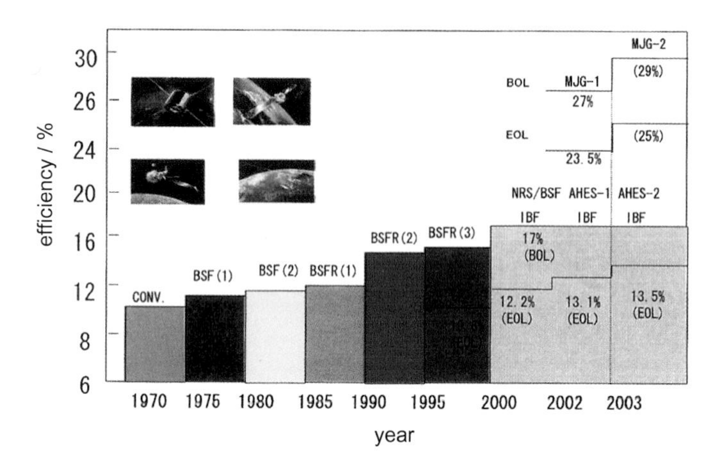

Figure 7.24 Sharp space solar cell conversion efficiency heritage. CONV., BSF, BSFR, NRS/BSF (IBF) show conventional *p-n* junction, back-surface-field, back-surface-field-reflector, non-reflective surface/back-surface field (improved back-surface field) structures for Si space solar cells. AHES shows advanced high-efficiency solar cell structures for space cells. BOL and EOL show beginning-of-life and end-of-life, respectively. All efficiencies for 1 Sun AM 0 under standard conditions.

can be channelled into the lattice vibration mode which drives the defect motion: $E_I = E_A - E_R$.

7.6.3 *Space applications of InGaP/GaAs/Ge three-junction solar cells*

Advanced technologies for high-efficiency cells and the discovery of the superior radiation resistance of InGaP-based materials are contributing to the industrialisation of InGaP-based MJ space solar cells in Japan; InGaP/GaAs/Ge three-junction cells have been commercialised for space use in Japan since 2002 (Takamoto *et al.*, 2005; Imaizumi *et al.*, 2005). Figure 7.24 shows the Sharp space solar cell conversion efficiency heritage up to 2003.

7.7 Multijunction solar cells: recent results

InGaP/GaAs-based MJ solar cells have drawn increased attention for terrestrial applications because concentrator operation of MJ cells has great potential for providing high-performance, low-cost solar cell modules. For concentrator applications, the cell contact grid structure should be designed to reduce the energy loss due to series resistance. Efficiencies have steadily increased: Sharp demonstrated 38.9% efficiency at 489 Suns (Takamoto *et al.*, 2005). Fraunhofer ISE achieved

41.1% efficiency under 454 Suns with $In_{0.65}Ga_{0.35}P/In_{0.17}GaAs_{0.83}/Ge$ 3-J cells (Bett *et al.*, 2009), and Spectrolab reached 41.6% efficiency under 364 Suns with lattice-matched InGaP/InGaAs/Ge 3-J cells (King *et al.*, 2009). More recently, Spire achieved 42.3% efficiency under 406 Suns with bifacial epitaxially grown InGaP/GaAs/InGaAs three-junction cells (Wojtczuk *et al.*, 2010) and Solar Junction (Solar Junction, 2011; Sabnis, 2012) reached 43.5% efficiency under 418 Suns with lattice-matched InGaP/GaAs/GaInNAs 3-J cells. At the time of writing, the world record is held by Sharp with their InGaP/GaAs/InGaAs 3-J solar cells, which are 44.4% efficient at 302 Suns (Sharp, 2013; Sasaki, 2013; Yamaguchi and Luque, 2013; Green *et al.*, 2013).

The 1 Sun efficiency of triple-junction solar cells is also improving. A world-record efficiency (37.9%) at 1 Sun (AM 1.5 G) has recently been realised with inverted epitaxially grown InGaP/GaAs/InGaAs three-junction cells by Sharp (Sharp, 2013; Sasaki, 2013; Yamaguchi and Luque, 2013; Green *et al.*, 2013). Figure 7.25 shows the fabrication process of this cell (Takamoto *et al.*, 2010; Yoshida *et al.*, 2011) and Fig. 7.26 shows its current–voltage curve. Figure 7.27 shows the chronological improvement in the conversion efficiencies of III–V compound MJ solar cells under 1 Sun and concentrator conditions.

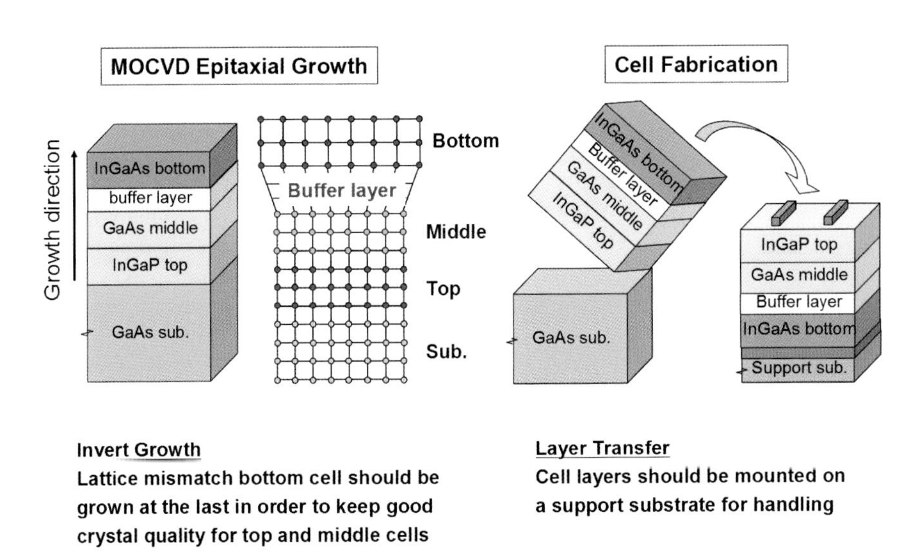

Figure 7.25 Fabrication process of inverted epitaxially grown InGaP/GaAs/InGaAs 3-J solar cells (Takamoto *et al.*, 2010; Yoshida *et al.*, 2011).

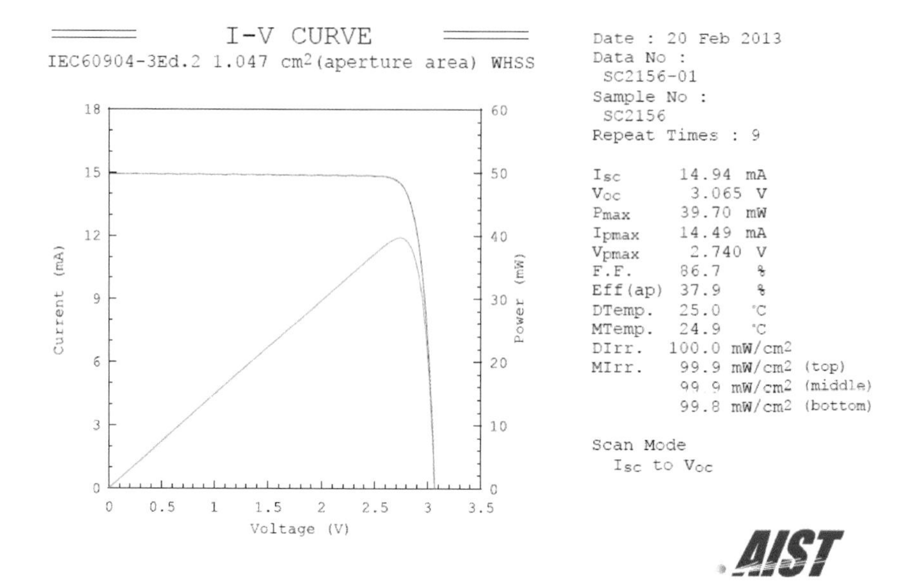

Figure 7.26 Current–voltage curve of world-record efficiency InGaP/GaAs/InGaAs cell as measured by the AIST (Sharp, 2013; Sasaki, 2013; Yamaguchi and Luque, 2013).

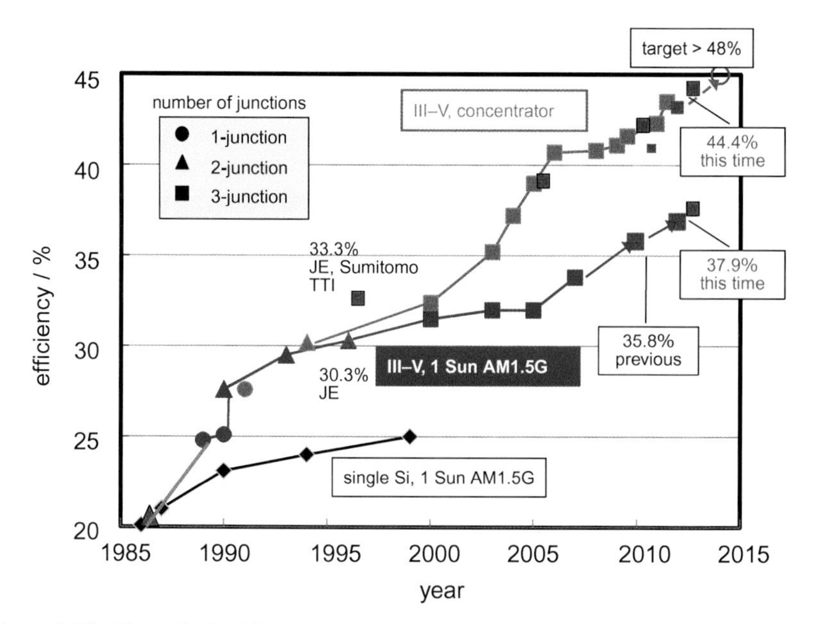

Figure 7.27 Chronological improvements in conversion efficiencies of III–V compound multijunction solar cells under 1 Sun and concentrator conditions (Yamaguchi and Luque, 2013).

7.8 Future directions

Multijunction solar cells will be widely used in space because of their high conversion efficiency and good radiation resistance. However, in order to apply super-high-efficiency cells widely on Earth, it will be necessary to improve their conversion efficiency and reduce their cost. As described in Chapter 11, concentrator PV (CPV) systems with several times more annual power-generation capability than conventional crystalline silicon flat-plate systems will open a new market for apartment or building rooftop applications. Other interesting applications are in agriculture and large-scale PV power plants.

Currently, efficiencies of over 40% are achieved by concentrator MJ solar cells, as shown in Fig. 7.28. Concentrator four-junction or five-junction solar cells have great potential for realising super-high efficiencies of over 50% (Goetzberger *et al.*, 2001; Yamaguchi, 2004; updated by using Solar Efficiency Tables [Green *et al.*, 2011]). Lattice-mismatched and III–V–N compounds are thought to be promising materials for realising more than 50% efficiency. As a three-junction combination, InGaP/InGaAs/Ge cells on Ge substrates will be widely used because this system has already been developed. The four-junction combination of a top cell with a bandgap energy $E_g = 2.0$ eV, a GaAs second-layer cell, a third-layer cell with a material of bandgap 1.05 eV and a Ge bottom cell would be lattice-matched to Ge

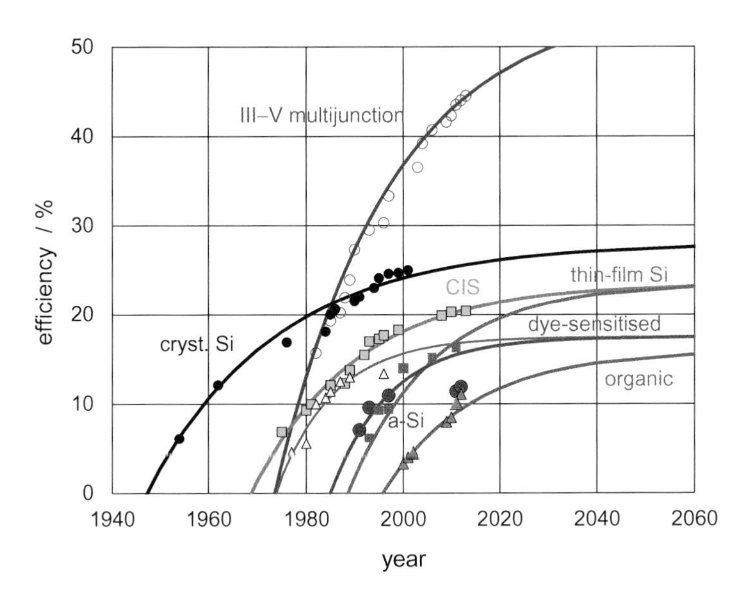

Figure 7.28 Predictions of future solar cell efficiencies (Goetzberger *et al.*, 2001; Yamaguchi, 2004; updated by using Solar Cell Efficiency Tables [Green *et al.*, 2013]).

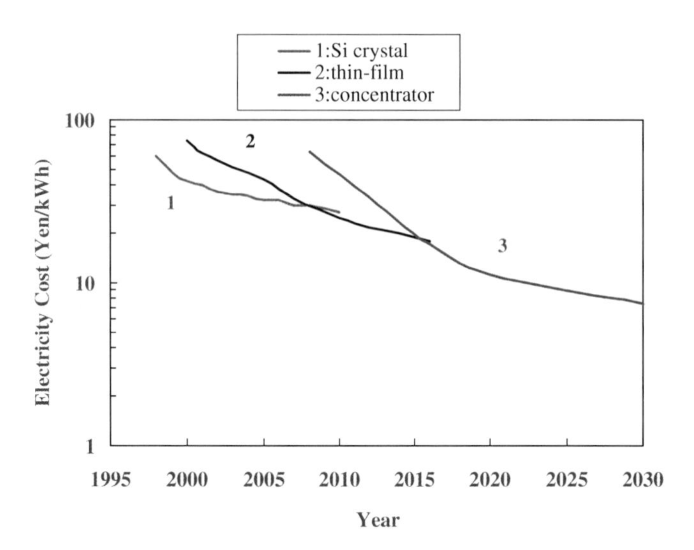

Figure 7.29 Scenario of electricity cost reduction by developing concentrator solar cells (Yamaguchi, 2003).

substrates and would have a theoretical efficiency of about 42% under 1 Sun AM 0, and over 47% under 500 Suns AM 1.5 G (Kurtz *et al.*, 1997).

In conclusion, as shown in Fig. 7.29 (Yamaguchi, 2004), we would like to contribute to commercialisation of CPV technologies as the third-generation PV technologies in succession to the first-generation crystalline Si PV and the second-generation thin-film PV technologies.

Acknowledgements

This work was partially supported by the Japanese New Energy and Industrial Technology Development Organization (NEDO) under METI (Ministry of Economy, Trade and Industry). The author thanks members of the Toyota Institute, Sharp, Daido Steel, JAXA, JAEA, the University of Tokyo, Meijo University, Kyushu University and Miyazaki University for their collaboration and cooperation.

References

Ahrenkiel R. K., Keyes B. M., Durbin S. M. and Gray J. L. (1993), 'Recombination lifetime and performance of III–V compound photovoltaic devices', *Proc. 23rd IEEE Photovoltaic Specialists Conf.*, IEEE, New York, NY, pp. 42–51.

Amano C., Sugiura H., Yamaguchi M. and Hane K. (1989), 'Fabrication and numerical analysis of AlGaAs/GaAs tandem solar cells with tunnel interconnections', *IEEE Trans. Electron Devices* **ED–36**, 1026–1035.

Ando K., Amano C., Sugiura H., Yamaguchi M. and Saletes A. (1987), 'Non-radiative *e-h* recombination characteristics of mid-gap electron trap in $Al_xGa_{1-x}As$ ($x = 0.4$) grown by molecular beam epitaxy', *Jpn. J. Appl. Phys.* **26**, L266–L269.

Araki K., Uozumi H., Egami T., Hiramatsu M., Miyazaki Y., Kemmoku Y., Akisawa A., Ekins-Daukes N. J., Lee H.-S. and Yamaguchi M. (2005), 'Development of concentrator modules with dome-shaped Fresnel lenses and triple-junction concentrator cells', *Progr. Photovoltaics* **13**, 513–527.

Bertness K. A., Kurtz S. R., Friedman D. J., Kibbler A. E., Kramer C. and Olson J. M. (1994), '29.5%-Efficiency GaInP/GaAs tandem solar cells', *Appl. Phys. Lett.* **65**, 989–991.

Bett A. W., Dimroth F., Guter W., Hoheisel R., Oliva E., Philips S. P., Schone J., Siefer G., Steiner M., Wekkeli A., Welser E., Meusel M., Kostler W. and Strobl G. (2009), 'Highest efficiency multijunction solar cell for terrestrial and space applications', *Proc. 24th European Photovoltaic Solar Energy Conf.*, WIP, Munich, pp. 1–6.

Bowler D. L. and Wolf M. (1980), 'Interactions of efficiency and material requirements for terrestrial silicon solar cells', *IEEE T Compon. Hybr.*, **CHMT–3**, 464–472.

Brown M. R., Goldhammer L. J., Goodelle G. S., Lortz C. U., Perron J. N., Powe J. S. and Schwartz J. A. (1997), 'Characterization testing of dual junction $GaInP_2$/GaAs/Ge solar cell assemblies', *Proc. 26th IEEE Photovoltaic Specialists Conf.*, IEEE, New York, NY, pp. 805–810.

Carlin J. A., Hudait M. K., Ringel S. S., Wilt D. M., Clark E. B., Leitz C. W., Currie M., Langdo T. and Fitzgerald E. A. (2000), 'High-efficiency GaAs-on-Si solar cells with high Voc using graded GeSi bufers', *Proc. 28th IEEE Photovoltaic Specialists Conf.*, IEEE, New York, NY, pp. 1006–1011.

Chung B.-C., Virshup G. F., Hikido S. and Kaminar N. R. (1989), '27.6% efficiency (1 Sun, air mass 1.5) monolithic $Al_{0.37}Ga_{0.63}$ As/GaAs two-junction cascade solar cell with prismatic cover glass', *Appl. Phys. Lett.* **55**, 1741–1743.

Fan J. C. C., Tsaur B. Y. and Palm B. J. (1982), 'Optical design of high-efficiency tandem cells', *Proc. 16th IEEE Photovoltaic Specialists Conf.*, IEEE, New York, NY, pp. 692–701.

Flores C. (1983), 'A three-terminal double junction GaAs/GaAlAs cascade solar cells', *IEEE Electr. Device L.* **EDL–4**, 96–99.

Fraas L. M., Avery J. E., Martin J., Sundaram V. S., Girard G., Dinh V. T., Davenport T. M., Yerkes J. W. and O'Neill M. J. (1990), 'Over 35% efficient GaAs/GaSb tandem solar cells', *IEEE Trans. Electr. Devices* **37**, 443–449.

Goetzberger G., Luther J. and Willeke G. (2002), 'Solar cells: past, present, future', *Solar Energy Mater. Solar Cells*, **74**, 1–11.

Green M., Emery K., Hishikawa Y., Warta W. and Dunlop E. D. (2013), 'Solar efficiency tables (Version 42)', *Progr. Photovoltaics* **21**, 827–837.

Hutchby J. A., Markunas R. J., Timmons M. L., Chiang P. K. and Bedair S. M. (1985), 'A review of multijunction concentrator solar cells', *Proc.18th IEEE Photovoltaic Specialists Conf.*, IEEE, New York, NY, pp. 20–27.

Imaizumi M., Matsuda S., Kawakita S., Sumita T., Takamoto T., Ohshima T. and Yamaguchi M. (2005), 'Activity and current status of R&D on space solar cells in Japan', *Progr. Photovoltaics* **13**, 529–543.

Jackson E. D. (1955), 'Areas for improving of the semiconductor solar energy converter', *Trans. Conf. on the Use of Solar Energy* **5**, University of Arizona Press, Tucson (1958), pp. 122–126.

Khan A., Yamaguchi M., Bourgoin J. C. and Takamoto T. (2000), 'Room-temperature minority-carrier injection-enhanced recovery of radiation-induced defects in *p*-InGaP and solar cells', *Appl. Phys. Lett.* **76**, 2559–2561.

King R. R., Boca A., Hong W., Liu X.-Q., Bhusari D., Larrabee D., Edmondson K. M., Law D. C., Fetzer C. M., Mesropian S. and Karam N. H. (2009), 'Bandgap-engineered architectures for high-efficiency multijunction concentrator solar cells', *Proc. 24th European Photovoltaic Solar Energy Conf.*, WIP, Munich, pp. 55–61.

Kurtz S. R., Myers D. and Olson J. M. (1997), 'Projected performance of three-and four-junction device using GaAs and GaInP', *Proc. 26th IEEE Photovoltaic Specialists Conf.*, IEEE, New York, NY, pp. 875–878.

Lamorte M. F. and Abbott D. H. (1980), 'Computer modeling of a two-junction monolithic cascade solar cell', *IEEE Trans. Electr. Devices* **ED–25**, 831–840.

Lang D. V., Kimerling L. C. and Leung S. Y. (1976), 'Recombination-enhanced annealing of the E1 and E2 defect levels in 1-MeV-electron-irradiated *n*-GaAs', *J. Appl. Phys.* **47**, 3587–3591.

Letay G. and Bett A. W. (2001), 'Theoretical investigation III–V multijunction solar cells', *Proc. 17th European Photovoltaic Solar Energy Conf.*, WIP, Munich, pp. 178–181.

Loferski J. J. (1976), 'Tandem photovoltaic solar cells and increased energy conversion efficiency', *Conf. Record 12th IEEE Photovoltaic Specialists Conf.*, Baton Rouge. IEEE Press, Piscataway, NJ, pp. 957–961.

Ludowise M. J., LaRue R. A., Borden P. G., Gregory P. E. and Dietz W. T. (1982), 'High-efficiency organometallic vapor phase epitaxy AlGaAs/GaAs monolithic cascade solar cell using metal interconnects', *Appl. Phys. Lett.* **41**, 550–552.

Mitchell K. W. (1981), 'High efficiency concentrator cells', *Conf. Record 15th IEEE Photovoltaic Specialists Conf.*, Kissimmee. IEEE Press, Piscataway, NJ, pp. 142–146.

Nell M. E. and Barnett A. M. (1987), 'The spectral *p-n* junction model for tandem solar-cell design, *IEEE Trans. Electr. Devices* **ED–34**, 257–266.

Olson J. M., Kurtz S. R. and Kibbler A. E. (1990), 'A 27.3% efficient $Ga_{0.5}In_{0.5}P$/GaAs tandem solar cell', *Appl. Phys. Lett.* **56**, 623–625.

Sabnis V., Yuen H. and Wiemer M. (2012), 'High-efficiency multijunction solar cells employing dilute nitrides', *Proc. 9th Int. Conf. on Concentrating Photovoltaics Systems*, Toledo, Spain.

Sasaki K., Agui T., Nakaido K., Takahashi N., Onitsuka R. and Takamoto T. (2013), 'Development of InGaP/GaAs/InGaAs inverted triple junction concentrator solar cells', *Proc. 9th Int. Conf. on Concentrating Photovoltaics Systems*, Miyazaki, Japan.

Sharp (2013), http://www.sharp.co.jp.

Solar Junction (2011), http://www.sj-solar.com.

Sugiura H., Amano C., Yamamoto A. and Yamaguchi M. (1988), 'Double heterostructure GaAs tunnel junction for AlGaAs/GaAs tandem solar cells', *Jpn. J. Appl. Phys.* **27**, 269–272.

Takamoto T., Agui T., Ikeda M. and Kurita E. (2000), 'High-efficiency InGaP/GaAs tandem solar cells on Ge substrates', *Proc. 28th IEEE Photovoltaic Specialists Conf.*, IEEE, New York, NY, pp. 976–981.

Takamoto T., Agui T., Kamimura K., Kaneiwa M., Imaizumi M., Matsuda S. and Yamaguchi M. (2003), 'Multijunction solar cell technologies — high efficiency, radiation resistance and concentrator applications', *Proc. 3rd World Conf. Photovoltaic Energy Conversion*, WCPEC-3, Osaka, pp. 581–586.

Takamoto T., Agui T., Yoshida A., Nakaido K., Juso H., Sasaki K., Nakamora K., Yamaguchi H., Kodama T., Washio H., Imaizumi M. and Takahashi M. (2010), 'World's highest efficiency triple-junction solar cells fabricated by inverted layers transfer process', *Proc. 35th IEEE Photovoltaic Specialists Conf.*, IEEE, New York, NY, pp. 412–417.

Takamoto T., Ikeda E., Kurita H. and Ohmori M. (1997a), 'Over 30% efficient InGaP/GaAs tandem solar cells', *Appl. Phys. Lett.* **70**, 381–383.

Takamoto T., Ikeda E., Kurita H., Ohmori M. and Yamaguchi M. (1997b), 'Two-terminal monolithic InGaP/GaAs tandem solar cells with a high conversion efficiency of over 30%', *Jpn. J. Appl. Phys.* **36**, 6215–6220.

Takamoto T., Kaneiwa M., Imaizumi M. and Yamaguchi M. (2005), 'InGaP/GaAs-based multijunction solar cells', *Progr. Photovoltaics* **13**, 495–511.

Takamoto T., Yamaguchi M., Ikeda E., Agui T., Kurita H. and Al-Jassim M. (1999), 'Mechanism of Zn and Si diffusion from highly doped tunnel junction for InGaP/GaAs tandem solar cells', *J. Appl. Phys.* **85**, 1481–1485.

WBGU (German Advisory Council on Global Change) (2003), *World in Transition — Towards Sustainable Energy Systems*, Earthscan, London, http://wbgu.de.

Wojtczuk S., Chiu P., Zhang X., Derkacs D., Harris C., Pulver D. and Timmons M. (2010), 'InGaP/GaAs/InGaAs 41% concentrator cells under bi-facial epigrowth', *Proc. 35th IEEE Photovoltaic Specialists Conf.*, IEEE, New York, NY, pp. 1259–1264.

Wolf M. (1960), 'Limitations and possibilities for improvement of photovoltaic solar energy converters', *Proc. Inst. Radio Engineers* **48**, 1246–1263.

Yamaguchi M. (2003), 'III–V compound multijunction solar cells: present and future', *Solar Energy Mater. Solar Cells*, **75**, 261–269.

Yamaguchi M. (2004), 'Toward 50% efficiency III–V compound multijunction and concentrator solar cells', *Proc. 19th European Photovoltaic Solar Energy Conf.*, WIP, Munich, xl–xlii.

Yamaguchi M. and Amano C. (1985), 'Efficiency calculations of thin film GaAs solar cells on Si substrates', *J. Appl. Phys.* **58**, 3601–3606.

Yamaguchi M., Amano C., Sugiura H. and Yamamoto A. (1987), 'High-efficiency AlGaAs/GaAs tandem solar cells', *Proc. 19th IEEE Photovoltaic Specialists Conf.*, IEEE, New York, NY, pp. 1484–1485.

Yamaguchi M. and Luque, A. (1999), 'High efficiency and high concentration in photovoltaics', *IEEE Trans. Electron Devices* **46**, 2139–2144.

Yamaguchi M. and Luque A. (2013), 'Outline of Europe–Japan collaborative research on concentrator photovoltaics', *Proc. 39th IEEE Photovoltaic Specialists Conf.*, IEEE, New York, NY.

Yamaguchi M., Ohmachi Y., Oh'hara T., Kadota Y., Imaizumi M. and Matsuda S., (2001), 'GaAs solar cells grown on Si substrates for space use', *Progr. Photovoltaics* **9**, 191–201.

Yamaguchi M., Okuda T., Taylor S. J., Takamoto T., Ikeda E. and Kurita H. (1997), 'Superior radiation-resistant properties of InGaP/GaAs tandem solar cells', *Appl. Phys. Lett.* **70**, 1566–1568.

Yamaguchi M., Warabisako T. and Sugiura H. (1994), 'CBE as a breakthrough technology for PV solar energy applications', *J. Crystal Growth* **136**, 29–36.

Yang M.-J., Yamaguchi M., Takamoto T., Ikeda E., Kurita H. and Ohmori M. (1997), 'Photoluminescence analysis of InGaP top cells for high-efficiency multijunction solar cells', *Solar Energy Mater. Solar Cells* **45**, 331–339.

Yoshida A., Agui T., Nakaido K., Murasawa K., Juso H., Sasaki K. and Takamoto T. (2011), 'Development of InGaP/GaAs/InGaAs inverted triple junction solar cells for concentrator application', *Extended Abstracts of 21st International Photovoltaic Science and Engineering Conf.*, Fukuoka, Japan, 28 November–2 December, 4B–4O–01.

CHAPTER 8

ORGANIC PHOTOVOLTAICS

DAN CREDGINGTON
Cavendish Laboratory, JJ Thomson Avenue
Cambridge CB3 0HE, UK
djnc3@cam.ac.uk

One day carpenters will fashion plastic trees,
And artists will paint their nylon leaves.
Anonymous.

8.1 Introduction

In this chapter, we move our focus away from the numerous elemental and crystalline semiconductors which dominate the existing technological landscape of photovoltaics. For these materials, it is the specific arrangement of elements in a (usually) crystalline lattice which gives rise to delocalised electronic states and resulting band structure. The quality of the band structure therefore depends rather sensitively on how such lattices are formed, and the necessary delocalisation of electrons leads to materials that are highly sensitive to defects — including both impurities and crystallite boundaries. The creation of high-performance materials by this approach is an intrinsically intolerant, and therefore expensive, process. As is detailed elsewhere in this volume, the primary brake on uptake of photovoltaics for the last several decades has been the relatively high cost of silicon crystal, which must be processed from a purified melt at extremely high temperature. While the price of silicon crystal has reduced considerably as the demand for solar energy has grown, routes that circumvent high-tolerance manufacturing may lead to even cheaper photovoltaic devices.

An alternative approach is to utilise materials in which semiconducting behaviour is intrinsic, rather than an emergent property of lattice formation. Molecular semiconductors are such a class of materials, whereby pre-synthesised molecules with desirable electronic structure are used to form the light-absorbing layer of the photovoltaic device. Interactions between molecules lead to broadening of the molecular orbitals, but this is a relatively small perturbation, which does not typically lead to three-dimensional extended electronic states. The advantage of such an approach is that the electronic properties of molecular semiconductors are largely 'hard-wired' into the molecule, and do not rely on a particular crystal structure being formed. This renders them significantly more tolerant to impurities, packing defects or grain boundary effects and so allows far greater scope for low-cost processing. Decoupling electronic and physical structure also casts bandgap engineering as a problem of molecular design, rather than crystal composition, and thus renders it accessible to the vast field of synthetic chemistry. In particular, the tremendous flexibility of organic (carbon-based) chemistry has led to the dominance of organic molecules within molecular semiconductor technology.

In this chapter, we examine the nature of semiconducting behaviour in organic molecules, the challenges and advantages associated with their use in photovoltaic devices, and the progress made in recent years to bring this technology to fruition. This chapter is split into two main themes. Sections 8.2 and 8.3 provide an overview of organic semiconductors, their operation in photovoltaic devices and the current state of the art. These sections are intended to provide a broad introduction to the technology. Sections 8.4 and 8.5 examine in more detail the processes governing organic photovoltaic (OPV) operation, in terms of current understanding and empirical observation. These sections are intended to provide a more in-depth description of current OPV research, and highlight some of the questions that remain within the field.

8.2 Basic concepts in organic semiconductors

8.2.1 *Chemistry and conjugation*

Semiconducting and metallic organic materials are based on chains and sheets of sp^2-hybridised carbon. Neutral carbon contains six electrons, two tightly bound in 1s orbitals and four valence electrons in a $2s^2$, $2p^2$ configuration. This configuration is maintained only in atomic carbon; in a carbon-based molecule the total bond energy may be reduced by the formation of '*sp*' hybrid states between the filled $2s$- and partially filled $2p$-shells. Three primary hybrid states are possible: (i) tetrahedrally aligned sp^3 orbitals as found in diamond and saturated polymers; (ii) planar hexagonal sp^2 orbitals as found in graphite and all organic semiconductors; and (iii) linear *sp* orbitals as found in linear acetylenic carbon (carbyne) — a highly reactive allotrope which is usually synthesised in short lengths capped with other organic groups. Figure 8.1 shows schematics of these configurations.

Since only three of the four valence electrons hybridise in the sp^2 configuration, the final electron occupies a perpendicular $2p_z$ orbital. The sharing of *sp* orbitals between adjacent carbons leads to strong, highly directional 'σ-bonding', which allows the formation of the stable carbon chains, sheets and branched structures

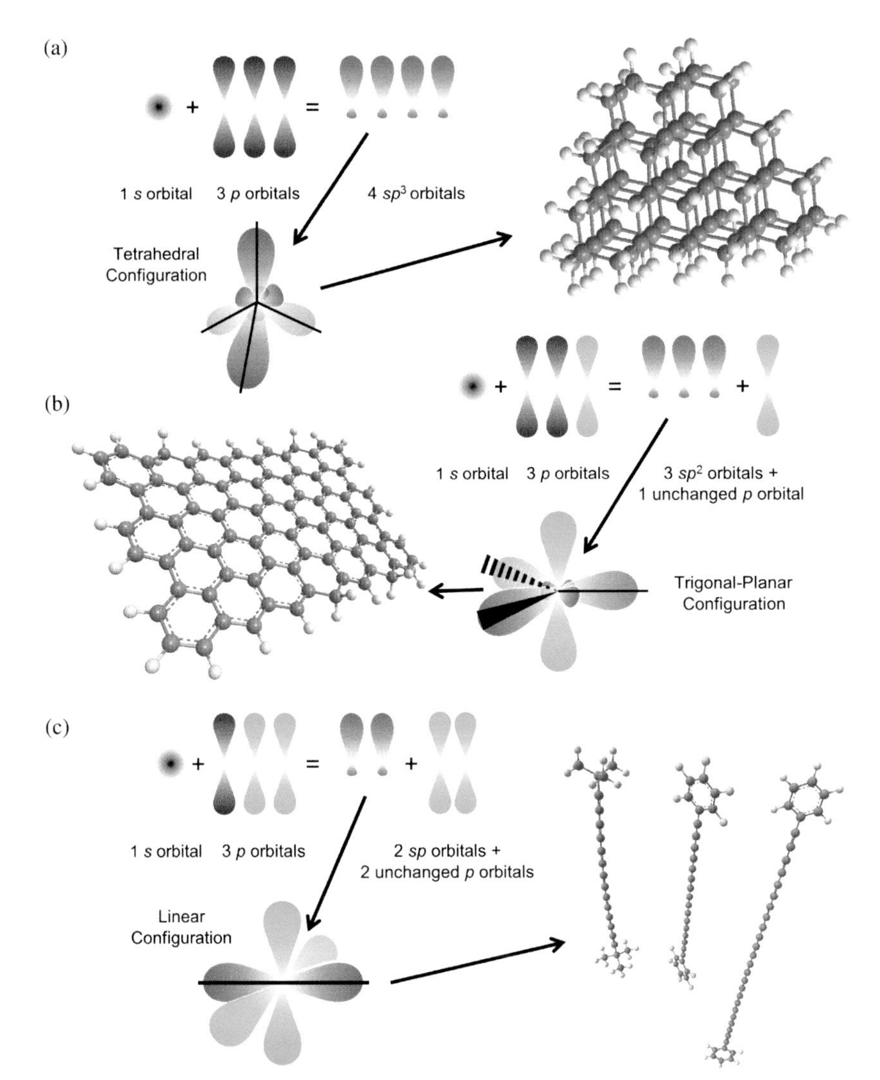

Figure 8.1 Schematics of the electronic orbitals of a) sp^3 hybridisation; b) sp^2 hybridisation; and c) sp hybridisation, with examples showing the a) diamond, b) graphene and c) heptayne, octayne and dodecayne linear acetylenic carbon structures, with various capping groups.

which form the basis of the rich field of synthetic organic chemistry, examples of which are shown in Fig. 8.2.

Hybridisation between the remaining $2p_z$ orbitals gives rise to weaker π (bonding) or π^* (anti-bonding) orbitals, the former arising from a symmetric

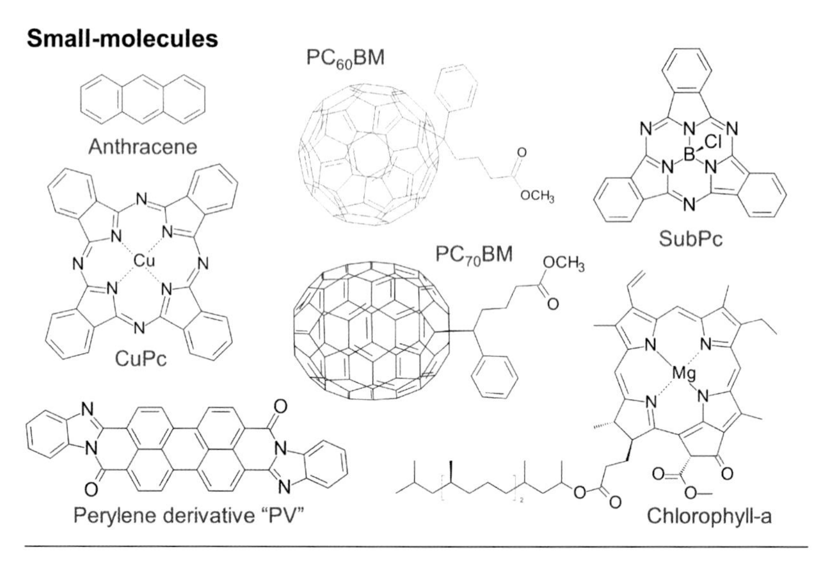

Figure 8.2 Chemical structures for a range of commonly used small-molecule and polymer semiconductors. CuPc = Copper phthalocyanine; SubPc = boron subphthalocyanine chloride.

superposition of $2p_z$ orbitals, the latter from an anti-symmetric superposition. Such hybridisation typically encourages a preferred rotational orientation between the main π-bonded units, which induces planarisation of the molecule and consequently allows delocalisation of π/π^*-states over many carbon atoms — a process known as *conjugation*. In the symmetric (π) configuration, significant electron density is present between neighbouring carbon atoms, stabilising this configuration compared with the anti-symmetric (π^*) one. In a simple two-carbon system, the two unpaired $2p_z$ electrons will therefore tend to occupy the two spin-allowed π states, leaving the unoccupied π^* states above a forbidden energy gap E_g. Increasing the conjugation length has the effect of broadening the ensemble of hybrid states into a quasi-continuous 'band' and narrowing this gap: for example, controlling the degree of polymerisation of ethylene can vary E_g from 6.7 eV for the monomer to around 1.5 eV for stretch-oriented polyacetylene (with effective conjugation over more than 20 carbon atoms) as illustrated in Fig. 8.3 (Townsend *et al.*, 1985). In the solid state, a similar effect is observed when the π orbitals of neighbouring molecules overlap to allow delocalisation over several molecular

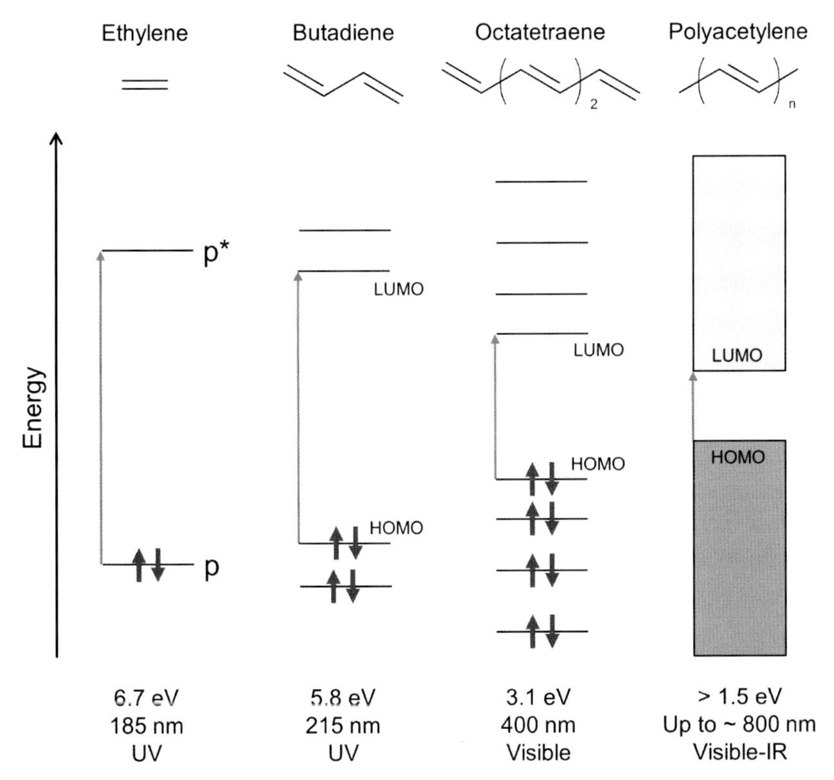

Figure 8.3 Schematic of energy-level splitting and absorption in alkenes with increasing conjugation length, highlighting the lowest-energy optical transitions from HOMO to LUMO. The bandgap of polyacetylene depends on both the number of monomers n and the effective conjugation length in the polymer. Arrows represent spin-paired electrons.

layers. Such '$\pi - \pi$ stacking' is important for transport (see Section 8.4.3) and also leads to a lower optical bandgap. A combination of intra- and inter-molecular delocalisation can therefore lead to molecules with absorption in the visible spectral range.

Since thermal excitation from π to π^* is negligible for visible-light transitions, such materials behave as intrinsic semiconductors with the 'bandgap' defined by E_g, the difference between the highest occupied molecular orbital (HOMO) and lowest unoccupied molecular orbital (LUMO). Doping of the molecule, or injection of external charge, is therefore necessary to initiate and exploit conduction. In addition, each conjugated region will not necessarily extend over the whole molecule (particularly in the case of larger polymers), limiting the practical extent of delocalisation. However, as we shall see in Section 8.4.4, numerous alternative routes exist for engineering the HOMO–LUMO offset, and a key advantage of these materials is that E_g is a molecular property, so may be defined before the material is incorporated into a photovoltaic device. In addition, most organic molecules are direct-bandgap materials with low symmetry, and thus exhibit extremely high absorption coefficients — a film of organic material 100–200 nm thick is usually sufficient to achieve complete absorption at the primary $\pi - \pi^*$ transition. However, the molecular (rather than band-like) nature of absorption usually means that the optical density is peaked rather than step-like at this transition — ideal for dyes, but not for photovoltaic absorbers.

8.2.2 *Excited states*

Optical excitation in conjugated organic molecules occurs by absorption of light with photon energy greater than E_g, promoting an electron from the ground state to the vibronic manifold of the first electronic excited state, and leaving behind a hole. In general, this electron-hole pair will remain coulombically bound, forming a neutral molecular excited state referred to as an *exciton*. The behaviour of excited states in organic semiconductors is distinct from that of inorganic materials, for the reasons outlined below.

- Organic semiconductors are low dielectric constant (ϵ) materials

For example, $\epsilon_{\text{pentacene}} \sim 4$, $\epsilon_{\text{PPV}} \sim 2$ (Martens *et al.*, 1999) whereas $\epsilon_{\text{Si}} \sim 12$, $\epsilon_{\text{GaAs}} \sim 13$. In materials with high ϵ, dielectric screening reduces the effective Coulomb attraction of the excited electron-hole pair, leading to very low binding energies and characteristic radii greater than the material's lattice spacing — i.e. electrons and holes are delocalised over many unit cells. This subset of excitons, described by and named after Gregory Wannier (Wannier, 1937), are characterised by binding energies of a few kT at room temperature, and thus may undergo spontaneous dissociation into free charges via interactions with lattice phonons.

In materials with low ϵ, the coulomb attraction between the electron and hole is not strongly screened, leading to tightly bound Frenkel-type excitons (named after Jakov Frenkel [Frenkel, 1931]). These have binding energies of order 0.1–1 eV and characteristic radii of the same order as the unit cell (for crystals) or individual monomers/molecular units (for molecular materials). Dissociation of these excitons into free charge carriers is not typically spontaneous at room temperature, and must be achieved by other means, which we discuss in Section 8.3.1. If they do not dissociate, excitons in organic materials typically exhibit luminescence lifetimes of a few nanoseconds. Since one of the most obvious effects of encouraging exciton dissociation is to quench this luminescence, the term 'exciton quenching' is often used to describe processes leading to exciton dissociation, whether or not the material in question is intrinsically luminescent.

- Organic materials are electronically 'soft'

Charge carriers moving through an organic material are usually localised to a few carbon atoms, and so act as large perturbations to the local electronic structure. Polarisation of the local medium (enhanced by low ϵ) and the breaking of local conjugation by the occupancy of the π^* anti-bonding state leads to geometric relaxation of the molecule. The fundamental charge carrier in organic materials is thus a quasi-particle, comprising the charge self-localised by its associated molecular distortion: this entity is called a *polaron*. The 'binding' energy of the polaron (as compared with the energy of a free charge without distortion) is estimated to be of order 100 meV, and is determined by the competition between the energy gained by relaxation of the surrounding medium around the charged excitation and the energy cost of decreasing delocalisation. A consequence of this self-localisation is that the migration of polarons between polymer chains or conjugated regions involves significant reorganisation energy, and is limited by the electronic coupling between conjugated regions. This relaxation also produces a significant Stokes shift between absorption and emission spectra, which is particularly beneficial for reducing self-absorption in organic light-emitting diodes (LEDs), but represents a potential loss of free energy in a photovoltaic. The extent of relaxation is governed by the physical properties of the particular molecule; thus physically 'stiff' materials tend to exhibit lower polaron binding energies and faster charge transport.

- Transport of carriers is activation-limited

The polaronic nature of charge carriers leads to self-localisation, and to move from one conjugated molecule (or conjugated segment) to another therefore requires additional thermal energy. This energy is associated with the geometric relaxation of the 'acceptor' molecule combined with the geometric rebound of the 'donor' molecule. In addition, the probability for a carrier to transfer is also dependent on the wave-function overlap between donor and acceptor states. Since organic molecules (and particularly polymers) typically exhibit low symmetry, transport can be highly anisotropic. For a stiff conjugated polymer, intra-chain transport will be faster than transport in the $\pi - \pi$ stacking direction, while lateral transport through saturated side groups can be very poor. For example, intra-chain hole mobilities as high as 600 cm^2/Vs have been reported for ladder-type poly(p-phenylene) (Prins *et al.*, 2006), whereas measured mobilities are of order 1–10 cm^2/Vs (Li *et al.*, 2012) in solid films, and depend strongly on polymer orientation (Lee *et al.*, 2011). Such transport also depends on temperature and electronic disorder, which we will discuss in more detail in Section 8.4.2.

- Transport of excitons is disorder-limited

The migration of excitons through the semiconductor occurs via transfer of the bound polaron pair from one conjugated molecule/chain segment to another through either Förster or Dexter energy transfer. Förster transfer (Förster, 1948) involves short-range dipole–dipole interactions equivalent to the exchange of a virtual photon; Dexter transfer (Dexter, 1953) involves direct exchange of electrons. Requiring only coupling between dipoles, Förster transfer of singlet excitons is the dominant process in photovoltaics and is typically restricted to a range of 5–20 nm over the lifetime of the exciton (Halls *et al.*, 1996; Stubinger and Brutting, 2001), depending on the material in question. This range is limited by the degree of overlap between the material's absorption and emission spectra (reduced by polaronic relaxation), and by disorder in the energetic landscape accessible to the exciton. Since Dexter transfer requires wave-function overlap between the donating and accepting species, it is usually dominant only in situations where optical transitions are forbidden, for example in the migration of spin-triplet excitons.

8.2.3 *Thin-film processing*

Small organic molecules may be sublimed under vacuum at relatively low temperature, enabling the creation of well-defined films and layer structures without the need for high-temperature, high-purity processing. However, the scalability and cost of such an approach must be balanced against its benefits, as we outline in Section 8.5.6. An alternative approach is possible since organic semiconductors are usually synthesised using wet chemistry from soluble precursors. Solubility can therefore be specifically designed into the molecule without significantly affecting its electronic structure (as illustrated in Fig. 8.4). Since conjugated molecules

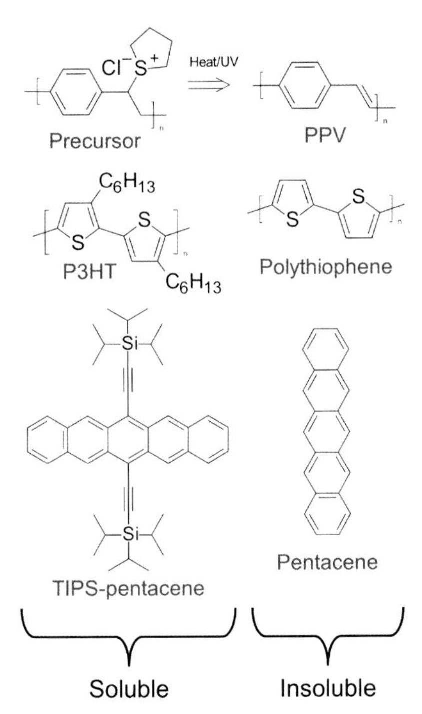

Figure 8.4 Typical routes to solubilisation. Insoluble poly(p-phenylenevinylene) (PPV) is formed *in situ* from a soluble precursor via a leaving-group reaction. Insoluble polythiophene and pentacene are rendered soluble by the inclusion of flexible side chains.

are usually quite stiff due to the planarising effect of their π-bonding, solubilisation is usually achieved by attaching flexible saturated side groups along the main backbone, whose chemistry determines the range of compatible solvents. For simplicity, we will here refer to both the conjugated segments of a polymer chain, and the conjugated carbon scaffold of organic small molecules as the 'backbone' of the structure.

Many organic molecules may therefore be formed into inks and processed from solution. This approach is of particular importance for depositing large semiconducting molecules such as conjugated polymers, for which vacuum sublimation is impossible. As solid films, these polymeric semiconductors are flexible and highly resistant to fracture, lending the collection of technologies which exploit them the umbrella term 'plastic electronics'.

In solution, the freedom of organic molecules to explore a range of configurations is enhanced by the presence of the solvent. Deposition from solution is therefore thermodynamically analogous to a rapid quench, as the evaporation of solvent molecules quickly restricts this freedom. Depending on the details of the deposition (and particularly on the evaporation rate of the solvent), the resulting material is typically left in a metastable configuration, with a mix of glassy and more ordered regions. This effect is even more prevalent in systems comprising mixtures of multiple materials. Since electronic structure is influenced by both local conjugation length and intermolecular interactions, solution deposition leads unavoidably to significant spatial and energetic disorder. This intrinsic disorder has implications for many photovoltaic properties, as we discuss in Section 8.4.2.

Deposition of organic semiconductors from solution can take advantage of the wide range of thin-film coating technologies already employed by existing industries. These include inkjet and screen-printing for individual substrates and extend to continuous roll-to-roll manufacture, for example using slot-die, gravure printing or doctor-blading. The existence of mature, low-cost, large-area manufacturing routes means that scaling-up of organic semiconductor technology is widely expected to be a relatively cheap process, if the materials set can also reach maturity. We discuss the progress made to date in large-area OPV manufacture in Section 8.5.6.

8.2.4 *Device structure*

The short optical absorption lengths ($\sim$100–200 nm) associated with organic semiconductors, combined with the ease with which they may be deposited over large areas, renders them highly suited for use in thin-film devices. When employed as photovoltaics, the conventional arrangement is a 'sandwich' structure with the photoactive layer deposited on top of one electrode, and capped with a second. The low carrier mobilities typical of organic molecules generally preclude the use of laterally separated electrodes in efficient devices. To allow light to enter the absorbing layer, at least one electrode must be optically transparent. Indium tin oxide (ITO) is commonly used in this case, and is usually deposited as the 'bottom' (though sun-facing in the completed device) electrode of the sandwich onto a glass or plastic substrate. It is also common to choose a highly reflective metal such as Al or Ag as the 'top' electrode to allow transmitted photons a second pass through the device. It has become conventional to refer to as 'normal' or 'non-inverted' a device where the transparent electrode is primarily used to collect hole current and the reflective electrode used to collect electron current. Where this assignment is reversed, the device is referred to as 'inverted'. While the inverted structure is a relatively recent development, it is generally thought that equivalent efficiencies are achievable in both arrangements, but that inverted structures may allow for more stable devices (see Section 8.5.1). Figure 8.5 shows conventional and inverted device structures.

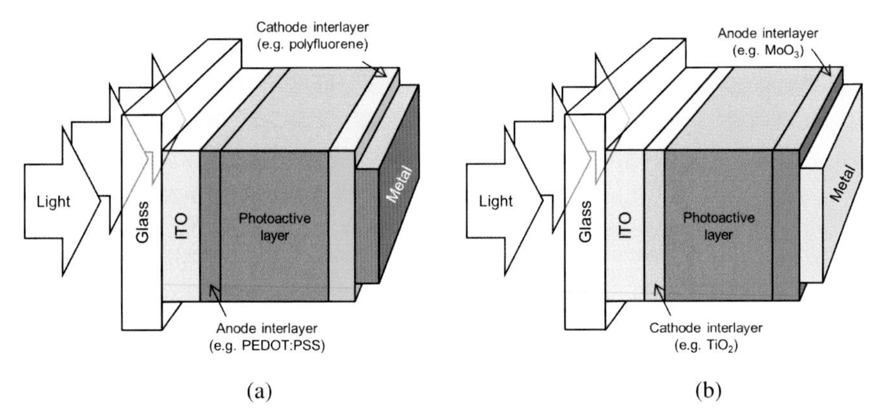

Figure 8.5 Conventional (a) and inverted (b) device structures, indicating stack order and interlayer positions.

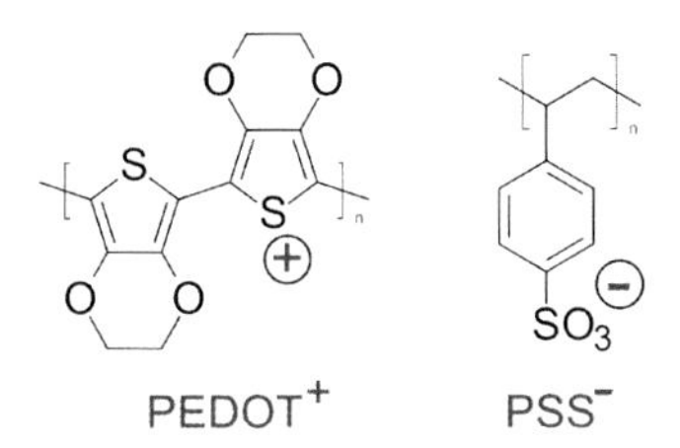

Figure 8.6 Poly(3,4-ethylenedioxythiophene) and poly(styrenesulphonate) — synthesised as a water-based colloidal suspension to form PEDOT:PSS.

8.2.5 *Interface control*

Whichever the orientation, the device electrodes have to be made selective for the appropriate carrier, or significant current may be lost as carriers diffuse to the 'wrong' electrode. For silicon solar cells, selectivity is imposed by heavy *n*- or *p*-doping close to the contact. Since direct doping of organic semiconductors remains difficult, particularly from solution, introducing such selectivity is usually encouraged by introducing thin interfacial layers ('interlayers') between the electrode and the photoactive layer. Interlayers may comprise both organic and inorganic materials, and may fulfil additional functions such as preventing exciton quenching at the electrodes and modifying the electrode work function (see Section 8.3.2). The most common interlayer material is the water-soluble composite poly(3,4-ethylenedioxythiophene): poly(styrenesulphonate) (PEDOT:PSS), a mixture of hole-conducting PEDOT doped by PSS, illustrated in Fig. 8.6.

8.3 Overview of organic photovoltaic devices

Organic photovoltaics aim to take advantage of the properties of organic semiconductors described above, while mitigating their pitfalls. In particular, the aim is to use the flexibility of organic chemistry to design direct-bandgap semiconductors with absorption tuned to closely match the solar spectrum. By choosing soluble organic molecules, large-area, low-temperature production techniques become available, and by utilising materials with high absorption coefficients, materials usage and module weight may be significantly reduced. Combined, these factors have the potential to considerably reduce the cost of photovoltaic power generation.

However, a number of challenges must be overcome to achieve these goals. Photogeneration of dissociated charge carriers from excitons is not spontaneous, transport of dissociated carriers is characterised by low mobilities as a result of polaronic self-localisation and electronic disorder, individual materials typically exhibit narrow absorption features, and the molecular origin of their semiconducting properties means that performance is sensitive to chemical degradation.

In the following sections, we briefly illustrate how these issues have been addressed and give an overview of the state of the art, starting with the process of exciton dissociation, which represented the most significant obstacle to the field.

8.3.1 *Exciton dissociation*

Homogeneous devices

The first classes of organic molecules to be considered seriously as photovoltaic materials were the porphyrins and phthalocyanines (shown in Fig. 8.2). Chlorophyll, the familiar green pigment responsible for the collection of solar energy in plants, is a dihydroporphyrin (a chlorin), so the prospect of emulating the highly evolved natural process of photosynthesis in an artificial solar cell inspired considerable investigation. The use of thin films of chlorophyll (Meilanov *et al.*, 1970a; Meilanov *et al.*, 1970b; Tang and Albrecht, 1975) and related 'natural' molecules (Delacote *et al.*, 1964; Kearns and Calvin, 1958) in photovoltaic cells was initially disappointing, with external quantum efficiencies (EQEs) less than 0.1%, and power conversion efficiencies (PCEs) throughout the 1960s and 1970s generally less than 0.01%. Parallel investigation on 'artificial' semiconductors such as anthracene (Kallmann and Pope, 1959) and pentacene (Silinsh *et al.*, 1974) yielded similar results. Several useful concepts were introduced, the most important being the use of electrodes with dissimilar work functions to break the symmetry of the device. This asymmetry generates a 'built-in' electric field which drives oppositely charged carriers in opposite directions, and allows photovoltages in the 200–500 mV range to be developed. A sufficiently high internal field could also be used directly to split excitons, which was found to be consistent with the electric-field-induced ionisation mechanism proposed by Onsager, and later refined by Braun (Braun, 1984; Onsager, 1934). However, the primary limit on efficiency remained the competition between the short exciton lifetime (a few ns) and the low rate of spontaneous exciton dissociation under normal operating conditions.

Planar heterojunctions

A number of pieces of evidence pointed to a possible solution. In single-material devices, dissociation was found to be more likely at surfaces or impurities, and in the presence of water or oxygen. It was also discovered that electron transfer from organic pigments to other semiconductors — such as from chlorophyll to ZnO — could be efficient (Tributsch and Calvin, 1971). The implication of these observations was that exciton dissociation could be encouraged by allowing charges to transfer to a nearby heterogeneous molecule. This was confirmed by Tang and co-workers (Tang, 1986), who showed that the photocurrent of an organic photovoltaic could be increased by orders of magnitude if two different semiconducting molecules were combined to form a bilayer, or 'planar' heterojunction (shown in Fig. 8.7a). Rather than relying on thermal energy or an electric field to ionise the excitons, this structure drives dissociation by using the gain in free energy associated with the transfer of an electron from the LUMO of the 'donor' semiconductor to the LUMO of the 'acceptor' semiconductor (or the transfer of a hole from acceptor HOMO to donor HOMO). In the language of crystalline semiconductors, a type II heterojunction is required. Figure 8.8 illustrates this process.

To a first approximation, the band-edge offset must be greater than or equal to the exciton binding energy, a point we shall discuss in more detail in Section 8.4.4.

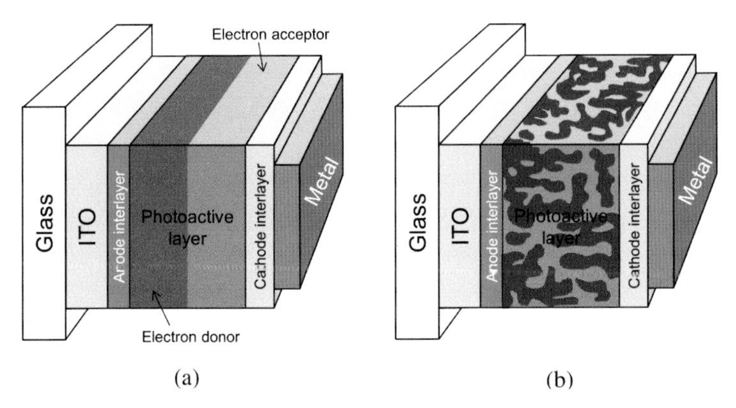

Figure 8.7 Heterojunction structures, showing schematics of: a) planar heterojunction; and b) bulk heterojunction photoactive layers in a conventional structure.

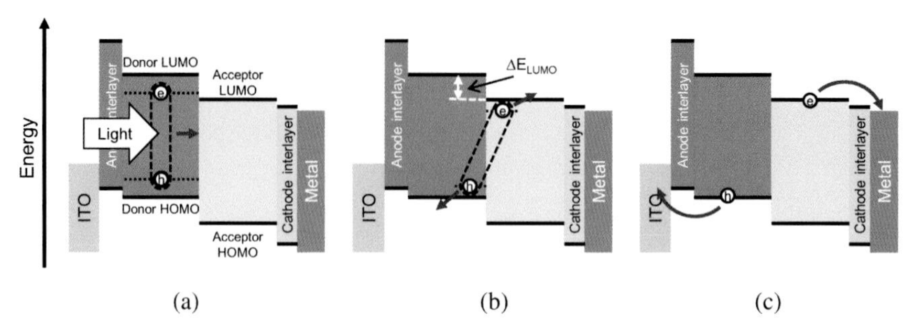

Figure 8.8 Photogeneration in a planar heterojunction device, illustrating: a) exciton formation and diffusion in the donor; b) formation and dissociation of a 'charge-transfer' state across the heterojunction; c) collection of dissociated charges at the electrodes. Solid lines represent single-carrier molecular energy levels, dotted lines represent (induced) excitonic energy levels, and dashed lines indicate a coulombically bound state. Interlayer energy levels have been chosen so as to be selective for electrons or holes and the LUMO:LUMO offset is highlighted.

By careful choice of materials — CuPc and the perylene derivative 'PV' in this case (shown in Fig. 8.2) — devices with PCEs of around 1% were realised. In addition, the use of two semiconductors opened the possibility of combining materials with complementary absorption spectra in a single device.

Bulk heterojunctions

While a significant advance, the EQE of planar heterojunction devices peaked at around 15%, since only those excitons able to diffuse to the heterojunction could be dissociated. With the diffusion length of singlet excitons limited to approximately 10 nm, and absorption lengths of order 100 nm, we must often accept either incomplete light harvesting or incomplete exciton harvesting when using such a bilayer configuration.

This problem was solved in the mid-1990s with the introduction of the 'bulk' heterojunction (BHJ) structure (Halls *et al.*, 1995; Yu *et al.*, 1995), the philosophy of which is to 'bring the heterojunction to the exciton'. By blending the donor and acceptor materials together in a single layer, it is possible to produce a distributed heterojunction extending throughout the device, such that every newly generated exciton is within one diffusion length of an interface (Fig. 8.7b).

The discovery of this approach coincided with the development of highly effective soluble fullerene derivatives (small molecules based on buckyballs) as acceptors (Sariciftci *et al.*, 1992). These favoured the use of organic polymers as donors, since the entangled polymer chains can prevent regions of the film becoming electrically isolated, and can also encourage de-mixing (discussed in

Section 8.4.1). By combining the polymer poly[2-methoxy-5-(2-ethyl-hexyloxy)-1,4-phenylene vinylene] (MEH-PPV) and C_{60}, PCEs of nearly 3% were realised, with peak EQE of 30% (Yu *et al.*, 1995). The recent dominance of the organic photovoltaics field by polymer:fullerene devices largely stems from this result. It was subsequently discovered that blends of the fullerene derivative [6,6]-phenyl-C_{61}-butyric acid methyl ester ($PC_{60}BM$) with the soluble semicrystalline thiophene donor poly(3-hexylthiophene) (P3HT) could achieve PCEs approaching 5% after a low temperature (140°C) anneal (Ma *et al.*, 2005; Padinger *et al.*, 2003). Such high efficiencies, particularly compared with those achieved in previous decades, suggested that commercial exploitation of OPV was a genuine possibility, and confirmed fullerenes as the acceptor of choice. Since fullerenes are usually weakly absorbing owing to their spherical symmetry, it has also become common to focus on the donor as the primary absorber, and the LUMO–LUMO offset as the main driving force for charge separation.

The BHJ structure has disadvantages, however. While continuous single-component pathways from the interface to the electrodes are always present in a bilayer heterojunction, in a BHJ this is not guaranteed, meaning isolated 'islands' of donor or acceptor may form. In addition, both electrons and holes remain in close proximity to the heterojunction as they travel to the electrodes, allowing dissociated carriers to recombine across an interface. The ideal bulk heterojunction is therefore likely to be an interpenetrating network of de-mixed donor and acceptor phases on a number of length scales, but since microstructure forms spontaneously during deposition this is very difficult to control. In addition, since a solution-cast microstructure is usually metastable, its long-term stability is always of concern. Direct nanoscale patterning of the semiconducting materials could solve this issue, but to date no low-cost scalable technology has achieved comparable results to simple thermodynamically mediated self-assembly. Because of its importance, influencing BHJ microstructure by controlling the thermodynamics of film formation still remains at the forefront of device engineering after nearly 20 years (Campoy-Quiles *et al.*, 2008; Chirvase *et al.*, 2004; Lee *et al.*, 2008; Li *et al.*, 2005; Moulé and Meerholz, 2008; Nguyen *et al.*, 2007; Peet *et al.*, 2007; Shaheen *et al.*, 2001; Xin *et al.*, 2010). We discuss this topic further in Section 8.4.1.

Despite the inherent complexity, the advantages of the BHJ continue to outweigh the disadvantages. By utilising absorption from low-bandgap polymers and asymmetric (and so more strongly absorbing) fullerenes, device currents in polymer:fullerene cells can exceed $15\,\mathrm{mA\,cm^{-2}}$ under solar illumination, e.g. utilising PDPP-TTT and PTB7 (Bronstein *et al.*, 2011; He *et al.*, 2012), and internal quantum efficiencies can reach nearly 100%, e.g. utilising PCDTBT (Park *et al.*, 2009), the structures of which are shown in Fig. 8.2. This means that the technology has advanced from quantum yields of 10^{-3} to a point where nearly every exciton can reach the heterointerface and dissociate, and under short-circuit conditions the resulting carriers escape the device without loss.

8.3.2 *Photovoltage*

For any photovoltaic material, the absorption of photons with energy greater than the semiconductor bandgap leads to thermalisation losses as electrons/holes relax to the conduction/valence band edges (or LUMO/HOMO in this case). The maximum free energy difference per carrier that remains after transport to the electrodes defines the open-circuit voltage (V_{oc}) of the cell. This is generally limited by either the electrodes or the materials, as we now discuss.

Electrode-limited regime

After the development of the bulk heterojunction solved the problem of exciton dissociation, several authors observed that a correlation existed between V_{oc} and the work function difference of the cell's electrodes, $\Delta\Phi$ (Brabec *et al.*, 2002; Mihailetchi *et al.*, 2003; Mihailetchi *et al.*, 2004). The implication was that the energy difference between the donor HOMO and acceptor LUMO (often referred to as the heterojunction bandgap, E_{HJ}) is reduced to $\Delta\Phi$ when the carrier transfers to the electrodes. This picture became known as the metal–insulator–metal (MIM) model, and drove the development of materials to achieve increasingly large $\Delta\Phi$. It is now common to use interlayers comprising soluble conducting polymers (Aernouts *et al.*, 2004), *p*-type oxides (Irwin *et al.*, 2008; Zilberberg *et al.*, 2012) and surface treatments to increase the work function of the hole-collecting electrode, and to use salts (Brabec *et al.*, 2002), polymers (Zhou *et al.*, 2012), *n*-type oxides (Kim *et al.*, 2006) and low work-function metals as interlayers to decrease the work function of the electron-collecting electrode. These layers can provide an additional benefit if they act to increase the 'selectivity' of the appropriate contact — i.e. to block the transport of electrons diffusing to the hole-collecting contact, and vice versa.

Materials-limited regime

The drive to increase $\Delta\Phi$, and the development of lower-bandgap absorbers, has revealed that where $\Delta\Phi$ is greater than E_{HJ}, it is the latter that limits the available free energy (see Fig. 8.9a). Outside the MIM regime, applicable where $\Delta\Phi < E_{HJ}$, numerous studies have shown correlations between the achievable V_{oc} and E_{HJ}, the most well-known being the study by Scharber and co-workers (Scharber *et al.*,

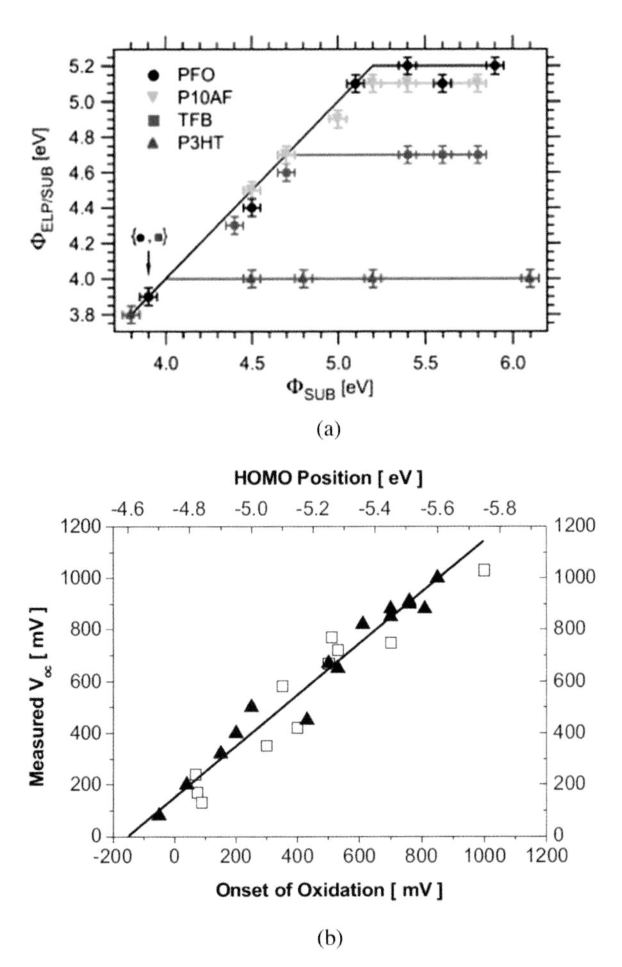

Figure 8.9 a) Dependence of the work function of polymer-coated substrate ($\Phi_{ELP/SUB}$) on the work function of bare substrate (Φ_{SUB}), for four polymers: P3HT, TFB, P10AF, and PFO. The transition from electrode-limited (1:1 linear correlation) to materials-limited (no dependence on Φ_{SUB}) performance is clear. Reproduced from Tengstedt *et al.* (2006). b) Experimentally observed correlation between V_{oc} of different bulk-heterojunction solar cells plotted versus the oxidation potential/HOMO position for the donor polymer used in each device. The solid line is a linear fit with slope of one. Reproduced from Scharber *et al.* (2006).

2006) reproduced in Fig. 8.9b. The correlation is not exact — in all cases the measured difference between donor HOMO and acceptor LUMO overestimates the achievable V_{oc} — but it provides an excellent design rule. In Section 8.4.5 we discuss in more detail the origin of this discrepancy, and how the switch between electrode-limited and materials-limited performance may be interpreted. For now, we note that maximising E_{HJ} has allowed V_{oc} in excess of 1 V to be achieved in single-layer devices (Najari *et al.*, 2012), while cells exhibiting efficient photogeneration can achieve 0.9 V (Park *et al.*, 2009).

8.3.3 *Maximum efficiency*

Since absorber bandgap, LUMO–LUMO (or HOMO–HOMO) offset and heterojunction bandgap are interdependent, it is not possible to simultaneously optimise energetic structure for high V_{oc}, broad absorption and efficient exciton dissociation. Numerous studies exist which aim derive an equivalent of the Shockley–Queisser limit for the maximum PCE of a practical organic photovoltaic, balancing carrier generation with free energy loss and helping to guide the synthesis of new molecules with more favourable energetics. Estimates for maximum achievable PCE range from 15–20% (Dennler *et al.*, 2009; Giebink *et al.*, 2011), though tend to be revised upward year on year as limiting processes are circumvented or better understood. Such calculations are typically based on the physical efficiency limits described in Chapter 2 of this volume, combined with additional constraints such as the free energy lost during exciton dissociation and the finite bandwidth of molecular absorbers (Deibel and Dyakonov, 2010).

8.3.4 *State of the art*

The rapid progress of materials design and solar cell performance over the last decade means that any examples of 'top' efficiencies described herein will rapidly become out of date. However, it is instructive to compare the relative performance of different cell types and provide a snapshot of the field. Until recently, most organic photovoltaics were produced over small areas in university research laboratories, with relative flexibility regarding testing conditions. Robust certification of large-area ($>1\ cm^2$) devices under standard temperature, humidity and spectrum (Reese *et al.*, 2011) is therefore still rare for all but the most eye-catching 'headline' cells. Because of this, throughout this chapter we make no distinction between self-reported and certified performance, nor do we exclude reports on small-area ($<1\ cm^2$) devices. The reader should therefore not give undue weight to incremental differences between performance metrics, which may arise simply from slightly different testing conditions. Figure 8.10 shows the progress in certified solar cell efficiency since 2000, collated by the US National Renewable Energy Laboratory (NREL). At the time of writing (July 2013), the world record was held by Heliatek's tandem evaporated small molecule OPV (Heliatek, 2013).

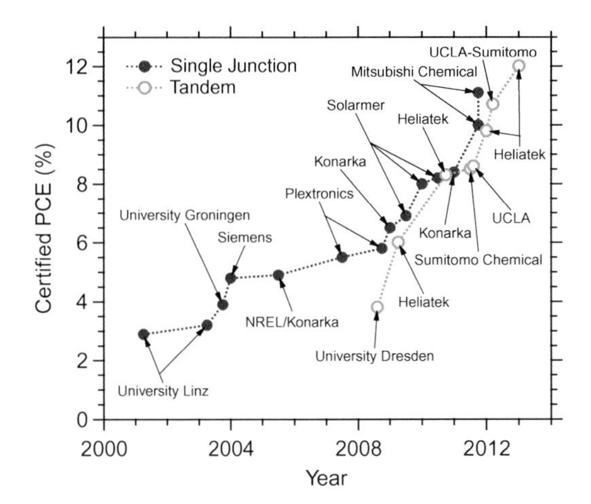

Figure 8.10 Progress of certified organic photovoltaic efficiencies over time since 2000, including both single-junction and tandem devices. Labels indicate institution/company of origin. Data from Green *et al.* (2013).

Bilayer heterojunctions

Bilayers are mainly of interest for studying interface physics in a well-defined system, meaning that thermally evaporated small molecules are particularly well suited to this architecture. Solution-processed bilayers require either strict solvent orthogonality (i.e. two materials with mutually exclusive solubility) or a means of rendering one layer insoluble before deposition of the next — for instance by cross-linking (Png *et al.*, 2010) or removal of solubilising groups (Bradley, 1987). Failure to do so risks rapid mixing of the bilayer components. Even where this is initially prevented, inter-diffusion of materials over time is often unavoidable (Treat *et al.*, 2011).

The best small-molecule absorbers remain the phthalocyanine and sub-phthalocyanine derivatives (see Fig. 8.2), which achieve good light harvesting in exceptionally thin (~10–15 nm) films. When combined with the acceptor C_{60}, PCEs of order 3–4% are achievable (Lin *et al.*, 2012; Peumans and Forrest, 2001).

Small-molecule bulk heterojunctions

Small organic molecules may be processed from solution, or co-evaporated under vacuum, to produce a mixed film. However, the as-deposited film is typically highly amorphous owing to the ease of intermixing, and requires further treatment to encourage the coarsening of donor and acceptor domains through crystallisation. The highest efficiency for small-molecule cells with published structures has been achieved using a solution-processed oligothiophene donor with a fullerene acceptor, reaching 6.7% PCE (Sun *et al.*, 2012). However, it is important to point out the exceptional performance of small-molecule devices reported by developers in the commercial sector, detailed below.

All-polymer bulk heterojunctions

Blends of conjugated polymers form the archetypical bulk heterojunction: two de-mixed interpenetrating networks of relatively pure material, with domains easily distinguishable using high-resolution microscopy. We consider the mechanism for this de-mixing in more detail in Section 8.4.1. The development of polymer:polymer blends has historically been hampered by the difficulty of producing acceptor polymers with high electron mobility, leading to excitons remaining trapped at the heterointerface. Recent development of new high-mobility polymer acceptors containing naphthalene diimide electron-accepting units have led to significant improvements in the current generation of devices, with efficiencies increasing from 1–2% up to 6.4% (Polyera, 2013).

Polymer:small-molecule bulk heterojunctions

Since the discovery of the efficient P3HT:PCBM system, the development of new conjugated polymer donors in combination with soluble fullerene acceptors has been the main focus of the organic photovoltaics field, while the optoelectronic and thermodynamic complexity of such systems has been responsible for the majority of organic semiconductor research output. The most efficient systems to date are also the most efficient lab-scale organic devices yet reported, with PCE now exceeding 9% for single-junction devices comprising alternating 'push-pull' copolymers with $PC_{70}BM$ (He *et al.*, 2012), and approaching 9.5% for multijunction devices (Li *et al.*, 2013).

Commercial developers

For the last two decades, research into new materials and device optimisation has been undertaken mainly within university research laboratories. It is worth noting that within the last few years, this trend has changed: the most efficient organic photovoltaics now reported are produced almost exclusively by commercial developers. As noted in connection with Fig. 8.10, the record for any OPV device currently lies with Heliatek (2013) and their 12%-efficient evaporated small-molecule tandem solar cell.

Mitsubishi Chemical have also achieved certified PCEs of 11% with a solution-processed small-molecule heterojunction, which is reported to be a single-junction device (Green *et al.*, 2013). However, in all cases the materials and device structure are not known. It is therefore not clear in which direction the future of the OPV field lies, but it is certain that the materials on which it is based are still far from maturity.

8.4 Device physics of organic photovoltaics

In the following sections, we will revisit the key operating principles of organic photovoltaics and give an overview of the current state of understanding, beginning with the process of thin-film formation, then focussing on the links between emergent device properties and molecular/morphological design.

8.4.1 *Morphology of the photoactive layer*

Thermodynamics and self-assembly

Many of the organic photovoltaic devices described throughout this chapter rely on a two-phase blend of semiconducting polymers or small molecules to form their photoactive layer. Homogeneous mixtures are also common, and may likewise display useful properties absent in the component parts. PEDOT:PSS is one such example, the two component polymers exchanging charge to create a soluble, conductive composite.

The mixing process for dissimilar materials can be understood within a framework set out in two papers published simultaneously by Flory (1941) and Huggins (1941). The Flory–Huggins approach gives a basic understanding of the possible behaviours and phases in a two-component blend by modifying the expression for the Gibbs free energy of a mixture to include molecules of different molecular weight.

The Gibbs free energy of a particular system is given by the difference between its enthalpy, H, and TS, a measure of the system's disorder, where T is temperature and S the entropy of the system:

$$G = H - TS \tag{8.1}$$

This quantity estimates the energy currently available to a system to do work, or to allow transitions to other states. The change in free energy, ΔG, associated with moving from one state to another is therefore an indication of the nature of the transition: only transitions for which $\Delta G < 0$ are thermodynamically favourable. For a mixture, an initial state of a pure A phase and a pure B phase and a final state of a mixed AB phase are considered, with the change in free energy on mixing given by ΔG_{mix}, which in the Flory–Huggins approximation is considered as a sum of two terms:

$$\Delta G_{\text{mix}} = G_{\text{AB}} - (G_{\text{A}} + G_{\text{B}}) = \Delta G_{\text{local}} - T\Delta S_{\text{trans}} \tag{8.2}$$

where G_{X} is the free energy associated with phase X, ΔG_{local} captures the changes to the local chemistry and motions of the mixed units, including any changes to the local entropy due to volume changes, and ΔS_{trans} captures the change in translational entropy of the system. By assuming that ΔG_{local} arises primarily from enthalpic interactions between a given molecule and the mean field originating from the rest of the mixture, it may be shown that

$$\Delta G_{\text{mix}} = RTV\left(\chi\phi_{\text{A}}\phi_{\text{B}} + \frac{\phi_{\text{A}}}{N_{\text{A}}}\ln\phi_{\text{A}} + \frac{\phi_{\text{B}}}{N_{\text{B}}}\ln\phi_{\text{B}}\right) \tag{8.3}$$

where V is the volume of the system, R the molar gas constant, ϕ_{A} and ϕ_{B} the volume fractions of component A and B, and N_{A} and N_{B} their degrees of polymerisation. χ is the dimensionless 'Flory–Huggins' parameter, which represents the change in local free energy per reference unit (monomer, in this case). The first term represents local enthalpy change, while the second and third terms represent entropy change and will be negative. Homogeneous mixing will occur only if $\Delta G_{\text{mix}} < 0$, so compatible molecules ($\chi < 0$) will always mix, but incompatible molecules ($\chi > 0$) will only do so for sufficiently small N. This implies that a homogeneous mix of dissimilar polymers can usually lower its free energy via spinodal decomposition into a phase-separated morphology, with little entropic penalty. Figure 8.11a shows an example of such a morphology.

Real blend systems are always more complicated, as illustrated in Fig. 8.11c, in particular because it has been observed empirically that at least one semicrystalline component is required for a device to be efficient. If phase separation can also be driven by crystallisation, then even small molecules may spontaneously de-mix, given sufficient thermodynamic freedom. For example, the small size of fullerene acceptors means they are usually miscible in amorphous conjugated polymers (or within the amorphous fraction of semicrystalline polymers) up to around 50% by volume, as might be expected from a simple Flory–Huggins description. Beyond this miscibility threshold, purer fullerene domains develop which may subsequently crystallise, adding significant complexity to the system.

For a two-component semicrystalline blend at least five phases are therefore expected: two pure phases per component, one crystalline, one amorphous, plus a mixed amorphous phase. The widely studied P3HT:PCBM system exhibits such behaviour, with the final phase structure and domain size depending crucially on the thermodynamic trajectory. The somewhat unusual system of poly(2,5-bis(3-hexadecylthiophen-2-yl)thieno[3,2-b]thiophene) (PBTTT):PCBM adds a sixth

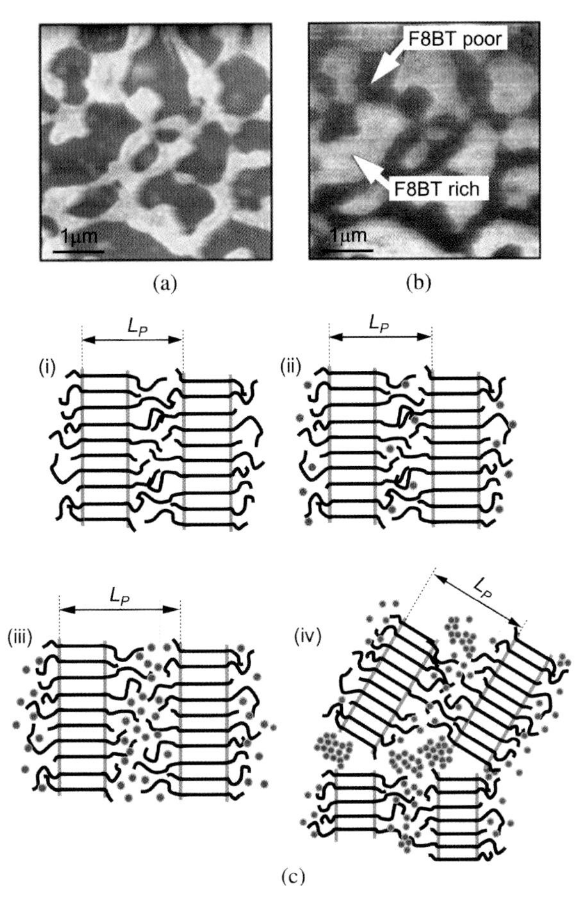

Figure 8.11 a) Topography (vertical range 81 nm) of a (1:1) PFO/F8BT thin film; b) F8BT fluorescence (arbitrary scale) from the same region of the blend, illustrating the classic phase-separated structure of a de-mixed polymer:polymer blend. Adapted from Chappell *et al.* (2003). c) Sketch illustrating the much finer microstructure formed by the blend of semicrystalline P3HT (lines) and PCBM (dots) with increasing fullerene loading: (i) with no PCBM; (ii) with PCBM dissolved in the amorphous fraction of the P3HT film without disrupting the typical crystallite spacing (L_P, of order 10 nm); (iii) with sufficient PCBM inclusion to swell the amorphous phase and increase L_P; (iv) saturation of amorphous-phase swelling and the onset of PCBM aggregation between P3HT crystallites. Adapted from Kohn *et al.* (2013).

phase, a co-crystal of PBTTT and PCBM, which is strongly implicated in exciton dissociation (see Fig. 8.12). The result in both cases is a hierarchical bulk heterojunction, with highly intermixed regions in quasi-equilibrium with purer phases and crystalline regions, the effects of which are discussed below.

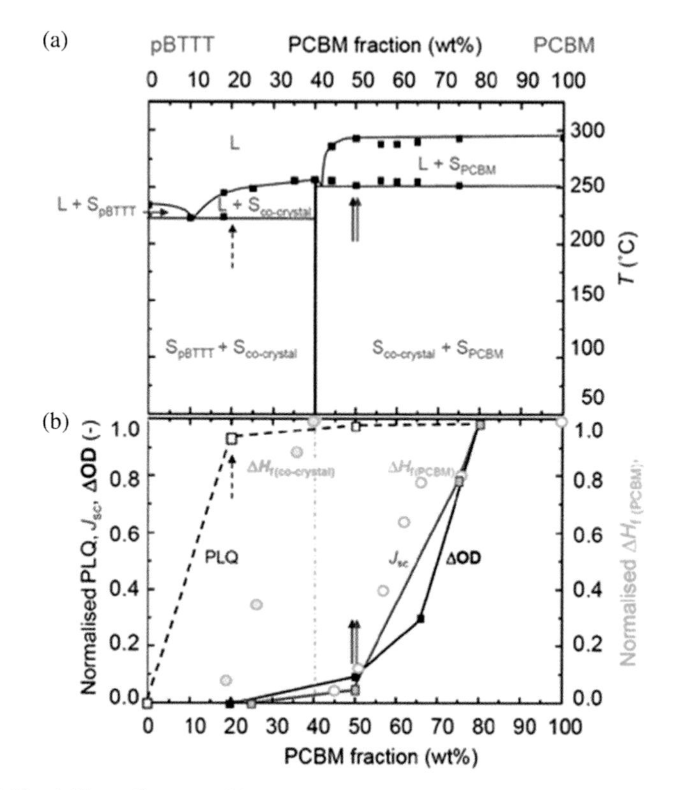

Figure 8.12 a) Phase diagram of the pBTTT/PCBM/co-crystal system, showing two eutectics: at ~1 wt% PCBM (pBTTT/co-crystal binary) and 43 wt% PCBM (co-crystal/PCBM binary; b Photoluminescence quenching (PLQ) is correlated with the onset of co-crystal formation beyond the first eutectic, as indicated by $\Delta H_{t(co-crystal)}$, whereas long-lived charge generation (ΔOD) and collection (J_{SC}) are correlated with the formation of crystalline PCBM domains beyond the second eutectic. Reprinted from Jamieson *et al.* (2012).

Ensuring a bulk heterojunction with beneficial morphology develops spontaneously following deposition requires a well-controlled thermodynamic trajectory, which in turn requires control of solution composition, temperature and evaporation rate as well as the interactions between component molecules. Techniques to improve device morphology therefore include:

- Varying the relative solubility of the blended materials — either by altering the molecules themselves (Park *et al.*, 2006), changing the type or temperature of the solvent (Liu *et al.*, 2001), or adding co-solvents (Lee *et al.*, 2008).

- Increasing the thermodynamic freedom of the blend to allow coarsening — either by annealing above its glass transition temperature (Ma *et al.*, 2005), by swelling the blend using a solvent vapour (Li *et al.*, 2007), or simply by restricting the solvent evaporation rate during deposition.
- Altering the interaction between molecules, for instance by encouraging/discouraging planarisation, enhancing/suppressing stacking or by affecting nucleation and growth of crystallites (Ballantyne *et al.*, 2007; Kim *et al.*, 2006).

Top-down patterning

The difficulty of generating a beneficial heterojunction structure using thermodynamics alone has led to interest in enforcing a particular nanostructure on the blend. To achieve this, a huge variety of techniques have been explored for patterning organic semiconductors (Brédas *et al.*, 2009). These include:

- Techniques which utilise pre-patterned substrates or stamps to guide or define the final structures, typically over large areas, such as site-selective chemical vapour deposition on patterned precursors (Cacialli and Bruschi, 1996), micromoulding in capillaries (Rogers *et al.*, 1998), micro-contact printing (Rogers *et al.*, 1999) and nanoimprint lithography (Chou *et al.*, 1996; Dong *et al.*, 2006).
- Techniques which rely on self-assembly to pre-form or form *in situ* a beneficial morphology, for instance using electrodeposition (Yun *et al.*, 2004), electrospinning (Kameoka *et al.*, 2004) and all forms of pre-aggregation in solution, including colloid, nanowhisker or vesicle formation. This may extend to templating using self-assembled diblock copolymer structures (Chappell *et al.*, 2003) and inorganic scaffolds (Martens *et al.*, 1999).
- Techniques which pattern directly to an arbitrary design, typically over relatively small areas, including electron-beam lithography (Stabile *et al.*, 2007), as well as scanning-probe technologies such as dip-pen nanolithography (via electrostatic or electrochemical deposition), scanning tunnelling microscopy (Granström, 1998), nanothermal lithography (Fenwick *et al.*, 2009) and scanning near-field optical lithography (Credgington *et al.*, 2010).

Each technique has particular strengths, which may be in terms of flexibility, resolution, or the amount of damage inflicted on the active polymer, though most are not easily scalable and few are able to achieve three-dimensional structures.

The ideal bulk heterojunction

A more fundamental difficulty when employing top-down approaches is that the ideal bulk heterojunction structure is not currently known. However, it must have some general properties:

1. A region of pure donor material adjacent to the hole-extracting contact to provide a barrier to the extraction of electrons.
2. A region of pure acceptor material adjacent to the electron-extracting contact to provide a barrier to the extraction of holes.
3. Continuous percolation pathways through high-mobility donor/acceptor phases from every interface to the electrodes.
4. Transport pathways aligned with the internal field, to enhance the contribution of carrier drift.
5. Interfaces distributed throughout the photoactive layer, such that there is a high probability of every photogenerated exciton reaching the heterojunction.
6. Thickness greater than the absorption depth of above-bandgap photons

Criteria 1 and 2 above improve the selectivity of the contacts, and reduce the need for interlayer materials to be carrier-blocking. Criteria 3 and 5 are particularly difficult to achieve simultaneously, since even in semicrystalline materials excitons rarely diffuse much further than 10 nm, while criterion 6 requires a film thickness of $\sim$200 nm. In principle, these requirements dictate an interdigitated structure of pure donor and acceptor material with a very high aspect ratio, which has been attempted by a variety of means using the techniques outlined above.

However, recent improvements in polymer:fullerene solar cell performance have shown that neither pure donor and acceptor phases, nor well-structured heterojunction networks, are necessary for efficient operation. Regions of intermixed donor and acceptor relax the need for single-phase percolation pathways, and separate excitons very efficiently. There is also evidence, discussed in Sections 8.4.4 and 8.4.5, that a hierarchical structure may provide an energetic driving force to move carriers away from the interface. The assumption that a pure two-component bulk heterojunction is ideal is therefore likely to be too simplistic, and the inclusion of mixed or amorphous phases may be necessary for efficient operation.

8.4.2 *Electronic disorder*

One of the benefits of using molecular semiconductors is that removal of defects is a relatively simple chemical purification process. While separation of very similar molecules is difficult — for instance C_{60} and C_{70}, or polymers with similar molecular weights — the electronically active backbone of the organic semiconductor is usually defect-free. This feature has several implications:

- Impurity tolerance. Since there is no lattice to disrupt, electronically inactive impurities have little effect on electronic structure.
- Passive interfaces. Junctions between molecular layers are not associated with broken bonds or lattice mismatch, and so are not detrimental to performance. This underpins the ability of organic heterojunctions to function, and allows relatively facile interface engineering by the insertion of interlayers.
- Both semicrystalline and amorphous materials may be employed, since the formation of a well-ordered lattice is not necessary.

Organic semiconductors are nevertheless usually highly disordered systems, despite their low intrinsic molecular disorder. This is because spatial and energetic disorder can be significant, as we outline below.

Configurational disorder

Although all small organic molecules, or monomers, of a given material are structurally identical, electronic interactions between two neighbouring molecules are strongly dependent on their relative position (Linares *et al.*, 2010). Since as-deposited films are usually amorphous, or at most semicrystalline, the relative position and orientation of neighbouring molecules varies significantly throughout the film. This spatial disorder leads to highly anisotropic electronic coupling between molecules, which impacts on charge transport. Similarly, since coupling strengths between molecules affect the energetics of intermolecular and delocalised states, this spatial disorder may contribute to energetic disorder, as each molecule experiences a slightly different local environment.

Conformational disorder

For flexible molecules, variation in conformation also leads to spatial and energetic disorder. This is particularly applicable to semiconducting polymers, where charge delocalisation is dependent on conjugation length. Kinks or twists in the polymer chain will break conjugation, limiting the extent of delocalisation and narrowing the molecular 'bands'. Chain-extended polymers are therefore optically distinct from coiled polymers within the same sample (Clark *et al.*, 2009), meaning that variations in processing history and polydispersity, as well as random variation in conjugation length, result in additional energetic disorder.

Dynamic disorder

The mechanisms above describe 'static' disorder, which will be valid at low temperatures where molecular positions are fixed. At elevated temperature, dynamic spatial disorder is induced by the presence of phonons. Since organic semiconductors are typically flexible, and adjacent molecules not strongly bound together, this thermally induced disorder can be large enough to perturb electronic coupling between molecules significantly — affecting both electronic transitions and charge transport. Since couplings strengths vary in time, modelling such behaviour is extremely complex, meaning dynamic effects are often approximated as an additional temperature-dependent contribution to static disorder (Bässler and Kohler, 2012).

8.4.3 *Charge transport in a disordered material*

Charge transport in organic semiconductors is quite different from that of their inorganic counterparts, where a three-dimensional periodic lattice structure often exists. The charges in the latter are strongly delocalised and their movement is interrupted only by scattering with phonons or impurities; therefore their mobility decreases with increasing temperature. In polymer semiconductors, spatial and energetic disorder prevents long-range band-like transport. Charge transport in organic semiconductors is instead described by quantum-mechanical tunnelling between localised states, and so depends on the ability of the charge carriers to access states in their immediate vicinity. As we will discuss below, the availability of accessible states is dependent on temperature, local electric field and carrier density. We note that most models for carrier transport assume a rigid lattice structure, and so implicitly ignore the effects of dynamic disorder and polaronic self-localisation, which will act to reduce carrier mobility. In general this simplification is only important where quantitative predictions are sought. In addition, we will not consider here the effects of anisotropic coupling between molecules (which will lead to anisotropic mobility), or include the contribution of intra-molecular transport, which we assume will only serve to reduce the per-state localisation.

Miller–Abrahams and Mott variable-range hopping

The Miller–Abrahams model was originally developed to describe conduction through impurities in inorganic semiconductors (Miller and Abrahams, 1960). Transport is assumed to occur in two steps — an initial phonon-assisted hop ‘up’, to conserve energy, followed by lateral tunnelling to the new site. The charge transfer rate W_{ij} for hopping from site i (with energy E_i) to site j (with energy E_j) is given by

$$W_{ij} = \nu_0 \exp(-R_{ij}/\Lambda) \begin{cases} \exp(-(E_j - E_i)/k_B T), & E_j - E_i \geq 0 \\ 1, & E_j - E_i < 0 \end{cases} \tag{8.4}$$

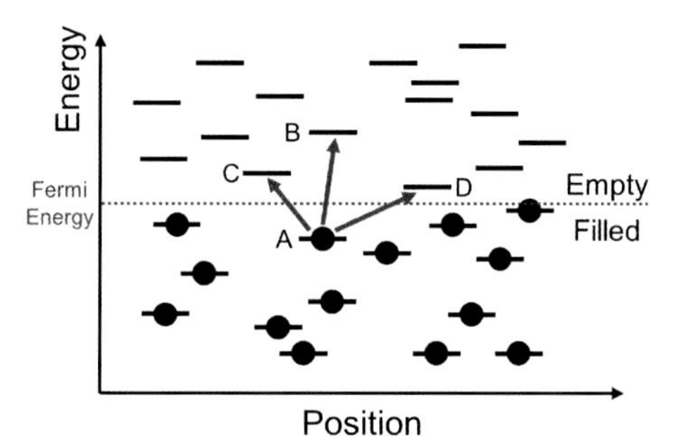

Figure 8.13 Schematic illustration of charge (circles) hopping between localised states (lines). Hopping from site A to site B represents nearest-neighbour hopping, while hopping to sites C or D represents variable-range hopping, with lower energetic activation competing with lower tunnelling probability.

where ν_0 is the hopping attempt frequency, Λ is the localisation length, and R_{ij} is the distance between sites i and j. The first exponential term describes the tunnelling probability, while the second exponential term accounts for thermal activation, with the assumption that there is no activation barrier for 'downhill' hops. Such a system is illustrated schematically in Fig. 8.13.

The Miller–Abrahams model was used by Mott to describe conduction in a d-dimensional system with uncorrelated spatial and energetic disorder and a constant density of states (DoS). By integrating *Wij* over available states in $d + 1$ dimensions (d spatial and one energetic), the conductivity σ of such a system was found to follow the expression

$$\sigma \propto \exp[-(T_0/T)^{\frac{1}{d+1}}] \tag{8.5}$$

where T_0 is a parameter proportional to the reciprocal of the DoS at the Fermi energy. This expression is therefore valid for variable-range hopping close to the Fermi energy, where the assumption of constant DoS holds. The key prediction of this model, and those below which derive from it, is that transport in organic semiconductor devices is faster at elevated temperatures.

Gaussian disorder model

A more realistic treatment of the effect of energetic disorder suggests several modifications to the constant DoS model. Uncorrelated disorder around a mean value is likely to give rise to a Gaussian 'band' of localised states, as illustrated in Fig. 8.14a. If instead disorder arises from a single species, the localised states may exist in a separate 'impurity band' within the forbidden energy gap — the original target for Miller–Abrahams theory. The scheme shown in Fig. 8.14b represents a band

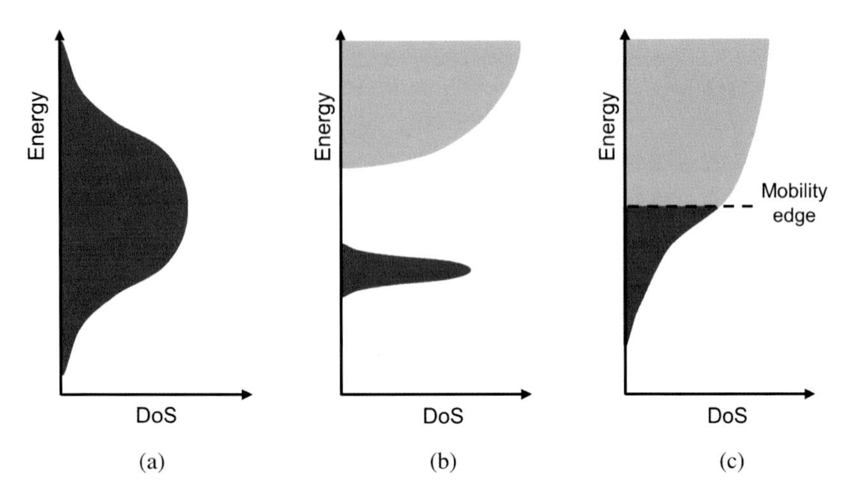

Figure 8.14 Models for transport in a disordered (electron) density of states. Localised states are indicated in red, delocalised states in grey. a) Fully localised Gaussian DoS; b) Band-like DoS with localised donor-like states within the bandgap; c) Transition from localised to delocalised states within a continuous DoS, demarcated by an abrupt mobility edge.

of lower-lying 'donor' states, separated from delocalised states of the conduction band. Figure 8.14c presents the situation where the presence of disorder induces a tail of localised states at the bottom of the conduction band. The threshold between localised and delocalised states is termed the *mobility edge* (discussed below).

By assuming Miller–Abrahams hopping rates between localised states, and employing Monte Carlo simulations, Bässler derived an expression for the mobility μ of a material described by a distribution of localised states characterised by two Gaussian distributions: one spatial, with width Σ and one energetic, with width σ

$$\mu(\mathcal{E}, T) = \mu_0 \exp\left(-\frac{4}{9}\hat{\sigma}^2\right) \begin{cases} \exp[C(\hat{\sigma}^2 - \Sigma^2)\sqrt{\mathcal{E}}], & \Sigma \geq 1.5 \\ \exp[C(\hat{\sigma}^2 - 2.25)\sqrt{\mathcal{E}}], & \Sigma < 1.5 \end{cases} \tag{8.6}$$

where μ_0 is the zero-field mobility at infinite temperature, $\hat{\sigma} \equiv \sigma/k_B T$, C is a constant related to the typical lattice spacing and $\mathcal{E}$ is electric field (Bässler, 1993).

This Gaussian disorder model (GDM) predicts that the mobility is dependent on the field by a factor of $\exp(\mathcal{E}^{1/2})$, which arises from the increased number of 'downhill' hops made available by the applied field. It was later found that the agreement with experiment could be improved by taking the correlation between spatial and energetic disorder into consideration — the correlated disorder model (CDM) (Dunlap *et al.*, 1996). Further detail may be added by considering the effect of state filling as the carrier density increases. Low-lying states with low transition probabilities will be filled first, implying that the mobility of subsequent carriers progressively increases. Such a modification was developed by Pasveer and co-workers (Pasveer *et al.*, 2005).

Multiple trapping and release

Since the number of states at a given energy is likely to correlate with their localisation, we may also consider a model whereby sparse, localised 'tail states' give way to increasingly delocalised 'band-like' states as energies approach the unperturbed HOMO/LUMO level. This tail is typically treated either as a Gaussian or, more simply, as a decaying exponential. Extracting analytical expressions for temperature-, field- and density-dependent mobility is difficult in this case, and statistical methods such as Monte Carlo simulations are required to understand charge transport. The behaviour of carriers in such a regime is usually simplified by defining a mobility edge, as illustrated in Fig. 8.14c. States below this edge are treated as localised, while states above the mobility edge are treated as delocalised and mobile. Carriers in localised states are therefore trapped until thermally released to the extended states, while free carriers in the extended states can be trapped to the localised states (Nelson, 2003). Transport of charge carriers in this model depends on the spatial and energetic distribution of localised sites, which may be included as a parameter, or extracted if sufficiently detailed data are available (MacKenzie *et al.*, 2012).

Numerical device simulations

Based on such models of carrier mobility, simulations of the current–voltage curves of complete devices are possible. The literature on such simulations is extensive, but we note here three important concepts. At low internal field, transport is primarily diffusive, with the diffusion coefficient D assumed to be given by either the Einstein relation (Eq. (8.7)) or the generalised Einstein relation (Eq. (8.8)) (Reese *et al.*, 2010):

$$D/\mu = kT/q \tag{8.7}$$

$$D/\mu = g(n, T)kT/q \tag{8.8}$$

where μ is the zero-field mobility and g is a function of the carrier density (n) and temperature which attempts to account for the disordered nature of the DoS. The need for such an 'enhancement factor' in practical devices is contested, however (Wetzelaer *et al.*, 2011a).

Since organic photovoltaics are thin-film ($\sim$100 nm) devices with low dielectric constants, they are subject to significant internal fields under operation, of the order $10^7\,\mathrm{V\,m^{-1}}$. Carrier drift is therefore a significant process and usually dominates even at modest applied voltage, such that forward-biased organic diodes are usually found to be space-charge limited (Mihailetchi *et al.*, 2005).

Therefore, complete drift–diffusion simulations are required to capture the main features of device operation in the power-generating quadrant, solving (in the 1D case)

$$i_n = i_{\mathrm{diffusion}} + i_{\mathrm{drift}} = qD\frac{\mathrm{d}n(x)}{\mathrm{d}x} + q\mu_n\mathcal{E}(x)n(x) \tag{8.9}$$

under the continuity condition

$$\frac{\mathrm{d}i_n(x)}{\mathrm{d}x} + G(x) - R(x) = 0 \tag{8.10}$$

where i_n is the current density due to electrons, μ_n is the electron mobility, G and R are local generation and recombination rates for electrons, and we assume equivalent expressions for holes with a requirement for charge neutrality. It is usual to base such a model on an 'effective medium' representing the combined contribution of each blend component, which can be made more realistic by including localised variations in the DoS or separate contributions from different thermodynamic phases.

In practice, obtaining sufficiently detailed data over a range of time and temperature scales to isolate all of the internal processes contributing to charge transport is extremely difficult (Bässler and Kohler, 2012).

8.4.4 *Carrier generation*

Optical bandgap

In order to generate large current densities, it is first necessary for a solar cell to absorb a large fraction of the solar spectrum, which requires a semiconductor with a low optical bandgap (of order 1–1.5 eV). However, as outlined in Section 8.2.1, most small organic molecules and polymers are found to have relatively large HOMO–LUMO offsets. Figure 8.15 illustrates this with the prototypical homopolymer P3HT, which has an absorption edge at around 1.9 eV (650 nm). Even assuming complete absorption of all photons with energies greater than this, such a large bandgap covers only 20% of the AM1.5G solar photon flux. This is compounded by the well-defined (rather than band-like) electronic transitions in molecular materials, which lead to relatively low absorption bandwidths.

The simplest method to collect additional photons is to combine donor and acceptor species with complementary absorption. For polymer:polymer cells, this is possible, but for the more efficient polymer:fullerene (or small-molecule:fullerene)

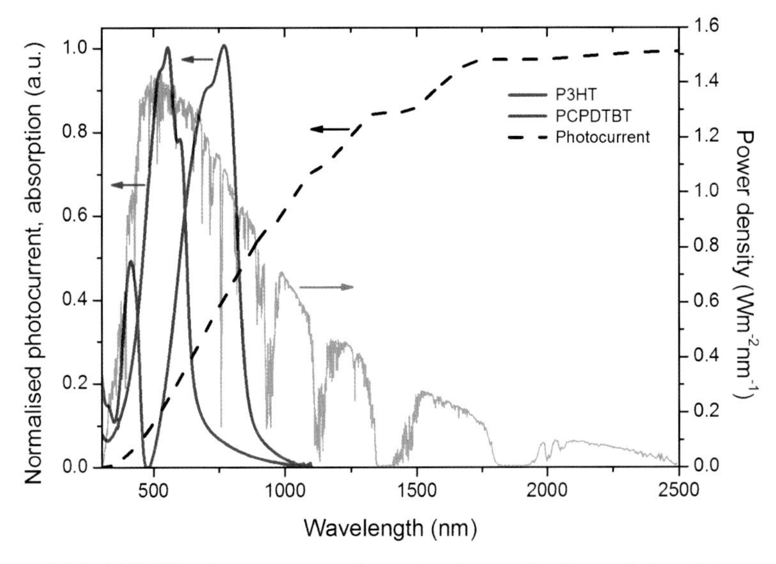

Figure 8.15 AM1.5G solar spectrum and associated normalised cumulative photocurrent, assuming that all photons below the given wavelength are converted to carriers and collected. The absorption spectra of P3HT and PCPDTBT thin films are included for comparison.

systems, changing the acceptor is more difficult. Since C_{60} and its soluble derivatives have spherical symmetry, their extinction coefficients are very low in the visible range. Substituting C_{60} by the prolate C_{70} fullerene significantly increases absorption in the visible and ultraviolet (UV), but at higher materials cost. To date, it has not been possible to replace fullerenes successfully with more highly absorbing acceptor materials.

The primary responsibility for absorption thus lies with the donor material. Increasing conjugation lengths to reduce the HOMO–LUMO gap can only be taken so far before molecules become too large or too stiff to deposit. Equally, including heteroatoms within the (primarily hydrocarbon) organic backbone shifts both HOMO and LUMO energies, but not necessarily their offset (Son *et al.*, 2011). This problem has been solved by coupling together electron-rich and electron-withdrawing units in a 'push-pull' (or, confusingly 'donor-acceptor') configuration, which can lead to the molecular HOMO states resting primarily on the electron-rich unit, while the molecular LUMO states reside primarily on the electron-withdrawing unit.

Figure 8.16 shows an example of this behaviour, illustrating the shift in electron density for a 3CDTBT oligomer (analogous to the backbone of PCDTBT) from the cyclopentadithiophene 'push' units to the benzothiadiazole 'pull' units. The result

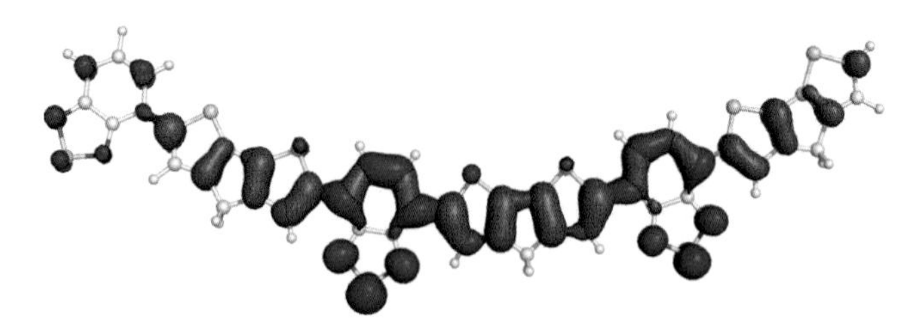

Figure 8.16 Shift in electron density on a 3CDTBT oligomer on excitation from the ground (HOMO) to the first excited state (LUMO), calculated using time-dependent density functional theory. Electron density moves from red to blue regions. Image courtesy of Sheridan Few.

has been a class of push-pull small molecules and polymers with absorption now extending to beyond 900 nm (Bronstein *et al.*, 2011; Mühlbacher *et al.*, 2006; Zhong *et al.*, 2013). An additional benefit is that the higher-lying electronic transitions are still accessible, which can broaden the absorption profile of such materials over much of the visible range. The best performing polymer:fullerene cells utilise the push-pull polymer PTB7 in combination with $PC_{70}BM$ to achieve >70% EQE from 400 to 750 nm, resulting in short-circuit current densities in excess of 17 mA cm^{-2} (He *et al.*, 2012).

Exciton dissociation

While absorption of photons is a necessary precursor to current generation, the early work on single-material devices showed that it was not sufficient. The necessity for a type 2 (staggered) heterojunction indicates that a driving force is required to separate charge, which represents a loss of available free energy. For example, in the P3HT:PCBM system, around 800 meV of free energy is lost per carrier during dissociation. Considerable effort has been expended to understand how this energy loss may be minimised, while maintaining good separation efficiency (Clarke and Durrant, 2010). We illustrate the main processes in Fig. 8.17, and refer to the rates for those processes below.

When excitons reach the heterointerface, charge transfer from donor to acceptor is exceptionally fast, with the process complete within 50 fs — i.e. $k_{CT} \gg k_{PL}$. Studies of blend photoluminescence show that only 5–10 wt% of a good acceptor mixed homogeneously with a donor is necessary to quench over 90% of excitons. However, the result of such quenching is not necessarily dissociated charges. The low dielectric constant of organic semiconductors means that the exciton may remain bound across the heterointerface in a charge-transfer (CT) state. The spatial separation necessitated by the interface means the binding energy of the CT state

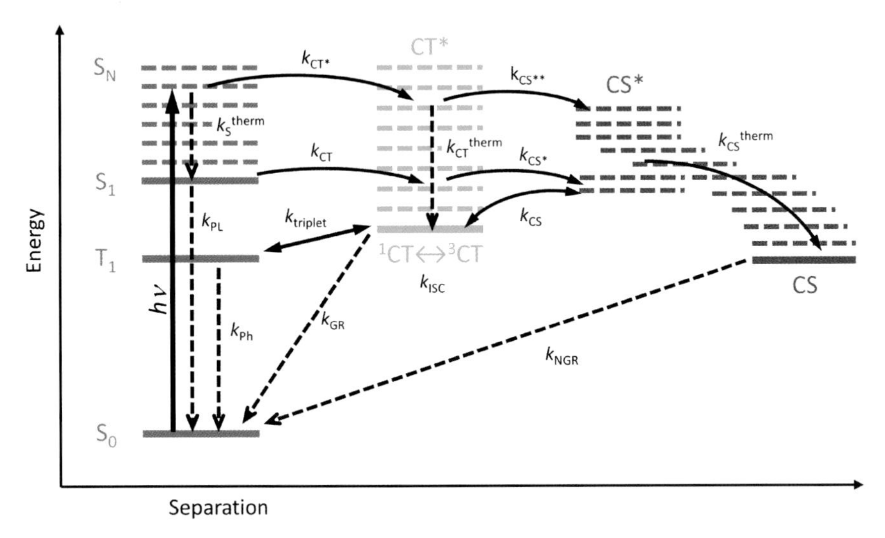

Figure 8.17 Energy-level diagram illustrating the competing processes (with rates k_i) involved in photogeneration following absorption of a photon with energy $h\nu$. Exciton diffusion and intersystem crossing of the donor singlet are ignored. Solid arrows represent processes increasing charge separation; dashed arrows represent relaxation towards the ground state S_0. k_{CT}: Dissociation of relaxed (or hot, k_{CT^*}) singlet excitons (S) to form hot interfacial charge-transfer (CT) states. k_{CS}: Dissociation of relaxed (or hot, k_{CS^*} and $k_{CS^{**}}$) CT states to form fully dissociated charge-separated (CS) states. k_{ISC}: Intersystem crossing between singlet and triplet CT states. $k_{triplet}$: Generation of molecular triplet excitons from triplet CT states, including possible re-dissociation at the heterojunction. $k_{PL/Ph}$: Relaxation of singlet/triplet excitons to the ground state. k_{GR}: Geminate recombination of CT states. k_{NGR}: Non-geminate recombination of dissociated carriers, either directly to the ground state or, more likely, via re-formation of CT states at the heterointerface. k^{therm}: Thermal relaxation to the corresponding lowest energy excited state. Since the CT state involves partially separated carriers, there is unlikely to be a significant difference in exchange interaction between ^{1}CT and ^{3}CT. Adapted from Clarke and Durrant (2010).

will be less than that of the molecular exciton, but still greater than the available thermal energy; estimates for CT binding energies vary from around 0.1 to 0.5 eV depending on the assumed separation and local polarisation. Excitons occupying singlet CT states are observed to decay to the ground state within a few nanoseconds (a process usually termed *geminate recombination*, k_{GR} in Fig. 8.17). While the majority of CT states decay non-radiatively, some couple radiatively to the ground state and produce CT luminescence, which provides a useful tool for their study (Tvingstedt *et al.*, 2009; Vandewal *et al.*, 2009). The mechanism by which CT excitons avoid geminate recombination and dissociate remains unclear, but several routes have been proposed, as we now discuss.

Field-assisted ionisation

In many organic systems, macroscopic electric fields can distort the local energy landscape and enhance separation efficiency. This process, observed in single-component systems (Dick *et al.*, 1994; Silinsh *et al.*, 1974), polymer:polymer systems (Gonzalez-Rabade *et al.*, 2009), and some systems with fullerene acceptors (Credgington *et al.*, 2012; Veldman *et al.*, 2008), is usually understood in terms of Onsager–Braun theory. This describes the separation of electrons and holes in a dielectric medium by calculating the probability that a Brownian random walk will separate an electron-hole pair by more than their Coulomb capture radius — defined by Onsager as the distance at which the Coulomb binding energy equals the thermal energy kT. Excess energy provided by the incident photon (or heterojunction) would generate an initially 'hot' carrier, and thus increase the chance of exceeding this radius. Any macroscopic electric field will also lower the Coulomb potential in the direction of the field, and so aid dissociation. Within this theory, the rate of dissociation for a geminate pair k_{CS} as a function of electric field $\mathcal{E}$ was derived by Braun (1984) as:

$$k_{CS}(\mathcal{E}) = \frac{3\langle\mu\rangle q}{4\pi\langle\epsilon_s\rangle a^3}\exp(-\Delta E/kT)\left[1 + b + \frac{b^2}{3} + \frac{b^3}{18} + \cdots\right] \quad (8.11)$$

where $\langle\mu\rangle$ is the spatially averaged sum of the electron and hole mobilities, $\langle\epsilon_s\rangle$ is the spatially averaged dielectric constant, ΔE is the Coulomb attraction of the electron-hole pair after thermalisation at a distance a and $b = q^3\mathcal{E}/8\pi\langle\epsilon_s\rangle(kT)^2$, where the final terms arise from a power-series approximation to a first-order Bessel function.

The efficiency of dissociation thus depends on the cumulative dissociation probability over the CT state lifetime. While effective at describing spontaneous dissociation in single-component systems, applying this theory to predict quantitative yields of dissociated carriers in heterojunction systems remains problematic. In these systems, carrier generation is often observed to be independent of temperature, and the CT lifetimes (or local mobilities) which must be assumed are usually unphysical. Dissociation is empirically observed to be far more efficient (and thus far less dependent on macroscopic electric field) than might be expected — i.e. $k_{CS} > k_{GR}$. Part of this difficulty arises from the assumed Coulomb capture radius, which for a material with $\epsilon_s \sim$3–4 is 10–20 nm. This ignores the significant influence of static and dynamic energetic disorder, which can change the energy landscape experienced by a carrier by $\sim$100 meV over this length scale. In addition, the local electric field at the heterojunction will be influenced by the orientation of the junction with respect to the macroscopic field, by space charge in the photoactive layer and by the presence of interfacial dipoles/polarisation (Rogers *et al.*, 1998).

The physical mechanisms allowing efficient exciton dissociation at organic heterojunctions (and particularly in fullerene-based systems) are thus not well understood. Several empirical observations help to shape our picture of the process, however.

Generation from CT states

As we have described, CT states present a paradox. When long-lived, they do not ionise efficiently and so are considered to be dead ends in the generation of dissociated carriers. However, since the CT state is often radiatively coupled (albeit weakly) to the ground state, it is possibly to optically excite these states directly. A number of studies have examined the quantum efficiency of polaron generation from such sub-bandgap absorption, and concluded that $k_{CS} = k_{CS^*}$: i.e. that there is no difference in quantum efficiency for polaron generation from CT excitons as compared with those generated from molecular excitons (Lee *et al.*, 2010).

This apparent contradiction suggests that the behaviour of CT states when they are first generated plays a critical role in their subsequent dissociation. The initially excited CT state is not geometrically relaxed, and thus corresponds to a relatively delocalised excitation. Subsequent geometric relaxation of those CT states which do not dissociate leads to a significant increase in their binding energy. The difference between CT absorption and emission spectra indicates this increase is of order 0.3 eV, due to both reorganisation of the molecular structure and increased coulomb attraction (Gélinas *et al.*, 2011). Equilibrated CT states are therefore tightly bound, and unlikely to dissociate, consistent with observation. However, non-equilibrium CT species — which are generated either by direct photon absorption or by charge-transfer from a molecular exciton — can achieve high dissociation yields. For instance, Bakulin and co-workers (Bakulin *et al.*, 2012) showed that exciting long-lived relaxed CT states with a low energy light pulse recreates these more delocalised species, which subsequently separate into dissociated carriers.

Separation of carriers therefore requires that more energy is available than is needed to form a relaxed CT exciton. Since this excess represents a loss of free energy available to do electrical work (as discussed in Section 8.4.5), there is considerable interest in establishing the minimum excess energy needed to drive polaron dissociation. Since we are mainly concerned with systems with fullerene acceptors, this is usually described in terms of the minimum LUMO–LUMO offset.

Empirical observations suggest that for many blend systems, a correlation exists between LUMO–LUMO offset and the yield of long-lived dissociated polarons (Shoaee *et al.*, 2010; Shoaee *et al.*, 2013). However, different classes of material exhibit different 'minimum' driving energies, suggesting that molecular structure and blend morphology are significant complicating factors in this process, and may obscure the exact role of energetics. It is also speculated that the dipole moment associated with transitions to the excited state may play a part, and has been cited as a possible reason why devices based on modern push-pull donors achieve excellent dissociation efficiency with offsets of only 100–200 meV (Carsten *et al.*, 2011). While the underlying mechanism remains uncertain, a relatively safe empirical design rule is that around 300 meV of driving energy is usually sufficient to achieve electron (or hole) transfer and polaron dissociation in a well-optimised organic blend.

Hot exciton effects

In some crystalline organic systems, including the fullerenes, absorption of a photon with energy well above the bandgap can lead to spontaneous generation of polarons — in accordance with the Onsager–Braun model of exciton dissociation. In a heterojunction system, and where absorption occurs away from the heterointerface, this effect is expected to be small: since excitons diffusing to the interface on picosecond to nanosecond timescales are likely to be thermally relaxed, they will have no memory of the initial photon energy. However, if absorption occurs directly at the interface, dissociation of the 'hot' exciton may compete with thermalisation ($k_S^{therm} \approx k_{CT^*}$), since both can be femtosecond processes. In bulk heterojunction systems with highly intermixed donor and acceptor phases, absorption directly at the heterointerface is common, and the additional thermal energy has been shown to assist the dissociation process. In particular, it has been observed that 'cold' excitons lead to the population of lower-lying CT states, which may dissociate less efficiently than higher-lying CT states populated by 'hot' excitons ($k_{CS^{**}} > k_{CS^*}$) (Dimitrov *et al.*, 2012; Grancini *et al.*, 2013).

Suppression of geminate recombination

Geminate recombination is a severely limiting process for organic solar cells, and while current evidence suggests that it can be suppressed in empirically optimised materials by a sufficiently large band offset at the heterojunction, this has a significant cost in free energy. However, the generation of CT states (if not their dissociation) appears to be a highly overdriven reaction — once an exciton reaches the heterojunction, the quantum yield of charge transfer is unity. This suggests that much of the free energy of charge transfer could be better spent moving the geminate pair further apart. Inspiration for this approach comes from biological photosynthesis, which employs an energetic cascade to drive electrons away from the chlorophyll absorbers. Each step costs energy, but decreases the associated back-recombination rate. Considerable effort has therefore been directed towards including spacer molecules or groups at the heterointerface to increase the electron-hole separation of the CT state and thus increase its lifetime. This necessarily reduces the rate of CT state formation, but since there are no competing processes on the timescales involved, this need not be restrictive.

A related approach may already have been implemented empirically in hierarchical polymer:fullerene bulk heterojunctions, where amorphous and crystalline regions exist together. Since crystalline phases typically exhibit narrower bandgaps than more amorphous phases, the existence of a crystalline region adjacent to the more amorphous interface region may also drive carriers away from the interface and thus reduce the geminate recombination rate (Jamieson *et al.*, 2012).

8.4.5 *Free energy per carrier*

When a photon with energy E_{ph} is absorbed by the photoactive layer and generates carriers, only the free energy of those carriers reaching the contacts can be used to do work. The remainder is lost, generally to phonons or emitted photons. The maximum free energy available is therefore qV_{oc}, since at open circuit we require negligible carrier flow. Maximising V_{oc} therefore involves minimising the free energy lost by carriers as they transit the device. We have already discussed the losses of free energy associated with carrier formation, which are:

1. Relaxation of the initially hot exciton back to the band edges. This is unavoidable for absorbers with a single bandgap, and sets a limit on how small that bandgap can be made.
2. The excess energy required to drive charge separation at the heterojunction.

We now discuss a third, which is common to all photovoltaic devices. The free energy available at open circuit is the difference in the chemical potentials (or quasi-Fermi energies) of electrons and holes at their respective extracting contacts, which incorporates both the enthalpy associated with the occupancy of excited states and the entropic cost of maintaining thermochemical equilibrium between the solar cell and light source. In the Shockley–Queisser treatment of an idealised solar cell, this equilibrium is maintained only by the radiative recombination of carriers. As discussed in Chapter 2 of this volume, for a device illuminated by the Sun, this gives a voltage loss of around 0.3 V arising primarily from the mismatch between the solid angle subtended by the Sun at the Earth and the solid angle of re-radiation from the device. Explicit in this analysis is the scope for excitons and photons to exchange energy multiple times before the photon finally escapes or electrons and holes are collected at their electrodes. However, any non-radiative recombination processes will necessarily increase the recombination rate, and so reduce the available voltage.

Within this framework, we may understand the empirically observed correlations between device energetics and V_{oc}. We begin by assuming that the heterojunction can be approximated as an effective medium with a single, well-defined pair of HOMO and LUMO states and no disorder, and that the reduction in free energy due to recombination may be treated as an empirical offset parameterised as ΔV. We will also assume for now that illumination intensity is fixed.

Since most of these recombination processes are a function of carrier density or cell voltage (or both), they increase with the filling fraction of the DoS. The point at which the total recombination current i_{rec} balances the rate of carrier generation i_{gen} sets the equilibrium filling fraction, and thus the maximum quasi-Fermi energy splitting. Organic solar cells therefore experience a voltage loss ΔV beyond the 0.3 V predicted by the Shockley–Queisser analysis, owing to both non-radiative recombination and disorder.

Metal–insulator–metal regime

In the metal–insulator–metal (MIM) regime, the electrode work functions lie within the heterojunction bandgap. Since organic semiconductors are usually intrinsic, equilibration between the electrodes is likely to deplete the entire semiconductor thickness, preserving the mismatch between the heterojunction energy levels and the electrode work functions. Carriers reaching an electrode must therefore thermalise to the work function, reducing their enthalpy. As such, the quasi-Fermi energies in the bulk are limited by $\Delta\Phi$, giving

$$V_{oc} = \frac{1}{q}(\Phi_h - \Phi_e) - \Delta V \tag{8.12}$$

where $\Phi_{e/h}$ are the work functions of the electron/hole extracting contacts, respectively.

Materials-limited regime

If the electrode work functions lie outside the heterojunction bandgap, charge may be transferred from the electrodes to empty states in the semiconductor, causing sharp band bending. This allows equilibration between the electrodes and the HOMO/LUMO levels of the heterojunction without the build-up of space charge in the majority of the semiconductor (provided the offset is not too great). In this regime, the quasi-Fermi energies at the contacts are 'pinned' to states in the semiconductor (Crispin *et al.*, 2006). For well-defined (donor) HOMO and (acceptor)

LUMO levels, V_{oc} will be primarily governed by the relative difference between these states, which we have termed E_{HJ}. These energies are usually estimated from pure donor or acceptor films using electrochemical or spectroscopic techniques such as cyclic voltammetry or ultraviolet photoelectron spectroscopy. As such, it is common for the measured quantities Ionisation Potential (IP) and Electron Affinity (EA) to be quoted instead of E_{HOMO} and E_{LUMO}. The latter are not generally known, since blending donor and acceptor together may induce a change in energetics, either through changes in molecular packing or the introduction of interfacial dipoles. Where these changes may be ignored, however, we expect

$$V_{oc} = \frac{1}{q}(\mathrm{IP}_{\mathrm{donor}} - \mathrm{EA}_{\mathrm{acceptor}}) - \Delta V \tag{8.13}$$

which is a widely observed correlation. A more accurate measure of the HOMO–LUMO gap may be obtained by using the weak luminescence of the CT state as an *in situ* probe of E_{HJ}, thereby implicitly including the effects of blending. The CT energy (E_{CT}) may be estimated by measuring either the electroluminescence of the blend (which will be dominated by CT emission), or by isolating sub-bandgap absorption features arising from CT states. The most accurate correlation between energetics and open-circuit voltage is therefore given by (Faist *et al.*, 2012; Vandewal *et al.*, 2008)

$$V_{oc} = \frac{1}{q}E_{CT} - \Delta V \tag{8.14}$$

where ΔV in equation (8.14) is reduced by any contribution from the coulomb binding energy of the CT state.

Disorder and recombination

All of the approaches above give good empirical correlation with the measured open-circuit voltage within the limits of their assumptions. However, all require an empirical offset ΔV of order 300–500 mV, which varies between systems (and depends on measurement technique). This is because we have not yet accounted for the influence of recombination and disorder on the quasi-Fermi energy positions.

If the molecular HOMO and LUMO bands are broadened by disorder, then the experimentally determined CT state energy, or IP and EA, will correspond to some average energy located mid-band, as will any reasonable definition of E_{HJ}. Photogenerated carriers will therefore be able to thermalise into the tail of empty states below this energy, with an associated loss of enthalpy. Where the characteristic disorder width is greater than the width of the Fermi distribution, such that the quasi-Fermi energies lie within the disordered DoS, this enthalpic

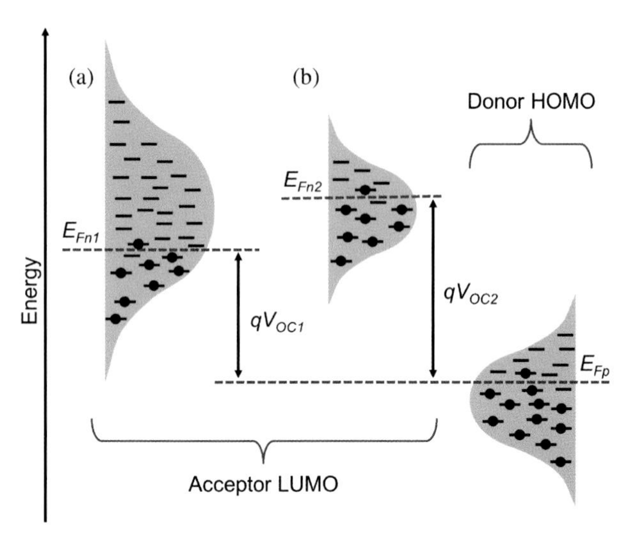

Figure 8.18 Schematic illustrating the effect of DoS width on achievable V_{OC}. For a fixed number of photoexcited carriers, the wider LUMO DoS (a) exhibits a lower electron quasi-Fermi energy E_{Fn1} than the narrower DoS (b) with quasi-Fermi energy E_{Fn2}. Both LUMO distributions share a common centre.

loss will dominate ΔV (Blakesley and Neher, 2011). At constant carrier density, ΔV will therefore be determined by the extent of disorder, as illustrated in Fig. 8.18.

The carrier density in a solar cell under operation is not a constant, however. At open circuit, where no net current flows out of the device, the internal carrier density will be set by the equilibrium between carrier generation and carrier recombination. Within a real solar cell, numerous additional recombination pathways exist beyond the radiative recombination described in the Shockley–Queisser approach. These include geminate recombination of CT excitons, collection of carriers at the 'wrong' electrode, non-radiative recombination between dissociated polarons (non-geminate recombination), recombination through trap sites and direct leakage currents, all of which can be important in organic solar cells and all of which can vary in severity between devices (Credgington and Durrant, 2012).

A general expression for the open-circuit voltage must therefore include not only the average heterojunction bandgap, but also the characteristic disorder width and the severity of recombination losses. For a device limited by both geminate and Langevin-type recombination (see below) and with fixed DoS N (assumed to be the same for electrons and holes), Koster *et al.* (2005) derived the expression

$$V_{\mathrm{oc}} = \frac{E_{\mathrm{HJ}}}{q} - \frac{\mathrm{kT}}{q} \ln\left(\frac{(1-P)\gamma N^2}{PG}\right) \tag{8.15}$$

where G is the generation rate of excitons per unit volume, P their dissociation probability and γ a measure of the likelihood of Langevin recombination. Including the effects of a general, disordered DoS and more complex recombination mechanisms is not straightforward (Kirchartz *et al.*, 2011), but in the case where the dominant recombination process gives rise to a rate which varies as a power law in carrier density, and the DoS varies exponentially with energy, the open-circuit voltage may be expressed as

$$V_{oc} = \frac{E_{HJ}}{q} - \frac{\eta kT}{q}\ln\left(\frac{J_{BI}}{J_{gen}}\right) \tag{8.16}$$

where η is an experimentally determined function of the DoS shape and recombination order, and J_{BI} is related to the recombination rate at $V = E_{HJ}/q$ (Credgington and Durrant, 2012).

Optimising open-circuit voltage

Routes to optimise V_{oc} rely on minimising the free-energy losses described above, and illustrated in Fig. 8.19. Minimising thermalisation losses requires a multijunction solar cell, with each junction absorbing progressively lower energy photons. The free-energy loss per photon is thus considerably reduced. Table 8.1 gives the thermodynamic limits for various junction arrangements, assuming only radiative losses, spontaneous exciton dissociation and that the solar spectrum can be approximated to a black body at 6000 K (Vos, 1980). The continuous multijunction is the limit of an infinite series of thin solar cells with bandgaps varying continuously from infinity to zero.

Minimising the loss at the LUMO–LUMO offset requires minimising the required driving force for exciton dissociation. This is limited by the need to overcome the CT-state binding energy, but there are indications that a driving force of 100–200 meV may be sufficient for some materials systems.

Minimising disorder widths is more difficult without employing more expensive fabrication methods. The two most practical routes are reducing impurities/unwanted isomorphs during synthesis (including reducing the polydispersity of polymers) and increasing the tendency of materials to crystallise. This must be weighed against the requirement for thermal stability — a high driving force for crystallisation may lead to Ostwald ripening over time (see Section 8.5.1).

Since the Shockley–Queisser approach considers only radiative recombination, it is often commented that a good solar cell is also a good LED. This must be understood to mean that if carriers can be made to recombine radiatively (which is a relatively slow process) then all other non-radiative recombination pathways have implicitly been eliminated and the cell can operate closer to its maximum

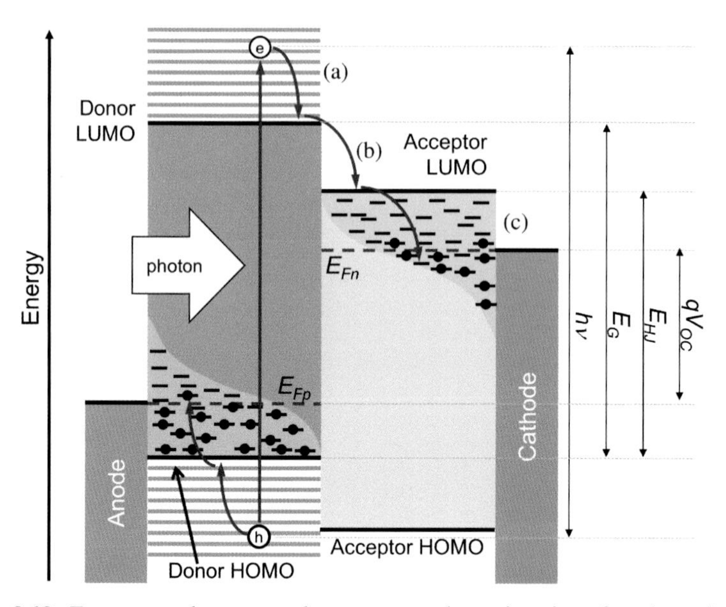

Figure 8.19 Free energy losses at a donor-acceptor heterojunction after absorption of a photon with energy $h\nu$. a) Thermal relaxation to the donor bandgap E_g; b) exciton separation at the heterojunction; c) relaxation into the disordered DoS, with electron/hole quasi-Fermi energies $E_{Fn,p}$. For simplicity, energetic disorder in the donor LUMO and acceptor HOMO is not considered, nor is relaxation of the exciton into the disordered DoS. Equilibration between the electrode work function and semiconductor Fermi energies is assumed (i.e. the system is not within the MIM regime).

Table 8.1 Theoretical efficiency limits for single and multijunction solar cells under 1 Sun and maximally concentrated sunlight (in brackets).

Solar cell type	**Maximum PCE under 1:1 (45,900:1) concentration**
Single junction	30% (40%)
Tandem	42% (55%)
Triple junction	49% (63%)
Continuous multijunction	68% (86%)

thermodynamic efficiency. Non-radiative recombination losses can most simply be reduced by improving the integrity of the device stack, to reduce leakage through pinholes and spikes, and by engineering the electrode interfaces to reduce surface recombination. However, methods to reduce the more important Langevin

(Eq. (8.17)) or Shockley–Read–Hall (SRH) (Eq. (8.18)) recombination currents are in their relative infancy, since the rate-limiting steps for these processes are not yet well understood. These recombination currents are given by

$$i_L = \frac{q(np - n_i^2)}{\epsilon A}\left(\frac{\mu_n + \mu_p}{2}\right) \tag{8.17}$$

$$i_{SRH} = \frac{qN_t C_n C_p (np - n_i^2)}{A[C_n(n + n_1) + C_p(p + p_1)]} \tag{8.18}$$

where A is cell area, n, p are the density of electrons and holes in LUMO/HOMO bands and μ_n, μ_p their mobilities, n_i is the intrinsic electron density, N_t is the trap density, C_n, C_p are the electron and hole capture cross-sections and n_1, p_1 are auxillary parameters such that $n_1 p_1 = n_i^2$. Note that Eq. (8.17) tends towards Eq. (8.18) as the mobility of either carrier drops towards zero, and that Eq. (8.18) simplifies considerably if one carrier is in significant excess.

It is known that certain materials systems (particularly those comprising semicrystalline polythiophenes and fullerenes) exhibit recombination rates which are significantly suppressed with respect to those expected from a simple mobility-limited Langevin recombination model (Credgington and Durrant, 2012). Whether this is due to morphological or energetic effects remains contentious, but it suggests that there is scope to reduce recombination rates in other materials systems.

Finally, note that we have not yet considered the influence of light intensity on recombination rate. The dominant recombination mechanism sets the ideality factor η for a device, defined as (Wetzelaer *et al.*, 2011b)

$$\eta = \left(\frac{kT}{q}\frac{\mathrm{d}\ln i_{\mathrm{ph}}}{\mathrm{d}V_{oc}}\right)^{-1} \tag{8.19}$$

where i_{ph} is the photogenerated current density (often replaced simply with illumination intensity). Recombination between mobile carrier populations (Langevin-type) leads to an ideality of 1, whereas recombination through traps (SRH-type) and at interfaces, or where one carrier is in significant excess, leads to a characteristic ideality > 1. Suppressing non-Langevin recombination processes therefore has the additional benefit that efficiency under low-light conditions can more easily be maintained. Considerable emphasis has therefore been placed on balancing carrier mobilities, balancing extraction efficiencies at the electrodes and removing energetic traps.

8.4.6 *Carrier collection efficiency*

While in Section 8.4.4 we considered the maximum yield of dissociated carriers, and in Section 8.4.5 the maximum free energy available per carrier pair, neither uniquely sets the efficiency of an OPV at its maximum power point. This is determined by how EQE varies with internal field — i.e. the fill factor of the cell.

As we have outlined, the losses within any heterojunction solar cell arise from either: (i) Relaxation of excitons which fail to diffuse to, and separate at, a heterojunction; (ii) Recombination of geminate pairs formed at the heterojunction which fail to fully dissociate; or (iii) Recombination of fully dissociated carriers via surface recombination, Langevin-type bulk recombination and SRH-type recombination.

Field-dependent exciton lifetime

In most efficient bulk heterojunction solar cells, the quenching efficiency of excitons on the donor molecule is close to unity, implying that the relaxation of excitons during diffusion is not a strongly limiting process (Jamieson *et al.*, 2012; Morana *et al.*, 2010). Due to their significant binding energy, the intrinsic lifetime of the exciton is only affected by internal fields well in excess of those experienced during normal operation.

Field-dependent geminate recombination

The extent to which the dissociation of geminate pairs is affected by macroscopic electric fields within the photoactive layer is currently the subject of significant uncertainty. The mechanism for this process is assumed to be a variation on the Onsager–Braun model of geminate pair dissociation, such that the Coulomb barrier to dissociation is lowered in the direction of the internal field. Device models that include field-dependent geminate recombination losses are widely employed to analyse the output curves of OPV devices, and have provided support for the importance of such losses (Andersson *et al.*, 2011; Mandoc *et al.*, 2007; Ooi *et al.*, 2008; Peumans and Forrest, 2004; Veldman *et al.*, 2008).

In polymer:polymer bulk heterojunctions, there is good experimental evidence that geminate pair separation efficiency can be enhanced by the presence of a macroscopic electric field (Gonzalez-Rabade *et al.*, 2009).

For heterojunctions employing a fullerene acceptor, the evidence that geminate pair separation efficiency is influenced by macroscopic electric fields is much less clear-cut, and appears to be dependent on both materials selection and processing regime. Direct experimental measurements of field-dependent geminate losses have often observed these to vary under strong reverse bias — important for devices used as photodetectors, but of less significance for the operation of photovoltaic cells (Tvingstedt *et al.*, 2010). Measurements in the power-generating quadrant give mixed results — some authors conclude that the variation of geminate recombination rate with internal electric field is sufficient to significantly impact the fill factor (Albrecht *et al.*, 2012; Credgington *et al.*, 2012), others that the effect is negligible (Etzold *et al.*, 2011; Kniepert *et al.*, 2011; Shuttle *et al.*, 2010; Street *et al.*, 2010). The fundamental mechanism that determines whether a given device suffers from this loss process therefore remains unclear. Nevertheless, the highest efficiency bulk heterojunction OPVs can achieve fill factors of greater than 70% (He *et al.*, 2012), implying that the dependence of geminate recombination rate on internal field can be suppressed, albeit empirically.

Field-dependent non-geminate recombination

Recombination of dissociated carriers will lead to further losses in extraction efficiency. In a device with highly carrier-selective electrodes, such that surface recombination is minimised, recombination of carriers back across the heterojunction will be the dominant loss pathway. While the mechanisms of geminate and non-geminate recombination are physically related, since both proceed via an intermediate CT-like state, they are usually treated as separable due to their differing timescales. Geminate recombination is usually a nanosecond process, and non-geminate recombination lifetimes can extend to microseconds or even milliseconds, meaning that the non-geminate recombination rate is more strongly affected by local carrier density.

The non-geminate recombination rate in an OPV device is determined by several field-dependent processes. Under constant illumination, increasing internal field leads to shorter carrier dwell times and therefore lower equilibrium carrier densities, reducing recombination rate. In addition, since non-geminate recombination is likely to proceed via CT-state formation, high internal fields may act to dissociate some of these CT states and regenerate free carriers — particularly since dissociated carriers may form longer-lived triplet CT excitons (with reference to Fig. 8.17, if k_{triplet} is low, k_{GR} may be suppressed for triplet CT states). However, observing such an effect is difficult since under high field, carrier densities, and so recombination rates, are already low. Increasing internal field also alters the equilibrium carrier distribution, affecting the spatial overlap between electron and hole populations. Depending on the zero-field carrier distribution within a given device, this may either increase or decrease the recombination rate as internal field varies.

The low mobilities and high internal fields common in OPV materials mean that carrier distributions are also dependent on illumination intensity. Therefore the equilibrium carrier distribution at a given applied bias in the dark differs from that present at the same bias under illumination (Shuttle *et al.*, 2010), and so the principle of superposition does not normally hold for OPVs. That is, the current–voltage curve of an illuminated cell cannot usually be separated into dark current and photocurrent components, making the study of non-geminate recombination rates difficult. As such, analyses of recombination rates typically have to rely on time-resolved, rather than steady state, measurements (Albrecht *et al.*, 2012; Credgington and Durrant, 2012; Marsh *et al.*, 2010).

While control of non-geminate recombination rates remains largely empirical, there is now a growing library of experimental techniques capable of quantifying both geminate and non-geminate recombination processes, and their variation across the current–voltage curve. Reductions in recombination rate are correlated with improved carrier collection efficiency and higher fill factors, as might be expected, though it is not yet clear which methods for achieving this will emerge as the most fruitful. As noted above, the fill factor can exceed 70% in the highest efficiency devices, meaning that suppression of recombination losses is clearly possible.

8.5 Beyond the state of the art

In this section, we detail directions in which the OPV field is moving beyond the current state of the art, both in terms of focus, new technology and new physics.

8.5.1 *Stability*

Device lifetime is a critical concern for any commercial photovoltaic technology. With the development of lab-scale OPVs with PCEs exceeding 10%, significant attention has been focussed on the stability of organic devices. To compete directly with existing technology, lifetimes approaching 100,000 hours under illumination are sought. As with device performance, the rate of progress in this area has been rapid, meaning that the state-of-the art described here will quickly become out-of-date. We therefore outline the recent progress in developing stable devices, and refer the reader to the series of reviews by Krebs and co-workers for a more comprehensive outlook (Jørgensen *et al.*, 2012). In particular, we do not consider the interplay between stability and manufacturing process and the evolution of encapsulation methods, which are not unique to organic photovoltaics.

In the following, we divide our discussion by degradation target, beginning with the photoactive layer and extending to the electrodes, including the impact of interfacial layers.

Photoactive layer: Photostability

The function of organic semiconductors is determined by their molecular structure. Chemical changes to that structure are usually detrimental to device performance, so the main concern for the long-term stability in OPVs is the ability of the organic layers to endure prolonged exposure to illumination and elevated temperature without chemical degradation. This is of particular concern in luminescent organic materials, where excited states can persist for relatively long times and photon energies (particularly in the UV) may be comparable to the strengths of intra-molecular bonds. In this respect, experience with organic LEDs (OLEDs) is extremely encouraging. The lifetimes of solution-processed OLED materials can now exceed 1,000,000 hours, suggesting there is no practical limit to the intrinsic stability of organic semiconductors (Dupont, 2009).

Nevertheless, current OPV materials systems fall short of the stability of OLEDs. One of the most important reactions limiting lifetime are the interactions between excited organic molecules and ambient oxygen. The prototypical poly(phenylene vinylene)-related polymer semiconductors (PPVs) were observed to degrade within minutes on exposure to light, arising from attack by oxygen on both the solubilising side chains and the main vinylene backbone units (Chambon *et al.*, 2007).

With the growing library of semiconducting polymers and experience with OLEDs and OPV devices, certain design rules have emerged for the design of stable materials — which has until recently been of secondary concern to efficiency. For example, by estimating the degradation rate of individual monomer units under a given photon flux, Manceau and co-workers developed an empirical ranking of donor and acceptor units in terms of their intrinsic photostability, reproduced in Fig. 8.20 (Manceau *et al.*, 2011). The least stable, fluorene, is widely observed to undergo oxidation to fluorenone regardless of the side-chain configuration (Grisorio *et al.*, 2011). This was particularly detrimental to the blue OLEDs in which it was first used, since fluorenone is a green emitter and acts as a trap for charge transport.

The continuous addition of new monomer units to the available library of molecular design means that optimised stability remains something of a moving target. However, studies such as these suggest that certain units, such as fluorene, should be avoided in the design of stable materials, and that stability may be enhanced by the substitution of silicon in place of carbon where possible. In addition, even for the most stable units, the addition of solubilising side chains provides a general route for photodegradation. For example, significantly longer lifetimes are observed for the PPV-replacing poly(alkyl-thiophenes) (Manceau *et al.*, 2010a), which was one reason for the P3HT:PCBM system being developed as a possible target for scale-up and large-area processing. Nevertheless, photodegradation is still observed, and is linked to attack by oxygen on the solubilising hexyl groups (Manceau *et al.*, 2009).

Since side chains are almost always required to render an organic semiconductor soluble, removing them entirely to improve stability is not possible.

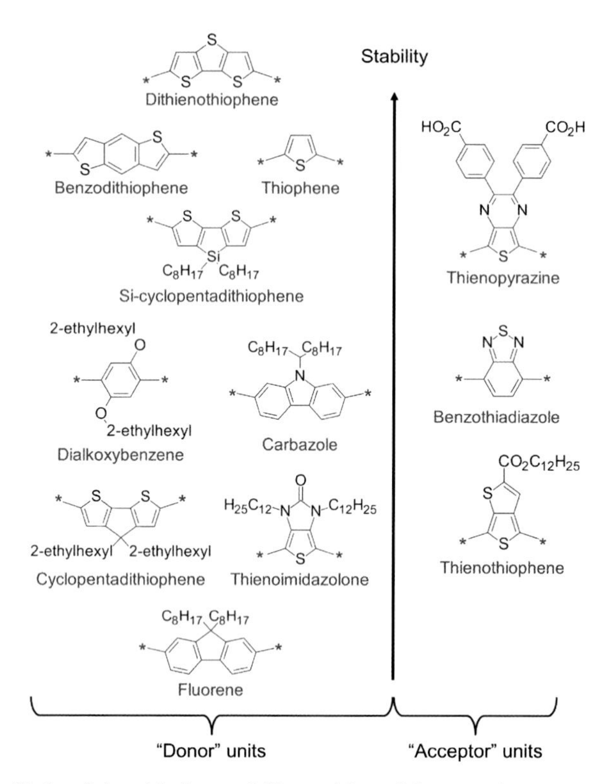

Figure 8.20 'Rule-of-thumb' photostability ranking of donor and acceptor monomers for OPV polymers. Adapted from Manceau *et al.* (2011).

One option to circumvent this problem is to include thermocleavable solubilising groups, which may be removed post-deposition by heating. Such approaches are not yet widespread, but have been shown to improve photostability (Manceau *et al.*, 2010b).

Since the degradation of organic semiconductors is linked to the presence of both an excited molecule and oxygen (degradation rates for material kept in the dark are very slow), the removal of either improves stability significantly. Removing excited states is relatively straightforward — the addition of an electron or hole acceptor rapidly quenches excitons. The photostability of blended heterojunction films is therefore significantly higher than that of pristine materials (Neugebauer *et al.*, 2000), although PCBM itself may also photodegrade (Reese *et al.*, 2010). Removal of oxygen relies on good encapsulation combined with oxygen-scavenging 'getter' materials.

Photoactive layer: Morphological stability

As we outlined in Section 8.4.1, the multiphase structure formed during the controlled deposition of a bulk heterojunction is thermodynamically unstable. Domains comprising equilibrium phases will Ostwald-ripen at the expense of non-equilibrium structures, and both donor and acceptor materials may form crystallites. These can reduce the heterointerface area, reducing carrier generation, and cut percolation pathways for carrier transport. The timescale over which these processes happen will be determined by operating temperature and the size of the molecules in question. In particular, small fullerene acceptors are extremely mobile within amorphous polymer domains at elevated temperature (Treat *et al.*, 2011), meaning that formation of fullerene crystallites is a significant degradation process within polymer:fullerene photovoltaics.

Stabilising morphology has been attempted by a number of routes. The first is to tether donor and acceptor molecules together as block-copolymers, which may be tailored to form a thermodynamically stable interpenetrating network (Segalman *et al.*, 2009). However, these devices have yet to achieve PCEs beyond 2%, meaning that morphological stability in such systems currently comes at too high a cost.

An alternative is to constrain the diffusion of blend components post-deposition by means of cross-linking between polymer chains using the solubilising side groups. This may be achieved using a photo- or thermally curable cross-linking agent as a tertiary component in the blend (Liu *et al.*, 2012; Png *et al.*, 2010), although this must be carefully targeted to avoid damaging the semi-conducting backbone. It has also been possible to chemically engineer polymers with side groups that are inherently cross-linkable (Charas *et al.*, 2006). Neither technique has yet been explored fully in the highest efficiency polymer:fullerene systems, but these approaches can in principle be applied universally. However, cross-linking the polymer may not entirely prevent the much smaller fullerene from diffusing through the blend and aggregating. Employing cross-linkable fullerenes is one route to address this problem (Drees *et al.*, 2005). Another is to remove the side groups altogether, since these are responsible for a significant proportion of the blend's free volume, and thus facilitate both polymer reptation and fullerene diffusion.

Thermo-cleavable side groups, as outlined above, can therefore also aid morphological stability, as well as photostability. P3HT-derivatives incorporating thermo-cleavable carboxylic esters, blended with PCBM, exhibit significantly enhanced stability after cleaving above 200°C (Krebs and Spanggaard, 2005). However, while P3HT:PCBM typically requires an annealing step at around 150°C to achieve the good efficiency, annealing has been shown to be detrimental for many better performing systems. It therefore remains to be seen whether thermo-cleaving can be achieved at temperatures low enough to avoid this process.

An advantage of all of the approaches outlined above is that they also render the blend insoluble, allowing for multiple thin-film stacks to be deposited without the need for orthogonal solvents. This ability is crucial for developing multijunction cells, as described in Section 8.5.2.

Electron-collecting electrode

The photoactive layer is not the only device component susceptible to degradation. Historically, the electron-collecting electrode has been formed of an evaporated metal. The requirement for this metal to have a low work function (see Section 8.4.5) means that highly reactive materials such as calcium and aluminium are needed. Such metals react in the presence of water or oxygen, creating an insulating metal oxide layer which grows from pinholes and the edges of the electrode, eventually spreading over the entire device. As the oxide increases in thickness, the transport barrier created eventually impedes the extraction of electrons from the entire active layer (Glatthaar *et al.*, 2007).

One method to circumvent this problem is to use a less reactive, high-work-function metal such as Ag or Au. These must be combined with a second material, such as a conducting metal oxide, to achieve a low work function at the organic:electrode interface. An alternative is to use an inverted structure, such that Ag or Au, sometimes combined with a high-work-function interlayer, acts as the hole-collecting electrode while a stable conductive oxide such as TiO_2 is used to extract electrons. The use of oxides as low-work-function electrodes has driven interest in inverted device structures, since they are intrinsically stable to oxygen degradation. In addition, while somewhat less conductive than TiO_2, solution-processed titanium sub-oxide (TiO_x) has been shown to act as an oxygen 'getter' which also improves the photostability of the active layer (Li *et al.*, 2011).

Hole-collecting electrode

While interfaces between ITO and organic materials have been shown to impact stability, its relatively low work function means that it is unusual to use ITO without a high-work-function hole-transporting layer — PEDOT:PSS (Fig. 8.6) being by far the most widely used. However, PEDOT:PSS has been implicated in device degradation through several mechanisms.

Since the conductivity of PEDOT:PSS relies on *p*-doping of the PEDOT chain by the charged PSS component, phase separation, structural degradation and electrochemical reactions that interfere with this charge exchange reduce the conductivity of the PEDOT:PSS layer over time (Vitoratos *et al.*, 2009). Commercial PEDOT:PSS formulations can be highly acidic, such that ambient humidity can significantly increase the rate of corrosion of other material components — particularly low-work-function metals, but also ITO (de Jong *et al.*, 2000). PEDOT:PSS is also highly hygroscopic, enhancing this effect even when minimal atmospheric water is present.

As a result, there have been significant efforts to try and replace PEDOT:PSS with a stable, transparent, high-work-function hole conductor. Examples of possible replacements include vacuum-deposited NiO (Irwin *et al.*, 2008), and solution- or vacuum-deposited MoO_3 (Zilberberg *et al.*, 2012). Both have been shown to significantly enhance the stability of devices, particularly under humid conditions.

As in the previous section, an alternative approach is to switch to an inverted structure such that a transparent electron-collecting electrode is used, and holes are collected through a stable, high-work-function metal (Ag or Au, potentially with an additional interlayer).

A huge number of other electron and hole-transporting interlayers have been examined in the OPV and OLED fields, though many have not been assessed for stability. For details, we refer the reader to the existing literature (Steim *et al.*, 2010).

8.5.2 *Multijunction organic photovoltaics*

As discussed in Section 8.4.5, thermalisation losses and narrow spectral response both limit the performance of OPVs. For any solar cell technology, a common route to reducing thermalisation losses and broadening spectral response is to absorb high-energy photons in a wide-bandgap semiconductor before absorbing low-energy photons in a low-bandgap semiconductor — i.e. to construct a tandem cell. A detailed analysis of tandem cells is beyond the scope of this chapter; instead we outline current progress towards realising such cells using organic materials.

The basic stacked tandem structure is shown in Fig. 8.21, and requires precise control over each layer of the stack. Evaporated small-molecule devices are therefore particularly well suited for combining into the tandem architecture, and the highest-efficiency organic solar cells to date are produced in this way. In particular, controlled doping of individual layers using molecular dopants is possible (Lüssem *et al.*, 2013), and the optical field structure can be precisely controlled by introducing transparent spacer layers to ensure efficient absorption by the photoactive components (Riede *et al.*, 2011). This flexibility also means that there is no technological barrier to producing evaporated multijunction organic cells (Sullivan *et al.*, 2013).

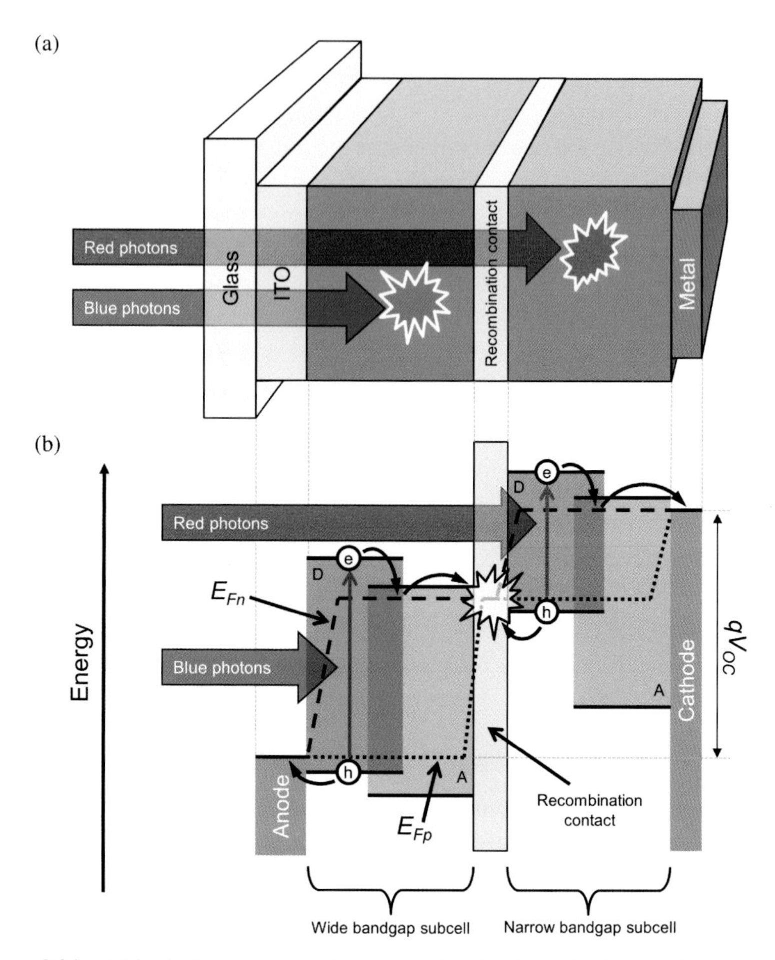

Figure 8.21 a) Physical structure; and b) energetic structure of a tandem heterojunction solar cell. High-energy (blue) photons are absorbed by the donor (D) of the front, wide-bandgap sub-cell while low-energy (red) photons pass through to the narrow-bandgap donor of the second sub-cell. Optimum performance requires lossless recombination (i.e. matched electron/hole Fermi energies ($E_{Fn,p}$)) and matched electron/hole currents at the common recombination contact. In this example, identical acceptors (A) are assumed.

Achieving solution-deposited tandem cells is complicated by the need to prevent subsequent layers dissolving those beneath them. This can be achieved by rendering each layer insoluble before deposition of the next (through cross-linking or similar), by choosing orthogonal solvents for the deposition of each layer, or by depositing one or more layers separately, then combining the resulting films using soft lithography.

In either approach, of critical importance is the choice of material for the recombination contact, which must allow holes from the top sub-cell to recombine with electrons from the bottom sub-cell, or vice versa. In early attempts at tandem cells, a metallic interlayer or metal nanoparticle layer was used (Dennler *et al.*, 2006); metals provide an optimum recombination contact since electron and hole quasi-Fermi energies are identical, meaning that no free energy is lost during recombination.

However, their broad parasitic absorption is a significant disadvantage, as is the need to interrupt solution processing to deposit thermally evaporated metals. More recently, evaporated devices have employed a heavily doped organic *p-n* junction to fulfil this role (Timmreck *et al.*, 2010), whereas solution-processed devices have utilised transparent metal oxides such as TiO_x and ZnO in combination with PEDOT:PSS as the recombination layer (Li *et al.*, 2013); both routes exhibit negligible free-energy losses, and also provide interconnects which are transparent in the visible range. In a solution-processed inverted structure, the oxide layer can also act to 'reset' the solubility of the stack, allowing repeat use of solvents for each sub-cell.

The complexity of tandem fabrication means that research-scale structures remain relatively rare, and developing tandems with efficiencies approaching the sum of their sub-cells is a significant challenge. However, there are now sufficient examples of success to confirm that such an approach can bear fruit. As a result, the discovery of new low-bandgap tandem-ready materials has proceeded at pace (Bronstein *et al.*, 2011; Zhong *et al.*, 2013), with the understanding that methods of tandem cell fabrication can be developed in parallel, and await a mature materials set.

8.5.3 *Hybrid photovoltaics*

One of the most significant functional problems for organic solar cells is their low dielectric constant. This enhances geminate recombination in almost all materials systems except those incorporating fullerene-based electron acceptors. However, as we have already noted, fullerenes are poorly absorbing and can be expensive to produce. One promising approach is therefore to replace fullerenes with a high-dielectric constant inorganic acceptor, creating a 'hybrid' solar cell, which combines the advantages of a highly absorbing organic donor with a stable and highly conductive acceptor. This approach is closely related to that taken by dye-sensitised solar cells (Hagfeldt *et al.*, 2010).

The commonly used acceptors are TiO_2 and ZnO, which may be deposited from a precursor or grown *in situ*. Since an organic donor is still required, a BHJ architecture must be used to ensure efficient exciton harvesting. It is not usually possible to form a hybrid heterojunction by *in situ* self-assembly, as it is for organic systems, though routes do exist (Dowland *et al.*, 2011). The inorganic material is therefore either pre-formed into particles which are deposited within an organic matrix (Beek *et al.*, 2004), or deposited first to form a mesoporous or nanostructured scaffold around which a soluble organic is deposited (Peiro *et al.*, 2006). The latter process renders hybrid devices very well suited for top-down heterojunction patterning.

While stable and relatively easy to produce, oxide semiconductors absorb primarily in the UV and so do not improve the spectral response of the solar cell. An alternative is to use semiconducting nanocrystals, the absorption of which can be tuned by controlling both the nanocrystal composition and size. In combination with polymers, CdSe is the most commonly studied nanocrystal-forming material owing to its controllable and relatively straightforward synthesis. The first polymer:nanocrystal hybrid solar cell, investigated in 1996, used CdSe as the acceptor, although the maximum EQE achieved was only 12% (Greenham *et al.*, 1996). There are two main difficulties when blending isotropic nanocrystals (also referred to as 'quantum dots') and polymers. The first is that as-synthesised nanocrystals are 'capped' with long, solubilising ligands, which insulate the particle electrically. These need to be replaced with shorter, more conducting groups (or removed entirely) to allow charge to tunnel between particles. The second is that a balance must be struck between nanoparticle aggregation, to create pathways for charge transport, and dispersion, to maintain a high heterojunction area. Thus the ability to control particle shape offers an additional route to tune device performance. Using anisotropic rod- and tetrapod-shaped nanocrystals can therefore allow simultaneous formation of a charge-transporting network while achieving efficient exciton quenching (Huynh *et al.*, 2002; Sun *et al.*, 2003).

In addition to CdSe, silicon nanocrystals have attracted considerable attention in hybrid devices (Liu *et al.*, 2008), and efforts to absorb more infrared light have spurred interest in low-bandgap materials combinations such as CdTe and PbS (Kumar and Nann, 2004; Watt *et al.*, 2005).

8.5.4 *Singlet fission*

One option for reducing thermalisation losses when absorbing high-energy photons is to combine wide and narrow bandgap absorbers in a tandem (or multijunction) solar cell, as described in Section 8.5.2. However, this necessarily increases fabrication complexity. An emerging strategy is to employ an organic donor material that undergoes spontaneous singlet fission — whereby a singlet exciton donates some of its energy to a neighbouring chromophore, converting both to triplets, and where it does so more rapidly than the singlet quenches at a heterojunction or emits. This strategy is particularly appropriate for organic semiconductors, since their tightly confined excitons necessarily lead to an increase in exchange energies between singlets and triplets. If both triplets can be ionised, quantum efficiencies of up to 200% are possible, and in combination with a low-bandgap absorber, theoretical efficiencies can reach ~50%, exceeding the Shockley–Queisser single-junction limit (Smith and Michl, 2010).

Singlet fission was first discovered in the 1960s in polyacenes, primarily via indirect observations, and considered a scientific curiosity (Singh *et al.*, 1965). More recently, singlet fission has been thought of as a means to enhance solar cell performance, and modern ultrafast spectroscopy has allowed direct observation of singlet fission and its kinetics. In the five-ringed polyacene pentacene, singlet fission occurs with near unity probability, since it proceeds on timescales of around 80 fs (Wilson *et al.*, 2011), out-competing all other decay channels including exciton thermalisation and charge transfer.

Harvesting these triplet excitons requires a heterojunction, since they are still tightly bound, with the associated heterojunction bandgap chosen such that $E_{\text{triplet}} > E_{\text{CT}}$. The large optical bandgap of pentacene (around 1.8 eV) requires that it be paired with a low-bandgap absorber, typically acting as acceptor. Tuneable low-bandgap quantum-dot acceptors such as PbS and PbSe are particularly suited in this role, allowing absorption into the infrared and leading to efficient hybrid devices (Ehrler *et al.*, 2012b; Ehrler *et al.*, 2012c). The implementation of such a hybrid singlet-fission device is illustrated in Fig. 8.22, which bears similarities with a tandem device. By analogy with the requirement for current matching between sub-cells in a tandem device, here we require voltage matching between the triplet energy level of the singlet fission material with the CT energy of the heterojunction.

While photons absorbed by pentacene do contribute to the photocurrent of such devices, internal quantum efficiencies currently remain below those of pentacene:fullerene devices and well below the maximum possible efficiencies, emphasising that this approach is in its relative infancy.

Singlet fission also suggests a way to increase the performance of existing low-bandgap solar cells, such as those made from silicon, as it allows for better utilisation of high-energy photons. The first efforts towards dissociating triplets generated via singlet fission in this way have recently been achieved using amorphous silicon (Ehrler *et al.*, 2012a).

8.5.5 *Multi-component systems*

The complexity of binary blends has been sufficient to drive OPV research for the last two decades, and may yet result in the optimum materials system. However,

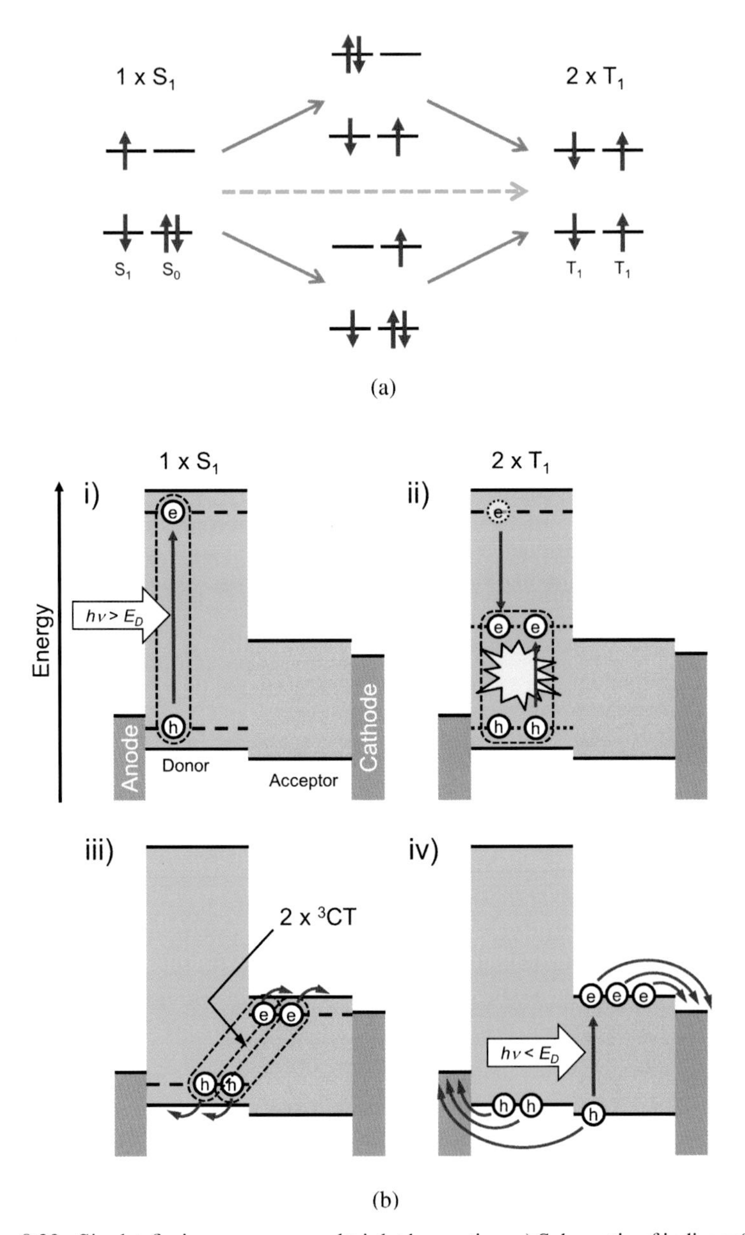

Figure 8.22 Singlet-fission processes and triplet harvesting. a) Schematic of indirect (blue) and direct (green) pathways for the fission of one singlet into two triplets, which are arranged on neighbouring chromophores in a pure singlet configuration. Adapted from Smith and Michl (2010). b) Photogeneration in a singlet fission device: i) singlet exciton formation

Figure 8.22 (*continued from caption on previous page*) in the donor following absorption of a Singlet-fission processes and triplet harvesting. a) Schematic of indirect (blue) and direct (green) pathways for the fission of one singlet into two triplets, which are arranged on neighbouring chromophores in a pure singlet configuration. Adapted from Smith and Michl (2010). b) Photogeneration in a singlet fission device: i) singlet exciton formation in the donor following absorption of a photon with energy greater than the donor bandgap E_D; ii) singlet fission resulting in two coupled triplet excitons; iii) diffusion of triplet excitons to the heterojunction forming two triplet CT states, which subsequently dissociate; iv) collection of dissociated charges arising from the dissociated triplets and absorption of low-energy photons in the acceptor. Free-carrier generation is typically spontaneous in inorganic acceptors. Solid lines represent single-carrier molecular energy levels. Dotted lines represent excitonic 'virtual' energy levels where the exciton binding energy has been schematically included. Dashed lines indicate coulombically bound carriers. Interlayers and competing processes have been ignored for clarity.

there is in principle no additional processing complexity involved in preparing an ink with three or more functional components. Early work on small-molecule additives was initially directed towards incorporating these materials into the heterointerface, though they are now usually seen primarily as co-solvents. Nevertheless, such a process is still viable and can produce ternary systems.

Incorporation of block copolymers into a two-component blend has also been attempted as a means of interface control. Spontaneous formation of polymer-rich 'skin' layers is commonly observed during film deposition, suggesting that modification of the electrode interfaces may be possible by incorporating materials that preferentially wet the electrode surface. Even more simply, combining components of different regioregularity or molecular weight offers a route to controlling the relative crystallinity within a blend (Campoy-Quiles *et al.*, 2009). More so than any other sub-field of OPV research, the parameter space for multi-component systems remains largely unexplored.

8.5.6 *Scale-up*

The increasing activity of industrial research groups in the OPV field is paralleled by increasing focus on applying scalable thin-film coating techniques to their manufacture, and developing design rules for taking lab-scale devices to large scale production. While much of this activity occurs within the commercial sector, meaning that limited information is publicly available, scaling up OPV technology requires several key transitions. We briefly outline these here, and refer the reader to the growing literature on OPV scale-up for further details (Krebs, 2009; Krebs *et al.*, 2009).

Scalable deposition

The first such transition is to move away from small-area serial production techniques such as spin-casting. While suitable for lab-scale devices, this requires relatively high solution wastage. In addition, variations in device properties result from changes in evaporation rate, wet-film thickness, shear rates and nucleation surfaces for crystallisation, which in turn depend on the spin rate, wetting behaviour and solution viscosity. Spin-casting can be very reproducible within a given laboratory environment for a given material batch, but it is very difficult to know the exact parameters of a deposition such that it can be reproduced in a different location.

Large-area, roll-to-roll (R2R) compatible techniques are therefore preferred for scalable and reproducible manufacture, as illustrated in Fig. 8.23. Numerous processes have been explored, including screen-printing, doctor blading, wire-bar coating, inkjet printing, knife-over-edge coating, slot-die coating, gravure printing and spray coating (Krebs, 2009). Roll-to-roll is particularly appealing for solution-based techniques since in-line assays can be used to correct for variations in film formation caused by changing ambient conditions and solution composition.

In terms of reproducibility, it may be that the advantage lies with all-vacuum R2R processing of thermally evaporated small molecules, as has been developed by Heliatek (2013). While thermal evaporation under vacuum is a more technically intensive route than solution processing, provided that all processing steps can be

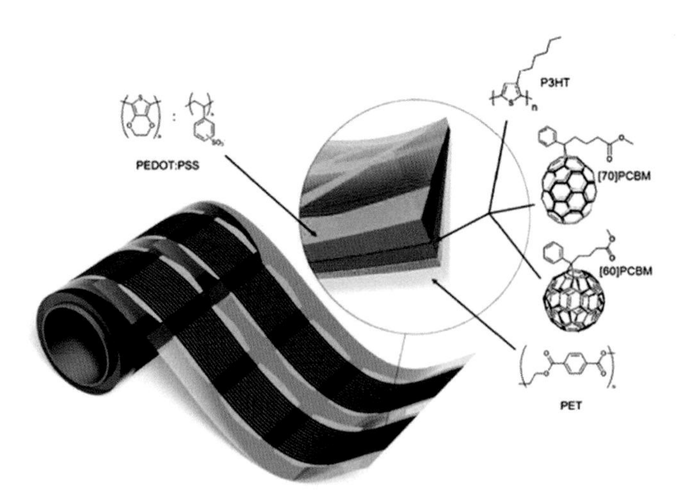

Figure 8.23 A roll of printed OPV modules produced in a continuous R2R process. The inset shows the layer stack, comprising PET-ITO-ZnO-P3HT:PCBM-PEDOT:PSS-printed silver paint and encapsulated using R2R lamination post-production. Reproduced from Krebs *et al.* (2010).

achieved without breaking vacuum there is little fundamental difference between the two methods. Solution processing allows for more complex thermodynamic control of active layer deposition, and a wider range of available materials, whereas thermal evaporation allows for tighter control of multilayers and higher material purity. Furthermore, lessons learned from lab-scale evaporated devices may be directly translated to R2R, whereas the need to limit the use of environmentally damaging heavy metal catalysts and halogenated solvents may lead to a more difficult translation from lab-scale solution processing.

In addition to re-optimising device production for new deposition processes, a transition to flexible substrates with flexible electrodes and encapsulation is required. Current lab-scale devices tend to employ glass-on-glass encapsulation, which offers exceptionally low permeability to oxygen and water. Flexible barrier layers of similar quality and low enough cost remain elusive, and are seen as a significant challenge for achieving long module (rather than just material) lifetimes.

Large-area devices

The primary limit on efficient scale-up of OPV module size is series resistance. Since organic semiconductors are low-mobility materials, it is not currently feasible to employ laterally spaced high-conductivity electrodes, which are used in crystalline solar cells to collect carriers while allowing the passage of light. It is thus necessary to employ at least one transparent electrode with complete coverage of the photoactive layer. The only material yet found to combine high optical transmission ($>80\%$) with acceptable sheet resistance (<15ohm/square) and a scalable deposition route is ITO.

The relatively high sheet resistance of ITO requires that modules are formed of multiple strips of photoactive material connected in series. Roll-to-roll printing techniques can reliably ensure electrode spacing of 1–2 mm between strips, and strip widths of $\sim$15 mm are empirically found to optimise between the fill factor and light-harvesting efficiency. For this reason, lab-scale devices with area >1 cm $\times$ 1 cm are considered to give a reasonable indication of the achievable performance of a larger-area module. This architecture has disadvantages, however, since the intercell gap represents dead module area and shading of a single strip can lead to a significant drop in module performance unless additional bypass diodes are used.

Because solution-processed ITO of sufficient quality is not currently available, ITO is typically pre-sputtered onto PET film and the composite used as the web for R2R printing. Such an approach has significant drawbacks since indium is expensive and sputtering is a relatively costly deposition process; ITO/PET is currently estimated to comprise half of the cost of large-area OPV manufacture.

By comparison, the polymers and fullerene photoactive components account for only around 10% of module cost, despite the lack of a sufficiently mature materials set to drive a reduction in their manufacturing costs (Nielsen *et al.*, 2010).

As a result, there is a significant drive to find cheap ITO replacements for transparent electrodes. Promising candidates include conducting polymers similar to PEDOT:PSS, graphene, carbon nanotubes, composites of highly conductive metallic nanowire networks dispersed in a lower-conductivity matrix and alternative transparent conducting oxides.

8.6 Conclusions

In this chapter we have traced the progress of organic photovoltaics from their early development, through the various routes found to increase their performance, to the state-of-the-art. We have set out the current scientific understanding and engineering progress in the use of organic semiconductors in thin-film solar cells, and highlighted both where there is broad consensus on the electronic processes involved in light absorption, charge generation, transport and collection, and where uncertainty still remains. This growing understanding and continued technological achievement indicate that excellent performance is achievable (towards that of current crystalline silicon technology). This will require continued engineering development to achieve long-lasting, highly efficient, mass-produced devices, but recent rapid advances give cause for great optimism. The most promising outlook comes from the sheer size of the space still to explore. Crystalline silicon has been a stable materials platform for four decades, yet the main driver of increased OPV performance has been the continual discovery of new organic semiconductors. There is no indication yet that the ideal OPV materials system has been found, and until it has, the process of true engineering optimisation remains ahead. Twenty years from their first discovery, there is every reason to believe that organic photovoltaics have a bright future.

Acknowledgement

Compiling this chapter would not have been possible without many fruitful discussions between myself and Professor Sir Richard Friend, who has provided invaluable insights from his long experience working in the organic electronics field, and whom I acknowledge unreservedly.

References

Aernouts T., Vanlaeke P., Geens W., Poortmans J., Heremans P., Borghs S., Mertens R., Andriessen R. and Leenders L. (2004), 'Printable anodes for flexible organic solar cell modules', *Thin Solid Films* **451**, 22–25.

Albrecht S., Schindler W., Kurpiers J., Kniepert J., Blakesley J. C., Dumsch I., Allard S., Fostiropoulos K., Scherf U. and Neher D. (2012), 'On the field dependence of free charge carrier generation and recombination in blends of pcpdtbt/PC70BM: influence of solvent additives', *J. Phys. Chem. Lett.* **3**, 640–645.

Andersson L. M., Müller C., Badada B. H., Zhang F., Würfel U. and Inganäs O. (2011), 'Mobility and fill factor correlation in geminate recombination limited solar cells', *J. Appl. Phys.* **110**, 024509.

Bakulin A. A., Rao A., Pavelyev V. G., van Loosdrecht P. H., Pshenichnikov M. S., Niedzialek D., Cornil J., Beljonne D. and Friend R. H. (2012), 'The role of driving energy and delocalized states for charge separation in organic semiconductors', *Science* **335**, 1340–1344.

Ballantyne A. M., Chen L. C., Nelson J., Bradley D. D. C., Astuti Y., Maurano A., Shuttle C. G., Durrant J. R., Heeney M., Duffy W. and McCulloch I. (2007), 'Studies of highly regioregular poly(3-hexylselenophene) for photovoltaic applications', *Adv. Mater.* **19**, 4544–4547.

Bässler H. (1993), 'Charge transport in disordered organic photoconductors a Monte Carlo simulation study', *Phys. Status Solidi B* **175**, 15–56.

Bässler H. and Kohler A. (2012), 'Charge transport in organic semiconductors', *Top. Curr. Chem.* **312**, 1–65.

Beek W. J. E., Wienk M. M. and Janssen R. A. J. (2004), 'Efficient hybrid solar cells from zinc oxide nanoparticles and a conjugated polymer', *Adv. Mater.* **16**, 1009–1013.

Blakesley J. C. and Neher D. (2011), 'Relationship between energetic disorder and open-circuit voltage in bulk heterojunction organic solar cells', *Phys. Rev. B* **84**, 075210.

Brabec C. J., Shaheen S. E., Winder C., Sariciftci N. S. and Denk P. (2002), 'Effect of LiF/metal electrodes on the performance of plastic solar cells', *App. Phys. Lett.* **80**, 1288–1290.

Bradley D. D. C. (1987), 'Precursor-route poly(*p*-phenylenevinylene): Polymer characterisation and control of electronic properties', *J. Phys. D: Appl. Phys.* **20**, 1389–1410.

Braun C. L. (1984), 'Electric field assisted dissociation of charge transfer states as a mechanism of photocarrier production', *J. Chem. Phys.* **80**, 4157–4161.

Brédas J.-L., Norton J. E., Cornil J. and Coropceanu V. (2009), 'Molecular understanding of organic solar cells: the challenges', *Acc. Chem. Res.* **42**, 1691–1699.

Bronstein H., Chen Z., Ashraf R. S., Zhang W., Du J., Durrant J. R., Shakya Tuladhar P., Song K., Watkins S. E., Geerts Y., Wienk M. M., Janssen R. A. J., Anthopoulos T., Sirringhaus H., Heeney M. and McCulloch I. (2011), 'Thieno[3,2-b]thiophene–diketopyrrolopyrrole-containing polymers for high-performance organic field-effect transistors and organic photovoltaic devices', *J. Amer. Chem. Soc.* **133**, 3272–3275.

Cacialli F. and Bruschi P. (1996), 'Site-selective chemical-vapor-deposition of submicron-wide conducting polypyrrole films: morphological investigations with the scanning electron and the atomic force microscope', *J. Appl. Phys.* **80**, 70–75.

Campoy-Quiles M., Ferenczi T., Agostinelli T., Etchegoin P. G., Kim Y., Anthopoulos T. D., Stavrinou P. N., Bradley D. D. C. and Nelson J. (2008), 'Morphology evolution via self-organization and lateral and vertical diffusion in polymer:fullerene solar cell blends', *Nat. Mater.* **7**, 158–164.

Campoy-Quiles M., Kanai Y., El-Basaty A., Sakai H. and Murata H. (2009), 'Ternary mixing: a simple method to tailor the morphology of organic solar cells', *Org. Electronics* **10**, 1120–1132.

Carsten B., Szarko J. M., Son H. J., Wang W., Lu L., He F., Rolczynski B. S., Lou S. J., Chen L. X. and Yu L. (2011), 'Examining the effect of the dipole moment on charge separation in donor–acceptor polymers for organic photovoltaic applications', *J. Amer. Chem. Soc.* **133**, 20468–20475.

Chambon S., Rivaton A., Gardette J.-L., Firon M. and Lutsen L. (2007), 'Aging of a donor conjugated polymer: photochemical studies of the degradation of poly[2-methoxy-5-(3′, 7′-dimethyloctyloxy)-1,4-phenylenevinylene]', *J. Polymer Sci. Part A: Polymer Chem.* **45**, 317–331.

Chappell J., Lidzey D. G., Jukes P. C., Higgins A. M., Thompson R. L., O'Connor S., Grizzi I., Fletcher R., O'Brien J., Geoghegan M. and Jones R. A. L. (2003), 'Correlating structure with fluorescence emission in phase-separated conjugated-polymer blends', *Nat. Mater.* **2**, 616–621.

Charas A., Alves H., Alcacer L. and Morgado J. (2006), 'Use of cross-linkable polyfluorene in the fabrication of multilayer polyfluorene-based light-emitting diodes with improved efficiency', *Appl. Phys. Lett.* **89**, 143519.

Chirvase D., Parisi J., Hummelen J. C. and Dyakonov V. (2004), 'Influence of nanomorphology on the photovoltaic action of polymer-fullerene composites', *Nanotechnology* **15**, 1317.

Chou S. Y., Krauss P. R. and Renstrom P. J. (1996), 'Imprint lithography with 25-nanometer resolution', *Science* **272**, 85–87.

Clark J., Chang J. F., Spano F. C., Friend R. H. and Silva C. (2009), 'Determining exciton bandwidth and film microstructure in polythiophene films using linear absorption spectroscopy', *Appl. Phys. Lett.* **94**, 163306.

Clarke T. M. and Durrant J. R. (2010), 'Charge photogeneration in organic solar cells', *Chem. Rev.* **110**, 6736–6767.

Credgington D. and Durrant J. R. (2012), 'Insights from transient optoelectronic analyses on the open-circuit voltage of organic solar cells', *J. Phys. Chem. Lett.* **3**, 1465–1478.

Credgington D., Fenwick O., Charas A., Morgado J., Suhling K. and Cacialli F. (2010), 'High-resolution scanning near-field optical lithography of conjugated polymers', *Adv. Funct. Mater.* **20**, 2842–2847.

Credgington D., Jamieson F. C., Walker B., Nguyen T.-Q. and Durrant J. R. (2012), 'Quantification of geminate and non-geminate recombination losses within a solution-processed small-molecule bulk heterojunction solar cell', *Adv. Mater.* **24**, 2135–2141.

Crispin A., Crispin X., Fahlman M., Berggren M. and Salaneck W. R. (2006), 'Transition between energy level alignment regimes at a low band gap polymer-electrode interfaces', *Appl. Phys. Lett.* **89**, 213503.

de Jong M. P., van Ijzendoorn L. J. and de Voigt M. J. A. (2000), 'Stability of the interface between indium-tin-oxide and poly(3,4-ethylenedioxythiophene)/poly (styrenesulfonate) in polymer light-emitting diodes', *Appl. Phys. Lett.* **77**, 2255–2257.

Deibel C. and Dyakonov V. (2010), 'Polymer:fullerene bulk heterojunction solar cells', *Rep. Progr. Phys.* **73**, 096401.

Delacote G. M., Fillard J. P. and Marco F. J. (1964), 'Electron injection in thin films of copper phtalocyanine', *Solid State Commun.* **2**, 373–376.

Dennler G., Prall H.-J., Koeppe R., Egginger M., Autengruber R. and Sariciftci N. S. (2006), 'Enhanced spectral coverage in tandem organic solar cells', *Appl. Phys. Lett.* **89**, 073502.

Dennler G., Scharber M. C. and Brabec C. J. (2009), 'Polymer-fullerene bulk-heterojunction solar cells', *Adv. Mater.* **21**, 1323–1338.

Dexter D. L. (1953), 'A theory of sensitized luminescence in solids', *J. Chem. Phys.* **21**, 836–850.

Dick D., Wei X., Jeglinski S., Benner R. E., Vardeny Z. V., Moses D., Srdanov V. I. and Wudl F. (1994), 'Transient spectroscopy of excitons and polarons in C_{60} films from femtoseconds to milliseconds', *Phys. Rev. Lett.* **73**, 2760–2763.

Dimitrov S. D., Nielsen C. B., Shoaee S., Shakya Tuladhar P., Du J., McCulloch I. and Durrant J. R. (2012), 'Efficient charge photogeneration by the dissociation of PC70BM excitons in polymer/fullerene solar cells', *J. Phys. Chem. Lett.* **3**, 140–144.

Dong B., Lu N., Zelsmann M., Kehagias N., Fuchs H., Sotomayor Torres C. M. and Chi L. F. (2006), 'Fabrication of high-density, large-area conducting-polymer nanostructures', *Adv. Funct. Mater.* **16**, 1937–1942.

Dowland S., Lutz T., Ward A., King S. P., Sudlow A., Hill M. S., Molloy K. C. and Haque S. A. (2011), 'Direct growth of metal sulfide nanoparticle networks in solid-state polymer films for hybrid inorganic–organic solar cells', *Adv. Mater.* **23**, 2739–2744.

Drees M., Hoppe H., Winder C., Neugebauer H., Sariciftci N. S., Schwinger W., Schaffler F., Topf C., Scharber M. C., Zhu Z. and Gaudiana R. (2005), 'Stabilization of the nanomorphology of polymer-fullerene "bulk heterojunction" blends using a novel polymerizable fullerene derivative', *J. Mater. Chem.* **15**, 5158–5163.

Dunlap D. H., Parris P. E. and Kenkre V. M. (1996), 'Charge-dipole model for the universal field dependence of mobilities in molecularly doped polymers', *Phys. Rev. Lett.* **77**, 542–545.

Dupont (2009), 'DuPont displays surpasses million hour milestone for lifetime of new OLED material', 27 May 2009, www2.dupont.com.

Ehrler B., Musselman K. P., Bohm M. L., Friend R. H. and Greenham N. C. (2012a), 'Hybrid pentacene/a-silicon solar cells utilizing multiple carrier generation via singlet exciton fission', *Appl. Phys. Lett.* **101**, 153507.

Ehrler B., Walker B. J., Böhm M. L., Wilson M. W. B., Vaynzof Y., Friend R. H. and Greenham N. C. (2012b), '*In situ* measurement of exciton energy in hybrid singlet-fission solar cells', *Nature Commun.* **3**, 1019.

Ehrler B., Wilson M. W., Rao A., Friend R. H. and Greenham N. C. (2012c), 'Singlet exciton fission-sensitized infrared quantum dot solar cells', *Nano Letters* **12**, 1053–1057.

Etzold F., Howard I. A., Mauer R., Meister M., Kim T.-D., Lee K.-S., Baek N. S. and Laquai F. (2011), 'Ultrafast exciton dissociation followed by nongeminate charge recombination in PCDTBT:PCBM photovoltaic blends', *J. Amer. Chem. Soc.* **133**, 9469–9479.

Faist M. A., Kirchartz T., Gong W., Ashraf R. S., McCulloch I., de Mello J. C., Ekins-Daukes N. J., Bradley D. D. C. and Nelson J. (2012), 'Competition between the charge transfer state and the singlet states of donor or acceptor limiting the efficiency in polymer:fullerene solar cells', *J. Amer. Chem. Soc.* **134**, 685–692.

Fenwick O., Bozec L., Credgington D., Hammiche A., Lazzerini G. M., Silberberg Y. R. and Cacialli F. (2009), 'Thermochemical nanopatterning of organic semiconductors', *Nature Nanotechnology* **4**, 664–668.

Flory P. J. (1941), 'Thermodynamics of high polymer solutions', *J. Chem. Phys.* **9**, 660–661.

Förster T. (1948), 'Zwischenmolekulare energiewanderung und fluoreszenz', *Annalen der Physik* **437**, 55–75.

Frenkel J. (1931), 'On the transformation of light into heat in solids. I', *Phys. Rev.* **37**, 17–44.

Gélinas S., Paré-Labrosse O., Brosseau C.-N., Albert-Seifried S., McNeill C. R., Kirov K. R., Howard I. A., Leonelli R., Friend R. H. and Silva C. (2011), 'The binding energy of charge-transfer excitons localized at polymeric semiconductor heterojunctions', *J. Phys. Chem. C* **115**, 7114–7119.

Giebink N. C., Wiederrecht G. P., Wasielewski M. R. and Forrest S. R. (2011), 'Thermodynamic efficiency limit of excitonic solar cells', *Phys. Rev. B* **83**, 195326.

Glatthaar M., Riede M., Keegan N., Sylvester-Hvid K., Zimmermann B., Niggemann M., Hinsch A. and Gombert A. (2007), 'Efficiency limiting factors of organic bulk heterojunction solar cells identified by electrical impedance spectroscopy', *Solar Energy Mater. Solar Cells* **91**, 390–393.

Gonzalez-Rabade A., Morteani A. C. and Friend R. H. (2009), 'Correlation of heterojunction luminescence quenching and photocurrent in polymer-blend photovoltaic diodes', *Adv. Mater.* **21**, 3924–3927.

Grancini G., Maiuri M., Fazzi D., Petrozza A., Egelhaaf H. J., Brida D., Cerullo G. and Lanzani G. (2013), 'Hot exciton dissociation in polymer solar cells', *Nat. Mater.* **12**, 29–33.

Granström M. (1998), 'Micropatterned luminescent polymer films', *Acta Polymerica* **49**, 514–517.

Green M. A., Emery K., Hishikawa Y., Warta W. and Dunlop E. D. (2013), 'Solar cell efficiency tables (version 41)', *Progr. Photovoltaics* **21**, 1–11.

Greenham N. C., Peng X. and Alivisatos A. P. (1996), 'Charge separation and transport in conjugated-polymer/semiconductor-nanocrystal composites studied by photoluminescence quenching and photoconductivity', *Phys. Rev. B* **54**, 17628–17637.

Grisorio R., Allegretta G., Mastrorilli P. and Suranna G. P. (2011), 'On the degradation process involving polyfluorenes and the factors governing their spectral stability', *Macromolecules* **44**, 7977–7986.

Hagfeldt A., Boschloo G., Sun L., Kloo L. and Pettersson H. (2010), 'Dye-sensitized solar cells', *Chem. Rev.* **110**, 6595–6663.

Halls J. J. M., Pichler K., Friend R. H., Moratti S. C. and Holmes A. B. (1996), 'Exciton diffusion and dissociation in a poly(*p*-phenylenevinylene)/C_{60} heterojunction photovoltaic cell', *Appl. Phys. Lett.* **68**, 3120–3122.

Halls J. J. M., Walsh C. A., Greenham N. C., Marseglia E. A., Friend R. H., Moratti S. C. and Holmes A. B. (1995), 'Efficient photodiodes from interpenetrating polymer networks', *Nature* **376**, 498–500.

He Z., Zhong C., Su S., Xu M., Wu H. and Cao Y. (2012), 'Enhanced power-conversion efficiency in polymer solar cells using an inverted device structure', *Nat. Photonics* **6**, 591–595.

Heliatek (2013), 'Heliatek consolidates its technology leadership by establishing a new world record for organic solar technology with a cell efficiency of 12%', 16 January 2013, www.heliatek.com.

Huggins M. L. (1941), 'Solutions of long chain compounds', *J. Chem. Phys.* **9**, 440.

Huynh W. U., Dittmer J. J. and Alivisatos A. P. (2002), 'Hybrid nanorod-polymer solar cells', *Science* **295**, 2425–2427.

Irwin M. D., Buchholz B., Hains A. W., Chang R. P. H. and Marks T. J. (2008), '*P*-type semiconducting nickel oxide as an efficiency-enhancing anode interfacial layer in polymer bulk-heterojunction solar cells', *Proc. Natl. Acad. Sci. U S A* **105**, 2783–2787.

Jamieson F. C., Domingo E. B., McCarthy-Ward T., Heeney M., Stingelin N. and Durrant J. R. (2012), 'Fullerene crystallisation as a key driver of charge separation in polymer/fullerene bulk heterojunction solar cells', *Chem. Sci.* **3**, 485–492.

Jørgensen M., Norrman K., Gevorgyan S. A., Tromholt T., Andreasen B. and Krebs F. C. (2012), 'Stability of polymer solar cells', *Adv. Mater.* **24**, 580–612.

Kallmann H. and Pope M. (1959), 'Photovoltaic effect in organic crystals', *J. Chem. Phys.* **30**, 585–586.

Kameoka J., Czaplewski D., Liu H. and Craighead H. G. (2004), 'Polymeric nanowire architecture', *J. Mater. Chem.* **14**, 1503–1505.

Kearns D. and Calvin M. (1958), 'Photovoltaic effect and photoconductivity in laminated organic systems', *J. Chem. Phys.* **29**, 950–951.

Kim J. Y., Kim S. H., Lee H. H., Lee K., Ma W., Gong X. and Heeger A. J. (2006), 'New architecture for high-efficiency polymer photovoltaic cells using solution-based titanium oxide as an optical spacer', *Adv. Mater.* **18**, 572–576.

Kim Y., Cook S., Tuladhar S. M., Choulis S. A., Nelson J., Durrant J. R., Bradley D. D. C., Giles M., McCulloch I., Ha C.-S. and Ree M. (2006), 'A strong regioregularity effect in self-organizing conjugated polymer films and high-efficiency polythiophene:fullerene solar cells', *Nat. Mater.* **5**, 197–203.

Kirchartz T., Pieters B. E., Kirkpatrick J., Rau U. and Nelson J. (2011), 'Recombination via tail states in polythiophene:fullerene solar cells', *Phys. Rev. B* **83**, 115209.

Kniepert J., Schubert M., Blakesley J. C. and Neher D. (2011), 'Photogeneration and recombination in P3HT/PCBM solar cells probed by time-delayed collection field experiments', *J. Phys. Chem. Lett.* **2**, 700–705.

Kohn P., Rong Z., Scherer K. H., Sepe A., Sommer M., Müller-Buschbaum P., Friend R. H., Steiner U. and Hüttner S. (2013), 'Crystallization-induced 10-nm structure formation in P3HT/PCBM blends', *Macromolecules* **46**, 4002–4013.

Koster L. J. A., Mihailetchi V. D., Ramaker R. and Blom P. W. M. (2005), 'Light intensity dependence of open-circuit voltage of polymer:fullerene solar cells', *Appl. Phys. Lett.* **86**, 123509.

Krebs F. C. (2009), 'Fabrication and processing of polymer solar cells: a review of printing and coating techniques', *Solar Energy Mater. Solar Cells* **93**, 394–412.

Krebs F. C., Fyenbo J. and Jorgensen M. (2010), 'Product integration of compact roll-to-roll processed polymer solar cell modules: methods and manufacture using flexographic printing, slot-die coating and rotary screen printing', *J. Mater. Chem.* **20**, 8994–9001.

Krebs F. C., Gevorgyan S. A. and Alstrup J. (2009), 'A roll-to-roll process to flexible polymer solar cells: model studies, manufacture and operational stability studies', *J. Mater. Chem.* **19**, 5442–5451.

Krebs F. C. and Spanggaard H. (2005), 'Significant improvement of polymer solar cell stability', *Chemistry of Materials* **17**, 5235–5237.

Kumar S. and Nann T. (2004), 'First solar cells based on CdTe nanoparticle/MEH-PPV composites', *J. Mater. Res.* **19**, 1990–1994.

Lee J., Vandewal K., Yost S. R., Bahlke M. E., Goris L., Baldo M. A., Manca J. V. and Voorhis T. V. (2010), 'Charge transfer state versus hot exciton dissociation in polymer-fullerene blended solar cells', *J. Amer. Chem. Soc.* **132**, 11878–11880.

Lee J. K., Ma W. L., Brabec C. J., Yuen J., Moon J. S., Kim J. Y., Lee K., Bazan G. C. and Heeger A. J. (2008), 'Processing additives for improved efficiency from bulk heterojunction solar cells', *J. Amer. Chem. Soc.* **130**, 3619–3623.

Lee M. J., Gupta D., Zhao N., Heeney M., McCulloch I. and Sirringhaus H. (2011), 'Anisotropy of charge transport in a uniaxially aligned and chain-extended, high-mobility, conjugated polymer semiconductor', *Adv. Funct. Mater.* **21**, 932–940.

Li G., Shrotriya V., Yao Y. and Yang Y. (2005), 'Investigation of annealing effects and film thickness dependence of polymer solar cells based on poly(3-hexylthiophene)', *J. Appl. Phys.* **98**, 043704.

Li G., Yao Y., Yang H., Shrotriya V., Yang G. and Yang Y. (2007), '"Solvent annealing" effect in polymer solar cells based on poly(3-hexylthiophene) and methanofullerenes', *Adv. Funct. Mater.* **17**, 1636–1644.

Li J., Kim S., Edington S., Nedy J., Cho S., Lee K., Heeger A. J., Gupta M. C. and Yates Jr J. T. (2011), 'A study of stabilization of P3HT/PCBM organic solar cells by photochemical active TiO_X layer', *Solar Energy Mater. Solar Cells* **95**, 1123–1130.

Li J., Zhao Y., Tan H. S., Guo Y., Di C.-A., Yu G., Liu Y., Lin M., Lim S. H., Zhou Y., Su H. and Ong B. S. (2012), 'A stable solution-processed polymer semiconductor with record high-mobility for printed transistors', *Sci. Rep.* **2**, 754.

Li W., Furlan A., Hendriks K. H., Wienk M. M. and Janssen R. A. J. (2013), 'Efficient tandem and triple-junction polymer solar cells', *J. Amer. Chem. Soc.* **135**, 5529–5532.

Lin C.-F., Liu S.-W., Lee C.-C., Hunag J.-C., Su W.-C., Chiu T.-L., Chen C.-T. and Lee J.-H. (2012), 'Open-circuit voltage and efficiency improvement of subphthalocyanine-based organic photovoltaic device through deposition rate control', *Solar Energy Mater. Solar Cells* **103**, 69–75.

Linares M., Beljonne D., Cornil J. r. m., Lancaster K., Bré das J.-L., Verlaak S., Mityashin A., Heremans P., Fuchs A., Lennartz C., Idé J., Méreau R. L., Aurel P., Ducasse L. and Castet F. (2010), 'On the interface dipole at the pentacene–fullerene heterojunction: a theoretical study', *J. Phys. Chem. C* **114**, 3215–3224.

Liu B., Png R.-Q., Zhao L.-H., Chua L.-L., Friend R. H. and Ho P. K. H. (2012), 'High internal quantum efficiency in fullerene solar cells based on crosslinked polymer donor networks', *Nature Commun.* **3**, 1321.

Liu C.-Y., Holman Z. C. and Kortshagen U. R. (2008), 'Hybrid solar cells from P3HT and silicon nanocrystals', *Nano Letters* **9**, 449–452.

Liu J., Shi Y. J. and Yang Y. (2001), 'Solvation-induced morphology effects on the performance of polymer-based photovoltaic devices', *Adv. Funct. Mater.* **11**, 420–424.

Lüssem B., Riede M. and Leo K. (2013), 'Doping of organic semiconductors', *Phys. Stat. Solidi A* **210**, 9–43.

Ma W., Yang C., Gong X., Lee K. and Heeger A. J. (2005), 'Thermally stable, efficient polymer solar cells with nanoscale control of the interpenetrating network morphology', *Adv. Funct. Mater.* **15**, 1617–1622.

MacKenzie R. C. I., Shuttle C. G., Chabinyc M. L. and Nelson J. (2012), 'Extracting microscopic device parameters from transient photocurrent measurements of P3HT:PCBM solar cells', *Adv. Energy Mater.* **2**, 662–669.

Manceau M., Bundgaard E., Carle J. E., Hagemann O., Helgesen M., Sondergaard R., Jorgensen M. and Krebs F. C. (2011), 'Photochemical stability of small π-conjugated polymers for polymer solar cells: a rule of thumb', *J. Mater. Chem.* **21**, 4132–4141.

Manceau M., Chambon S., Rivaton A., Gardette J.-L., Guillerez S. and Lemaître N. (2010a), 'Effects of long-term uv-visible light irradiation in the absence of oxygen on P3HT and P3HT:PCBM blend', *Solar Energy Mater. Solar Cells* **94**, 1572–1577.

Manceau M., Helgesen M. and Krebs F. C. (2010b), 'Thermo-cleavable polymers: materials with enhanced photochemical stability', *Polym. Degrad. Stabil.* **95**, 2666–2669.

Manceau M., Rivaton A., Gardette J.-L., Guillerez S. and Lemaître N. (2009), 'The mechanism of photo- and therMoOxidation of poly(3-hexylthiophene) (P3HT) reconsidered', *Polym. Degrad. Stabil.* **94**, 898–907.

Mandoc M. M., Veurman W., Koster L. J. A., de Boer B. and Blom P. W. M. (2007), 'Origin of the reduced fill factor and photocurrent in MDMO-PPV:PCNEPV all-polymer solar cells', *Adv. Funct. Mater.* **17**, 2167–2173.

Marsh R. A., Hodgkiss J. M. and Friend R. H. (2010), 'Direct measurement of electric field-assisted charge separation in polymer:fullerene photovoltaic diodes', *Adv. Mater.* **22**, 3672–3676.

Martens H. C. F., Brom H. B. and Blom P. W. M. (1999), 'Frequency-dependent electrical response of holes in poly(*p*-phenylene vinylene)', *Phys. Rev. B* **60**, R8489–R8492.

Meilanov I. S., Benderskii V. A. and Bliumenfel'd L. A. (1970a), 'Photoelectric properties of chlorophyll a and b layers. I. Photocurrents during constant illumination', *Biofizika* **15**, 822–827.

Meilanov I. S., Benderskii V. A. and Bliumenfel'd L. A. (1970b), 'Photoelectric properties of chlorophyll a and b layers. II. Photocurrents during impulse illumination', *Biofizika* **15**, 958–964.

Mihailetchi V. D., Blom P. W. M., Hummelen J. C. and Rispens M. T. (2003), 'Cathode dependence of the open-circuit voltage of polymer:fullerene bulk heterojunction solar cells', *J. Appl. Phys.* **94**, 6849–6854.

Mihailetchi V. D., Koster L. J. A. and Blom P. W. M. (2004), 'Effect of metal electrodes on the performance of polymer:fullerene bulk heterojunction solar cells', *Appl. Phys. Lett.* **85**, 970–972.

Mihailetchi V. D., Wildeman J. and Blom P. W. M. (2005), 'Space-charge limited photocurrent', *Phys. Rev. Lett.* **94**, 126602.

Miller A. and Abrahams E. (1960), 'Impurity conduction at low concentrations', *Phys. Rev.* **120**, 745–755.

Morana M., Azimi H., Dennler G., Egelhaaf H.-J., Scharber M., Forberich K., Hauch J., Gaudiana R., Waller D., Zhu Z., Hingerl K., van Bavel S. S., Loos J. and Brabec C. J. (2010), 'Nanomorphology and charge generation in bulk heterojunctions based on low-bandgap dithiophene polymers with different bridging atoms', *Adv. Funct. Mater.* **20**, 1180–1188.

Moulé A. J. and Meerholz K. (2008), 'Controlling morphology in polymer–fullerene mixtures', *Adv. Mater.* **20**, 240–245.

Mühlbacher D., Scharber M., Morana M., Zhu Z., Waller D., Gaudiana R. and Brabec C. (2006), 'High photovoltaic performance of a low-bandgap polymer', *Adv. Mater.* **18**, 2884–2889.

Najari A., Berrouard P., Ottone C., Boivin M., Zou Y., Gendron D., Caron W.-O., Legros P., Allen C. N., Sadki S. and Leclerc M. (2012), 'High open-circuit voltage solar cells based on new thieno[3,4-c]pyrrole-4,6-dione and 2,7-carbazole copolymers', *Macromolecules* **45**, 1833–1838.

Nelson J. (2003), 'Diffusion-limited recombination in polymer-fullerene blends and its influence on photocurrent collection', *Phys. Rev. B* **67**, 155209.

Neugebauer H., Brabec C., Hummelen J. C. and Sariciftci N. S. (2000), 'Stability and photodegradation mechanisms of conjugated polymer/fullerene plastic solar cells', *Solar Energy Mater. Solar Cells* **61**, 35–42.

Nguyen L. H., Hoppe H., Erb T., Günes S., Gobsch G. and Sariciftci N. S. (2007), 'Effects of annealing on the nanomorphology and performance of poly(alkylthiophene):fullerene bulk-heterojunction solar cells', *Adv. Funct. Mater.* **17**, 1071–1078.

Nielsen T. D., Cruickshank C., Foged S., Thorsen J. and Krebs F. C. (2010), 'Business, market and intellectual property analysis of polymer solar cells', *Solar Energy Mater. Solar Cells* **94**, 1553–1571.

Onsager L. (1934), 'Deviations from Ohm's law in weak electrolytes', *J. Chem. Phys.* **2**, 599–615.

Ooi Z. E., Tam T. L., Sellinger A. and deMello J. C. (2008), 'Field-dependent carrier generation in bulk heterojunction solar cells', *Energy Environ. Sci.* **1**, 300–309.

Padinger F., Rittberger R. S. and Sariciftci N. S. (2003), 'Effects of postproduction treatment on plastic solar cells', *Adv. Funct. Mater.* **13**, 85–88.

Park S. H., Roy A., Beaupre S., Cho S., Coates N., Moon J. S., Moses D., Leclerc M., Lee K. and Heeger A. J. (2009), 'Bulk heterojunction solar cells with internal quantum efficiency approaching 100%', *Nat. Photonics* **3**, 297–302.

Park Y. D., Kim D. H., Jang Y., Cho J. H., Hwang M., Lee H. S., Lim J. A. and Cho K. (2006), 'Effect of side chain length on molecular ordering and field-effect mobility in poly(3-alkylthiophene) transistors', *Org. Electronics* **7**, 514–520.

Pasveer W. F., Cottaar J., Tanase C., Coehoorn R., Bobbert P. A., Blom P. W. M., de Leeuw D. M. and Michels M. A. J. (2005), 'Unified description of charge-carrier mobilities in disordered semiconducting polymers', *Phys. Rev. Lett.* **94**, 206601.

Peet J., Kim J. Y., Coates N. E., Ma W. L., Moses D., Heeger A. J. and Bazan G. C. (2007), 'Efficiency enhancement in low-bandgap polymer solar cells by processing with alkane dithiols', *Nat. Mater.* **6**, 497–500.

Peiro A. M., Ravirajan P., Govender K., Boyle D. S., O'Brien P., Bradley D. D. C., Nelson J. and Durrant J. R. (2006), 'Hybrid polymer/metal oxide solar cells based on ZnO columnar structures', *J. Mater. Chem.* **16**, 2088–2096.

Peumans P. and Forrest S. R. (2001), 'Very-high-efficiency double-heterostructure copper phthalocyanine/C_{60} photovoltaic cells', *Appl. Phys. Lett.* **79**, 126–128.

Peumans P. and Forrest S. R. (2004), 'Separation of geminate charge-pairs at donor–acceptor interfaces in disordered solids', *Chem. Phys. Lett.* **398**, 27–31.

Png R.-Q., Chia P.-J., Tang J.-C., Liu B., Sivaramakrishnan S., Zhou M., Khong S.-H., Chan H. S. O., Burroughes J. H., Chua L.-L., Friend R. H. and Ho P. K. H. (2010), 'High-performance polymer semiconducting heterostructure devices by nitrene-mediated photocrosslinking of alkyl side chains', *Nat. Mater.* **9**, 152–158.

Polyera (2013), 'Polyera achieves 6.4% all-polymer organic solar cells', 12 April 2013, www.polyera.com.

Prins P., Grozema F. C., Schins J. M., Patil S., Scherf U. and Siebbeles L. D. A. (2006), 'High intrachain hole mobility on molecular wires of ladder-type poly(*p*-phenylenes)', *Phys. Rev. Lett.* **96**, 146601.

Reese M. O., Gevorgyan S. A., Jørgensen M., Bundgaard E., Kurtz S. R., Ginley D. S., Olson D. C., Lloyd M. T., Morvillo P., Katz E. A., Elschner A., Haillant O., Currier T. R., Shrotriya V., Hermenau M., Riede M., R. Kirov K., Trimmel G., Rath T., Inganäs O., Zhang F., Andersson M., Tvingstedt K., Lira-Cantu M., Laird D., McGuiness C., Gowrisanker S., Pannone M., Xiao M., Hauch J., Steim R., DeLongchamp D. M., Rösch R., Hoppe H., Espinosa N., Urbina A., Yaman-Uzunoglu G., Bonekamp J.-B., van Breemen A. J. J. M., Girotto C., Voroshazi E. and Krebs F. C. (2011), 'Consensus stability testing protocols for organic photovoltaic materials and devices', *Solar Energy Mater. Solar Cells* **95**, 1253–1267.

Reese M. O., Nardes A. M., Rupert B. L., Larsen R. E., Olson D. C., Lloyd M. T., Shaheen S. E., Ginley D. S., Rumbles G. and Kopidakis N. (2010), 'Photoinduced degradation of polymer and polymer–fullerene active layers: experiment and theory', *Adv. Funct. Mater.* **20**, 3476–3483.

Riede M., Uhrich C., Widmer J., Timmreck R., Wynands D., Schwartz G., Gnehr W.-M., Hildebrandt D., Weiss A., Hwang J., Sundarraj S., Erk P., Pfeiffer M. and Leo K. (2011), 'Efficient organic tandem solar cells based on small molecules', *Adv. Funct. Mater.* **21**, 3019–3028.

Rogers J. A., Bao Z., Makhija A. and Braun P. (1999), 'Printing process suitable for reel-to-reel production of high-performance organic transistors and circuits', *Adv. Mater.* **11**, 741–745.

Rogers J. A., Bao Z. and Raju V. R. (1998), 'Nonphotolithographic fabrication of organic transistors with micron feature sizes', *Appl. Phys. Lett.* **72**, 2716–2718.

Sariciftci N. S., Smilowitz L., Heeger A. J. and Wudl F. (1992), 'Photoinduced electron transfer from a conducting polymer to buckminsterfullerene', *Science* **258**, 1474–1476.

Scharber M. C., Mühlbacher D., Koppe M., Denk P., Waldauf C., Heeger A. J. and Brabec C. J. (2006), 'Design rules for donors in bulk-heterojunction solar cells — towards 10% energy-conversion efficiency', *Adv. Mater.* **18**, 789–794.

Segalman R. A., McCulloch B., Kirmayer S. and Urban J. J. (2009), 'Block copolymers for organic optoelectronics', *Macromolecules* **42**, 9205–9216.

Shaheen S. E., Brabec C. J., Sariciftci N. S., Padinger F., Fromherz T. and Hummelen J. C. (2001), '2.5% efficient organic plastic solar cells', *Appl. Phys. Lett.* **78**, 841–843.

Shoaee S., Clarke T. M., Huang C., Barlow S., Marder S. R., Heeney M., McCulloch I. and Durrant J. R. (2010), 'Acceptor energy level control of charge photogeneration in organic donor/acceptor blends', *J. Amer. Chem. Soc.* **132**, 12919–12926.

Shoaee S., Subramaniyan S., Xin H., Keiderling C., Tuladhar P. S., Jamieson F., Jenekhe S. A. and Durrant J. R. (2013), 'Charge photogeneration for a series of thiazolo-thiazole donor polymers blended with the fullerene electron acceptors PCBM and ICBA', *Adv. Funct. Mater.* **23**, 3286–3298.

Shuttle C. G., Hamilton R., O'Regan B. C., Nelson J. and Durrant J. R. (2010), 'Charge-density-based analysis of the current–voltage response of polythiophene/fullerene pho tovoltaic devices', *Proc. Natl. Acad. Sci. USA* **107**, 16448–16452.

Silinsh E. A., Belkind A. I., Balode D. R., Biseniece A. J., Grechov V. V., Taure L. F., Kurik M. V., Vertzymacha J. I. and Bok I. (1974), 'Photoelectrical properties, energy level spectra, and photogeneration mechanisms of pentacene', *Phys. Stat. Solidi A* **25**, 339–347.

Singh S., Jones W. J., Siebrand W., Stoicheff B. P. and Schneider W. G. (1965), 'Laser generation of excitons and fluorescence in anthracene crystals', *J. Chem. Phys.* **42**, 330–342.

Smith M. B. and Michl J. (2010), 'Singlet fission', *Chem. Rev.* **110**, 6891–6936.

Son H. J., Wang W., Xu T., Liang Y., Wu Y., Li G. and Yu L. (2011), 'Synthesis of fluorinated polythienothiophene-co-benzodithiophenes and effect of fluorination on the photovoltaic properties', *J. Amer. Chem. Soc.* **133**, 1885–1894.

Stabile R., Camposeo A., Persano L., Tavazzi S., Cingolani R. and Pisignano D. (2007), 'Organic-based distributed feedback lasers by direct electron-beam lithography on conjugated polymers', *Appl. Phys. Lett.* **91**, 101110.

Steim R., Kogler F. R. and Brabec C. J. (2010), 'Interface materials for organic solar cells', *J. Mater. Chem.* **20**, 2499–2512.

Street R. A., Cowan S. and Heeger A. J. (2010), 'Experimental test for geminate recombination applied to organic solar cells', *Phys. Rev. B* **82**, 121301.

Stubinger T. and Brutting W. (2001), 'Exciton diffusion and optical interference in organic donor–acceptor photovoltaic cells', *J. Appl. Phys.* **90**, 3632–3641.

Sullivan P., Schumann S., Da Campo R., Howells T., Duraud A., Shipman M., Hatton R. A. and Jones T. S. (2013), 'Ultra-high voltage multijunction organic solar cells for low-power electronic applications', *Adv. Energy Mater.* **3**, 239–244.

Sun B., Marx E. and Greenham N. C. (2003), 'Photovoltaic devices using blends of branched CdSe nanoparticles and conjugated polymers', *Nano Letters* **3**, 961–963.

Sun Y., Welch G. C., Leong W. L., Takacs C. J., Bazan G. C. and Heeger A. J. (2012), 'Solution-processed small-molecule solar cells with 6.7% efficiency', *Nat. Mater.* **11**, 44–48.

Tang C. W. (1986), 'Two-layer organic photovoltaic cell', *Appl. Phys. Lett.* **48**, 183–185.

Tang C. W. and Albrecht A. C. (1975), 'Photovoltaic effects of metal–chlorophyll-*a*–metal sandwich cells', *J. Chem. Phys.* **62**, 2139–2149.

Tengstedt C., Osikowicz W., Salaneck W. R., Parker I. D., Hsu C.-H. and Fahlman M. (2006), 'Fermi-level pinning at conjugated polymer interfaces', *Appl. Phys. Lett.* **88**, 053502.

Timmreck R., Olthof S., Leo K. and Riede M. K. (2010), 'Highly doped layers as efficient electron-hole recombination contacts for tandem organic solar cells', *J. Appl. Phys.* **108**, 033108.

Townsend P. D., Pereira C. M., Bradley D. D. C., Horton M. E. and Friend R. H. (1985), 'Increase in chain conjugation length in highly oriented Durham-route polyacetylene', *J. Phys. C: Solid State Physics* **18**, L283–L289.

Treat N. D., Brady M. A., Smith G., Toney M. F., Kramer E. J., Hawker C. J. and Chabinyc M. L. (2011), 'Interdiffusion of PCBM and P3HT reveals miscibility in a photovoltaically active blend', *Adv. Energy Mater.* **1**, 82–89.

Tributsch H. and Calvin M. (1971), 'Electrochemistry of excited molecules: photo-electrochemical reactions of chlorophylls*', *Photochem. Photobiol.* **14**, 95–112.

Tvingstedt K., Vandewal K., Gadisa A., Zhang F., Manca J. and Inganäs O. (2009), 'Electroluminescence from charge transfer states in polymer solar cells', *J. Amer. Chem. Soc.* **131**, 11819–11824.

Tvingstedt K., Vandewal K., Zhang F. and Inganás O. (2010), 'On the dissociation efficiency of charge transfer excitons and Frenkel excitons in organic solar cells: A luminescence quenching study', *J. Phys. Chem C* **114**, 21824–21832.

Vandewal K., Gadisa A., Oosterbaan W. D., Bertho S., Banishoeib F., Van Severen I., Lutsen L., Cleij T. J., Vanderzande D. and Manca J. V. (2008), 'The relation between open-circuit voltage and the onset of photocurrent generation by charge-transfer absorption in polymer:fullerene bulk heterojunction solar cells', *Adv. Funct. Mater.* **18**, 2064–2070.

Vandewal K., Tvingstedt K., Gadisa A., Inganäs O. and Manca J. V. (2009), 'On the origin of the open-circuit voltage of polymer-fullerene solar cells', *Nat. Mater.* **8**, 904–909.

Veldman D., İpek Ö., Meskers S. C. J., Sweelssen J., Koetse M. M., Veenstra S. C., Kroon J. M., van Bavel S. S., Loos J. and Janssen R. A. J. (2008), 'Compositional and electric field dependence of the dissociation of charge transfer excitons in alternating polyfluorene copolymer/fullerene blends', *J. Amer. Chem. Soc.* **130**, 7721–7735.

Vitoratos E., Sakkopoulos S., Dalas E., Paliatsas N., Karageorgopoulos D., Petraki F., Kennou S. and Choulis S. A. (2009), 'Thermal degradation mechanisms of PEDOT:PSS', *Org. Electronics* **10**, 61–66.

Vos A. D. (1980), 'Detailed balance limit of the efficiency of tandem solar cells', *J. Phys. D: Appl. Phys.* **13**, 839–846.

Wannier G. H. (1937), 'The structure of electronic excitation levels in insulating crystals', *Phys. Rev.* **52**, 191–197.

Watt A., Blake D., Warner J. H., Thomsen E. A., Tavenner E. L., Rubinsztein-Dunlop H. and Meredith P. (2005), 'Lead sulfide nanocrystal: conducting polymer solar cells', *J. Phys. D: Appl. Phys.* **38**, 2006–2012.

Wetzelaer G. A. H., Koster L. J. A. and Blom P. W. M. (2011a), 'Validity of the Einstein relation in disordered organic semiconductors', *Phys. Rev. Lett.* **107**, 066605.

Wetzelaer G. A. H., Kuik M., Lenes M. and Blom P. W. M. (2011b), 'Origin of the dark-current ideality factor in polymer:fullerene bulk heterojunction solar cells', *Appl. Phys. Lett.* **99**, 153506.

Wilson M. W. B., Rao A., Clark J., Kumar R. S. S., Brida D., Cerullo G. and Friend R. H. (2011), 'Ultrafast dynamics of exciton fission in polycrystalline pentacene', *J. Amer. Chem. Soc.* **133**, 11830–11833.

Xin H., Reid O. G., Ren G., Kim F. S., Ginger D. S. and Jenekhe S. A. (2010), 'Polymer nanowire/fullerene bulk heterojunction solar cells: how nanostructure determines photovoltaic properties', *ACS Nano* **4**, 1861–1872.

Yu G., Gao J., Hummelen J. C., Wudl F. and Heeger A. J. (1995), 'Polymer photovoltaic cells: enhanced efficiencies via a network of internal donor-acceptor heterojunctions', *Science* **270**, 1789–1791.

Yun M., Myung N. V., Vasquez R. P., Lee C., Menke E. and Penner R. M. (2004), Electrochemically grown wires for individually addressable sensor arrays', *Nano Letters* **4**, 419–422.

Zhong H., Li Z., Deledalle F., Fregoso E. C., Shahid M., Fei Z., Nielsen C. B., Yaacobi-Gross N., Rossbauer S., Anthopoulos T. D., Durrant J. R. and Heeney M. (2013), 'Fused dithienogermolodithiophene low band gap polymers for high-performance organic solar cells without processing additives', *J. Amer. Chem. Soc.* **135**, 2040–2043.

Zhou Y., Fuentes-Hernandez C., Shim J., Meyer J., Giordano A. J., Li H., Winget P., Papadopoulos T., Cheun H., Kim J., Fenoll M., Dindar A., Haske W., Najafabadi E., Khan T. M., Sojoudi H., Barlow S., Graham S., Brédas J.-L., Marder S. R., Kahn A.

and Kippelen B. (2012), 'A universal method to produce low-work function electrodes for organic electronics', *Science* **336**, 327–332.

Zilberberg K., Gharbi H., Behrendt A., Trost S. and Riedl T. (2012), 'Low-temperature, solution-processed MoO_x for efficient and stable organic solar cells', *ACS Appl. Mater. Interfaces* **4**, 1164–1168.

CHAPTER 9

DYE- AND PEROVSKITE-SENSITISED MESOSCOPIC SOLAR CELLS

MICHAEL GRÄTZEL
Laboratory for Photonics and Interfaces
Swiss Federal Institute of Technology
CH–1015 Lausanne, Switzerland
michael.graetzel@epfl.ch

and

JAMES R. DURRANT
Department of Chemistry, Imperial College London SW7 2AZ, UK
Materials Science Centre,
College of Engineering, University of Swansea,
Swansea SA12 7Ax, UK
j.durrant@imperial.ac.uk

On the arid lands there will spring up industrial colonies without smoke and without smokestacks; forests of glass tubes will extend over the plains and glass buildings will rise everywhere; inside of these will take place the photochemical processes that hitherto have been the guarded secret of the plants, but that will have been mastered by human industry which will know how to make them bear even more abundant fruit than nature, for nature is not in a hurry and mankind is. And if in a distant future the supply of coal becomes completely exhausted, civilisation will not be checked by that, for life and civilisation will continue as long as the Sun shines!

Giacomo Ciamician, *Eighth International Congress of Applied Chemistry*, Washington and New York, September 1912.

9.1 Introduction

Photovoltaic (PV) devices are based on charge separation at an interface of two materials of different conduction type or mechanism. To date, this field has been dominated by solid-state junction devices, usually made of silicon, and profiting from the experience and material availability resulting from the semiconductor industry. The dominance of the photovoltaic field by inorganic solid-state junction devices is now being challenged by the emergence of a range of new device concepts, including devices based on nanocrystalline and conducting polymer films. These devices offer the prospect of very low-cost fabrication and present a range of attractive features that will facilitate market entry. It is now possible to depart completely from the classical solid-state junction device by replacing the phase contacting the semiconductor by an electrolyte, liquid, gel or solid, thereby forming a photoelectrochemical cell. The phenomenal progress realised recently in the fabrication and characterisation of nanocrystalline materials has opened up vast new opportunities for these systems. Contrary to expectation, devices based on interpenetrating networks of mesoscopic semiconductors have shown strikingly high conversion efficiencies, competing with those of conventional devices. The prototype of this family of devices is the dye-sensitised solar cell (DSSC), which realises the optical absorption and charge-separation processes by the association of a sensitiser as light-absorbing material with a wide-bandgap semiconductor of nanocrystalline morphology (O'Regan and Grätzel, 1991).

Recently, solid-state embodiments of mesoscopic solar have emerged where the sensitiser is replaced by a semiconductor quantum dot (Grätzel *et al.*, 2012) or a perovskite pigment. Solid-state cells based on perovskites as light harvesters have stunned the PV community by their amazing performance, certified power conversion efficiencies having reached 14.14% only within one year after their inception (Burschka *et al.*, 2013).

9.2 Historical background

The history of the sensitisation of semiconductors to light of wavelength longer than that corresponding to the semiconductor bandgap has been presented elsewhere (McEvoy and Grätzel, 1994; Hagfeldt and Grätzel, 2000). It represents an interesting convergence of photography and photoelectrochemistry, both of which rely on photoinduced charge separation at a liquid–solid interface. The silver halides used in photography have bandgaps of the order of 2.7–3.2 eV, and are therefore insensitive to much of the visible spectrum, as are the metal oxide films now used in DSSCs.

The first panchromatic film able to render the image of a scene realistically into black and white followed on the work of Vogel in Berlin after 1873 (West, 1874), in which he associated dyes with the halide semiconductor grains. The first sensitisation of a photoelectrode followed shortly thereafter, using similar chemistry (Moser, 1887). However, the clear recognition of the parallels between the two procedures — realisation that the same dyes can in principle function in both systems (Namba and Hishiki, 1965) and verification that their operating mechanism is by injection of electrons from photoexcited dye molecules into the conduction band of the $n-$type semiconductor substrates (Gerischer and Tributsch, 1968) — dates only to the 1960s. In the years that followed it became recognised that the dye could function most efficiently if chemisorbed on the surface of the semiconductor (Tsubomura *et al.*, 1976; Anderson *et al.*, 1979; Dare-Edwards *et al.*, 1980). The concept of using dispersed particles to provide a sufficient interface area then emerged (Dung *et al.*, 1984), and was subsequently employed for photoelectrodes (Desilvestro *et al.*, 1985).

Titanium dioxide (TiO_2) quickly became the semiconductor of choice for the photoelectrode on account of its many advantages for sensitised photochemistry and photoelectron chemistry: it is a low-cost, widely available, non-toxic and biocompatible material, and as such is even used in healthcare products as well as industrial applications such as paint pigmentation. Initial studies of the TiO_2-based DSSC employed tris-bipyridyl ruthenium(II) dyes, which are paradigm sensitisers for photochemical studies, functionalised by the addition of carboxylate groups to attach the chromophore to the oxide substrate by chemisorption. Progress thereafter was incremental, a synergy of structure, substrate roughness and morphology, dye photophysics and electrolyte redox chemistry, until the announcement in 1991 (O'Regan and Grätzel, 1991) of a sensitised electrochemical photovoltaic device with an energy conversion efficiency of 7.1% under solar illumination. That evolution has continued progressively since then, with certified efficiencies now over 12%. These advances in efficiency have been complemented by significant advances in processability, cost and stability which have greatly enhanced the commercial viability of this technology.

9.3 Mode of function of dye-sensitised solar cells

9.3.1 *Device configuration*

Figure 9.1 is a schematic of the components of a DSSC. At the heart of the system is a mesoporous oxide layer composed of nanometre-sized particles that have been sintered together to allow electronic conduction to take place. The material of choice has been TiO_2 (anatase) although alternative wide-gap oxides such as ZnO, SnO_2 and Nb_2O_5 have also been investigated (Hoyer and Weller, 1995; Ferrere *et al.*, 1997; Rensmo *et al.*, 1997; Sayama *et al.*, 1998; Green *et al.*, 2005). Attached to the surface of the nanocrystalline film is a monolayer of a sensitiser dye. Photoexcitation of the latter results in the injection of an electron into the conduction band of the oxide, generating the dye cation. The original state of the dye is subsequently restored by electron donation from the electrolyte; this step is often referred to as the regeneration reaction. The electrolyte usually comprises

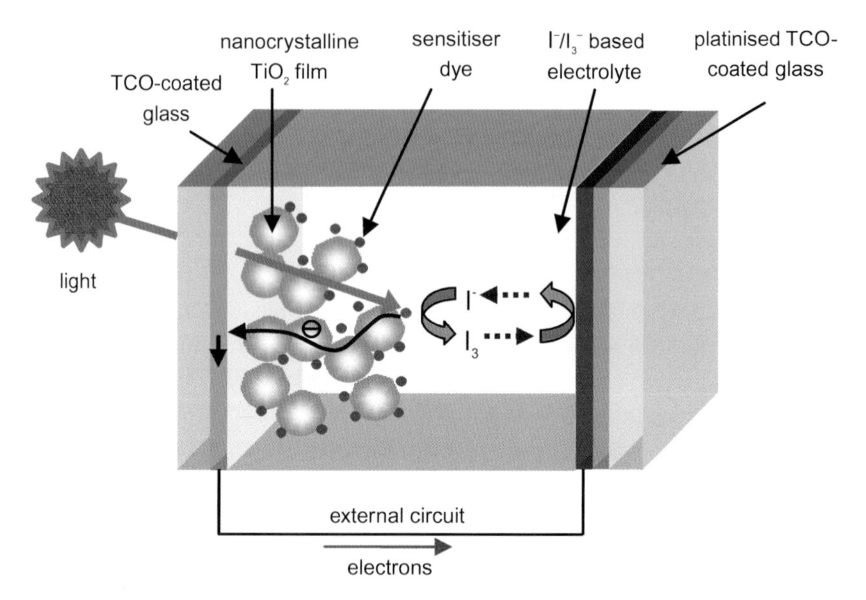

Figure 9.1 Schematic of a liquid electrolyte dye-sensitised solar cell. Photoexcitation of the sensitiser dye is followed by electron injection into the conduction band of the mesoporous oxide semiconductor, and electron transport through the metal oxide film to the TCO-coated, glass, working electrode. The dye molecule is regenerated by the redox system, which is itself regenerated at the platinised counter electrode by electrons passed through the external circuit.

an iodide/triiodide redox couple dissolved in a liquid organic solvent, although attention is increasingly focusing on alternatives for the solvent, including ionic liquids, gelled electrolytes and polymer electrolytes (Stathatos *et al.*, 2003; Wang *et al.*, 2003b; Nogueira *et al.*, 2004; Durr *et al.*, 2005) as well as alternatives to the iodide/iodine redox couple which are now yielding very exciting device efficiencies (Yella *et al.*, 2011). The regeneration of the sensitiser by iodide intercepts the recapture of the injected electron by the oxidised dye. The iodide is in turn regenerated by the reduction of triiodide at the counter electrode, with the electrical circuit being completed via electron migration through the external load.

The high surface area of the mesoporous metal oxide film is critical to efficient device performance as it allows strong absorption of solar irradiation to be achieved by only a monolayer of adsorbed sensitiser dye. Whereas a dye monolayer absorbed on a flat interface exhibits only negligible light absorption (the optical absorption cross-sectional areas for molecular dyes being typically two to three orders of magnitude smaller than their physical cross-sections), the use of a mesoporous film dramatically enhances the interfacial surface area over the geometric surface area, by up to a 1000-fold for a 10 μm thick film, leading to high visible-light absorbance from the many successive monolayers of adsorbed dye in the optical path. Another advantage of the use of a dye monolayer is that there is no requirement for exciton diffusion to the dye/metal oxide interface, and also the non-radiative quenching of excited states often associated with thicker molecular films is avoided. The high surface area of such mesoporous films does, however, have a significant downside, as it also enhances interfacial charge-recombination losses, a topic we return to in more detail below.

A recent alternative embodiment of the DSSC concept is the replacement of the redox electrolyte with a solid-state hole conductor, which may be either inorganic (Tennakone *et al.*, 1995; O'Regan and Schwartz, 1998) or organic (Bach *et al.*, 1998), thereby avoiding the use of a redox electrolyte. Such solid-state sensitised heterojunctions can be regarded as functionally intermediate between redox electrolyte-based photoelectrochemical DSSCs and the organic bulk heterojunctions described in Chapter 6. For more than a decade, power conversion efficiencies (PCEs) measured under AM 1.5 standard reporting conditions for such solid-state DSSCs were below those of liquid electrolyte-based devices, the highest reported PCE value for the former being 7.2% (Burschka *et al.*, 2011) as compared with 12.3% for the latter (Yella *et al.*, 2011). However, the situation has changed dramatically with the advent of perovskite light harvesters, which now reach efficiencies of 15% in a solid hole-conductor-based device.

9.3.2 *Device fabrication*

Dye-sensitised solar cells are typically fabricated on transparent conducting oxide (TCO) glass substrates, enabling light irradiance through this substrate under photovoltaic operation. The conductive coating typically used is fluorine-doped SnO_2 (FTO), preferred over its indium-doped analogue (ITO) for reasons of lower cost and enhanced stability. Prior to deposition of the mesoporous TiO_2 film, a dense TiO_2 film may be deposited to act as a hole-blocking layer, preventing recombination (shunt resistance) losses between electrons in the FTO and oxidised redox couple.

The TiO_2 nanoparticles are typically fabricated by the aqueous hydrolysis of titanium alkoxide precursors, followed by autoclaving at temperatures up to 240°C to achieved the desired nanoparticle dimensions and crystallinity (anatase) (Barbe *et al.*, 1997). The nanoparticles are deposited as a colloidal suspension by screen-printing or by spreading with a doctor blade, followed by sintering at ~450°C to ensure good interparticle connectivity. The film porosity is controlled by the addition of an organic filler such as carbowax to the suspension prior to deposition; this filler is subsequently burnt off during the sintering step. Figure 9.2

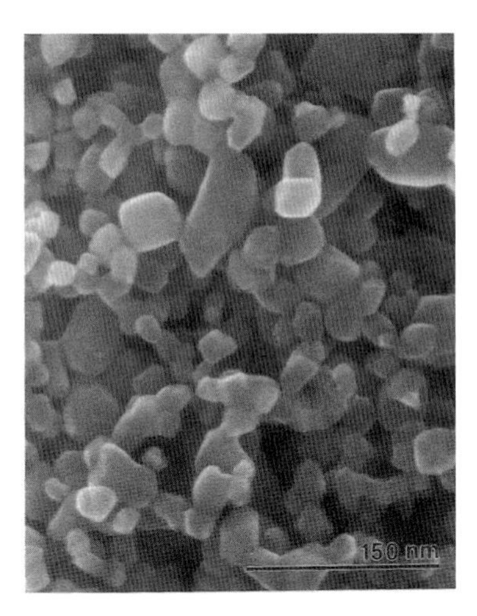

Figure 9.2 Scanning electron micrograph of a typical mesoscopic TiO2 film employed in DSSCs. Note the bipyramidal shape of the particles, with (101) oriented facets exposed. The average particle size is 20 nm.

shows a scanning electron micrograph of a typical mesoporous TiO_2 film. Typical film thicknesses are 5–20 μm, with TiO_2 mass 1–4 mg cm^{-2}, film porosity 50–65%, average pore size 15 nm and particle diameter 15–20 nm.

The classic sensitiser dye employed in DSSCs is a ruthenium(II) bipyridyl dye, cis-bis(isothiocyanato)-bis(2,2′-bipyridyl-4,4′-dicarboxylato)-ruthenium(II), which is usually referred to as 'N3', or in its partially deprotonated form (a di-tetra butylammonium salt) as 'N719' (Nazeeruddin *et al.*, 1993). Figure 9.3 shows the structure of this dye. The incorporation of carboxylate groups allows ligation to the film surface via the formation of bidendate and ester linkages, whilst the (–NCS) groups enhance the absorption of visible light. Adsorption of the dye to the mesoporous film is achieved by simple immersion of the TiO_2 film in a solution of dye, which results in the conformal adsorption of a self-assembled dye monolayer to the film surface. The counter electrode is fabricated from FTO-coated glass, with the addition of a Pt catalyst to catalyse the reduction of the redox electrolyte at this electrode. Electrical contact between working and counter electrodes is achieved by the redox electrolyte, with capillary forces being sufficient to ensure the electrolyte efficiently penetrates the film pores.

9.3.3 *Energetics of operation*

Figure 9.4 illustrates the energetics of DSSC operation. In contrast to silicon and organic devices, the high concentration (~0.5 M) of mobile ions in the electrolyte

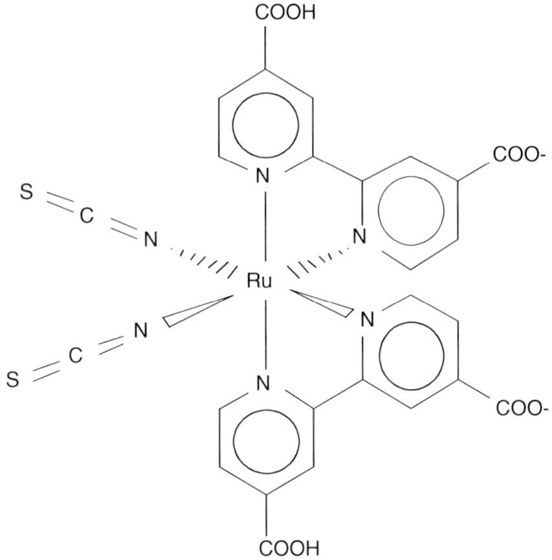

Figure 9.3 Chemical structure of the N719 ruthenium complex used as a charge-transfer sensitiser in dye-sensitised solar cells. The N3 dye has the same structure, except that all four carboxylates are protonated.

effectively screens out any macroscopic electric fields. Charge separation is therefore primarily driven by the inherent energetics (oxidation/reduction potentials) of the different species at the TiO_2/dye/electrolyte interface rather than by the presence of macroscopic electrostatic potential energy gradients. Similarly, charge-transport processes are primarily diffusive, driven by concentration gradients in the device generated by the photoinduced charge separation. Photoinduced charge separation occurs at the dye-sensitised TiO_2/electrolyte interface. Electron injection requires the dye excited state to be more reducing than the TiO_2 conduction band edge, enabling transfer of an electron from photoexcited dye into the metal oxide. This energetic requirement is equivalent to the dye excited-state lowest unoccupied molecular orbital (LUMO) having a lower electron affinity than the electrode conduction-band edge. Similarly, regeneration of the dye ground state by the redox couple requires the dye cation to be more oxidising than the iodide/triiodide redox couple. Charge separation in DSSCs can be regarded as a two-step redox cascade, resulting in the injection of electrons into the TiO_2 electrode and the concomitant oxidation of the redox electrolyte. Figure 9.4 shows typical values for the interfacial energetics in DSSCs. Both charge-separation reactions are thermodynamically downhill and can be achieved with near unity quantum efficiency in efficient devices.

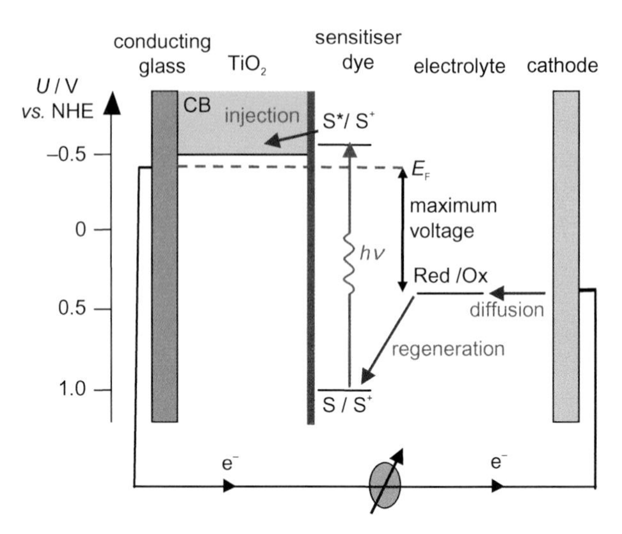

Figure 9.4 Energetics of operation of DSSCs. The primary free-energy losses are associated with electron injection from the excited sensitiser into the TiO_2 conduction band and regeneration of the dye by the redox couple. The voltage output of the device is approximately given by the splitting between the TiO_2 Fermi level (dashed line) and the redox potential of the redox electrolyte.

Following charge separation, charge collection from the device requires transport of the photogenerated charges to their respective electrodes. For an efficient DSSC under AM 1.5 solar irradiance, these charge fluxes are of the order of $20\,mA\,cm^{-2}$. The high ionic concentrations in the device effectively screen out any macroscopic electric fields, thereby removing any significant drift component of these transport processes. The transport of both electrons and redox ions is therefore primarily driven by diffusion processes resulting from concentration gradients. Under optimum conditions (e.g. good TiO_2 nanoparticle interconnections and a low-viscosity electrolyte), these charge-transport processes (electrons towards the FTO working electrode and the oxidised component of redox couple towards the counter electrode) can be efficiently driven with only modest concentration gradients, and therefore only small free-energy losses ($<50\,meV$). At the counter electrode, the oxidised redox couple is re-reduced, a platinum catalyst typically being used to enable this reaction to proceed with minimal overpotential (again $<50\,meV$). A similarly ohmic contact is achieved at the TiO_2/FTO interface. It is apparent that the energetics at the TiO_2/dye/electrolyte interface are of primary importance in determining the overall device energetics.

Power output from the DSSC requires not only efficient charge collection by the electrodes but the generation of a photovoltage, corresponding to a free-energy difference between the working and counter electrodes. In the dark at equilibrium, the Fermi energy of the TiO_2 electrode (corresponding to the free energy of electrons in this film after thermalisation) equilibrates with the mid-point potential of the redox couple, resulting in zero output voltage. Under these conditions, the TiO_2 Fermi level lies deep within the bandgap of the semiconductor, and the film is effectively insulating, with a negligible electron density in the TiO_2 conduction band. Photoexcitation results in electron injection into the TiO_2 conduction band and concomitant hole injection into (that is, oxidation of) the redox electrolyte. The high concentrations of oxidised and reduced redox couple present in the electrolyte in the dark mean that this photooxidation process does not result in a significant change in redox potential of the electrolyte, which remains effectively fixed at its dark, resting value. In contrast, electron injection into the TiO_2 conduction band results a dramatic increase in electron density (from approx. $\leq 10^{13}\,cm^{-3}$ to $\geq 10^{18}\,cm^{-3}$), raising the TiO_2 Fermi level towards the conduction-band edge, as illustrated in Fig. 9.4, and allowing the film to become conducting. This shift of the TiO_2 Fermi level under irradiation increases the free energy of injected electrons and is responsible for the generation of the photovoltage in the external circuit.

The mid-point potential of the redox couple is given by the Nernst equation, and is therefore dependent on the relative concentrations of iodide and iodine. The concentrations of these species required for efficient device function are in turn constrained by kinetic requirements of dye regeneration at the working electrode, and iodide regeneration at the counter electrode, as discussed below. Typical concentrations of these species for the iodine/iodide couple are in the range 0.1–0.7 M iodide and 10–200 mM iodine, constraining the mid-point potential of this electrolyte to ~0.4 V vs. NHE. It should furthermore be noted that in the presence of excess iodide, the iodine is primarily present in the form I_3^-, resulting in this electrolyte often being referred to as the iodide/triiodide redox couple.

Determining the energetics of the TiO_2 conduction band is more complex. As with most oxides, the surface of TiO_2 may be more or less protonated depending on the pH of the surrounding medium. The resultant changes in surface charge cause the surface potential to exhibit a Nernstian dependence on effective pH, shifting by 60 mV per pH unit (Rothenberger *et al.*, 1992). In bulk metal oxides, surface charge can result in significant bending of the conduction and valence bands adjacent to the surface. However, in the mesoporous TiO_2 films employed in DSSCs, the nanoparticles are too small to support significant band bending. As a consequence, the whole conduction band of such mesoporous films shifts with the surface potential. At pH 1 in aqueous solutions, the conduction band edge for TiO_2 has been reported at −0.06 V vs. NHE, shifting negatively as the pH is increased (Hengerer *et al.*, 2000). Dye-sensitised solar cells typically employ organic rather than aqueous electrolytes, complicating quantification of the effective pH. Nevertheless, studies in organic solvents have demonstrated shifts of the conduction band of mesoporous TiO_2 films of up to 1 V depending on the concentration of potential-determining ions (primarily small cations such as protons or lithium cations) in the electrolyte (Redmond and Fitzmaurice, 1993). For this reason, the concentration of such potential-determining ions in the electrolyte plays a key role in determining the energetics of the dye-sensitised interface, and thereby device performance. Additives added to the electrolyte to determine such energetics include Li^+, guanadinium ions, N-methylbenzimidazol and *t*-butyl pyridine (which functions as a base). Further influence on the interfacial energetics can be achieved by variation of the extent of protonation of the sensitiser dye. A further consideration is the presence of a tail of sub-bandgap states in the TiO_2 which function as localised electron trap states, as discussed elsewhere (O'Regan and Durrant, 2009).

The choice of suitable sensitiser dye energetics is essential to achieve suitable matching to the metal oxide and redox couple. The excited-state oxidation potential ($U_m(S^+/S^*)$) must be sufficiently negative to achieve efficient electron injection into the TiO_2 conduction band, whilst the ground-state oxidation potential must be sufficiently positive to oxidise the redox couple. The redox properties of adsorbed sensitiser dyes may differ significantly from those measured in solution, mainly due to the high surface-charge densities and dipoles present at this interface (Zaban *et al.*, 1998). Notwithstanding this, typical dye redox potentials empirically found to be compatible with efficient device function are U_m $(S^+/S) > 0.6$ V *vs.* NHE.

9.3.4 *Kinetics of operation*

Figure 9.5 is a photochemical view of the function of a DSSC, illustrating the sequence of electron transfer and charge-transport processes which result in photovoltaic device function. In addition to the forward electron transfer and transport processes, this figure also illustrates several competing loss pathways, shown as black arrows. These loss pathways include decay of the dye excited state to ground, and charge recombination of injected electrons with dye cations and with the redox couple. Each charge-transfer step results in an increased spatial separation of electrons and holes, increasing the lifetime of the charge-separated state, but at the expense of reducing the free energy stored in this state. This functionality exhibits a close parallel to function of photosynthetic reaction centres. As in natural photosynthesis, kinetic competitions between the various forward and loss pathways are critical to determining the quantum efficiencies of charge separation and collection, and are therefore key factors determining energy conversion efficiency.

The theory of interfacial electron transfer, and the dependence of electron-transfer rate on interfacial energetics, are discussed in Volume 3 of this book series,

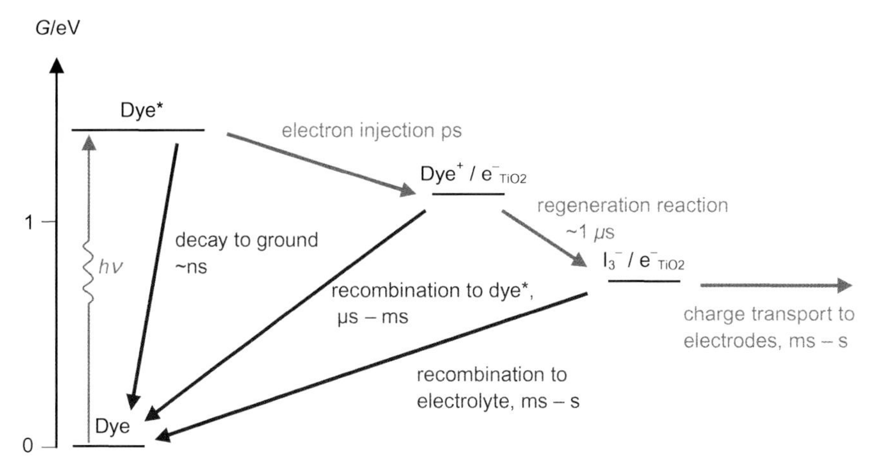

Figure 9.5 State diagram representation of the kinetics of DSSC function. Forward processes of light absorption, electron injection, dye regeneration and charge transport are indicated by red arrows. The competing loss pathways of excited-state decay to ground and electron recombination with dye cations and oxidised redox couple are shown in black. The vertical scale corresponds to the free energy G stored in the charge-separated states. Note the free energy of injected electrons is determined by the Fermi level of the TiO_2; the figure is drawn assuming a TiO_2 Fermi level 0.8 V above the chemical potential of the redox electrolyte, corresponding to the Fermi level of a typical iodide/iodine-based DSSC at open circuit. It should be noted that at maximum power point, the Fermi level is lower due to charge transport to the electrodes (also shown in diagram).

and the dynamics of charge transport in mesoscopic electrodes are addressed in Chapter 11 of that volume. In this chapter we will therefore confine ourselves only to consideration of the impact of these dynamics on the performance of DSSCs. More detailed reviews of these dynamics have been reported elsewhere (Hagfeldt *et al.*, 2010; Listorti *et al.*, 2011).

Efficient electron injection requires the rate of electron injection to exceed excited-state decay to ground. Typical rates of dye excited-state decay to ground are in the range 10^7–$10^{10}\,s^{-1}$. The rate of electron injection depends on the electronic coupling between the dye excited-state LUMO orbital and accepting states in the TiO_2, and on the relative energetics of these states. In model system studies of dye-sensitised metal oxide films, electron injection rates of $>10^{12}\,s^{-1}$ have been reported for a range of sensitiser dyes, consistent with efficient electron injection (Anderson and Lian, 2005; Durrant *et al.*, 2006). However it should be noted that fast electron-injection dynamics require both strong electronic coupling of the dye LUMO orbital to the metal oxide conduction-band states, and a sufficient free-energy difference to drive the reaction (Schmickler, 1996). As such, electron-injection dynamics are dependent on the energetics of the TiO_2 conduction band, and therefore on the concentration of potential-determining ions (e.g. Li^+) in the electrolyte. Omission of such ions from the electrolyte can result in an insufficient energetic driving force, reducing the quantum yield of charge injection, and thereby reducing device photocurrent.

It should be noted that the heavy metal ion in the ruthenium bipyridyl dyes widely employed in DSSCs results in an ultrafast (up to $10^{13}\,s^{-1}$) intersystem crossing from the initially formed singlet excited state to a triplet state. This relaxation process reduces the excited-state free energy by ~400 meV. While several studies have reported sub-picosecond electron injection from the singlet excited state of these dyes, electron injection from the triplet state is significantly slower (10^{10}–$10^{11}\,s^{-1}$), consistent with the lower free energy of this excited state, but still fast compared with decay of the triplet state to ground (10^7–$10^8\,s^{-1}$) (Anderson and Lian, 2005; Durrant *et al.*, 2006). Studies of complete devices have suggested that this relatively slow electron injection from the dye triplet state may be the dominating injection pathway in efficient DSSCs (Haque *et al.*, 2005; Koops *et al.*, 2009).

Efficient dye regeneration requires the rate of re-reduction of the dye cation by the redox couple to exceed that of the charge recombination of injected electrons with these dye cations. This recombination reaction has been shown to be strongly dependent on the electron density in the TiO_2 electrode (and therefore light intensity and cell voltage), accelerating by at least an order of magnitude between short-circuit and open-circuit conditions (Durrant *et al.*, 2006). It is furthermore dependent on the spatial separation of the dye cation (highest occupied molecular orbital [HOMO]) orbital from the metal oxide surface, with the rate constant decaying exponentially with distance, consistent with electron tunneling theory. The regeneration reaction is dependent on the iodide concentration, electrolyte viscosity and dye structure. For the N719 sensitiser dye, and employing a low-viscosity electrolyte such as acetonitrile, the regeneration reaction has a half-time of ~1 μs, sufficiently fast to compete effectively with the competing recombination reaction and ensuring that the regeneration reaction can be achieved with unit quantum efficiency (Green *et al.*, 2005).

Efficient charge collection by the external circuit requires the time constant for electron transport within the TiO_2 matrix to be faster than charge recombination of injected electrons with the redox couple. Electron transport is a diffusive process, strongly influenced by electron trapping in localised sub-bandgap states, resulting in the dynamics being strongly dependent on the position of the TiO_2 electron Fermi level: raising the Fermi level towards the conduction-band edge results in increased trap filling. Typical electron-transport times under solar irradiation are of the order of milliseconds (Peter and Wijayantha, 2000; Frank *et al.*, 2004). If a low-viscosity solvent such as acetonitrile is used, transport of the oxidised redox couple to the counter electrode is not rate-limiting. However, the use of higher viscosity solvents more suitable for practical device applications (for reasons of device stability) can result in significantly lower ionic diffusion constants, with the resultant iodide/triiodide concentration gradients causing significant free-energy (series resistance) losses, and also potentially accelerating interfacial charge recombination.

Given the relatively slow timescale for charge transport in DSSCs compared with most other photovoltaic devices, and the extensive interfacial area available for charge recombination in the device (due to its mesoscopic structure), it is remarkable that the quantum efficiency of charge collection can approach unity. The key factor enabling this high efficiency is the slow rate constant for the interfacial charge recombination of injected electrons with the oxidised redox couple. This reaction is a multi-electron reaction, most simply being described by the equation

$$I_3^- + 2e^- \rightarrow 3I^- \tag{9.1}$$

which must therefore proceed via one or more intermediate states. The mechanism of this reaction has been extensively studied, and while the details remain somewhat controversial, it is apparent that without a suitable catalyst such as platinum, one or more of the intermediate steps exhibits a significant activation barrier, resulting in a slow overall rate constant for this reaction. The low rate constant for this recombination reaction on TiO_2, contrasting to the facile electrochemistry of this redox couple on the platinised counter electrode, is a key factor behind the remarkable efficiencies achieved to date for DSSCs. Similarly, interface engineering through sensitiser design to minimise this recombination rate constant has been key to recent successes with alternative one electron couples (Feldt *et al.*, 2010). Nevertheless, Eq. (9.1) is the primary recombination pathway in DSSCs. The flux of this recombination pathway increases with increasing electron density in the TiO_2 electrode (and therefore with the TiO_2 Fermi level or cell voltage). In the dark it is responsible for the diode-like leakage current observed in current–voltage scans, while under illumination it is the primary factor limiting the voltage output of the device.

The kinetic competition between charge transport and recombination in DSSCs has been analysed in terms of an effective carrier diffusion length L_n, given by

$$L_n = \sqrt{D_{\text{eff}}\,\tau} \tag{9.2}$$

where D_{eff} is the effective electron diffusion coefficient, and τ the electron lifetime due to the charge-recombination reaction given by Eq. (9.1) (Peter and Wijayantha, 2000). D_{eff} increases with light intensity (due to the increased electron density in the TiO_2 film) whilst τ shows a proportional decrease, resulting in L_n being largely independent of light intensity. Typical values for L_n are 5–20 μm, and even longer near the optimum power point for cells with $> 10\%$ conversion efficiency, consistent with the high carrier-collection efficiencies observed in efficient DSSCs.

It is important to emphasise that the energetics and kinetics of DSSC function are not independent considerations. The kinetics of the interfacial electron-transfer dynamics depend strongly on the energetics of the TiO_2/dye/electrolyte interface and on the density of electrons in the TiO_2 (i.e. the TiO_2 Fermi level). Raising the energy of the TiO_2 conduction band reduces recombination losses (as for a given TiO_2 Fermi level, the electron density in the TiO_2 film will be lower), and therefore may give a high cell output voltage, but at the expense of a lower free energy driving force for charge separation, which may result in a lower quantum efficiency for charge generation and therefore a lower output current. In practice, modulation of these energetics and kinetics to achieve optimum device performance remains one of the key challenges in DSSC research and development.

9.4 DSSC research and development

9.4.1 *Panchromatic sensitisers*

The ideal sensitiser for a single-junction photovoltaic cell converting standard global AM 1.5 sunlight to electricity should absorb all light below a threshold wavelength of about 920 nm. In addition, it must also carry attachment groups such as carboxylate or phosphonate to firmly graft it to the semiconductor oxide surface. On excitation, it should inject electrons into the solid with a quantum yield of unity. The energy level of the excited state should be well matched to the lower bound of the conduction band of the oxide to minimise energetic losses during the electron-transfer reaction. Its redox potential should be sufficiently high that it can be regenerated via electron donation from the redox electrolyte or the hole conductor. Finally, it should be stable enough to sustain about 100 million turnover cycles, corresponding to about twenty years of exposure to natural light. A single-junction device with such a sensitiser could reach a maximum conversion efficiency of 32% in global AM 1.5 sunlight.

Much of the research in DSSC dye chemistry is devoted to the identification and synthesis of sensitisers matching these requirements, while retaining stability in the photoelectrochemical and photovoltaic environment. A number of recent reviews reflects the impressive scientific thrust of this research (Imahori *et al.*,

2009; Hagfeldt *et al.*, 2010; Mishra *et al.*, 2010; Ning *et al.*, 2010; Walter *et al.*, 2010; Qin and Peng, 2012; McEvoy and Grätzel, 2013). The attachment group of the dye ensures that it spontaneously assembles as a molecular layer on exposing the oxide film to the dye solution. This molecular dispersion ensures a high probability that, once a photon is absorbed, the excited state of the dye molecule will relax by electron injection into the semiconductor conduction band. However, the optical absorption of a single monolayer of dye is weak, a fact which originally was cited as ruling out the possibility of high-efficiency sensitised devices, as it was assumed that smooth substrate surfaces would be imperative in order to avoid the recombination loss mechanism associated with rough or polycrystalline structures in solid-state photovoltaics. This objection was invalidated by recognising that the injection process places electrons in the semiconductor lattice, spatially separated from the positive charge carriers in the electrolyte or a hole carrier. Hence, mastering the interface is key to slow down the rate of charge recombination. The dye molecules themselves, if judiciously designed, can block the access of the redox shuttle to the interface providing a barrier for charge recombination. By now, the use of nanocrystalline thin-film structures with a roughness factor of >1000 has become standard practice.

Polypyridyl complexes of ruthenium and osmium have over many years maintained a lead in photovoltaic performance in terms of both conversion yield and long-term stability. Initially, interest focused on sensitisers having the general structure $ML_2(X)_2$ where L stands for 2,2′-bipyridyl-4,4′-dicarboxylic acid, M is Ru or Os, and X represents a halide, cyanide, thiocyanate, acetyl acetonate, thiocarbamate or water substituent. In particular, the ruthenium complex *cis*-$RuL_2(NCS)_2$, known as N3 dye, became the paradigm heterogeneous charge-transfer sensitiser for mesoporous solar cells (Nazeeruddin *et al.*, 1993). The absorption spectrum of fully protonated N3 has maxima at 518 and 380 nm, with extinction coefficients of $1.3 \times 10^4\,M^{-1}cm^{-1}$ and $1.33 \times 10^4\,M^{-1}cm^{-1}$, respectively. The complex emits at 750 nm, the excited-state lifetime being 60 ns. The optical transition has MLCT (metal-to-ligand charge-transfer) character: excitation of the dye involves transfer of an electron from the metal to the π^* orbital of the surface-anchoring carboxylated bipyridyl ligand, from where it is released in a timescale of femtoseconds to picoseconds into the conduction band of TiO_2, generating electric charges with near unit quantum yield.

Discovered in 1993, the photovoltaic performance of N3 was unmatched for eight years by virtually hundreds of other complexes that have been synthesised and tested. However, in 2001 the 'black dye' tri(cyanato)-2,2′2″-terpyridyl-4,4′4″-9-tricarboxylate)Ru(II) achieved 10.4% AM1.5 solar-to-power conversion efficiency in full sunlight (Nazeeruddin *et al.*, 2001). Conversion efficiencies have meanwhile

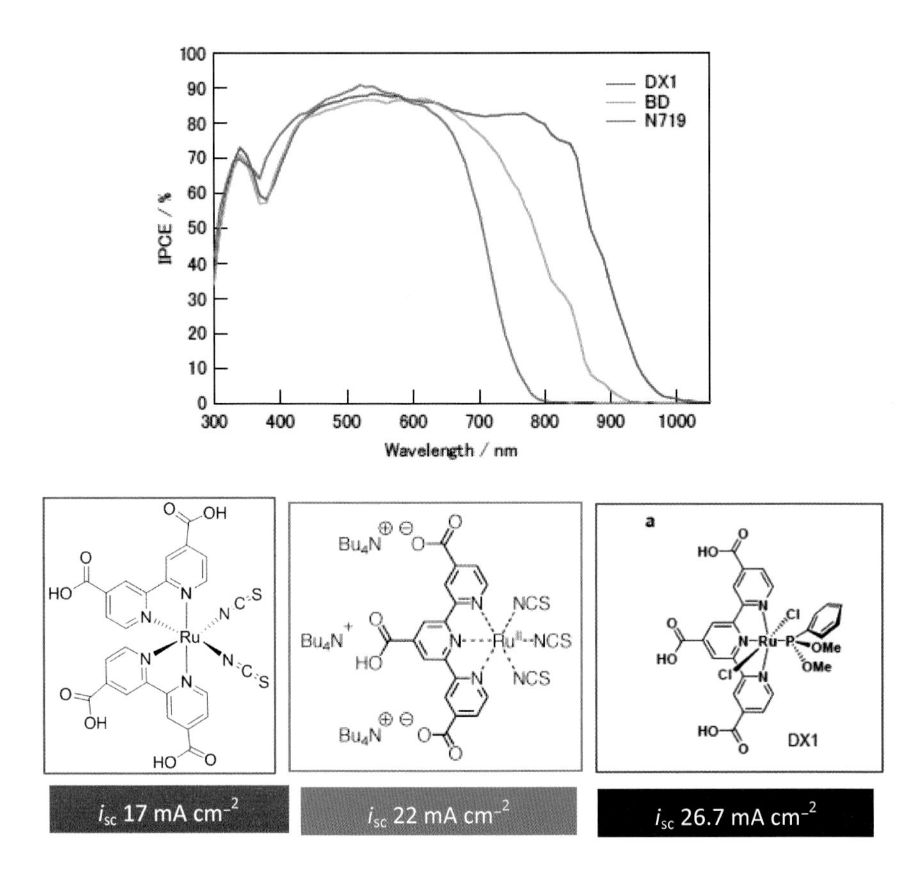

Figure 9.6 Spectral response of the photocurrent for the N3, black dye and DX1 sensitisers. The incident photon-to-current conversion efficiency (IPCE) is plotted as a function of excitation wavelength. The structure of the sensitisers and their short-circuit photocurrent is shown below the graph.

been improved further, the current record validated by an accredited laboratory being 11.9% (Han *et al.*, 2012). In the latter case an organic co-sensitiser was employed to enhance the light harvesting in the blue spectral region.

Figure 9.6 compares the spectral response of the photocurrent observed with the N3 dye, the black dye and the DX1 sensitisers. The incident photon-to-current conversion efficiency (IPCE), also referred to as the external quantum efficiency (EQE) of the DSSC, is plotted as a function of excitation wavelength. Both chromophores show very high IPCE values in the visible range. However, the response of the black dye extends 100 nm further into the infrared (IR) than that of N3. The photocurrent onset is close to 920 nm, near the optimal threshold for single-junction converters. From there the IPCE rises gradually until at 700 nm it reaches a plateau of ~80%. If one accounts for reflection and absorption losses in the conducting glass, the conversion of incident photons to electric current is practically quantitative over the whole visible domain. From the overlap integral of the curves in Fig. 9.6 with the AM 1.5 solar spectrum, the short-circuit photocurrents of the N3, black dye and DX1 sensitisers cells are predicted to be 17 mA cm^{-2}, 22 mA cm^{-2} and 27.6 mA cm^{-2} under standard AM 1.5 G reporting conditions, respectively, in agreement with experimental observations.

Thus, over the last decade a whole new family of judiciously engineered ruthenium complexes has emerged that show excellent photovoltaic performance (Wang *et al.*, 2005a; Reddy *et al.*, 2007; Grätzel, 2009; Kalyanasundaram and Grätzel, 2010). The spectral response of the photocurrent for of the DX1 dye, shown by the blue curve in Fig. 9.6, illustrates the amazing advance in this area (Kinoshita *et al.*, 2013). By replacing the three thiocyanate groups in the black dye with two chloride ions and one phosphine ligand, a new panchromatic sensitiser, coded DX1, was fashioned that harvests sunlight over the whole visible and near-IR range up to 1000 nm, closely matching the spectral response of silicon photovoltaic cells and generating a short circuit photocurrent of 27.6 mA cm^{-2}.

Recently, new donor acceptor porphyrins have been developed, in particular the YD2-o-C8 zinc complex, whose PCE exceeds that of the black dye when used in conjunction with Co((II/III)(bipy)$_3$ complexes as a redox couple (Yella *et al.*, 2011). The Nernst potential of this redox mediator is more positive than that of the iodide/triiodide couple, yielding substantial gains in open-circuit potential of the cells. The performance of YD2-o-C8 is enhanced by judicious substitution of two meso-positions at the tetrapyrrole ring with a donor and acceptor moiety and the two remaining meso-positions bearing phenyl substituents with two bulky alkoxy groups in ortho-positions. This increases the light-harvesting capacity of the porphyrin and blocks the unwanted interfacial electron back transfer. Further gain in efficiency was obtained via the use of a co-sensitiser, coded Y123, absorbing strongly in the green spectral region. The structure of the two sensitisers is shown in Fig. 9.7, which presents a plot of the photocurrent density versus voltage for a mixture of the two dyes, adsorbed at the surface of the nanocrystalline titania film and the IPCE spectra for this. The i_{sc}, V_{oc} and fill factor values under standard air mass (AM 1.5) reporting conditions were 17.8 mA cm^{-2}, 0.935 V and 0.74, respectively, yielding a power conversion efficiency of 12.3%.

9.4.2 *Present status of DSSC performance*

The solar-to-electric power conversion efficiency (PCE) of PV cells is calculated from the short-circuit photocurrent density (i_{sc}), the open-circuit photovoltage

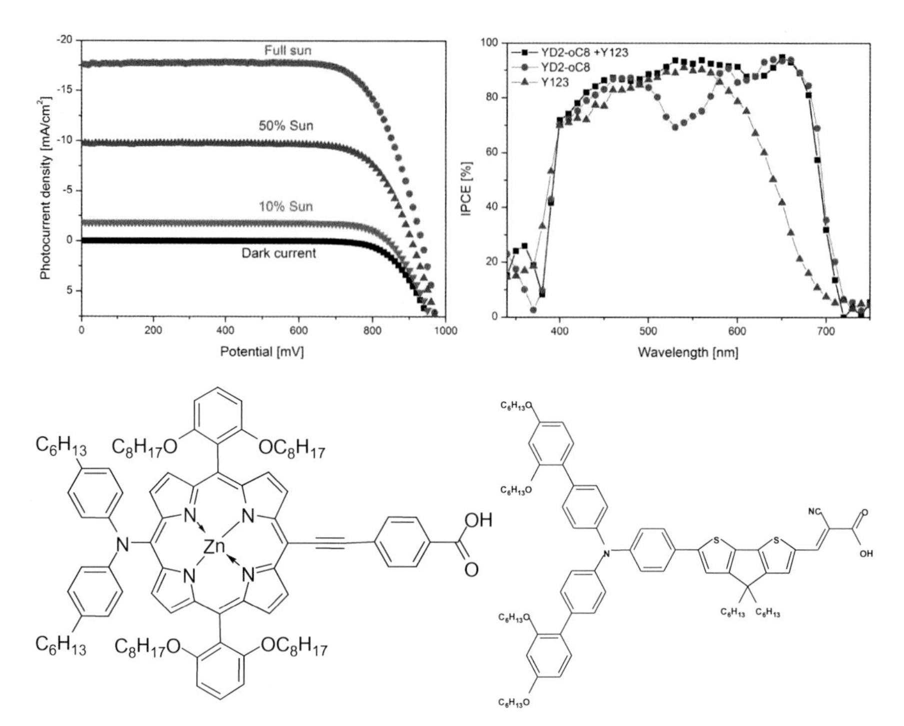

Figure 9.7 Top left: current–voltage characteristics of a YD2-*o*-C8/Y123 co-sensitised DSSC, measured under standard AM 1.5 G sunlight under various light intensities. The molar YD2-*o*-C8/Y123 ratio in the dye solution was 7. Top right: spectral response of the incident photon-to-electric current conversion efficiency (IPCE) for YD2-*o*-C8 (red dots), Y123 (blue triangles) and YD2-*o*-C8/Y123 co-sensitised nanocrystalline TiO_2 films (black squares). Bottom left: chemical structure of the sensitiser YD2-o-C8. Bottom right: chemical structure of the sensitiser Y123.

(V_{oc}), the fill factor (η_{fill}) and the standard incident solar irradiance (E_s^o = 1000 W m^{-2}) as

$$\mathrm{PCE} = i_{sc} \times V_{oc} \times \eta_{fill}/E_s^o \tag{9.3}$$

The PCE value of 12.3% achieved with the YD2-o-C8/Y123 co-sensitised titania films present a new record for laboratory cells based on redox electrolytes. However, solid-state DSSC embodiments based on perovskite light harvesters achieve currently a PCE of 15% (Burschka *et al.*, 2013). Due to significant industrial upscaling efforts, the conversion efficiency of DSSC modules has also been rapidly rising over the last few years. Using the black ruthenium dye along with an organic co-sensitiser, Sony Corporation has reached a certified PCE efficiency under AM 1.5 standard conditions of 9.9% (Green *et al.*, 2011). This implies that scaling up dye-sensitised laboratory cells by a factor of at least 100 does not entail a significant decrease in photovoltaic performance, the loss being mainly due to interconnects or collector grids reducing the active device area.

Further substantial gains in the PCE of the DSSC can be expected in the near term by implementing the substantial increase in the open-circuit photovoltage that can be achieved with cobalt complexes rather than iodide/triiodide as redox electrolyte. For example, a very high V_{OC} of 1.1 V was recently reached with the cobalt bisbipyridyl pyrazol complex (Yum *et al.*, 2012) whose redox potential matches well that of the sensitiser ground state, reducing the free-energy loss in the regeneration step from 0.6 eV for I^-/I_3^- to 0.2 eV, as shown by the energy level diagram in Fig. 9.8.

9.4.3 *Tandem cells*

A feature that makes the DSSC particularly attractive for tandem cell application is that its optical transmission and short-circuit photocurrent can be readily adjusted by changing the film thickness, pore size, nature of the dye, and dye loading. This, along with the ease of forming layered structures (for example, by producing the mesoscopic oxide films using screen-printing or doctor blading methods), renders the DSSC particularly well suited for the fabrication of tandem solar cell structures that capture the solar spectrum in an optimal fashion. Note that for a two-level tandem cell a conversion efficiency of up to 45% can be reached, the optimal bandgaps for the top and bottom cells being around 1.65 eV and 1 eV, respectively.

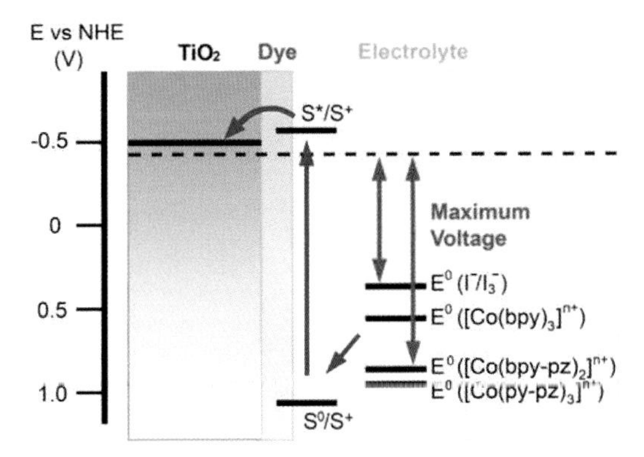

Figure 9.8 Energy-level diagram of a DSSC. Shown are the ground and excited-state redox potentials for a typical sensitiser as well as for the iodide/triiodide redox couple and a series of cobalt complexes; bpy-pz and py-pz represent bipyridylpyrazol and pyridylpyrazol, respectively.

Several publications have dealt with the use of stacked DSSC configurations in which two dyes absorbing different parts of the solar spectrum serve as sensitisers (Durr *et al.*, 2004; Kubo *et al.*, 2004). We have demonstrated that a tandem device comprising a DSSC as a top cell for high-energy photons and a copper indium gallium selenide (CIGS) thin-film bottom cell, capturing the red and near-IR solar emission respectively, produces AM 1.5 solar-to-electric conversion efficiencies greater than 15% (Liska *et al.*, 2006).

Figure 9.9 shows the I–V curve and other characteristics obtained under AM 1.5 insolation for one of the first embodiments of this two-terminal tandem device. The performance of the tandem was clearly superior to that of the individual cells, despite the fact that the short-circuit currents of the two cells were not perfectly matched, as evidenced by the shoulder in the I–V curve. Likewise, no effort was made to minimise the optical losses. This leaves no doubt that further efficiency gains reaching well beyond the 20% mark can be expected from the fructuous marriage of these two thin-film PV technologies (Liska *et al.*, 2006). Combining a relatively low-cost thin-film CIGS substrate cell with a DSSC superstrate cell may be a cheaper method of achieving efficiencies above 15% than use of a high-efficiency CIGS cell alone.

However, in order to minimise optical and ohmic losses it is mandatory to integrate the two cells in a monolithic device in which they are series-connected

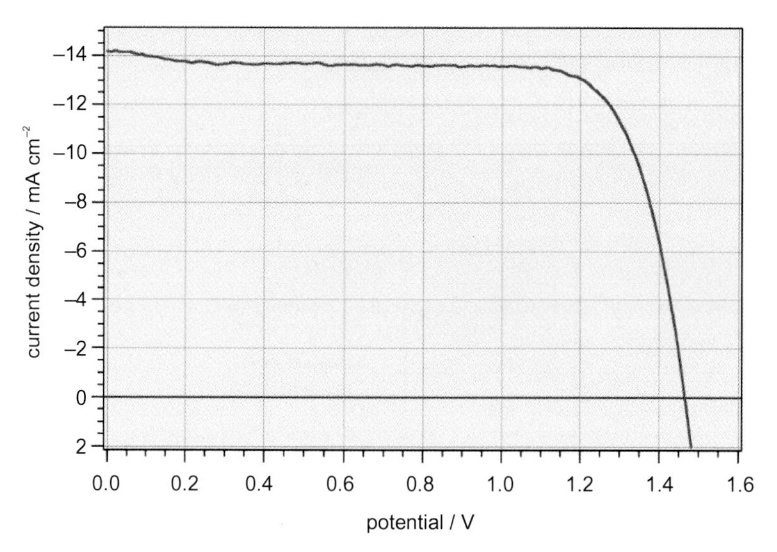

Figure 9.9 Photocurrent density–voltage characteristics under AM 1.5 full sunlight (1000 mW cm^{-2}) for a two-terminal tandem DSSC/CIGS cell. The PCE is 15.8%. Adapted from Liska *et al.* (2006).

through a transparent recombination layer. Attempts in this direction have been made (Wenger *et al.*, 2009) but the ZnO front contact of the CIGS cell was unstable in the presence of the iodide/triidode-based electrolyte. The advent of perovskite-sensitised solid-state cells has rekindled interest in such two-cell tandem structures, and several groups are actively working on devices of this type. Indeed, perovskite photovoltaics are very attractive candidates for the top cell of a two-cell tandem with CIGS or silicon as a bottom cell because their bandgap can be adjusted to a value between 2.1 and 1.5 eV, for example by using mixtures of iodide and bromide as anions (Noh *et al.*, 2013). Moreover, the open-circuit voltages of perovskite cells are amazingly high, reaching 1.1 V for the 1.5 eV bandgap, close to the theoretical maximum of 1.2 V. Realising tandem cells with PCEs >20% has therefore come within close reach.

9.4.4 *Stability*

Unlike amorphous silicon, which suffers from degradation due to the well-known Staebler–Wronski effect, the intrinsic stability of the DSSC has been confirmed by extensive accelerated light-soaking tests carried out over the last fifteen years. One major issue that has been settled during this period is that the sensitisers employed in the current DSSC embodiments can sustain twenty years of outdoor service without significant degradation. However, as new and more advanced dye structures emerge, and in order to avoid repeating these lengthy tests every time the sensitiser is modified, kinetic criteria have been developed to allow prediction of long-term sensitiser performance.

Figure 9.10 illustrates the catalytic cycle that the sensitiser undergoes during cell operation. Critical for stability are side reactions that occur from the excited state (S*) or the oxidised state of the dye (S^+), which would compete with electron injection from the excited dye into the conduction band of the mesoscopic oxide

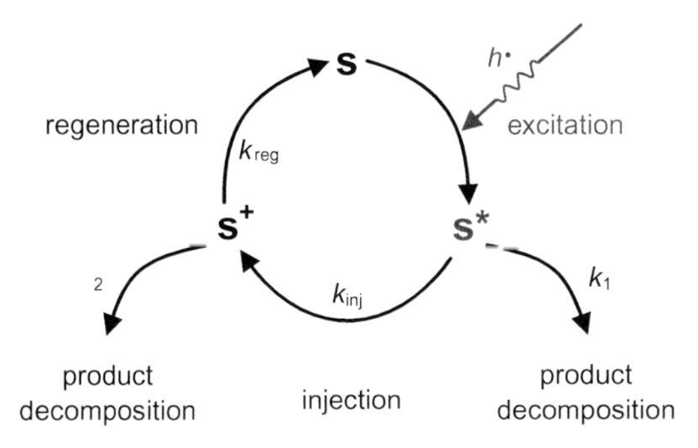

Figure 9.10 The catalytic cycle of the sensitiser during DSSC operation.

and with the regeneration of the sensitiser. These destructive channels are assumed to follow first or pseudo-first-order kinetics and are assigned the rate constants k_1 and k_2. Introducing the two branching ratios $P_1 = k_{inj}/(k_1 + k_{inj})$ and $P_2 = k_{reg}/(k_2 + k_{reg})$, where k_{inj} and k_{reg} are the first-order or pseudo-first-order rate constants for the injection and regeneration process, respectively, the fraction of the sensitiser molecules that survives n cycles is given by the product $(P_1 \times P_2)^n$. A simple calculation (Grätzel, 2006) shows that the sum of the branching ratios for the two bleeding channels should not exceed 1×10^{-8} in order for the sensitiser to turn over 100 million times, corresponding to a lifetime of at least twenty years. The turnover frequency of the dye in the working DSSC, averaged diurnally and seasonally, is about $0.16\,s^{-1}$.

For most of the common sensitisers, the rate constant for electron injection from the excited state of the dye to the conduction band of the TiO_2 particles is in the picosecond or femtosecond range. Assuming $k_{inj} = 10^{10}$–$10^{13}\,s^{-1}$, any destructive side reaction should have $k_1 < 10^2\,s^{-1}$. Ruthenium sensitisers of the N719 or C106 type readily satisfy this condition as the decomposition from the excited-state level occurs at a much lower rate than the $10^2\,s^{-1}$ limit. Precise kinetic information has also been gathered for the second destructive channel involving the oxidised state of the sensitiser, the key parameter being the ratio k_2/k_{reg} of the rate constants for the degradation of the oxidised form of the sensitiser and its regeneration. The S^+ state of the sensitiser can readily be produced by chemical or electrochemical oxidation and its lifetime can be independently determined by absorption spectroscopy. A typical value of k_2 is around $10^{-4}\ s^{-1}$ while the regeneration rate constant is at least in the $10^5\,s^{-1}$ range. Hence the branching ratio is well below the limit of 10^{-8}, which can be tolerated to achieve the 100 million turnovers and a twenty-year lifetime for the sensitiser. Apart from the photoinduced degradation of the sensitiser during the catalytic cycle, its chemical reaction with constituents of the electrolyte or its desorption or its mere reorganisation on the surface to a non-productive conformation could possibly harm cell performance in the long term. An appropriate way to rule out the existence of any problems of this type is to carry out accelerated tests under heat stress.

Many long-term tests that have been performed with the N3-type ruthenium complexes have confirmed the extraordinary stability of these charge-transfer sensitisers. For example, a European consortium supported by the Joule program (Hinsch *et al.*, 2001) confirmed cell photocurrent stability during 8,500 hours of light soaking at 2.5 Suns, corresponding to about 56 million turnovers of the dye without any significant degradation. More recently, DSSC lifetimes of over 20 years have been projected from continuous light-soaking tests performed over

20,000 hours (Harakirisun and Desilvestro, 2011). These results corroborate the projections from the kinetic considerations made above.

A more difficult task has been to achieve stability under prolonged stress at higher temperatures, i.e. 80–85°C. The introduction of hydrophobic sensitisers has been particularly rewarding in allowing DSSCs to meet, for the first time, the specifications laid out for outdoor applications of silicon photovoltaic cells (Wang *et al.*, 2003a). In addition these dyes show enhanced extinction coefficients owing to the extension of the π-conjugation onto one of the bipy ligands by styrene moieties. Taking advantage of these properties and using a novel robust electrolyte formulation, a $\geq$8% efficient DSSC that shows strikingly stable performance under both prolonged thermal stress and light soaking was realised (Wang *et al.*, 2005b; Sauvage *et al.*, 2011). Dyesol recently announced completion of a high-temperature test carried out with a new electrolyte formulation based on sulfone, a polar non-toxic high boiling point solvent and using the above mentioned Y123 dye as a sensitiser. Impressively, stable performance was maintained over 3000 hours, establishing beyond any doubt that the DSSC can survive long-term heat exposure without any significant degradation.

While impressive progress has been made in the development of stable, non-volatile electrolyte formulations, the efficiencies of these systems are presently in the 7–10% range (Sauvage *et al.*, 2011), below the 11.9% reached with Ru-based sensitisers and volatile solvents. Future research efforts will be dedicated to reducing the performance gap between these systems. The focus will be on hole conductors and solvent-free electrolytes, in particular ionic liquids. The latter are a particularly attractive choice for the first commercial modules, owing to their high stability, negligible vapour pressure and excellent compatibility with the environment. Fujikura (Arakawa *et al.*, 2009) has announced excellent stability data with cells based on solid (gel) 'nanocomposite' electrolytes based on a mixture of ionic liquids and silica nanoparticles. Modules showing 8% PCE passed all accelerated stability tests required by the IEC 61646 protocol for outdoor PV panels including exposure to 85% humidity at 85°C for 1000 hours, long-term light soaking and cycling the temperature of the modules between −40°C and +90°C. These results confirm the extraordinary robustness of dye-sensitised solar cells when employed in conjunction with appropriately designed electrolytes, showing their aptness for outdoor deployment.

In keeping with the encouraging results obtained by Fujikura, the Israeli company 3GSolar announced that its DSSC modules have operated on the company's rooftop continuously for two years and they continue to perform to the same standard as on the day they were placed outdoors (3GSolar, 2011). Importantly, 3GSolar and ECN (DeWild-Scholten and Veltkamp, 2007) have also performed life-cycle analysis, which shows the energy payback time for the DSSC to be less than one year in southern European climates as compared with about three years for silicon solar cells at that time; a more recent analysis shows the latter has improved to 1.3–1.5 years (Mann *et al.*, 2013).

9.4.5 *Organic dyes*

Over the last two decades, ruthenium complexes endowed with appropriate ligands and anchoring groups have by far been the preferred choice of charge-transfer sensitisers for mesoscopic solar cells. In recent years, however, there has been a surge of interest in organic donor–acceptor dyes (Qin *et al.*, 2008). The groups of Arakawa and Uchida have made remarkable advances in the development of organic dyes for use in DSSCs (Horiuchi *et al.*, 2004; Hara *et al.*, 2005). The majority of these sensitisers are of the 'push-pull' type comprising a donor and acceptor group bridged by a π-conducting moiety. Coumarin or polyene-type sensitisers have attained PCE values of up to 9.2% in full sunlight (Ito *et al.*, 2006), increasing to 9.5% for the indoline dye D205 (Ito *et al.*, 2008). Figure 9.11 shows examples of typical structural elements of such D–π–A dyes. Numerous representatives of this class of compounds have been synthesised (Yum *et al.*, 2009), and PCE values up to 9.8% have been reported with iodide-based electrolytes (Zhang *et al.*, 2009). This rapid development has led to the discovery of a new class of D–π–A dyes where the

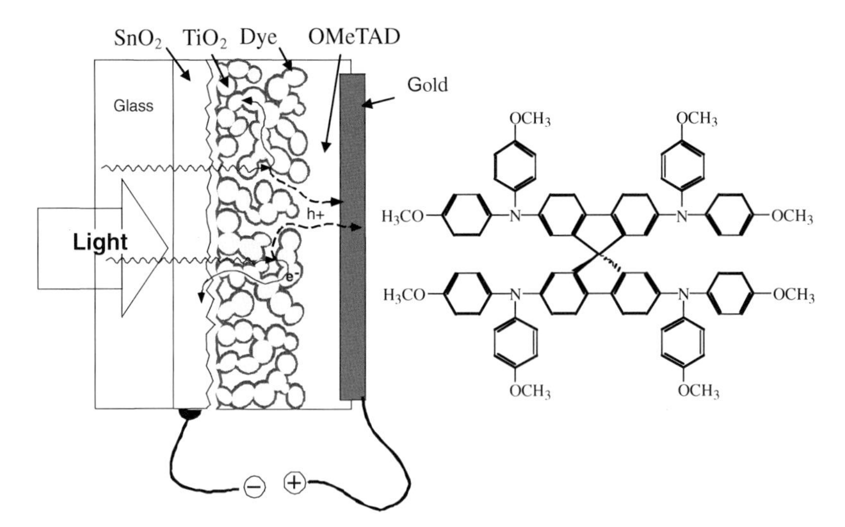

Figure 9.11 Cross-sectional view of a solid-state DSSC containing the hole conductor *spiro*-OMeTAD, the structure of which is indicated on the right. Left-hand drawing courtesy of B. O'Regan.

π-bridge is constituted by a porphyrin moiety (Hsieh *et al.*, 2010; Bessho *et al.*, 2010).

The number of suitable options for D–π–A dye structures being very large, state-of-the-art theoretical chemical calculations are being employed as a guide for the selection of the most promising candidates for synthesis. The recently developed scaled opposite-spin configuration interaction singles-doubles technique, abbreviated as SOS-CIS(D) (Rhee and Head-Gordon, 2007), appears to offer great accuracy in calculating the UV–vis spectra of novel D–π–A structures (Casanova *et al.*, 2010) This new method has advantages over time-dependent density functional theory in computing the energy of charge transfer type optical excitations and will assist the experimentalists in the judicious selection of molecular components to engineer the best-performing push charge-transfer sensitisers.

When considering organic dyes for use in DSSCs, porphyrins and phthalocyanines have attracted particular attention, the former because of the analogy with natural photosynthetic processes, the latter because of their photochemical and phototherapeutic applications. Through judicious molecular engineering, porphyrins have become the leading photosensitiser with PCE values reaching 12.3%, as shown in Fig. 9.7 (Yella *et al.*, 2011). Phthalocyanines show intense absorption bands in the 600–800 nm spectral region. Previous problems with aggregation and the unsuitable energetic position of the LUMO level, which was too low for electron transfer to the TiO_2 conduction band to occur efficiently, have now been largely overcome (Ince *et al.*, 2012), rendering this class of sensitisers also very attractive for DSSC applications in buildings where the attractive blue-green tinge of the glass panels is very much in demand by architects.

9.4.6 *Quantum dots as sensitisers*

Semiconductor quantum dots (QDs) are another attractive option for panchromatic sensitisers. These are II–VI and III–V type semiconductor particles whose size is small enough to produce quantum confinement effects. The absorption spectrum of such quantum dots can be adjusted by changing the particle size. Thus, the bandgap of materials such as InAs and PbS can be adapted to attain the value of 1.35 eV, which is ideal for a single-junction solar quantum converter. During the last two decades a wealth of information has been gathered on the physical properties of QD materials and research is being pursued very actively. With liquid electrolytes, photocorrosion of the quantum dots is likely to happen. However, quantum dots are expected to display higher stability in solid-state heterojunction devices (Plass *et al.*, 2002). Hence, recent work has focused on solid-state cells where a layer of quantum dots deposited on a mesoscopic oxide film such as ZnO or TiO_2 forms a depleted heterojunction (Pattantyus-Abraham *et al.*, 2010; Barkhouse *et al.*, 2011;

Sargent, 2012). These simple devices function surprisingly well, exhibiting high short-circuit photocurrents that exceed 20 mA cm^{-2} under full sunlight, the highest certified PCE being 7.5%.

Multiple exciton generation (MEG) can be obtained from the absorption of a single photon by a quantum dot if the photon energy is at least two times higher than its bandgap (Nozik, 2003; Schaller and Klimov, 2004). The challenge is to find ways to collect the excitons before they recombine. As recombination occurs on a picosecond time scale, the use of mesoporous oxides as electron collectors presents a promising strategy, because the electron transfer from the quantum dot to the conduction band of the oxide electrode occurs within femtoseconds (Plass *et al.*, 2002). This opens up research avenues that ultimately may lead to photon converters reaching external quantum efficiencies over 100%. Recent results on depleted heterojunctions of PbSe(ZnO) confirm this expectation (Semonin *et al.*, 2011). A calculation based on Henry's model (Henry, 1980) shows that the maximum conversion efficiency of a single-junction cell could be increased from 34% to 44% by exploiting MEG effects.

9.4.7 *Mesoporous oxide film development*

When the dye-sensitised nanocrystalline solar cell was first presented, perhaps the most puzzling phenomenon was the highly efficient charge transport through the nanocrystalline TiO_2 layer. Mesoporous electrodes differ greatly from their compact analogues because: 1) the inherent conductivity of the film is very low; 2) the small size of the nanocrystalline particles does not support a built-in electrical field; and 3) the electrolyte penetrates the porous film all the way to the back contact, making the semiconductor/electrolyte interface essentially three-dimensional. The mechanism of charge transport in mesoporous systems is still under debate and several interpretations based on the Montrol–Scher model for random displacement of charge carriers in disordered solids have been advanced (Nelson and Chandler, 2004). However, the 'effective' electron diffusion coefficient is expected to depend on a number of factors such as trap filling and space-charge compensation by ionic motion in the electrolyte. The factors controlling the rate of charge-carrier percolation across the nanocrystalline film are presently under intense scrutiny. Intensity-modulated impedance spectroscopy has proved to be an elegant and powerful tool (Bisquert, 2002; Kubo *et al.*, 2002; Cass *et al.*, 2003; Frank *et al.*, 2004; Peter, 2009; Fabregat-Santiago *et al.*, 2005) to address these and other important questions related to the characteristic time constants for charge-carrier transport and reaction dynamics in dye-sensitised nanocrystalline solar cells.

On the materials science side, future research has been directed towards synthesising structures with a higher degree of order than a random assembly of nanoparticles. A desirable morphology of the films would have the mesoporous channels or nanorods aligned in parallel to each other and vertically with respect to the TCO glass current collector. This would facilitate pore diffusion, give easier access to the film surface, avoid grain boundaries and allow the junction to be formed under better control. One approach to fabricate such oxide structures is based on surfactant template-assisted preparation of TiO_2 nanotubes, as described by Jiu *et al.* (2005), another the use of mesoporous TiO_2 single crystals (Crossland *et al.*, 2013). These and the hybrid nanorod–polymer composite cells developed by Alivisatos and co-workers (Huynh *et al.*, 2002) have confirmed the superior photovoltaic performance of such films as compared with random-particle networks.

9.4.8 *Molecular engineering of the interface*

The large contact area of the junction in nanocrystalline solar cells renders mandatory the grasp and control of interfacial effects for future improvement of cell performance. The nature of the exposed surface planes of the oxide and the mode of interaction with the dye is the first important information to gather. As far as the adsorption of the N3 dye on TiO_2 is concerned, this is now well understood. The prevalent orientation of the anatase surface planes is (101) and the sensitiser is adsorbed through two of the four carboxylate groups, at least one of them being anchored via a bidentate configuration bridging two adjacent titanium sites (Nazeeruddin *et al.*, 2000). Molecular dynamic calculations employing a classical force field have been carried out to predict the equilibrium geometry of the adsorbed sensitiser state (Burnside *et al.*, 1998; Shklover *et al.*, 1998). More sophisticated first-principle density functional calculations have also been undertaken (Vittadini *et al.*, 1998) to model the surface interactions of TiO_2 with simple adsorbates as well as the surface reconstruction effects resulting from the adsorption. The latter approach is particularly promising and will provide an important tool for future theoretical investigations.

Synthetic efforts focus on the molecular engineering of sensitisers that enhance charge separation at the oxide solution interface. The structural features of the dye should match the requirements for current rectification: by analogy to the photofield effect in transistors, the gate for unidirectional electron flow from the electrolyte through the junction and into the oxide is opened by the photoexcitation of the sensitiser. The reverse charge flow, i.e. recapture of the electron by the electrolyte, could be impaired by judicious design of the sensitiser, which is an insulator in the ground state. The latter should form a tightly packed monolayer blocking the dark current. The gain in open-circuit voltage can be calculated from the diode equation

$$V_{oc} = (nRT/F)\ln[(i_{sc}/i_o) - 1] \tag{9.4}$$

where n is the ideality factor, whose value is between 1 and 2 for DSSCs, and i_0 is the reverse saturation current. Thus for each order of magnitude decrease in the dark current, the gain in V^{oc} would be between 59 and 118 mV at room temperature. Work in this direction is indispensable to raise the efficiency of the DSSC significantly over the 15% limit with the currently employed redox electrolytes.

9.5 Solid-state mesoscopic cells based on molecular dyes or perovskite pigments as sensitisers

9.5.1 *Mesoscopic photovoltaics based on molecular sensitisers*

Research on solid-state DSSCs has gained considerable momentum in recent years as this embodiment is attractive for realising flexible PV cells in a roll-to-roll production. The most successful *p*-type organic conductor employed to date is *spiro*-OMeTAD, which has a work function of $\sim$5.1 eV and hole mobility of $2 \times 10^{-4}\,\text{cm}^2\,\text{s}^{-1}$ (see Fig. 9.11 for the structure of this material). First reported in 1998, conversion efficiencies of solid-state cells incorporating this hole conductor have increased dramatically over the last few years, from a fraction of a percent (Bach *et al.*, 1998) to over 4% (Schmidt-Mende *et al.*, 2005a), reaching presently 7.2% (Burschka *et al.*, 2011). The main challenge posed by these cells has been fast interfacial electron-hole recombination, reducing the electron diffusion length to a few microns (Kruger *et al.*, 2003) as compared with 20–100 μm for electrolyte-based DSSCs. This restricts the film thickness employed in these cells to only 2 μm, which is insufficient for the adsorbed sensitiser to harvest all incident visible sunlight, thus reducing the photocurrent. The dye monolayer can block this back reaction to some extent because it is electrically insulating (Snaith *et al.*, 2005). Hence, current efforts are directed towards molecular engineering of the interface to improve the compactness and order of the monolayer and prevent charge carriers from recombining.

Another difficulty that has been encountered has been the filling of the porous network with the hole conductor. This impediment may be overcome by developing oxide films having regular mesoporous channels aligned perpendicular to the current collector. On the other hand, the V_{oc} values obtained with solid-state DSSCs are high, reaching nearly 1 V, owing to a better match of the hole conductor work function with the redox potential of the sensitiser than that of the electrolyte. The future of these solid hole-conductor systems thus looks very bright if the recombination and pore-filling problems can be solved.

9.5.2 *Mesoscopic photovoltaics based on metal halide perovskites*

The recent introduction of lead (Kojima *et al.*, 2009) and tin halide (Chung *et al.*, 2012) perovskites as light harvesters and solid hole conductors in mesoscopic solar cells has triggered one of the most exciting developments in the history of photovoltaics. While these compounds have been known for 120 years (Wells, 1893) and have been extensively studied for optoelectronic applications (Hirasawa *et al.*, 1994) in, for example, light-emitting diodes (Chondroudis and Mitzi, 1999) and thin-film transistors (Kagan *et al.*, 1999), surprisingly their photovoltaic properties have not been examined until very recently, even though it was remarked in 1958 that Cs_3PbI_3 is a semiconductor exhibiting photoconductivity (Moller, 1958).

Photovoltaic investigations on metal halide perovskites started only in 2009 when methylammonium lead halides, of the composition $CH_3NH_3PbX_3$ (X = I, Br), were introduced as replacement for the dye sensitiser in liquid electrolyte-based DSSCs. The PCE was initially 3.8% and device stability was poor. By changing the electrolyte Im *et al.* (2011) improved the performance and stability, attaining a PCE of 6.5%. Importantly it became apparent that the $CH_3NH_3PbI_3$ perovskite is a better light harvester than the molecular N719 ruthenium sensitisers, enabling the thickness of the mesoscopic titania film to be reduced significantly.

A key advance was subsequently made by replacing the liquid electrolyte with the solid-state hole conductor *spiro*-OMeTAD (Kim *et al.,* 2012; Lee *et al.*, 2012). Not only did the conversion efficiencies practically double but the cell stability was vastly improved by avoiding the liquid solvent which dissolves the perovskite. The studies by Lee *et al.* (2012) and Etgar *et al.* (2012) revealed that the perovskite itself acts as an ambipolar semiconductor, obviating the need to employ a TiO_2 scaffold or organic hole conductor. This prompted further intensive investigations, which yielded a wealth of new information on these exciting systems. Today the best-performing cells reach a PCE of 15% (Burschka *et al.*, 2013), which is more than four times higher than the PCE value attained in 2009 — an unprecedented rapid rise in efficiency for a photovoltaic. Figure 9.12 shows a scanning electron microscope (SEM) view of the cross-section of the cell and a photocurrent–voltage curve for the device. The field is presently in a very dynamic state and reports on further improvements in performance are expected in the near future.

9.6 Pilot production of modules, field tests and commercial DSSC development

Prototypes of monolithic Z-type interconnected DSSC modules, fabricated by Aisin Seiki in Japan, use carbon as a back contact to cut costs. Comparative field

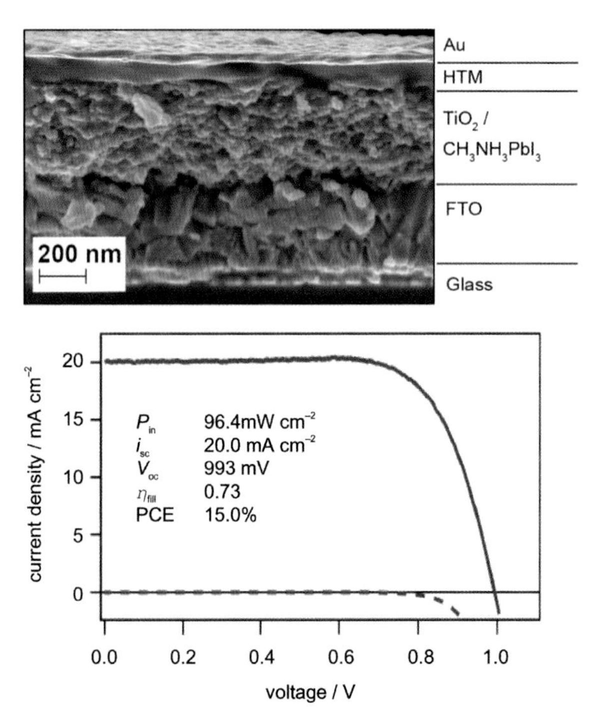

Figure 9.12 Top: Cross-sectional SEM photograph of a mesoscopic solar cell using $CH_3NH_3PI_3$ as light harvester. The perovskite is infiltrated into the mesopores of the TiO_2 scaffold. FTO and HTM signify fluorine-doped tin dioxide and hole-transport material. Bottom: i–V curve of the best-performing device.

tests of these modules and polycrystalline silicon (pc-Si) have been running for several years (Figure 9.13 shows a photograph of the test station). The test results revealed the advantages of the DSSC as compared with silicon modules under realistic outdoor conditions: for equal rating under standard test conditions (STC), the DSSC modules produced 20–30% more energy than pc-Si modules (Toyoda, 2006). Recent data presented by Sony (Noda, 2012) show that today's DSSC module, having a standard AM 1.5 conversion efficiency of 10%, is expected to produce on average over one year the same electricity as a 15% standard-rated pc Si module. The superior performance of the DSSC can be ascribed to the following factors:

- The DSSC efficiency is practically temperature-independent in the range 25–65°C while that of monocrystalline and pc-Si declines by ~20% over the same range.
- Outdoor measurements indicate that light capture by the DSSC is less sensitive to the angle of incidence, although this needs to be further assessed. The DSSC

Figure 9.13 Outdoor field tests of DSSC modules by Fujikura Corporation near Tokyo, Japan.

is more efficient than pc-Si in diffuse light or cloudy conditions. This advantage is further enhanced for glass modules, which collect light in a bifacial way from all angles.

While it is up to the commercial supplier to set the final price for such modules, it is clear that the DSSC shares the cost advantages of all thin-film devices. In addition it uses no high-vacuum, cost-intensive steps in manufacture and only cheap and readily available materials. Although it might be thought that the ruthenium-based sensitiser adds high material cost, its contribution to cost is < US$ 0.01 W_p^{-1} because of the small amount used. Also, the present organic sensitisers match or even surpass the performance of Ru complexes. Given these advantages at comparable conversion efficiency, module costs below €0.5 W_p^{-1} are realistic targets even for production plants having well below GW capacity. The DSSC has thus become a viable contender for large-scale solar energy conversion systems on the basis of cost, efficiency, stability and availability, as well as environmental compatibility. The British company G24i has started commercial sales of a flexible DSSC model. Today the production rate is in the MW per year range.

9.7 Outlook

Using a principle derived from natural photosynthesis, mesoscopic injection solar cells and in particular the DSSC, have become a credible alternative to solid-state *p-n* junction devices. Conversion efficiencies of 12.3% and 15% have already been obtained with single-junction liquid and solid-state mesoscopic solar cells, respectively, on the laboratory scale, but there is ample room for further improvement. Future research will focus on improving short-circuit current by extending the light response of the sensitisers in the near-IR spectral region. Substantial gains in the open-circuit voltage are expected from introducing ordered oxide mesostructures and controlling the interfacial charge recombination by judicious engineering on the molecular level. Hybrid cells based on solid-state inorganic and organic hole conductors are an attractive option, in particular for the flexible DSSC embodiment. Solar cells based on perovskite light absorbers are showing very promising efficiencies and are rapidly developing into a particularly exciting new research area.

Mesoscopic dye-sensitised cells are well suited for a whole realm of applications, ranging from the low-power market to large-scale applications. Their excellent performance in diffuse light gives them a competitive edge over silicon in providing electric power for stand-alone electronic equipment both indoors and outdoors. Dye-sensitized solar cells are already being applied in building-integrated PV and this will become a fertile field of future commercial development.

As the epigraph at the chapter opening shows, at the 1912 IUPAC conference in Washington a hundred years ago, the famous Italian photochemist Professor Giacomo Ciamician predicted that mankind would unravel the secrets of photosynthesis and apply the principles used by plants to harvest solar energy in glass buildings. His visionary thoughts now appear close to becoming a reality.

Acknowledgements

We gratefully acknowledge the insights and support from the many co-workers with whom we have had the pleasure of working on these devices, and particularly thank Dr Brian O'Regan for his help in preparing this manuscript.

References

3GSolar (2011), '3GSolar announces record outdoor reliability of third generation photovoltaic technology', http:// www.3Gsolar.com.

Anderson N. A. and Lian T. Q. (2005), 'Ultrafast electron transfer at the molecule-semiconductor nanoparticle interface', *Annu. Rev. Phys. Chem.* **56**, 491–519.

Anderson S., Constable E. C., Dare-Edwards M. P., Goodenough J. B., Hamnett A. and Seddon K. R. (1979), 'Chemical modification of a titanium(IV) oxide electrode to give stable dye sensitization without a supersensitiser', *Nature* **280**, 571–573.

Arakawa H., Yamaguchi T., Okada K., Matsui K., Kitamura T. and Tanabe N. (2009), 'Highly durable dye-sensitised solar cells', *Fujikura Tech. Rev.* **2009**, 55–59.

Bach U., Lupo D., Comte P., Moser J. E., Weissortel F. and Salbeck J. (1998), 'Solid-state dye-sensitised mesoporous TiO_2 solar cells with high photon-to-electron conversion efficiencies', *Nature* **395**, 583–585.

Barbe C. J., Arendse F., Comte P., Jirousek M., Lenzmann F. and Shklover V. (1997), 'Nanocrystalline titanium oxide electrodes for photovoltaic applications', *J. Amer. Ceramic Soc.* **80**, 3157–3171.

Barkhouse D. A. R., Debnath R., Kramer, I. J., Zhitomirsky D., Pattantyus-Abraham A. G., Levina L., Etgar L., Grätzel M. and Sargent E. H. (2011), 'Depleted bulk heterojunction colloidal quantum dot photovoltaics', *Adv. Mater.* **23**, 3134–3140.

Bessho T., Zakeeruddin S. M., Yeh C.-Y., Diau E. W. G. and Grätzel M. (2010), 'Highly efficient mesoscopic dye-sensitised solar cells based on donor-acceptor-substituted porphyrins', *Angew. Chem. Int. Ed.* **49**, 6646–6649.

Bisquert J. (2002), 'Theory of the impedance of electron diffusion and recombination in a thin layer', *J. Phys. Chem. B* **106**, 325–333.

Burnside S. D., Shklover V., Barbe C., Comte P., Arendse F. and Brooks K. (1998), 'Self-organisation of TiO_2 nanoparticles in thin films', *Chem. Mater.* **10**, 2419–2425.

Burschka A., Dualeh F., Kessler F., Baranoff E., Cevey-Ha N.-L., Yi Md. C., Nazeeruddin M. K. and Grätzel M. (2011), 'Tris(2-(1H-pyrazol-1-yl) pyridine) cobalt(III) as *p*-type dopant for organic semiconductors and its application in highly efficient solid-state dye-sensitised solar cells', *J. Amer. Chem. Soc.* **133**, 18042–18045.

Burschka J., Pellet N., Moon S. J., Humphry-Baker R., Gao P., Nazeeruddin M. K. and Grätzel M. (2013), 'Sequential deposition as a route to high-performance perovskite-sensitized solar cells', *Nature* **499**, 316–319.

Casanova D., Rotzinger F. P. and Grätzel M. (2010), 'Computational study of promising organic dyes for high-performance sensitized solar cells', *J. Chem. Theory Comput.* **6**, 1219–1227.

Cass M. J., Qiu F. L., Walker A. B., Fisher A. C. and Peter L. M. (2003), 'Influence of grain morphology on electron transport in dye sensitised nanocrystalline solar cells', *J. Phys. Chem. B* **107**, 113–119.

Chondroudis K. and Mitzi D. B. (1999), 'Electroluminescence from an organic-inorganic perovskite incorporating a quaterthiophene dye within lead halide perovskite layers', *Chem. Mater.* **11**, 3028–3030.

Chung I., Lee B., He J., Chang R. P. H. and Kanatzidis M. G. (2012), 'All-solid-state dye-sensitized solar cells with high efficiency', *Nature* **485**, 486–489.

Crossland E. J. W., Nakita N., Sivaram V., Leijtens T., Alexander-Webber J. A. and Snaith H. J. (2013), 'Mesoporous TiO_2 single crystals delivering enhanced mobility and opto-electronic device performance', *Nature* **495**, 215–219.

Dare-Edwards M. P., Goodenough J. B., Hamnett A., Seddon K. R. and Wright R. D. (1980), 'Sensitization of semiconducting electrodes with ruthenium-based dyes', *Faraday Discussions* **70**, 285–298.

Desilvestro J., Grätzel M., Kavan L., Moser J. and Augustynski J. (1985), 'Highly efficient sensitization of titanium dioxide', *J. Amer. Chem. Soc.* **107**, 2988–2990.

De Wild-Scholten M. J. and Veltkamp A. C. (2007), 'Environmental life cycle analysis of dye sensitised solar devices: status and outlook', 22nd EU-PVSEC, https://www.ecn.nl/publicaties/ECN-M-07-081.

Dung D. H., Serpone N. and Grätzel M. (1984), 'Integrated systems for water cleavage by visible-light sensitization of TiO_2 particles by surface derivatization with ruthenium complexes', *Helvetica Chim. Acta* **67**, 1012–1018.

Durr M., Bamedi A., Yasuda A. and Nelles G. (2004), 'Tandem dye-sensitised solar cell for improved power conversion efficiencies', *Appl. Phys. Lett.* **84**, 3397–3399.

Durr M., Schmid A., Obermaier M., Rosselli S., Yasuda A. and Nelles G. (2005), 'Low-temperature fabrication of dye-sensitised solar cells by transfer of composite porous layers', *Nature Materials* **4**, 607–611.

Durrant J. R., Haque S. A. and Palomares E. (2006), 'Photochemical energy conversion: from molecular dyads to solar cells', *Chem. Comm.* **31**, 3279–3289.

Dyesol (2013), 'Dyesol exceeds stringent PV durability test by 400%', May 28, www.dyesol.com/posts/cat/corporate-news/post/dyesol-exceeds-stringent-pv-durability-test-by-400-per-cent/.

Etgar L., Gao P., Xue Z., Peng Q., Chandiran A. K., Liu B., Nazeeruddin M. K. and Grätzel M. (2012), 'Mesoscopic $CH_3NH_3PbI_3/TiO_2$ heterojunction solar cells', *J. Amer. Chem. Soc.* **134**, 17396–17399.

Fabregat-Santiago F., Bisquert J., Garcia-Belmonte G., Boschloo G. and Hagfeldt A. (2005), 'Influence of electrolyte in transport and recombination in dye-sensitised solar cells studied by impedance spectroscopy', *Solar Energy Mater. Solar Cells* **87**, 117–131.

Feldt S. M., Gibson E. A., Gabrielsson E., Sun L., Boschloo G. and Hagfeldt A. (2010), 'Design of organic dyes and cobalt polypyridine redox mediators for high-efficiency dye-sensitized solar cells', *J. Amer. Chem. Soc.* **132**, 16714–16724.

Ferrere S., Zaban A. and Gregg B. A. (1997), 'Dye sensitization of nanocrystalline tin oxide by perylene derivatives', *J. Phys. Chem. B* **101**, 4490–4493.

Frank A. J., Kopidakis N. and van de Lagemaat J. (2004), 'Electrons in nanostructured TiO_2 solar cells: transport, recombination and photovoltaic properties', *Coordination Chem. Rev.* **248**, 1165–1179.

Gerischer H. and Tributsch H. (1968), 'Elektrochemische Untersuchungen über den Mechanismus der Sensibilisierung und Übersensibilisierung an ZnO-Einkristallen', *Ber. Bunsenges. Phys. Chem.* **72**, 437–445.

Grätzel M. (2006), 'Photovoltaic performance and long-term stability of dye-sensitised mesosocopic solar cells', *Comptes Rend. Chimie* **9**, 578–583.

Grätzel M. (2009), 'Recent advances in sensitised mesoscopic solar cells', *Acc. Chem. Res.* **42**, 1788–1798.

Grätzel M., Janssen R. A. J., Mitzi D. B. and Sargent E. H. (2012), 'Materials interface engineering for solution-processed photovoltaics', *Nature* **488**, 304–312.

Green A. N. M., Palomares E., Haque S. A., Kroon J. M. and Durrant J. R. (2005), 'Charge transport versus recombination in dye-sensitised solar cells employing nanocrystalline TiO_2 and SnO_2 films', *J. Phys. Chem. B* **109**, 12525–12533.

Green M. A., Emery K., Hishikawa Y. and Warta W. (2011), 'Solar efficiency tables (Version 37)', *Progr. Photovoltaics* **19**, 84–92.

Hagfeldt A., Boschloo G., Sun L., Kloo L. and Pettersson H. (2010), 'Dye-sensitized solar cells', *Chem. Rev.* **110**, 6595–6663.

Hagfeldt A. and Grätzel M. (2000), 'Molecular photovoltaics', *Acc. Chem. Res.* **33**, 269–277.

Han L., Islam A., Chen H., Malapaka C., Chiranjeevi B., Zhang S., Yang X. and Yanagida M. (2012), 'High-efficiency dye-sensitised solar cell with a novel co-adsorbent', *Energy Env. Science* **5**, 6057–6060.

Haque S. A., Palomares E., Cho B. M., Green A. N. M., Hirata N. and Klug D. R. (2005), 'Charge separation versus recombination in dye-sensitised nanocrystalline solar cells: the minimization of kinetic redundancy', *J. Amer. Chem. Soc.* **127**, 3456–3462.

Hara K., Sato T., Katoh R., Furube A., Yoshihara T. and Murai M. (2005), 'Novel conjugated organic dyes for efficient dye-sensitised solar cells', *Adv. Functional Mat.* **15**, 246–252.

Harikisun R. and Desilvestro H. (2011), 'Long-term stability of dye solar cells', *Solar Energy* **85**, 1179–1188.

Hengerer R., Kavan L., Krtil P. and Grätzel M. (2000), 'Orientation dependence of charge-transfer processes on TiO_2 (anatase) single crystals', *J. Electrochem. Soc.* **147**, 1467–1472.

Henry C. H. (1980), 'Limiting efficiencies of ideal single and multiple energy-gap terrestrial solar cells', *J. Appl. Phys.* **51**, 4494–4500.

Hinsch A., Kroon J. M., Kern R., Uhlendorf I., Holzbock J. and Meyer A. (2001), 'Long-term stability of dye-sensitised solar cells', *Progr. Photovoltaics* **9**, 425–438.

Hirasawa M., Ishihara T., Goto T., Uchida K. and Miura N. (1994), 'Magnetoabsorption of the lower exciton in perovskite–type compound $CH_3NH_3PbI_3$', *Physica B* **291**, 427–430.

Horiuchi T., Miura H., Sumioka K. and Uchida S. (2004), 'High efficiency of dye-sensitised solar cells based on metal-free indoline dyes', *J. Amer. Chem. Soc.* **126**, 12218–12219.

Hoyer P. and Weller H. (1995), 'Potential-dependent electron injection in nanoporous colloidal ZnO films', *J. Phys. Chem.* **99**, 14096–14100.

Hsieh C.-P., Lu H.-P., Chiu C.-L., Lee C.-W., Chuang S.-H., Mai C.-L., Yen W.-N., Hsu S.-J., Diau E. W.-G. and Yeh C.-Y. (2010), 'Synthesis and characterization of porphyrin sensitizers with various electron-donating substituents for highly efficient dye-sensitized solar cells', *J. Mater. Chem.* **20**, 1127–1134.

Huynh W. U., Dittmer J. J. and Alivisatos A. P. (2002), 'Hybrid nanorod-polymer solar cells', *Science* **295**, 2425–2427.

Im J.-H., Lee C. R., Lee J. W., Park S. W. and Park N. G. (2011), '6.5% efficient perovskite quantum-dot-sensitized solar cell', *Nanoscale* **3**, 4088–4093.

Imahori H., Umeyama T. and Ito S. (2009), 'Large π-aromatic molecules as potential sensitizers for highly efficient dye-sensitized solar cells', *Acc. Chem. Res.* **42**, 1809–1818.

Ince M., Cardinali F., Raugossi M., Yum J., Gouloumis A., Martinez-Diaz M., Torre G., Nazeeruddin M., Grätzel M. and Torres T. (2012), 'Phthalocyanines for molecular photovoltaics', Abstracts of Papers, 243rd ACS National Meeting and Exposition, San Diego, CA, 25–29 March 2012, FUEL-522.

Ito S., Miura H., Uchida S., Takata M., Sumioka K., Liska P., Comte P., Pechy P. and Gratzel M. (2008), 'High conversion efficiency organic dye sensitised solar cells with a novel indoline dye', *Chem. Commun.* **41**, 5194–5196.

Ito S., Zakeeruddin S. M., Humphry-Baker R., Liska P., Charvet R. and Comte P. (2006), 'High-efficiency organic-dye-sensitised solar cells controlled by nanocrystalline TiO_2 electrode thickness', *Adv. Mater.* **18**, 1202–1205.

Jiu J. T., Wang F. M., Isoda S. and Adachi M. (2005), 'Highly efficient dye-sensitised solar cells based on single crystalline TiO_2 nanorod film', *Chem. Lett.* **34**, 1506–1507.

Kagan C. R., Mitzi D. B. and Dimitrakopoulos C. D. (1999), 'Organic-inorganic hybrid materials as semiconducting channels in thin-film field-effect transistors', *Science* **286**, 945–947.

Kalyanasundaram K. and Graetzel M. (2010), 'Artificial photosynthesis: biomimetic approaches to solar energy conversion and storage', *Current Opinion Biotechnol.* **21**, 298–310.

Kim H. S., Lee C. R., Im J. H., Lee K. B., Moehl T., Marchioro A., Moon S. J., Humphry-Baker R., Yum J. H., Moser J. E., Grätzel M. and Park N. G. (2012), 'Lead iodide perovskite sensitized all-solid-state submicron thin film mesoscopic solar cell with efficiency exceeding 9%', *Nature Sci. Reports* **2**, 591.

Kinoshita,T., Dy J. T., Uchida S., Kubo T. and Segawa H. (2013), 'Wideband dye-sensitized solar cells employing a phosphine-coordinated ruthenium sensitizer', *Nat. Photonics* **7**, 535–539.

Kinoshita T., Dy J. T., Uchida S., Kubo T. and Segawa H. (2013), 'Wideband dye-sensitized solar cells employing a phosphine-coordinated ruthenium sensitizer', *Nature Photonics* **7**, 535–539.

Kojima A., Teshima K., Shirai Y. and Miyasaka T. (2009), 'Organometal halide perovskites as visible-light sensitizers for photovoltaic cells', *J. Amer. Chem. Soc.* **131**, 6050–6051.

Koops S. E., O'Regan B. C., Piers R. F., Barnes P. R. F. and Durrant J. R. (2009), 'Parameters influencing the efficiency of electron injection in dye-sensitised solar cells', *J. Amer. Chem. Soc.* **131**, 4808–4818.

Kruger J., Plass R., Grätzel M., Cameron P. J. and Peter L. M. (2003), 'Charge transport and back reaction in solid-state dye-sensitised solar cells: a study using intensity-modulated photovoltage and photocurrent spectroscopy', *J. Phys. Chem. B* **107**, 7536–7539.

Kubo W., Kitamura T., Hanabusa K., Wada Y. and Yanagida S. (2002), 'Quasi-solid-state dye-sensitised solar cells using room temperature molten salts and a low molecular weight gelator', *Chem. Comm.* **4**, 374–375.

Kubo W., Sakamoto A., Kitamura T., Wada Y. and Yanagida S. (2004), 'Dye-sensitised solar cells: improvement of spectral response by tandem structure', *J. Photochem. Photobiol. A — Chemistry* **164**, 33–39.

Lee M. M., Teuscher J., Miyasaka T., Murakami T. N. and Snaith H. J. (2012), 'Efficient hybrid solar cells based on meso-superstructured organometal halide perovskites', *Science* **338**, 643–647.

Liska P., Thampi K. R., Grätzel M., Bremaud D., Rudmann D. and Upadhyaya H. M. (2006), 'Nanocrystalline dye-sensitised solar cell/copper indium gallium selenide thin-film tandem showing greater than 15% conversion efficiency', *Appl. Phys. Lett.* **88**, 203103–203106.

Listorti A., O'Regan B. and Durrant J. R. (2011), 'Electron transfer dynamics in dye-sensitized solar cells', *Chem. Mater.* **23**, 3381–3399.

Mann S. A., de Wild-Scholten M. J., Fthenakis V. M., van Sark W. G. J. H. M. and Sinke W. C. (2013), 'The energy payback time of advanced crystalline PV modules in 2020: a prospective study', *Progr. Photovoltaics*, http://onlinelibrary.wiley.com/doi/10.1002/pip.2363/full.

McEvoy A. J. and Grätzel M. (1994), 'Sensitization in photochemistry and photovoltaics', *Solar Energy Mat. Solar Cells* **32**, 221–227.

McEvoy A. J. and Grätzel M. (2013), 'Nanotechnology in dye sensitized solar cells', in Garcia-Martinez J. (ed.), *Nanotechnology for Energy Challenges*, Wiley, Weinheim.

Mishra A., Fischer M. K. R. and Bauerle P. (2009), 'Metal-free organic dyes for dye-sensitized solar cells: from structure:property relationships to design rules', *Angew. Chemie Int. Ed.* **48**, 2474–2499.

Moller C. K. (1958), 'Crystal structure and photoconductivity of caesium plumbohalides', *Nature* **182**, 1436.

Moser J. (1887), 'Notiz über Verstärkung photoelectrischer Ströme durch optische Sensibilisierung', *Monatsch. Chem.* **8**, 373.

Namba S. and Hishiki Y. (1965), 'Color sensitization of zinc oxide with cyanine dyes', *J. Phys. Chem.* **69**, 774–779.

Nanu M., Schoonman J. and Goossens A. (2005), 'Solar energy conversion in TiO_2/ $CuInS_2$ nanocomposites', *Adv. Functional Mater.* **15**, 95–100.

Nazeeruddin M. K., Amirnasr M., Comte P., Mackay J. R., McQuillan A. J. and Houriet R. (2000), 'Adsorption studies of counterions carried by the sensitiser cis-dithiocyanato (2,2′-bipyridyl-4,4′-dicarboxylate) ruthenium(II) on nanocrystaline TiO_2 films', *Langmuir* **16**, 8525–8528.

Nazeeruddin M. K., Kay A., Rodicio I., Humphry-Baker R., Muller E. and Liska P. (1993), 'Conversion of light to electricity by cis-X_2-bis(2,2′-bipyridyl-4,4′-dicarboxylate) ruthenium(II) charge-transfer sensitisers ($X = Cl^-$, Br^-, I^-, CN^- and SCN^-) on nanocrystalline TiO_2 electrodes', *J. Amer. Chem. Soc.* **115**, 6382–6390.

Nazeeruddin M. K., Pechy P., Renouard T., Zakeeruddin S. M., Humphry-Baker R. and Comte P. (2001), 'Engineering of efficient panchromatic sensitisers for nanocrystalline TiO_2-based solar cells', *J. Amer. Chem. Soc.* **123**, 1613–1624.

Nelson J. and Chandler R. E. (2004), 'Random walk models of charge transfer and transport in dye-sensitised systems', *Coordination Chem. Rev.* **248**, 1181–1194.

Ning, Z. J., Fu, Y. and Tian, H. (2010), 'Improvement of dye-sensitized solar cells: what we know and what we need to know', *Energy Envir. Sci.* **3**, 1170–1181.

Noda K. (2012), Lecture, 'Dye-sensitized Solar Cell Development at Sony', PV Expo, Tokyo, 29 February, Japan.

Nogueira A. F., Longo C. and de Paoli M. A. (2004), 'Polymers in dye-sensitised solar cells: overview and perspectives', *Coordination Chem. Rev.* **248**, 1455–1468.

Noh J. H., Im S. H., Heo J. H., Mandal T. N. and Seok S. I. (2013), 'Chemical management for colorful, efficient, and stable inorganic-organic hybrid nanostructured solar cells', *Nano Lett.* **13**, 1764–1769.

Nozik A. J. (2003), 'Quantum dot solar cells', in Marti A. and Luque A. (eds), *Next Generation Photovoltaics: High Efficiency Through Full Spectrum Utilization*, CRC Press, Boca Raton.

O'Regan B. and Grätzel M. (1991), 'A low-cost, high-efficiency solar-cell based on dye-sensitised colloidal TiO_2 films', *Nature* **353**, 737–740.

O'Regan B. and Schwartz D. T. (1998), 'Large enhancement in photocurrent efficiency caused by UV illumination of the dye-sensitised heterojunction TiO_2/RuLL'NCS/ CuSCN: initiation and potential mechanisms', *Chem. Mater.* **10**, 1501–1509.

O'Regan B. C. and Durrant J. R. (2009), 'Kinetic and energetic paradigms for dye-sensitized solar cells: moving from the ideal to the real', *Acc. Chem. Res.* **42**, 1799–1808.

Pattantyus-Abraham A. G., Kramer I. J., Barkhouse A. R., Wang Y., Konstantatos G., Debnath R., Levina L., Raabe I., Nazeeruddin M. K., Grätzel M. and Sargent E. H. (2010), 'Depleted-heterojunction colloidal quantum dot solar cells', *ACS-Nano* **6**, 3374–3380.

Peter L. (2009) "'Sticky electrons" transport and interfacial transfer of electrons in the dye-sensitized solar cell', *Accounts Chem. Res.* **42**, 1839–1847.

Peter L. M. and Wijayantha K. G. U. (2000), 'Electron transport and back reaction in dye-sensitised nanocrystalline photovoltaic cells', *Electrochim. Acta* **45**, 4543–4551.

Plass R., Pelet S., Krueger J., Grätzel M. and Bach U. (2002), 'Quantum dot sensitization of organic–inorganic hybrid solar cells', *J. Phys. Chem. B* **106**, 7578–7580.

Qin H., Wenger S., Xu M., Gao F., Jing X., Wang P., Zakeeruddin S.-M. and Grätzel M. (2008), 'An organic sensitiser with a fused dithienothiophene unit for efficient and stable dye-sensitised solar cells', *J. Amer. Chem. Soc.* **130**, 9202–9203.

Qin, Y. C. and Peng, Q. (2012), 'Ruthenium sensitizers and their applications in dye sensitized solar cells', *Int. J. Photoenergy*, http://www.hindawi.com/journals/ijp/2012/291579/.

Reddy P. Y., Giribabu L., Lyness C., Snaith H. J., Vijaykumar C., Chandrasekharam M., Lakshmikantam M., Yum J.-H., Kalyanasundaram K., Grätzel M. and Nazeeruddin M. K. (2007), 'Efficient sensitization of nanocrystalline TiO_2 films by a near-IR-absorbing unsymmetrical zinc phthalocyanine', *Angewandte Chem. Int. Ed.* **46**, 373–376.

Redmond G. and Fitzmaurice D. (1993), 'Spectroscopic determination of flat-band potentials for polycrystalline TiO_2 electrodes in nonaqueous solvents', *J. Phys. Chem.* **97**, 1426–1430.

Rensmo H., Keis K., Lindstrom H., Sodergren S., Solbrand A. and Hagfeldt A. (1997), 'High light-to-energy conversion efficiencies for solar cells based on nanostructured ZnO electrodes', *J. Phys. Chem. B* **101**, 2598–2601.

Rhee Y. M. and Head-Gordon M. (2007), 'Scaled second-order perturbation correction to configuration interaction singles: efficient and reliable excitation energy methods', *J. Phys. Chem.* **111**, 5314–5326.

Rothenberger G., Fitzmaurice D. and Grätzel M. (1992), 'Optical electrochemistry. 3. Spectroscopy of conduction-band electrons in transparent metal-oxide semiconductor films — optical determination of the flat-band potential of colloidal titanium dioxide films', *J. Phys. Chem.* **96**, 5983–5986.

Sargent E. H. (2012), 'Colloidal quantum dot solar cells', *Nat. Photonics* **6,** 133–135.

Sauvage F., Chhor S., Marchioro A., Moser J.-E. and Grätzel M. (2011), 'Butyronitrile-based electrolyte for dye-sensitised solar cells', *J. Amer. Chem. Soc.* **133**, 13103–13109.

Sayama K., Sugihara H. and Arakawa H. (1998), 'Photoelectrochemical properties of a porous Nb_2O_5 electrode sensitised by a ruthenium dye', *Chem. Mater.* **10**, 3825–3832.

Schaller R. D. and Klimov V. I. (2004), 'High efficiency carrier multiplication in PbSe nanocrystals: implications for solar energy conversion', *Phys. Rev. Lett.* **92**, 86601–86604.

Schmidt-Mende L., Bach U., Humphry-Baker R., Horiuchi T., Miura H. and Ito S. (2005a), 'Organic dye for highly efficient solid-state dye-sensitised solar cells', *Adv. Materials* **17**, 813–815.

Schmidt-Mende L., Kroeze J. E., Durrant J. R., Nazeeruddin M. K. and Grätzel M. (2005b), 'Effect of hydrocarbon chain length of amphiphilic ruthenium dyes on solid-state dye-sensitised photovoltaics', *Nano Lett.* **5**, 1315–1320.

Semonin O. E., Luther J. M., Choi S., Chen H.-Y., Gao J., Nozik A. J. and Beard M. C. (2011), 'Peak external quantum efficiency exceeding 100% via MEG in a quantum dot solar cell', *Science* **334**, 1530–1533.

Schmickler W. (1996), *Interfacial Electrochemistry*, Oxford University Press, Oxford.

Shklover V., Ovchinnikov Y. E., Braginsky L. S., Zakeeruddin S. M. and Grätzel M. (1998), 'Structure of organic/inorganic interface in assembled materials comprising molecular components. Crystal structure of the sensitiser bis[(4,4′-carboxy-2,2′-bipyridine)(thiocyanato)]ruthenium(II)', *Chem. Mater.* **10**, 2533–2541.

Snaith H. J., Moule A. J., Klein C., Meerholz K., Friend R. H. and Grätzel M. (2007), 'Efficiency enhancements in solid-state hybrid solar cells via reduced charge recombination and increased light capture', *Nano Letters* **7**, 3372–3376.

Snaith H. J., Zakeeruddin S. M., Schmidt-Mende L., Klein C. and Grätzel M. (2005), 'Ion-coordinating sensitiser in solid-state hybrid solar cells', *Angew. Chem. Int. Ed.* **44**, 6413–6417.

Stathatos E., Lianos R., Zakeeruddin S. M., Liska P. and Grätzel M. (2003), 'A quasi-solid-state dye-sensitised solar cell based on a sol–gel nanocomposite electrolyte containing ionic liquid', *Chem. Mater.* **15**, 1825–1829.

Tennakone K., Kumara G. R. R. A., Kumarasinghe A. R., Wijayantha K. G. U. and Sirimanne P. M. (1995), 'A dye-sensitised nanoporous solid-state photovoltaic cell', *Semiconductor Sci. Technol.* **10**, 1689–1693.

Toyoda T. (2006), 'Outdoor tests of large area dye-sensitised solar cell modules by Aisin Seiki, Inc.', *Re2009 Renewable Energy Congress*, Makura, Japan, October.

Tsubomura H., Matsumura M., Nomura Y. and Amamiya T. (1976), 'Dye-sensitised zinc oxide–aqueous electrolyte–platinum photocell', *Nature* **261**, 402–403.

Vittadini A., Selloni A., Rotzinger F. P. and Grätzel M. (1998), 'Structure and energetics of water adsorbed at TiO_2 anatase (101) and (001) surfaces', *Phys. Rev. Lett.* **81**, 2954–2957.

Wang P., Klein C., Humphry-Baker R., Zakeeruddin S. M. and Grätzel M. (2005a), 'A high molar extinction coefficient sensitiser for stable dye-sensitised solar cells', *J. Amer. Chem. Soc.* **127**, 808–809.

Wang P., Klein C., Humphry-Baker R., Zakeeruddin S. M. and Grätzel M. (2005b), 'Stable 8% efficient nanocrystalline dye-sensitised solar cell based on an electrolyte of low volatility', *Appl. Phys. Lett.* **86**, 1235081–1235083.

Wang P., Zakeeruddin S. M., Humphry-Baker R., Moser J. E. and Grätzel M. (2003a), 'Molecular-scale interface engineering of TiO_2 nanocrystals: improving the efficiency and stability of dye-sensitised solar cells', *Adv. Mater.* **15**, 2101–2103.

Wang P., Zakeeruddin S. M., Moser J. E., Nazeeruddin M. K., Sekiguchi T. and Grätzel M. (2003b), 'A stable quasi-solid-state dye-sensitised solar cell with an amphiphilic ruthenium sensitiser and polymer gel electrolyte', *Nat. Materials* **2**, 402–407.

Walter M. G., Rudine A. B. and Wamser C. C. (2010), 'Porphyrins and phthalocyanines in solar photovoltaic cells', *J. Porphyrins and Phthalocyanines* **14**, 759–792.

Wells H. L. (1893), 'Uber die Casium- und Kalium-Blei halogenide', *Z. Anorg. Chem.* **3**, 195–210.

West W. (1874), *Vogel Centennial Symp., Photogr. Sci. Eng.* **18**, 35.

Wenger S., Seyrling S., Tiwari A. N. and Graetzel M. (2009), Fabrication and performance of a monolithic dye-sensitized TiO_2/Cu(In,Ga)Se_2 thin film tandem solar cell, *Appl. Phys. Lett.* **94**, 173508/1–173508/3.

Yella A., Lee H.-W., Tsao H. N, Yi C., Chandiran A. K., Nazeeruddin M. K., Diau E. W.-G., Yeh C.-Y., Zakeeruddin S. M. and Grätzel M. (2011), 'Porphyrin-sensitised solar cells with cobalt (II/III)-based redox electrolyte exceed 12% efficiency', *Science* **334**, 629–634.

Yum J.-H., Baranoff E., Kessler F., Moehl T., Ahmad S., Bessho T., Marchioro A., Ghadiri E., Moser J.-E., Yi C., Nazeeruddin M. K. and Grätzel M. (2012), 'A cobalt complex redox shuttle for dye-sensitised solar cells with high open circuit potentials', *Nature Commun.* **3**, Art. 631.

Yum J.-H., Hagberg D. P., Moon S.-J., Karlsson K. M., Marinado T., Sun L., Hagfeldt A., Nazeeruddin M. K. and Grätzel M. (2009), 'A light-resistant organic sensitiser for solar-cell applications', *Angew. Chem. Int. Ed.* **48**, 1576–1580.

Zaban A., Ferrere S. and Gregg B. A. (1998), 'Relative energetics at the semiconductor sensitizing dye electrolyte interface', *J. Phys. Chem. B* **102**, 452–460.

Zhang G., Bala H., Cheng Y., Shi D., Lv X., Yu Q. and Wang P. (2009), 'High efficiency and stable dye-sensitised solar cells with an organic chromophore featuring a binary π-conjugated spacer', *Chem. Comm.* **16**, 2198–2200.

CHAPTER 10

QUANTUM WELL SOLAR CELLS

JENNY NELSON and NED EKINS-DAUKES
Blackett Laboratory, Imperial College London, SW7 2BZ, UK
jenny.nelson@ic.ac.uk

No steam or gas ever drives anything until it is confined. No Niagara is ever turned into light and power until it is tunneled.
Harry Emerson Fosdick.

10.1 Introduction

The quantum well solar cell (QWSC) was proposed by Barnham and Duggan (1990) as a new type of multiple-bandgap, single-junction solar cell. The principle is similar to that of the monolithic tandem cell: several bandgaps are used to absorb different spectral ranges efficiently. However, rather than using two junctions made from different semiconductors, the QWSC uses ultra-thin layers of different materials in a monolithic, two-terminal arrangement. The technology of quantum wells (QWs) is borrowed from the optoelectronics industry. The 1980s and 1990s saw developments in the use of low-dimensional semiconductor structures in optical modulators, photodetectors, lasers and other devices (Weisbuch and Vinter 1991; Zory, 1993). These applications exploit the special optical and electronic properties available with highly confined carrier populations. Before 1989, low-dimensional systems had been considered in connection with solar cells only in the context of improved carrier transport through the neutral regions. In the QWSC, QWs are used in the active region of the solar cell to enhance solar photon absorption and boost the photocurrent. The QWs can be considered approximately as an incremental current source in parallel with a conventional homojunction solar cell.

The use of low-dimensional semiconductor structures in solar cells was first proposed in the early 1980s, where QW and superlattice arrays were introduced into the doped regions of a tandem device based on *p-n* junctions (Wanlass and Blakeslee 1982; Chaffin *et al.*, 1984; Goradia *et al.*, 1985). However, poor collection due to inefficient carrier diffusion within a doped QW structure led to performances inferior to those of bulk alloys, and the idea was abandoned.

The usual and most effective geometry is to insert an array of QWs called a multiple QW (MQW) into the depletion region (rather than the doped region) of a *p-n-* or *p-i-n*-junction solar cell. Quantum wells absorb more solar photons and so enhance the photocurrent. However, because of their lower bandgap, QWs also enhance the dark recombination current that opposes the photocurrent. If the increase in photocurrent exceeds the increase in dark current under operating conditions, then the power-conversion efficiency of the cell will be increased.

The possibility of a higher limiting efficiency than is possible with a simple single-junction cell is the main reason for interest in the QWSC, but QW structures have also been considered as a means of improving the performance of practical solar cells. For instance, the ability to control the bandgap through QW width is useful in various applications, and the response of QW cells to increases in temperature makes them attractive for solar concentration. The properties of QWSCs have been studied, theoretically and experimentally, by a number of research groups. Significant milestones in the development of the QWSC include the demonstration of increased photocurrent from photogeneration in the quantum wells of a QWSC (Barnham *et al.*, 1991; Freundlich *et al.*, 1994) and an understanding of the carrier escape and current transport processes in forward bias (Nelson *et al.*, 1993; Kitatani *et al.*, 1995). In parallel, the opportunity for a fundamental efficiency advantage was debated (Corkish and Green, 1993; Araujo *et al.*, 1994; Ragay *et al.*, 1994; Luque *et al.*, 2001) and full device models were developed (Renaud *et al.*, 1994; Anderson, 1995; Nelson, 1995; Ramey and Khoie, 2003). More recently, developments in so-called 'strain-balanced' QW structures enabled high efficiency single-junction and tandem devices. This development made QWSC technology potentially viable for applications in space and terrestrial concentrator systems. Quantum well solar cell technology is now being commercialised by JDS Uniphase Corporation. Progress in epitaxial growth of highly strained layers has enabled strain-balanced superlattice solar cells to be achieved (Wang*et al.*, 2012).

This chapter covers the following areas: QW technology and materials issues; the physics of QWs and how they differ from bulk materials in the processes important for solar energy conversion; experimental results showing how these properties affect the performance of the solar cell; the question of the limits to efficiency; and practical applications of QWSCs.

10.2 Device design, materials and technology

10.2.1 *Device design*

The typical QWSC is a *p-i-n* junction containing a number of QWs in the intrinsic (*i*) region, as shown schematically in Fig. 10.1. The *i* region acts as a spacer layer

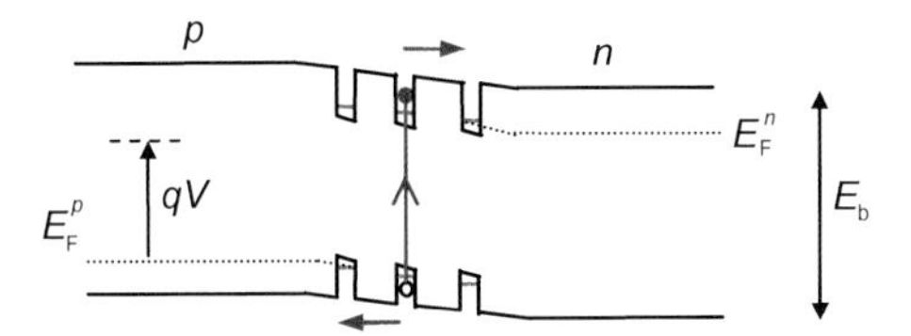

Figure 10.1 Schematic diagram of energy band vs. distance for a p-i-n QWSC at forward bias V. Quantum wells are placed in the undoped i-layer, where the electric field helps to separate photogenerated electron–hole pairs. Under illumination, the charges are separated by the electric field and the electron and hole quasi-Fermi levels E_F^p and E_F^n are split by qV, as shown.

to extend the field-bearing region. The rationale is that QWs should be placed in this region where charge separation and collection is most efficient.

The QWs are thin layers of a second, narrower gap semiconductor between barrier layers of the host material. III–V semiconductors are normally used for both QW and host material. Typically the QWs are 60–150 Å wide, separated by barriers of 50 Å or more. Some 50 QWs can be placed in an i-region 0.5–1 μm thick. In practice the i-layer thickness is limited by the background level of charged impurities, since these can cause the electric field to fall to zero within the i-layer and render the remaining, neutral part of the junction useless (Zachariou *et al.*, 1996; Barnham *et al.*, 1991).

The remaining parts of the cell — the p-layer, n-layer, substrate, window layers, contacts and optical coatings — are modelled on conventional III–V solar cell designs. The emitter (top layer) may be somewhat narrower in order to admit as much light as possible to the active i-region. Otherwise, the same design considerations apply: choice of polarity (p-n or n-p); techniques to enhance minority-carrier collection such as surface passivation, window layers and graded emitters; choice of materials for dopant, antireflective (AR) coat and substrate. The original goal was to use the incremental photocurrent provided by the QWs to enhance the efficiency of a well-designed homojunction cell.

An additional advantage is the possibility of controlling the QW width to tune the absorption edge of one or more cells in a multijunction device.

10.2.2 *Materials*

Quantum well solar cells have been studied in several III–V materials combinations. A short summary of some key examples follows, expressed as Barrier/QW material: AlGaAs/GaAs (Barnham *et al.*, 1991), GaAs/InGaAs (Barnes *et al.*, 1996), InP/InGaAs (Zachariou *et al.*, 1998), InGaP/GaAs (Barnham *et al.*, 1996), InP/InAsP (Monier *et al.*, 1998), InGaAsP/InGaAs (Rohr *et al.*, 2006) and

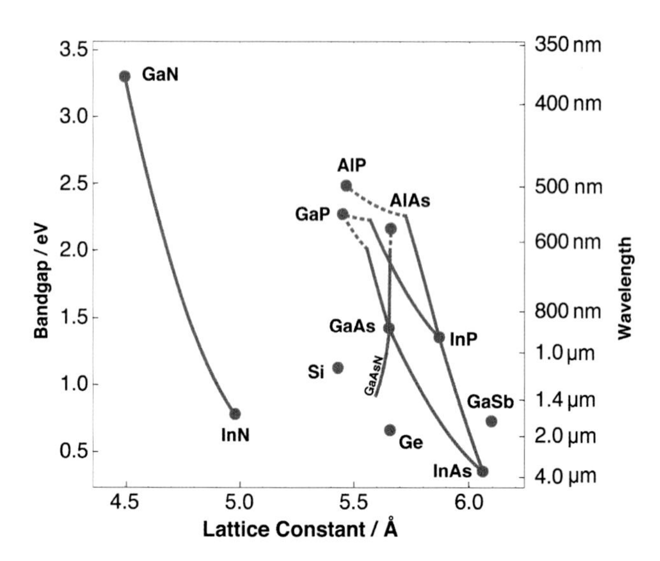

Figure 10.2 Plot of bandgap vs. lattice constant for III–V semiconductors.

GaAs/GaAsN (Freundlich *et al.*, 2007). In choosing a compatible pair of materials the lattice constant of the two should match approximately with each other and with a suitable substrate. In addition, some strain-balanced combinations of QW and barrier material where a lattice mismatch is accommodated through regions of alternating strain have been investigated, notably GaAsP/InGaAs (Ekins-Daukes *et al.*, 1999) and In_xGaAs/In_yGaAs (Connolly and Rohr, 2003).

As shown in Fig. 10.2, a wide range of lattice constants and bandgaps, spanning the infrared (IR) to ultraviolet (UV) parts of the spectrum, can be achieved using alloys of III–V semiconductor material. The AlGaAs/GaAs system has been most used as a test system to investigate the basic physical principles of the QWSC, because this combination has been widely used in optoelectronics and is well understood. AlGaAs/GaAs QW solar cells are not in themselves interesting as high-efficiency devices, since the bandgap of AlGaAs is higher than is optimal for solar energy conversion, and performance can always be improved simply by reducing the amount of Al.

The technologically relevant systems are gallium arsenide (GaAs, $E_g = 1.42\,\text{eV}$) or indium phosphide (InP, $E_g = 1.34\,\text{eV}$) *p-n* cells to which QWs have been added. Both materials have bandgaps close to the optimum for solar energy conversion. GaAs is currently a suitable material for high-efficiency multijunction PV systems, since an efficient, high-bandgap (1.9 eV) cell composed of InGaP can be grown lattice-matched to GaAs. These semiconductors can form the basis for the industry-standard InGaP/GaAs/Ge triple-junction solar cell that is presently used in terrestrial concentrator systems and to power space satellites. However, this material system is constrained by the bandgap of Ge, which is too low. The optimal would be 1 eV but there is presently a lack of high-quality compound semiconductors with this bandgap lattice that is matched to GaAs.

InP is attractive on account of its radiation hardness and, unlike GaAs, it has a number of alloys suitable for forming low-gap solar cells. However, it suffers from a relative lack of the lattice-matched high-gap alloys that are necessary for a triple-junction solar cell.

The obvious material for QWs in GaAs is the alloy $In_xGa_{1-x}As$, where $0 < x < 1$. GaAs/$In_xGa_{1-x}As$ structures are widely used in optoelectronics in IR photodetectors and lasers. However, they tend to suffer from strain-induced dislocations arising from the difference in the lattice constants of GaAs and $In_xGa_{1-x}As$, which increases with increasing indium fraction x. These dislocations degrade the device quality and impose a limit on the number and depth of QWs that may be added to a *p-i-n* structure (Griffin *et al.*, 1996). One means to overcome this limitation is to compensate the compressive strain introduced by the QWs with a tensile barrier material that is different from the *p* and *n* host material. While each bilayer remains strained, the QWs exert no net force on overlying layers, enabling thick stacks of strained semiconductor material to be grown (Ekins-Daukes *et al.*, 2002). The GaAsP/InGaAs material system is one such strain-balanced alloy combination that enables materials with absorption thresholds below that of GaAs to be achieved, using strained semiconductors yet without introducing dislocations (Ekins-Daukes *et al.*, 1999). The value of strain-balanced structures was also recognised by early pioneers of QW superlattice structures in solar cells (Chaffin *et al.*, 1984).

To achieve an efficient four-junction solar cell, a 1 eV threshold absorber is required and achieving this on a GaAs substrate requires the use of dilute nitride materials. In this respect, GaAs/GaAsN QW structures have been shown to achieve an absorption threshold of 1.1 eV (Freundlich *et al.*, 2007).

For InP, the best QW material is $In_{0.53}Ga_{0.47}As$. At this composition In_xGa_{1-x} As is lattice-matched to InP, eliminating strain problems. Since $In_{0.53}Ga_{0.47}As$ has a much narrower bandgap than InP, the QWs are deep, extending absorption out to around 1600 nm, so they offer significant photocurrent enhancements by utilising IR solar radiation. In addition InP/$In_xAs_{1-x}P$ QW structures have shown radiation resistance equal or superior to that of an InP homojunction (Walters *et al.*, 2000).

For thermophotovoltaics, where a lower bandgap is needed because of the lower source temperature, InGaAsP/$In_xGa_{1-x}As$ QW cells have been studied (see Section 10.6), and InGaAsSb/GaSb is likely to be a promising system.

Also, QWSCs have been fabricated in the GaN/InGaN material system, extending the spectral response of a GaN-based solar cell to 520 nm (Farrell *et al.*, 2011).

10.2.3 *Fabrication technology*

III–V QW structures are grown by epitaxial techniques such as molecular beam epitaxy (MBE) and metal organic vapour phase epitaxy (MOVPE). Layers of controlled composition are deposited, one atomic layer at a time, on a monocrystalline substrate of the same or another lattice-matched III–V material. These methods are also used to grow compound heterostructures for windowed and tandem solar cells and can reach growth rates of several microns per hour. When growing precise QW heterostructures, the growth rate is slowed, but this is the only significant difference relative to the growth of conventional III–V solar cells. The manufacturing cost of III–V concentrator solar cells is spread roughly equally between the substrate cost, epitaxy cost and cell-processing cost. Growth on inexpensive substrates (e.g. silicon) or epitaxial lift-off and wafer reuse (Bauhuis *et al.*, 2010) can help reduce the substrate cost. Epitaxy and processing costs may be reduced through high growth rate deposition methods to be developed specifically for III–V photovoltaic devices, as well as efficient, high-throughput metallisation technologies. In parallel with the microelectronics industry, when the market grows sufficiently large, transitioning to larger substrate sizes (6-inch GaAs or 12-inch Si substrates) will help reduce all these manufacturing costs. Since cell efficiency directly impacts the levelised cost of electricity, surprisingly small increases in efficiency can offset higher manufacturing cost. A quantum well cell technology must therefore offer ample efficiency premium to become commercially attractive.

10.2.4 *Other materials*

The possibilities of using QW structures for photoconversion in materials other than III–V semiconductors have been, to date, largely unexplored. The first attempt to grow a II–IV QWSC using the materials combination CdTe/CdMnTe (Scott *et al.*, 1997) was hindered by problems with series resistance. II–IV quantum dot (QD) structures, developed mainly for their electroluminescent properties, have been widely applied as *n*-type components of hybrid organic/inorganic bulk heterojunction solar cells, but these do not exploit the quantum properties of the semiconductor nanoparticles and are reviewed elsewhere (Moulé *et al.*, 2012; McGehee, 2009). More recently, chalcopyrite QWSCs have been considered and while interdiffusion of elements has been observed, this is considered not to be a major limitation for a functioning device (Afshar *et al.*, 2011).

Si-based QWs and superlattices have also been investigated in an attempt to fabricate an all-silicon tandem solar cell (Green, 2003), exploiting the 1.1 eV bulk silicon junction as a bottom cell. Three-dimensional confinement is required to achieve a 2 eV gap in Si for the top junction. For this purpose, arrays of Si QDs confined by SiO and SiN barriers have been fabricated with an absorption threshold approaching this value, although in common with II–VI QD structures, efficient current transport through the QD array remains a challenging problem (Conibeer *et al.*, 2010).

10.3 Physics of QWs

Quantum wells possess a distinct electronic structure due to the confinement of charge carriers into thin semiconductor layers. This special electronic structure affects the physical processes of photon absorption, carrier relaxation and transport, which are of central importance to photovoltaic energy conversion. In this section, we will discuss the electronic structure of the QW and its effect on each of these microscopic processes. In Section 10.4, we will relate the basic physics of QWs to the performance of the photovoltaic device.

10.3.1 *The quantum well*

A QW is formed when a layer of a narrow-gap semiconductor a few nanometres thick is grown epitaxially between layers of the wider-gap semiconductor. It is important that the two materials be *electronically* similar (i.e. that the atomic part of the crystal wave function is similar, and that the two materials have the same crystal symmetry), that the interface be clean and that the same crystal planes form the interface in either material. Then the different potential of the QW may be treated as a perturbation to the crystal potential and delocalised carrier states can extend across the two materials (Bastard, 1988).

If the conduction band edge is lower in energy and the valence band higher in energy in the well material than in the barrier, then electrons and holes are both confined in the well material. This is known as a Type I QW (Fig. 10.3). If only one carrier type is confined in the well, the QW is Type II. Many QWs together form an MQW, and if the barriers are thin enough for neighbouring wells to be electronically coupled the structure is known as a superlattice (SL). Only Type I QWs have so far been studied for solar cells, although SLs have been proposed as a

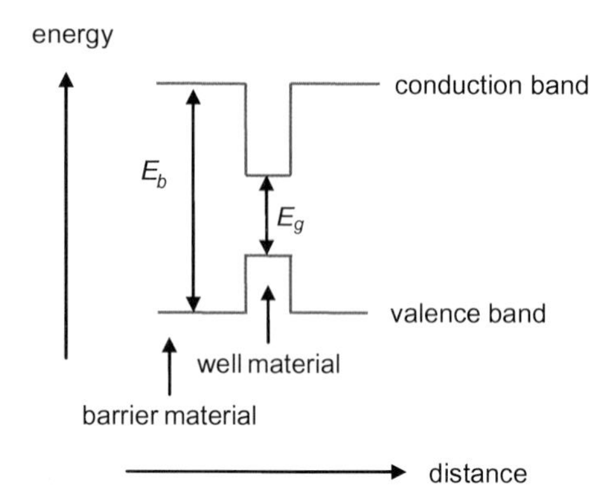

Figure 10.3 Band profile of a Type I quantum well.

means of improving carrier transport in high-resistivity InP solar cells (Varonides and Berger, 1997).

10.3.2 *Density of states*

The QW forms a quasi-two-dimensional system. Confinement of electrons and holes in the growth (say, z) direction leads to quantisation of the z component of their momentum and kinetic energy. The quantised energy E_n of the nth level is related to the z component of the wave vector k_n through

$$E_n = \frac{\hbar^2 k_n^2}{2m^*}$$

where m^* is the effective mass of the carrier in the growth direction. The carriers are confined to a set of sub-bands of minimum energy U_n, but are free to move in the (xy) plane of the well where the symmetry of the crystal is maintained. Hence a carrier in the nth sub-band has total energy

$$E(\mathbf{k}) = E_n + \frac{\hbar^2 k_\parallel^2}{2m_\parallel^*} \tag{10.1}$$

where $\mathbf{k}$ is the total wave vector, $k_\parallel$ is the component in the xy plane (such that $\mathbf{k^2} = k_\parallel^2 + k_n^2$), and $m_\parallel^*$ is the effective mass of the carrier in this plane.

In the envelope function approximation, the shift $V(z)$ in the conduction or valence band edge due to the QW is considered as a perturbation to the periodic crystal potential, and the wave functions as crystal eigenfunctions modulated by an '*envelope* function'. The confined state energies E_n and envelope functions $F_n(z)$ are solutions to an 'effective mass' equation, which resembles Schrödinger's equation for a one-dimensional potential well. They are analogous to the energy levels and wave functions of a one-dimensional quantum system. For a QW of width L and depth V

$$-\frac{\hbar^2}{2m^*}\frac{d^2F_n(z)}{dz^2} + E(z)F_n(z) = E_nF_n(z)$$

where

$$\begin{aligned} E(z) &= 0, \quad -L/2 \le z \le L/2 \\ E(z) &= V, \quad z > |L/2| \end{aligned} \tag{10.2}$$

This equation holds for both electrons in the conduction band and holes in the valence band, but with different values of m^* and V. Energies E_n are measured up from the bottom of the QW in the conduction band for electrons, and down from the top of the valence band for holes. The well depth V depends on the composition of the barrier and well materials and on how the difference in bandgap is divided between the valence and conduction bands. The effective mass m^* for each carrier type is in general different for well and barrier.

In III–V semiconductors two different types of hole, *heavy* and *light*, need to be considered. In the bulk crystal, heavy and light holes are carriers with different effective mass associated with two degenerate crystal bands. For a QW in unstrained material, heavy and light holes occupy the same potential well in the valence band, but with different sets of confined-state energies on account of their different effective masses. In a strained QW, the well depths for heavy and light holes can be different. The number N of confined states contained in the QW for each carrier type is given by

$$N = \text{int}\left(\frac{L\sqrt{2m^*V}}{\pi\hbar}\right) + 1 \tag{10.3}$$

where int(x) means the integer part of x. N increases with increasing well width and depth, and carrier effective mass. The well is normally narrow enough to admit only a few confined states. At energies $E > V$ the carriers are no longer confined and a continuum of states becomes available, as in the bulk material. These continuum states will not be considered here.

In accordance with the uncertainty principle, the lowest energy level is always shifted away from the bottom of the well, by an amount that increases with increasing quantum confinement. This means that the ground-state energy, and hence the absorption edge, can be controlled simply by varying the well width.

The corresponding envelope functions have well-defined parity and penetrate further into the barrier as energy is increased. Energy levels and envelope functions

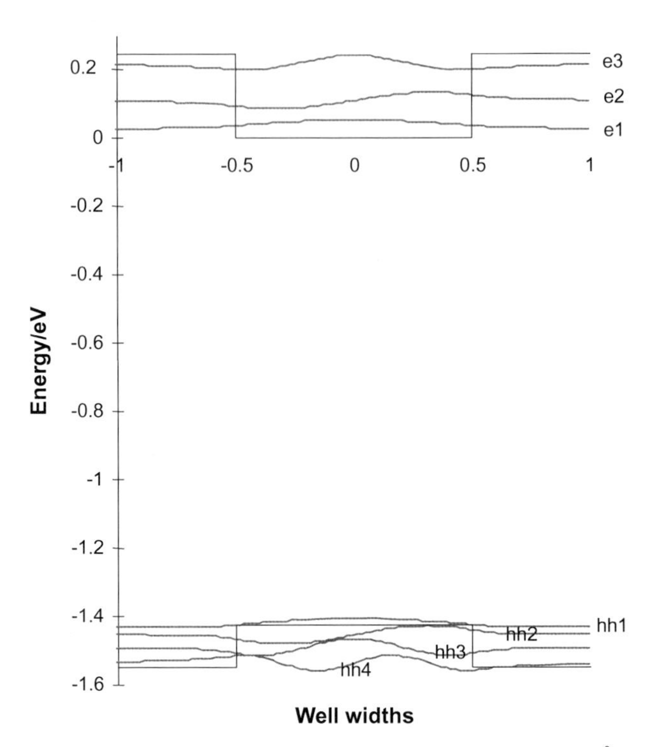

Figure 10.4 Calculated energy levels and envelope functions for a 100 Å GaAs QW in Al0.3Ga0.7As. The relative energies of confined states and bandgaps are to scale, and the bottom of the conduction band is taken as the zero of energy. Quantum number is measured up from the bottom of the well for electrons, and down from the top of the well for holes.

for a typical $Al_xGa_{1-x}As$ /GaAs QW are shown in Fig. 10.4. In the SL configuration, neighbouring wells are coupled, extended-state envelope functions span the entire SL, and the previously discrete energy levels of the QW broaden into bands. These effects improve carrier transport in the growth direction.

The density of states function can be constructed from the energy spectrum in the usual way. For a QW of width L the density of states per unit volume V is given by

$$D(E) = \frac{2}{V}\sum_{\mathbf{k}} \delta[E - E(\mathbf{k})] = \frac{m^*}{\pi \hbar^2 L}\sum_{n=1}^{N} \Theta(E - E_n) \qquad (10.4)$$

where δ is the Dirac delta function and Θ is the Heaviside function. As illustrated in Fig. 10.5, $D(E)$ has the staircase structure characteristic of quasi-two-dimensional systems.

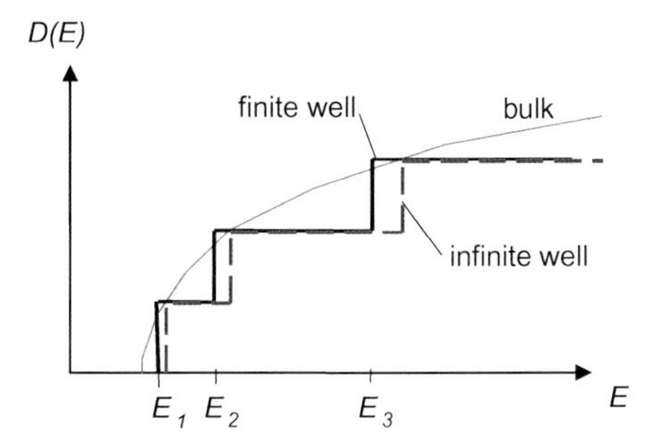

Figure 10.5 Schematic density of states function $D(E)$ for a finite QW, compared with that for an infinitely deep QW and for the well material in the bulk. The first three confined state energies, E_1, E_2 and E_3, are shown.

This allows us to calculate the concentrations n (of electrons) and p (of holes) in the QW, assuming a local quasi-thermal equilibrium. For electrons with density of states function $D_{\mathrm{cb}}(E)$ in the conduction band

$$n = \int_{E_{cb}}^{\infty} D_{cb}(E(\mathbf{k})) f_{FD}(E(\mathbf{k}), T, \widetilde{\mu}_e) dE \tag{10.5}$$

where $f_{\mathrm{FD}}(E)$ is the Fermi–Dirac distribution function, E_{c} is the conduction band-edge energy, $\widetilde{\mu}_e$ the quasi-Fermi level for electrons in the QW, and T the effective electron temperature. When $E_{\mathrm{cb}} \gg \widetilde{\mu}_e$, n is given by

$$n = n_{iq} \exp[(\widetilde{\mu}_e - E_i + \theta_n)/kT] \tag{10.6}$$

where n_{iq} is the intrinsic carrier density of the QW material in the bulk, E_i is the intrinsic potential energy — the level at which the Fermi level would lie in a perfectly intrinsic material — and θ_n is a measure of the shift in n due to quantum confinement. This is analogous to the expression for n, namely $n_i \exp[(\widetilde{\mu}_e - E_i)/kT]$, in a non-degenerate bulk semiconductor.

For the remainder of the discussion, we will assume that the QW is described by a quasi-two-dimensional density of states and by a local quasi-Fermi level that is not necessarily continuous with that in the barrier material.

In operating conditions, QWs placed in the space-charge region of a *p-n* junction will be subject to a (small) electric field. The field tilts the QW band profile, as shown in Fig. 10.6, distorts the confined-state functions and shifts their energies.The energy of the lowest confined state is reduced. Strictly speaking, in the presence of the field these wave functions are no longer confined — carriers penetrate further into the barrier on the side of decreasing potential energy and can tunnel out. For solar cells, the electric field is small enough for the 'flat-band'

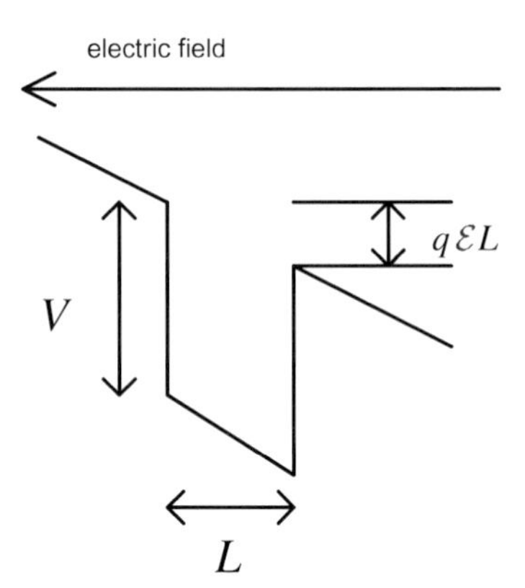

Figure 10.6 Band profile for a QW subject to an electric field in the growth direction. As the field is increased the right-hand barrier is reduced, increasing the probability of electron escape by thermionic emission or tunnelling.

approximation to the band structure to be adequate. However, it is relevant that tunnelling through the barrier is possible.

10.3.3 *Photogeneration*

In a solar cell, photon absorption across the bandgap is important. Fermi's golden rule gives the absorption coefficient α in terms of the confined-state energies and overlap integrals. For transitions between an initial valence-band state $|i\rangle$, of energy E_i and a final, conduction-band state $|f\rangle$, of energy E_f, under the influence of an electromagnetic field of angular frequency ω and polarisation **e** we have (Bastard, 1988)

$$\alpha(E) = \frac{A}{\omega}\sum_{i,f} |\langle f|\mathbf{e}.\mathbf{p}|i\rangle|^2 \delta[E_f - E_i - E](f_{FD}(E_i) - f_{FD}(E_f)) \qquad (10.7)$$

where E is the photon energy $\hbar\omega$, $\mathbf{p}$ is the quantum mechanical momentum operator and A is a sample-dependent optical constant. In the usual case where the light is incident normal to the plane of the QW, the matrix element is proportional to the overlap integral M_{lm} between the initial valence sub-band, l, and final conduction band, m, envelope functions

$$M_{lm} = \int F_{el}(z)F_{hm}(z)\mathrm{d}z \qquad (10.8)$$

This means that optical transitions are allowed only between sub-bands of the same parity (l and m both even or both odd), and are strong only when $l = m$.

In addition, coulombic bound states (excitons) are formed at an energy just below the minimum for each optically allowed sub-band-to-sub-band transition. The excitons appear as strong peaks in the spectrum, even at room temperature, because of their higher binding energy in two-dimensional systems. Including only the principal ($1s$) exciton and summing Eq. (10.7) over initial and final state energies for the lth electron – mth hole sub-band pair, we have

$$\alpha_{lm}(E) = \alpha_{lh/hh} M_{lm} [f_{lm}\delta(E - E_{lm} - B_{lm}) + \Theta(E - E_{lm})] \tag{10.9}$$

where E_{lm} is the electron-hole transition energy before Coulombic effects are included, B_{lm} and f_{lm} are the exciton binding energy and oscillator strength, respectively, and the constants $\alpha_{lh/hh}$ represent the absorption coefficient on the first step edge. In III–V semiconductors, optical transitions occur between both electron–heavy hole (hh) and electron–light hole (lh) states. The total absorption is the sum of contributions from all such transitions.

$$\alpha(E) = \sum_{l,m} \alpha_{e_l hh_m}(E) + \sum_{l,m} \alpha_{e_l lh_m}(E) \tag{10.10}$$

where each electron–hole sub-band pair contributes a step function and a set of excitons to the total absorption spectrum. The absorption coefficient for a typical $Al_xGa_{1-x}As$ /GaAs QW is shown in Fig. 10.7.

The QW absorption spectrum thus reflects the step-like form of the density of states, modified by strong excitonic peaks. (Because of the strong exciton, the QW spectrum may have a steeper absorption edge than the equivalent bulk alloy,

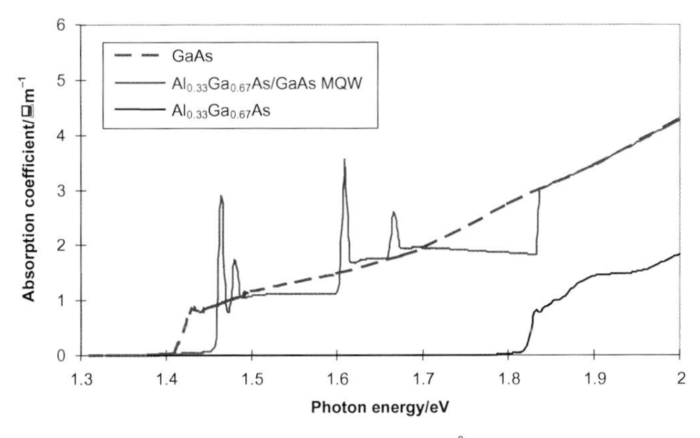

Figure 10.7 Calculated absorption coefficient for a 100 Å $Al_{0.33}Ga_{0.67}As$/GaAs QW compared with the absorption of bulk GaAs and bulk $Al_{0.33}Ga_{0.67}As$. (For the QW, the absorption coefficient is per unit thickness of *well* material, not including barrier thickness.)

which could be useful for certain PV applications.) The absorption edge or effective bandgap E_a is blue-shifted from the absorption edge E_g of the well material *in the bulk* by the joint confinement energies E_{11h} of the lowest electron and heavy hole sub-bands less the corresponding exciton binding energy B_{11h}.

$$E_a = E_g + E_{11h} - B_{11h} \tag{10.11}$$

The effective bandgap E_a is most strongly influenced by QW width and varies from the bandgap E_g of the well material for very wide wells, to the bandgap E_b of the barrier — or host — material for very narrow wells. This tunability of the absorption edge is one of the most important features of the QWSC.

At photon energies above E_b, photogenerated carriers are no longer confined in the QW and the simple quantum mechanical model of absorption becomes unhelpful. In this range the absorption spectrum of the QW begins to resemble that of the bulk material.

10.3.4 *Transport and recombination*

As in any semiconductor device, the electron and hole, once excited, may be transported away from the point of creation, or recombine with each other or with trap states in the bandgap. In the steady state, these processes are described by the continuity equations for electrons and holes:

$$\frac{1}{q}\frac{\mathrm{d}i_e}{\mathrm{d}z} + g - r = 0$$

and

$$-\frac{1}{q}\frac{\mathrm{d}i_h}{\mathrm{d}z} + g - r = 0$$

where r is the volume recombination rate, g the volume generation rate, i_e the electron current density and i_h the hole current density. The materials parameters that are normally used to quantify these processes in a bulk crystalline semiconductor device — the recombination lifetimes and diffusion constants — are properties of the bulk material and only have meaning in a material many times the thickness of a QW. Level quantisation affects not only the generation term through the QW absorption, discussed above, it also affects the rate of recombination and the mechanism of transport in the direction of the built-in field. We shall discuss these effects next.

10.3.5 *Recombination*

The processes that govern recombination in bulk materials apply to QWs. For III–Vs the most important, in practice, is non-radiative recombination through traps. For a single trap state in the bandgap, the Shockley–Read–Hall recombination rate is given by

$$r_{nr} = \frac{np - n_i^2}{\tau_e(p + p_t) + \tau_h(n + n_t)} \tag{10.12}$$

where p_t, n_t are the equilibrium populations of trap states occupied by holes and electrons, and τ_p, τ_n are the respective carrier trapping times. This formulation should be appropriate to a QW provided that n and p are defined using the quasi-Fermi level of the carriers in the QW (Eq. (10.6)). The lifetime parameters are properties of the material and so as a first approximation to a QW we may take the same values as for the well material in the bulk. However, the accumulation of defects at the QW interface may affect the location and density of trap states, and quantum confinement may reduce the trapping times.

In the limit of ideal material, radiative recombination is the process that determines solar cell efficiency. The excess radiative recombination in the biased device (i.e. in addition to the recombination that balances thermal generation in equilibrium) then constitutes the dark current. In any volume element δV the radiative recombination rate r_{rad} depends on the local absorption spectrum $\alpha(E)$ and the local quasi-Fermi level separation $\Delta\tilde{\mu}_{\text{F}}$, according to

$$r_{\text{rad}}\delta V = \int \alpha(E) j(E, T, \Delta\tilde{\mu}_F) \mathrm{d}E\delta V \tag{10.13}$$

where the emitted flux density j is given by the generalised Planck equation (Würfel, 1982; Tiedje *et al.*, 1984)

$$j(E, T, \Delta\tilde{\mu}_F) = \frac{2n_r^2}{h^3c^2} \frac{E^2}{e^{(E-\Delta\tilde{\mu}_F)/kT} - 1} \tag{10.14}$$

and n_r is the local refractive index, h is Planck's constant and c the speed of light.

10.3.6 *Transport in the growth direction*

In a homojunction solar cell, electron and hole currents are normally described by the drift–diffusion equations. The electron current is given by (Sze, 1981; Hovel, 1975)

$$i_e = i_{\text{diff}} + i_{\text{drift}} = qD_e\frac{dn}{dz} + qu_e n\mathcal{E} \tag{10.15}$$

where D_e is the electron diffusion coefficient, $\mathcal{E}$ the electric field strength and u_n the electron mobility. In the region of a QW, carriers with energy above the barrier band edge will be scattered into and out of the QW by much the same processes as in the bulk. One approximation to transport through the QW is to apply the drift–diffusion equations to the composite structure and represent the QW by a region of increased carrier population. In this approach it is assumed that the QW and host are in quasi-thermal equilibrium and that the quasi-Fermi level is continuous across the interfaces of the QW.

However, this treatment neglects the fact that carriers with energies below the barrier band edge are essentially trapped in the QW. These carriers may escape from the QW by quantum mechanical tunnelling or thermally assisted tunnelling when the QW is subject to an electric field in the growth direction, and other carriers may be captured into the QW by scattering from higher energy states. Both of these processes are quantum mechanical and cannot effectively be described with the drift–diffusion equations. An alternative approach is to consider two populations for each carrier: a mobile population with energy greater than the barrier band edge, and a confined population in the QW. The confined population does not actually contribute to the current, but it affects the current by influencing the rate at which carriers are added to the mobile population through escape, or removed from it through capture. Equation (10.15) for i_e should therefore include terms for escape and capture:

$$i_e = i_{\text{diff}} + i_{\text{drift}} + i_{\text{esc}} - i_{\text{capture}} \tag{10.16}$$

The escape current is important whenever carriers are generated in the QW and an electric field is present perpendicular to the plane of the QW. It can be calculated semiclassically from

$$i_{\text{esc}} = \int D_{cb}(E(\mathbf{k})) f_{FD}(E(\mathbf{k}), T, \widetilde{\mu}_e) Tr(k_z) \frac{\hbar k_z}{m_e^*} d^3\mathbf{k} \tag{10.17}$$

where $Tr(k_z)$ is the probability of transmission through the barrier at wave vector k_z. The integral should be carried out with respect to wave vector rather than energy, because of the anisotropy of the density of states. For shallow QWs in short-circuit and low-forward-bias conditions, the escape current appears to be equivalent to the sheet-generation rate in the QW (Nelson *et al.*, 1993). In the operating conditions of a QW, when barrier states are also populated, Eq. (10.17) should be modified to allow for a finite probability of population of the final states, and a term added for capture of carriers into the QW from states above the barrier. This is a notoriously difficult problem, involving both localised and continuum states, and has not at the time of writing been resolved for the QWSC.

Once given the expressions for carrier-pair generation, recombination and current, carrier continuity may be invoked to complete the set of equations for the QW device. Discretising the continuity equations for electrons in a QW of width L

we have, in the steady state

$$\frac{\Delta i_e}{L} = q(g - r) \tag{10.18}$$

where Δi_e represents the change in electron current across the QW, and g and r the volume rates of pair generation and recombination in the QW. At each interface, i_e should be continuous. Thus a set of current and continuity equations can, in principle, be constructed to describe the behaviour of the QWSC, much as for conventional solar cells. To do this, the correct descriptions of escape and capture must be known so that the carrier current can be related to the quasi-Fermi levels in the QW.

10.4 Performance characteristics of QWSCs

10.4.1 *Quantum efficiency*

Quantum wells enhance the photocurrent by extending absorption. The relevant quantity for PV is the spectral response (SR) or zero-bias external quantum efficiency of the cell, which is the probability that, for each incident photon of given wavelength, an electron will be collected at the contacts. Figure 10.8 shows how QWs extend the SR to longer wavelengths for an AlGaAs *p-i-n* cell with and without GaAs QWs. The SR at photon energies above the baseline cell bandgap E_b is barely affected by the QWs. This shows that QWs do not degrade the collection of minority carriers. Indeed, the SR is improved slightly at energies close to E_b on account of the higher absorption of the QW material at these wavelengths.

The SR at photon energies below E_b is entirely due to absorption in the QWs and reflects features of the QW absorption spectrum. This behaviour is straightforward to model. For QWs placed in the depletion region of the cell, we can ignore diffusion and make the usual assumption that all minority carriers reaching the edges of the depletion region by drift are collected. Then for N QWs of width L, the SR is related to absorption through

$$\mathrm{SR}(E) = (1 - r)\eta_{\mathrm{esc}}(1 - \exp[-N\alpha_{\mathrm{QW}}(E)L]) \quad E < E_b \tag{10.19}$$

where r is the front-surface reflectivity and η_{esc} the probability of carrier escape from the QW. The fraction $(1 - \eta_{\mathrm{esc}})$ of the carriers that do not escape can be assumed to recombine in the QW. In fact, for QWs in any of the systems mentioned at room temperature, η_{esc} is unity. This has been confirmed by experimental studies of the dependence on the field and temperature of the photocurrent from single QW devices (Nelson *et al.*, 1993; Barnes, 1994; Zachariou *et al.*, 1998), which show that carriers escape from the QWs at room temperature by thermally assisted

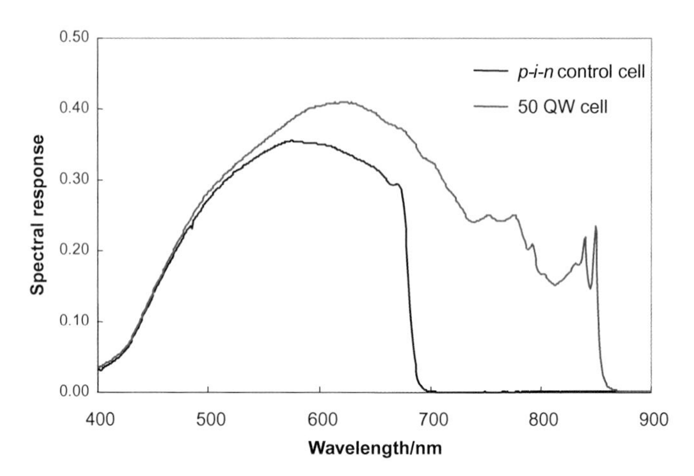

Figure 10.8 Measured spectral response (external quantum efficiency) for an $Al_{0.33}Ga_{0.67}$ As *p-i-n* test device with and without 50 87 Å QWs in the i-region. These devices were not AR coated.

tunnelling. Equation 10.19 reproduces the short-circuit SR fairly reliably for a well-characterised QW (Paxman *et al.*, 1993; Barnes *et al.*, 1996; Connolly *et al.*, 1995). It is implicit in Eq. 10.19 that, once a carrier has escaped from the QW, it will be collected.

The incremental photocurrent due to the QWs in an incident spectrum of photon flux $j_{\text{inc}}(E)$ is thus

$$\Delta i_{\text{ph}} = \int \Delta SR(E) j_{\text{inc}}(E)\, \mathrm{d}E \tag{10.20}$$

where $\Delta\text{SR}(E)$ is the incremental SR due to the QWs, which is given by Eq. (10.19) for $E < E_b$ and the difference between the SR of well and host material for $E > E_b$.

To maximise the effect of the QWs on photocurrent we need an SR which is as high as possible for $E > E_a$. In principle this can be achieved by increasing the number of QWs and reducing the optical depth of the top layer (the *p*-region in a *p-i-n* structure) by reducing its thickness or compositional grading, as shown in Fig. 10.9.

In practice it is not always straightforward to increase the number of QWs. A wide *i*-region requires a relatively low level of charged impurities ($<10^{15}$ cm^{-3}) in order to maintain the electric field (Barnham *et al.*, 1991; Zachariou *et al.*, 1996). Figure 10.10 shows how the QW SR is affected by high background doping, in this case arising from diffusion of the zinc dopant into the *i*-region.

For GaAs/$In_xGa_{1-x}As$ QWs, strain limits the spacing of the QWs. The GaAs barrier layers need to exceed a critical thickness several times the $In_xGa_{1-x}As$

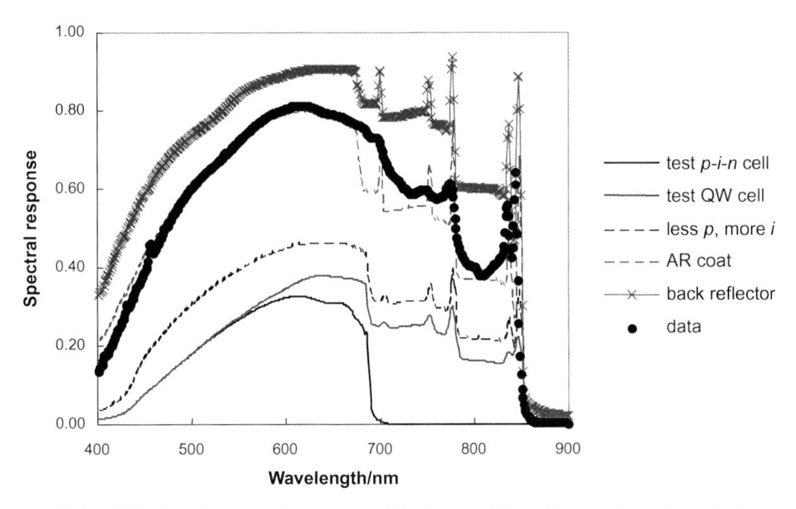

Figure 10.9 Calculated spectral response SR for an $Al_{0.33}Ga_{0.67}As$ *p-i-n* device and a series of $Al_{0.33}Ga_{0.67}As/GaAs$ QW devices, showing the effect of: a) reducing the *p*-layer thickness and increasing the *i*-layer thickness; b) adding an AR coat; c) adding a front-surface window and back-surface reflector. The measured SR for a cell of design (b) is compared with the calculation.

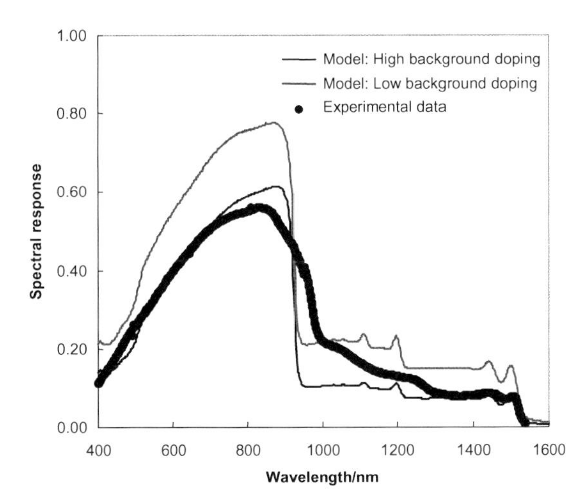

Figure 10.10 Measured and calculated SR for an InP *p-i-n* device containing 20 $In_{0.53}Ga_{0.47}As$ QWs. The calculated curves show that an unintentional background doping in the *i*−layer of around $2 \times 10^{16}\ cm^{-3}$ is required to explain the low SR. Diffusion of zinc dopants increases the space charge in the *i*-region so that at zero bias the electric field falls to zero near the centre of the *i*-region and only around half of the carriers generated in the QWs are collected.

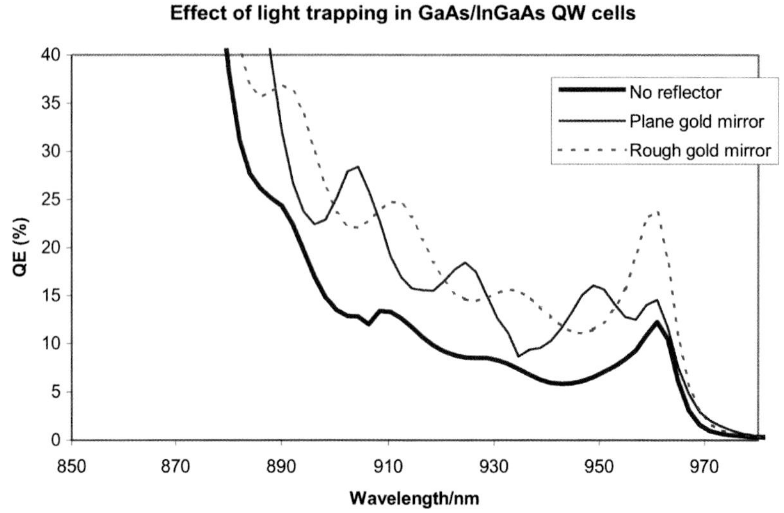

Effect of light trapping in GaAs/InGaAs QW cells

QE (%) — Wavelength/nm — No reflector, Plane gold mirror, Rough gold mirror

Figure 10.11 Measured SR for a series of GaAs/$In_xGa_{1-x}As$ devices, each with 10 QW, processed in different ways. The low SR with no light trapping is substantially increased by adding a plane or a rough gold mirror at the back surface. The rough mirror, a gold epoxy resin, achieves light trapping effectively through large-angle internal scattering.

QW width, in order to accommodate the lattice mismatch without dislocations (Barnes *et al.*, 1996; Griffin *et al.*, 1996). In this system, light trapping is essential to increase the SR to a useful level. Figure 10.11 illustrates the effect of light trapping on the SR.

An alternative strategy is to use $In_xGa_{1-x}As$ of low indium content rather than pure GaAs in the n- and p-regions, so partly compensating the strain introduced by the QWs (Mazzer, 1997). This reduces the critical spacing, and allows more QWs to be inserted into an i-region of given width.

10.4.2 *Dark current density*

The dark current $i_{\text{Dk}}(V)$ of a solar cell at forward bias V is the sum of contributions from radiative and non-radiative processes:

$$i_{\text{Dk}}(V) = i_{\text{nr}}(V) + i_{\text{rad}}(V) \tag{10.21}$$

Since the generation rate $g = 0$ in the dark, i_{Dk} is found by integrating all the contributions to the recombination rate r over the volume of the cell. Radiative recombination within the interior of the device is partly cancelled by absorption, and the radiative current density i_{rad} is given by q times the net photon flux escaping through the device surface. Thus i_{rad} is obtained by integrating Eq. (10.13) over surface elements S and solid angle Ω

$$i_{\text{rad}} = q \int_0^{\infty} \int_S^a (E, \theta, \mathbf{s}) j(E, T, \Delta\widetilde{\mu}_F) \mathrm{d}\Omega . \mathrm{d}S . \mathrm{d}E$$

where $a(E, \theta, \mathbf{s})$ is the probability that a photon of energy E will be emitted from the point $\mathbf{s}$ on the surface at an angle θ. This equation should apply to the radiative currents from a QW device provided that the appropriate values of $a(E)$ and $\Delta\widetilde{\mu}_{\text{F}}$ are used. In an ideal solar cell, radiative processes are dominant and $\Delta\widetilde{\mu}_{\text{F}}$ is assumed to be constant and equal to the cell voltage V.

In real solar cells at typical operating biases, non-radiative processes are dominant, and the non-radiative current density $i_{\text{nr}}(V)$ is found by integrating the non-radiative recombination rate $r_{\text{nr}}(z)$ over the volume of the cell:

$$i_{\text{nr}}(V) = \int r_{nr}(V, z) \mathrm{d}z \tag{10.22}$$

In III–V devices, r_{nr} is well modelled by the Shockley–Read–Hall expression (Eq. (10.12)), and will be greatest in the depleted space-charge layer (SCL) of the junction, where n and p are similar. Calculating i_{nr} in the depletion approximation (also known as the exhaustion layer approximation), we find for a *p-n* or *p-i-n* homojunction cell

$$i_{\text{nr}}(V) = \frac{q n_i W}{\tau_{nr}} \frac{\sinh(qV/2kT)}{(V_{bi} - V)q/kT} \frac{\pi}{2} \tag{10.23}$$

where W is the width of the SCL, τ_{nr} the ambipolar carrier lifetime and V_{bi} the built-in bias of the junction (Sah *et al.*, 1957).

Since QWs are placed in the SCL and have a much higher intrinsic carrier density than the host material, we may expect them to dominate the dark current. We can approximate the incremental recombination current Δi_{nr} by treating the QWSC in the depletion approximation in the same way as the *p-i-n* cell above. (The depletion approximation appears to be valid so long as the quasi-Fermi levels energy remain below the QW band-edges.) Then, for a single QW cell, we find (Nelson *et al.*, 1995)

$$\Delta i_{\text{nr}} = \frac{q n_{\text{iq}} L}{\tau_{\text{nrq}}} \frac{\sinh(qV/2kT)}{\cosh(q\varphi_q/kT)} \tag{10.24}$$

where n_{iq} is the intrinsic carrier density in the QW, τ_{nrq} is the mean non-radiative recombination time in the QW and ϕ_{q} the value of the intrinsic potential at the QW relative to the mid-point of the quasi-Fermi potentials for electrons and holes, which depends on the location of the QW relative to the point where $n = p$ and vanishes for a QW located in the centre of a symmetrically doped junction. In this case, Eq. (10.24) reaches its maximum value and $\Delta i_{\rm nr}$ takes the usual ideality factor of 2 for recombination–generation currents. Comparing Eqs (10.23) and (10.24) we could expect the dark current to increase by a factor of $(n_{\rm iq}L/n_i W)$ for a single QW added to the centre of the i-region. Typically, this factor may be one or two orders of magnitude, which could reduce the open-circuit voltage by 10–20%.

As Fig. 10.12 shows, dark currents for QW and p-i-n solar cells in a range of materials show that $i_{\rm Dk}$ values do indeed increase when QWs are added. However, when compared to theoretical calculations, the measured $i_{\rm Dk}$ is smaller than predicted by around one order of magnitude, and has a different bias dependence or ideality factor (Nelson *et al.*, 1994, 1995). This means that the open-circuit voltage is slightly larger than expected. Figure 10.13 shows dark currents calculated by solving the carrier transport equations self-consistently to determine the n and p profiles, and then integrating the SRH recombination rate across the device. A local quasi-thermal equilibrium is assumed, so that electron and hole quasi-Fermi levels are continuous across the QW interfaces.

Several possible explanations for the discrepancy are suggested by Eq. (10.24). One is that the recombination lifetime for carriers in the QW is longer than for the well material in the bulk. This is rather unlikely since material quality should, if anything, be worse in the QW configuration on account of the interfaces. Another is that background doping could shift the point where non-radiative recombination

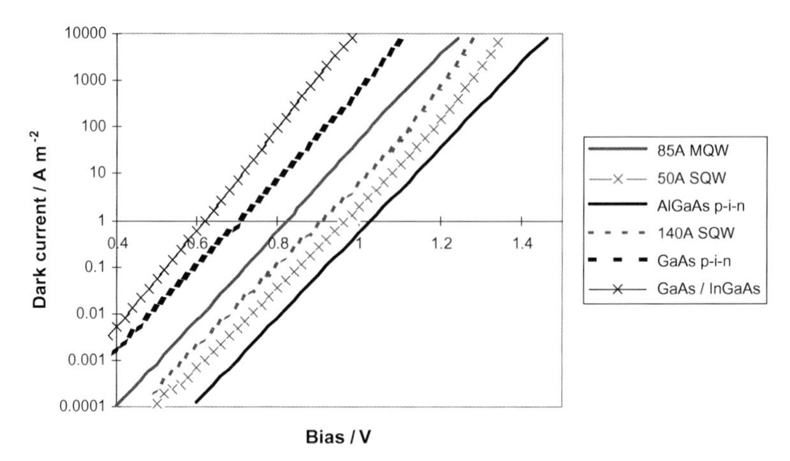

Figure 10.12 Measured dark current densities for a series of QW and homogenous p-i-n devices. The curves show that adding GaAs QWs to an $Al_xGa_{1-x}As$ p-i-n cell increases the dark current, as does adding $In_xGa_{1-x}As$ QWs to a GaAs device. The effect on the dark current is larger for wider QWs, and for a greater number of QWs.

is maximal away from the centre of the i-region, and so reduce the recombination rate at a centred QW. This would also explain the change in ideality factor, as shown in Nelson *et al.* (1995, 1999). But it would not explain why the dark current is lower than expected in *multi*-quantum well devices, where we would expect the regular MQW array to sample all n/p ratios equally.

A possible explanation is that n and p are smaller than expected in the QW, i.e. the quasi-Fermi levels are closer to the centre of the bandgap than in the neighbouring host material. This is supported by complementary measurements of the radiative recombination current in single QW devices (Nelson *et al.*, 1997, 1998), which show that the emission spectrum from a single QW is some tens of meV smaller (and the radiative current a factor of 2–4 smaller) than expected if the quasi-Fermi level separation were constant across the well/barrier interface.

This explanation suggests that the QWs are not in quasi-thermal equilibrium with the barrier material. Quasi-equilibrium is appropriate when current is dominated by low-field carrier drift and diffusion. These conditions are usually satisfied for a homojunction cell, where variations in carrier population are smooth. In QWSCs in the region of the QW interfaces, as discussed above, current may be dominated by carrier escape from the QWs through thermionic emission and thermally assisted tunnelling. These currents may greatly exceed drift–diffusion currents in the direction of increasing kinetic energy, and therefore the notion of quasi-thermal equilibrium may not be valid here.

This explanation was proposed by Corkish and Honsberg (1997). They draw on studies of bulk heterojunctions which show that a step in the minority-carrier Fermi level may result at a heteroface in conditions where transport is limited by thermionic emission or transport across a space-charge region. They show that a moderately high level of background doping in the QW could give rise to reduced quasi-Fermi level separation and lower dark currents. This is a promising idea and more detailed modelling incorporating carrier escape and capture may well provide a quantitative explanation of the observed dark currents.

Optimisation of the QWSC clearly requires minimisation of the incremental recombination current. One way to do this would be to choose a material for the QW with long non-radiative recombination times. In GaAs, for example, minority carrier lifetimes are long compared to those in $Al_xGa_{1-x}As$, and it is possible that the recombination current from an $Al_xGa_{1-x}As$ /GaAs MQW may be lower than that from the $Al_xGa_{1-x}As$ alloy of equivalent effective bandgap. Also the dependence of recombination rate on QW position means that it should be possible to optimise the design of QWSCs by locating QWs away from the centre of the SCR, in the regions where $n \gg p$ or $p \gg n$. (Nelson *et al.*, 1999). This effect results from the asymmetry in carrier populations and the form of the Shockley–Read–Hall recombination rate; it does not require that the quasi-Fermi level in the QW be reduced.

10.4.3 *Current–voltage characteristic*

As explained in Section 10.3, QWs are expected to increase both the photocurrent and the dark current of a single-bandgap solar cell. Making the assumption that the dark current is not changed by illumination (the superposition assumption), it is usual to write the current–voltage characteristic as

$$i(V) = i_{sc} - i_{Dk}(V) \tag{10.25}$$

where i_{sc} is the short-circuit photocurrent density. The efficiency is given by the ratio of the maximum of the current–voltage product, $i(V_{mp}) \times V_{mp}$, to the incident light power. For a QWSC:

$$i(V) = i_{sc} + \Delta i_{sc} - i_{Dk}(V) - \Delta i_{Dk}(V) \tag{10.26}$$

Therefore, QWs can benefit the power conversion efficiency only when the increase in photocurrent Δi_{sc} exceeds the increase in recombination current Δi_{Dk}

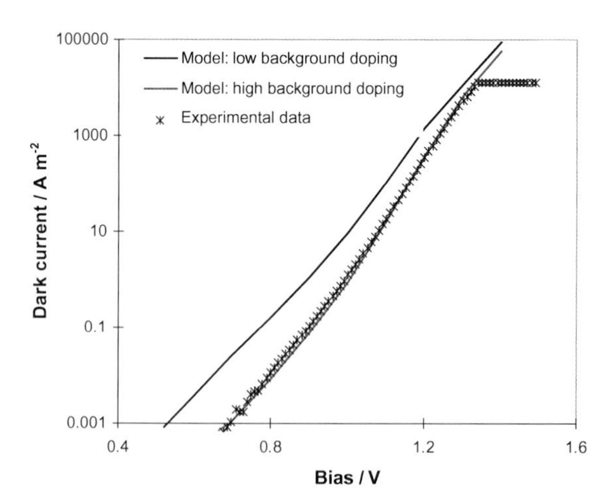

Figure 10.13 Measured and modelled dark currents for an $Al_{0.34}Ga_{0.66}As$/GaAs single QW device. The black line shows the dark current expected when the background doping level N_i in the i region is small ($\sim 10^{14}$ cm^{-3}) and the QW is located close to the plane where electron and hole densities are equal. For high background doping ($N_i \approx 2 \times 10^{16}$ cm^{-3}), $p \gg n$ at the QW and non-radiative recombination is reduced. The lower ideality factor reflects the asymmetry in the electron and hole densities.

at the operating voltage. For small changes this means

$$\Delta i_{\mathrm{sc}} - \Delta i_{\mathrm{Dk}}(V) > 0 \tag{10.27}$$

The advantage increases with the number of QWs since, while Δi_{sc} increases approximately linearly with the number of QWs, the decrease in open-circuit voltage due to Δi_{Dk} changes only logarithmically.

Figures 10.14 and 10.15 show I–V characteristics for an $Al_{0.3}Ga_{0.7}As$ *p-i-n* cell with and without 30 GaAs QWs and for a GaAs *p-i-n* cell with and without 10

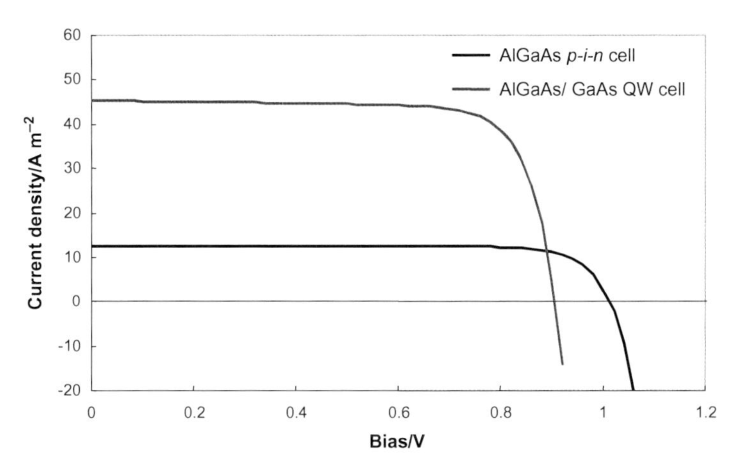

Figure 10.14 Measured current–voltage characteristics for an $Al_{0.3}Ga_{0.7}As$ *p-i-n* device with and without 30 GaAs QWs, in a white light source approximating to a 3200 K black-body spectrum. Note that these devices were not AR-coated, hence the low current densities.

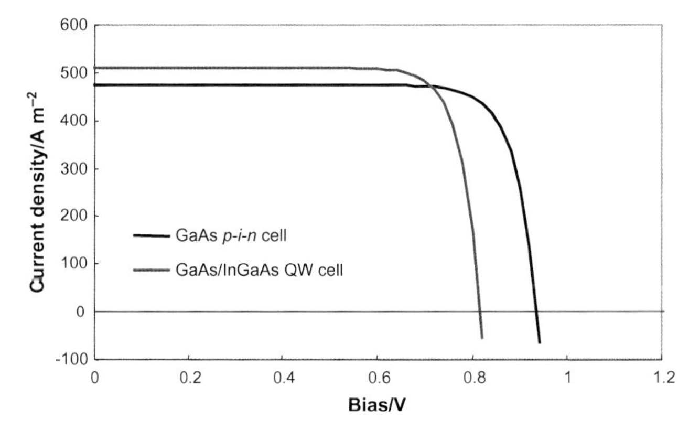

Figure 10.15 Measured current–voltage characteristics for a GaAs *p-i-n* cell with and without 10 InGaAs QWs, using the same light source as for Fig. 10.14.

$In_{0.16}Ga_{0.84}As$ QWs. In both cases, introducing the QWs has increased i_{sc} and reduced V_{oc}. The latter effect results from the increased dark current, which is evident from Eq. (10.25).

In the case of the $Al_{0.3}Ga_{0.7}As$ host cell, where the host bandgap exceeds the optimum for solar energy conversion, the net effect of QWs is to increase the cell efficiency. This is as expected, since QWs added to a wide-gap host cell reduce its effective bandgap towards the optimum. In the case of the GaAs host, the efficiency decreases, which is again the result expected simply from arguments about the optimum bandgap for photoconversion: the addition of a lower-bandgap material to GaAs will reduce the effective bandgap for absorption, and from detailed balance arguments this is expected to reduce the efficiency of the solar cell.

However, the results for the GaAs and InP host materials are complicated by material quality issues. In GaAs the number of QWs that may be added before strain degrades the device quality is too few to produce an adequate photocurrent enhancement. In InP, problems of high background doping have made it impossible to prepare good quality devices for comparison. However, $In_xGa_{1-x}As$ QWs have been observed to increase the efficiency of a less than perfect InP p-i-n cell (Zachariou *et al.*, 1996). Strain-balanced GaAsP/InGaAs represents the cleanest material system in which QWSCs have been grown showing a linear increase in photocurrent with approximately constant voltage (Bushnell, 2005). As discussed further in Section 10.5, this material becomes dominated by radiative recombination at high solar concentration and for a short period set a world-record power-conversion efficiency, discussed further in Section 10.7.2. However, the record was subsequently reclaimed by a GaAs homojunction (Green, 2011), so it is therefore not yet possible to decide on the effect, in practice, of QWs on solar cell efficiency.

However, we can learn something from the effect of QWs on the open-circuit voltage V_{oc} of test devices. Detailed balance arguments (below) imply that V_{oc} should be controlled by the absorption edge E_a. Therefore we would expect a decrease ΔE_a in the absorption edge to cause a decrease in V_{oc} of the same magnitude, and we have seen above that the decrease in E_a caused by the introduction of QWs is accompanied by a reduction in V_{oc}. However, measurements (Fig. 10.16) show that V_{oc} is less sensitive than expected to the effective bandgap E_a of the well material (Barnham *et al.*, 1996). This is reasonable since it is the *host* material that controls carrier injection currents, and hence the population of carriers available for recombination.

10.5 Developments in QWSC design and performance

In most applications, the aim of adding quantum wells is to increase the photocurrent of the host homostructure cell. In most cases it is not possible to grow

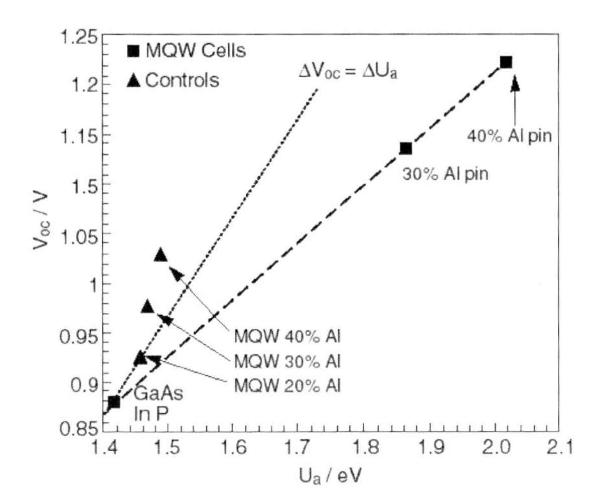

Figure 10.16 Open-circuit voltage against effective bandgap for a series of $Al_xGa_{1-x}As$/GaAs QWSCs and $Al_xGa_{1-x}As$ *p-i-n* cells of different Al fraction. Note how, for the QW devices, V_{OC} is higher than expected either from the measured dependence of V_{OC} on Al fraction for $Al_xGa_{1-x}As$ devices (dashed line) or from the theoretical dependence of V_{OC} on the absorption edge of the QW, the *effective* bandgap, (dotted line) expected from detailed balance arguments.

an optically thick QW stack, since background impurity levels limit the width of depleted i-region that can be grown. As a result, rear surface reflectors, such as a distributed Bragg reflector (DBR), have been employed to increase the optical path length in the material (Bushnell *et al.*, 2003). These have resulted in quantum efficiency levels in excess of 50% in the QW region.

Apart from increasing photogeneration, the effect of introducing a high-quality reflector into the solar cell enables re-absorption of radiative recombination to take place, an effect known as photon recycling (Marti *et al.*, 1997). In a high-quality material such as strain-balanced GaAsP/InGaAs, recombination is dominated by radiative processes at high current levels (equivalent to 200× solar concentration), and the introduction of reflecting stacks in devices made from such materials was observed to reduce the electrical dark current of the solar cell (Johnson *et al.*, 2007). The observed reduction in recombination rate corresponds to a 0.3% increase in absolute efficiency, yet in principle a gain of 2% in absolute efficiency is possible through photon recycling.

In addition to trapping light inside the QWSC, it is also possible to manipulate the emission profile for efficiency gain. In a bulk semiconductor and unstrained quantum well, radiative recombination is roughly isotropic, but under compressive biaxial strain, the radiative recombination becomes directional, favouring the transverse electric (TE) polarised direction propagating normal to the QW plane (Adams, 1986). Since a solar cell need only emit light into the optical modes through which it receives light (Markvart, 2008), manipulating the radiation pattern of radiative recombination can lead to substantial efficiency gains of 9% absolute (Hirst *et al.*, 2011). However, in practice the incomplete suppression of transverse propagating light in QW samples reduces the size of the effect, delivering a modest efficiency enhancement of 0.12% absolute for a compressively strained InGaAs/GaAsP QW solar cell (Adams *et al.*, 2011).

In a multijunction solar cell there is less scope for photon recycling since photons cannot easily be confined in a single junction, but downward-propagating radiative emission can be absorbed by an underlying junction. This radiative coupling of sub-cells can lead to a significant recovery of efficiency in a multijunction solar cell with sub-optimal threshold energies (Marti *et al.*, 1996). In most bulk cells, radiative coupling efficiencies have been low, on account of re-absorption losses in the base region of the cell, but in a QW device the base is transparent to the QW emission, resulting in an efficient transfer of energy from one sub-cell to the underlying device. Recently, photon coupling efficiencies in excess of 50% have been measured between sub-cells (Lee *et al.*, 2012).

10.6 Limits to efficiency

A generalised detailed balance theory due to Araujo and Marti (Araujo and Marti, 1994; Araujo *et al.*, 1994) concludes that no solar cell can achieve a higher limiting efficiency than a homojunction cell of optimum bandgap. This conclusion rests on two important assumptions: that the quasi-Fermi level separation is constant throughout the device and equal to the applied bias V, and that each absorbed photon delivers exactly one electron to the external circuit. The first condition means that, according to detailed balance, the probability of photon emission from the device is equal to the probability of photon absorption. The second means that both probabilities have value unity for photon energies greater than the bandgap. Then for the homogenous cell of bandgap E_b, the current–voltage characteristic is given by

$$i(V) = F_\Omega \int_{E_g}^{\infty} j_{\text{inc}}(E, T_s, 0)\mathrm{d}E - F_e \int_{E_g}^{\infty} j(E, T_c, qV)\mathrm{d}E \qquad (10.28)$$

where j_{inc} is the incident photon flux from the Sun, radiating at temperature T_s, j is the radiative flux within the cell at temperature T_c with quasi-Fermi level separation qV (Eq. (10.14)), and F_Ω and F_e are geometrical factors giving the solid angles for solar photon absorption and emission. For a black-body Sun at 5800 K, the available power $i(V)V$ has a maximum at a bandgap of around 1.3 eV. When applied to the case of a QWSC of effective bandgap E_a this approach gives for the incremental photocurrent and dark current

$$\Delta i_{\mathrm{ph}} = F_{\Omega} \int_{E_a}^{E_g} b(E, T_s) a(E) \, \mathrm{d}E$$

$$\Delta i_{\mathrm{Dk}}(V) = F_e \int_{E_a}^{E_g} j(E, T_c, \Delta\widetilde{\mu}_{\mathrm{F}}) a(E) \, \mathrm{d}E \qquad (10.29)$$

where $a(E)$ is the probability of photon absorption in the MQW, i.e. its spectral response. Now for the optimum QWSC $a(E) = 1$ for all $E > E_a$ and, if the quasi-Fermi level separation $\Delta\widetilde{\mu}_{\mathrm{F}}$ in the QW is equal to qV, then Eq. (10.26) becomes identical to Eq. (10.28), and the optimum QWSC will be identical to the optimum single-bandgap homojunction cell.

There has been some debate about whether the detailed balance theory applies to the QWSC in practice (Corkish and Honsberg, 1997; Anderson, 1995; Araujo *et al.*, 1994). Measurements of radiative recombination currents from biased single-QW test cells suggest that $\Delta\widetilde{\mu}_{\mathrm{F}}$ is smaller in the QW than in the surrounding host material. Irreversible carrier escape from the QW under the small electric field present at the operating point has been suggested as a reason for this (Nelson *et al.*, 1995; Corkish and Honsberg, 1997). Interestingly, the same effect persists under illuminated conditions (Ekins-Daukes *et al.*, 2003), lending weight to the hypothesis of the breakdown of quasi-thermal equilibrium at quantum well length scales. Microscopic non-equilibrium theories for the QWSC are beginning to reveal the carrier dynamics present in these systems (Aeberhard, 2010).

Another interesting idea is the possibility of exploiting 'hot' carrier effects (Ross and Nozik, 1982) in QWSCs. At high carrier densities the relaxation of excited carriers to the band edge can be slowed down by quantum confinement in a QW. The carrier populations then appear to have a higher effective temperature than the lattice, and recombination is reduced. Retarded relaxation has already been observed in QW photoelectrodes (Rosenwaks *et al.*, 1993), with the potential to reach efficiencies in excess of 50% (Le Bris and Guillemoles, 2010). Some attempts have been made to design hot-carrier superlattice solar cells (Hanna *et al.*, 1997).

10.7 Applications

Because QWSCs are as costly to produce as high-efficiency III–V homojunction cells, we may expect them to be interesting only in those applications where III–Vs are preferred. At the present time that means space, concentrator and thermophotovoltaic systems. Finally, we mention certain applications where QWSCs are particularly promising.

10.7.1 *Tandem cells*

The efficiency of a monolithic tandem cell is highly sensitive to the combination of bandgaps and to the requirement of current matching between the wide and narrow bandgap components. Although the bandgap of bulk ternary semiconductors such as $Al_xGa_{1-x}As$ can be adjusted simply by varying the composition (e.g. the Al fraction x), non-radiative recombination increases rapidly with increasing x and degrades collection efficiency. $Al_xGa_{1-x}As$/GaAs QWSCs offer the alternative possibility of controlling the bandgap through the width of the GaAs QWs. Compared with wide-gap bulk alloys such as $Al_xGa_{1-x}As$ and InGaP, QW structures in $Al_xGa_{1-x}As$/GaAs, InGaP/GaAs and other combinations offer the advantages of: (i) bandgap tunability through the QW width and (ii) control of the current through the number of QWs.

At the time when the *p-i-n* AlGaAs/GaAs QW configuration was successfully demonstrated (Barnham *et al.*, 1991), the AlGaAs/GaAs tandem solar cell marked the highest efficiency attained for photovoltaic power conversion. Recognising that recombination will occur primarily in the lower-bandgap GaAs QWs, where recombination lifetimes are longer than in $Al_xGa_{1-x}As$, it was considered that a QWSC would have superior practical performance to the $Al_xGa_{1-x}As$ homojunction cell of the same effective bandgap (Connolly, 1998).

Around the same time, the InGaP/GaAs multijunction solar cell was being developed and constrained by the relatively high bandgap of GaAs. The use of GaAs/InGaAs strained QWs provided a means for lowering the bandgap and therefore increasing the current from the sub-cell but with a loss in voltage (Freundlich *et al.*, 1998). The advent of strain-balanced GaAsP/InGaAs enabled much of the sub-cell voltage to be preserved (Ekins-Daukes *et al.*, 1999).

Later, when triple-junction tandem cells started to outperform double-junction tandems, QW structures became of interest for the high-gap cell in structures such as the (high-gap semiconductor)/GaAs/Ge device.

10.7.2 *Concentrator cells*

In a homojunction cell, efficiency decreases at high light concentrations when the increased temperature causes the bandgap to shrink and the open-circuit voltage, which is directly related to the bandgap, to fall. In a QWSC, although the bandgaps of the well and host material still reduce with increasing temperature, the effect on V_{oc} is less marked. Figure 10.17 compares the efficiency and temperature dependence of V_{oc} for a pair of QWSCs and homojunction cells. Although the mechanism

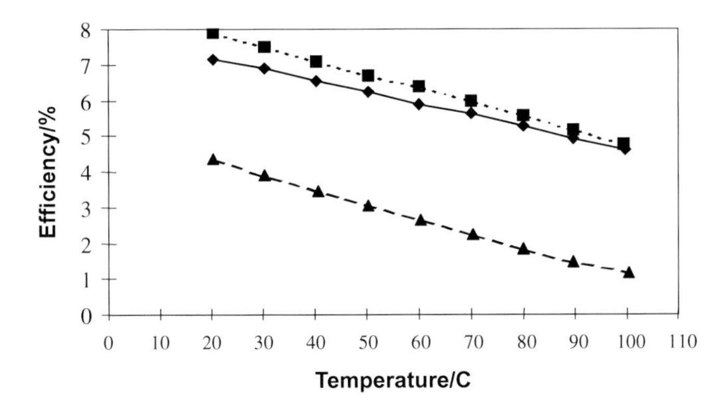

Figure 10.17 Temperature dependence of the efficiency of an $InP/In_xGa_{1-x}As$ QW cell (full line) in comparison with a homogenous InP *p-i-n* device (squares) and an $InP/In_xGa_{1-x}As$ heterostructure device with an $In_xGa_{1-x}As$ *i*-region (triangles). The measurements were made in a 3000 K black-body spectrum and scaled by correcting the photocurrent to the standard terrestrial AM 1.5 spectrum using the measured spectral response.

is not fully understood, clearly the efficiency of carrier escape from the QWs will increase, or remain at unity, as T is increased. Faster carrier escape is likely to reduce the probability of recombination in the QWs, and so offset the effect of the decreasing QW bandgap.

Very high peak performance can be obtained from QWSCs under concentrated sunlight. Figure 10.18 shows the increase in efficiency of a GaAsP/InGaAs quantum well solar cell measured under concentrated sunlight, retaining the temperature stable at 25°C. A peak efficiency of 28.3% at 535 Suns was measured under AM 1.5 D.

10.7.3 *Thermophotovoltaics*

In thermophotovoltaics, low-bandgap photovoltaic cells are used to produce electricity from the long-wavelength radiation emitted by a hot (2000–3000°C) source. The source is usually provided through fossil fuel combustion in a combined heat and power system. Often a selective emitter is used to reabsorb the very low energy photons and reemit them at higher energies to prevent heating. The reshaped spectrum is concentrated around certain bands characteristic of the emitter. For such a spectrum, control of the bandgap of the PV cell is essential for good power conversion efficiency. The flexibility of bandgap makes QWSCs of great interest for TPV. It is also possible that Auger recombination, a longstanding problem in low-bandgap solar cells, is suppressed in the QW device.

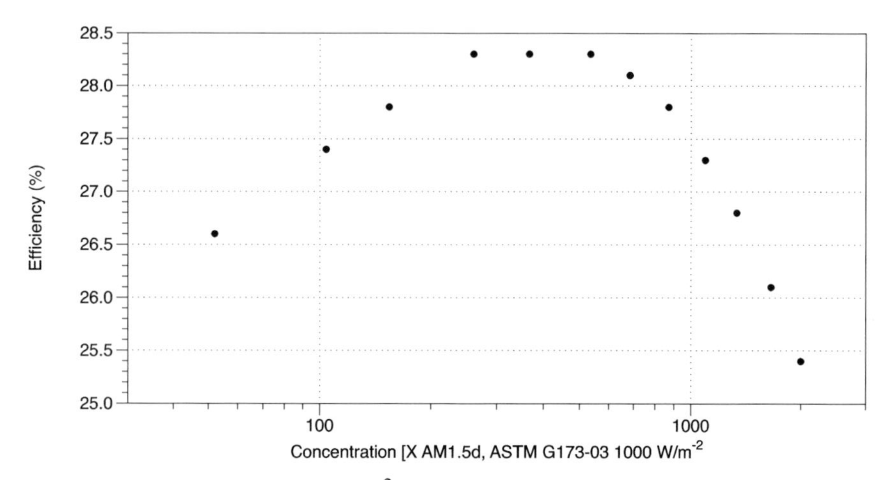

Figure 10.18 Single-junction 1.2 mm^2 GaAsP/InGaAs quantum well solar cell efficiency measured as a function of concentration. Data courtesy of Quantasol Ltd, measured at the Fraunhofer Institute for Solar Energy Systems, May 2009.

Quantum well solar cells in InGaAsP/$In_xGa_{1-x}As$ have already been shown to produce a higher V_{oc} than the comparable $In_xGa_{1-x}As$ homojunction cell (Griffin *et al.*, 1997; Connolly and Rohr, 2003). For low-temperature emitters, an In_xGaAs/In_yGaAs strain-balanced combination grown on InP has demonstrated absorption thresholds up to 1.95 μm (Rohr *et al.*, 2006).

10.8 Conclusions

We have reviewed the use of novel QW semiconductor heterostructures in solar cells. Quantum well structures are of interest as a means of enhancing the photocurrent and efficiency of crystalline solar cells. Photocurrent enhancement has been demonstrated in a range of materials and is well understood. Efficiency enhancement has been observed in materials whose bandgap is larger than the optimum for solar energy conversion. In materials of bandgap close to the optimum, experimental tests on QW cells of equivalent quality to homojunction cells have not yet been possible. Nevertheless there is some evidence that the effect of QWs in increasing recombination within the device is smaller than expected from arguments based on a quasi-thermal equilibrium distribution of carriers. If this is true under operating conditions, then higher efficiencies may also be available with optimum-bandgap cells.

Quantum well structures have the advantages over homojunction cells that the effective bandgap can be controlled by tuning the width of the QW, rather than by varying the composition of a bulk alloy. This means that QWs may produce better cells of better material quality than bulk alloys when particular bandgaps are required. This is relevant for high-efficiency tandem cells and for thermophotovoltaic cells, and QW structures are being researched for both these applications. A further important advantage is that QW structures have a better response to temperature and consequently are expected to perform better under concentrated light.

Some of the major challenges that remain are: to find and verify a theoretical explanation for the observed dark currents and V_{oc} behaviour; to establish whether the suppressed recombination behaviour observed in the dark occurs under solar cell operating conditions; and to prepare an optimum-bandgap QWSC of equivalent quality and superior efficiency to a GaAs homojunction solar cell. More generally, work on QW structures has stimulated a range of new ideas about the role of quantum nanostructures in photovoltaics and the limits to efficiency of solar cells.

References

Aeberhard E. (2010), 'Spectral properties of photogenerated carriers in quantum well solar cells', *Solar Energy Mater. Solar Cells* **94**, 1897–1902.

Adams A. (1986), 'Band-structure engineering for low-threshold high-efficiency semiconductor lasers', *Electron Lett.* **22**, 249–250.

Adams J. G. J., Browne B. C., Ballard I. M., Connolly J. P., Chan N. L. A., Ioannides A., Elder W., Stavrinou P., Barnham K. W. J. and Ekins-Daukes N. J. (2011), 'Recent results for single-junction and tandem quantum well solar cells', *Progr. Photovoltaics* **19**, 865–877.

Afshar M., Sadewasser S., Albert J., Lehmann S., Abou-Ras D., Marron D. F., Rockett A. A., Räsänen E. and Lux-Steiner M. C. (2011), 'Chalcopyrite semiconductors for quantum well solar cells', *Adv. Energy Mater.* **1**, 1109–1115.

Anderson N. G. (1995), 'Ideal theory of quantum-well solar cells', *J. Appl. Phys.* **78**, 1850–1861.

Araujo G. L. and Marti A. (1994), 'Absolute limiting efficiencies for photovoltaic energy conversion', *Solar Energy Mater. Solar Cells* **33**, 213–240.

Araujo G. L., Marti A., Ragay F. W. and Wolter J. H. (1994), 'Efficiency of multiple quantum well solar cells', *Proc. 12th. European Photovoltaic Solar Energy Conf.*, H. S. Stephens & Associates, Bedford, pp. 1481–1484.

Barnes J. M. (1994), 'An experimental and theoretical study of GaAs/InGaAs quantum well solar cells and carrier escape from quantum wells', PhD Thesis, University of London.

Barnes J., Nelson J., Barnham K. W. J., Roberts J. S., Pate M. A., Grey R., Dosanjh S. S., Mazzer M. and Ghiraldo F. (1996), 'Characterization of GaAs/ InGaAs quantum wells using photocurrent spectroscopy', *J. Appl. Phys.* **79**, 7775–7777.

Barnham K., Ballard I., Barnes J., Connolly J., Griffin P., Kluftinger B., Nelson J., Tsui E. and Zachariou A. (1997), 'Quantum well solar cells', *Appl. Surf. Sci.* **113/114**, 722–733.

Barnham K., Connolly J., Griffin P., Haarpaintner G., Nelson J., Tsui E., Zachariou A., Osborne J., Button C., Hill G., Hopkinson M., Pate M., Roberts J. and Foxon T. (1996), 'Voltage enhancement in quantum well solar cells', *J. Appl. Phys.* **80**, 1201–1206.

Barnham K. W. J., Braun B., Nelson J., Paxman M., Button C., Roberts J. S. and Foxon C. T. (1991), 'Short-circuit current and energy efficiency enhancement in a low-dimensional structure photovoltaic device', *Appl. Phys. Lett.* **59**, 135–137.

Barnham K. W. J. and Duggan G. (1990), 'A new approach to high-efficiency multi-bandgap solar cells', *J. Appl. Phys.* **67**, 3490–3493.

Bastard G. (1988), *Wave Mechanics Applied to Semiconductor Heterostructures*, Editions de Physique, Les Ulis.

Bauhuis G. J., Mulder P., Haverkamp E. J., Schermer J. J., Bongers E., Oomen G., Koestler W. and Strobl G. (2010), 'Wafer reuse for repeated growth of III–V solar cells', *Progr. Photovoltaics* **18**, 155–159.

Bushnell D. B., Ekins-Daukes N. J., Barnham K. W. J., Connolly J. P., Roberts J. S., Hill G., Airey R. and Mazzer M. (2003), 'Short-circuit current enhancement in Bragg stack multi-quantum-well solar cells for multijunction space cell applications', *Solar Energy Mater. Solar Cells* **75**, 299–305.

Bushnell D. B., Tibbits T. N. D., Barnham K. W. J., Connolly J. P., Mazzer M., Ekins-Daukes N. J., Roberts J. S., Hill G. and Airey R. (2005), 'Effect of well number on the performance of quantum-well solar cells', *J. Appl. Phys.* **97**, 124908–124908-4.

Chaffin R. J., Osbourn G. C., Dawson L. R. and Biefeld R. M. (1984), 'Strained superlattice quantum well multijunction photovoltaic cell', *Proc. IEEE Photovoltaic Specialists Conf.*, IEEE, New York, NY, pp. 743–746.

Conibeer G., Green M. A., Konig D., Perez-Wurfl I., Huang S., Hao X., Di D., Shi L., Shrestha S., Puthen-Veetil B., So Y., Zhang B. and Wan Z. (2010), *Progr. Photovoltaics* **19**, 813–824.

Connolly J. P., Barnham K. W. J., Nelson J., Griffin P., Haarpaintner G., Roberts C., Pate M. and Roberts J. S. (1995), 'Optimisation of high efficiency $Al_xGa_{1-x}As$ MQW solar cells', *Proc. Int. Solar Energy Society 1995 Solar World Congress*, Harare, Zimbabwe, http://www.researchgate.net/publication/235764284_Optimisation_of_high_efficiency_AlxGa1-xAs_MQW_solar_cells.

Connolly J. and Rohr C. (2003), 'Quantum well cells for thermophotovoltaics', *Semicond Sci. Tech.* **18**, S216–S220.

Connolly J. P. (1998), Private communication.

Corkish R. and Green M. (1993), 'Recombination of carriers in quantum-well solar-cells', *Conf. Record 23rd. IEEE Photovoltaic Specialists Conf.*, IEEE, New York, pp. 675–680.

Corkish R. and Honsberg C. B. (1997), 'Dark currents in double-heterostructure and quantum-well solar cells', *Conf. Record 26th. IEEE Photovoltaic Specialists Conf.*, IEEE, New York, pp. 923–926.

Ekins-Daukes N. J., Ballard I., Calder C. D. J., Barnham K. W. J., Hill G. and Roberts J. S. (2003), 'Photovoltaic efficiency enhancement through thermal up-conversion', *Appl. Phys. Lett.* **82**, 1974–1976.

Ekins-Daukes N. J., Barnham K. W. J., Connolly J. P., Roberts J. S., Clark J. C., Hill G. and Mazzer M. (1999), 'Strain-balanced GaAsP/InGaAs quantum well solar cells', *Appl. Phys. Lett.* **75**, 4195–4197.

Ekins-Daukes N. J., Kawaguchi K. and Zhang J. (2002), 'Strain-balanced criteria for multiple quantum well structures and its signature in X-ray rocking curves', *Cryst. Growth Des.* **2**, 287–292.

Farrell R. M., Neufeld C. J., Cruz S. C., Lang J. R., Iza M., Keller S., Nakamura S., Denbaars S. P., Mishra E. K. and Speck J. S. (2011), 'High quantum efficiency InGaN/GaN multiple quantum well solar cells with spectral response extending out to 520 nm', *Appl. Phys. Lett.* **98**, 201107–201110.

Freundlich A., Fotkatzikis A., Bhusal L., Williams L., Alemu A., Zhu W., Coaquira J. A. H., Feltrin A. and Radhakrishnan G. (2007), 'Chemical beam epitaxy of GaAsN/GaAs multiquantum well solar cell', *J. Vac. Sci. Technol. B* **25**, 987–990.

Freundlich A., Rossignol V., Vilela M. F. and Renaud P. (1994), 'InP-based quantum well solar cells grown by chemical beam epitaxy', *Conf. Record IEEE First World Conf. on Photovoltaic Energy Conversion*, IEEE, New York, pp. 1886–1889.

Freundlich A. and Serdiukova I. (1998), 'Multi-quantum well tandem solar cells with efficiencies exceeding 30% AM0', *Proc. 2nd World Conf. Photovoltaic Energy Conversion*, Vienna, p. 3707.

Goradia C., Clark R. and Brinker D. (1985), 'A proposed GaAs based superlattice solar cell structure with high efficiency and high radiation tolerance', *Proc. IEEE Photovoltaic Specialists Conf.*, IEEE, New York, NY, pp. 776–781.

Green M. A. (2003), *Third Generation Photovoltaics*, Springer, New York.

Green M. A., Emery K., Hishikawa Y. and Warta W. (2011), 'Solar cell efficiency tables version 37', *Progr. Photovoltaics* **19**, 84–92.

Griffin P., Ballard I., Barnham K., Nelson J. and Zachariou A. (1997), 'Advantages of quantum well solar cells for TPV', in Coutts T. J., Allman C. S. and Benner J. P., (eds), *Thermophotovoltaic Generation of Electricity*, American Institute of Physics, New York.

Griffin P., Barnes J., Barnham K. W. J., Haarpaintner G., Mazzer M., Zanotti-Fregonara C., Grunbaum E., Olson C., Rohr C., David J. P. R., Roberts J. S., Grey R. and Pate M. A. (1996), 'Effect of strain relaxation on forward bias dark currents in GaAs/InGaAs multiquantum well *p-i-n* diodes', *J. Appl. Phys.* **80**, 5815–5820.

Hanna M. C., Lu Z. H. and Nozik A. J. (1997), 'Hot carrier solar cells', in *Future Generation Photovoltaic Technologies — First NREL Conf.*, McConnell R. D. (ed.), American Institute of Physics, New York, pp. 309–316.

Hirst L. C. and Ekins-Daukes N. J. (2011), 'Fundamental losses in solar cells', *Progr. Photovoltaics* **19**, 286–293.

Hovel H. J. (1975), *Semiconductor and Semimetals, Volume 11 — Solar Cells*, Academic Press, London.

Johnson D. C., Ballard I. M., Barnham K. W. J., Connolly J. P., Mazzer M., Bessiere A., Calder C., Hill G. and Roberts J. S. (2007), 'Observation of photon recycling in strain-balanced quantum well solar cells', *Appl. Phys. Lett.* **90**, 213505–213508.

Le Bris A. and Guillemoles J.-F. (2010), 'Hot carrier solar cells: achievable efficiency accounting for heat losses in the absorber and through contacts', *Appl. Phys. Lett.* **97**, 113506–113506-3.

Lee K.-H., Barnham K. W. J., Connolly J. P., Browne B. C., Airey R. J., Roberts J. S., Führer M., Tibbits T. N. D. and Ekins-Daukes N. J. (2012), 'Demonstration of photon coupling in dual multiple-quantum-well solar cells', *IEEE J. Photovoltaics* **2**, 68–74.

Luque A., Marti A. and Cuadra L. (2001), 'Thermodynamic consistency of sub-bandgap absorbing solar cell proposals', *IEEE Trans. Electron Dev.* **48**, 2118–2124.

Kitatani T., Yazawa Y., Minemura J. and Tamura K. (1995), 'Vertical transport-properties of photogenerated carrier in InGaAs/GaAs strained multiple-quantum wells', *Jpn. J. Appl. Phys.* **34**, 1358–1361.

Markvart T. (2008), 'The thermodynamics of optical étendue', *J. Opt. A: Pure Appl. Opt.* **10**, 015008.

Marti A. and Araujo G. L. (1996), 'Limiting efficiencies for photovoltaic energy conversion in multigap systems', *Solar Energy Mater. Solar Cells* **43**, 203–222.

Marti A., Balenzategui J. L. and Reyna R. F. (1997), 'Photon recycling and Shockley's diode equation', *J. Appl. Phys.* **82**, 4067–4075.

Mazzer M. (1997), Private communication.

McGehee M. D. (2009), 'Nanostructured organic–inorganic hybrid solar cells', *MRS Bulletin* **34**, 95–100.

Monier C., Vilela M., Serdiukova I. and Freundlich A. (1998), 'Photocurrent and photoluminescence spectroscopy of $InAs_xP_{1-x}$/InP strained quantum wells grown by chemical beam epitaxy', *J. Cryst. Growth* **188**, 332–337.

Moulé A. J., Chang L., Thambidurai C., Vidu R. and Stroeve P. (2012), 'Hybrid solar cells: basic principles and the role of ligands', *J. Mater. Chem.* **22**, 2351–2368.

Nelson J. (1995), 'Multiple quantum well structures for photovoltaic energy conversion', *Physics of Thin Films* **21**, Francombe M. H. and Vossen J. L. (eds), pp. 311–368.

Nelson J., Barnes J., Ekins-Daukes N., Barnham K. W. J., Kluftinger B., Tsui E. S.-M., Foxon C. T., Cheng T. S. and Roberts J. S. (1998), 'Reduced radiative currents from GaAs/InGaAs and AlGaAs/GaAs *p-i-n* quantum well devices', in *Conf. Record 24th. IEEE Int. Symposium on Compound Semiconductors*, IEEE, New York, pp. 413–416.

Nelson J., Barnes J., Ekins-Daukes N., Kluftinger B., Tsui E., Barnham K., Foxon C. T., Cheng T. and Roberts J. S. (1997), 'Observation of suppressed radiative recombination in single quantum well *p-i-n* photodiodes', *J. Appl. Phys.* **82**, 6240–6246.

Nelson J., Barnham K., Ballard I., Connolly J. P., Roberts J. S. and Pate M. (1999), 'Effect of QW location on quantum well photodiode dark currents', *J. Appl. Phys.* **86**, 5898–5905.

Nelson J., Barnham K., Connolly J. and Haarpaintner G. (1994), 'Quantum well solar cell dark currents — a theoretical approach', *Proc. 12th. European Photovoltaic Solar Energy Conf.*, H. S. Stephens & Associates, Bedford, pp. 1370–1373.

Nelson J., Kluftinger B., Tsui E. and Barnham K. (1995), 'Quasi-Fermi level separation in quantum well solar cells', *Proc. 13th. European Photovoltaic Solar Energy Conf.*, H. S. Stephens & Associates, Bedford, pp. 150–153.

Nelson J., Paxman M., Barnham K. W. J., Roberts J. S. and Button C. (1993), 'Steady state carrier escape from single quantum wells', *IEEE J. Quantum Electron.* **29**, 1460–1467.

Paxman M., Nelson J., Barnham K. W. J., Braun B., Connolly J. P., Button C., Roberts J. S. and Foxon C. T. (1993), 'Modelling the spectral response of the quantum well solar cell', *J. Appl. Phys.* **74**, 614–621.

Pearsall T. P. (1989), 'Optical properties of Ge–Si alloys and superlattices', *J. Luminescence* **44**, 367–380.

Ragay F. W., Wolter J. H., Marti A. and Araujo G. L. (1994), 'Experimental analysis of the efficiency of multiple quantum well solar cells', *Proc. 12th. European Photovoltaic Solar Energy Conf.*, H. S. Stephens & Associates, Bedford, pp. 1429–1433.

Ramey S. and Khoie R. (2003), 'Modeling of multiple-quantum-well solar cells including capture, escape, and recombination of photoexcited carriers in quantum wells', *IEEE Trans. Electron Devices* **50**, 1179–1188.

Renaud P., Vilela M. F., Freundlich A., Bensaoula A. and Medelci N. (1994), 'Modeling *p-i* (multi quantum well)-*n* solar cells: a contribution for a near optimum design', *Proc. 1994 IEEE 1st. World Conf. Photovoltaic Energy Conversion*, IEEE, New York, NY, pp. 1787–1790.

Rohr C., Abbott P., Ballard I., Connolly J. P., Barnham K. W. J., Mazzer M., Button C., Nasi L., Hill G., Roberts J. S., Clarke G., and Ginige R. (2006), 'InP-based lattice-matched InGaAsP and strain-compensated InGaAs/InGaAs quantum well cells for thermophotovoltaic applications', *J. App. Phys.* **100**, 114510.

Rosenwaks Y., Hanna M. C., Levi D. H., Szmyd D. M., Ahrenkiel R. K. and Nozik A. J. (1993), 'Hot-carrier cooling in GaAs — quantum wells versus bulk', *Phys. Rev. B.* **48**, 14675–14678.

Ross R. T. and Nozik A. J. (1982), 'Efficiency of hot-carrier solar-energy converters', *J. Appl. Phys.* **53**, 3813–3818.

Sah C.-T., Noyce R. N. and Shockley W. (1957), 'Carrier generation and recombination in *p-n* junctions and *p-n* junction characteristics', *Proc. Inst. Radio Engineers* **45**, 1228–1243.

Scott C. G., Sands D., Yousaf M., Abolhassani N., Ashenford D. E., Aperathitis E., Hatzopoulos Z. and Panayotatos P. (1997), '$P-i-n$ solar cell efficiency enhancement by use of MQW structures in the i-layer', *Proc. 14th. European Photovoltaic Solar Energy Conf.*, H. S. Stephens & Associates, Bedford, pp. 2499–2502.

Sze S. M. (1981), *Physics of Semiconductor Devices*, Wiley, New York.

Tiedje T., Yablonovitch E., Cody G. D. and Brooks B. G. (1984), 'Limiting efficiency of silicon solar cells', *IEEE Trans. Electron Devices* **31**, 711–716.

Varonides A. C. and Berger A. W. (1997), *Proc. 14th. European Photovoltaic Solar Energy Conf.*, H. S. Stephens & Associates, Bedford, pp. 1712–1715.

Wang Y. P., Ma S. J., Watanabe K., Sugiyama M. and Nakano Y. (2012), 'Management of highly strained heterointerface in InGaAs/GaAsP strain-balanced superlattice for photovoltaic application', *J. Cryst. Growth* **352**, 194–198.

Walters R. J., Summers G. P., Messenger S. R., Freundlich A., Monier C. and Newman F. (2000), 'Radiation hard multi-quantum well InP/InAsP solar cells for space applications', *Progr. Photovoltaics* **8**, 349–354.

Wanlass M. W. and Blakeslee A. E. (1982), 'Superlative cascade solar cell', *Proc. IEEE Photovoltaic Specialists Conf.*, IEEE, New York, NY, p. 584.

Weisbuch C. and Vinter B. (1991), *Quantum Semiconductor Structures*, Academic Press, San Diego.

Würfel P. (1982), 'The chemical potential of radiation', *J. Phys. C.* **15**, 3967–3985.

Zachariou A., Barnes J., Barnham K. W. J., Nelson J., Tsui E. S.-M., Epler J. and Pate M. (1998), 'A carrier escape study from InP/InGaAs single quantum well solar cells', *J. Appl. Phys.* **83**, 877–881.

Zachariou A., Barnham K. W. J., Griffin P., Nelson J., Button C., Hopkinson M., Pate M. and Epler J. (1996), 'A new approach to *p*-doping and the observation of efficiency enhancement in InP/InGaAs quantum well solar cells', *Conf. Record 25th. IEEE Photovoltaic Specialists Conf.*, IEEE, New York, pp. 113–117.

Zory P. S. (ed., 1993), *Quantum Well Lasers*, Academic Press, London.

CHAPTER 11

CONCENTRATOR SYSTEMS

IGNACIO LUQUE-HEREDIA
Compañía Española de Alta Eficiencia Fotovoltaica BSQ Solar, SL
Madrid 28040, Spain
iluque@bsqsolar.com

and

ANTONIO LUQUE
Instituto de Energía Solar, Universidad Politécnica de Madrid
Madrid 28040, Spain
luque@ies-def.upm.es

Porque el que no sabe mas que las palabras sin saber el fundamento que la regla tiene: siguen se le muchos daños/y hallandose en ellos/no sabe ni alcança de donde le vienen.
(Because much harm will come to him who only knows the words; without knowing the foundation of the rule, he does not know and cannot reach where they come from.)

Pedro de Medina, *Regimiento de Navegación*, fo. 26, 1563.

11.1 Introduction

Historically, solar cells have been considered to be expensive. A potential way of reducing their cost is casting onto them a higher light intensity than is available naturally. For this solar concentrators are used. Concentrators are optical elements that collect the Sun's energy in a certain area and redirect it onto the solar cells. Obviously the collecting optical element has to be cheaper per unit area than the solar cell, a necessary although not sufficient condition to render the concentrated light system less expensive than an unconcentrated one.

The rapid reduction of the cost of silicon solar cells and the irruption of thin-film cells in the market has weakened the convincing strengths of this approach. However, another factor that has revitalised the interest in concentrated photovoltaics (CPV) has appeared, namely the fabulous potential of multijunction (MJ) cells for high efficiency under concentration. This fact is perspicuous in the very widely used National Renewable Energy Laboratory (NREL) chart in Fig. 11.1. In this we see that the efficiency of triple-junction cells under concentration is not

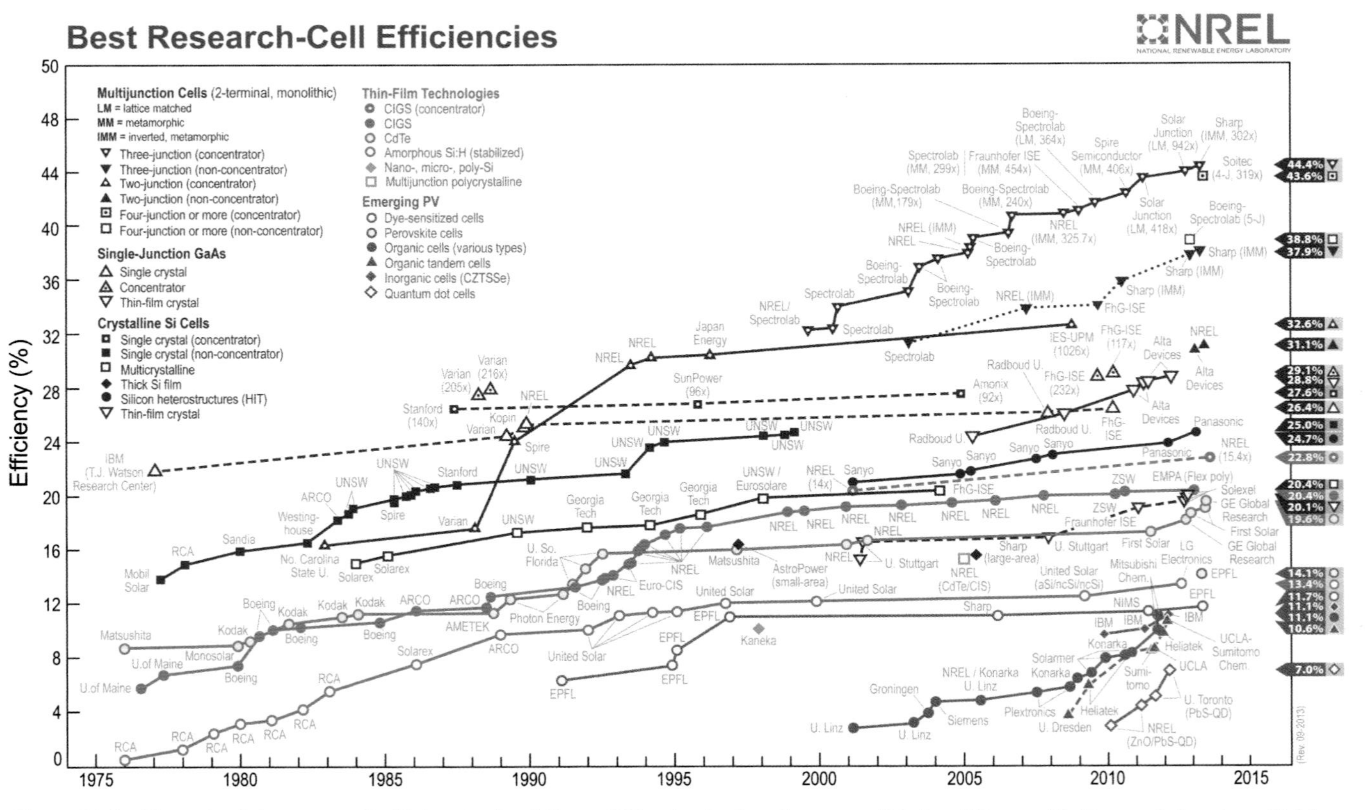

Figure 11.1 Champion laboratory cell efficiencies for different PV technologies. Courtesy of National Renewable Energy Laboratory, Golden, CO (www.nrel.gov/ncpv/images/ efficiency_chart.jpg).

matched by any other photovoltaic technology. In the last decade their efficiency has increased at the rate of about 1% (absolute) per year.

The increase of efficiency by concentration has solid theoretical grounds that are explained later in this chapter. One of the authors has expressed his opinion that 50% cell efficiency will eventually be achievable (Luque, 2011). This promising potential overcomes the fact that CPV only uses the direct sunbeam and not the diffuse component of the radiation, which is harnessed by conventional photovoltaics (PV). At the same time the earlier motivation of concentration — reducing the number and hence the cost of the cells required to produce a given power output — becomes convincing in this context. Only under concentrated sunlight can the sophisticated and expensive multijunction cells be exploited. But we think that these high efficiencies have the potential of becoming cheaper than any other PV technology and, furthermore, than any other energy technology.

The level of irradiance (luminous power flux) at which concentrating cells operate is very variable. Recalling that for the purpose of solar cell rating, the *standard* solar irradiance at the Earth's surface is 1 kW m^{-2}, the level of irradiance in static concentrator cells is in the range 1.5–5 kW m^{-2}, while silicon tracking concentrators range today from 10 to 500 kW m^{-2}, and for multijunction cells irradiances between 100 and 1500 kW m^{-2} are used or envisaged. It is very common to refer to the irradiance level in 'Suns', meaning the number of times the actual irradiance is higher than the standard solar irradiance. Thus a cell operating at 1500 kW m^{-2} is said to operate at 1500 Suns.

In this chapter, we shall look briefly first at the early history of CPV, then at the basic operation of solar cells under concentration to explain the grounds of the increase of the efficiency with concentration and also the limits of this: why concentration cannot be increased indefinitely and lead to more efficient operation. Then we outline the modern multijunction structures and their behaviour.

As regards concentrators, we start with their description. Then we examine methods for their optical design. The theoretical grounds of concentrator optics are also presented and the limits on increasing the concentration factor are described. Following that, we analyse how concentrator cells are mounted and cooled, including a quantitative analysis of the cooling.

Concentrating systems usually (but not invariably) need tracking mechanisms to keep the sunlight focussed on the cells. These constitute an important part of the cost of a CPV system, so we look at them in some detail in this chapter. Static concentrators also exist, although they permit only very moderate concentration. They also collect diffuse sunlight, to different extents. Such systems have not so far been commercialised and will not be treated in this chapter. For more information, the reader is directed to Luque (1986 and 1989).

Figure 11.2 Photovoltaic concentrator panel fabricated in the late 1970s at Sandia National Laboratories, Albuquerque, NM.

Next we consider the performance and cost of CPV and the potential for cost reduction. Finally we discuss what future we foresee for tracking concentrator technology.

11.2 The early development of CPV

Figure 11.2 shows the first modern photovoltaic concentrating panel, developed at Sandia National Laboratories in the late 1970s (Burgess and Pritchard, 1978). Each elementary concentrator is formed by a point focus Fresnel lens (see Section 11.4.1) that casts the radiation onto a circular cell of about 5 cm in diameter. The concentration ratio is about 40× and total rated power about 1 kWp. A set of 15 × 8 such elementary concentrators is attached to a beam that rotates around a horizontal axis (elevation tracking). This beam is placed on the top of a pedestal and also rotates as a whole in azimuth. Jointly, this is the azimuth–elevation two-axis tracking mechanism. We shall refer to this tracking configuration as the pedestal type. The tracking electronics of this system were based on a sensor that provided an error signal when the Sun moved out of aim. Since the panel movement is very slow — one turn per day — the power spent tracking the Sun is negligible, usually less than 1% of the energy produced. In this first prototype, the cooling was provided by cool water fed into the cell holder.

This concept has inspired several research prototypes, among them the Ramón Areces panel, installed in the late 1970s at the Instituto de Energía Solar of the Universidad Politécnica de Madrid (IES-UPM) (Sala *et al.*, 1979). In later prototypes (including the Ramón Areces), cooling of the cells was achieved by an extruded aluminium multifin heat sink. A closed housing was also provided to protect the cells from the environment. In some cases, *parquets* of several lenses were fabricated to facilitate the assembling.

The passive-cooling modified Sandia Labs design was adopted by several manufacturers in those days, Martin Marietta being the most remarkable. This organisation installed about 350 kW in Saudi Arabia in the early 1980s (Salim and Eugenio, 1990). Some other US companies, for instance Alpha Solarco and Amonix (Garboushian *et al.*, 1996) also continued this concept to develop large panels in the 15–25 kW range.

While many other designs were considered at that time, the next original concept that came into reality was the ENTECH concentrator. A plant of 300 kW and several other smaller ENTECH arrays were installed (O'Neill *et al.*, 1991). In this concentrator, still commercialised today (mainly for space applications), the cells are series-connected in a linear row located under an arched Fresnel lens of linear focus, as shown in Fig. 11.3. The performance of such arched lenses is strikingly insensitive to their position. The concentration ratio can be up to 20×, and in some cases screen-printed cells are used. Cooling is again effected with extruded heat sinks, and a 'housing' contains the whole module.

The concentrator array comprises a set of linear cell/lens modules in an elevated east–west oriented frame. Each module can rotate separately to follow the hour angle, and the frame as a whole can rotate in elevation on two supporting poles situated at its east and west ends. This is the declination–hour angle tracking mechanism. Its tracking control is similar to that of the Sandia prototype.

A concept developed jointly by BP Solar and IES-UPM, in an EU joint project in which the University of Reading and ZSW-Stuttgart also participated, was the EUCLIDES concentrator (Sala *et al.*, 1996) shown in Fig. 11.4. The rationale

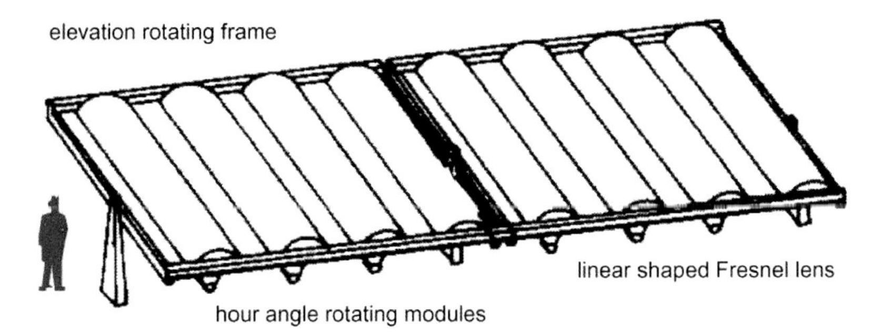

Figure 11.3 ENTECH concentrator array. This has axial focus curved lenses, tracking the hour angle in a frame that tracks the Sun's elevation angle.

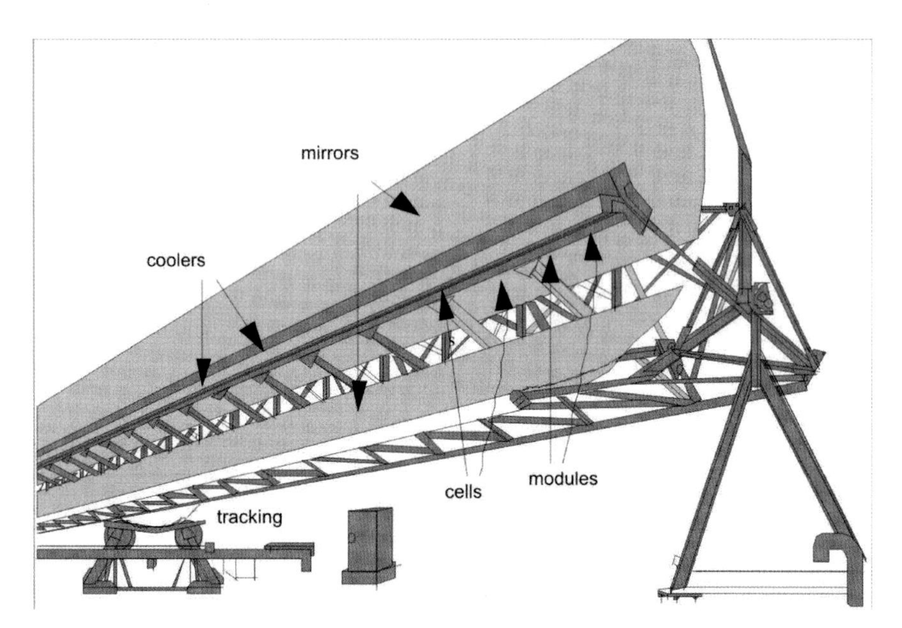

Figure 11.4 View of the one-axis tracking EUCLIDES concentrator. The cells, encapsulated inside modules, are the darker strip that can be seen in the figure (there is another strip underneath). There are two separated asymmetric parabolic mirrors illuminating two rows of cells.

behind this design was to have a high voltage output, able to feed an inverter without the use of an intermediate transformer. The output needed for this purpose was 600–650 V DC (direct current), this requiring many cells ($\sim$1400) to be connected in series, which resulted in very long arrays (84 m in the commercial version). The arrays should therefore be horizontal and have a one-axis horizontal tracking. This implies linear mirror optics, because linear lenses change their focal distance when the Sun is at inclined angles.

Subject to these specifications, the concept of cell housing was no longer valid. Instead, cells were encapsulated in receiving modules inspired by flat module encapsulation, which provided excellent environmental protection. The receiver module was formed of an aluminium tray, on which the cells were stuck on a layer of material that had to be both an electric insulator and a good thermal conductor. The cells were electrically connected in series and covered with glass. The inside of this module was filled with a transparent resin.

The concentrating optics were made of mirrors instead of the Fresnel lenses used previously in all (successful) photovoltaic concentration designs. The mirrors were parabolic in outline, and their profiles were optimised using non-imaging optics (Luque, 1980), so as to allow for the highest manufacturing errors for a given level of concentration (geometrical concentration of 32×). The focal distance, position of the axis, useful arch within the infinite parabola, and the receiver (cell) angle, were also optimised.

Since this mirror profile was not available in the market, and because of the precise profile that was required (which we were not sure could be achieved with glass), a new fabrication technology was developed using a weather-protected 3M silvered layer laminated onto a thin aluminium sheet that was subsequently formed to the desired shape with high precision and then stuck to a supporting aluminium frame.

Cell cooling was provided in the EUCLIDES module by means of an aluminium-finned heat sink, the aluminium sheets being held together by a core stuck onto the aluminium tray of the modules. Cooling by natural convection is more effective in vertical fins than in the inclined fins used in the two-axis tracking configuration, so less aluminium is required for the same cooling performance. The structure, which held two rows of cells and mirrors as shown in Fig. 11.4, consisted of a long horizontal reticulated beam rotating on a large centrally placed wheel resting on two smaller wheels. This configuration also provided the tracking mechanisms. Two additional passive supports provided the vertical reaction to the weight at the beam ends. The tracking control was provided by a microprocessor that calculated the correct aiming of the system at any moment of the year, based on astronomic data. However, in order to account for inaccurate module positioning, a feedback based on system power output maximisation was occasionally used, generating a table of corrections for the astronomic calculations.

The EUCLIDES concentrator used crystalline silicon cells of the laser-grooved buried contact (LGBG) type, a concept that had been developed by the University of New South Wales. These cells (also called SATURN, after the name of the industrialisation project of BP Solar) had proven high efficiencies. These cells were, at the time, very convenient for use in concentrator systems because, while their 1 Sun version was more expensive than those used in conventional c-Si modules, when used as $125 \times 125\ \text{mm}^2$ concentrator cells and designed for 30×, they could be sold at €10–12 each, thus becoming economically very attractive to the manufacturer. In their concentrator version, LGBG cells showed a highly homogeneous voltage along their metallisation and a low surface recombination, which allowed them to reach an average of 18.5% efficiency at 30× and up to 20% in small ($1\ \text{cm}^2$) cells at 100× (Bruton *et al.*, 1994).

The first prototype of this concentrator technology was installed in Madrid at IES-UPM in 1995. It proved to have an efficiency of 10.8% close to noon on a typical summer day which, extrapolated to the then-proposed concentrator standard

Figure 11.5 Seven of the fourteen arrays constituting the EUCLIDES demonstration plant installed in Tenerife by the ITER, the IES and BP Solar.

test conditions — direct radiation of 800 W m^{-2} and cell temperature of 25°C — resulted in an efficiency of 14.4% (Luque *et al.*, 1997). This was achieved at a lower cost than a typical flat-plate PV power plant of the same rating at that time.

Following the Madrid prototype, a 480 kW EUCLIDES demonstration plant was built in Tenerife (see Fig. 11.5) through the joint initiative of BP Solar, the Instituto de Tecnología y Energías Renovables (ITER) in Tenerife and IES-UPM. This became the biggest CPV plant of its time. However, several problems that appeared in the series production of the receiver, some overestimation of the concentrator benefits, and finally the merger of BP with Amoco (which resulted in the consolidation of their respective solar subsidiaries BP Solar and Solarex), led to the abandonment of concentrator projects by these companies.

Other smaller CPV prototypes were developed using combinations of the above designs. For instance, a point-focus concentrator in a declination–hour angle frame configuration like that of ENTECH was developed by Midway Laboratories. Other less fully developed ideas have also been published. For instance, among the most original suggested was a large reticulated platform resting on wheels or even floating on water to provide the azimuth tracking. On this platform, rows of arrays would be installed, each with its own elevation tracking mechanism (Alarcón *et al.*, 1982).

These developments and others, which appeared in the last two decades of the past century, constituted the 'first wave' of photovoltaic concentrators, setting the foundations of CPV and paving the way for current CPV start-ups and developments.

After this overview of the historical background of CPV, which has also introduced the different elements of a concentrator system by describing several real examples, we proceed in the next sections to further analyse them one by one and in greater detail.

11.3 Concentrator solar cells

We devote this section to a theoretical overview of single-junction concentrator solar cells. They are easier to understand and contain the fundamentals of any concentrator solar cell of today. Actually, MJ solar cells are just series-connected stacks of several single-junction solar cells. Many of the statements in this section are therefore valid for MJ solar cells, which are discussed by Masafumi Yamaguchi in Chapter 7.

Concentrator solar cells differ from conventional solar cells in that they must be able to extract more current per unit area. Furthermore, as they must be no more than a small part of the concentrator cost, they can afford to be more expensive in order to be more efficient.

Under increasing luminous flux, the current from a solar cell usually increases proportionally. The open-circuit voltage will increase according to the formula $V_{oc} = V_{th} \ln(I_{th}/I_0$, derived from the single-diode model of the solar cell. In this formula I_{ph} is the photogenerated current, which is very close to the short-circuit one; I_0 is the reverse-bias saturation current; and $V_{th} = \beta kT/q$ is the thermal voltage. Here β is a quality factor used in the single-diode model. This tends to be one in concentrator cells so that $V_{th} = 0.0257$ at 25°C. In consequence V_{oc} increases logarithmically with the current and thus with the luminous flux. Some smaller deviations may be observed at very high fluxes due to inhomogeneous distribution of the current, but they are not relevant at the operation conditions.

The fill factor can be written as (Luque, 1989):

$$\eta_{fill} = \left(1 - \frac{V_{th}}{V_{oc}}\right)\left[1 - \frac{V_{th}}{V_{oc}} \ln\left(\frac{V_{oc}}{V_{th}}\right) - \frac{I_{ph} R_S}{V_{oc}}\right] \quad (11.1)$$

Excluding the term in R — the series resistance — η_{fill} slowly increases with V_{oc} and therefore with the flux intensity. However, at high currents, when the ohmic drop increases, the fill factor decreases and this becomes the main factor causing an efficiency decrease. Obviously this implies that the series resistance must be

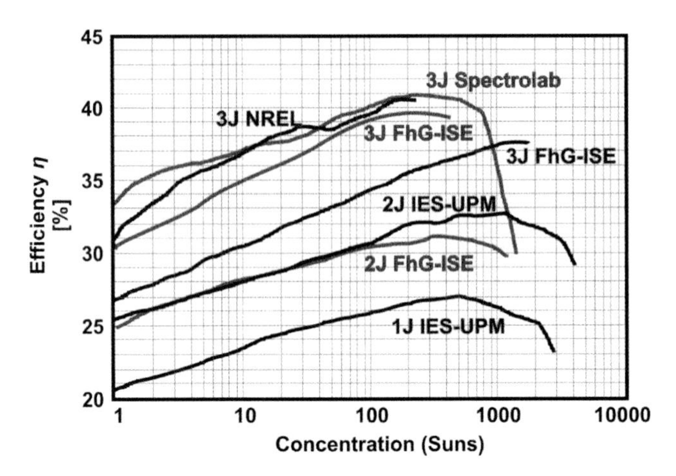

Figure 11.6 Evolution of efficiency with concentration factor in recent record-breaking multijunction cells. Labels give information on the research lab or company developing the cell, and the number of junctions; all of them are monolithic lattice-matched. The typical evolution of efficiency with concentration that increases with flux intensity until loss due to ohmic drop becomes significant and can be seen, particularly in the 3J Spectrolab cell. Courtesy of Prof. Algora, personal communication.

very small in concentrator cells. This is perhaps the most important difference as compared with the non-concentrator solar cells.

As a consequence of all this, the solar cell efficiency η_{mp}

$$\eta_{\mathrm{mp}} = \left(\frac{I_{\mathrm{ph}}}{P_{\mathrm{L}}}\right) V_{\mathrm{oc}}\eta_{\mathrm{fill}} = S_{\mathrm{L}} V_{\mathrm{oc}}\eta_{\mathrm{fill}} \tag{11.2}$$

increases with concentration until the moment when the ohmic drop causes a reduction, as seen in Fig. 11.6 (single-junction cell labelled 1J IES-UPM). Here S_{L} is the cell photosensitivity (approximately constant) and P_{L} is the luminous flux. A very easy rule is that the highest efficiency is achieved when $I_{\mathrm{ph}} R_{\mathrm{S}} \approx V_{\mathrm{th}}$ (Luque, 1989). This behaviour, although ruled by a more complex expression, is also present (for the same fundamental reasons) in MJ solar cells, as can also be seen in Fig. 11.6.

A very important feature of the concentrator solar cells is their thermal behaviour. At higher temperatures the bandgaps decrease and therefore the current tends to increase, but this is a very small effect — the relative increase of the short-circuit current in silicon is only $3 \times 10^{-4}\,\mathrm{K}^{-1}$. The most important thermal effect in a solar cell is the voltage reduction, which is governed by the expression

$$\frac{\mathrm{d}V}{\mathrm{d}T} = -\frac{E_{\mathrm{g}}/q - V - I R_{\mathrm{S}}}{T} \tag{11.3}$$

where E_g is the bandgap. In Si the temperature coefficient of the voltage drop is about $-2.1\,\mathrm{mV\,K^{-1}}$ whereas in GaAs it is only $-1.4\,\mathrm{mV\,K^{-1}}$. This means that cells experience a serious performance reduction when they operate at high temperatures, which is smaller under concentration when the voltages are bigger.

Most concentrator cells have a classical structure, with an extended *p-n* junction on the front face. The series resistance is reduced by using a carefully designed grid. For the same surface-covering fraction, the series resistance is decreased if the fingers are deep and narrow (less than $10\,\mu$m). That is why lift-off photolithography is used to delineate the grid of concentration cells. The optimum grid-covering factor is higher than in non-concentrator cells. Furthermore, the ohmic losses also decrease if the cell is of small size, but in some cases, in particular in silicon cells, going below 1 cm^2 leads to reduced efficiency owing to increased edge recombination. In GaAs cells — of much shorter carrier lifetime — cells of 1 mm^2 are more effective.

In general, screen-printing technology — used for most non-concentrator silicon cells — works poorly at concentrations above 10×. On the contrary, the LGBG cell technology chosen for the EUCLIDES concept could be operated at concentrations of 40×, as has been explained above.

Nevertheless, in cells of classical structure there is a fundamental limitation on the series resistance that is independent of the grid and of the cell size: this is the base series resistance. This component can only be reduced by reducing the base resistivity and thickness. In silicon cells of conventional structure, this limits the concentration factor to some 120× (Terrón *et al.*, 1992). In III–V semiconductors the base is usually very narrow and is situated on a substrate that is, to a large extent, electrically inactive. This allows it to be of very low resistivity and therefore these cells can operate properly at much higher concentrations, may be at 5000×, but certainly at 1000×. It was at IES-UPM that the ability of III–V cells to operate efficiently at more than 1000× was first proved (Algora *et al.*, 2001).

Swanson and co-workers developed a new type of silicon cell (Sinton *et al.*, 1985) — the back point contact (BPC) cell — with a low-doped base of high lifetime and p^+ and n^+ dot-shaped emitters that are the positive and the negative contacting areas of the solar cells, respectively, all located closely intercalated at the cell back face. In such cells no current flows from top to bottom, as in conventional ones (but carriers of both types do, annihilating the overall current), and because of this such cells do not have a base component of the series resistance. Thus they can operate at higher concentration. These cells do not have any grid on the front face and all the electric contacts are made to the rear face through interdigitated dense grids.

In BPC cells the limit to the concentration efficiency is not determined by series resistance but by the super-linear increase of Auger recombination at high luminous flux caused by the steep profile of carriers produced by intense illumination. This leads to a sub-linear current–irradiance characteristic that precludes operation at very high concentration. Best efficiencies are at some 200×. New cells have been developed based on the same principle, with closely intercalated p^+ and n^+ emitters in the front face, covered by interdigitated grids in wedge form so that they deflect the light towards the uncovered silicon areas. Such cells are in principle able to operate at higher concentration, over 500× (Luque *et al.*, 2004a). However, the fabricated cells were imperfect and did not lead to the expected high efficiencies (Luque *et al.*, 2004b).

Despite Swanson's cell invention originally being targeted at integration in concentrator systems (and in fact it was used until 2010 by well-known CPV manufacturers such as Amonix in the USA and Guascor Fotón in Spain), the BPC cell, after having been reengineered to make it cost-effective in flat PV modules, has been successfully commercialised as such by SunPower.

11.3.1 *Beyond the Shockley–Queisser limit: multijunction solar cells*

One of the main reasons behind today's interest in CPV has been the striking increase in efficiency over the past 20 years of MJ cells based on III–V semiconductors. These cells were originally devised for space applications — powering of satellites — but proposals for their use in large-scale terrestrial power concentrator plants are about as old. The potential of high-efficiency photovoltaic devices to make the promise of concentrators to provide cheap solar electricity real, has triggered the quest for other concepts that, independently or complementing MJ cells, are today being developed in the laboratory, and may become tomorrow's concentrator cells. All these new concepts, treated in detail by Martin Green in Chapter 2 of this book, are sometimes grouped together as third-generation cells, which have as their common property their capacity to circumvent the so-called Shockley–Queisser efficiency limit (Luque and Martí, 2011).

In 1961, William Shockley and Hans Queisser (SQ) published an elegant paper establishing the efficiency limit of a solar cell (Shockley and Queisser, 1961). They considered the solar cell as a system of two levels, the valence band (VB) and the conduction band (CB), and assumed that each photon with energy above the bandgap pumps one electron from the valence band to the conduction band. They also assumed infinite mobility so that the CB and VB quasi-Fermi levels are horizontal and their difference everywhere is the cell voltage (times the electron charge). They also assumed that selective contacts (heavily doped n and p regions) are formed so that it is possible to extract electrons and holes separately. They further assumed that all non-radiative recombination processes are suppressed and that radiative recombination is only produced by radiation that escapes through the cell surfaces. Thus they established a balance equation of the photons entering and escaping the semiconductor and the electron-hole pairs produced and extracted. No mention is made of the properties of the semiconductors — excepting the bandgap — or of the need of any *p-n* junction.

The calculations were repeated by Araújo and Martí (1994) for the case of solar radiation impinging on the solar cell isotropically rather than within the natural cone of direct solar radiation. As we will see in the next section, this condition corresponds to the maximum possible sunlight concentration at the Earth's surface. In this case the maximum possible efficiency is 40.7% for black-body radiation at 6000/300 K (Sun/ambient) temperatures, as shown in Fig. 11.7, together with cases for other spectra (direct and global) and concentration levels.

However, if the design principles of the solar cells are not considered, and only thermodynamic considerations are taken into account, the limiting efficiency of a solar energy converter was calculated by Landsberg and Tonge (1980) and is

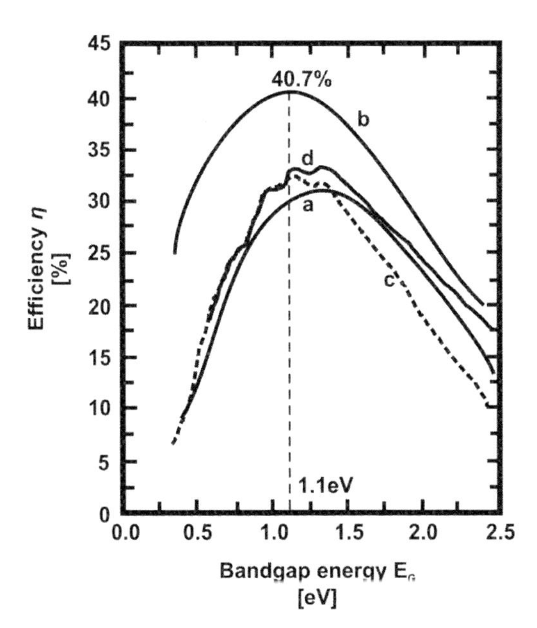

Figure 11.7 Shockley–Queisser efficiency limit for an ideal solar cell versus bandgap energy for: a) unconcentrated 6000 K black-body radiation (1595.9 Wm^{-2}); b) fully concentrated 6000 K black-body radiation (7349.0 × 104 $W\,m^{-2}$); c) unconcentrated AM 1.5-Direct13 (767.2 $W\,m^{-2}$); d) AM 1.5 Global13 (962.5 $W\,m^{-2}$). Reprinted with permission from Araújo and Martí (1994), © Elsevier.

given by

$$\eta_{\text{Landsberg}} = 1 - \frac{4}{3}\left(\frac{T_{\text{amb}}}{T_{\text{Sun}}}\right) + \frac{1}{3}\left(\frac{T_{\text{amb}}}{T_{\text{Sun}}}\right)^4 \tag{11.4}$$

Assuming the Sun to be a reservoir of photons at 6000 K (T_{Sun}) and the Earth a cold reservoir at 300 K (T_{amb}), this results in a maximum 93.33% maximum conversion efficiency (Section 2.4 of this book gives a detailed treatment).

The main reason a solar cell, even if ideal as in the SQ assumptions, is so far from the Landsberg efficiency is that it does not use the whole solar spectrum but only those photons of energy higher than the bandgap. A second reason is that entropy is generated by the cooling of the electron-hole pairs produced by photons with energy higher than the bandgap.

The straightforward way to overcome the fundamental limitation of the SQ single-bandgap cell is to use several solar cells of different bandgaps to convert photons of different energies. A simple configuration to achieve this is vertical stacking of the cells so the uppermost cell has the highest bandgap, and lets the photons with energy less than its bandgap pass through to the cells below. The last cell in the stack is the one with the narrowest bandgap. In this way the entire solar spectrum can be used and the upper cells can act as a filter for the lower ones, each being illuminated by a narrower range of photon energies, thus reducing the production of entropy. For an infinite number of ideal cells, as per the SQ definition, the maximum efficiency of this arrangement is 86.8% under maximum concentration, for a 6000/300 K Sun/ambient temperature.

The most common practical realisation of this concept is the so-called monolithic MJ cell, consisting in a stack of solar cells of different semiconductor materials, along with tunnel diodes to interconnect them in series, all grown on a single substrate. III–V compound semiconductors of a variety of materials can be grown epitaxially with very high quality, if all the materials have a similar lattice constant (although the so-called metamorphic cells allow for some lattice mismatch to achieve a better bandgap combination). Figure 11.8 shows a schematic of a stack of three cells built on a Ge substrate; the figure caption shows the roles of the different layers. The so-called window layers are intended to prevent recombination at the surface or in the tunnel junctions. The fraction of Group III elements in each layer is adjusted to obtain the proper lattice constant and, if possible, the desired bandgap. As noted in the figure, the data are illustrative and each manufacturer has their own recipe.

As noted above, the different cells in a monolithic MJ device are series-connected, and this introduces an additional constraint, because the photogenerated current of the stack will be the smallest of the currents generated by each cell.

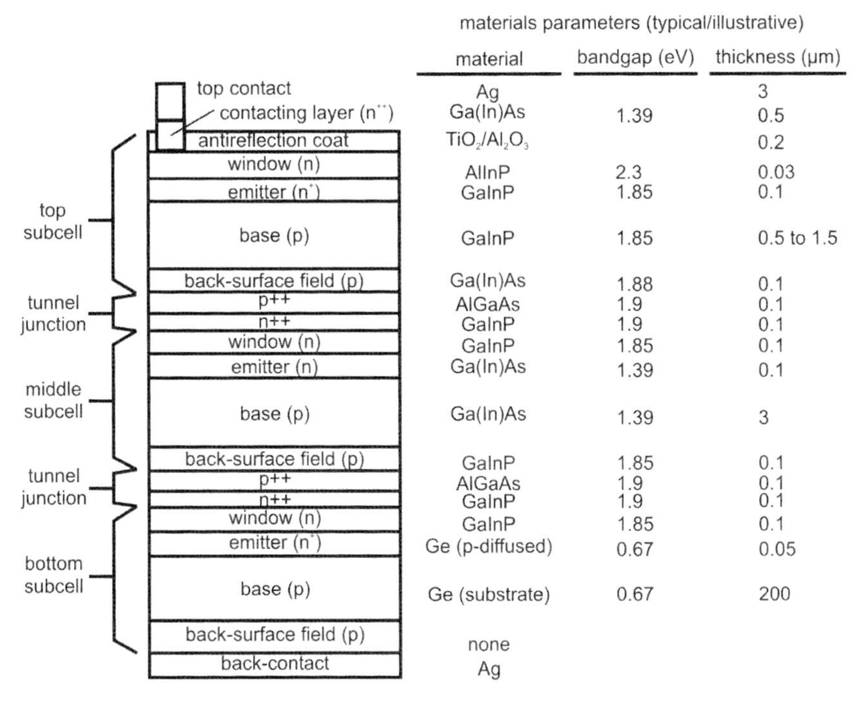

Figure 11.8 Schematic cross-section of a monolithic two-terminal series-connected three-junction solar cell. An n-on-p configuration is illustrated. Doping indicated by n^{++}, n^{+} and n (or p^{++}, p^{+}, p) corresponds to electron (or hole) concentrations of the order of $10^{19}-10^{20}$, 10^{18} and 10^{17}, respectively. Typical materials, bandgaps and layer thicknesses for the realisation of this device structure as a GaInP/GaAs/Ge cell are indicated. Note that not all layers in an actual device (e.g. tunnel-junction cladding layers) are included in the illustration. The figure is not to scale. Reprinted with permission from Friedman *et al.* (2011), © Wiley.

The best situation is when all the cells produce the same photogenerated current. This restricts the efficiency of a series-connected stack with respect to that of an unconstrained stack. For an infinite number of cells, the efficiency limit is the same (Tobías and Luque, 2002). However, in the case of three cells (the configuration that is commercially available today), the efficiency limit for the unconstrained stack is 63.6% compared with 63.0% for the series-connected stack, where conditions of maximum concentration (about 46000 Suns) and Sun/ambient temperatures of 5762 K/298.15 K are assumed in both cases. The Sun temperature considered here is that deriving from the extra-terrestrial spectrum and the ambient temperature is 25°C. These are some of the commonly used standard measuring conditions.

Optimal bandgaps for the series-connected case are 1.75, 1.10, 0.58 eV, but optimal bandgaps at lower concentration are higher, and efficiency is lower.

Two main strategies are followed for the growth of the epitaxial cells: the lattice-matched (LM) approach and the metamorphic (MM) approach. In the first case, the lattices of the different cells are as closely matched as possible. That is why a small proportion of In (1%) is added to the GaAs, allowing close matching with the Ge substrate. The drawback is that this procedure does not admit optimal bandgaps for ternaries of the elements under consideration (Ga, In, P and As). This is evident in Fig. 11.9 (Guter *et al.*, 2009), where, for the lattice-matched (LM) cell, the bandgaps of the two upper cells are a bit too large. In this scheme, the upper III–V cells give too little current as compared with the Ge cell.

In the metamorphic approach the bandgaps are better adjusted, but at the expense of permitting the formation of threading dislocations, as shown in

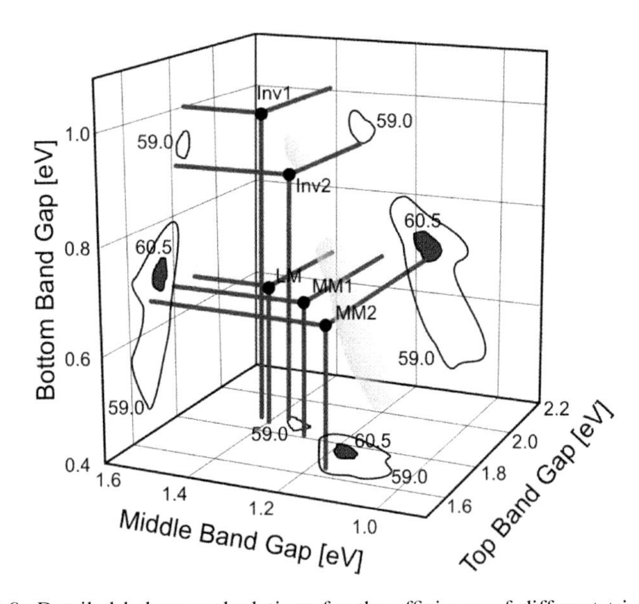

Figure 11.9 Detailed balance calculations for the efficiency of different triple-junction solar cell structures under the AM 1.5D ASTM G173–03 spectrum at 500 kW m^{-2} and 298 K. The grey haze, whose contours are projected in the three Cartesian planes, represents bandgap combinations, which allow efficiencies from 60.5% to 59.0% and hence mark the optimum. Five specific triple-junction solar cell structures are shown: the lattice-matched $Ga_{0.5}In_{0.5}P/Ga_{0.99}In_{0.01}As/Ge$ (LM), two metamorphic GaInP/GaInAs/Ge (1.8, 1.29, 0.66 eV for MM1) and (1.67, 1.18, 0.66 eV for MM2), as well as two inverted metamorphic GaInP/GaInAs/GaInAs (1.83, 1.40, 1.00 eV for Inv1) and (1.83, 1.34, 0.89 eV for Inv2) devices. Reprinted with permission from Guter *et al.* (2009), © AIP.

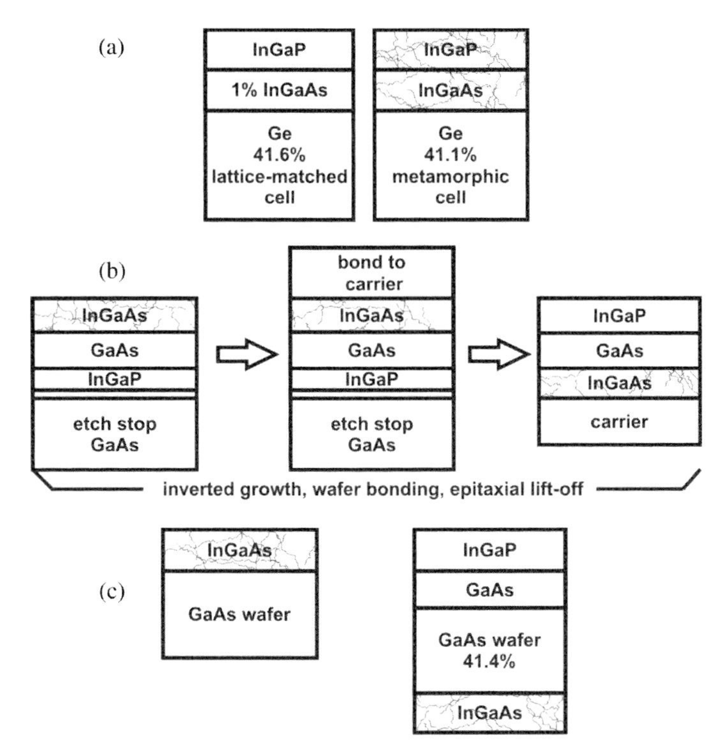

Figure 11.10 a) Lattice-matched and metamorphic three-junction cells grown on Ge. Dislocations are shown in the stressed layers; b) inverted metamorphic cell grown on GaAs, detached and bonded to carrier; c) bifacial epitaxy metamorphic cell. Reprinted with permission from Wojtczuk *et al.* (2010), © IEEE.

Fig. 11.10. This reduces the cell quality and may balance out the advantages of better bandgap matching.

Very good results have been obtained with the present structures, as detailed by Yamaguchi in Chapter 7. Several three-junction cells have reached efficiencies of well over 40%; in late 2012, the world record was held by Solar Junction's LM three-junction InGaP/GaAs/GaInNAs cell, which is 44% efficient at 947 Suns. As for four-junction cells, in autumn 2013 the record was held by Fraunhofer/Helmholtz Centre's cell, reported as 44.7% efficient under 297 Suns. These efficiency measurements have been made at certified laboratories. However, it is worth noting that uncertainty in the measurements is given as ±2.5 absolute percentage points (Green *et al.*, 2010).

Further increase in the efficiency of MJ cells may come through the use of quaternary compounds that provide an additional degree of freedom to tune lattice match and bandgap simultaneously, or through the use of nanotechnology to tune the semiconductor bandgap, such as in quantum well (QW) structures, which are discussed in Chapter 10. The most obvious way of increasing the potential MJ cell efficiency is to add more junctions. This is not straightforward, but there are several studies on how to proceed. The work by King and co-workers of Spectrolab (King *et al.*, 2009) provides recommended reading on the subject. They have, in fact, already presented a five-junction cell, that achieved an efficiency record under no concentration of 37.8% (Optics.org, 2013).

Even if their detailed analysis is beyond the scope of this chapter, it has already been mentioned that other high-efficiency concepts have been proposed that, having the potential of exceeding the SQ limit, could also lead the way to cells that could be stacked as multijunctions, further boosting the overall efficiency. Basically these new concepts seek both a better matching of bandgaps and photon energies, as well as the reduction of entropy production from excess photonic energy over the bandgap. Three of these new concepts were qualified as 'Revolutionary Photovoltaic Devices: 50% Efficient Solar Cells' in a book edited by the US Department of Energy (Lewis *et al.*, 2005), these being the multiple exciton generation (MEG) solar cell (Kolodinski *et al.*, 1993), the intermediate-band (IB) solar cell (Luque and Martí, 1997; 2012) and the hot-carrier solar cell (Ross and Nozik, 1982). Their maximum theoretical efficiencies under the SQ assumptions would be, respectively, 85.4%, 63.2% and 85.4% (Luque and Martí, 2011) for 6000 K/300 K Sun/ambient temperature. Martin Green discusses these and other high-efficiency concepts in Chapter 2.

11.4 Optics for photovoltaic concentrators

Introducing the theoretical framework of the optics of the photovoltaic concentrator, as well as its relatively modern recent design toolbox, is a good first step before going on to describe the most common practical realisations.

Geometric optics has usually been applied to the design of image-forming optical systems, and in this ambit it has achieved a very high degree of sophistication. The problem of collection and concentration of luminous power was also traditionally one of the fields of application of geometric optics. However, it was not until the early 1970s that specific design methods were developed for this type of problem. These methods demonstrated that the most efficient light concentration systems could be obtained when image formation is not considered as a design constraint, allowing thus for an extra degree of freedom.

Geometric optics results from the solutions obtained in Maxwell's equations for the electromagnetic field when wavelength tends to zero, i.e. spatial variations of the field are big compared with the wavelength. Then the concept of the light ray is defined as the curve that is always normal to the wavefronts. The light ray (light path) between any two points will have to satisfy Fermat's principle, this being a variational formulation stating that the length of the optical path followed will be stationary ($\mathrm{d}L = 0$), i.e. a maximal, a minimal or a saddle point, where the optical path length between two points is defined as

$$L = \int_B^A n(x, y, z)\mathrm{d}l \tag{11.5}$$

where $n(x, y, z)$ is the refractive index of the medium at point (x, y, z) and $\mathrm{d}l$ is the differential length along the light's path between points A and B. All the laws of geometric optics can be derived from Fermat's principle, including Snell's laws of reflection and refraction. In fact, minimal path length is the case when a light ray passes from one medium into another (refraction), or is reflected by a planar mirror (this was already known by the Greeks, who formulated the reflection law exactly through Fermat's principle). However, it can also be a maximum, as occurs in gravitational lensing.

An alternative formulation of geometrical optics results from the Hamiltonian equations. A ray, specified as passing through a point (x, y, z) with a direction given by the unitary vector $\bar{v}$ of its direction cosines, is represented as a six-vector (x, y, z, p, q, r) where $(p, q, r) = n(x, y, z)\bar{v}$ is the vector of the so-called *optical direction cosines* of the ray. The Hamiltonian formulation states that the rays follow trajectories that are solutions of the following system of first-order differential equations

$$\begin{aligned}
\frac{\mathrm{d}x}{\mathrm{d}s} &= \frac{\partial H}{\partial p} & \frac{\mathrm{d}p}{\mathrm{d}s} &= -\frac{\partial H}{\partial x} \\
\frac{\mathrm{d}y}{\mathrm{d}s} &= \frac{\partial H}{\partial q} & \frac{\mathrm{d}q}{\mathrm{d}s} &= -\frac{\partial H}{\partial y} \\
\frac{\mathrm{d}z}{\mathrm{d}s} &= \frac{\partial H}{\partial r} & \frac{\mathrm{d}r}{\mathrm{d}s} &= -\frac{\partial H}{\partial z}
\end{aligned} \tag{11.6}$$

where $H = n^2(x, y, z) - p^2 - q^2 - r^2$ is the Hamiltonian function, and s is a parameter with no specific physical meaning. Valid solutions must result in $H = 0$ as required by the definition of (p, q, r) above. The Hamiltonian formulation can be derived from Fermat's variational principle through its equivalent in the form of Lagrange equations, and then applying the Legendre transformation.

This Hamiltonian formulation invites us to consider geometric optics' rays as points in the six-dimensional space (x, y, z, p, q, r), where rays are curves belonging to the five-dimension variety $p^2 + q^2 + r^2 = n^2(x, y, z)$. This five-dimension space is called the *extended phase space*. If M is a set of points in the extended phase space, and $\Im(M)$ is the set of the (points contained in the) rays passing through the points in M, then M is a ray manifold if there are no two points in M that belong to the same ray, i.e. two points that represent the same ray in two different time instants. It is usual to define a ray manifold M through its intersection with a reference surface Σ_R of the space $x - y - z$ that must intersect the rays of M only once. This reference surface will then define a four-dimension variety termed the *phase space*.

If M_{4D} is a four (dimensional) parameter ray manifold then the *etendue* of the manifold is defined as the integral:

$$E(M_{4D}) = \int_{M_{4D}(\Sigma_R)} \mathrm{d}x\,\mathrm{d}y\,\mathrm{d}p\,\mathrm{d}q + \mathrm{d}x\,\mathrm{d}z\,\mathrm{d}p\,\mathrm{d}r + \mathrm{d}y\,\mathrm{d}z\,\mathrm{d}q\,\mathrm{d}r \tag{11.7}$$

which measures how many rays a manifold has. The theorem of conservation of the etendue is fundamental to non-imaging optics design and it states that the etendue is an invariant of any ray manifold as it propagates through an optical system. In other words, the etendue is independent of the reference surface used for its calculation.

In concentrator optics design, the bundle of rays that come from a certain light source, the Sun in the case of solar energy, and impinge on the entry surface of the optical concentrator is called the *input manifold* and is denoted by M_i. The manifold of rays that traverse from the exit surface of the concentrator to the photovoltaic cell, performing as receiver is denoted by M_o. The collected manifold M_c of rays is defined as those rays that are connected by the concentrator and belong both to the input and exit manifold. Also M_o is a subset of the manifold M_R composed by all the rays that can reach the receiver.

Design problems in non-imaging optics can be of several different types. If the design is intended to couple the input and output manifolds perfectly, the concentrator obtained will be such that $\Im(M_i) = \Im(M_o) = \Im(M_c)$, and it is then said to be ideal. If $\Im(M_c) = \Im(M_o) = \Im(M_R)$ then the concentrator is said to be maximal provided it is able to illuminate the receiver isotropically. A concentrator design with the properties of being both ideal and maximal is known as optimum. On other occasions the target may be to just have M_i included in M_o, $\Im(M_c) = \Im(M_i) \subset \Im(M_o)$, in which case we can impose an additional design condition, which could be that M_c produces a prescribed irradiance distribution on the receiver, e.g. homogeneous irradiance in photovoltaic cells improves their conversion efficiency.

There are some ray manifolds that are especially useful. For example, the so-called finite source is that of the rays that connect two parallel concentric planar discs, both within the same homogeneous optical medium. When one of the discs is moved away from the other towards infinity, then we talk of an infinite source and the manifold consists of all the rays that impinge on the other disc with an incidence angle (with respect to the entry surface normal) less or equal to a certain constant angle α, named the source's *acceptance angle*. It can be proved that the etendue of an infinite source E_{inf} is

$$E_{\mathrm{inf}} = A\pi n^2 \sin^2 \alpha \tag{11.8}$$

where A is the disc's surface and n is the refractive index of the surrounding medium.

The *geometrical concentration* is defined as the ratio of the area of the entry surface of the concentrating system to that of the receiver. As we have already been doing, in a concentrator design the geometric concentration is usually written followed by an $\times$, e.g. a $1000\times$ design is a concentrator with an entry surface 1,000 times bigger than the surface of the cell.

In order to better quantify geometrical concentration, let M_i be the ray manifold of an infinite source with acceptance angle α, impinging on the entry surface of a concentrator system surrounded by a medium of refractive index n. Also let M_o be the output manifold, characterised by a homogeneous angular spread β, and n_o the refractive index of the medium surrounding the receiver. Then if the entry and receiver surface areas are A_{E} and A_{R} respectively, due to the conservation of etendue the input and output manifolds, E_{M_i} and E_{M_o}, must be equal if the concentrator perfectly couples these two manifolds (i.e. it is an ideal concentrator):

$$E_{M_i} = A_{\mathrm{E}} n_i^2 \pi \sin^2 \alpha = A_{\mathrm{R}} n_o^2 \pi \sin^2 \beta = E_{M_o} \le E_{M_R} = A_{\mathrm{R}} n_o^2 \pi \tag{11.9}$$

where the last inequality relates the input and output etendue to the maximum possible etendue at the receiver when this is illuminated isotropically ($\beta = \pi/2$) (i.e. it is a maximal concentrator), from which a maximum limit for geometric concentration C_g can be derived:

$$C_g = \frac{A_{\mathrm{E}}}{A_{\mathrm{R}}} \le \frac{n_o^2}{n_i^2 \sin^2 \alpha} \tag{11.10}$$

This is usually referred to as the thermodynamic limit of concentration, because it can also be derived by equalling the temperature of an ideal black absorber to that of the Sun considered a black-body radiator with $T_{\mathrm{S}} = 5777\,\mathrm{K}$ (this is closer to reality than the 6000 K often used for mnemonics) in accordance with the second law of thermodynamics. In the most common case of the entry surface to the concentrator being surrounded by air ($n_i = 1$), Eq. (11.10) turns into

$$C_g \leq \frac{n_o^2}{\sin^2 \alpha} \tag{11.11}$$

This can also be expressed as

$$\text{CAP} = C_g \sin^2 \alpha \leq n_o^2 \tag{11.12}$$

where the square of the refractive index of the medium surrounding the receiver becomes the upper bound to the *concentration acceptance product* (CAP). This parameter is frequently used today as a figure of merit of concentrator designs, showing how close they get to ideality.

If we have a four-parametric ray manifold M, then δM is defined as its edge-ray manifold and is the boundary, in topological terms, of that manifold. δM is itself a three-dimensional subset of manifold M. One of the most powerful tools of non-imaging optics design is the so-called *edge-ray theorem*, stating that if we want to perfectly couple two manifolds M_i and M_o then it is enough to match their δM_i and δM_o subsets. This can be expressed as $\Im(M_i) = \Im(M_o) \Leftrightarrow \Im(M_i) = \Im(M_o)$ as the inverse relationship is also true: if M_i and M_o are coupled by an optical system then their edge-ray manifolds will also be coupled. This means that the design process can focus just on edge-ray manifolds, thus reducing by one the dimension of the problem of coupling four-dimensional manifolds. Even though this theorem was assumed to be true and has been in use since the beginning of non-imaging optics, it was not to be proved till the mid-1980s by IES-UPM's Miñano (1985a, b and c).

The problem of coupling two four-dimensional manifolds is usually too complex, and in order to make it more attainable the usual simplification consists in imposing an axis or plane of symmetry, on both the ray manifolds and the optical system developed, and then working out the design in one of the planar sections. Then it is a couple of two-dimensional manifolds, m_i and m_o subsets of M_i and M_o, that are to be matched. For example, in the case of assuming a rotationally symmetric optical system the design is carried out in one of its meridian planes — those that contain the axis of symmetry — where ideal solutions perfectly matching m_i and m_o can be obtained. Afterwards the three-dimensional optical system will be obtained by rotating the two-dimensional resulting design around the symmetry axis. However, there will be no guarantee that non-meridian rays in the M_i and M_o manifolds will also be ideally coupled, and the assessment of the behaviour of these other rays usually is completed using ray-tracing software applications. Ray-tracing CAD tools, essentially able to compute Snell's laws for very large sets of light rays as they proceed through different optical media, are very much used in non-imaging optics to simulate and characterise the performance of resulting designs.

An important characterisation parameter of CPV optical designs is the *total transmission* ratio, defined as

$$\eta = \frac{P(M_c)}{P(M_i)} \tag{11.13}$$

where $P(M)$ is the power transmitted by the M ray-manifold. Then the *optical concentration* C_o of the system is defined as

$$C_o = C_g \eta \tag{11.14}$$

When designing for infinite sources with a certain acceptance angle, *transmission* can be plotted as a function of the rays' input angle θ so that we obtain the transmission angle curve

$$T(\theta) = \frac{\mathrm{d}P[M_c(\theta, \theta + d\theta)]}{\mathrm{d}P[M_i(\theta, \theta + d\theta)]} \tag{11.15}$$

where $\mathrm{d}P[M(\theta, \theta + d\theta)]$ is the infinitesimal power transmitted by rays in manifold M in the angular differential interval $(\theta, \theta + d\theta)$. The maximum of $T(\theta)$ is termed as the *optical efficiency*, η_o, usually occurring at normal incidence, i.e. $\theta = 0$.

Several different design methods have been developed in the field of non-imaging optics. The earliest was proposed by Welford and Winston (1978) and used by them to develop all the family of the so-called CPCs (compound parabolic concentrators) based on the application of the edge-ray theorem in two dimensions (Fig. 11.11). The primary prototype of this family of concentrators consists of two symmetric parabolic mirrors, each focusing edge rays entering the concentrator with the maximum acceptance angle at the other's lower rim, which coincides with the receiver's edge. In its two-dimensional version, this design performs as an optimum concentrator. Several other versions of the CPC were later developed, such as the dielectric-filled CPC (Ning *et al.*, 1987) using lossless total internal reflection instead of mirrors (which incur losses of ~10%), or CPC designs for finite sources or for non-maximal concentrations.

Welford and Winston attempted the extension of their initial edge-ray design method to three-dimensional systems with their flow-line method, the most remarkable product of which was the *hyperboloid concentrator*, also called the *trumpet concentrator* (Winston and Welford, 1979a, b). Later, the Poisson bracket method was developed by Miñano (1985a, b). While this method was specifically conceived for the design of three-dimensional concentrators, it usually requires the

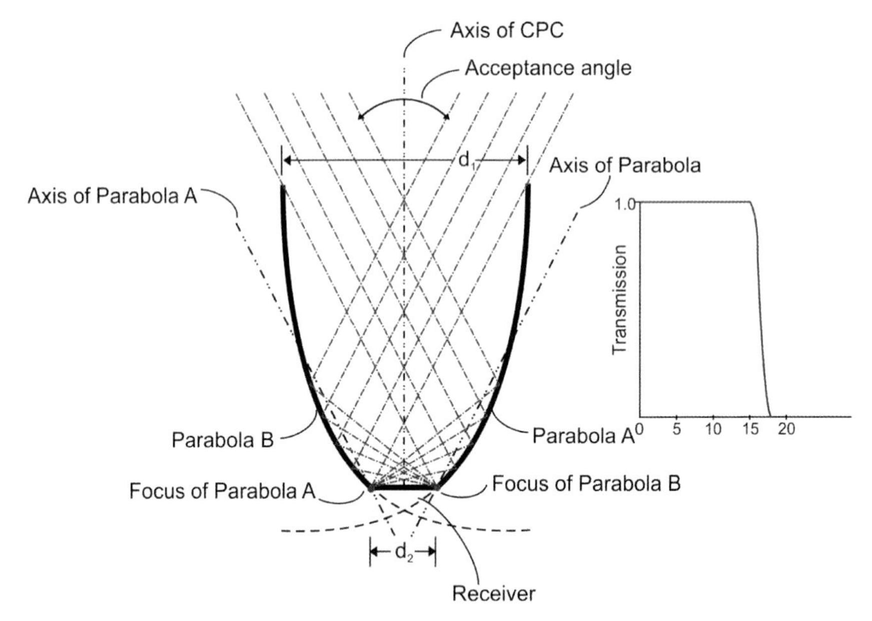

Figure 11.11 Winston's CPC consisting of two symmetric parabolas with an inner mirrored surface, and with their focal points at each end of the receiver. While the two-dimensional version performs as an optimum concentrator, the plot on the right shows the transmission performance of a three-dimensional rotationally symmetric design, with a design acceptance angle of 16° where transmission is brought down to zero in approximately 1°.

use of variable refractive index media, this being impractical in most cases. The main interest of the Poisson bracket method is that it is able to produce ideal three-dimensional concentrators, thus ultimately proving their theoretical existence. It also produced a family of ideal two-dimensional concentrators, the compound triangular concentrator (CTC), with two different regions of constant refractive index (Miñano, 1985c) (see Fig. 11.12).

The Simultaneous Multiple Surface (SMS) method was developed in the early 1990s at IES-UPM (Miñano and González, 1992) as a two-dimensional method, and has to date produced a long list of different designs. Quite curiously, it was born while trying to solve a conjecture. It was known that a single refractive surface can exactly focus a plane wavefront into a point; these sort of surfaces are the so-called Cartesian ovals. The conjecture, based once again on the edge-ray theorem, was that two plane wavefronts could be exactly focused into two points, the receiver edges, by means of consecutive refractions or reflections in two optical surfaces. In fact the conjecture can be extended into the imaging optics world, by rephrasing it as a number n of spherical wavefronts generated at object points in the near field

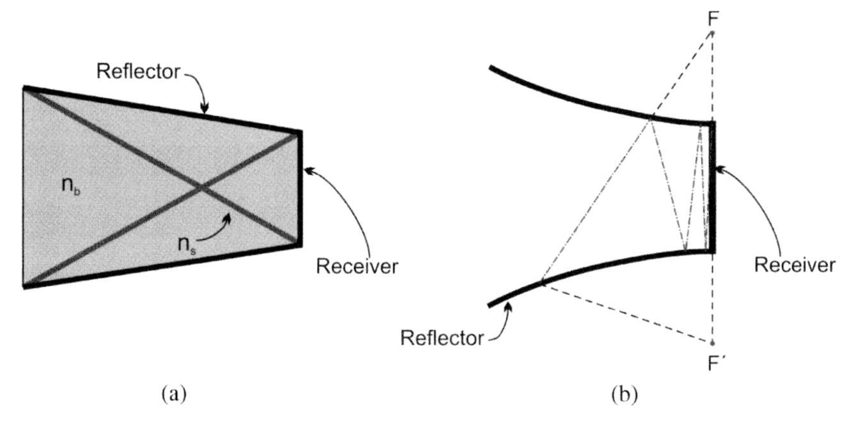

Figure 11.12 a) Miñano's compound triangular two-dimensional concentrator designed using the Poisson bracket method, which requires no curved lines. It is filled with a medium of refractive index $n_b > 1$ plus narrow strips of refractive index $n_s > n_b$. These strips are transparent or act as mirrors depending on the angle of incidence of the rays, and the direction of the collected rays can be varied by modifying the refractive index of the strips; b) Winston's trumpet concentrator.

(or again plane wavefronts for points at infinity) that are to be focused into n image points, again by sequential refractions or reflections in n optical interfaces.

No formal proof of the two-surface conjecture has yet been produced, but the SMS method was developed as an iterative method by which these two sought optical surfaces could be generated. Several designs have been produced by this method, generally grouped by families named RR, XR, RX, XX, RXI, where the R stands for refraction, X for reflection, and I for total internal refraction, which are listed in the same order as incoming light rays will encounter their two constituent optical interfaces (Fig. 11.13).

11.4.1 *State-of-the-art of concentrator optics*

Some of the canonical non-imaging optics developments described above have important practical limitations, starting for example with the famous CPC. Despite its optimality in two dimensions, and high optical efficiency in three dimensions (over 90%), if designed for low acceptance angles, as required by Sun-tracking concentration systems, its length-to-width ratio becomes unfeasibly high (e.g. 920 for an acceptance angle equal to the Sun's half-subtended angle).

Despite the powerful theoretical design toolbox developed, in many cases the optics used in photovoltaic concentration are simply derived from elementary geometric optics, basically using the refraction and reflection laws. Design, when not trivial, is undertaken by ray-tracing codes based on these laws.

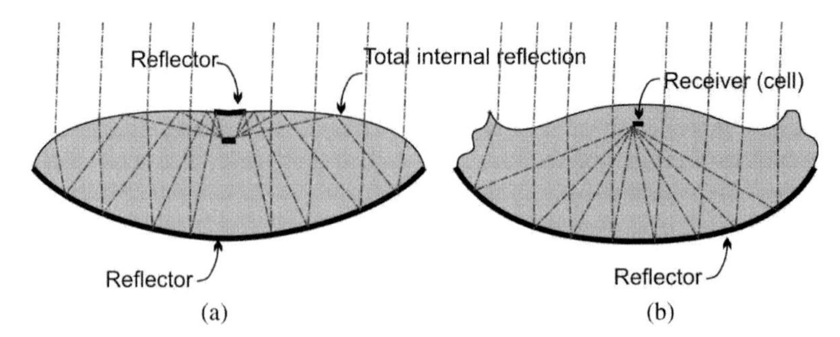

Figure 11.13 a) The RXI (Refraction + Reflection & Total Internal Reflection) concentrator designed using the Simultaneous Multiple Surface (SMS) method. These light rays refract at the front surface, then reflect in the mirrored back surface and are finally reflected towards the cell in the inner side of the front surface. This reflection is mostly total internal reflection, except for the central part in which a reflective spot is placed to redirect rays arriving at high angles; b) represents an RX SMS design. In a) and b), the entering edge rays arrive at both ends of the receiver and at the receiver; thus obeying the edge-ray theorem.

In particular, among the simple options, it is known from elementary geometry that, in two dimensions, a parabola focuses all incident rays parallel to its axis onto its focus. Where the rays will fall when they come from small angles off the axis due to optical aberrations is also a simple matter to determine. In general, a receiver at the focus is designed to collect all these rays, and this puts a limit on the maximum achievable concentration. Lenses may also be used in an elementary way, although thick lenses have usually (but not always) been avoided in concentrator optics because of their high cost.

Fresnel lenses are generally used instead of bulk lenses. These are usually made of highly transparent plastics such as polymethyl methacrylate (PMMA) that are manufactured by moulding or similar techniques. They consist of a set of small prisms that deflect incident rays according to Snell's laws. In thin continuous lenses, only the tangent orientations (derivatives) at the entering and exiting faces of the lens are relevant to the optical design. The same is true of Fresnel lenses, so the derivatives of the continuous lens may be replicated. However, in the case of continuous lenses there is a link between position and derivative, while in the case of Fresnel lenses every prism may be designed independently, thus providing an extra degree of freedom (see Fig. 11.14). In photovoltaic applications, this extra degree of freedom can be used, for example, to improve the homogeneity of the irradiance incident on the cell, which as we have said is beneficial in photovoltaic conversion efficiency. In fact, Fresnel lens designs for solar applications do also commonly embody the principles of non-imaging optics (Leutz and Suzuki, 2001), with precursor designs dating back to the early 1980s (Lorenzo and Luque, 1981).

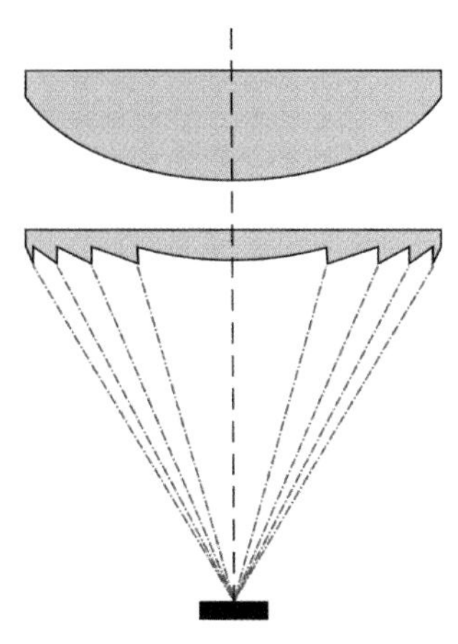

Figure 11.14 Schematic of a Fresnel lens showing its facets. The slopes of a plano-convex lens (above) are projected into a thinner plate.

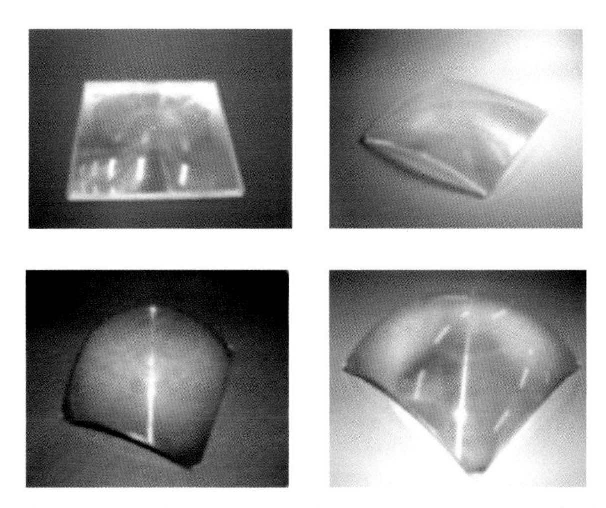

Figure 11.15 Generations of injection-moulded Fresnel lenses from CPV manufacturer Daido Steel. Top left: First-generation flat design (400×, 77.3% of peak efficiency, 2001). Top right: Second-generation half-dome design (400×, 81.5% of peak efficiency, 2002). Bottom left: Third-generation full-dome design made by collapsible moulding die (400×, 85.4% of peak efficiency). Bottom right: Fourth-generation full-dome design made by collapsible moulding die (556×, 91% of theoretical efficiency).

For instance, dome- shaped Fresnel lenses as shown in Fig. 11.15 exhibit much better characteristics.

What we usually desire of the optical design is to achieve the highest possible concentration with the highest optical efficiency. In addition we wish this efficiency to be maintained even under aiming errors or manufacturing imperfections, i.e. we want the lens to have the widest possible acceptance angle. Thus we would be seeking a high CAP concentrator according to the definition of Eq. (11.12). Moreover, we usually want the concentrator to produce uniform illumination of the cell, i.e. homogeneous on it and zero outside. The latter is a hard condition that can be only partly met.

The theorem of conservation of the etendue explained above establishes that the highest concentration can be achieved only if the cell is isotropically illuminated by the concentrator. Going back to Eq. (11.11) this concentration is

$$C_g = \frac{n_o}{\sin^2 \phi_s} \approx 46747 n_o \tag{11.16}$$

where n_o is the index of refraction of the medium surrounding the cell and ϕ_S is the Sun's angular semi-diameter (about 0.265°).

Even theoretically such concentrations cannot be achieved with the lenses or mirrors we have described above, because they are not able to provide the necessary isotropic illumination of the receiver. This isotropy can be increased through the use of two-stage optics, with a primary optics element (POE) that directly receives sunlight at its entry surface, and a secondary optics element (SOE), that receives the concentrated light from the primary, and further conducts it into the PV cell sited at its exit surface. Apart from increasing concentration in the receiver by enabling light isotropy gain, secondaries are usually designed to increase homogeneity of the light flux in the PV cell or to increase the acceptance angle of the overall concentrating system. In the design terms discussed previously, the theoretical requirement of a secondary is that it is able to cast ray manifolds from a finite source, the output of the primary, into the receiver.

While not practically feasible as a POE — where small design acceptance angles are involved — non-imaging CPCs or their derivatives, such as the dielectric-filled CPCs or trumpet concentrators, can be used as efficient secondary optics as acceptance angles for this stage are much bigger. Other simpler devices, such as the truncated cone (for round cells) or truncated pyramid (for square cells), have also been used. In these devices the inner walls are reflective — either made from aluminium or covered with some reflective film — and redirect into the cell those rays that enter and would fall beyond the cell's perimeter, and increasing concentration and acceptance angle. However, solid dielectric secondaries can achieve higher concentrations for the same acceptance angles thanks to the n_0^2 factor in Eq. (11.16) being higher than unity. Another successful secondary design along these lines is the dielectric truncated pyramid, where lossless total internal reflection is used to provide two to four times additional concentration and at the same time, through the so-called kaleidoscopic effect, produces high homogeneity in the light flux in the cell.

One of the ways of providing some tolerance in manufacturing, assembly and Sun-tracking is to design for the collection of rays from a region of the sky larger than the solar disk, i.e. to increase the optical design's acceptance angle. The maximum concentration achievable by this means can be calculated from Eq. (11.16) using an angle greater than the Sun's semi-diameter for ϕ_s. Table 11.1 (Luque, 1989) shows the limit of concentration for several families of concentrators for the case of collecting only the Sun's semi-diameter, 0.265°, and for the more practical case of an allowance of 1°. The concentrator's primary entry surface (also called its aperture) is specified through the f-number, which is defined as the ratio of aperture diameter to the focal distance. This table can be used to estimate the angular allowance required by a certain design. For instance, the above-mentioned ENTECH concentrator with arched Fresnel lenses had a geometrical concentration of 20× (see Table 11.1) and the value in Table 11.1 for an angular tolerance of 1° is 26×. Simple proportion gives us an approximate angular allowance for this concentrator of 1.3°. In the table we can observe that linear concentrators achieve much lower concentrations than point-focus ones. This is because there is no concentrating effect in one of the dimensions. For two-dimensional concentrators, the maximum thermodynamic concentration is only the square root of what appears in Eq. (11.16). Note that the highest concentration for the Sun's angular spread is 54× (limited by chromatic aberration) for arched Fresnel lenses, 108× for linear parabolic mirrors and 376× for point-focus Fresnel lenses. In some cases concentrations higher than those shown in Table 11.1 with high efficiency have been announced; and this is surprising to us as the results in this table are the result of fundamental limitations.

Table 11.1 also shows how the highest concentrations are achieved through the integration of secondaries. However, it must be noted that full isotropy in the receiver is not practical, if incidence angles beyond 65° Fresnel reflection on the cells is regarded as too high. So then again applying the conservation of etendue principle, the maximum practical gain of a secondary is:

$$C_{g\,\mathrm{sec}} \leq \frac{n_o^2 \sin 65^\circ}{\sin^2 \alpha_{\mathrm{sec}}} \tag{11.17}$$

where α_{sec} is the incidence angle at the entry surface of the secondary, which can be expressed as a function of the f-number as $\tan \alpha_{\mathrm{sec}} = 1/2\mathrm{f}$. Thus we obtain

$$C_{g\,\mathrm{sec}} \leq 0.9 n_o^2 (1 + 4f^2) \tag{11.18}$$

Table 11.1 Maximum concentration for concentrators accepting all incident rays.

Concentrator type	Angular acceptance $C_g(C_{gsec})$	0.265° Aperture	Angular acceptance $C_g(C_{gsec})$	1° Aperture
Circular reflective parabolic dish	11678	f/0.50	821	f/0.50
Parabolic dish with secondary ($n_o = 1.49$)	67666 (16.1)	f/1.51	4752 (16.3)	f/1.51
Square flat Fresnel lens–square cell	376	f/1.35	73	f/1.19
Square flat Fresnel lens with axisymmetric secondary ($n_o = 1.49$)	12356 (42.1)	f/2.12	1856 (38.4)	f/1.98
Linear, flat Fresnel lens	22	f/1.74	10	f/1.38
Linear, arched Fresnel lens	54		26	
Linear, parabolic reflector	108	f/0.50	29	f/0.50

Source: Luque (1989). C_g is the overall geometrical concentration (primary optics aperture-to-cell area ratio); C_{gsec} is the secondary geometrical concentration (aperture of the secondary-to-cell area ratio), where the refractive index of the secondary is 1.49, which is that of PMMA. Aperture diameter or width (for the primary) is the (freely selected) focal distance divided by the number in the table (f-number).

so that, for example in the case of the square flat lens with a secondary with $n_{\rm o} = 1.49$ in Table 11.1, we find $C_{g,sec} \leq 37.9$, i.e. lower than that achieving maximum overall concentration.

Figure 11.16 shows the results of the analysis of the transmission angle curve, as defined above, of different common secondaries when placed below a Fresnel lens with f/1 (Victoria *et al.*, 2009). It can be seen that the rotational CPC provides the highest acceptance angle.

Another important requirement for concentrator optics is homogeneity of illumination. Strongly inhomogeneous illumination produces much higher than average local concentrations, leading to losses in the fill factor of the cell and reducing efficiency. One solution is the above-mentioned kaleidoscope SOE, consisting of a truncated glass pyramid in which the irradiance is expected to be homogenised by multiple total internal reflection of the entering rays. However, this solution substantially reduces the acceptance angle. An interesting solution has recently been presented by Benítez *et al.* (2010): the Fresnel–Koehler (FK) concentrator, which uses the principle of designing the SOE to image the POE onto the cell (see Fig. 11.17). As the POE is homogeneously illuminated by the Sun it produces a homogeneous light profile. The POE can be accurately imaged onto the cell with the SOE. This produces a nice squared homogeneous illuminated area that fits with the square shape of the cell. The FK SOE has the aspect of a dome and presents an

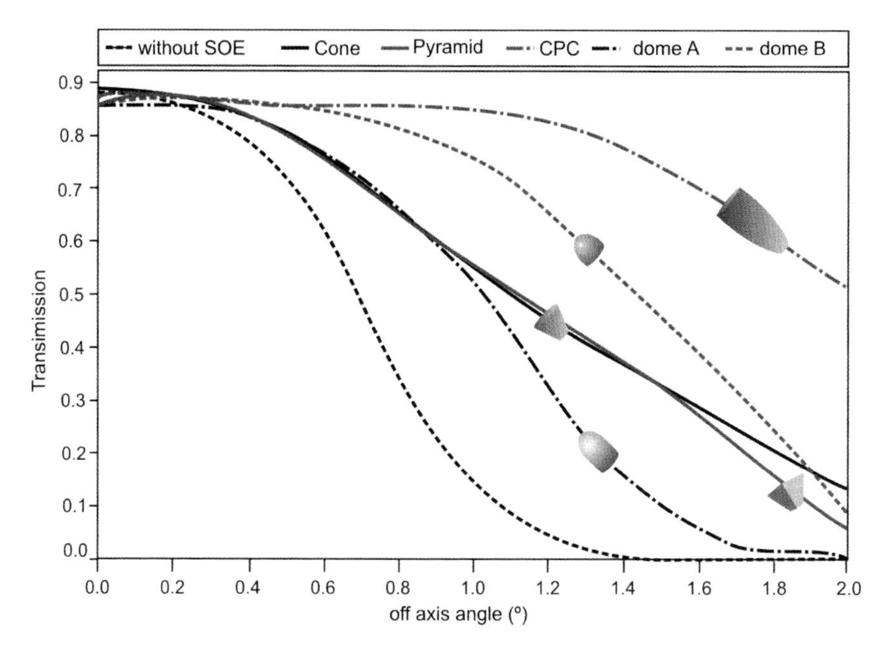

Figure 11.16 Angular transmission curves for the different SOE (secondary optics element) studied. Lens-to-cell geometrical concentration 1000× and f/1. Reproduced with permission from Victoria *et al.* (2009), © OSA.

inactive optical region near the cell that is helpful when sticking the concentrator to the cell (if this zone is active it gives rise to high losses due to the brimming glue). As matter of fact in the figure, the FK concentrator is formed of four independent POE–SOE couples, all four manufactured together, so this design aspect is irrelevant for the user. Their acceptance angle is reasonably high, in the range of 1.1° for 1000×.

The optics introduces changes in the spectrum impinging on the cell so that its optimum design will not be that used today in champion cells designed for a standard spectrum. Furthermore, chromatic aberrations may cause the illumination profile at the cell to be different for different spectral ranges. An additional advantage of the FK concentrator is that it is almost free of chromatic aberrations.

Reflective optics is also used, although much less often. Figure 11.18 shows the solution of the pioneering CPV company SolFocus, based on non-imaging optics principles (Winston and Gordon, 2005). Its acceptance angle is rather high (~1.4° for 500 Suns) but the efficiency is relatively low owing to metal reflections (governed by the metal absorption) and two Fresnel reflections. Note that the element over the cell is a homogenising kaleidoscope. Mirrors can bend rays more

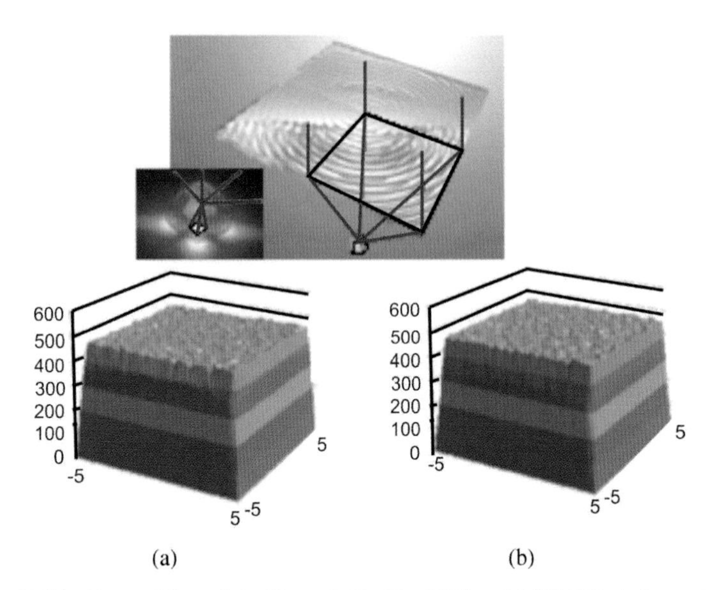

Figure 11.17 Above: View of the Fresnel–Koehler POE and SOE. This is formed of four POE in a single Fresnel lens, each one imaged onto the cell by the four SOE elements. The four SOE elements are cast in the same block. Below: Irradiance distribution on the cell for the FK concentrator with $C_g = 625\times$, f/1, no antireflective (AR) coating on the SOE, when the Sun is on axis and the solar spectrum is restricted to: a) the top sub-cell range (360–690 nm); and b) the middle sub-cell range (690–900 nm). Reproduced with permission from Benítez *et al.* (2010), © OSA.

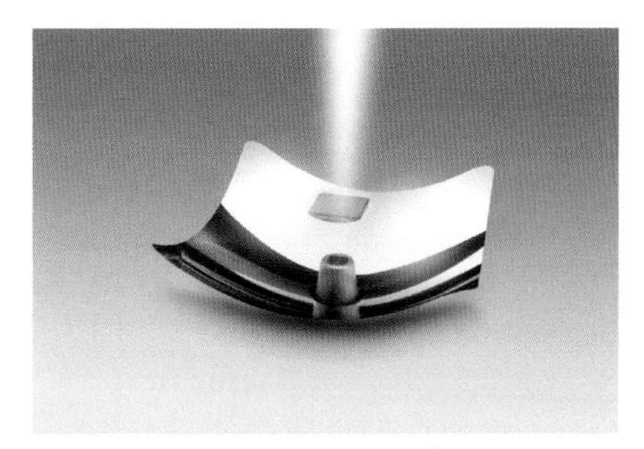

Figure 11.18 Cassegrainian optics in the SolFocus CPV module. Two reflections, first in a parabolic dish and then in a hyperbolic mirror with same focus, sends light down to the receiver block where a TIR kaleidoscope homogenises light reaching the cell below. Courtesy of SolFocus, Inc.

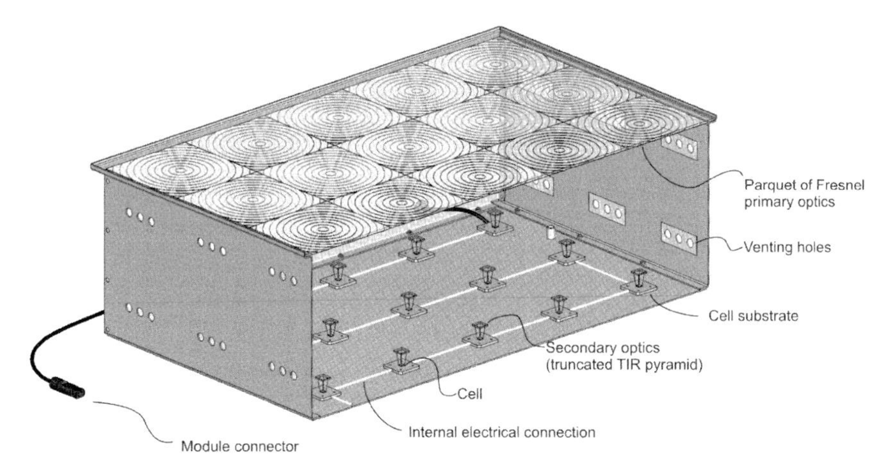

Figure 11.19 Side-cut view of a common concentrator module design, based on a parquet of Fresnel lenses as primaries.

than lenses; therefore this concentrator is about three times more compact than a refractive system for the same POE entry aperture.

11.5 Photovoltaic concentration modules

In their most common configuration, cells and optics are put in a box to form a concentrator module, usually made of metal except for the front cover, the module's entry surface, which will be a covering glass. Figure 11.19 shows the main parts of a generic concentrator module. Primary plastic lenses can then be stuck to the inner face of the covering glass. Parquets of plastic Fresnel lenses can be supplied by some specialised manufacturers such as 3M, and directly assembled in place of a covering glass, but care must be taken over surface scratching, especially in sandy and windy environments. A common alternative to this possible problem is to have the Fresnel lenses made of transparent silicone rubber polymerised directly on the inner face of the covering glass, the so-called silicone-on-glass technique that was first used in the Ramón Areces concentrator referred to above (Lorenzo and Sala, 1979).

The cells have to be carefully attached to the supporting structure in such a way that the wires issuing from the solar cell can be easily connected to external wires that connect to the next cell. The arrangement must provide electric insulation of all the cells, which must then be floating with respect to the module box, thus enabling its grounding for safety purposes, and at the same time it should provide a good thermal connection to the module's back plate. To this end, an etched copper layer on an electrical insulating substrate, possibly alumina or AlN with high thermal conductivity, is used. This must be stuck close to the metal plate that closes the module at the back. Furthermore, a bypass diode is connected to the cell in an antiparallel configuration to enable current flow to bypass a particular cell which may for any reason be shadowed. This ensures that the module will operate even under partial shading and prevents the appearance of strong reverse bias in the cells, which is potentially destructive.

For single-substrate structures, removing heat from the cells is not a difficult problem if the cells are relatively small. In today's high-concentration modules cooling is usually passive and cooling fins are seldom used. The amount of heat to be removed is the fraction of the power incident on the module that is not converted into electricity. This heat is partially radiated by the two faces of the module and partially removed by convection, as in the flat module case. The temperature drop in the thin insulating layer and in the metal layers is relatively small. The main concern is the removal of heat from the cell itself where the power density is high (although at 1000 Suns it is still ten times less that in today's high-power light-emitting diodes [LEDs]). The typical operating temperature of the cells does not exceed 80–90°C, that is, about 50°C over the ambient. This is roughly twice the increase in a flat module.

Cells in modules are usually interconnected in different ways depending on the cell size, which varies between 1 mm^2 and 1 cm^2. Usually, cells are grouped in parallel (if the cells are small) and then these groups are connected in series. Greater paralleling in the cell interconnection will result in a higher acceptance angle.

Something has to be done to keep the cells aligned with the primary optics, given the difference in the expansion coefficients of glass and most metals. For example, in order to address this problem, the so-called all-glass technology was developed by engineers of the Ioffe Institute in St. Petersburg (Andreev *et al.*, 2003). In this design the module's front cover was of the silicone-on-glass type while the back plane was also made of sheet glass. This technology was later transferred to German CPV start-up Concentrix, which successfully industrialised it (Fig. 11.20). Beside the problem of optical misalignment, due to differences in thermal expansion coefficients thermal stress is to be prevented and in this respect expansion joints should be allowed and sealants must be chosen as sufficiently elastic to avoid mechanical stress.

Moreover, the inside of the module must remain clean for 20 years without requiring maintenance or any sort of disassembling. This may get somewhat complicated when some sort of ventilation holes may be needed in the module enclosure to help dew to dry up in the morning. Different solutions are used by

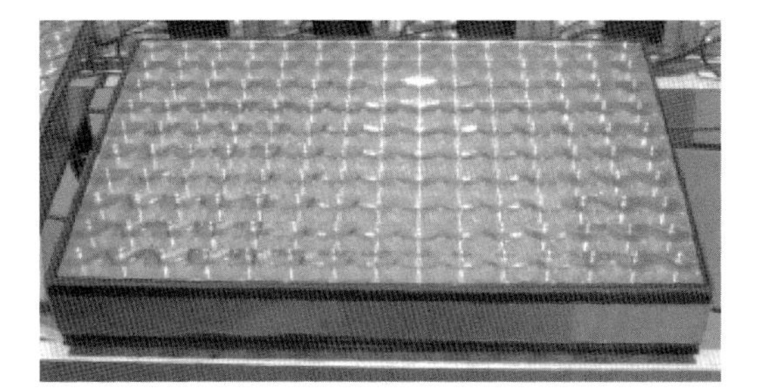

Figure 11.20 Concentrix's 500× module, following the all-glass design principle (initially the side panels were also made of glass and one could look inside the module and check if it was roughly focused). On the right is an inside view of the module where interconnected cell substrates can be seen receiving the concentrated sunspot.

every manufacturer and details are usually kept secret. Some have opted for making a vacuum inside the enclosure, which is hermetically sealed, while others go for an open approach in which venting holes are only made dustproof through metal cloth coverings, and rain water is allowed inside provided the cell encapsulation and wiring is made waterproof. In between these two approaches, manufacturers frequently choose venting solutions based on holes covered with breathable waterproof Gore-Tex membranes.

Finally, it is interesting to describe, following Sala's analysis (Sala and Antón, 2011), the impact of cell size on several other module design parameters and what the range of its optimum value is. For the most usual case of Fresnel primaries, optical efficiency decreases with f-number and usually this value is set above unity. Assuming a minimum f-number of one, then we can establish an approximate relation between cell size and focal distance

$$A_{\mathrm{C}} = \frac{A_{\mathrm{E}}}{C_g} = \frac{F^2}{C_g} \tag{11.19}$$

where A_{C} is the cell size; A_{E} is the primary lens surface or, in optical design terms, the entry aperture surface; F the focal distance; and C_g the geometric concentration. In other words, the smaller the cell the less the module volume, this meaning a reduction in the cost of the housing materials and in the module's requirements for storage and transportation space, and frequently in the cost of the tracker components as well (not only because of a possible reduction in module weight, but also because the centre of gravity of the array of modules stays closer to the tracking drive, reducing its load rating). Also, smaller cell size will result in lesser series resistance and consequently I^2R ohmic losses, and as will be seen below a smaller size also enables better heat dissipation. Furthermore, the smaller cells attain a

Figure 11.21 Side view of one of the early designs of Daido Steel, a 550× module integrating domed Fresnel primaries where one enclosure side panel has been removed to show the module's inside.

higher wafer-filling factor (i.e. the ratio of the total surface of cells made in a wafer to the total wafer surface). But there are also some disadvantages in going to too-small cells, for example a lesser fraction of active area per cell, or higher material losses when dicing the cell wafers, and usually higher complexity in the assembly process. Figure 11.21 shows a side inner view of one of the early module designs of the Japanese manufacturer Daido Steel, while Fig. 11.22 shows some examples of current and recent commercial high concentration PV (HCPV) modules.

Sala's costing exercise assumed a fixed geometric concentration of 1000× and 20% efficient modules with a design based on a flat Fresnel lens as primary, and a silo-type secondary. It focuses on two production regimes of 10 MW yr^{-1} and 30 MW yr^{-1}, which are considered likely in the start-up phase of a CPV production facility. By exploring cell sizes from 1 mm^2 to 100 mm^2 the curve in Fig. 11.23 is obtained, which shows a module cost minimum for cells of size between 10–20 mm^2 in area, beyond which module cost increases slowly due to decreasing efficiency deriving from higher thermal losses.

11.5.1 *Cooling of solar cells*

Despite the simplicity of cell cooling solutions in most of commercially produced CPV modules, the analysis of this deserves some attention in the design stage, given that proper cooling is essential to obtain good system performance. The heat sink is, in most cases, passive, meaning that the heat is dissipated by natural convection to the air. This may be achieved by multi-finned structures or, as happens in most of today's high-concentration modules, by a single flat aluminium plate placed at the back of the cells.

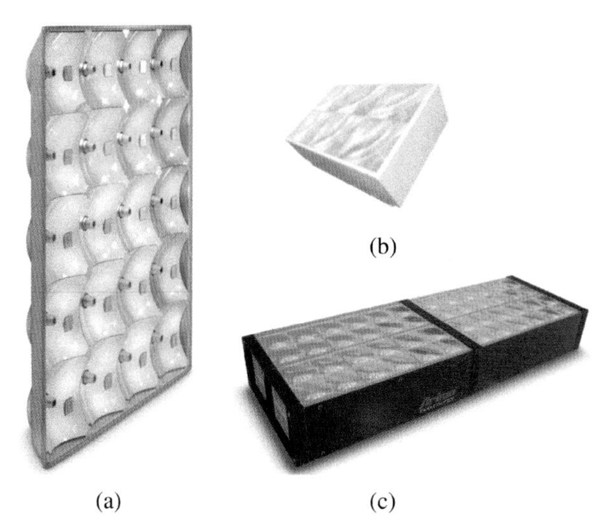

Figure 11.22 Different commercial CPV modules: a) the US's SolFocus; b) Spain's Isofotón; c) Taiwan's Arima Eco. Courtesy of SolFocus, Inc. Isofotón, SA, Arima Eco Energy Technologies Corp.

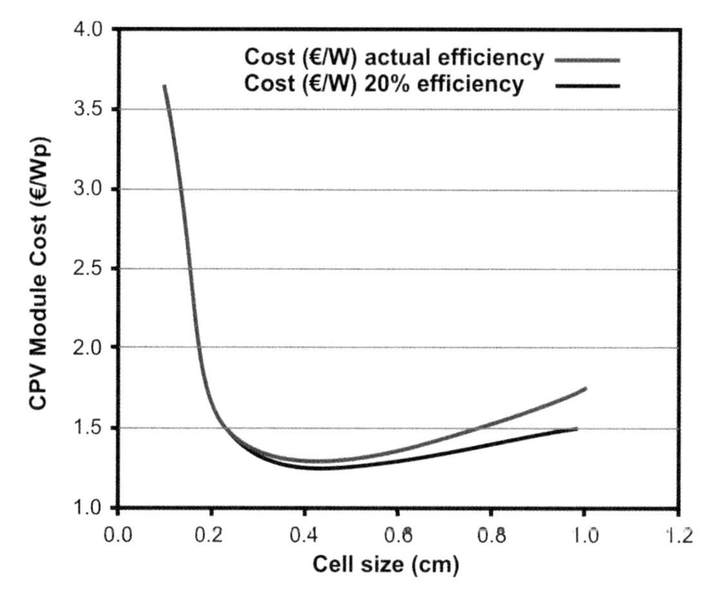

Figure 11.23 Estimated cost estimate of 1000× module based on Fresnel POE and silo SOE, rated at a DNI of 850 W m^{-2} in a manufacturing scenario of 30 MW yr^{-1}. The black curve assumes fixed efficiency while the red curve takes into account efficiency variations due to cell size, showing the impact of cooling degradation as the cells grow larger. Courtesy of Prof. Sala, personal communication.

The heat is produced at the cell itself, mainly where light absorption occurs. This tends to be at the cell front. Thus most of the heat goes through the cell volume to the heat sink or back plate, and then to the interface with air of this metallic part, where it is transferred to the ambient. Temperature gradients are the driving force for this heat flow, so some temperature rise over the ambient cannot be avoided. In many parts of the path, the heat is transferred by conduction so the heat flow density is proportional to the gradient of the temperature, the thermal conductivity κ being the constant of proportionality. This law, formally equivalent to Ohm's law of electric conduction, justifies the use of an electric equivalence between heat and current flow. In this equivalence the heat flow plays the role of the electric current and the temperature that of the electric potential. For a rod of cross-section A and thickness d the thermal resistance (K W^{-1}) is then

$$R_{th} = d/\kappa A \tag{11.20}$$

Table 11.2 gives the thermal conductivity of some common materials, and shows the importance of the encapsulation materials. Copper and aluminium are good conductors but other metals are poorer. However, there are metals (like iron, $\kappa = 0.5\,W\,cm^{-1}K^{-1}$) with a linear expansion coefficient better matched to that of solar cell semiconductors than Cu, and with a reasonable thermal conductivity so that they can be used as the stage on which the cell is bonded. Cells are either soldered to these metal substrates or attached with heat- and electricity-conducting glues. In order not to shunt the cell, it has to be electrically isolated from the metal substrate performing as heat sink by an insulator that should also be a good thermal conductor. In this respect several solutions are applied, usually coming from the power electronics world, such as the use of IMS (insulated metal substrate) or DCB (direct bonded copper) substrates that are essentially printed circuit boards (PCBs)

Table 11.2 Thermal conductivity of some common materials.

Material	**Thermal conductivity/ W cm^{-1} K^{-1}**	**Material**	**Thermal conductivity/ W cm^{-1} K^{-1}**
Si	1.45	Solder	0.5
GaAs	0.8	Air	0.000252
Cu	3.85	Glass	0.007
4Al	2.1	Al_2O_3	0.375
Fe	0.5	BeO	2.2
Epoxy (thermal)	0.027	AlN	1.8

Source: Sala, 1989. Reproduced with permission. © IOP.

having a ceramic dielectric layer in between which has good thermal-conducting properties. Alumina (Al_2O_3), aluminium nitride (AlN) and beryllium oxide (BeO) have been used as dielectrics, or in the case of an IMS the dielectric is usually some heat-conductive epoxy. Air has been included because it can form voids or gaps between mechanically attached parts, this having a negative impact due to a serious increase in thermal resistance.

Finally, the heat flow will find an interface with a fluid, usually air. This air may remove heat either by natural convection (the surrounding air being at rest), or by forced convection (when the air is in movement, perhaps because of natural wind). To model convection a heat transfer coefficient h ($\mathrm{W\,cm^{-2}\,K^{-1}}$) is used. The modelling of h is a complex matter. Its value varies from $5 \times 10^{-4}\,\mathrm{W\,cm^{-2}\,K^{-1}}$ for natural convection from a flat plate into air, to $(10-15) \times 10^{-4}\,\mathrm{W\,cm^{-2}\,K^{-1}}$ for the case of added vertical fins with wind speeds of 1–$2\,\mathrm{m\,s^{-1}}$.

Cooling by water is much more effective, with h values that may lie in the range $(400-3000) \times 10^{-4}\,\mathrm{W\,cm^{-2}\,K^{-1}}$. Water cooling is often used for experiments, but tends to be avoided in present-day concentrator designs, as it usually increases complexity and introduces reliability issues.

In the case of the back plate of the module housing acting as heat sink, solving the differential heat equation can prove that heat dissipation decreases with distance to the cell (Napoli *et al.*, 1977; Sala, 1989), to the point that for a certain value of plate thickness w, there is a maximum effective radius R_{max} beyond which more heat sink material is of no use:

$$R_{\mathrm{max}} = 0.66/\alpha = 0.66(w\kappa/h_{\mathrm{eq}})^{1/2} \tag{11.21}$$

where $h_{\mathrm{eq}} = h + 8\varepsilon\kappa_R T_{\mathrm{amb}}^3$ is the equivalent thermal conductance both due to convection and radiation at the back plate (radiation through two surfaces and convection through the outer one) with κ_R in the radiation term, the right addend, being Wien's constant and ε being the surface emissivity.

The temperature difference between the coldest point of the back plate and the receiver bottom is

$$T_i - T_o = -B(n)\eta_{op}\frac{A_E}{\pi w\kappa}\ln(\alpha R_c) \tag{11.22}$$

where $B(n)$ is the direct irradiance, R_c is the cell radius, and as above, A_E is the primary optics entry surface to the concentrator, and η_{op} is the optical efficiency. The thermal resistance of the back plate working as heat distributor will be $R_{\mathrm{th}} = \ln(\alpha R_c)/\pi w\kappa$, which decreases with cell size. In general, for a fixed temperature difference, the smaller the cell radius and the entry surface, the less the plate thickness that will be required to maintain it. In the case of the back plate acting as the only heat conductor and dissipater, its surface will be the same as that of the entry surface of the concentrator and in this respect we could impose that this surface is limited by the maximum effective radius as in Eq. (11.21), so that $A_E = \pi R_c^2$ and Eq. (11.22) becomes

$$T_i - T_o = -0.44B(n)\eta_{op}\frac{\ln(\alpha R_c)}{h_{\text{eq}}} \tag{11.23}$$

The cell temperature can be obtained simply by adding to the drop in Eq. (11.22) or its special case, Eq. (11.23), the contribution of the layers interspersed between the light absorption surface of the cell and the upper surface of the heat-sink back plate, e.g. the cell itself, the solder, the insulating layer, that act as conductors with their own specific thermal resistance. According to Eq. (11.20) its drop can be written as

$$T_i - T_o = \frac{0.44}{\alpha^2 R_c^2}B(n)\eta_{\text{op}}\left[\sum_i \frac{d_i}{\kappa_i} - \frac{R_c^2}{\kappa w}\ln(\alpha R_c)\right] \tag{11.24}$$

where κ_i is the thermal conductivity of the ith layer between cell and back plate and d_i its thicknesses. However, for well-manufactured receivers most of the temperature gap occurs in the back plate and the contribution of the upper layers is small, so that their thermal properties can be to some extent disregarded. The use of conductive glue instead of solder can create an exception.

A worked example with a real 500× module design allows us to see the importance of the different elements of the encapsulation. Assume we are using a triple-junction solar cell of area $0.5\,\text{cm}^2$, $150\,\mu$m thick, bonded with $50\,\mu$m of solder to a copper tab $500\,\mu$m thick, which is then laminated to a thermal conductive epoxy layer of $500\,\mu$m thickness, and then to 4 mm thick aluminium substrate slab which is screwed to the module's back plate, also made of aluminium and 2mm thick. The surface of the Al slab is $16\,\text{cm}^2$ and the Cu tabs and thermally conductive epoxy printed in its surface cover 40% of this surface. Finally, in-between the Al slab and the module's back plate, thermal conductivity is enhanced by a $50\,\mu$m thick layer of thermal grease. Taking $h = 5 \times 10^{-4}\,\text{W}\,\text{cm}^{-2}\,\text{K}^{-1}$ for the air–back-plate interface and the emissivity of Al as 0.1, we find using Eq. (11.21) that $R_{\text{max}} = 17.15\,\text{cm}$, which is greater than the radius of the $250\,\text{cm}^2$ back-plate dissipating surface, whether this is squared or circular. This means that all of is being effectively used for heat dissipation. Assuming a direct normal irradiance (DNI) value of $900\,\text{W}\,\text{m}^{-2}$, and an optical efficiency of 80%, we can calculate the temperature difference between ambient temperature and the receiver bottom using Eq. (11.22) as $T_i - T_o = 56\text{C}$.

Table 11.3 Thermal drops in layers between the cell and the heat-sink back plate.

Layer	Thermal drop (°C)
Cell	0.67
Solder	0.36
Copper tab	0.03
Thermal epoxy	2.67
Aluminium slab	0.21
Thermal grease	1.87

Also using the respective thermal conductivities, thicknesses and areas of the different layers, from the cell to the receiver's bottom in contact with the back plate, we would find the temperature jumps shown in Table 11.3.

The total thermal drop in the conducting layers adds up to 5.83°C, which added to the difference between the ambient temperature and the receiver makes a total of 61.83°C, which is the temperature above ambient reached by the cell. This example is good to show how most of thermal drop occurs in the module's back plate, performing as a heat sink. Care must be taken with the insulating layers — the epoxy layer in our example — to ensure the required breakdown voltage that is commonly specified in the 1–2 kV range is not exceeded. Here we have not taken into account the convection in the Al substrate slab, which also reduces the overall temperature drop somewhat.

In summary, cooling the cells is an important problem to tackle when facing CPV module design, in which a combination of electric insulation and good thermal contact has to be achieved. But in the end, it is helpful to know that simple passive cooling means can provide the heat dissipation needed by the cells, even under solar concentrations of 1000×.

11.6 Tracking systems for photovoltaic concentration

Most CPV systems use only direct solar radiation, and they must therefore permanently track the Sun's apparent daytime motion, and hence incorporate an automatic Sun-tracking structure able to mount and position the concentrator optics in such a way that direct sunlight is always focused on the cells. This Sun tracker is basically composed of a structure with a sunlight-collecting surface on which to attach concentrator modules or systems, which is in some way coupled to a one- or two-axis mechanical drive, and also some Sun-tracking control system which operates over the drive axes, and maintains optimum aim of the collecting surface or aperture towards the Sun.

Static mounts are only feasible today for low concentration factors below 5. In the long term, static concentrators with higher ratios making use of luminescence and photonic crystals might become available. However, all these issues are beyond the scope of this work.

Line-focus reflective concentrators, such as troughs (the already-mentioned mentioned EUCLIDES case), require only one-axis tracking to maintain the PV receiver along the focus line. However, due to the daily variations in the Sun's elevation, the incidence of sunlight on the tracker's aperture is usually somewhat oblique, reducing the intercepted energy and causing the Sun's image to move up and down within the focal axis, producing further losses whenever it oversails the receiver's ends. Line-focus refractive concentrators (for example, those based on linear Fresnel lenses, as in the ENTECH type) experience severe optical aberrations when light incidence is not normal, thus requiring two-axis sun tracking, and the same happens to most of the point-focus concentrators, excepting some low-concentration factor devices with sufficient acceptance angle to admit the Sun's altitude variations.

Nearly all PV concentrators that are already commercialised or currently under development use two-axis tracking, the so-called pedestal tracker, with its azimuth-elevation axes, being the most common configuration, followed by the tilt-roll tracker operating on the declination–hour angle axes shown in Fig. 11.24.

Strictly speaking, the main commitment to be fulfilled by a CPV Sun tracker is permanently to align the pointing axis of the supported concentration system with the local Sun vector, in this way producing maximum power output. The reasons for the decrease of Sun-tracking performance can be classified into two main types (Luque-Heredia et *al.*, 2006): (1) those purely related to the precise pointing of the

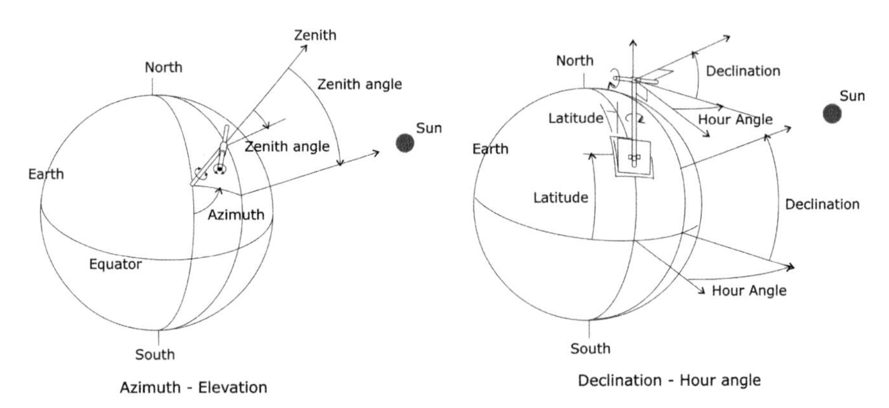

Figure 11.24 The two most common Sun-tracking axes geometries used in solar trackers: declination–hour angle (on the left) and azimuth-elevation tracking (on the right).

tracker at the Sun; and (2) those which provoke shrinkage of the overall acceptance angle of the concentrator system, thus indirectly increasing the tracking accuracy required. Among those related to the tracking accuracy are on the one hand the exactness of the Sun's positional coordinates generated by the control system, expressed in terms of rotation angles of the tracking axes, either by Sun ephemeris-based computations or derived from the feedback of Sun-pointing sensor readings, or a combination of both, which are affected by numerous error sources. On the other hand, there is the precision with which the tracker can be positioned at these dictated orientations, i.e. the positioning resolution of the tracking drive and its control system, which essentially depends on the performance of tracking speed control and on the mechanical backlash introduced by the drive's gearings.

Regarding acceptance angle losses caused by the tracking system, these are due to the accuracy, which can be attained in the mounting and alignment of the concentrator system atop of the tracker. This is basically a design problem having to do with the fixtures provided for this purpose, their accurate assembly and the means provided for in-field fine tuning, but also with the mounting protocols devised to carry out these tasks. Also resulting in acceptance angle cuts is the stiffness of the tracker, which is to say the deformations allowed in the different elements of its structure under service conditions.

In a first iteration, characterisation of the service conditions for a CPV tracker basically consist in determining the CPV array payload (modules' weight) and fixing the maximum wind load, i.e. wind speed, to be withstood during Sun-tracking operation without any effect on the concentrator's power output. To ensure service under these maximum permanent and variable loads, a maximum flexure deformation should be specified for the tracker structure.

As for the variable loads, the higher the maximum wind speed to be resisted while maintaining productive operation, the heavier and more expensive is the tracking structure required to keep deformation within the bounds required for accurate tracking. This behaviour can be seen in Fig. 11.25 for one of Inspira's early tracking designs, a 9 m^2 CPV pedestal tracker, in which normalised cost vs. maximum service wind speed for a maximum 0.1° flexure is plotted (Luque-Heredia *et al.*, 2003).

A cost-effective approach here is to determine the maximum wind load from the cross-correlation between wind speed and direct radiation, in the location or set of locations in which the trackers are to be installed. A case example of this type of analysis is presented in Fig. 11.26, worked out with one year of continuous wind speed and direct normal irradiation hourly data (assuming a two-axis tracker), for the Spanish city of Granada (Luque-Heredia *et al.*, 2007). Above this threshold stiffness specifications do not have to be met, service is not guaranteed, and the

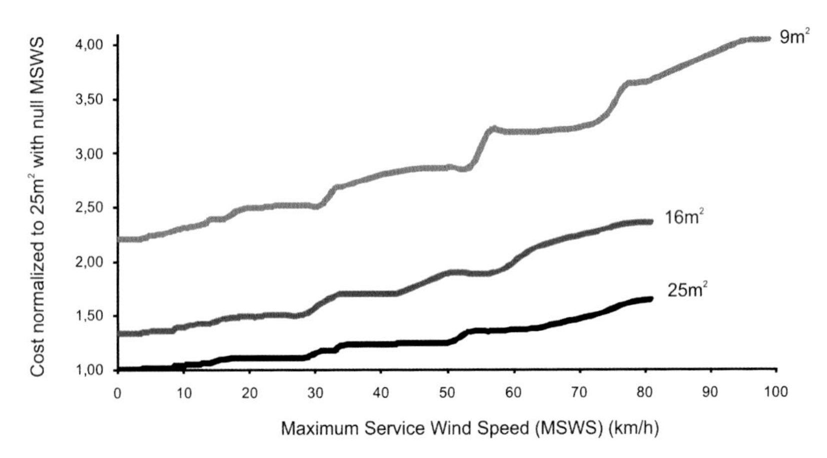

Figure 11.25 Tracker cost vs. maximum service wind speed for a specific tracker design dimensioned for three different aperture surfaces and 1000 units/yr production.

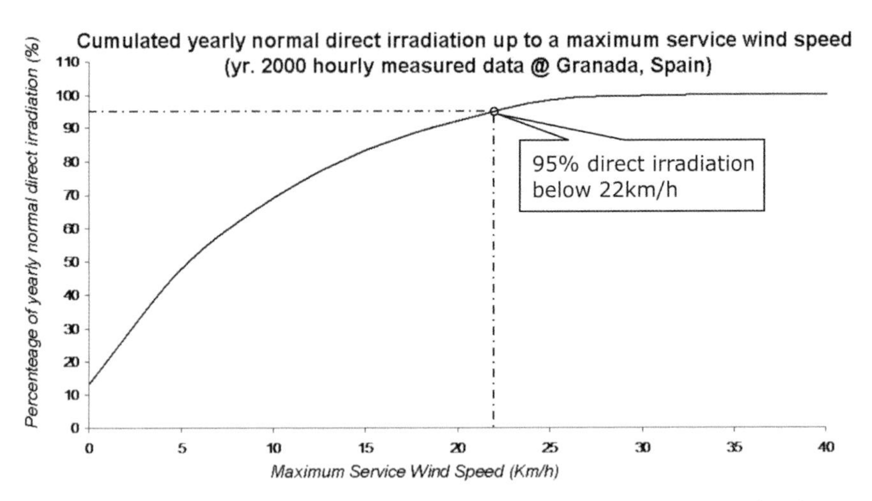

Figure 11.26 Wind speed vs. yearly DNI correlation for the determination of optimum service conditions applied to Granada.

tracker would be better to switch to some low wind profile stow position to decrease stress and increase operative lifetime. For example, as a general rule of thumb a $13\,\mathrm{m\,s^{-1}}$ maximum wind load both windward or leeward to the modules in any of the Sun-tracking orientations of the tracker can be considered a reasonably conservative value which has been proved to comprise a minimum of 95% of the direct radiation measured by the 26 weather stations of the SOLMET network distributed over the contiguous United States (Randall and Grandjean, 1982).

Regarding the CPV array payload, and again bearing in mind the targets set for a possible CPV module design team in order to decrease tracker cost, it would be worthwhile to decrease the module's weight as much as possible; the weights of present module technologies range from 40–140 g W^{-1}. Moreover, module sizing is also an important issue, and an optimum is to be sought in which the size of modules does not require redundant framing from the tracker just to hold them, but on the other hand in which module stiffness doesn't develop into an excess of self-weight.

The achievable tracking accuracy and the acceptance angle of the modules determine the maximum deformation that can occur in the tracker structure without affecting power output. The higher the tracking accuracy and the wider the nominal acceptance angle of a module, the bigger the deformation tolerance which can be set for the structure, and consequently the lower its cost.

Aside from the structural specifications derived from service conditions, basically determining the tracker's stiffness, its ultimate limit state and the loads at which this is reached is to be considered. This is usually a better-paved way in the sense that there is a good set of international standard building codes. Even if these do not yet consider the particular case of Sun-trackers, the CPV designer will be able to interpret and adapt them, dimensioning structural ultimate resistance for the recommended values of variable loads, namely wind and (depending on the location) snow and earthquakes. Along with these static loads, dynamic effects should also be taken into account in what respects the resonant frequencies of the structure.

For a flat-plate tracker, acceptance angle is meaningless and thus ultimate limit state dimensioning constitutes only structural specification; stiffness is only an issue if excess bending can somehow stress or damage the PV modules. However, in the case of CPV trackers it can easily happen that the stiffness specifications are more stringent than the ultimate resistance ones, so the fulfilment of the former implies that the latter are overridden. For this reason a CPV tracker, whichever its tracking axes configuration, is frequently heavier than its conventional flat-plate counterpart, and therefore somewhat more expensive. But on the other hand, thanks to the fact that HCPV modules are already more than double the average PV efficiency, tracker costs per unit Wp have the potential to outperform their flat-plate counterparts.

Further exploring the sensitivity of tracker cost to flexure specifications, Fig. 11.27 shows how for a 50 m^2 CPV pedestal tracker designed by Inspira, weight diminishes when maximum allowed flexure under maximum service conditions is increased, apparently following a potential law, to the point that below 0.4° we

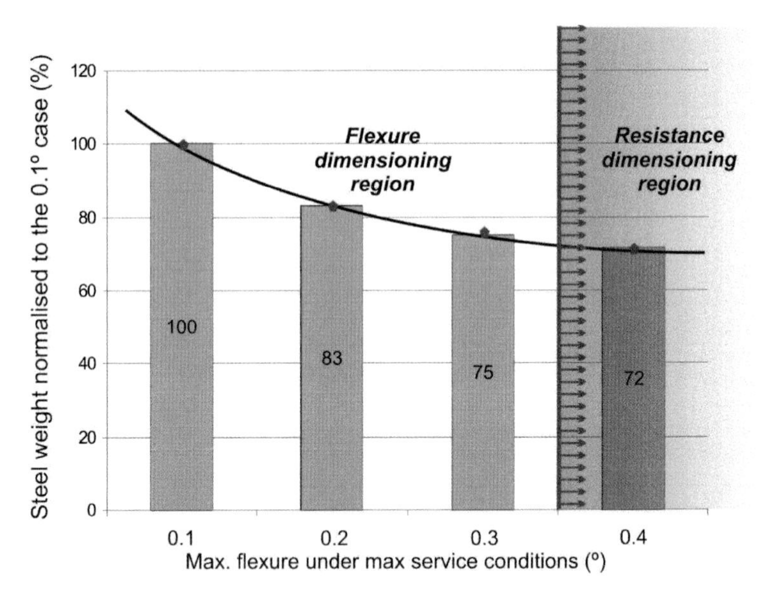

Figure 11.27 Steel weight vs. maximum flexure under maximum service conditions for a $50\,m^2$ pedestal CPV tracker design.

start to have weights which correspond to those of flat-plate trackers, flexure is no longer the dimensioning factor and the ultimate limit state starts to command.

11.6.1 *Sun-tracking accuracy*

Field experience with photovoltaic concentrators consistently indicates closed-loop, sensor-based tracking controllers as not being up to the stringent reliability standards required by low-maintenance plants, which are necessary to sustain the economic feasibility of these technologies.

Open-loop controllers, based on the computation of Sun ephemeris, implemented as digital embedded systems, appear as a reliable and almost maintenance-free alternative, able to attain the increasingly high accuracies demanded by the rising trends in concentration ratio, and offering the possibility of including added value content, such as remote control and monitoring capabilities, at reasonable cost when approaching industrial volumes. However, an open-loop controller, even if operating on the very precise Sun ephemeris equations available to date, is affected, once connected in the field to its concentrator, by many sources of error which can degrade its final tracking accuracy well below its ephemeris' nominal value, even to the point of missing the concentrator's specifications.

Among these error sources the most significant have a deterministic nature and mainly result from self-characterisation inaccuracies, i.e. tolerance deviations appearing in the controller's default assumptions regarding the tracker's installation or manufacturing. Drifts in the internal timing required for the computation of the Sun ephemeris are the other critical error source, which should be restrained. Second-order error sources, such as gravitational bending in wide-aperture trackers, the effect of mismatch in multi-secondary axis trackers, or even ephemeris inaccuracies due to the effect of local atmospheric refraction, might have to be considered as well.

To suppress the tracking errors caused by the above-mentioned sources, feedback has to be integrated in the control strategy. This open-loop core strategy blended with the fed-back closed-loop is sometimes referred to as the hybrid approach, which in the past has been mainly connected with control theory (Luque-Heredia *et al.*, 2005a). Related to this approach but relying on the principles of automated instrumentation calibration, and working in much the same way as the pointing control of big telescopes, Inspira developed the SunDog sun-tracking control unit (STCU) (Luque-Heredia *et al.*, 2005b), whose operation is based on the internal calculation of high-precision solar ephemeris (mean error $\approx 0.01°$). To convert the solar coordinates generated by the ephemeris into tracking axes rotation angles with respect to a specific reference orientation, the SunDog STCU contains a non-linear calibration model characterised by a series of parameters that determine the transformation in a precise manner. Those parameters are related to: (1) the relative position of the tracking axes with respect to the Earth's local surface; (2) the orientation of the axes of rotation references; and (3) the orientation of the pointing vector of the concentrator system with respect to the tracking axes. The parameters of the model are least squares fitted internally by means of global optimisation routines and from precise measurements of the Sun's position, determined by maximising the concentrator's power output. From then on the STCU operates as an open-loop controller, with proven long-term accuracies of better than 0.1° with 97% probability and below 0.05° with 60% probability, as shown in Fig. 11.28. This was measured with Inspira's SunSpear tracking accuracy sensor, which is based on solid-state image sensors (Luque-Heredia *et al.*, 2005c). This tracking control technique was first used, in a different context, by Arno Penzias, the discoverer of cosmic background radiation, for pointing early satellite receivers by fitting an error model with position measurements of a precisely located radio galaxy.

Also of the utmost importance to achieve Sun-tracking accuracy is the positioning resolution, i.e. the capacity to position the tracker at whatever command position is produced by the controller, for example in the case of a SunDog STCU

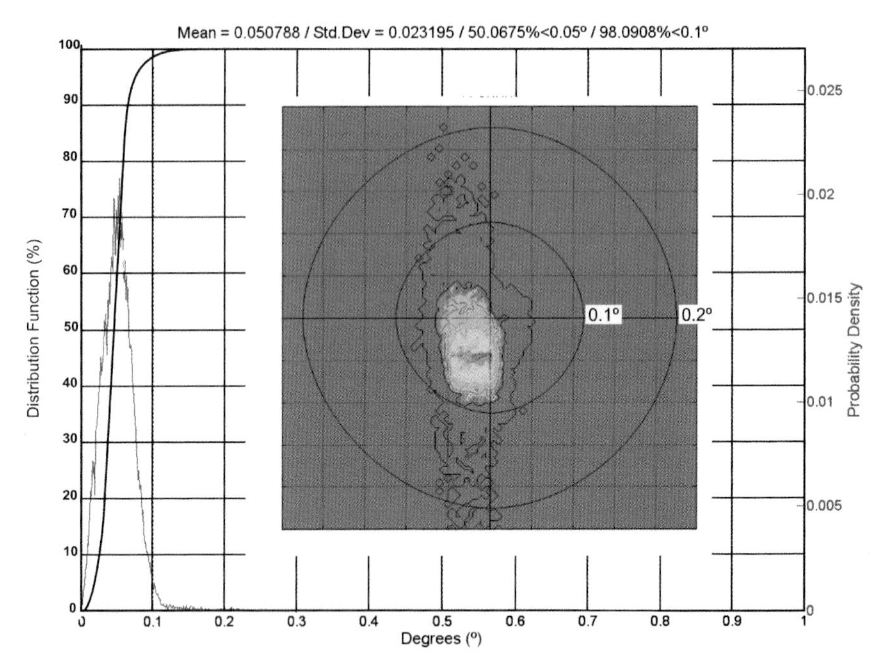

Figure 11.28 Example of daily Sun-tracking error statistics, probability density and distribution function, with the probability density of collimated sun beam over the tracking accuracy sensor surface superimposed, showing the 0.1° and 0.2° tracking error rings.

the position resulting from the Sun ephemeris computation and the later conversion of these Sun coordinates into tracking axes rotation angles by the calibration model. These high resolutions can usually be achieved by a mixture of versatile speed control and low backlash gearings, which is in the case of step-based tracking control schemes will permit a tracking step that is as small as possible, which in the last instance will serve to decrease tracking error variance.

11.7 High-concentration systems

After the overview of the technical elements behind CPV systems, we now continue reviewing the timeline of CPV development from the point we left it in Section 11.2. As we saw there, the origin of concentration as a means to reduce the costs of photovoltaic systems through a reduction of the necessary cell area can be traced back to the late 1970s, when the first 1 kW 40× prototype was produced by Sandia Laboratories in the US. Several other similar early prototypes were developed and tested in France, Italy and Spain, and since then the number of attempts to develop CPV systems has been connected to the pace of development of the photovoltaics industry. Different corporations were created in the United States in the mid-1980s around this concept, out of which, as we will see below, Amonix and Entech Solar are still operating in 2013. Also, some important demonstration projects were built, such as the above-mentioned EUCLIDES plant in Tenerife, which was until 2008 the biggest CPV plant in the world. However, concentration photovoltaic projects were hardly ever developed further than their prototype stage, even when they showed promising results in the laboratory, either because they did not reach a commercialisation stage, or if they did, they could not develop the minimum economies of scale necessary to demonstrate cost reductions.

The situation has changed radically since those early days of CPV. Because of the booming growth experienced in the period 2000–2010, the photovoltaics industry reached sufficient critical mass to boost business ideas based on alternative technologies such as concentration.

With this fertile scenario, the rapid increase of multijunction cell efficiency in recent years has played an important role to recall and reassert the promise of the high cost-reduction potential of concentration systems. Likewise, the fact that theoretically there is still plenty of room for efficiency increase by means of increasing the number of junctions, and also in the medium term through the introduction of other high- efficiency concepts now being researched, has taken concentration systems, formerly based on more or less customised silicon cells conceived to operate in the 20–100× range, to the so-called HCPV range. Here the concentration factors range from 500× to 1000×, which is the level at which these new high-efficiency cells can show a clear economic advantage. Using a simple example, at a concentration ratio of 1000, a triple-junction cell cost of 100,000 \$/m^2 falls to 100 \$/m^2, which at efficiency levels of 33% and normal direct irradiation of 1000 W/m^2, results in cell costs of 0.3 \$/W, as opposed to the 2.3 \$/W cost of conventional PV modules in 2011 (here, it must be noted, including also housing costs).

Concentrated photovoltaics has evolved to become the platform from which to launch high-efficiency cells onto the market, even though this market entry has significantly slowed down due to the drastic price reductions in flat-plate silicon modules that occurred in 2012/13. However, the tendency to achieve higher concentration ratios has been confronted with important technical challenges. As we have seen, the higher the concentration factor, the smaller is the maximum possible acceptance angle of the optics. At a system level, additional acceptance angle losses will also have to be subtracted from the nominal acceptance angles of single concentrator units, these deriving from misalignments in the mounting of such sub-systems in modules and subsequently of these on the tracking systems, and also from structural flexure in the tracking system due to its payload or service wind load. In spite of these difficulties, moving to high concentration factors also yields some advantages. For example, moving to the use of small cells of at most 1 cm^2 size helps to make heat dissipation easier and at some point also facilitates the adoption of the manufacturing processes used by optoelectronic industries that have achieved very high learning rates, i.e. extent of cost reductions as cumulative production increases.

In sum, a quickly growing PV industry, the potential foreseen in high-efficiency cells, and the knowledge gathered from past CPV prototypes and projects, all together favoured the development of concentration photovoltaics until the economic downturn of 2012–2013. And as a clear signal in this respect, more than 30 new CPV start-ups appeared during the first decade of the twenty-first century. In Table 11.4 we give a list of some of these start-ups, choosing mostly those that started some industrial production and have operating installations of a relevant size, or those that represent an example of a novel design.

11.7.1 *CPV's market entry*

As we will see in Section 11.9 when reviewing CPV's cost projections, the cumulative production figures now required to prove CPV's superior cost reduction rates are relatively small compared with the actual volume of the PV market, and CPV is therefore apparently more affordable. However, CPV's traditional chicken-and-egg problem still remains when attempting market entry. This in the sense that customers of conventional PV (flat-plate silicon modules), i.e. PV plant developers, the EPC (engineering, procurement and construction) companies that build these plants, and most importantly, the banks that finance their construction, view CPV as a new and still relatively expensive technology that lacks the necessary track record to enter the mainstream PV market. But on the other hand, CPV needs to enter the market and accumulate production volume, in order to progress along its expectedly fast-learning curve and demonstrate its cost-reduction promise.

One of the landmarks in the history of CPV is ISFOC, the Instituto de Sistemas Fotovoltaicos de Concentración, which was created in the La Mancha town of Puertollano, Spain in 2006 by the Consejería de Educación de Castilla La Mancha following a project developed and managed by IES-UPM. The main target of ISFOC was to provide new CPV companies with an instrument to bridge the gap to the market. Conscious that CPV systems technology was very much ready to enter the market, but that some entity, not subject to strict market rules, had to help as introducer, ISFOC is performing as a customer-like research centre that purchases and operates CPV plants, through international calls for tenders. The idea was to encourage CPV start-ups to get beyond prototypes, and act as suppliers for real plants of significant size, in order to create a track record audited by ISFOC,

Table 11.4 Main CPV start-ups and industrial projects worldwide.*

Company	Type[a]	C_g	kW[b]	Comments
Amonix (USA)	Point, Si, Fresnel (TP-R), 2-axis, Az&El, pedestal	500×	950	Redesign underway; moving to higher concentrations and III–V, MJ cells
Arima Eco (Taiwan)	Point, III–V, Fresnel (TP-R), 2-axis, Az&El, pedestal	476×	400	Subsidiary of Arima electronics company
BSQ Solar (Spain)	Point, III–V, Fresnel (TP-TIR), 2-axis, Az&El, pedestal	820×	148	
Concentrix (Germany)	Point, III–V, Fresnel (–), 2-axis, Az&El, pedestal	530×	1, 622	Acquired by Soitec (France) manufacturer of electronics substrates Redesign underway; introducing secondary
Daido Steel (Japan)	Point, III–V, Domed Fresnel (TP-TIR), 2-axis, Az&El, pedestal	550×	270	Business division within specialty steel manufacturer Daido Steel Redesign underway; moving to higher concentrations
Emcore (USA)	Point, III–V, Fresnel (TP-R), 2-axis, Az&El, pedestal	520×	830	Redesign underway; moving to 1090× concentration and two-axis tilt-roll tracker Joint venture Suncore with Chinese group San'an Optoelectronics
Entech Solar (USA)	Linear, Si, Arched Fresnel (-), 2-axis, Dec&Ha, tilt–roll	20×	640	Redesign underway; moving to two-axis pedestal tracker
Greenvolts (USA)	Point, III–V, Mirror-off axis parabolic, 2-axis, Az&El, turntable	625×	1, 000	Redesign underway; moving to Fresnel + secondary and two-axis tilt-roll tracker Acquired by Swiss group ABB

(Continued)

Table 11.4 (Continued)

Company	Type[a]	C_g	kW[b]	Comments
Guascor Fotón (Spain)	Point, Si, Fresnel (TP-R), 2 axis, Az&El, pedestal	400×	12, 550	Subsidiary of engine manufacturer Guascor Redesign underway; moving to higher concentrations and III–V, MJ cells Renamed as Foton HC Systems
Isofotón (Spain)	Point, III–V, TIR-R (specific secondary), 2 axis, Az&El, pedestal	1000×	400	Business division within flat-plate PV manufacturer Isofotón Redesign underway; moving to Fresnel + secondary and lower concentration
Opel (USA)	Point, III–V, Fresnel (TP-TIR), 2-axis, Az&El, pedestal	500×	470	Sold solar assets to Northern States Metals in Dec. 2012
Semprius (USA)	Point, III–V, Plano convex (glass ball), 2-axis, Az&El, pedestal	1110×	8	Siemens acquires minority share
Sharp (Japan)	Point, III–V, Fresnel (−), 2-axis, Az&El, pedestal	700×	100	
Skyline (USA)	Linear, III–V, Mirrored Trough, 1-axis, horizontal, NS	10×	27	
Silicon CPV (UK)	Point, Si, Parquet of prisms, 2-axis, Az&El, pedestal	120×	75	Also developing a III–V, MJ 960× system
Sol3g (Spain)	Point, III–V, Fresnel (TP-TIR), 2-axis, Az&El, pedestal	476×	1000	Acquired by engineering and construction company Abengoa (Spain) Redesign underway
Solar Systems (Australia)	Point, III–V, Mirrored Parabolic dish, 2-axis, Az&El, pedestal	500×	1, 200	Acquired by engineering company Silex Systems (Australia)

(Continued)

Table 11.4 (Continued)

Company	Type[a]	C_g	kW[b]	Comments
SolFocus (USA)	Point, III–V, Mirror Cassegrain (TP-TIR), 2-axis, Az&El, pedestal	650×	4,010	
Soliant (USA)	Point, III–V, Fresnel (TP-TIR), 2-axis, Dec&Ha, tilt-roll	500×	50	Acquired by Emcore (USA)
Suntrix (China)	Point III–V, Fresnel (–), 2-axis, Az.&El, pedestal	595×	270	
Whitfield (UK)	Point, Si, Fresnel (–), 2-axis, Dec&Ha, tilt-roll	70×	10	Redesign underway; moving to higher concentrations and III–V, MJ cells

[a]Focus geometry (linear, point); cell material (Si, III–V compound); type of primary optics; type of secondary optics (TP: Truncated pyramid, TIR: Total internal reflection, R: Refractive); one- or two-axis tracking; tracking geometry.
[b]Installed capacity in kW.
*This information was correct at the time of compilation (November 2011).

that could convince conventional customers of the excellence of their performance and their reliability. At the same time winning bidders would go through a process that would help them to refine their manufacturing, logistics, field works and plant operation, again with the help of a proactive customer willing to obtain and share the widest possible breadth of knowledge regarding the performance and operation of CPV plants. It is worth noting that, believing in the greater potential of high-concentration photovoltaic systems, ISFOC had as one of the eligibility conditions that bidders' technologies would implement concentration factors of over 250×.

Two calls for tenders were completed by ISFOC, resulting in 2.7 MW of CPV plants. In the first call, the winning contenders were SolFocus (500 kW), Concentrix (500 kW) and Isofotón (400 kW), and these were in fact the very first large-scale installations of these three CPV companies, which were completed in 2008 (see Fig. 11.29). Also awarded with contracts in ISFOC's second call for tenders were Arima Eco (300 kW), Emcore (300 kW), Renovalia (300 kW) and Sol3g (400 kW). Analyses of the first operating periods for these CPV plants were released by ISFOC, focusing on the performance and operations and maintenance (O&M) aspects, and also providing first recommendations regarding plant design and sizing

Figure 11.29 One of the ISFOC CPV plants at Puertollano, in which three different technologies can be seen. In the foreground is Soitec (formerly Concentrix), in the background SolFocus and on the right is Isofotón. Courtesy of ISFOC.

(Rubio *et al.*, 2010a; Gil *et al.*, 2010; Alamillo *et al.*, 2010). Some of these results will be reviewed in the next section.

One of CPV's problems has been the wide variety of different designs proposed, as usually happens with any nascent industry (recall the early days of aviation, for example). This lack of standardisation hindered the development of a set of specific CPV component providers (e.g. trackers, optics, etc.) that could attain cost reductions through aggregated market volume. However, in recent times, as can be inferred from the different system descriptions in Table 11.4, CPV system designs are converging towards a set of common features, firstly confirming the general trend towards higher concentration ratios, which is in most cases achieved through the use of Fresnel primary optics plus some kind of secondary, frequently a simple reflective truncated pyramid. Regarding the tracker, the two-axis azimuth-elevation pedestal configuration is the most-chosen option, unless when targeting the rooftop market, in which case the tilt-roll tracker has the advantage of a lower profile. Without debating the pros and cons of this coalescing standardisation, it did indeed help the appearance of dedicated suppliers to the CPV industry, whether start-ups or already-established companies creating product lines specifically to supply CPV manufacturers. Also a symptom of this standardisation is that design debates have become more specific. For example, given that a Fresnel primary is usually chosen, the question now is should it be made in PPMA acrylic plastic or should it be of the so-called silicone-on-glass (SOG) type, where the Fresnel profile is stamped in a layer of silicone behind a protecting glass.

11.7.2 *Crossing the valley of death*

Returning to the manufacturers' list in Table 11.4, it is interesting to focus our attention on the first two companies in terms of cumulative installations to date, and briefly explain some reasons behind their success. Way ahead of all the rest of CPV manufacturers is Guascor Fotón, a spin-off of Spain's manufacturer of diesel and gas engines Guascor. Guascor Fotón arose from a technology transfer agreement from US veteran CPV manufacturer Amonix. This led to a technology industrialisation project by Guascor Fotón, which ranged from all the tracker and module manufacturing and assembly down to the production of its high-efficiency silicon cells. The first Guascor Fotón 25 kW CPV prototype system was installed in the test field of IES-UPM, replacing its long-serving ancestor the 'Ramón Areces' 1 kW system already mentioned. This project was well timed in that Guascor Fotón had completed its manufacturing lines and was ready to start marketing the product during the rise of the Spanish PV market, thanks to generous FITs (feed-in tariffs). Guascor Fotón installed 11.5 MW of its CPV systems, all of them in Spain, between 2007 and 2008, which in 2011 still represented more than 50% of the global cumulative CPV power installed (see Fig. 11.30). A well-timed and structured project, a relatively well-proven technology, that of Amonix, even if not the newest of its time, and most importantly a strongly incentivised market in a high DNI country, which spurred such a high demand spike of PV as to admit also alternatives to conventional flat plate as CPV systems — all these contributed to the success of Guascor Fotón. The Spanish PV market boom (roughly 4 GW installed in two years) passed, but Guascor Fotón remains, now renamed Foton FC and working on a new generation of its product, in which silicon cells have been replaced by multijunction cells and concentration has been driven upwards. A 31.5% module efficiency had been achieved by the end of 2011.

The second in the list of top CPV installers in Table 11.4 is the Silicon Valley start-up SolFocus. If the Guascor success was mostly market-driven thanks to the Spanish national FIT — once successful technology transfer and industrialisation had been achieved — SolFocus is a good example of the thrust of venture capital (VC)-backed tech companies in one of the most dynamic entrepreneurial regions in the world: Silicon Valley. A typical US garage start-up founded in 2005 in Saratoga, California, by Gary Conley and Steve Horne, SolFocus was set to develop a CPV system built around a concentrating optics design developed in cooperation with Roland Winston, one of the inventors of non-imaging optics. By 2006, SolFocus had received its first \$3 million of VC seed funding and moved to the Xerox

Figure 11.30 Built in 2008 by Spanish developer Parques Solares de Navarra, the Villafranca plant with 10 MW of Guascor Fotón CPV systems was the largest CPV installation in the world until 2012, when the 30 MW Alamosa Solar Plant opened in Colorado. In the upper-left corner of the picture, the systems in a darker tone are 2 MW of flat PV mounted on two-axis trackers. Courtesy of Parques Solares de Navarra, SL.

Palo Alto Research Centre (PARC), a high-tech incubator with a distinguished reputation for its contributions to information technology and hardware systems. In just one year SolFocus left the PARC incubator and moved to headquarters in Mountain View and were able to raise $150 million, mostly coming from venture capital firm New Enterprise Associates (NEA), which will appoint the above-mentioned Nobel laureate Penzias as chairman of SolFocus' technical advisory panel. After having built a few prototypes in the US, it will be through ISFOC that SolFocus will build their first plant of a significant size, 500 kW distributed in two locations of La Mancha, in this way quickly moving from product development to manufacturing. United States VC funding, able to rapidly pour large amounts of money into technology start-ups, demands in exchange a relatively high and prompt return on its investment; thus SolFocus had to hurry up the closing of its product development stage, which in other CPV companies had taken several years, in order to get to the market as soon as possible. For example, conscious that tracking technology was a critical element in their CPV system, instead of spending time in its internal development, SolFocus opted to acquire the Spanish company Inspira, founded by one of the authors (Luque-Heredia), which had been developing

Figure 11.31 Victor Valley College 1 MW plant built by Solfocus in 2010 was the first one set by this company in the United States. Courtesy of Solfocus, Inc.

tracker technology specifically designed for CPV's accuracy specifications since its participation in the EUCLIDES project.

In 2011 (the time of writing), SolFocus continues to raise VC funding and has been able to build about 4 MW of CPV power plants (see Fig. 11.31), both in the US and Europe (Spain, Italy and Greece), in very competitive PV markets against a background of conventional PV achieving fast and pronounced cost reductions, and during the world's recent economic downturn, which has forced the reduction of PV incentives.[1]

11.8 Rating and performance

The first issue that appears when considering the performance of today's CPV systems is their power rating. Several approaches have been proposed, regarding both the standard test conditions (STC) used and the measurement procedures.

The STC under which flat PV modules are rated are well known: 1000 W m^{-2}, 25°C cell temperature and AM 1.5 G spectrum. These are not related to real

[1] Sadly, SolFocus later fell victim to the fall in price of flat-panel silicon PV and in September 2013 was reportedly in negotiations to sell its assets and intellectual property (IP).

operation conditions — they are overly optimistic with respect to what is normally found in real operation — but there is by now sufficient track record for PV plant planners to know what the energy output will be for an installation rated under standard conditions in a location with a known irradiance and temperature. There are also well-accepted formulae that have been developed to transform electrical output readings — current, voltage and power — from a PV module, made at whatever irradiance and cell temperature conditions, into those that would correspond to STC.

A set of STC must also be established to rate CPV modules or systems. Having those for flat PV modules as the market's well-established reference, it seemed reasonable to lower the irradiance value somewhat, given that CPV modules do not capture diffuse irradiance as tracking flat-plate PV modules would do. However, ultimately it is the energy production which is of most interest, and in order to remain close to the way things are in the flat-plate PV market, it seems a good approach to determine CPV's STC as those that will result in a similar energy output to that of an equivalent non-concentrating PV system, i.e. a two-axis tracking PV system. This was conceptually the approach taken by Luque in the specifications for ISFOC's first call for tenders in 2007, where he estimated the STC to be 850 W m^{-2} and 60°C cell temperature. Obviously this estimate is site-specific, mostly through the available DNI irradiance and the ambient temperature.

This approach was further analysed by ISFOC (Rubio *et al.*, 2010b), who showed that at Puertollano — the location of most of ISFOC's installed CPV power — STC defined as 850 W m^{-2} DNI and 60°C cell temperature resulted in yearly cumulative production (kWh/kW) of CPV which was about 5% higher than that of tracked PV in the same location. On the other hand, were we to choose STC similar to those of flat-plate PV, i.e. 1000 W m^{-2} DNI and again 60°C cell temperature, then the yearly cumulative CPV production would be almost 15% below that of tracked PV. The best match in ISFOC's analysis, in-between the cumulative productions of CPV and tracked PV, when varying the STC DNI, is obtained with 900 W m^{-2} DNI and 60°C cell temperature. Also a good match, better than 5%, occurs when varying cell temperature instead, for example using 850 W m^{-2} and 25°C cell temperature. While the rationale behind this analysis and approach to a definition of the CPV rating STC puts tracked PV and CPV on the same level in energy production terms, some CPV manufacturers have justifiably opted to raise DNI and lower cell temperature under which they rate their products, while there is still no generally accepted STC, in order to somewhat increase the nominal power of their products and therefore lower their €/W, which as has been said is still the market's main metric. The obvious trade-off is that going to more favourable STC would lower their kWh/kWp, eventually below those of tracked PV systems, but initial sales experiences showed this was less important in a marketing strategy. The issue of the appropriate STC for CPV has been debated by the International Electrotechnical Commission (IEC) when working on its IEC 62670 standard. Although a good case was made by the IEC TC82 Working Group 7 for 900 W m^{-2} and 25°C cell temperature to be the standard, the IEC has recently ruled that the STC for CPV will remain the same as those for flat-plate PV as regards irradiance and cell temperature (1000 W m^{-2} and 25°C) but that the standard solar spectrum will be AM 1.5 DNI rather than AM 1.5 G (Solar Industry, 2013).

11.8.1 *Rating procedures and instruments*

Moving now to measurement procedures, it is obviously not practical to rely on outdoor measurements when rating the module production of a manufacturing line, and since the early days of the PV industry, solar simulators have been used to simulate natural sunlight for repeatable and accurate indoor rating of PV modules. The same is needed for CPV modules and a specific solar simulator has been developed at IES-UPM for this technology by Domínguez *et al.* (2008). This simulator concept is gradually being adopted by many of today's CPV manufacturers. The main difference between a conventional solar simulator for PV modules and that designed for CPV is that the latter should produce collimated sunlight with a subtended angle as close as possible to that of the Sun (±0.275°), in order to reproduce DNI. This has been achieved in the IES-UPM design by placing a Xenon flash light at the focus of a large-area parabolic mirror (see Fig. 11.32).

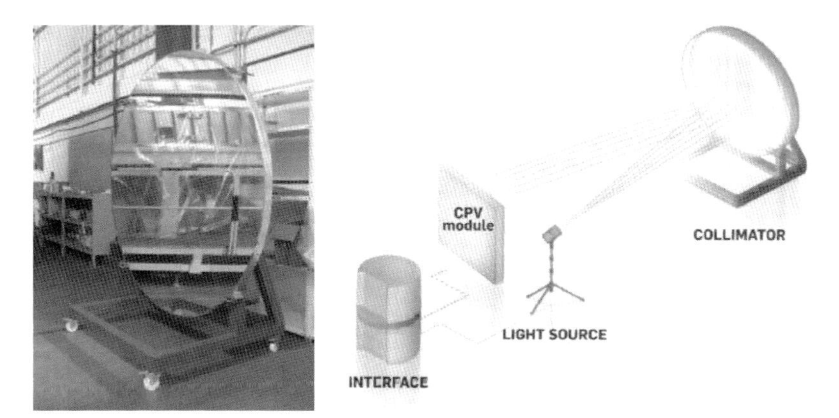

Figure 11.32 On the left is a drawing of the complete solar simulator developed by IES-UPM for the rating of CPV modules, providing an extended source of collimated light (with a subtended angle equivalent to the Sun's), through the reflection of the light of a Xenon flash lamp in a 2 m accurate parabolic mirror, shown in the picture on the right. Courtesy of Soldaduras Avanzadas, SL.

As is well known, one of the properties of a parabola is that every ray coming from its focus is reflected back parallel to its axis, so that by correctly sizing the diameter of the light source located at this focal point, the light reflected by the mirror will have a uniform angular spread. For example, in the currently available setup, a 2 m diameter parabolic mirror with a 6 m focal distance, at which a 6 cm diameter bulb is placed, will ideally provide a $\pm 0.275°$ spread beam in the area covered by the parabolic mirror. The simulator operates as a multi-flash testing system. The Xenon lamp is triggered to illuminate the CPV module that is biased at a different voltage at each flash pulse, recording different current–voltage pairs during the decay of the flash pulse. Each current measurement within a single flash corresponds to a certain irradiance level, which can be determined through a calibrated cell, so that at the end of the series of flashes, a family of I–V curves is obtained at varying concentration factors. A computer controls the whole process by triggering the lamp, setting the biasing voltage for each flash, and recording the current–voltage pairs.

One of the main difficulties in developing this type of solar simulator was to find a 2 m diameter parabolic dish, which due to the long focal distance required demanded extremely accurate optical quality. This could only be found at sky-high prices in the astronomical telescope industry. A parabolic profile was therefore machined out of a bulk aluminium piece, and its mirroring provided by an aluminium-based reflective coating. Errors in the profile are in the 10 μm range, which is higher than the tolerance permitted in astronomical telescopes but good enough for the CPV application. Scattering due to the mirror surface finishing has the effect of widening the angular spread of the collimated beam up to $\pm 0.43°$, i.e. the angular size of the Sun plus some circumsolar radiation, which is still good enough for most CPV modules.

The spectrum of the flash lamp varies over time during its discharge. This can be monitored by means of isotypes, which are triple-junction solar cells with a single cell electrically contacted. The result is that only one of the curves in the family described above has the desired spectrum. With faster electronics it is possible to draw the whole I–V curve in a single flash. Additional software can provide spectrum-corrected I–V curves at different irradiances.

We will also describe a procedure devised for the power rating of complete CPV plants, which is the one used by ISFOC in the acceptance tests it carries out on its contractors' plants and which constitute a contractual payment milestone (ISFOC, 2006). First, the plant's DC power is measured at the ISFOC-specified STC (850 W m^{-2}, 60°C cell temperature) by tracing a set of I–V curves, using a capacitive electronic load. This load essentially consists of a discharged capacitor that is connected to the CPV array, so that it will start to charge itself as it receives the array's current, thus gradually increasing its voltage and decreasing the current, sweeping in this way the I–V curve and recording a set of current and voltage pairs. This type of electronic load was specifically developed at IES-UPM for the measurement of high-power PV systems. Full 100 kW plant units can be measured with it, in a rather inexpensive way (Antón, 2004).

At least 15 I–V curves are required by the ISFOC procedure: this repetition is very important because the tracker is always slightly out of aim and the I–V curve may vary by about 5% depending on the moment of the measurement. These must be measured with DNIs over 700 W m^{-2} and wind speed values below 12 km/h. Before and after tracing each of these I–V curves, the temperature in the back plate of a few modules in the plant is measured and the average is recorded, together with the DNI. Each of these I–V curves will be translated from its registered measurement conditions of average back-plate temperature and DNI to STC, so that the STC DC power rating of the plant, P_{DC}, is the average of the maximum power values of all I–V translated curves. The translation equations used by the ISFOC procedure first calculate the cell temperature through the average temperature measured in the module's back plates. A thermal resistance parameter ρ_T is introduced, which is defined as dependent on the direct irradiance received by these modules:

$$\rho_T = \frac{T_{\mathrm{cell}} - T_{\mathrm{backplate}}}{B} \tag{11.25}$$

where T_{cell} and $T_{\mathrm{backplate}}$ are the cell and back-plate temperatures respectively, and B is the measured DNI. Translation of current–voltage pairs from measurement conditions to STC is obtained through

$$\begin{aligned} I_{\mathrm{STC}} &= I\frac{B_{\mathrm{STC}}}{B} \\ V_{\mathrm{STC}} &= V + N\frac{0.0257(T_{\mathrm{STC}} - T_{\mathrm{cell}})}{297}\ln\left[\frac{(I_{L1}-I)(I_{L2}-I)(I_{L3}-I)}{I_{L1}I_{L2}I_{L3}}\right] \\ &\quad + [N(E_{g1}+E_{g2}+E_{g3}) - V_{OC}]\left(1 - \frac{T_{\mathrm{STC}}}{T_{\mathrm{cell}}}\right) \end{aligned} \tag{11.26}$$

where I, V are respectively the measured current and voltage, I_{STC}, V_{STC} the current and voltage measured at STC, B_{STC}, T_{STC} the values for STC DNI and cell temperature (all temperatures in Kelvin), and N is the number of array cells connected in series. E_{g1}, E_{g2}, E_{g3} are the energy bandgaps, in eV, of each three junctions in the cell, and I_{L1}, I_{L2}, I_{L3} the short-circuit currents of each three cell junctions under the measurement conditions. When going through this rating procedure, ISFOC contractors could themselves provide the currents of the sub-cells (isotypes) measured at AM 1.5 D spectrum and duly validated, or by agreement with ISFOC at any other common spectrum, and it was allowed that the relative proportions among these currents would be maintained at any measurement conditions. If the contractor cannot provide these data, then $I_{sc} = I_{L1} = I_{L2} = I_{L3}/3$ is assumed, and also unless the contractor justifies other values, the bandgap energies are taken to be $E_{g1} = 1.85\,\text{eV}$, $E_{g2} = 1.42\,\text{eV}$ and $E_{g3} = 0.66\,\text{eV}$. These translation equations, proposed by Luque, are based on the simplification that all the CPV modules in the plant ultimately behave as its longest string of N cells that are all considered identical.

Once the DC rating is completed, the AC (alternating current) rating of the CPV system or plant can be obtained using its energy production, recorded by the counter at the grid connection point, for a period of at least seven days. During this period continuous DNI readings are stored together with the back-plate or heat-sink temperature of a randomly selected set of modules in the plant. Within this analysis time span, those periods in which the DNI goes below $600\,\text{W}\,\text{m}^{-2}$ will not be considered. For this analysis period, the initially obtained DC power rating of the plant will be translated to real conditions every five minutes using the DNI and the averaged module temperature readings and the translation equations, Eq. (11.26). Integration of these instantaneous power estimates will provide the expected lossless energy output at the inverter's input, so that we can then define F as the ratio of the energy output measured by the counter to the estimated DC energy output, and obtain the AC power rating, P_{AC}, from the equation

$$P_{\text{AC}} = P_{\text{DC}}\eta_{\text{inv}}F \tag{11.27}$$

where η_{inv} is the inverter efficiency.

Other technical specifications have been proposed to measure the power of a CPV plant, such as the American Standard Test Method for 'Rating Electrical Performance of Concentrator Terrestrial Photovoltaic Modules and Systems under Natural Sunlight', ASTM E 2527-06.

11.8.2 *Field performance of CPV plants*

Turning now to the performance of CPV systems, we will again resort to the data released by ISFOC regarding the operation of its plants (Rubio *et al.*, 2010b). Considering energy production, after one and a half years of data recording, the output of the CPV plants in Puertollano is very similar to that of tracked PV systems, which are also installed at the Puertollano site and provide real production data for comparison (see Fig. 11.33). This was to be expected given that the CPV systems are rated at $850\,\text{W}\,\text{m}^{-2}$ DNI and 25°C cell temperature, which as explained

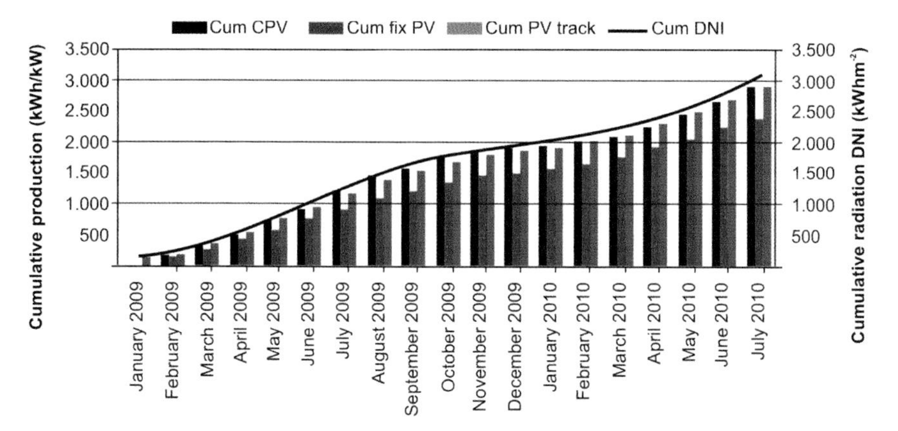

Figure 11.33 Comparison between cumulative energy of CPV, and tracked and fixed-mount conventional (non-concentrating) PV systems monitored by ISFOC for 1.5 years. In this case the CPV rating, 850 W m^{-2} and 25°C cell temperature, produces an energy output very similar to that of tracked PV and as can also be seen the cumulative DNI evolves very closely to the cumulative production, implying a performance ratio value for the case of CPV very close to unity. Reprinted with permission from Rubio *et al.* (2010b), © AIP.

above was the STC best option found by ISFOC. It is worth noting that this is the overall production of all the CPV plants at Puertollano, regardless of differences in their technologies, and without filtering any outage periods due to research and development (R&D) works, maintenance, etc.

It is also interesting to check energy production on a monthly basis. We can see in Fig. 11.34 that energy production in Puertollano during the summer months with the highest DNI is up to 25% better than that of tracked PV and as much as 65% better than fixed-mount PV. The good performance of CPV plants during the high DNI months is due not only to intrinsic module efficiency increase, but also to the more fluent operation of the inverters, which under clear skies do not suffer from intermittent connections and disconnections due to passing clouds. This can be seen more clearly in Fig. 11.35, where efficiency of CPV and tracked PV plants are plotted as a function of the DNI: while CPV's performance increases linearly with DNI, silicon PV's efficiency linearly worsens due to its higher temperature coefficient.

The inverter can also be a source of efficiency losses in the high DNI months if it is undersized, as can be seen in ISFOC's analysis of one of its concentrator systems that had a 6 kW inverter per tracker, which was found to be undersized during the summer, thus capping system power and therefore efficiency. Thinking engineering-wise this undersizing should be avoided; however, the sizing of the

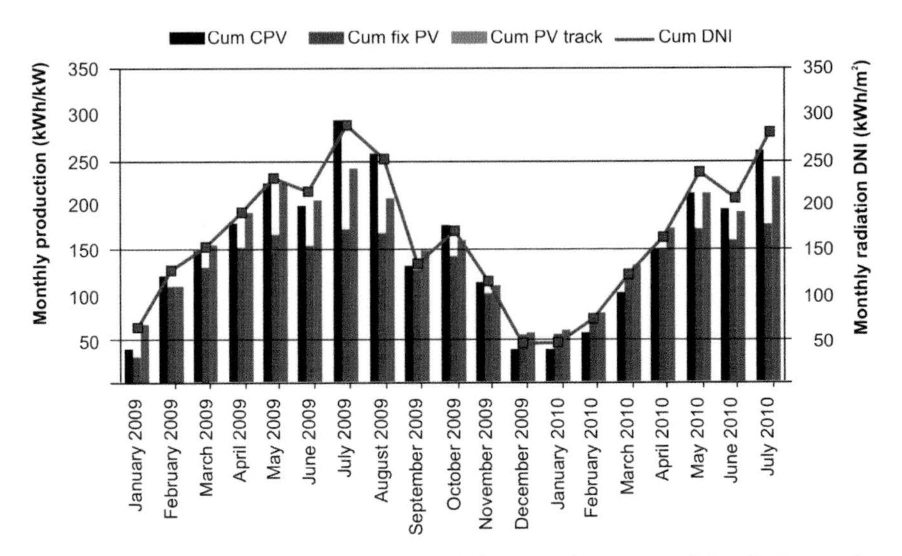

Figure 11.34 Monthly cumulative energy of the same three types of installations as in Fig. 11.33, showing how CPV outperforms tracked PV in the summer months. Reprinted with permission from Rubio *et al.* (2010b), © AIP.

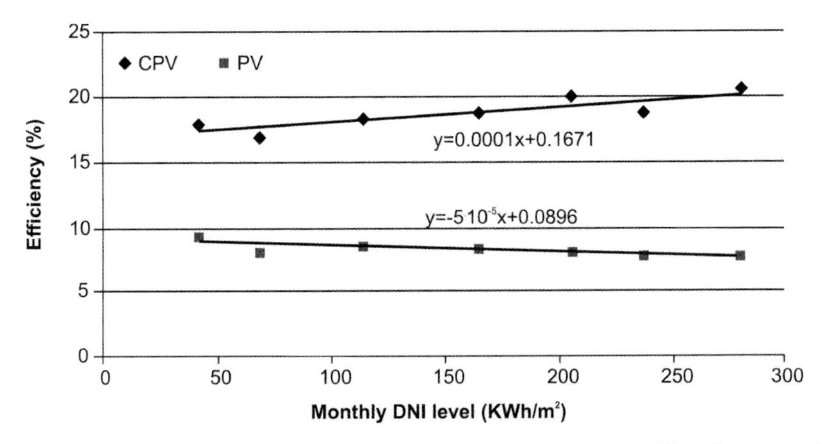

Figure 11.35 Monthly cumulative energy of the same three types of installations as in Fig. 11.34, showing how CPV outperforms tracked PV in the summer months. Reprinted with permission from Rubio *et al.* (2010b), © AIP.

inverter can get to be an important and rather subtle issue in a solar developer's financial spreadsheet, as certain market situations have proved to favour the so-called repowering of PV plants, i.e. inverters being undersized by as much as 30% (Franco de Saravia and Rossel, 2010). In CPV it has been a common practice from the beginning to install one inverter per concentrator system of power in the 5–10 kW range. However, there is no special advantage in doing this (Alamillo *et al.*, 2010) and it is a fact that centralised inverters — i.e. a single inverter servicing several concentrator systems of up to 500 kW — are as much as 35% cheaper than the small 5 kW ones (Knoll, 2010). Some inverter manufacturers have reported that power vs. operating voltage of concentrator fields, as these grow bigger, present several maxima, thus requiring MPPT (maximum power point tracking) algorithms that seek the global maximum. Usual P&O (peek & observe) algorithms will not then be good enough to lock the global maximum, and voltage sweeps should be programmed periodically.

Also among the ISFOC's first reports on its CPV plants were those dealing with reliability issues (Gil *et al.*, 2010). Classification and statistics of defects, detected during plant installation, prior to its commissioning and acceptance by ISFOC, were undertaken for 2 MW of total installed power, distributed in 20 plants of 100 kW, comprising five different CPV technologies. The mechanical tracker was reported as the most flaw-prone during commissioning, as can be seen in Table 11.5.

Note that the mechanical tracker is considered separately from the tracking controller. As regards the mechanical tracker, ISFOC's report also distinguishes between the tracker's steel structure, the tracking drive, and the foundation, finding that 76% of mechanical tracker problems correspond to the structure and 23% to the drive. Problems detected in the tracker structure are mainly due to faulty transportation, which has either bent the steel profiles or damaged their coating, with rust points appearing. Concentrated photovoltaics tracker transportation to the field and assembly and leveling of modules on site are possibly areas with much

Table 11.5 Distribution of installation problems at ISFOC.

CPV plant element	No. of reported flaws	(%)
Mechanical Tracker	905	36.1%
AC protections	530	21.1%
Wiring	524	20.9%
DC protections	253	10.1%
Earth connection	171	6.8%
Controller	90	3.6%
Modules	26	1.0%
Inverters	8	0.3%
Total	2507	100%

room for improvement. Regarding the drive problems, most of these seemed to be localised in the elevation axis, usually addressed by all tracker designs using power jacks. Regarding the second most important source of problems, in the wiring or electrical protection, these were usually related to defective layout and overlooking electrical safety codes.

In 2011, ISFOC reported in respect of degradation and O&M issues 2.5 years after the installation of its oldest plants (Rubio *et al.*, 2011). After a threefold analysis checking for efficiency reductions, they found no evidence of noticeable degradation having happened during this time.

To end, we have referred to the work being done towards the CPV rating norm within IEC, and it must also be mentioned that the first CPV-specific standard already approved and published is IEC 62108 (IEC, 2007). Essentially this is a set of accelerated aging tests devised to detect long-term reliability problems in the CPV modules (Muñoz *et al.*, 2010).

11.9 Cost considerations

Let us now talk about costs. In this respect even if CPV has got beyond its initial prototyping stage and an early industry has definitely appeared, its costs, and even its prices, are yet not well known, and manufacturing companies do not readily publish them except in their commercial quotations. In general, for all CPV manufacturers it is no big secret that their costs continue to be to a greater or lesser extent above the price of equivalent tracked flat-plate PV. This cannot be surprising given that flat-plate PV is well ahead of CPV in its learning curve, with Chinese flat-plate PV manufacturers having achieved yearly manufacturing volumes of several GW.

Probably more relevant than knowing the spread in the costs of the current early stage CPV industry is to review some of the published cost projections. Cost analysis and prognosis is usually a difficult task, highly sensitive to the initial assumptions. However, reduction of costs is frequently one of the main targets behind R&D projects, and cost projections produced in these, without pretending to attain very high accuracies, do serve as tools to compare different technology alternatives. Curiously, and possibly a symptom of the coming of age of CPV, the publication of cost projections, traditionally made in the past by academic researchers involved in CPV development, is gradually being handed over to the increasing number of PV industry research firms; in fact, CPV has already given birth to specific publications of this kind such as the British *CPV Today*. However, the cost analysis published by one of the authors a decade ago (Yamaguchi and Luque, 1999) and summarised in Table 11.6, still seems to retain useful information, and the way in

Table 11.6 HCPV system cost forecast by Yamaguchi and Luque (1999), reprinted with permission, © IEEE.

	Short-term		**Long-term**	
	1J[a]	**2J**	**No learning**	**Learning**
Substrate wafer ($ per cm^2wafer area)	8.50	8.50	8.50	2.37
Cells ($/$cm^{-2}$cell area)	13.4	15.85	15.85	4.43
Cells ($/$m^{-2}$module area)	134	159	159	44
Optics, heat sink & assembling ($/$m^{-2}$)	131	131	131	69
Module ($/$m^{-2}$)	265	290	290	113
Cell efficiency (%)	23.1	37	45	45
Module efficiency (%)	19.0	30.5	37.1	37.1
Module ($/W)	1.39	0.95	0.78	0.31
Area-related BOS ($ per m^2 aperture area)	114	114	114	60
Power-related BOS ($/W)	0.22	0.22	0.22	0.12
Plant price ($ per m^2 aperture area)	526	589	607	271
Madrid DNI (kWh $m^{-2}year^{-1}$)	1826	1826	1826	1826
Performance ratio*	0.606	0.606	0.606	0.606
Electricity costs in Madrid ($/kWh)	0.186	0.130	0.110	0.050
Electricity costs in EGL[b] ($/kWh)	0.131	0.091	0.077	0.035

[a] 1J: single-junction cell; 2J: double-junction cell.

which it was derived is clearer than that of present-day forecasts by market research firms, having at least a didactic value.

The cost analysis was produced for 1000× concentration modules that used III–V cells of 1 mm^2 surface area and integrated different, highly compact optical designs produced with the SMS method. This CPV technology was jointly developed by IES-UPM and Isofotón (Algora *et al.*, 2000). The module design followed the so-called 'LED-like approach' in that the assembly processes proposed resembled those in the optoelectronics industry (Algora *et al.*, 2006). The levelised cost of electricity (LCOE) for an amortisation period of 30 years was calculated for the average DNI of Madrid (1,826 kWh $m^{-2}yr^{-1}$) and also for that of an extremely good DNI location (EGL in Table 11.6 above) with 2,600 kWh m^{-2} yr^{-1}.

Single- and double-junction cells were the state of the art at the time (first and second column in Table 11.6, with header 1J and 2J) and the authors, assuming average prices in the wholesale electricity markets of 0.05 $/kWh, and customer electricity prices of 0.1 $/kWh, concluded that CPV electricity could become cost-effective for the customer at EGL locations. Overall system prices, including CPV module and area-related BOS (balance of system) (which were calculated based on the EUCLIDES plant data) envisaged for the 1J and 2J cell CPV systems, were respectively 1.62 \$/W and 1.17 \$/W and were calculated for a cumulative production of 100 MW, with a 25% mark-up applied on top to account for installation costs and commercial profit margin. The learning rate, LR, of a certain product or industry is the cost reduction achieved every time the cumulative production doubles (Nemet, 2005), and can be calculated from

$$C_t = C_0 \left(\frac{q_t}{q_0} \right)^{-b} \quad \text{where LR} = (1 - 2^{-b}) \tag{11.28}$$

with C_0 being the known cost of a product at some past time and C_t the projected cost at some future time, q_0 and q_t the cumulative productions achieved at these two times, and b the so-called learning coefficient; the value $b = 0.175$, appropriate for the PV industry of the time, was assumed.

The short-term projections when extended into the long term, supposing modules with four-junction tandem cells (third and fourth columns in Table 11.6) for which module efficiencies of 37% could be expected, are so reasonably given that present-day three-junction cells are now achieving module efficiencies in the 32–35% range. One of the authors of this chapter has written a paper in which his belief that 50% cells will eventually be achieved is substantiated (Luque, 2011). In the long-term case, Table 11.6 shows two options (third and fourth columns), firstly with no learning, i.e. no consideration of cost reduction due to production scaling, and secondly integrating the cost reduction achieved when arriving at a cumulative production of 1 GW with a 32% learning rate, which is that exhibited by electronics and optoelectronics. A higher learning rate than that achieved by conventional PV at that point, was used for this long-term estimate, considering on the one hand that CPV is much less mature than PV and therefore could achieve faster cost reductions in its initial industrialisation, and taking the optoelectronics reference as that closer to the LED-like design approach to HCPV modules. This last long-term forecast, assuming learning, is capable of matching wholesale electricity prices after a relatively low cumulative volume by today's standards.

An important positive surprise when looking at this table is the value assumed for the performance ratio (PR). This is defined as the ratio between the energy produced and the rated power installed times the 'equivalent hours' of operation of the modules. The latter are the energy falling into a unit module surface divided by the rating solar irradiance for the modules. In the easy case that the rating irradiance in the modules is 1 kW m^{-2} the equivalent hours are exactly the kWh of solar energy incident on 1 m^2 of concentrator aperture. In Table 11.6 this rating is used and the PR value is very low (0.606) as deduced from that measured in the EUCLIDES prototype. However, the data registered at ISFOC for three-junction solar cells are much higher, 0.82 vs. 0.72 for the silicon flat module plants in the area. This better behaviour is due to the smaller temperature effect in the three-junction solar cells as compared with the silicon cells and this also explains the low value (0.606) for concentrated silicon systems. Of course, the silicon tracking system collects global normal irradiance (GNI) while the CPV system collects DNI, which in this site is 76% of the former. That is why the concentrator system produced only 87% of the flat-panel's output if the rating is the same for both (1 kW m^{-2} and 25°C) but the ratio is not that of DNI/GNI but much better. With the measured PR (0.82) instead of the one derived from Si cells experiments (0.606) the cost of CPV electricity is substantially reduced.

11.9.1 *The value of the concentration factor*

Another interesting cost analysis exercise was undertaken by Algora (2007), in which the impact of the concentration factor and cell efficiency on CPV system costs was studied. The assumptions were very much the same of those of Yamaguchi and Luque above, as this study was also undertaken with the cooperation of IES-UPM and Isofotón in connection with the development of a 1000× concentration technology. Figure 11.36 shows the complete CPV system price at plant level as a function of cell efficiency, with the concentration factor as a variable parameter. The results of this analysis are presented for a cumulative production of 10 MW, and again estimating cost reduction after cumulative production of 1 GW with the same 32% learning rate as above. It can be seen how

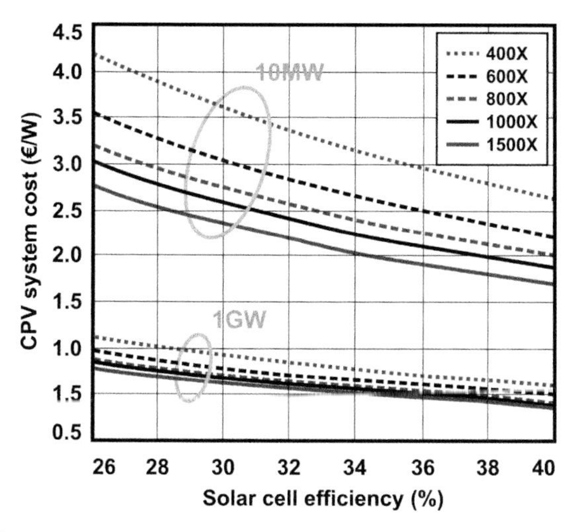

Figure 11.36 CPV system cost estimated by Algora (2007) as a function of the concentration factor and the cell efficiency for two different scenarios of cumulative production.

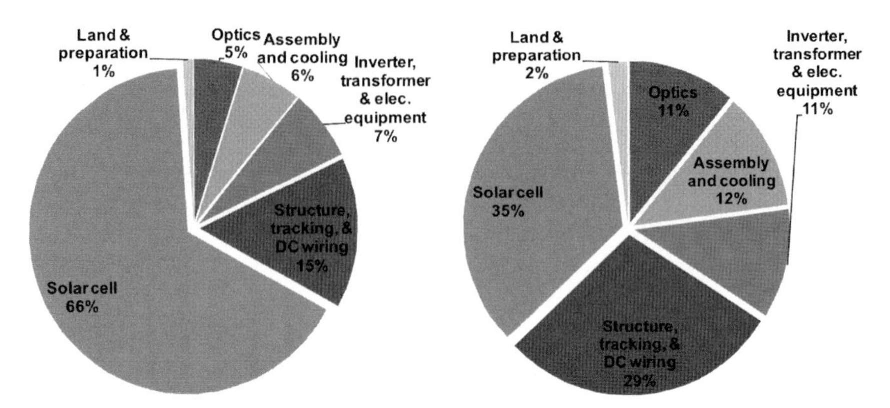

Figure 11.37 Breakdown of costs of a triple-junction concentrator panel after 10 MW of learning curve. Left: for 40% cells at 250 Suns, leading to an installed cost of 3.8 €/W. Right: for 33% cells at 1000 Suns, leading to an installed cost of 2.4 €/W. Reprinted with permission from Algora *et al.* (2008).

for example the estimated final price for an installation operating at 400× is about 1 €/W higher than if the installation operates at 100 Suns, almost independent of the cell efficiency after 10 MW of cumulative production.

Also in this analysis, to further stress the cost-reduction leverage of increasing the concentration factor, a system cost breakdown is provided for two different cases. The pie chart on the right in Fig. 11.37 represents the contribution of the different CPV plant cost elements, when using 40% three-junction cells operating at 250× (after 10 MW of cumulative production), resulting in a final cost of 3.8 €/W at plant level, mostly due to the big impact (66%) of the expensive solar cells, while on the left cheaper 33% two-junction solar cells operating at 1000× achieve, also after 10 MW of cumulative production, a much lower CPV plant cost of 2.4 €/W. In this way, Algora (2008) shows the importance of concentration factor as a CPV system cost reduction factor, in particular in the starting phase when mass production costs have not yet been achieved.

11.9.2 *Metrics and cost breakdown*

The two most common metrics that help to assess the financial feasibility of a PV project, which we have already seen used in the previous cost analyses, are on the one hand the installed cost in €/W and on the other the LCOE in €/kWh. The €/W is a relatively simple metric that is based solely on the production and installation costs of the system, divided by its power rated under some consensus irradiance and temperature conditions, usually STC. This is the main metric used by the PV market to date, basically because it can be easily verified and it will ultimately appear in the invoice issued by the PV manufacturer to the plant developer or the appointed EPC firm — i.e. it is what you will pay for the PV system if you want to build a plant, whether to operate it for its whole lifetime or to resell just after it is built.

The LCOE is defined as the cost of a unit of electricity in current monetary units. It incorporates the €/W of installed costs described above, and adds the operation and maintenance costs over the life cycle of the plant or system, also known as the total life-cycle costs (TLCC). These costs are divided by the energy generated, which is a function of the DNI irradiation in the plant's location, so that the overall expression is

$$\mathrm{LCOE} = \frac{\mathrm{TLCC}}{\sum_{n=1}^{N} \frac{Q_n}{(1+d)^n}} \tag{11.29}$$

where Q_n is the energy output in year n, N is the number of years of the analysis period, usually the total number of productive years, and d is the discount rate used to discount back future annual energy output to a present value. The discount rate is the rate of return that could be earned on an investment in the financial markets with similar risk.

The LCOE should be the most relevant metric for PV plant operators, for example the so-called IPPs (Independent Power Producers) that enter into PPAs (power purchase agreements) with electric utilities. This metric also has strong advocates among present CPV manufacturers, who are usually elusive in public regarding their installed costs (€/W), presumably because they yet do not bear favourable comparison with those of conventional flat-plate PV, but frequently present aggressive LCOE estimates and projections, and propose that LCOE is standardised as the main commercial metric, instead of the €/W of installed costs.

However, while it makes sense to base the financial assessment of investments in PV installations on their LCOE values, in general the PV market does not yet permit CPV manufacturers to base their marketing strategies on LCOE values, for two main reasons. The first is that there is still no confidence in the O&M cost figures and the long-term predicted power outputs of a new technology such as CPV, but it is these figures that are used to compute the LCOE; this problem is very much linked to that already mentioned: CPV's lack of track record. In other words, potential CPV customers frequently prefer to know about installed system prices in order to easily compare with other PV technologies, and afterwards, if they are in for a long-term investment, they try and make the LCOE calculation by themselves with all sorts of safety margins.

The second reason is that frequently, where PV plants are financed by banks through project finance, as happens in the FIT-based markets in Europe, there is only a small set of strong EPC companies that are considered eligible to receive funding (i.e. bankable) to build any plant. The EPC role will be to build a turnkey plant and hand it over to the developer — the one who has received the funding — for its operation, and therefore EPC's involvement in the plant ends after two or three years, and EPC companies are not too much concerned about metrics such as LCOE to assess the long-term profitability of the plant; their main target is to maximise profit in the plant construction (within quality margins) and for this purpose it is €/W of systems bought which is really of interest to them.

To get a feeling of HCPV costs in recent years, it is helpful to refer to market research done by specialised firms such as the above-mentioned *CPV Today* (Extance and Márquez, 2010). In this report, the main HCPV industry players (defining HCPV as geometric concentrations above 300×) were surveyed to determine current and projected installed costs, as well as the breakdown of costs among system components. As can be seen in Table 11.7, the surveyed HCPV companies were grouped into three categories, according to their installed costs, and averages were given — to keep each company's data anonymous — for each group. The lowest installed cost group offered quite a low average, falling in the range of flat PV modules. However, the authors noted that this group included a set of companies ranging from recent HCPV entrants who claim new approaches which can attain pronounced cost reductions (but which remain to be proven) to the most experienced HCPV vendors, who may be attaining the production volumes and experience necessary to bring costs down, but may be selling at a loss to enter the PV market and develop volume.

Probably more realistic is to take the survey's overall average of 3.7 €/W, which can be compared with the turnkey price of a tracked PV system which in summer 2011 could be assumed as around 2.8 €/W.

CPV Today also presented the averaged cost breakdown for HCPV systems, as shown in Fig. 11.38a, derived from their survey. The assembly costs in the pie

Table 11.7 Costs of HCPV (over 300×) according to *CPV Today's* market survey (Extance and Márquez, 2010).

Segment	**Installed cost 2010 (€/W, arithmetic mean)**	**Projected installed cost 2015 (€/W, arithmetic mean)**
HCPV overall	3.68	1.83
HCPV high installed cost tier	5.43	2.60
HCPV mid installed cost tier	3.41	1.92
HCPV low installed cost tier	2.28	1.17

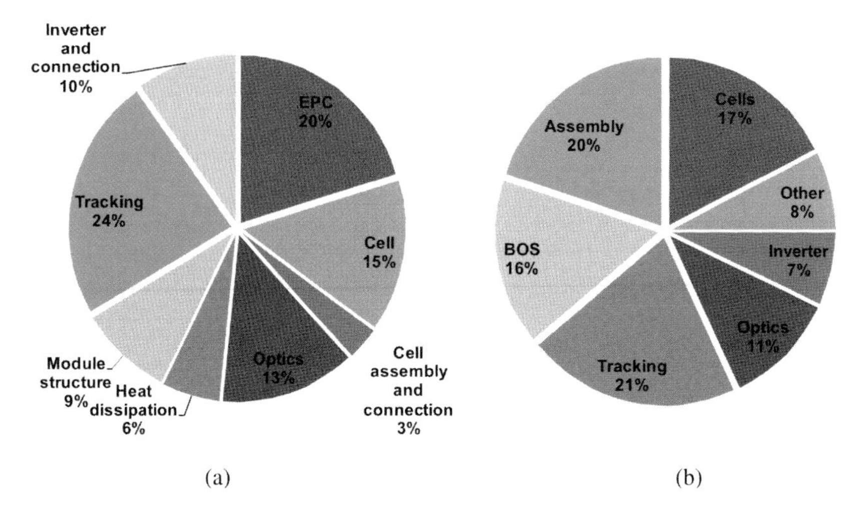

Figure 11.38 Two different cost breakdowns for CPV. a) An average one produced by *CPV Today* after polling a wide set of CPV manufacturers; b) that presented by Guascor Fotón derived from their experience for a 5 MW CPV plant with their technology. Reprinted with permission from Extance and Márquez (2010), © FC Business Intelligence. Courtesy of Guascor Fotón, SA.

chart were those related to the module integrating cells and optics. Figure 11.38b shows the CPV cost breakdown of Guascor's technology (Aurtenetxe, 2010) for a 5 MW PV plant, this company being the only that had built plants of this size at that time.

11.10 Conclusions

By the end of 2011, when this chapter was written, CPV technology had entered a phase of incipient commercialisation. Dozens of companies were involved in its development and the most promising of them, in our opinion, were the HCPV technologies, fuelled by the fantastic efficiency achievements of MJ solar cells. Most companies currently using high-efficiency silicon technology were moving to adapt MJ cells to their products. The rapid fall in flat-plate silicon modules in 2012 and 2013 has affected the commercial development of CPV, as it has for some of the other PV technologies described in this book.

It is predicted that the efficiency of MJ cells will continue to increase from their present 44% towards the 50% efficiency to be achieved with five or six junctions or perhaps with third-generation approaches, such as the intermediate-band solar cell (Luque and Martí, 1997).

Trends towards high concentration are underway from the 500 Suns or less of many present-day solutions towards the 1000 Suns or more to be expected in the near future. This increase will have a very important benefit to the cost of CPV systems in the short term, while the production volume of MJ cells is too small to provoke a reduction-by-experience of the cell cost.

Concentrator optics is well understood today and the main challenges are solved at least theoretically: good efficiency, large angular aperture and good illumination homogeneity. Challenges to come are maintaining these properties with the more demanding 1000× technology under way. In spite of these challenge, improvements are to be expected that will lead to practical optical efficiencies of 90% as compared with today's typical values of 80%.

Simple solutions suffice for heat management. Passive cooling without fins is generally adopted and, despite the high concentration, over 1000 Suns, the cell temperature is usually in the range of 80–90°C, which high-performance cells can easily cope with (Vázquez *et al.*, 2007; Núñez *et al.*, 2010).

Field experience is still small, but the ~10 MW commercial plant in Villafranca, (Spain), the largest in the world today (still with Si cells), and the 30 MW Alamosa Solar Plant in Colorado (with MJ cells) show that the time of no more than prototypes has gone. The establishment of the Institute for Photovoltaic Concentration Systems (ISFOC) in Spain by 2006, and the 3 MW international call-for-tenders launched there, have been important steps to help companies to develop MJ cell technology. Module manufacturing has matured, and from the ~20% efficiency at the module level by about 2008, when the plants for ISFOC were delivered, we are already at over 30% module efficiency at the time of writing (summer 2011). A module efficiency of 45% is foreseen (Luque, 2011).

Trackers are an important element of the systems cost, constituting maybe a third of the total. Therefore this has been carefully studied and the conditions for producing them at a cost per unit similar to those used for unconcentrated arrays with tracking are clear: excellent control and mechanisms and wide-aperture optics. Moreover, the high efficiency of MJ technology reduces by a factor of two to three the cost per watt of the tracker as compared with that for unconcentrated arrays.

Very good news is the high performance ratio of the MJ concentrator cells, over 82% as compared to 72% for unconcentrated silicon cells and about 60% for Si concentrated cells. This is attributable to the better temperature performance of MJ solar cells. This may have a very important influence on the LCOE for concentrators.

Ultimately CPV will develop if the prices are below those for unconcentrated approaches. For this they must climb the barrier of the low-cost conventional unconcentrated photovoltaics, fuelled by 50 GW of manufacturing experience. Yet academic studies of CPV in the late 1990s predicted very low costs in the long term, comparable or below most electricity sources, and by looking at these studies again today, we think these goals can be achieved, if the above-mentioned barrier can be overcome. If this is the case, CPV would become the leading technology for large PV plants, which in its turn is the largest and fastest growing PV application today.

References

Alamillo C., de la Rubia O., Gil E., Hipólito A., Martín A., Rubio F., Pachón J. L. and Banda P. (2010), 'Analysis of inverter configuration on CPV Plants', *Proc. 25th European Photovoltaic Solar Energy Conf.*, Valencia, pp. 4733–4736.

Alarcón J., Bassy A., Blazquez L., Luque A., Sala G. and Lorenzo E. (1982), 'Central fotovoltaica flotante', *Mundo Electrónico* **117**, 95–99.

Algora C. (2007), 'Very-high-concentration challenges of III–V multijunction solar cells', in Luque A. and Andreev V., *Concentrator Photovoltaics*, Springer Verlag, Berlin, pp. 89–112.

Algora C., Miñano J. C., Benítez P., Rey-Stolle I., Álvarez J. L., Díaz V., Hernández M., Ortiz E., Muñoz F., Peña R., Mohedano R., Luque A., Smekens G., de Villers T., Andreev V., Khostikov V., Rumiantsev V., Schvartz H., Nather H., Viehmann K. and Saveliev S. (2000), 'Ultra compact high flux GaAs cell photovoltaic concentrator', *Proc. 16th European Photovoltaic Solar Energy Conf.*, Glasgow, pp. 2241–2244.

Algora C., Ortiz E., Díaz V., Peña R., Andreev V. M., Khvostikov V. P. and Rumyantsev R. D. (2001), 'A GaAs solar cell with an efficiency of 26.2% at 1000 suns and 25.0% at 2000 suns', *IEEE Trans. Electron Dev.* **48**, 840–844.

Algora C., Rey-Stolle I., Galiana B., González J. R., Baudrit M. and García I. (2006), 'Strategic options for a LED-like Approach in III–V concentrator photovoltaics', *Conf. Record 4th World Conf. Photovoltaic Energy Conversion*, Waikoloa, Hawaii, pp. 741–744.

Algora C., Rey-Stolle I., García I., Galiana B., Baudrit M., Espinet P., Barrigón E., Datas A., González J. R. and Bautista J. (2008), 'A dual junction solar cell with an efficiency of 32.6% at 1000× and 31.0% at 3000× ', *Proc. 5th Int. Conf. on Solar Concentrators*, Palm Desert, CA.

Andreev V., Ionova E., Rumyantsev V., Sadchikov N. and Shvarts M. (2003), 'Concentrator PV modules of 'all-glass' design with modified structure', *Conf. Record 3rd World Conf. Photovoltaic Energy Conversion*, Osaka, pp. 803–876.

Antón I (2004), 'Métodos y equipos para la caracterización de sistemas fotovoltaicos de concentración', PhD Thesis, Universidad Politécnica de Madrid.

Araújo G. L. and Martí A. (1994), 'Absolute limiting efficiencies for photovoltaic energy conversion', *Solar Energy Mater. Solar Cells* **90**, 1068–1088.

Aurtenetxe A. (2010), 'Current and future HCPV competitiveness in utility scale installations', *3rd Concentrated Photovoltaics Summit*, Sevilla, 18 November.

Benítez P., Miñano J. C., Zamora P., Mohedano R., Cvetkovic A., Buljan M., Chaves J. and Hernández M. (2010), 'High performance Fresnel-based photovoltaic concentrator', *Optics Express* **18**, A25–A40.

Bruton T. M., Heasman K. C., Nagle. J. P., Cunningham D. W. and Mason N. B. (1994), 'Large area high efficiency silicon solar cells made by laser grooved buried grid process', *Proc. 12th European Photovoltaic Solar Energy Conf.*, Amsterdam. H. S. Stephens & Associates, Bedford, pp. 761–765.

Burgess E. L. and Pritchard D. A. (1978), 'Performance of a one kilowatt concentrator photovoltaic array utilizing active cooling', *Conf. Record 13th IEEE Photovoltaic Specialists Conf.*, Washington D.C. IEEE Press, Piscataway, NJ, pp. 1121–1124.

Domínguez C., Antón I. and Sala G. (2008), 'Solar simulator for concentrator photovoltaic systems', *Optics Express* **19**, 14894–14901.

Extance A. and Márquez C. (2010), *The Concentrated Photovoltaics Industry Report 2010*, FC Business Intelligence, London.

Franco de Saravia C. and Rossel A. D. (2010), 'Power Boost', *Photon International* **12**, 44–52.

Friedman D. J., Olson J. M. and Kurtz S. (2011), 'High efficiency III–V multijunction solar cells', in Luque A. and Hegedus S. (eds.), *Handbook of Photovoltaic Science and Engineering*, John Wiley & Sons, Chichester, UK.

Garboushian V., Roubideaux D. and Yoon S. (1996), 'An evaluation of integrated high-concentration photovoltaics for large-scale grid connected applications', *Conf. Record 25th IEEE Photovoltaic Specialists Conf.*, Washington D.C. IEEE Press, Piscataway, NJ, pp. 1373–1376.

Gil E., Sánchez D., de la Rubia O., Alamillo C., Rubio F. and Banda P. (2010). 'Field technical inspection of CPV power plants for the 25th EU PVSEC/ WCPEC-5', *Conf. Record 25th European Photovoltaic Solar Energy Conf.*, Valencia, pp. 4483–4486.

Green M. A., Emery K., King D. L., Hisikawa Y. and Warta W. (2010), 'Solar efficiency tables (version 36)', *Progr. Photovoltaics* **18**, 346–352.

Guter W., Schöne J., Philipps S. P., Steiner M., Siefer G., Wekkeli A., Welser E., Oliva E., Bett A. W. and Dimroth F. (2009), 'Current-matched triple-junction solar cell reaching 41.1% conversion efficiency under concentrated sunlight', *Appl. Phys. Lett.* **94**, 223504.

IEC (2007), *IEC 62108: 2007 Concentrator photovoltaic (CPV) modules and assemblies — design qualification and type approval*, International Electrotechnical Commission, Geneva, Switzerland.

ISFOC (2006), 'Specifications of general conditions for the call for tenders for concentration photovoltaic solar plants for the Institute of Concentration Photovoltaic Systems', Instituto de Sistemas Fotovoltaicos de Concentracion, Puertollano, Spain, pp. 34–39.

King R. R., Boca A., Hong W., Liu X.-Q., Bhusari D., Larrabee D., Edmondson K. M., Law D. C., Fetzer C. M., Mesropian S. and Karam N. H. (2009), *Proc. 13th European Photovoltaic Solar Energy Conf.*, Hamburg, pp. 55–61.

Knoll B. (2010), 'Still falling: inverter prices in the German spot market continue to get cheaper', *Photon International* **12**, 108–109.

Kolodinski S., Werner J. H., Wittchen T. and Queisser H. J. (1993), 'Quantum efficiencies exceeding unity due to impact ionization in silicon solar cells', *Appl. Phys. Lett.* **63**, 2405–2407.

Landsberg P. T. and Tonge G. (1980), 'Thermodynamic energy conversion efficiencies', *J. Appl. Phys.* **51**, R1–R20.

Lewis N. S., Crabtree G., Nozik A. J., Wasielewski M. R. and Alivisatos P. (2005), 'Basic research needs for solar energy utilization', *Office of Science — US Department of Energy*, http://science.energy.gov/~/media/bes/pdf/reports/files/seu_rpt.pdf.

Leutz R. and Suzuki A. (2001), *Non Imaging Fresnel Lenses*, Springer Verlag, Berlin/Heidelberg.

Lorenzo E. and Sala G. (1979), 'Hybrid silicone–glass Fresnel lens as a concentrator for photovoltaic applications', in *Sun II: Proc. Silver Jubilee Congress*, Pergamon Press, Atlanta, GA, pp. 536–539.

Lorenzo E. and Luque A. (1981) 'Fresnel lens analysis for solar energy applications', *Appl. Optics* **20**, 2941–2945.

Luque A. (1980), 'Quasi-optimum pseudo-lambertian reflecting concentrator: an analysis', *Appl. Optics* **19**, 2398–2402.

Luque A. (1986), 'Non-imaging optics in photovoltaic concentration', *Phys. in Technology* **17**, 118–124.

Luque A. (1989, ed.), *Solar Cells and Optics for Photovoltaic Concentration*, Adam Hilger, Bristol.

Luque A. (2011), 'Will we exceed 50% efficiency in photovoltaics?', *J. Appl. Phys.* **110**, 031301–031301-19.

Luque A., Gidon P., Pirot M., Anton I., Caballero L. J., Tobias I., Canizo C. del and Jausseaud C. (2004a), 'Performance of front contact silicon solar cells under concentration', *Progr. Photovoltaics* **12**, 517–528.

Luque A. and Martí A. (1997), 'Increasing the efficiency of ideal solar cells by photon induced transitions at intermediate levels', *Phys. Rev. Lett.* **78**, 5014–5017.

Luque A. and Martí A. (2011), 'Theoretical limits of photovoltaic conversion and new generation solar cell', in Luque A. and Hegedus S. (eds.), *Handbook of Photovoltaic Science and Engineering*, John Wiley & Sons, Chichester, UK.

Luque A. and Martí A. (2012), 'Understanding intermediate-band solar cells', *Nature Photonics* **6**, 146–152.

Luque A., Sala G., Arboiro J. C., Bruton T., Cunningham D. and Mason N. (1997), 'Some results of the EUCLIDES photovoltaic concentrator prototype', *Progr. Photovoltaics* **5**, 195–212.

Luque A., Tobias I., Gidon P., Pirot M., Canizo C. del, Anton I. and Jausseaud C. (2004b), 'Two-dimensional modeling of front contact silicon solar cells', *Progr. Photovoltaics* **12**, 503–516.

Luque-Heredia I., Cervantes R. and Quéméré G. (2005c), 'A sun tracking error monitor for photovoltaic concentrators', *Conf. Record 4th World Conf. on Photovoltaic Energy Conversion*, Hawaii, 706–709.

Luque-Heredia I., Martín C., Mañanes M. T., Moreno J. M., Auger J. L., Bodin V., Alonso J., Díaz V. and Sala G. (2003), 'A subdegree precision sun tracker for 1000× micro-concentrator modules', *Conf. Record 3rd World Conf. on Photovoltaic Energy Conversion*, Osaka, Japan, pp. 857–860.

Luque-Heredia I., Moreno J. M., Magalhães P. H., Cervantes R., Quéméré G. and Laurent O. (2007), 'Inspira's CPV sun tracking', in Luque A. and Andreev V. (eds), *Concentrator Photovoltaics*, Springer Verlag, Berlin, pp. 221–252.

Luque-Heredia I., Moreno J. M., Quéméré G., Cervantes R. and Magalhães P. H. (2005a), 'CPV Sun tracking at Inspira', *Conf. Record 3rd Int. Conf. on Solar Electric Concentrators for the Production of Electricity or Hydrogen*, Scottsdale, AZ.

Luque-Heredia I., Moreno J. M., Quéméré G., Cervantes R. and Magalhães P. H. (2005b), 'SunDog STCU TM generic sun tracking unit for concentration technologies', *Conf. Record 20th European Photovoltaic Solar Energy Conf.*, Barcelona, pp. 2047–2050.

Luque-Heredia I., Quéméré G., Magalhães P. H., Fraile de Lerma A., Hermanns L., de Alarcón E. and Luque A. (2006), 'Modelling structural flexure effects in CPV sun trackers', *Conf. Record 21st European Photovoltaic Solar Energy Conf. and Exhibition*, Dresden, Germany, pp. 2105–2110.

Miñano J. C. (1985a), 'Two-dimensional non-imaging concentrators with inhomogeneous media: a new look', *J. Opt. Soc. Amer. A* **2**, 1826–1831.

Miñano J. C. (1985b), 'Design of three-dimensional optical concentrators with inhomogeneous media', *J. Opt. Soc. Amer. A* **3**, 1345–1353.

Miñano J. C. (1985c), 'New family of 2-D non imaging concentrators: The compound triangular concentrators', *Appl. Optics* **24**, 3872–3876.

Miñano J. C. and González J. C. (1992), 'New method of designing of non-imaging concentrators', *Appl. Optics* **31**, 3051–3060.

Miñano J. C., González J. C. and Benítez P. (1995), 'RXI: a high gain, compact, non-imaging concentrator', *Appl. Optics* **34**, 7850–7856.

Muñoz E., Vidal P. G., Nofuentes G., Hontoria L., Pérez P., Terrados J., Almonacid G. and Aguilera J. (2010), 'CPV standardization: an overview', *Renewable and Sustainable Energy Rev.* **14**, 518–523.

Napoli L. S., Swartz G. A., Liu S. G., Klein N., Fairbanks D. and Tamatus D. (1977), 'High level concentration of sunlight on silicon solar cells', *RCA Review* **38**, 76–108.

Nemet G. F. (2005), 'Beyond the learning curve: factors influencing cost reductions in photovoltaics', *Energy Policy* **34**, 3218–3232.

Ning X., Winston R. and O'Gallagher J. (1987), 'Dielectric totally internally reflecting concentrators', *Appl. Optics* **26**, 300–305.

Núñez N., Vázquez M., Gonzalez J. R., Algora C. and Espinet P. (2010). 'Novel accelerated testing method for III–V concentrator solar cells', *Microelectronics Reliability* **50**, 1880–1883.

O'Neill M., McDanal A., Walters R. and Perry J. (1991), 'Recent development in linear Fresnel lens concentrator technology including the 300 kW 3M/Austin System and the 20 kW PVUSA system and the concentrator initiative', *Conf. Record 22nd. IEEE Photovoltaic Specialists Conf.*, Las Vegas. IEEE Press, Piscataway, NJ, pp. 523–528.

Optics.org (2013), 'Spectrolab beats unconcentrated PV efficiency record', http://optics.org/news/4/4/16, 15 April.

Randall D. E. and Grandjean N. R. (1982), 'Correlation of insolation and wind data for SOLMET stations', SAND82-0094, Sandia National Laboratories.

Ross R. T. and Nozik A. J. (1982), 'Efficiency of hot-carrier solar energy converters', *J. Appl. Phys.* **53**, 3813–3818.

Rubio F., Martínez M., Hipólito A., Martín A. and Banda P. (2010a), 'Status of CPV technology', *Proc. 25th European Photovoltaic Solar Energy Conf.*, Valencia, pp. 1008–1011.

Rubio F., Martínez M., Martín A., Hipólito A. and Banda, P. (2010b), 'Evaluation parameters for CPV production', *6th Int. Conf. on Concentrating Photovoltaic Systems*, American Institute of Physics, Freiburg, Germany, pp. 252–255.

Rubio F., Martínez M., Sánchez D., Aranda R. (2011), 'Results of three years of CPV demonstration plants in ISFOC', *Conf. Record 37th IEEE Photovoltaic Specialists Conf.*, Seattle> IEEE Press, Piscataway, NJ, pp. 3543–3546.

Sala G. (1989), 'Cooling of solar cells', Ch. 8 in Luque A., ed. *Solar Cells and Optics in Photovoltaic Concentration*, Adam Hilger, Bristol.

Sala G. and Anton I. (2011), 'Photovoltaic Concentrators', in Luque A. and Hegedus S. (eds), *Handbook of Photovoltaic Science and Engineering*, John Wiley & Sons, Chichester, UK.

Sala G., Araújo G. L., Luque A., Ruiz J. M., Coello A., Lorenzo E., Chenlo F., Sanz J. and Alonso A. (1979), 'The Ramón Areces concentration photovoltaic array', *Sun II: Proc. ISES International Solar Energy Society Silver Jubilee Congress*, Pergamon Press, Atlanta, GA, pp. 1737–1741.

Sala G., Arboiro J. C., Luque A., Zamorano J. C., Miñano J. C. and Dramsch C. (1996), 'The EUCLIDES prototype: an efficient parabolic trough for PV concentration', *Conf. Record 25th IEEE Photovoltaic Specialists Conf.*, Washington D. C. IEEE Press, Piscataway, NJ, pp. 1207–1210.

Salim A. and Eugenio N. (1990), 'A comprehensive report on the performance of the longest operating 350 kW concentrator photovoltaic power system', *Solar Cells* **29**, 1–24.

Sinton R., Kwark Y. and Swanson R. M. (1985), '23-percent efficient Si point contact concentrator solar-cell', *IEEE Trans. Electron Devices* **32**, 2553.

Shockley W. and Queisser H. J. (1961), 'Detailed balance limit of efficiency of *p-n* junction solar cells', *J. Appl. Phys.* **32**, 510–519.

Solar Industry (2013), 'CPV industry converges on standard rating conditions', 7 March.

Terrón M. J., Davies P. A., Oliván J., Alonso J. and Luque A. (1992), 'Deep emitter silicon solar cells for concentrated sunlight and operation with light confining cavities', *Proc. 11th European Photovoltaic Solar Energy Conf.*, Montreux, Switzerland. Harwood Academic Publishers, Chur, pp. 233–236.

Tobías I. and Luque A. (2002), 'Ideal efficiency of monolithic, series-connected multijunction solar cells', *Progr. Photovoltaics* **10**, 323–329.

Vázquez, M., Algora C., Rey-Stolle I. and González J. R. (2007), 'III–V concentrator solar cell reliability prediction based on quantitative LED reliability data', *Progr. Photovoltaics* **15**, 477–491.

Victoria M., Dominguez C., Antón I. and Sala G. (2009), 'Comparative analysis of different secondary optical elements for aspheric primary lenses', *Optics Express* **17**, 6487–6492.

Welford E. T. and Winston R. (1978), *The Optics of Non-Imaging Concentrators*, Academic Press, New York.

Winston R. and Gordon J. R. M. (2005), 'Planar concentrators near the étendue limit', *Optics Lett.* **30**, 2617–2619.

Winston R. and Welford E. T. (1979a), 'Geometrical vector flux and some new non-imaging concentrators', *J. Opt. Soc. Amer.* **69**, 532–536.

Winston R. and Welford E. T. (1979b), 'Ideal flux concentrators as shapes that do not disturb the geometrical vector flux field: a new derivation of the compound parabolic concentrator', *J. Opt. Soc. Amer.*, **69**, 536–539.

Wojtczuk S., Chiu P., Zhang X., Derkacs D., Harris C., Pulver D. and Timmons M. (2010), 'InGaP/GaAs/InGaAs 41% concentrator cells using bi-facial epigrowth', *35th IEEE Photovoltaic Specialists Conf.*, IEEE, Honolulu, Hawaii, pp.1259–1264.

Yamaguchi M. and Luque A. (1999), 'High efficiency and high concentration in photovoltaics', *IEEE Trans. Electron Devices* **46**, pp. 2139–2144.

CHAPTER 12

PHOTOVOLTAIC MODULES, SYSTEMS AND APPLICATIONS

NICOLA M. PEARSALL
Northumbria Photovoltaics Applications Centre
Northumbria University, Newcastle, NE1 8ST, UK
nicola.pearsall@northumbria.ac.uk

The best way to predict the future is to invent it.
Alan Kay, Apple Computers.

12.1 Introduction

The electricity from photovoltaic (PV) cells can be used for many different applications, from power supplies for small consumer products to large power stations feeding electricity into the grid. Previous chapters in this book have discussed different cell technologies and the optimisation of cell structures to achieve high efficiency of conversion from light to electricity. In this chapter, we will address the aspects that allow us to take those photovoltaic cells and incorporate them into a system delivering a required service.

The chapter concentrates on the use of the most common types of photovoltaic cells available commercially, described mainly in Chapters 3 to 6, and on typical system applications, including both stand-alone and grid-connected options. The stand-alone system can be defined as one where the PV array, together with other system components, is designed to be the only power source for the load. By contrast, the grid-connected PV system operates in parallel with the normal grid distribution system and feeds any electricity not required for local loads into the grid. This chapter considers the design and use of flat-plate PV systems. Readers are referred to Chapter 11 for a discussion of the issues involved in the design of PV systems incorporating concentration of sunlight.

In the next section, we discuss the construction and performance of photovoltaic modules. Individual solar cells must be connected to provide an appropriate electrical output and then encapsulated so as to protect the cells from environmental damage, particularly from moisture. The design of the module depends on the application for which it is to be used and an expansion of those applications in recent years has led to a range of alternative module designs, including the use of variable transparency and different electrical configurations. Finally, in Section 12.2, we discuss module testing, including the establishment of rated output and long-term performance.

In Section 12.3, we turn to the design of PV arrays, including electrical configuration, optimum tilt angle and orientation, protection from shading and mounting aspects. The variation in performance expected from different array configurations will be discussed.

The next Section 12.4 discusses the PV system, commencing with the rest of the system components, usually referred to as the balance-of-system (BOS) equipment. The BOS portion of the system differs substantially according to the application and the use of the electricity produced by the PV array. In this section we discuss the requirements of equipment to be included in a PV system, testing and standardisation, issues of power conditioning and sizing of the PV system to meet the required application. Finally, we will discuss applications of PV systems including current implementation, some aspects of integration into the electrical network and some environmental considerations.

12.2 Photovoltaic modules

In order to provide useful power for any application, the individual solar cells described in previous chapters must be connected together to give the appropriate current and voltage levels and they must also be protected from damage by the environment in which they operate. This electrically connected, environmentally protected unit is termed a photovoltaic module. It may be supplied with or without a metal frame that allows it to be fixed to an appropriate support. Figures 12.1a and b show simple schematic diagrams of typical module constructions for crystalline silicon and thin-film cells, respectively.

Due to the difference in fabrication process, module designs for crystalline and thin-film cells, whilst following the same basic principles in terms of electrical connections and environmental protection, differ in several aspects of module construction and design. Indeed, thin-film cells are usually fabricated in modular form, requiring only the encapsulation step after completion of the thin-film deposition processes. For simplicity, the crystalline silicon solar cell will be considered initially in each sub-section, since it currently represents the largest share of the PV market. Variations introduced by the use of thin-film cells will then be identified.

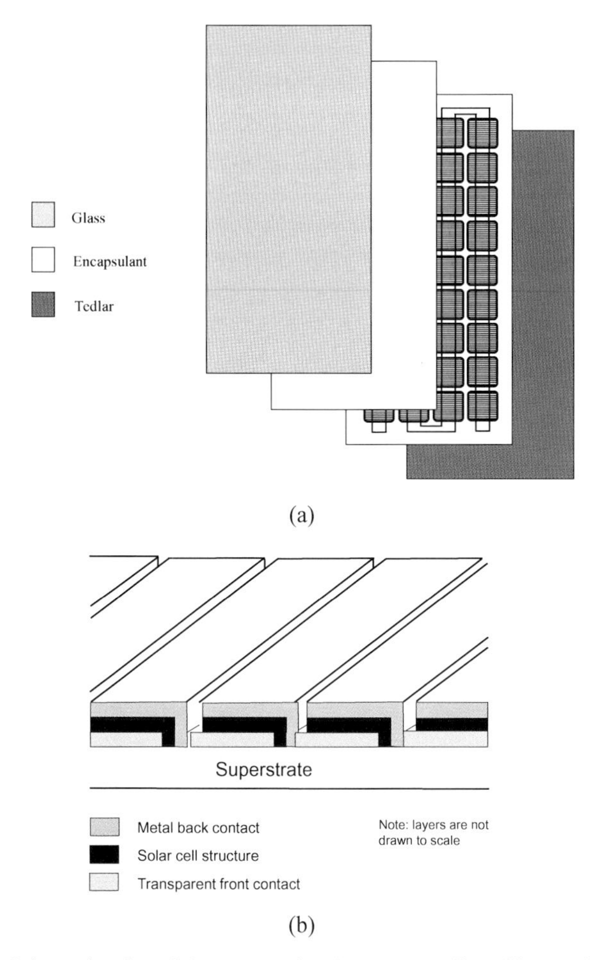

Figure 12.1 Schematic of module construction for: a) crystalline silicon cells — exploded view showing the different layers which make up the module; b) thin-film cells.

12.2.1 *Electrical connection of the cells*

The electrical output of a single cell is dependent on the design of the device and the semiconductor material(s) chosen, but is not usually large enough to meet most applications. In order to provide the appropriate quantity of electrical power, a number of cells must be electrically connected. There are two basic connection methods: series connection, in which the top contact of each cell is connected to the back contact of the next cell in the sequence, and parallel connection, in which

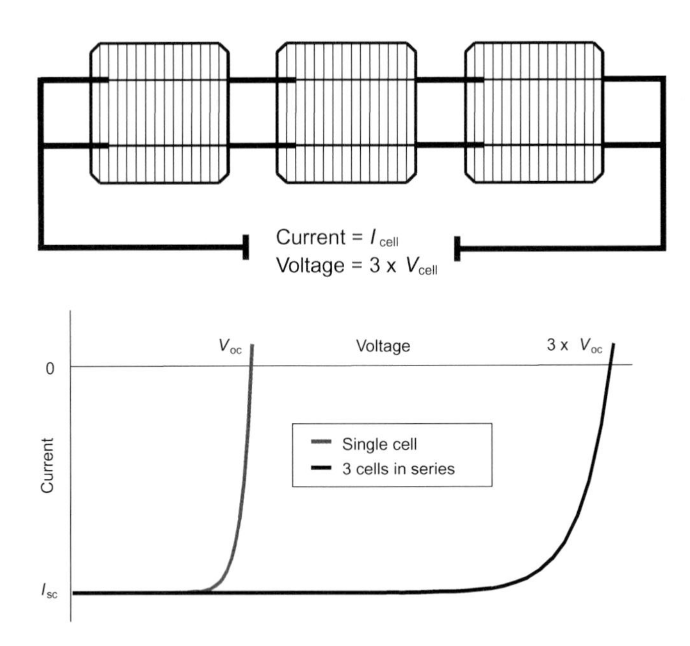

Figure 12.2 Series connection of cells, with resulting current–voltage characteristic.

all the top contacts are connected together, as are all the bottom contacts. In both cases, this results in just two electrical connection points for the group of cells.

Series connection

Figure 12.2 shows the series connection of three individual cells as an example; the resultant group of connected cells is commonly referred to as a series string. The current output of the string is equivalent to the current of a single cell and the voltage output is the sum of the voltages from all the cells in the string (in this example, since there are three cells connected, the voltage output is equal to $3\,V_{cell}$).

It is important to have well-matched cells in the series string, particularly with respect to current. If one cell produces a significantly lower current than the other cells (under the same illumination and temperature conditions), then the string will operate at that lower current level and the remaining cells will not be operating at their maximum power points. This could also happen in the case of partial shading of a string (further discussed in Section 12.3.5).

Parallel connection

Figure 12.3 shows the parallel connection of three individual cells as an example. In this case, the current from the cell group is equivalent to the sum of the

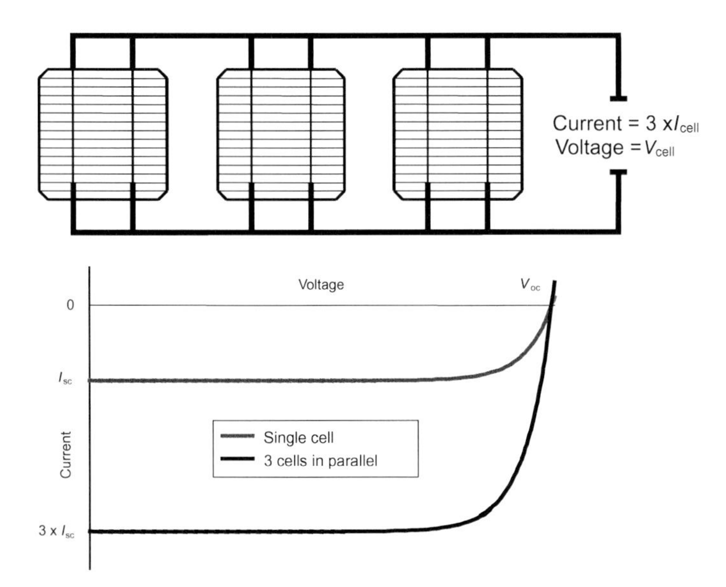

Figure 12.3 Parallel connection of cells, with resulting current–voltage characteristic.

current from each cell (here, $3I_{cell}$), but the voltage remains equivalent to that of a single cell.

As before, it is important to have the cells well matched in order to gain maximum output, but this time the voltage is the important parameter since all cells must be at the same operating voltage. If the voltage at the maximum power point is substantially different for one of the cells, then this will force all the cells to operate off their maximum power point and the output power will be reduced below the optimum.

Typical module configurations

The electrical connections within a module can be arranged in any desired combination of series and parallel connections, remembering the importance of the matching of the units in any series or parallel string. This means, for example, that parallel connection of series strings should be made using similar strings with the same number and type of cells. The series/parallel configuration will determine the current and voltage values obtained from the module under given illumination and load conditions.

The initial crystalline silicon module designs were for use in stand-alone applications for the charging of batteries and, thus, the electrical output was required to be appropriate for battery charging under a range of sunlight conditions. This was most readily achieved by the series connection of about 36 crystalline silicon cells. This produces a rated voltage at maximum power point voltage of around 18 V. Although the voltage reduces with reducing light intensity and increasing temperature, this configuration allows the module voltage to remain over 12 V, and therefore capable of charging a 12 V battery, for a wide range of operating conditions.

Many more PV modules are now used in grid-connected applications, where the output voltage is not constrained by battery requirements. However, most crystalline silicon power modules use between 36 and 72 series-connected cells since this provides a useful voltage and reasonable module size, which can be handled and transported. In some cases, a 72-cell module will be configured as two strings of 36 cells each, either then connected in parallel or with separate terminals so as to allow flexibility in module connection. In general, larger modules are used for large ground-mounted systems and architectural applications, whilst smaller modules are used for small systems to allow greater flexibility in electrical design.

The physical arrangement of cells in the module is generally designed to minimise the area, consistent with the minimum spacing between the cells required to prevent electrical interactions and at the edges of the module to provide environmental protection. Therefore, most modules are rectangular in shape and a typical configuration is four rows of nine cells per row to obtain a 36-cell module. Typical module sizes are 0.5–2 m^2 in area and the module should also be suitable for transportation and generally light enough to be lifted by one or two people for ease of installation.

In the case of thin-film modules, the same design principle is adopted in terms of voltage enhancement by series connection, although the number of cells in the string will differ since the voltage per cell is dependent on the cell technology. This is accomplished by the electrical connection of the cells during fabrication using a series of patterning steps. As new materials are developed, the module structure will be adapted for the characteristics of that material (see, for example, Chapter 9 for a discussion of dye-sensitised solar cells). Some modifications of the structure and manufacturing process can also be expected for developments of existing cell types (e.g. rear-contact silicon cells).

Module I–V characteristic

The typical I–V characteristic of a photovoltaic cell was described in Chapter 1. The module I–V characteristic is of a similar shape and can be described by the same equations, where now the parameters of open-circuit voltage, short-circuit current, fill factor, maximum power point, reverse saturation current, diode factor,

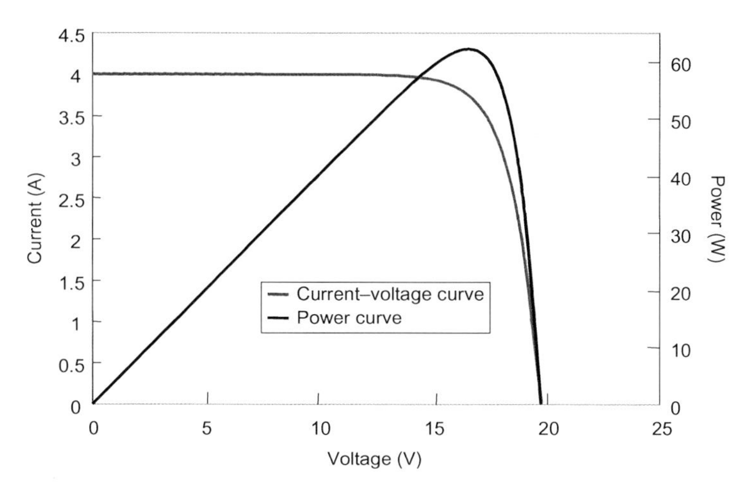

Figure 12.4 *I–V* characteristic of a crystalline silicon module, with the variation of power with voltage also shown. This illustrates the position of the maximum power point. In this case, the module has a peak power output of around 65 W.

series and shunt resistances refer to the whole module and are dependent on the type, number and electrical connection method of the cells.

Figure 12.4 shows an *I–V* characteristic together with the power curve, to illustrate the position of the maximum power point. Owing to mismatch between the characteristics of the component cells and to an increased overall series resistance, the module will typically have a reduced fill factor as compared with its constituent cells.

Module rating and efficiency

As with the individual cells, the module output varies with illumination and temperature conditions and therefore these must be defined when considering the power rating of the module. Module testing uses the same standard test conditions (STC) as are used for the measurement of cells, these being a light intensity of 1000 W m^{-2}, a spectral content corresponding to a standard AM 1.5 global spectrum and an operating temperature of 25°C (IEC, 2007).

In the ideal case, the module rating would simply be the sum of the rating of the individual cells but there are, of course, additional losses that must be taken into account. The most important of these is the mismatch between the cells, whereby differences in performance will mean that the maximum power point operation of the module as a whole does not coincide with the maximum power point operation of some or all of the cells in the module. The mismatch losses can vary depending on the operating conditions and whether differences in cell performance are light or temperature induced. Where possible — for example, for crystalline silicon cells — manufacturers usually sort their cells into batches according to their performance and use cells from the same batch to construct the modules. In this way, mismatch losses are minimised.

The module efficiency is related to the total area of the module in the same way that the efficiency of a cell is related to the total area of the cell. Because it is necessary to have the cells physically separated, the module area is always larger than the sum of the cell areas and therefore the module efficiency is always lower than the cell efficiency. The resulting reduction in efficiency depends on the configuration of the module and is defined by the packing density (ratio of cell area to module area). Typically, a fully packed crystalline silicon module will have a packing density of around 85% and so, if it uses 17% efficient cells, the module efficiency will be around 14.5%. While most modules have maximum packing density in order to produce the highest output possible per unit area, some architectural applications require the modules to have a higher level of transparency, for example to provide a combination of electrical generation, shading and daylight access. In this case, the cells are more widely separated and the resulting module efficiency is reduced accordingly.

For thin-film cells, the reduction in efficiency is lower because the strip cells are separated only by the contact strip. In this case, the mismatch between cell performances can be more important since it is not possible to sort and select the cells as for the crystalline devices. Since the mismatch arises from variations in the production process across the surface of the module, it is important to control the uniformity of all processes.

The performance of the module is also a function of its operating temperature and hence the rated efficiency is quoted at a standard temperature of 25°C. The module voltage reduces with increasing temperature and, although the current increases slightly, the overall effect is that the efficiency reduces as the temperature rises. The amount of the change depends on the cell type and structure, with crystalline silicon cells typically losing about 0.4–0.5% of their output per degree Celsius rise in temperature. Higher bandgap cells have a lower temperature coefficient; for example, thin-film silicon reduces by only about 0.2% per degree Celsius. However, thin-film silicon modules also exhibit a thermal dependence due to annealing of the light-induced (Staebler–Wronski) degradation and this acts in the opposite direction. So, their overall temperature coefficient can be zero or even slightly positive over some temperature ranges. This depends on cell structure and operating conditions.

The operating temperature varies as a function of the climatic conditions of ambient temperature and incident sunlight and also depends on the module design and mounting. Both these latter factors affect the ability of the module to lose heat and hence determine the operating temperature under given climatic conditions. A measure of the effect of module design is given by the nominal operating cell temperature (NOCT) of the module. This is the operating temperature of the cell junction under normal incidence sunlight at an intensity of 800 W m^{-2}, when the module is held at open circuit and mounted on an open rack, the ambient temperature is 20°C and the wind speed is 1 m s^{-1} (IEC, 2007).

12.2.2 *Module structure*

The structure of the PV module is dictated by several requirements. These include the electrical output (which determines the number of cells incorporated and the electrical connections), the transfer of as much light as possible to the cells, the cell temperature (which should generally be kept as low as possible) and the protection of the cells from exposure to the environment. The electrical connections have already been discussed, so in this section we will concentrate on physical protection from the environment and maintenance of cell operating conditions. Figure 12.5 shows a typical polycrystalline silicon PV module.

Figure 12.5 Typical polycrystalline silicon module (photograph courtesy of Sharp Electronics Europe GmbH). The square cells allow a high packing density.

In modern crystalline silicon modules, the front surface is usually glass, toughened to provide physical strength and with a low iron content to allow transmission of short wavelengths in the solar spectrum. The rear of the module can be made from a number of materials. One of the most common is Tedlar (see Fig. 12.1), although other plastic materials can also be used. If a level of transparency is required, then it is possible to use either a translucent Tedlar sheet or more commonly a second sheet of glass. The glass-glass structure is popular for architectural applications, especially for incorporation into glazed façades or roofs.

The glass-Tedlar module is usually fabricated by a lamination technique. The electrically connected cells are sandwiched between two sheets of encapsulant, for example EVA (ethylene vinyl acetate), and positioned on the glass sheet that will form the front surface of the module. The rear plastic sheet is then added and the whole structure is placed in the laminator. Air is removed and then reintroduced above a flexible sealing membrane to provide pressure on the module to expel air from the structure. The module is heated and the encapsulant flows and surrounds the cells. Additional encapsulant material is included at the module perimeter to ensure complete sealing of the module edges.

In the thin-film module, the glass substrate on which the cell is deposited is often used as the front surface of the finished module. The module is then laminated in the same way as for crystalline modules although only a single layer of encapsulant is required. Lower temperatures are often used to avoid damage to the cells. If the substrate used for the cell fabrication is positioned at the bottom of the module, a second glass layer is used to give the transparent front surface required. In some cases, the thin-film solar cells are deposited on a flexible substrate (metal foil or plastic). In all cases, particular care must be taken with edge sealing since all thin-film cells tend to be badly affected by the ingress of moisture. During manufacture, a clear gap must be left around the edge of the deposited cell area for proper sealing of the module.

The electrical connections to the module are made via a junction box, usually fixed to the rear of the module, or by flying leads. These typically exit the module through the rear Tedlar sheet. In glass-glass modules, the leads may exit through one edge of the module to avoid drilling holes in the glass sheet. The points at which the electrical connections are brought out of the module are sealed to prevent moisture ingress.

The module will exhibit the highest efficiency when the maximum amount of the light falling on the module is incident on the cells. Light which is incident on the spaces between cells or at the module edge is either reflected or converted to heat. Most power modules use the minimum cell spacing, which is 2–3 mm between the cell edges. This gap is to prevent any problems with electrical shorting between cells.

The shape of monocrystalline silicon cells is usually described as pseudo-square, where the cell is cut from a circular wafer and is square apart from the cut-off corners. Polycrystalline silicon cells are often truly square, depending on the manufacturing technique of the material. Thin-film cells are deposited in strips, usually of around 1 cm in width and running the length of the module, although dimensions can vary depending on cell properties.

In good sunlight conditions and with an ambient temperature of 25–30 °C, the module usually operates at temperatures in the region of 50–80 °C; the actual value depends on the module mounting. While these operating temperatures are not excessive, the difference in thermal expansion of the various components must be taken into account. Also, allowance must be made for the higher temperatures experienced during manufacture, albeit for a much shorter time. The cell stringing allows for some differential expansion in the length of ribbon between each cell. The electrical connection is also made in two or three places on each cell to allow for any problems with thermal expansion and other stresses during manufacture or operation.

The ideal module would also provide good heat transfer in order to keep the cell temperature as low as possible. However, the encapsulant is required to provide electrical isolation and physical protection, so a high heat transfer coefficient is not always possible. The operating temperature is also influenced by the module encapsulation: glass-glass structures usually run at a higher temperature than glass-Tedlar modules under similar conditions. The colour of the rear Tedlar film also has some influence. For example, a module with a white Tedlar backing will reject more heat than one with a black Tedlar backing, allowing it to operate at higher efficiency.

The module is often provided with a metal frame in order to make it straightforward to fix to a support structure, although this is less usual for building integrated applications, where use can be made of traditional mounting methods for glazing. The frame also provides some protection of the module edges in operation.

12.2.3 *Module testing*

The electrical output of the module is tested under STC as described earlier. The measurement under STC provides the module rating in peak watts (Wp) and defines the module efficiency. The testing method requires control of module temperature, light spectrum and illumination uniformity.

It is also important to assess the effectiveness of the module construction in protecting the cells from the environment, since this determines the lifetime of the module in operation. Testing conditions have been defined for accelerated life testing. These include thermal cycling, hail impact, humidity-freeze and electrical isolation tests and are detailed in IEC standard 61215 for crystalline silicon modules (IEC, 2005) and standard 61646 for thin-film modules (IEC, 2008). Whether a module meets the standard is determined by setting maximum limits for change in output and visual faults after each test. As the applications for photovoltaic systems change, the operating conditions and lifetime requirements also change. Therefore, the performance standards are regularly reviewed, updated and added to in order to meet the requirements of the evolving industry.

12.3 The photovoltaic array

A PV array consists of a number of PV modules, mounted in the same plane and electrically connected to give the required electrical output for the application. The PV array can be of any size from tens of watts to tens of megawatts or more, although the larger systems are often divided into several electrically independent sub-arrays each feeding into their own power conditioning system.

12.3.1 *Electrical connection of modules*

As with the connection of cells to form modules, a number of modules can be connected in a series string to increase the voltage output of the array, or in parallel to increase the current level, or in a combination of the two. The exact configuration depends on the current and voltage requirements of the load circuitry fed by the system output. Matching of interconnected modules in respect of their outputs can maximise the efficiency of the array, in the same way as matching cell output maximises module efficiency.

If there is a shaded module in a series-connected string of modules, it can act as a load to the string in the same way as a shaded cell does in an individual module. As with the cell, damage can occur due to heating by the current flowing through the module. The severity of the problem varies according to the number of modules in the string (and hence the potential voltage drop across the module) and the likelihood of partial shading of the string (which depends on system design and location). Where the shading situation may cause damage to the module, bypass diodes can be included. The bypass diode is connected in parallel with the module and, in the case of the module being shaded, current flows through the diode rather than through the module.

The use of bypass diodes adds some expense and reduces the output of the string by a small amount, because of the voltage that is dropped across the diode.

However, it is now common for bypass diodes to be incorporated into the module structure at the manufacturing stage. Several diodes may be used, each protecting different sections of the module, with the diode typically being connected across twelve to eighteen cells. This integration reduces the need for extra wiring, although it makes it more difficult to replace the diode if it fails. Where the likelihood of shading is very low and the system voltage is also low, modules without bypass diodes are sometimes used.

In systems where shading may reduce the output of one of the strings substantially below that of the others, it can also be advantageous to include a blocking diode connected in series with each string. This prevents the current from the remainder of the array being fed through the shaded string and causing damage.

12.3.2 *Mounting structure*

The main purpose of the mounting structure is to hold the modules in the required position without undue stress. The structure may also provide a route for the electrical wiring and may be freestanding or part of another structure (e.g. a building). At its simplest, the mounting structure is a metal framework, securely fixed into the ground. It must be capable of withstanding appropriate environmental stresses, such as wind loading, for the location. As well as the mechanical issues, the mounting has an influence on the operating temperature of the system, in determining how easily heat can be dissipated by the module.

12.3.3 *Tilt angle and orientation*

The orientation of the module with respect to the direction of the Sun determines the intensity of the sunlight falling on the module surface. Two main parameters are defined to describe this. The first is the *tilt angle*, which is the angle between the plane of the module and the horizontal plane. The second parameter is the *azimuth angle*, which is the angle between the normal to the plane of the module projected onto the horizontal plane and due south (or sometimes due north depending on the definition used). Correction of the direct normal irradiance to that on any surface can be determined using the cosine of the angle between the plane that is normal to the direct beam from the Sun and the module plane.

The optimum array orientation will depend on the latitude of the site, the prevailing weather conditions and the loads to be met. It is generally accepted that, for low latitudes, the maximum annual output is obtained when the array tilt angle is roughly equal to the latitude angle and the array faces due south (in the northern hemisphere) or due north (in the southern hemisphere). For higher latitudes, such as those in northern Europe, the maximum output is usually obtained for tilt angles of the latitude angle minus about 10–15°, due to the effect of both day length and sunlight levels in the summer. The optimum tilt angle is also affected by the proportion of diffuse radiation in the sunlight, since diffuse light is only weakly directional. For locations with a high proportion of diffuse sunlight, the effect of tilt angle is therefore reduced.

However, although this condition will give the maximum output over the year, there can be considerable variation in output with season. This is particularly true in high-latitude locations, where the day length varies significantly between summer and winter. Therefore, if a constant or reasonably constant load is to be met or, particularly, if the winter load is higher than the summer load, then the best tilt angle may be higher in order to boost winter output.

Prevailing weather conditions can also influence the optimisation of the array orientation if they affect the sunlight levels available at certain times of the day. Alternatively, the load to be met may also vary during the day and the array can be designed to match the output with this variable demand by varying the azimuth angle. For example, for air-conditioning loads peaking in the afternoon, it may be advantageous to orient the array to the west of south to boost the afternoon output.

Notwithstanding the ability to tailor the output profile by altering the tilt and azimuth angles, the overall array performance does not vary substantially for small differences in array orientation. Figure 12.6 shows the percentage variation in annual insolation levels for the location of London as tilt angle is varied between 0 and 90 degrees and azimuth angle is varied between −45° (south east) and +45° (south west). The maximum insolation level is obtained for a south-facing surface at a tilt angle of about 35°, as would be expected for a latitude angle of about 51°N. However, the insolation level varies by less than 10% with changing azimuth angle at this tilt angle. A similarly low variation is observed for south-facing surfaces for +/−30° from the optimum tilt angle.

The final point to consider when deciding on array orientation is incorporation into the support structure. For building-integrated applications, system orientation is also dictated by the nature of the roof or façade in which it is to be incorporated. It may be necessary to trade off the additional output from the optimum orientation against the additional costs incurred to accomplish this. Aesthetic issues must also be considered.

12.3.4 *Sun-tracking/concentrator systems*

The previous section assumed a fixed array with no change of orientation during operation. This is the usual configuration for a flat-plate array. However, some

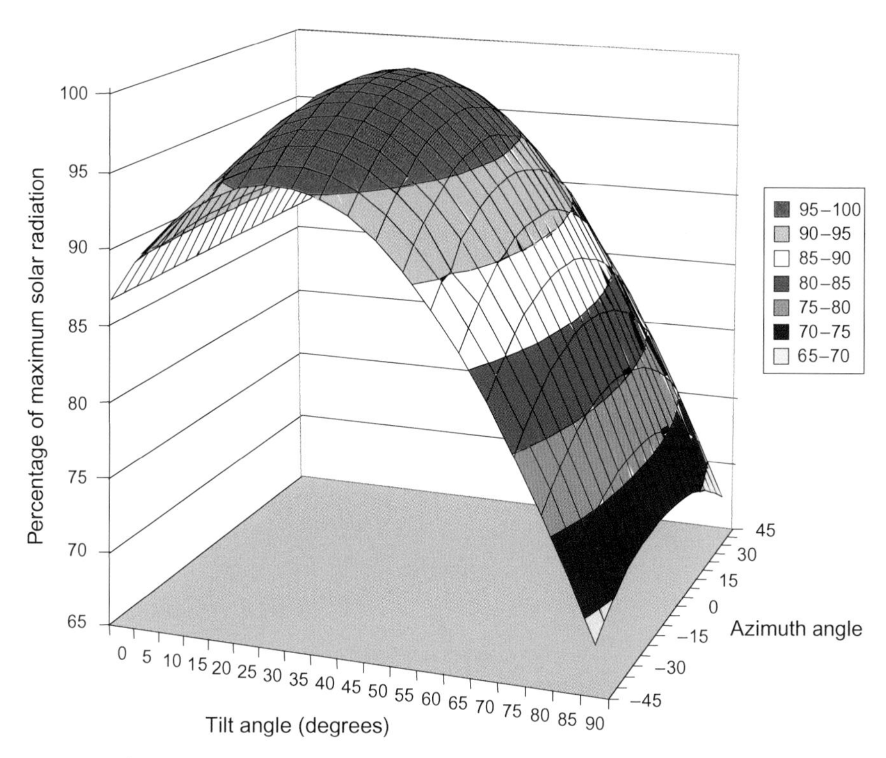

Figure 12.6 Percentage variation of annual global sunlight levels as a function of tilt angle and azimuth angle. The calculations were carried out for the location of London using Meteonorm Version 4.0.

flat-plate (non-concentrating) arrays are designed to track the path of the Sun. This can account fully for the Sun's movements during the day and by season by tracking in two axes, or partially by tracking only in one axis, from east to west.

The output gain achieved by tracking for a flat-plate system depends on the latitude, the climatic conditions (especially the ratio of direct to diffuse radiation), the type of photovoltaic module used and the local terrain. Because the tracked system receives more sunlight across the day, it will generally operate at a higher temperature than the fixed array and so the efficiency will be reduced by an amount determined by the array mounting and the temperature dependence of the type of module being used. However, this is usually more than offset by the increased light level received. In terms of climate, higher gains are seen where there is a high proportion of direct sunlight.

Huld *et al.* (2008) have predicted the gain from two-axis tracking for sites in Europe, compared with arrays fixed at the optimum angle for the site, in both cases based on crystalline silicon modules. They show that, on an annual basis, the potential gain is largest in northern latitudes (>60°N) and is in the region of 40–60%. This is because of the long days in the summer, when the movement of the Sun takes it behind the fixed-plate system at the beginning and end of the day. Gains of 30–40% are observed in southern Europe (latitudes <45°N) due to the high direct component in the sunlight at these locations. Lower gains of 20–30% would be expected in central Europe owing to higher proportions of diffuse sunlight. These values assume no restrictions due to the local terrain (e.g. mountains) preventing tracking throughout the day.

Of course, a tracking system incurs extra costs both for the initial installation and for maintenance, due to the required movement. Large installations will require multiple tracking pedestals, with the number dictated by the weight of the sub-array that can be accommodated by each pedestal. In general, a large installation of a two-axis tracked system would also require more land area than a fixed system of the same capacity, due to the need to avoid shading of the sub-arrays by neighbouring ones as they move. These extra costs need to be considered against the additional output in selecting whether to use a fixed or tracking approach for a flat-plate system.

For concentrator systems, described in Chapter 11, the system must track the Sun to maintain the concentrated light falling on the cell. The accuracy of tracking, and hence the cost of the tracking system, increases as the concentration ratio increases.

12.3.5 *Shading*

Shading of any part of the array will reduce its output, but this reduction will vary in magnitude depending on the electrical configuration of the array. Clearly, the output of any individual cell or module that is shaded will be reduced according to the reduction of light intensity falling on it. However, if this shaded cell or module is electrically connected to other cells and modules that are unshaded, their performance may also be reduced since this is essentially a mismatch situation. For example, if a single module in a series string is partially shaded, its current output will be reduced and this will then dictate the operating current of the whole string. If several modules are shaded, the string voltage may be reduced to the point where the open-circuit voltage of that string is below the operating point of the rest of the array, and then that string will not contribute to the array output. If this is likely to occur, it is often useful to include a blocking diode for string protection, as discussed earlier.

Thus, the reduction in output from shading of an array can be significantly greater than the reduction in illuminated area, since it results from the loss of output from:

- Shaded cells and modules.
- Illuminated modules in any severely shaded strings that cannot maintain operating voltage.
- The remainder of the array because the strings are not operating at their individual maximum power points.

For some systems, such as those in a city environment, it may be impossible to avoid all shading without severely restricting the size of the array and hence losing output at other times. In these cases, good system design, including the optimum interconnection of modules in relation to the shading pattern, the use of string or module inverters and, where appropriate, the use of protection devices such as blocking diodes, can minimise the reduction in system output for the most prevalent shading conditions.

12.4 The photovoltaic system

A PV system consists of a number of interconnected components designed to accomplish a desired task, which may be to feed electricity into the main distribution grid, to pump water from a well, to power a small calculator or one of many more possible uses of solar-generated electricity. The design of the system depends on the task it must perform and the location and other site conditions under which it must operate. In this section, we consider the components of a PV system, variations in design according to the purpose of the system, system sizing and aspects of system operation and maintenance.

12.4.1 *System design*

There are two main system configurations — *stand-alone* and *grid-connected*. As its name implies, the stand-alone PV system operates independently of any other power supply and it usually supplies electricity to a dedicated load or loads. It often includes a storage facility (e.g. battery bank) to allow electricity to be provided during the night or at times of poor sunlight levels. Stand-alone systems are also often referred to as autonomous systems since their operation is independent of other power sources. By contrast, the grid-connected PV system operates in parallel with the conventional electricity distribution system. It can be used to feed electricity into the grid distribution system or to power local loads, which can also be fed from the grid when there is insufficient power from the PV system.

It is also possible to add one or more alternative power supplies (e.g. diesel generator, wind turbine) to the system to meet some of the load requirements. These

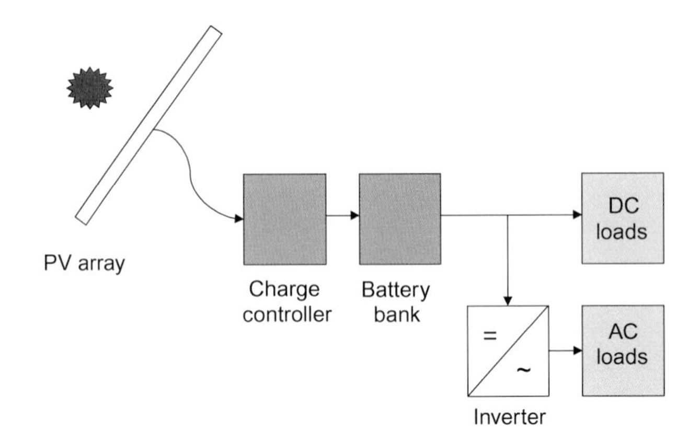

Figure 12.7 Schematic diagram of a stand-alone photovoltaic system.

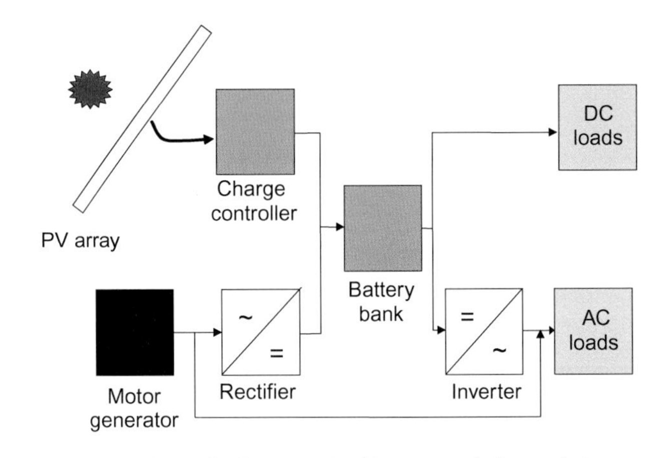

Figure 12.8 Schematic diagram of grid-connected photovoltaic system.

systems are then known as 'hybrid' systems. Hybrid systems can be used in both stand-alone and grid-connected applications but are more common in the former because, provided the power supplies have been chosen to be complementary, they allow reduction of the storage requirement without increased probability of not being able to meet the load, as discussed in Section 12.4.6. Figures 12.7 to 12.9 show schematic diagrams of the three main system types.

12.4.2 *System components*

The main system components are the PV array (which includes modules, wiring and mounting structure), power conditioning and control equipment, storage equipment

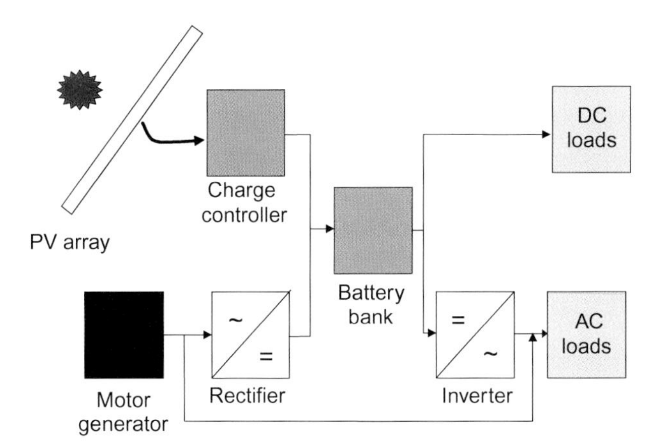

Figure 12.9 Schematic diagram of hybrid system incorporating a photovoltaic array and a motor generator (e.g. diesel or wind).

(if required) and load equipment. It is particularly important to include the load equipment for a stand-alone system because the system design and sizing must take the load into consideration. By convention, the array components are split into the photovoltaic part (the PV modules themselves) and the balance-of-system (BOS) components. In the remainder of this section, we briefly discuss the most common system components and their role in the system operation, with some examples of typical performance. There are many different options for BOS equipment, depending on the detail of the system, and it is only possible to give a general overview here.

The photovoltaic array

The PV array is made up of the PV modules themselves and the support structure required to position and protect the modules. The array has already been discussed in Section 12.3.

Power conditioning

Most systems incorporate electrical conditioning equipment to ensure that the system operates under optimum conditions. In the case of the array, the highest output is obtained for operation at the maximum power point. Since the voltage and current at maximum power point vary with both solar irradiance level and temperature, it is usual to include control equipment to follow the maximum power point of the array, commonly known as the maximum power point tracker (MPPT). The MPPT is an electrical circuit which can control the effective load resistance that the PV array sees and thus control the point on the I–V characteristic at which the system operates.

One of the most common tracking methods is the perturb and observe (P&O) approach. At regular intervals, the power level is checked for a small change in operating voltage on either side of the current operating point. If a gain in power is observed in one direction, then the MPPT moves the operating point in that direction until it reaches the maximum value. It is also possible to calculate the expected maximum power point based on measurement of open-circuit voltage or knowledge of the operating conditions of the module. These methods can achieve a more rapid determination of the maximum power point but require additional measurements to be made. For grid-connected systems, the MPPT is often incorporated into the inverter for ease of operation, although it is possible to obtain the MPPT as an independent unit.

Although many stand-alone systems use an MPPT approach, some systems operate at constant voltage, so as to allow battery charging or direct current (DC) loads to be met at a specific voltage level. It may be necessary to include a DC–DC converter to change the voltage level of the output of the array to that required for input to the load. For systems with storage batteries, it is also usual to include charge control circuitry, in order to control the rate of charge and so prevent damage to the batteries and extend their lifetime.

Inverter

If the PV system is to supply alternating current (AC) loads or feed power directly into the AC distribution grid, then an inverter must be included to convert the DC output of the PV array to the AC output required. As with PV systems, inverters can be broadly divided into two types, these being stand-alone and grid-connected (sometimes referred to as line-tied).

The stand-alone inverter is capable of operating independently from the utility grid and uses an internal frequency generator to obtain the correct output frequency (usually 50/60 Hz). By contrast, the grid-connected inverter must integrate smoothly with the electricity supplied by the grid in terms of both voltage and frequency and therefore uses the grid as a reference. Some modern inverters have the capability to change from stand-alone to grid-connected mode. The output voltage of the inverter is chosen according to the load requirements, e.g. 220–230 V single-phase for European domestic appliances. However, if the electricity from the PV system is to be fed directly into the supply of a large office building, for example, a 415 V three-phase output may be chosen. For large PV installations feeding directly into the distribution grid, higher voltages may also be chosen. The output voltage of the inverter is selected to match the grid voltage in the location at the network level required.

The input voltage to the inverter depends on the design of the PV array, the output characteristics required and the inverter type. Stand-alone systems commonly operate at 12, 24 or 48 V, since the system voltage is determined by the storage system, whereas grid-connected inverters usually operate at significantly higher voltages (between 150 and 1000 V, depending on the system size).

The shape of the output waveform is important because some loads can overheat or be damaged if a square wave output is used. True sine wave or quasi-sine wave (or modified sine wave) outputs are generally more costly but are much more widely applicable. Most modern stand-alone inverters provide a modified sine wave output, whilst grid-connected inverters should have a sine wave output with a very low harmonic content. Limits on the harmonic distortion are defined in the regulations for grid connection.

While most systems use an inverter connected to a group of modules, it is also possible to obtain modules with an integrated inverter, positioned on the rear of a module and converting the electrical output from that single module. This module–inverter combination is sometimes referred to as an AC module. There are some advantages in system design, such as the use of AC wiring for most of the power transmission and reduced losses for non-uniform systems (e.g. where there is shading) since each module has individual maximum power point tracking. Indeed, the system designer now has a wide choice of inverter configurations, from a single inverter per module via inverters connected to each string to a single central inverter per system. A recent development has been the use of individual maximum power point trackers, attached to one or two modules, with the output from a collection of modules then being sent to a single inverter for DC/AC conversion. The choice of system configuration depends on the uniformity of irradiance (particularly whether there are shading problems), the overall efficiency of the different options and the overall cost.

Inverters for PV systems are designed to have high conversion efficiency and, although this varies with inverter design, values in the range of 95–97% for maximum efficiency are now common. The efficiency varies with the operating point of the inverter, but usually reaches its maximum between 30 and 50% of rated capacity and shows only a small decrease as the power level increases. However, the efficiency generally reduces substantially at power levels below about 10% of full power. Because of the variation of efficiency with operating conditions, it is also usual for European inverter manufacturers to quote the Euro-efficiency value of their unit. This is calculated by considering the efficiency at several input levels and producing a weighted average according to a defined set of operating conditions.

Recently, it has been argued that this method should be refined to also take account of the maximum power point voltage range over which the inverter operates (Bletterie, 2011), although this modification has not yet been widely adopted.

In locations in the middle and north of Europe, the performance at low light levels (and hence low power levels) can have a significant effect on the overall system efficiency. Thus, it is usual to size the inverter at about 80% of the array capacity so that high inverter efficiencies are maintained at lower power levels. This means that the very high power levels are sacrificed since they are out of the range of operation of the inverter, but the balance of low and high power operation is usually such that it is more advantageous to use a reduced inverter size. This may not be the case for systems that experience a significant proportion of high power levels due to cold, clear weather conditions. Although all locations experience low light level conditions, at least for fixed flat plate systems, at the end of the day, southern European locations operate more often under high irradiance conditions and therefore it may be most beneficial to derate the inverter by 5–10% only. The nature of the installation should also be taken into account. For example, the solar irradiation received on an array mounted on a vertical façade will be less than that for the same array mounted at the optimum tilt angle in the same location. Therefore a smaller inverter would be advisable for the façade system.

When the inverter is grid-connected, it is essential to ensure that the system will not feed electricity back into the grid when there is a sustained fault on the grid distribution system. This problem is known as islanding, and safeguards are required in order to provide protection for equipment and personnel involved in the correction of the fault. Islanding is usually prevented by closing down the inverter when the supply from the grid is outside certain limits. The allowable limits vary from country to country but are usually around +/−2% in voltage and frequency. Requirements for prevention of islanding for systems are detailed in the connection regulations for each country.

In recent years, there has been a growing amount of embedded generation (i.e. electricity generation systems sited towards the consumer end of the distribution system) connected to the grid in some countries. This includes wind, photovoltaic and other systems. Problems can occur if all this generation turns off when there is a transient fault on the grid and in some countries grid operators are now requiring a fault ride-through capability on the inverter to address this issue. This means that the embedded generation will continue to operate for a short period to maintain continuity of supply. If the grid fault persists beyond a specified time, the inverter will then close down until the fault is cleared.

Although not yet fully realised, the inverter also has the capability to contribute to the safety and stability of the grid by a variety of means. Most inverters currently operate at unity power factor, but there is the possibility to supply reactive power when required by the grid, at the penalty of some loss of active power output. The inverter also often provides system monitoring and could be incorporated into a smart grid configuration, supplying information on the PV system status and output. Some inverters also supply integral storage to allow the exchange of electricity with the grid to be managed.

Storage

For many PV system applications, particularly stand-alone, electrical power is required from the system during hours of darkness or periods of poor weather. In this case, storage must be added to the system. Typically, this is in the form of a battery bank of an appropriate size to meet the demand when the PV array is unable to provide sufficient power.

The storage capacity must be chosen so as to provide sufficient security of supply for the load and this will vary depending on how critical it is to maintain the operation of the load. Of course, increasing storage capacity also increases the system cost and so an appropriate balance between expenditure and performance must be achieved.

There is a range of battery types and designs that can be used, although the most common type is a low maintenance lead-acid battery, designed for medium depth discharge. The design of lead-acid battery commonly used in vehicles is unsuitable for PV applications since this only allows shallow discharge if the lifetime of the battery is to be maintained. As previously mentioned, a charge controller should be used so as to control the rate of charging and discharging of the battery and to prevent complete discharge, which can seriously damage the battery. Attention must also be paid to the location of the battery, since extremes of temperature will also reduce operating efficiency and lifetime.

Generally, grid-connected systems do not include storage but this would be one method to offset the variability of output from the PV system. As mentioned in the previous section, some inverter manufacturers are beginning to offer integral storage as an option. This adds cost but may be advantageous if the value of deferring use of the electricity from the PV system is higher than that of exporting it to the network. It is likely that storage will become a more common feature of grid-connected systems as the level of penetration on the network increases.

Load equipment

The nature of the load equipment will determine the need for, and suitability of, the power-conditioning equipment and the capacity of both the PV system and the storage. The first consideration is whether the load or loads use DC or AC electricity. In the former case, the loads can be operated directly from the PV system or battery storage whereas AC loads will require an inverter to be included in the system.

Where the system is grid-connected, loads are almost always AC but for autonomous systems, a choice can be made. This choice will depend on the availability, cost and performance of the DC and AC versions of the load equipment. For example, it is possible to obtain high-efficiency DC fluorescent lighting which, by virtue of its superior performance compared with AC lighting, results in a smaller capacity requirement for the PV system and hence, usually, reduced costs. In the case of water pumping, the choice between DC and AC pumps depends on the nature of the water supply (e.g. deep borehole or surface pump).

The requirements of the load in terms of voltage and current input range will influence the type of power conditioning included in the system and the load profile will determine the relative sizes of the PV system and the storage, if used. System sizing in accordance with load details is discussed in more detail later in the chapter.

Cabling and switching equipment

The array cabling ensures that the electricity generated by the PV array is transferred efficiently to the load and it is important to make sure that this is specified correctly for the voltage and current levels which may be experienced. Since many systems operate at low voltages, the cabling on the DC side of the system should be as short as possible to minimise the voltage drop in the wiring. Switches and fuses used in the system should be rated for DC operation. In particular, DC sparks can be sustained for long periods, leading to possible fire risk if unsuitable components are used.

Because of their design and exposed situation, PV systems can be vulnerable to lightning strikes, with the probability depending on the prevalence of lightning at the location in question. This is usually addressed by ensuring correct grounding of the system and including surge protection of sensitive electrical components. In some cases (e.g. for arrays on buildings), lightning conductors can divert strikes away from the array although this does not prevent induced currents.

12.4.3 *System sizing*

It is important to determine the correct system size, in terms of both peak output and overall annual output, in order to ensure acceptable operation at minimum cost. If the system is too large, it will be more expensive than necessary without increasing performance levels substantially and therefore the system will be less cost-effective than it could be. However, if too small a system is installed, the availability of the system will be low and the customer will be dissatisfied with the equipment. Again, the cost-effectiveness is reduced.

Although many of the same principles are included in the sizing process, the approach differs somewhat for stand-alone and grid-connected systems. Considering stand-alone systems first, the initial step is to gather the relevant information on the location and purpose of the system. Location information includes:

- Latitude and longitude.
- Weather data — monthly average sunlight levels, ambient and maximum temperatures, rainfall, maximum wind speeds, other extreme weather conditions.
- Constraints on system installation — tilt angle, orientation, risk of shading.

Information on system purpose includes:

- Nature of load or loads.
- Likely load profile — daily, annual variation (if any).
- Required reliability — ability to cope with loss of load (for example, clinic lighting requires a higher level of reliability than a lighting system for a domestic house).
- Likelihood of increase of demand — many systems fail because they are sized for an existing load, but demand increases soon after provision of the PV supply.

If an autonomous system is required, the PV system must provide sufficient electricity to power the loads even under the worst conditions. Thus, system sizing is usually carried out for the month that represents the worst conditions in terms of the combination of high load levels and low sunlight conditions. This is not necessarily the month that has the lowest sunshine or the highest load, but that for which the combination represents the worst case.

For a given system design, the average electrical output in the sizing month can be calculated from the average daily insolation level (usually expressed in kWh m^{-2}) taking into account the number of modules, their rated efficiency, the efficiencies of all control and power conditioning equipment, the efficiency of any storage system, mismatch losses, wiring losses and the operating temperature. For an autonomous PV system, the average daily electrical output should match or exceed the average daily load. If this is not the case, then the PV array size must be increased.

The battery storage allows for variations in the load level during the day and the provision of power at night. The battery bank must be sized to accommodate the average daily need for electricity that cannot be directly supplied by the PV system and so that this results in only the allowable discharge of the batteries.

So far, we have considered only average values for load and sunlight levels. The daily sunlight levels can vary substantially and the battery storage must also allow for providing power in periods of unusually poor weather conditions. The length of the period to be allowed for is determined by consideration of local weather conditions (i.e. the probability of several days of poor weather) and the importance of maintaining power to the load. Clearly, if the system is used for medical purposes or communications, loss of power could have serious consequences, whereas for other situations, such as powering domestic TV or lighting, it is merely an inconvenience. Since an increase in the period for which supplies can be maintained involves an increase in the size of the PV array and/or battery bank and hence an increase in system cost, this aspect is an important part of the sizing exercise. Supply companies refer to this by many different terms, including reliability, availability and loss-of-load probability (LOLP).

Clearly, the sizes of the PV array and battery bank are linked, and an increase in the size of one can often allow a decrease in the size of the other. The sizing operation is usually an iteration of the problem to find the most cost-effective solution, taking into account the requirements and preferences of the user. A detailed sizing can be performed using system-sizing software, commercially available from a variety of sources, using detailed solar data for the location and allowing a more robust solution than using daily or monthly averages.

For a grid-connected system, it is not usually necessary to meet a particular load but only to contribute to the general electricity supply. Some systems are designed to feed all their output into the electricity grid whilst others (e.g. most building-integrated systems) are designed to meet some of the load in a local area with the rest of the requirement being supplied by the grid. These latter systems only feed power into the grid when their output exceeds the demand of the load. The system sizing is therefore not often governed by the size of the load, but by other constraints such as the area available for the system and the budget available for its purchase and installation. Therefore, most sizing packages are used to determine potential output and to compare different options of system location and design, rather than optimising system size as such. Modern sizing software includes options for different mounting configurations and for the inclusion of an allowance for shading based on details of the array surroundings.

The accuracy of the output of any simulation will depend on the accuracy of the input data, as with all such systems. However, since there is a natural variation in solar irradiation levels depending on climatic conditions, this must also be taken into account in the use of results from a simulation. If average insolation data are used, as is most common, then an average output will be obtained as a result. This is strictly speaking only the average value over the period represented by the input data rather than a prediction of what any future values will be. Thus it is possible to obtain practical results from a system, which are significantly different from the simulation results of the design process, simply because of normal climatic fluctuations.

12.4.4 *System operation*

The output of any PV system depends mainly on the sunlight conditions but can also be affected by temperature, shading and the accumulation of dirt on the modules. The overall system performance is usually represented by the efficiency, which is defined as the ratio of the electrical output delivered to the load (in kWh) to the sunlight energy input (also in kWh) over the surface of the array in the same period. In general, this overall efficiency results from several processes to which individual efficiency values can be assigned, e.g. the conversion of sunlight to DC electricity, the conversion of DC to AC by the inverter and so on.

The *system yield* is also a useful parameter. This expresses the annual output (or that over another defined period) normalised for the rated capacity of the system and is in units of kWh/kW. This allows comparison of systems in different locations. For example, based on average annual irradiation levels, system yields in southern Europe are typically in the range 1300–1500 kWh/kW, whereas in Germany a value of around 1000 kWh/kW might be expected. Since the yield does not explicitly include the sunlight level received over the period, account must be taken of whether the level was above or below average if a measured value is to be used for a critical assessment of system performance.

A useful parameter for assessing system performance is the *performance ratio*, which is either given as a percentage or as a number between zero and one. Essentially, this parameter expresses the performance of the system by comparison of the system output with that from a reference system of the same design and rating at the same location, but with no losses. Because the sunlight level is included in the assumption of the output of the reference system, the performance ratio is independent of sunlight conditions. Thus, it allows the comparison of system design and performance in different locations. The performance ratio is mainly affected by the operating characteristics of the inverter, including efficiency and reliability, the effect of temperature on the PV module efficiency, the effectiveness of the maximum power point tracking, electrical losses in the cabling, mismatch between modules, shading and dirt accumulation on the modules.

The performance ratio (PR) can be calculated from the following formula:

$$\begin{aligned}\text{PR} = {} & \text{system output over period/(average daily irradiance} \times \text{array rating} \\ & \times \text{number of days in period} \times \text{monitoring fraction)}\end{aligned}$$

where all parameters are values for the same period, the system output is in kWh, the average daily irradiance is in kWh m^{-2} and the array rating is in kW. The array rating is calculated as the sum of the power ratings of the individual modules making up the PV array. The monitoring fraction is the fraction of the period for which monitoring data are available and have been used to determine the values of the other parameters. The formula makes the assumption that average conditions are experienced for the time when data are not collected and so care must be taken with the use of PR values calculated for monitoring fractions less than 0.9.

Performance ratio values for grid-connected systems give a useful measure of system design and operating quality with a value of between 0.8 and 0.9 being the aim. This implies high inverter efficiency, minimum electrical losses between the modules and the inverter, good thermal transfer to keep module temperature to a reasonable level, and good system reliability. Performance ratio values cannot normally be used to indicate the quality of a stand-alone system since the array may not always be operating with maximum power point tracking and the requirement to supply the load fully under worst-case conditions means that some potential output is wasted under more favourable conditions. This can lead to a low PR value even for a well-designed system. A different approach must be used for the stand-alone system with availability of load (i.e. the provision of sufficient power when it is needed to operate the load) being the most important aspect.

12.4.5 *Operation and maintenance*

Because of the lack of moving parts and simple connections, a PV system generally requires little maintenance. However, it is necessary to ensure continued access to sunlight, by cleaning the panels at appropriate intervals, refraining from building any structures that could shade the panels and by cutting back any branches or other vegetation that could cover the system. The electrical connections should also be checked at regular intervals to eliminate any problems, such as corrosion or loose connections. If included in the system, the battery bank may need regular maintenance according to the type chosen.

The requirement for cleaning is often overestimated by those with little experience of PV systems. It is generally assumed that 3–5% of performance will be lost if the system is only cleaned annually, with up to half of that loss being experienced within a few weeks of cleaning. However, the losses incurred and thus the requirement for cleaning are very dependent on location and are best determined from practical applications operating under similar conditions. For example, if there is the possibility of dust or sandstorms causing accumulation on the modules, perhaps in a desert area, then more frequent cleaning will be required. This can also be the case for systems installed in industrial areas close to sources of airborne pollutants. For building integrated systems on houses in many parts of Europe, it may not actually be necessary to clean the systems regularly, since the action of rainwater on the inclined panels removes surface dust. A thorough cleaning every few years would then be carried out to remove persistent dirt.

Operational problems can occur as a result of poor maintenance of the BOS components (including loads and batteries) or allowing the array to become obscured or damaged. This latter problem indicates a lack of understanding of the operation of the system and there is a need for education of users to ensure that they operate the system correctly. This is also demonstrated by system failures arising from the addition of loads that were not included in the original system sizing for a stand-alone system. In this case, the combination of the PV and storage system cannot meet the increased demand and there is a danger of damage to the batteries from deep discharging.

The costs of operation and maintenance will vary with application, since they are dependent on the ease of access and the requirement for cleaning, the remoteness of the system and any replacements that may be required. However, they are generally not more than a few percent of the system cost per annum. For large applications, such as ground-mounted power plants, it is common to have a monitoring and maintenance contract that ensures minimum downtime on the plant operation and early identification of loss issues.

12.4.6 *Photovoltaic applications*

The wide range of applications in which photovoltaic systems are employed cannot be covered in depth in this chapter, but we will summarise some of the main uses for stand-alone systems before considering both building-integrated and ground-mounted systems that are connected to the grid.

Stand-alone PV systems

Most stand-alone PV systems are used to power loads in locations that are remote from the electricity grid supply, although they are also used in urban locations where it may be inconvenient or expensive to use the grid supply (e.g. for low-voltage lighting on bus shelters). Stand-alone systems generally range in size from a single module powering a solar home system (SHS) to a few kilowatts of PV supplying a local area grid network, although there are options for much larger stand-alone systems for applications such as rural electrification.

The systems are autonomous and so must include some type of energy storage to supply power in the absence of sunlight. Systems are usually categorised as critical, meaning that there are issues relating to health or safety, or non-critical, meaning that loss of load causes no more than inconvenience. In some cases, systems powering applications with a high economic value (e.g. communications) are also treated as critical applications from the design point of view. In general, a critical system will have a higher storage capacity to allow for continued operation through prolonged periods of low insolation compared with the average for that location. The economics of storage dictate that, for larger systems and for those where high reliability is paramount, a hybrid system is often used, allowing some of the energy storage to be in the form of fuel for an internal combustion engine. In locations where the seasonal availability of wind energy is complementary to that of the solar irradiance, it can also be cost-effective to include a wind turbine in the hybrid system. Other options would be biomass or small hydro-electricity systems, depending on availability at the location.

In a small, non-critical system, such as an SHS, one or two PV modules charge a battery during the day, and the power is used at night for a few high-efficiency lights and a radio or small TV. A charge controller ensures that the battery is not overcharged or deep-discharged, to provide as long a battery lifetime as possible. Battery capacity and lifetime can be substantially reduced by poor charging regimes, especially consistent discharging below the design level of charge. To keep costs as low as possible, standard systems are sold to all users, though there may a choice of the number of modules based on local needs and solar conditions. System reliability largely depends on users modifying their usage of the loads to ensure an appropriate battery charging regime, often using indicator lights on the charge controller.

For systems that are part of safety-critical networks, for example those in aircraft navigation aids or telecommunication systems, a very low LOLP must be guaranteed, perhaps as little as one or two hours per year on average. The stochastic variability of solar irradiance means that a large PV array feeding into a large battery storage capacity must be provided to ensure that almost all periods of poor weather can be accommodated, or additional charging must be provided from a small internal combustion engine, usually a diesel, with a fuel store sufficient to maintain this option with fuel deliveries under normal maintenance visits only. The PV array and battery system should be sized so that the engine is run at full power on a sufficiently regular basis to keep it in good condition, but so that the overall cost is minimised, taking into account the fuel and equipment costs. The design will vary with location in terms of both climate and cost of fuel.

Some stand-alone systems provide power for a local network for a small community. Again, the use of a hybrid system is common, although in this case the alternative power source may be another renewable energy technology such as wind or small hydropower to supplement the PV output. The optimum combination will take account of the required load profile and the generation profile of the various power sources. It is possible to commence a local network with a small PV or PV hybrid system and then add to it as consumer demand grows and finance allows.

The first commercial use of PV cells was in space, providing the electrical power required by satellites for their operation, and this continues to be an important application of the technology. These systems could be described as a special case of a stand-alone system, where the space environment must be considered in determining the system size needed to meet the load requirement. Particular attention must be paid to the PV array size and weight, because of launch constraints, and to the performance of the solar cells after exposure to irradiation in the space environment.

Stand-alone PV systems are used for a wide range of applications from lighting for rural homes to telecommunication repeater stations, from providing the power for parking meters to maintaining the vaccine cold chain by powering refrigerators in rural clinics, from water pumping to navigation buoys. Some systems have only the PV array as their power source, whilst others incorporate one or more other power supplies. Whilst the uses differ greatly, the design approach is the same. The load is carefully considered and then the system is sized so as to achieve the required probability that it will meet the load given the climatic conditions at the location. In this way, PV systems can make a remarkable contribution to providing power in difficult locations and to people who have no other source of electricity.

Grid-connected PV systems

The grid-connected system, as its name implies, is a PV system that is installed in a location where there is also grid electricity available and a direct connection is made to allow power to be fed to or taken from the grid. We can divide grid-connected PV systems into two main categories, which are generally described as *centralised* and *distributed*. Centralised systems feed electricity directly into the grid and are usually connected at medium voltage. These systems will be discussed later in this section. We will first consider distributed systems, which generally feed local loads first and then export any excess to the grid. These systems are usually connected at the level of the electricity consumer (i.e. at low voltage). The most common example of this kind of system is the one that is mounted on or integrated into a building.

Building-integrated photovoltaic systems

The building-integrated photovoltaic (BIPV) system is an excellent application for the use of photovoltaics in an urban environment since it takes advantage of the distributed nature of both sunlight and the electrical load. The potential benefits of the BIPV system can be summarised as follows:

- In common with other PV systems and most renewable energy technologies, it has a lower environmental impact than production of electricity from conventional fuels.
- The electricity is generated at the point of use, so reducing the impacts and costs of distribution.
- There is a possibility of offsetting some of the cost of the PV array by the amount that would have been paid for the building material it has replaced.
- The system does not require additional land area, since building surfaces are used to accommodate the array.

The PV modules can be integrated in several different ways, for example to replace roofing tiles, in place of façade materials or as sunshades. Figure 12.10

Figure 12.10 Example of façade integration of photovoltaics. The photograph shows the 40 kWp PV façade on the Northumberland Building at the University of Northumbria. The PV array is integrated into the rainscreen overcladding. This system was installed in 1994 and is an early example of façade integration. Photograph courtesy of University of Northumbria.

Figure 12.11 Integrated PV façade using copper indium gallium diselenide thin-film modules. Photograph courtesy of Würth Solar GmbH.

shows an early example of façade integration using crystalline silicon modules, but there are many different ways of including the PV array in the building design. Figure 12.11 shows a later example, this time with thin-film copper indium gallium diselenide modules. It can be seen that the visual aspect of the two buildings is significantly different.

A large proportion of current PV systems on buildings are not fully integrated, but rather the array is mounted on the roof or façade, leaving the original exterior of the building in place. This is generally less expensive than full integration, although it is often not as aesthetically pleasing and does not present as many opportunities for using the modules to perform other building-related functions.

The principle of the technical system design is similar to that for other PV applications, but there are some additional aspects to be taken into account. In contrast to the stand alone systems described in the previous section, the BIPV system is rarely sized to meet a particular load but usually makes a contribution to the electricity requirement of the building as a whole. It may be designed to match the general load profile or to provide higher output levels when, for instance, air conditioning is required, but it does not need to be an autonomous system since the building also has a grid supply.

The area available for the BIPV array may be constrained by building design, shading from surrounding structures or owner preference. Thus, the system size is often dictated by the building's structural details rather than its electrical loads. The visual aspect of the system is also important and this will influence the choice of module type, location and detailed integration method. Finally, the system design must take into account ease of installation, maintenance and operation, and compliance with local building regulations, both structural and electrical.

A fully integrated BIPV array performs at least two tasks, the generation of electricity for use in the building and the protective functions of the external building element, but arrays can also be designed to perform additional functions. The most common function is shading of internal spaces, by louvre systems on the exterior of the building, by designing the cladding so as to provide shading to the windows at high Sun positions or by the use of semitransparent PV elements for a roof or façade. Figure 12.12 shows an example of the use of semitransparent modules in a glazed façade, where the cells provide both visual stimulation by variation of the arrangement pattern and shading to reduce solar gain and glare.

In some cases the heat at the rear of the modules can also be used. For all except the highest efficiency modules, less than 20% of the light falling on the module is converted to electricity and, whilst a few percent is reflected, the rest is absorbed as heat. This results in a module operating temperature that can be

Figure 12.12 Example of the use of photovoltaic modules to influence indoor lighting patterns. The Solar Office at the Doxford International Business Park in Sunderland, UK, has a 73 kWp array formed from semitransparent PV modules. The cell spacing is varied to create the light effects in the inner atrium. Photograph courtesy of Akeler Developments Ltd.

25–50 °C above ambient, depending on the details of integration. Reducing the operating temperature by removing some of the heat is advantageous in terms of increasing system efficiency and a double benefit can be obtained if the heat is useful for another purpose.

Because of the rather large area of the module and the relatively modest temperature differential between the module and ambient temperatures, it is not usually cost-effective to use forced air or fluid flow to extract the heat unless there is a direct use for that heated air or fluid. However, the heat can be used to assist natural ventilation within the building by taking in cold air at the bottom of the building. As this air is heated behind the PV façade, it rises and pulls in more cold air to replace it. One of the earliest examples of this type of application was the Mataró Library in Spain (Lloret *et al.*, 1997).

Even for a system where no use is made of the heat, care must be taken to ensure that the PV array operating temperature remains at an acceptable level. For ground-mounted PV arrays, there is free air movement around the array and so some cooling is effected. This is not the case for a BIPV system, which forms part of the building fabric. The design must include adequate ventilation around the modules if significant losses in efficiency are to be avoided. It is somewhat easier to ensure adequate ventilation for PV arrays added to the building rather than fully integrated. In this case, a small gap (typically a few cm) is provided between the existing roof or façade and the rear of the PV array so as to allow some natural airflow.

One aspect of installation that must be carefully considered is the possibility of shading of the modules, from surrounding buildings, trees and other vegetation and, in some cases, parts of the same building (e.g. by ventilation outlets, offset roofs, etc.). It can often be difficult to avoid shading completely and therefore the design of the system needs to minimise the effect on the output by good choice of string connections and splitting the system into sub-systems using a number of inverters.

The PV system is connected in parallel with the local grid, with the conventional electricity supply meeting any shortfall between the system electrical output and the building demand. The system must conform to safety regulations for connection and those relating to voltage, frequency and cut-off in the case of grid faults, as discussed previously. Arrangements can usually be made to sell back any excess production from the BIPV system to the electricity supply company. There is a wide range of tariffs offered for this electricity, ranging from the replacement generation cost (i.e. the cost for production of the same amount of electricity by the electricity company, not including distribution costs and overheads) to several times the normal electricity rate, where a scheme to promote BIPV exists.

The wide range of possible designs and the variation in the cost of the building material that is being replaced, and hence the offset of the PV system cost that this provides, makes it difficult to generalise about the cost of BIPV systems. Some systems are less costly than the material being replaced and so the electricity produced is essentially free except for any small additional maintenance costs. Most systems are replacing material that would be less costly than the PV array and so there is some residual cost to assign to the electricity produced. Nevertheless, the reducing costs of PV modules and other system equipment means that these distributed systems are becoming economically attractive across a wide range of locations, especially as the costs of conventionally generated electricity rise.

Centralised grid-connected PV systems

Since the inception of strong financial support schemes for PV across several European countries and in the US and Asia, there has been increased interest in centralised PV systems and the capacity of such systems has grown considerably. These systems are generally ground-mounted (although there are a few that are included on the roofs of large buildings or building complexes) and they are predominantly designed to feed output into the grid rather than meeting local loads. In the last five years, the maximum system capacity has grown from around 50 MW to around 250 MW, with even larger systems currently under construction.

The system design is modular in nature, with the overall system being divided up into a number of similar sub-systems, typically of 0.1–1 MW in size, depending on the overall capacity of the system. This approach has several advantages. From the electrical point of view, this allows operation at suitable voltages for reduction of losses, usually between 500–1000 V, and means that multiple identical inverters can be used. These can be installed in a distributed manner across the plant site to reduce cabling losses. From an installation point of view, the use of sub-systems allows different configurations to be used in different parts of the array field to allow for any physical constraints of the site. Finally, the modular nature allows the plant to be built in sections and, if permitted, to connect each section to the grid as it is completed. This reduces the time between the start of plant construction and the first sale of exported electricity. The centralised PV system is usually connected to the grid at medium voltage levels (typically tens of kV depending on the location), rather than at the low voltage level typical of distributed systems.

The modules are usually mounted on open rack support structures to allow maximum ventilation and therefore minimum operating temperature. The strength of the support structures must take account of wind or snow loading likely at the location. The most common arrangement is to have rows of modules, tilted at the

Figure 12.13 Ground-mounted PV system at Douneika, Greece. The system is rated at 2 MW and consists of 9,274 polycrystalline silicon modules, tilted at 27°. The physical and electrical layout takes account of the topography of the site. Photograph courtesy of Phoenix Solar Greece.

optimum angle for solar capture, and spaced so that there is a low level of shading between the rows. This requires a trade-off between the available land area and the amount of shading at low solar elevations and the spacing can be expressed by a parameter known as the Ground Cover Ratio. This is the ratio of the total area of PV modules to the total land area of the site and is typically around 0.5–0.6, although it can vary depending on the latitude and the details of the site. In some cases, the spacing, tilt angle and even azimuth angle can be varied on different parts of the site to take account of different topographies, the aim being to generate the highest energy yield from the available area. Figure 12.13 shows an example of a ground-mounted centralised PV system.

While most of the large grid-connected PV systems are fixed, some flat-plate systems employ Sun-tracking on either one or two axes. For tracked systems, the most common configuration is to mount a group of modules on a pedestal, to give the required freedom of movement, similar to the configuration illustrated in Fig. 11.29 in Chapter 11 for concentrator systems. The array field is then made up of a multiple of pedestals up to the required system capacity. Tracking both increases the output of the modules and modifies the output profile, keeping it high across a longer portion of the day. As a disadvantage, a system using tracking will require a larger area of land so as to keep the shading between the modules to an acceptable level. The support structure and tracking system also make the installation more costly. The performance advantage for a tracked system depends on the climatic details at the location and is more pronounced for sites with a high fraction of clear days across the year. The choice of whether or not to use flat-plate tracking is therefore location-dependent. Some of the large centralised systems also use concentrator PV technology (see Chapter 11), where the climatic conditions are favourable.

Environmental impact

Whereas the stand-alone system is designed to meet a specific load, the motivation for the grid-connected system is ultimately to reduce consumer energy costs and to reduce the environmental impact of electricity production. There is an environmental impact from the manufacture and disposal of PV modules and the rest of the system components, but this is substantially lower per unit of electricity across the system lifetime than for fossil-fuel-based generation (IPCC, 2012). Furthermore, it can be expected to reduce consistently over time, due to reducing energy inputs in manufacturing and reducing carbon content in relation to those inputs, as more renewable energy technologies are incorporated into the grid. The comparison of impact between PV systems and other renewable technologies depends on assumptions made about the system design and location, so would require a fuller treatment than is possible here. It can be concluded that PV systems, as part of a portfolio of renewable technologies, can contribute significantly to the reduction of carbon emissions resulting from electricity provision.

As the use of PV systems increases, it is also necessary to address the issue of how to deal with the system components at the end of the system life. Photovoltaic systems are generally straightforward to decommission and the industry, especially in Europe, has now begun to put in place schemes for the collection and recycling of modules at the end of their life. Within Europe, PV equipment comes under the Waste Electrical and Electronic Equipment Directive (European Union, 2012) and this requires the suppliers to provide a route for customers to return their used equipment. It has been demonstrated that a large proportion of the PV module can be recycled, with materials either being reused for manufacturing new modules (e.g. refurbished crystalline silicon cells) or in other sectors (e.g. glass).

Network development

As the cost of PV electricity reduces towards the cost of conventionally generated electricity in several countries, one of the technical challenges to be faced is how to deal with large amounts of PV electricity being fed into the electricity network, whether from distributed or centralised systems. The output from the PV system is variable, depending on weather conditions. The output profile can be modified by introducing storage, whether local or central, and, if sufficient capacity is installed, by curtailment of output power to match requirements. The use of the inverter to provide both storage and reactive power has been discussed earlier in the chapter. Clearly, the network also has to include generation capacity to meet demand when there is no solar output.

The ability to control the supply of electricity via the grid where there is a high penetration level of variable renewable technologies (*e.g.* wind, PV) requires knowledge of likely outputs. This is leading to advances in forecasting techniques for system output over short and medium timescales and in the communication systems and models required to make use of those forecasts.

12.5 Conclusions

Photovoltaic cells have social and commercial value when they are used in a system to provide a service, whether that is the provision of electricity to meet a specific load or to contribute to the power network serving a multitude of loads. This chapter has given a brief overview of the technical considerations that allow the cells to provide such a service, and of the current applications of PV systems.

Photovoltaic cells may be incorporated directly into a product and add value to that product to the extent that their use is commercially viable. In most cases, however, the cells are contained in a PV module, interconnected to give a required output depending on the application. The module structure also protects the cells from damage in transportation, installation and use. This chapter has described the construction of PV modules and their quality assurance testing, designed to provide a product with an assured output, reliability and lifetime when operating in varied climatic conditions. It is these developments in module performance that have provided the basis for the expanding market for PV throughout the world. While the current module construction is similar in concept to that developed over 30 years ago, consistent developments in manufacturing have reduced the material and energy requirements in manufacture and reduced the electrical losses in moving from the individual cell to the module. Modules have also become larger and more varied to address different user requirements. As new cell types develop and new applications are found, PV modules will surely continue to evolve. Nevertheless, the requirements for lifetime and durability will remain.

This chapter has also discussed the range of other equipment needed in PV systems to give optimal operation in both stand-alone and grid-connected systems.

Here, it was possible to give only a brief overview of the equipment and the design criteria, especially in relation to the development of inverters. As the applications of PV systems become more widespread, especially in relation to connection to the grid, so the BOS components are also being developed to address specific market needs such as grid support. The interested reader is referred to the specialist journal and conference papers for the latest information, as well as to manufacturer information on the latest commercial products.

Photovoltaic systems are clearly well suited to the provision of electrical power at locations that are remote from the electricity grid and where reliability is a key requirement. This is especially true where only small amounts of power are needed to bring major improvements in quality of life, e.g. for rural medical applications. The main issues are the correct design of the system, the education of the user to ensure the system is operated correctly, and the provision of financial mechanisms to allow the purchase of the systems (which is beyond the scope of this chapter).

The great majority of the installed capacity of PV systems around the world is grid-connected, with a market share estimated to be $>98\%$ at the end of 2011 (IEA Photovoltaic Power Systems Programme, 2012). The reduction in PV module price over the last few years and the simplicity of their installation and operation has meant that small, distributed systems have become popular. The market promotion schemes introduced by several countries have also encouraged the installation of centralised PV systems and capacity has increased year on year, as described in Chapter 13.

While PV systems only provide a very small percentage of networked electricity at the moment, there are a few locations, particularly in southern Germany, where there is a high penetration of systems. Under the correct conditions, it has been demonstrated that the load can be met solely from a combination of wind and solar electricity. This situation brings challenges in relation to control and variability of supply, but also provides the stimulus for the developments in system hardware and output forecasting that will eventually allow PV to be a major contributor to the world's electricity supply.

References

Bletterie, B., Bründlinger R. and Lauss G. (2011), 'On the characterisation of PV inverters' efficiency — introduction to the concept of achievable efficiency', *Progr. Photovoltaics* **19**, 423–435.

European Union (2012), *Directive 2012/19/EU of the European Parliament and of the Council on Waste Electrical and Electronic Equipment (WEEE), Official Journal of the European Union*, 24 July.

Huld T., Šúri M. and Dunlop E. D. (2008), 'Comparison of potential solar electricity output from fixed-inclined and two-axis tracking photovoltaic systems in Europe', *Progr. Photovoltaics* **16**, 47–59.

IEA Photovoltaic Power Systems Programme (2012), *Trends in Photovoltaic Applications: Survey Report of Selected IEA Countries between 1992 and 2011*, Report IEA-PVPS T1-21:2012.

IEC (2007), *Solar Photovoltaic Energy Systems — Terms, Definitions and Symbols*, IEC/TS 61836:2007.

IEC (2005), *Crystalline Silicon Terrestrial Photovoltaic (PV) Modules — Design Qualification and Type Approval*, IEC 61215:2005.

IEC (2008), *Thin-film Terrestrial Photovoltaic (PV) Modules — Design Qualification and Type Approval*, IEC 61646: 2008.

IPCC (2012), *Renewable Energy Sources and Climate Change Mitigation*, Special Report of the Intergovernmental Panel on Climate Change, Cambridge University Press, p. 732.

Lloret A., Aceves O., Andreu J., Merten J., Puigdollers J., Chantant M., Eicker U. and Sabata L. (1997), 'Lessons learned in the electrical system design, installation and operation of the Mataró Public Library', *Proc. 14th European Photovoltaic Solar Energy Conf.*, Barcelona, Spain. H. S. Stephens & Associates, Bedford, pp. 1659–1664.

CHAPTER 13

THE PHOTOVOLTAIC BUSINESS: MANUFACTURERS AND MARKETS

ARNULF JÄGER-WALDAU
Senior Scientist, Renewable Energy Unit
Institute for Energy and Transport
EU Joint Research Centre, Ispra, Italy
Arnulf.Jaeger-Waldau@ec.europa.eu

Viewed in the light of the world's growing power needs, these gadgets are toys. But so was the first flea-power motor built by Michael Faraday over a century ago — and it sired the whole gigantic electrical industry.

From *Prospects for Solar Power*, Harland Manchester, Reader's Digest, June 1955.

13.1 Introduction

Over the last decade photovoltaics has been one of the fastest-growing industries, with a compound annual growth rate (CAGR) of over 50%. Solar cell and module production have increased by more than an order of magnitude during this decade and despite the negative impacts of the on-going economic crisis since 2009, cumulative installed capacity of photovoltaic (PV) systems surpassed 100 GW[1] at the end of 2012. In 2013, new installed photovoltaic capacity could for the first time surpass new installed wind-power capacity. This is mainly driven by rapid market expansion in Asia, which will be the largest market, surpassing Europe. Photovoltaics is on the right track to become a major electricity source within the next decade.

13.1.1 *Global production data*

Estimates of global cell production[2] in 2013 vary between 35 GW and 42 GW. The uncertainty in these data is due to the highly competitive market environment, as

[1] All production and installation figures cited in W in this chapter are actually Wp.

[2] Solar-cell production capacities mean: in the case of wafer silicon-based solar cells, only the cells; in the case of thin-films, the complete integrated module. Only those companies that actually produce the active circuits (solar cells) are counted; companies that purchase these circuits and make cells are not counted.

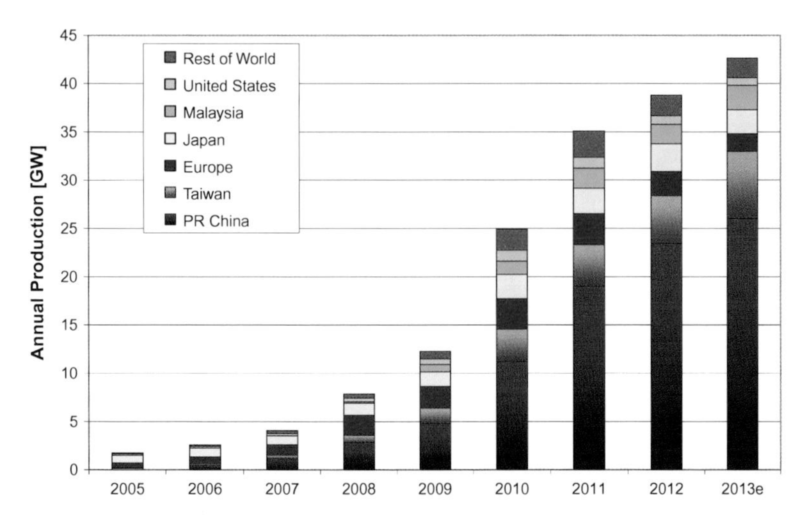

Figure 13.1 World PV cell/module production from 2005 to 2012, and 2013 (estimated). Data source: *Photon International* (2012), *RTS Corporation* (2013), *PV News* (2013) and the author's analysis.

well as the fact that some companies report shipment figures, while others report sales and others still production figures. The data presented, collected from stock market reports of listed companies, market reports and colleagues, were compared to various data sources and this led to an estimate of 38.5 GW (Fig. 13.1), representing a moderate increase of about 10% compared with 2012; as manufacturing plant utilisation began to improve in the second quarter of 2013 another moderate increase of about 10% is expected for 2013.

Total PV production has increased almost by two orders of magnitude since 2000, and the CAGR over the last decade has been above 50%. The most rapid growth in annual production over the last five years has occurred in Asia, where China and Taiwan together now account for more than 70% of worldwide production.

Publicly traded companies manufacturing products along the value chain, installing photovoltaic electricity systems or offering related services, have attracted a growing number of private and institutional investors. In 2012, worldwide new investments in clean energy decreased by 11% compared with 2011 but were still very considerable at $269 billion (€207 billion[3]), including $30.5 billion (€23.5 billion) corporate and government research and development spending (Bloomberg New Energy Finance, 2013a; PEW Charitable Trusts, 2013). In 2012,

[3]Exchange rate: 1€ = 1.30 US$.

clean energy markets outside the Group of 20 (G20) grew by more than 50% to surpass $20 billion (€15.4 billion), while investments in the G20 countries dropped by 16% to $218 billion (€167.7 billion). Despite the overall decline in investments, the decrease of renewable energy system prices more than compensated this and allowed these investments to install a record 88 GW of new clean-energy generation capacity, bringing the total to 648 GW, capable of producing more than 1500 TWh of electricity per annum. This corresponds to 64% of the electricity generated per annum by nuclear power plants worldwide.

For the third year in a row, solar power attracted the largest amount of new investment into renewable energies. Despite a 9% decline in solar-energy investments, solar power attracted $137.7 billion (€105.9 billion), which is 57.7% of all new renewable energy investments (Bloomberg New Energy Finance, 2013b).

In contrast to Europe and the Americas, where new investments in renewable energy decreased, new investments continued to increase in Asia/Oceania, reaching $101 billion (€77.7 billion) in 2012. Europe took the second place with $62.1 billion (€47.8 billion), followed by the Americas with $50.3 billion (€38.7 billion) (PEW Charitable Trusts, 2013). With a 20% increase, China became the largest investor in renewable energy with $65.1 billion (€50.1 billion) followed by the USA with $35.6 billion (€27.4 billion) and Germany with $22.8 billion (€17.5 billion). The country with the biggest change in 2012 was South Africa, which moved up to ninth place with $5.5 billion (€4.2 billion).

13.1.2 *Prices*

The existing overcapacity in the solar industry has led to continuous downward price pressure along the value chain and resulted in a reduction of spot market prices for polysilicon, solar wafers and cells, as well as solar modules. Photovoltaic module prices have decreased in price by 80% since 2008, and by 20% in 2012 alone (Bloomberg New Energy Finance, 2013b). These rapid price declines are putting all solar companies under enormous pressure and access to fresh capital is key to survival. It is believed that this situation will continue until at least 2015, when the global PV market should exceed 50 GW of new installations. The increase of polysilicon spot prices and the levelling of module prices since the beginning of 2013 indicate that some production capacity has been taken out, and that prices may stabilise for some time until they are back on the learning-curve slope. It should be noted that PV system hardware costs are priced more or less the same worldwide, and the so-called soft costs, which consist mainly of financing, permit and labour costs, and installer/system integrator profits, are the main reason for the significant differences which are still observed.

The continuation of the difficult worldwide financial situation and the fact that support schemes are changed with little notice, undermining long-term investor confidence, make project financing more difficult as risk premiums are added. On the other hand, the declining module and system prices have already opened new markets that offer perspectives for further growth of the industry — at least for those companies with the capability to expand and reduce their costs at the same pace.

Despite the problems of individual companies, business analysts are confident that the industry fundamentals, as a whole, remain strong and that the overall photovoltaics sector will continue to experience significant long-term growth. For example, in its second Medium-Term Renewable Energy Market Report, published in July 2013, the IEA increased its predicted installed capacity to 300 GW of cumulative PV installations by 2018 (IEA, 2013a).

Market predictions for new PV system installations in 2013 have been upgraded regularly during the year. They range between 31 GW from Solarbuzz, 33.1 GW in the Bloomberg conservative scenario, >35 GW from IHS research and 39.6 GW in the Bloomberg optimistic scenario (Bloomberg New Energy Finance, 2012; IHS Global Demand Tracker, 2013; Solarbuzz, 2013). The wide spread of analyst numbers derives from the different methodologies, and whether the given numbers represent constructed/installed systems or actual grid-connected systems. For 2014, analysts expect a further increase to over 40 GW, mainly driven by growing Asian markets. Even in the optimistic forecasts, massive overcapacities in cell and module manufacturing will continue to exist. Depending how capacities are calculated, overcapacity estimates for 2013 range between 60 and 70 GW.

Despite a number of bankruptcies and companies idling production lines, or even permanent closure of production lines, the number of new entrants into the field, notably large semiconductor or energy-related companies, has overcompensated this in the past. However, the rapid changes in the sector and the difficult financing situation make a reasonable forecast of future capacity developments very speculative. The consequence is the continuation of price pressure in an oversupplied market. This will accelerate the consolidation of the photovoltaics industry and spur even more mergers and acquisitions.

The existing overcapacity is a result of very ambitious investments dating back to 2010, triggered by the more than 150% growth of the PV market in that year. This led to a peak in capital spend on manufacturing equipment of about $14 billion (€10.8 billion) in 2011. Since then, equipment spending has declined dramatically and will probably hit the bottom with around $1–2 billion (€0.77–1.54 billion) invested in 2013 before a moderate recovery from 2014 onwards will be possible. This development has had a serious effect on equipment manufacturers, all of which now need a new strategy for the PV industry. Companies with no significant business segment outside the PV supply chain were hit the hardest, and some of them are struggling to survive the slump until the predicted recovery kicks in.

13.1.3 *Market share by cell/module type*

Current solar-cell technologies are well established and provide a reliable product, with sufficient efficiency and energy output for at least 25 years of lifetime. This reliability, and the increasing likelihood of electricity interruption from grid overloads, as well as the rise in electricity prices from conventional energy sources, add to the attractiveness of photovoltaic systems.

With over 85% of current production, wafer-based crystalline silicon remains the dominant market technology, discussed by Martin Green in Chapter 3. Projected silicon production in 2013 ranges between 290,000 metric tons (Bloomberg New Energy Finance, 2013b) and 409,690 metric tons (Ikki, 2013a). It is estimated that about 27,000 metric tons will be used by the electronics industry. The resulting solar-cell production will depend on the amount of material used per Wp; the current worldwide average is about 5.6 g/Wp.

Estimates of the thin-film share of the market by different analysts and market institutes range from 9 to 12%. This is shared by the three main thin-film technologies: 3.5–4% a-Si/μ-Si, ~5% CdTe and ~2.5% CIGS, described in Chapters 4, 5 and 6, respectively.

As in other areas of technology, new products will enter the market, enabling further cost reduction. Concentrated photovoltaics (CPV), described in Chapter 11, is an emerging market. There are two main CPV options — either high concentration >300 Suns (HCPV), or low to medium concentration with concentration factors of 2 to about 300. In order to maximise the benefits of CPV, high direct normal irradiation (DNI) is required and this is only found in a limited geographical range — the Sun Belt of the Earth. Dye cells, discussed in Chapter 9, are entering the market as well. The development of these technologies is accelerated by the growth of the PV market as a whole, but economic conditions for their market entry are getting more and more challenging as silicon module and system prices are continuing to decrease.

It can be concluded that, in order to maintain the extremely high growth rate of the photovoltaic industry, different technology pathways have to be pursued at the same time. As the cost share of solar modules in a PV system has declined to below 40% in a residential system and below 50% in a commercial system, the non-technology costs and 'soft costs' have to be targeted for further significant cost reductions.

With an increasing share of PV electricity in the grid, the economics of integration become more and more important and it is urgent to focus attention on issues such as:

- Development of new business models for the collection, sale and distribution of photovoltaic electricity, e.g. development of bidding pools at electricity exchanges, virtual power plants with other renewable power producers and storage capacities.
- Adaptation of regulatory and legal procedures to ensure fair and guaranteed access to the electricity grid and market.

The cost of PV-generated electricity has dropped to less than 0.05 €/kWh, and the main cost component is to transport the power from the module to where it is needed. Therefore innovative and cost-effective overall electricity system solutions for the integration of PV electricity are needed to realise the vision of PV as a major electricity source for everybody everywhere. More effort is needed to optimise the non-PV costs and greater public support, especially as regards regulatory measures, is crucial.

13.2 The photovoltaic market

After the worldwide photovoltaic market more than doubled in 2010, the market grew again by almost 30% in 2011, and then another 11% in 2012, despite difficult economic conditions. The 2010 market volume of 20.9 GW includes those systems in Italy, which were reported under the second '*conto energia*' and installed, but connected only in 2011. There is uncertainty about the actual installation figures in China. The Chinese National Energy Administration reported a cumulative installed capacity of 7 GW at the end of 2012, whereas most other market analysts cite figures between 8 and 8.5 GW (National Energy Administration, 2013). The stronger than expected market in Germany and the marked increase of installations in Asia and the USA resulted in a new installed capacity of about 30 GW in 2012; increases to about 37 GW and 45 GW are expected for 2013 and 2014, respectively, as shown in Fig. 13.2. This comprises mostly the grid-connected photovoltaic market.

To what extent the off-grid and consumer product markets are included is not clear, but it is believed that a substantial part of these markets are not accounted for as it is very difficult to track them. Figure 13.3 shows the cumulative capacity installed in the period 2000 to 2014. With a cumulative installed capacity of over 69 GW, the European Union (Rest of Europe + Italy + Spain + Germany) had the

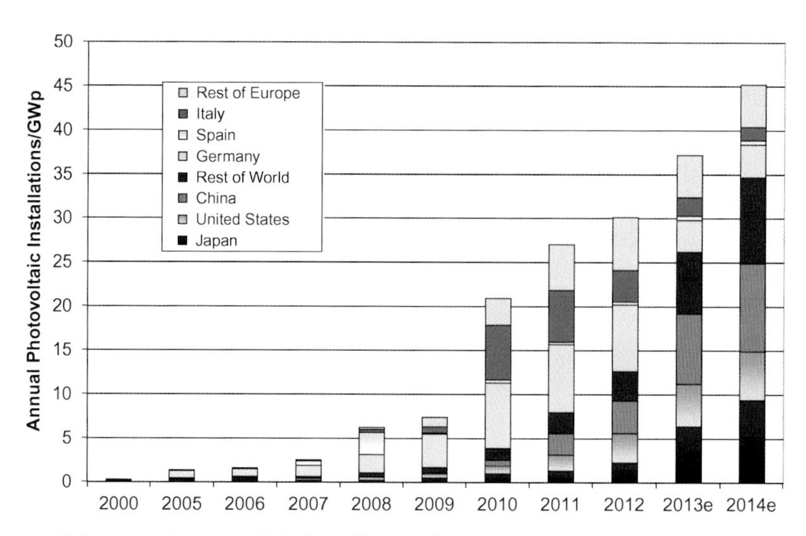

Figure 13.2 Annual photovoltaic installations from 2000 to 2012 and estimates for 2013 and 2014. Data sources: European Photovoltaic Industry Association, 2013; National Energy Administration, 2013; Systèmes Solaires, 2013, and author's analysis.

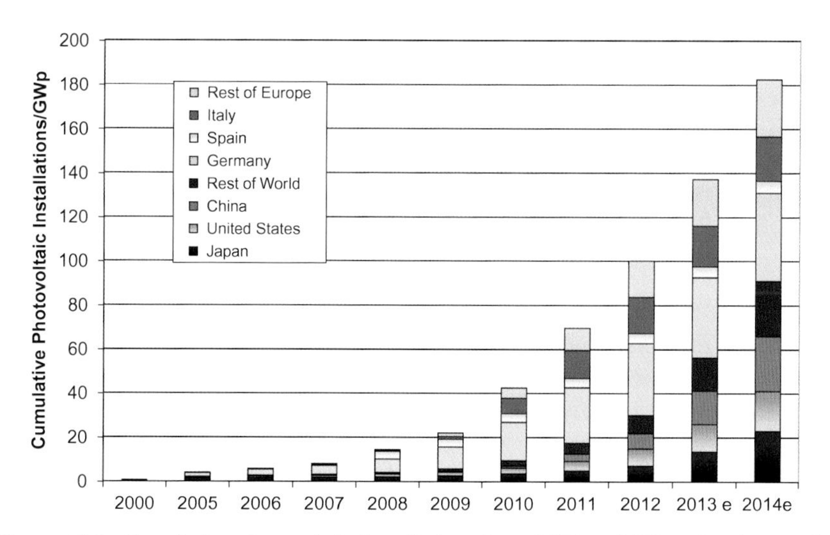

Figure 13.3 Cumulative photovoltaic installations from 2000 to 2012 and estimates for 2013 and 2014. Data sources: European Photovoltaic Industry Association, 2013; National Energy Administration, 2013; Systèmes Solaires, 2013, and author's analysis.

largest portion of the total worldwide 100 GW of installed capacity at the end of 2012.

13.2.1 *Asia and Pacific Region*

The Asia and Pacific Region continued its upward trend in PV installations. There are a number of reasons for this, ranging from declining system prices, heightened awareness and favourable policies to the sustained use of solar power for rural electrification projects. Countries such as Australia, China, India, Indonesia, Japan, Malaysia, South Korea, Taiwan, Thailand, the Philippines and Vietnam show a very positive upward trend, thanks to increasing government commitment to the promotion of solar energy and the creation of sustainable cities. The introduction or expansion of feed-in tariffs (FiTs) is expected to be an additional big stimulant for on-grid solar PV system installations for both distributed and centralised solar power plants in countries such as Australia, Japan, Malaysia, Thailand, Taiwan and South Korea.

In 2012, about 8.5 GW of new PV electricity generation systems was installed in the region, which corresponds to a 60% growth as compared with 2011. The largest market was China with 3.7 GW, followed by Japan with 2.3 GW and Australia with over 1 GW. For 2013, a market increase to about 20 GW is expected, driven by the major market growth in China ($\sim$6–8 GW), India ($>$1 GW), Japan (6.9–9.4 GW), and Malaysia and Thailand. For 2014, the market expectations exceed 25 GW.

Australia

In 2012, slightly more than 1 GW of new solar photovoltaic electricity systems were installed in Australia, bringing the cumulative installed capacity of grid-connected PV systems to 2.45 GW. Photovoltaic electricity systems accounted for 36% of all new electricity generation capacity installed in 2011. As in 2011, the 2012 market was dominated by grid-connected residential systems, which accounted for more than 90% of the total installed. The average PV system price for a grid-connected system fell from 6 AUD/Wp (4.29 €/Wp[4]) in 2010 to 3.9 AUD/Wp (3 €/Wp[5]) in 2011 and 3.0 AUD/Wp (2.15 €/Wp[4]) in early 2013. Due to this, the cost of PV-generated electricity has reached, or is even below, the average residential electricity price of 0.27 AUD/kWh (0.19 €/kWh).

[4] Average exchange rate 2010 and 2013: 1€ = 1.40 AUD.
[5] Average exchange rate 2011: 1 €= 1.30 AUD.

In 2012, PV electricity systems generated 2.37 TWh, which was about 1% of total Australian electricity supply. The total renewable electricity share was 13.34% and this should increase to 20% by 2020. For 2013, a PV market of 750 MW is forecast.

Within two years, Australia has installed almost 2 GW of PV capacity, and about 10% of residential buildings now have a PV system. At first, most installations took advantage of the incentives under the Australian Government's Renewable Energy Target (RET) mechanisms and feed-in tariffs in some states or territories. At the beginning of 2011, eight out of the eleven Australian federal states and territories had introduced eleven different kinds of FiT schemes, mainly for systems under 10 kWp. All except three of these schemes had built-in caps, which were partly reached that year, so that in 2012 only six schemes were available for new installations.

India

Market estimates for PV systems in India in 2012 vary between 750 and 1000 MW, because some statistics cite the financial year and others the calendar year. According to the Ministry of New and Renewable Energy (MNRE), total capacity at the end of financial year (FY) 2012/13[6] was 1.9 GW grid-connected and 125 MW off-grid capacity. The Indian Jawaharlal Nehru National Solar Mission (JNNSM) was launched in January 2010, and it was hoped that this would give impetus to the grid-connected market. This Mission aims to make India a global leader in solar energy and envisages an installed solar generation capacity of 20 GW by 2020, 100 GW by 2030 and 200 GW by 2050. After only a few MW had been installed in 2010, installations in 2011 slowly picked up, but the majority of the JNNSM projects will come online from 2015 onwards. Market expectations for 2013 vary between 1.00 and 1.35 GW (Bridge to India, 2013; Mercom Capital Group, 2013).

Israel

Three years after the introduction of an FiT in 2008, Israel's grid-connected PV market saw about 60 MW of capacity newly connected in 2012. One of the main drives behind the development of solar energy is energy security, and Israel plans to have about 1 GW of solar PV systems installed by the end of 2014. In August 2012, about 215 MW had been built and a further 300 MW approved (Bloomberg New Energy Finance, 2012). The FiTs depend on the system size segment and have individual caps. Market expectations for 2013 are in the range 150–200 MW.

[6]The Indian financial year ends 31 March.

Japan

The Japanese market experienced significant growth in 2012, increasing its domestic shipments to 2.47 GW. Cumulative installed capacity increased by about 1.7 GW to reach 6.6 GW at the end of 2012 (IEA, 2013b). By the end of February 2013, more than 5.5 GW had received approval under the new national feed-in scheme, which started in July 2012. The market outlook for Japan has been raised and is now between 6.9 and 9.4 GW new installed capacity by the end of 2013 (Bloomberg New Energy Finance, 2013d).

The consequence of the accident at the Fukujima Daiichi Nuclear Power Plant in March 2011 was the reshaping of the country's energy strategy. For PV power, an official target of 28 GW was set for 2020. In July 2012, a Ministry for Economy, Trade and Industry (METI) panel published its long-awaited plan to reform the country's power market. This aims to increase the share of renewable power supply from 11% in 2011 to 25–35% by 2030.

Until 2010, residential rooftop PV systems represented about 95% of the Japanese market. In 2011, due to a change in permit policy, large ground-mounted systems, as well as large commercial and industrial rooftop systems, increased their market share to about 20%.

In June 2012, METI finally issued the Ministerial Ordinances for the new FiTs for renewable energy sources and these were adjusted in March 2013. The tariff for commercial installations (total generated power) larger than 10kWp is ¥36 per kWh for 20 years and ¥38 per kWh[7] for residential installations (surplus power) smaller than 10 kWp for 10 years, starting from 1 April 2013 (Ikki, 2013a).

People's Republic of China

In 2012, the Chinese PV market grew to 3.7 GW, bringing the cumulative installed capacity to about 7 GW (NEA, 2013). About 3.3 GW of this capacity was connected to the grid at the end of 2012 (SERC, 2013). This is a 600% growth compared with 2010, but still represents only 16% of total Chinese PV production. In January 2013, the Chinese National Energy Administration (NEA) announced an ambitious target of 10 GW of new domestic PV installations during 2013 (NEA, 2013). A further increase can be expected due to the recent change of the solar energy target for 2015, which was increased to 41 GW for PV and solar thermal electricity combined by the NEA (Bloomberg New Energy Finance, 2013a).

According to the 12th Five-Year Plan, which was adopted on 14 March 2011, China intends to cut its carbon footprint and become more energy efficient. The

[7] Exchange rate: 1€ = 130 ¥.

Plan targets are 17% lower carbon dioxide emissions and 16% lower energy consumption per unit of gross domestic product (GDP) by 2015. The total investment in the power sector under the Plan is expected to reach $803 billion (€618 billion), divided into $416 billion (€320 billion), or 52%, for power generation, and $386 billion (€298billion) to construct new transmission lines and other improvements to China's electrical grid.

In August 2012, the NEA released China's new renewable energy five-year plan for 2011–2015 (NEA, 2012). The new goal of the NEA calls for renewable energy to supply 11.4% of the national energy mix by 2015. To achieve this goal, renewable power generation capacity has to be increased to 424 GW. Hydro-power is planned to be the main source, with 290 GW including 30 GW pumped storage, followed by wind with 100 GW, solar with 21 GW (this target was later increased to 41 GW) and biomass with 13 GW.

The plan estimated new investments in renewable energy of CNY 1.8 trillion (€222 billion) between 2011 and 2015. China aimed to add a total of 160 GW of new renewable energy capacity during the period 2011–15, in the form of 61 GW hydro, 70 GW wind, 21 GW solar[8] (10 GW small distributed PV, 10 GW utility scale PV and 1 GW solar thermal power), and 7.5 GW biomass. For 2020, the targets were set as 200 GW for wind, 50 GW for solar (27 GW small distributed PV, 20 GW utility-scale PV and 3 GW solar thermal power) and 30 GW for biomass.

These required investment figures to be in line with a World Bank report stating that China needs additional investment of $64 billion (€49.2 billion) annually over the next two decades to implement an energy-smart growth strategy (World Bank, 2010). However, according to the report, the reductions in fuel costs through energy savings could largely pay for these additional investment costs. At a discount rate of 10%, the annual net present value (NPV) of the fuel cost savings from 2010 to 2030 would amount to $145 billion (€111.5 billion), which is about $70 billion (€53.8 billion) more than the annual NPV of the additional investment costs required.

On 24 February 2012, the Chinese Ministry of Industry and Information Technology published its industrial restructuring and upgrading plan (2011–2015) for the photovoltaic industry (MIIT, 2012). In this document, the Ministry stated that by 2015 it expects to support only backbone enterprises, which should produce a minimum of 50000 MT polysilicon, or 5 GW of solar cell or module production. The Plan also projects a cost reduction of electricity generated with PV systems to 0.8 CNY/kWh (0.098 €/kWh[9]) by 2015 and 0.06 CNY/kWh (0.074 €/kWh) by 2020.

[8]This has been updated in the meantime.

[9]Exchange rate 2012: 1€ = 8.1 CNY.

South Korea

In 2012, about 250 MW of new PV systems were installed in South Korea, bringing the cumulative capacity to a total of 981 MW (IEA, 2013c). Since January 2012, Korea's Renewable Portfolio Standard (RPS) officially replaced its earlier FiTs. For 2013, the RPS set-aside quota for PV was set at 450 MW, and this should increase to 1.2 GW in 2016. The result is an annual target of 230 MW in 2013, 240 MW in 2014, 250 MW in 2015 and 260 MW in 2016. Under the RPS, income for power generated from renewable energy sources is a combination of the wholesale system marginal electricity price and the sale of Renewable Energy Certificates (RECs); certificate sales in the second half of 2012 were around 40000 KRW/MWh (0.026–0.035 €/MWh[10]). These RECs are multiplied by an REC multiplier, varying between 0.7 for ground-mounted free-field systems to 1.5 for building-adapted systems, to reflect the different costs of the different systems.

The new RPS programme obliges power companies with at least 500 MW of generating capacity to increase their renewable energy mix from not less than 2% in 2012 to 10% by 2022. The renewable energy mix in the Korean RPS is defined as the proportion of renewable electricity generation to the total non-renewable electricity generation.

Taiwan

In June 2009, the Taiwan Legislative Yuan gave its final approval to a Renewable Energy Development Act to bolster the development of Taiwan's green energy industry. The goal was to increase Taiwan's renewable energy generation capacity by 6.5 GW to a total of 10 GW within 20 years. Promotion of all types of renewable energy was expanded: total installed capacity of 9952 MW (accounting for 14.8% of total power generation installed capacity), with new installed capacity of 6600 MW, has been planned by 2025 so that the goal set by the Act can be achieved five years early. By 2030, the target will be further increased to 12.5 GW, accounting for 16.1% of total power generation installed capacity and capable of generating 35.6 billion kWh of electricity. This is equivalent to the annual electricity consumption of 8.9 million households (accounting for 78% of the number of households consuming electricity nationwide).

Between 2009 and 2012, a total capacity of 194 MW was installed, bringing the total installed capacity to 222 MW at the end of 2012 (Ministry of Economic Affairs, 2013). The FiTs in the first half of 2013 for rooftop systems were 8.4

[10]Exchange rate: 1€ = 1, 420 KRW.

NT$/kWh (0.215 €/kWh[11]) for systems up to 10 kW, 7.54 NT$/kWh (0.193 €/kWh) for systems between 10 and 100 kW, 7.12 NT$/kWh (0.183 €/kWh) for systems between 100 and 500 kW and 6.33 NT$/kWh (0.162 €/kWh) for systems larger than 500 kW. Ground-mounted systems had a tariff of 5.98 NT$/kWh (0.153 €/kWh). For the second half of 2013, a tariff reduction of 2.5% to 6.1% was foreseen to reflect declining system prices.

The installation targets for 2013 were increased twice and are currently at 175 MW. This is in line with Taiwan's new Million Solar Rooftop Programme, which aims to achieve installed capacity of 610 MW by 2015 and 3.1 GW by 2030. However, the increased installation target for 2013 represents only 3.5% of Taiwan's total 2012 PV production volume, reflecting the country's strength as a PV exporter.

Thailand

Thailand enacted a 15-year Renewable Energy Development Plan (REDP) in early 2009, setting the target of increasing the renewable energy share of the final energy consumption of the country to 20% by 2022. Besides a range of tax incentives, PV systems are eligible for a feed-in premium or 'adder' for a period of ten years. The original 8 THB/kWh[12] (0.182 €/kWh) 'adder' facilities in the three southern provinces, and those replacing diesel systems, are eligible for an additional 1.5 THB/kWh (0.034 €/kWh. This was reduced to 6.5 THB/kWh (0.148 €/kWh) for those projects not approved before 28 June 2010. The original cap of 500 MW was increased to 2 GW at the beginning of 2012, due to the high oversubscription of the original target. In addition to the 'adder' programme, projects are now being developed with power purchase agreements (PPAs).

At the end of 2012, grid-connected PV systems of about 360 MW total capacity were operational, of which 210 MW were installed in that year (IEA, 2013c). In September 2012, projects with around 1.8 GW capacity had signed PPAs, projects with 76 MW already had a letter of intent (LOI) and another 925 MW of projects were waiting for an LOI (Kruangam, 2013).

13.2.2 *Emerging markets*

Bangladesh

In 1997, the government of Bangladesh established the Infrastructure Development Company Limited (IDCOL) to promote economic development in Bangladesh.

[11] Exchange rate: 1€ = 39 NT$.
[12] Exchange rate: 1€ = 44 THB.

In 2003, IDCOL started its Solar Energy Programme to promote the dissemination of solar home systems (SHSs) in the remote rural areas of Bangladesh, with financial support from the World Bank, the Global Environment Facility (GEF), the German Kreditanstalt für Wiederaufbau (KfW), the German Technical Cooperation (GTZ), the Asian Development Bank and the Islamic Development Bank. By April 2013, the programme had seen more than 2 million SHSs installed in Bangladesh (World Bank, 2013). At the time of writing, the installation rate is more than 60,000 units per month. In 2011, the Asian Development Bank agreed to provide financial support to Bangladesh for implementing the installation of 500 MW within the framework of the Asian Solar Energy Initiative (The Daily Star, 2011; UNB Connect, 2011).

Indonesia

The development of renewable energy in Indonesia is regulated in the context of national energy policy by Presidential Regulation No. 5/2006 (The President of the Republic of Indonesia, 2006). This decree states that 11% of the national primary energy mix in 2025 should come from renewable energy sources and the target for solar PV was set at 1000 MW by 2025. By the end of 2011, about 20 MW of solar PV systems had been installed, mainly for rural electrification purposes. In 2012, the Indonesian Ministry of Energy and Mineral Resources (ESDM) drafted a Roadmap that foresaw installations of 220 MW between 2012 and 2015 (ESDM, 2012). A new policy to promote solar energy through auction mechanisms was published in June 2013 (Bloomberg New Energy Finance, 2013a); how this new policy will influence the market remains to be seen.

Kazakhstan

The development of renewable energy was one of the priorities of Kazakhstan's State Programme for Accelerated Industrial and Innovative Development for 2010–2014. The main goal of this programme is to develop a new and viable economic sector for growth, innovation and job creation. In addition, it drives the development of renewable energy sources for the electricity sector in Kazakhstan and is regulated by the Law on Support to the Use of Renewable Energy Sources, adopted in 2009 (CIS Countries Legislation Database, 2009). In February 2013, the Kazakh government decided to install at least 77 MW of PV by 2020 (Government of Kazakhstan, 2013). In 2011, JSC NAC Kazatomprom, jointly with a French consortium headed by Commissariat à l'Energie Atomique et aux Energies Alternatives (CEA), started the project Kaz PV with the aim of producing PV modules based on Kazakh silicon (Kaz Silicon, 2011). The first project step was concluded in January 2013, when a new 60 W PV module production plant was opened in Kazakhstan's capital city, Astana. In May 2013, Zhambilskie Electricheskie Seti LLP signed a Memorandum of Understanding with NanoWin Thin Film Tech to build a 60 MW CIGS factory and a 5 MW solar plant in the country (Nanowin, 2013).

Malaysia

The Malaysia Building Integrated Photovoltaic (BIPV) Technology Application Project was initiated in 2000, and by the end of 2009, a cumulative capacity of about 1 MW of grid-connected PV systems had been installed under this programme. The Malaysian government officially launched their GREEN Technology Policy to encourage and promote the use of renewable energy for Malaysia's future sustainable development in 2009. By 2015, about 1 GW is planned to come from renewable energy sources, according to the Ministry of Energy, Green Technology and Water.

In April 2011, renewable energy FiTs were approved by the Malaysian Parliament, with a target of 1.25 GW installed by 2020. For the period December 2011–June 2014, a total quota of 125 MW was allocated for PV. The 2013 tariffs were set by the Sustainable Energy Development Authority between 0.782 and 1.555 MRY/kWh (0.195 to 0.389 €/kWh[13]), depending on the type and system size. In addition, there is a small bonus for local module or inverter use. An annual reduction in these tariffs is foreseen. As of 30 April 2013, 28.92 MW of PV systems were operational under the new FiT scheme and another 141.58 MW had received approval, and were in various stages of project planning or installation (SEDA, 2013). First Solar (USA), Hanwha Q Cells (Korea/Germany), Sunpower (USA), and recently Panasonic (Japan), have set up manufacturing plants in Malaysia, with more than 3.8 GW of total production capacity.

The Philippines

The Philippines' Renewable Energy Law was passed in December 2008 (Philippine Department of Energy, 2008). Under this law, the country had to double the energy it derives from renewable energy sources within ten years. In June 2011, the Energy Secretary unveiled a new Renewable Energy Roadmap, which aimed to increase the share of renewables to 50% of total energy consumption by 2030, and to boost renewable energy capacity from the 2011 level of 5.4 GW to 15.4 GW by 2030. In early 2011, the country's Energy Regulator National Renewable Energy Board recommended a target of 100 MW of solar installations to be constructed in the

[13]Exchange rate: 1€ = 4.0 MRY.

country over the following three years. An FiT of 17.95 PHP/kWh (0.299 €/kWh)[14] was suggested, to be paid from January 2012 onwards. For 2013 and 2014, an annual degression of 6% was foreseen. The initial period of the programme is scheduled to end on 31 December 2014. In July 2012, the Energy Regulatory Commission decided to lower the tariff in view of lower system prices to 9.68 PHP/kWh (0.183 €/kWh[15]) and confirmed the degression rate. By the end of 2012, about 2 MW of the 20 MW of installed PV systems were grid-connected. SunPower had two cell-manufacturing plants outside of Manila, but closed down Fab. No 1 in early 2012. Fab. No 2, with a nameplate capacity of 575 MW, remains in action.

Vietnam

The National Energy Development Strategy of Vietnam was approved in December 2007. This gave priority to the development of renewable energy and included the following targets: increase the share of renewable energies from negligible to about 3% (58.6 GJ) of the total commercial primary energy in 2010, to 5% in 2020, 8% (376.8 GJ) in 2025, and 11% (1.5 TJ) in 2050. By the end of 2011, about 5 MW of PV systems had been installed, mainly in off-grid applications. In May 2011, the Indo-Chinese Energy Company broke ground in the central coastal Province of Quang Nam for the construction of a thin-film solar panel factory with an initial capacity of 30 MW and a final capacity of 120 MW. However, the company applied for permission to delay the project in June 2012, with no new launch date set. In January 2013, WorldTech Transfer Investment and the UAE-based Global Sphere began work on a solar-panel manufacturing plant in the central province of Thua Thien-Hue (Global Sphere, 2013). This plant is located in the Phong Dien Industrial Park and the first phase of the project (60 MW) should be operational by June 2015. The capacity is planned to be increased to 250 MW in a second phase.

13.2.3 *Europe and Turkey*

Market conditions for PV vary substantially from country to country across Europe. This is due to different energy policies and public support programmes for renewable energies, and especially for PV, as well as the varying degrees of liberalisation of domestic electricity markets. As shown in Fig. 13.4, overall growth in PV capacity has been impressive, increasing 373-fold from 185 MW in 2000 to 69 GW at the end of 2012 (European Photovoltaic Industry Association, 2013; Systèmes Solaires, 2013).

[14]Exchange rate 2011: 1€ = 60 PHP.

[15]Exchange rate 2012: 1€ = 53 PHP.

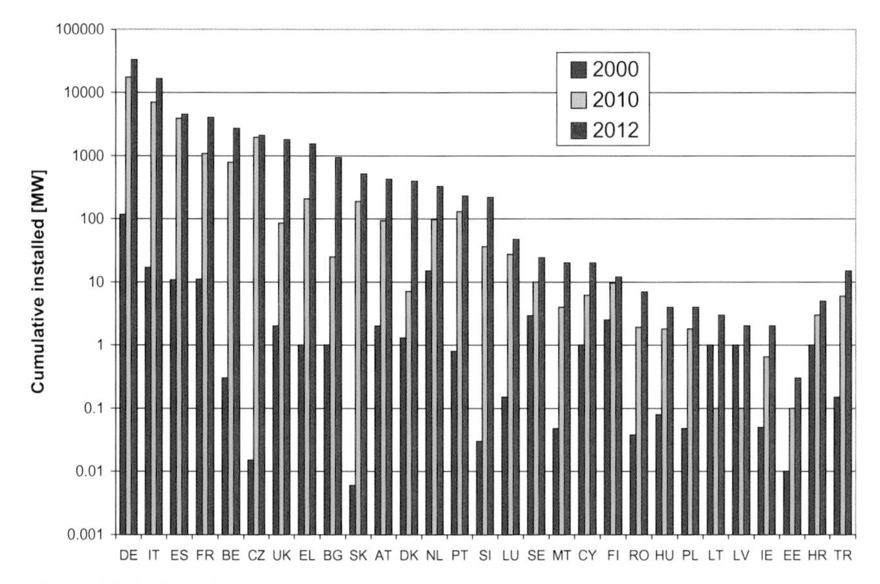

Figure 13.4 Cumulative installed grid-connected PV capacity in the European Union + Candidate Countries (CC). Note that the installed capacities do not correlate with solar resources.

The net growth of all renewable energy power generation capacity in Europe and Turkey between 2000 and 2012 was 178 GW. This compares with 121 GW of new gas-fired capacity and a reduction in capacity of generation from coal (−12.7 GW), oil (−7.4 GW) and nuclear (−14.7 GW). Wind (96.7 GW) and PV (69 GW) accounted for more than 93% of the renewable capacity. This net growth of 178 GW to a total of 316 GW of renewable capacity increased the total share of renewable power capacity in Europe and Turkey from 22.5% in 2000 to 33.9% in 2012.

A total of about 45 GW of new power capacity was connected in the European Union (EU) in 2012 and 12.5 GW was decommissioned, resulting in 32.5 GW of new net capacity, as shown in Fig. 13.5 (European Wind Association, 2013; Systèmes Solaires, 2013). Photovoltaic electricity generation capacity accounted for 16.8 GW, or 51.7%, of the new net capacity. Wind power was second with 11.7 GW (36%), followed by 5 GW (15.4%) gas-fired power stations, 1.3 GW (4%) biomass, 0.8 GW (2.5%) solar thermal power plants, 266 MW (0.8%) hydro and 61 MW (0.2%) other sources. The net installation capacity for coal-fired, oil-fired and nuclear power plants was negative, with decreases of 2.3 GW, 3.2 GW and 1.2 GW, respectively. The renewable share of new power installations was more than 69% and more than 95% of new net capacity in 2012.

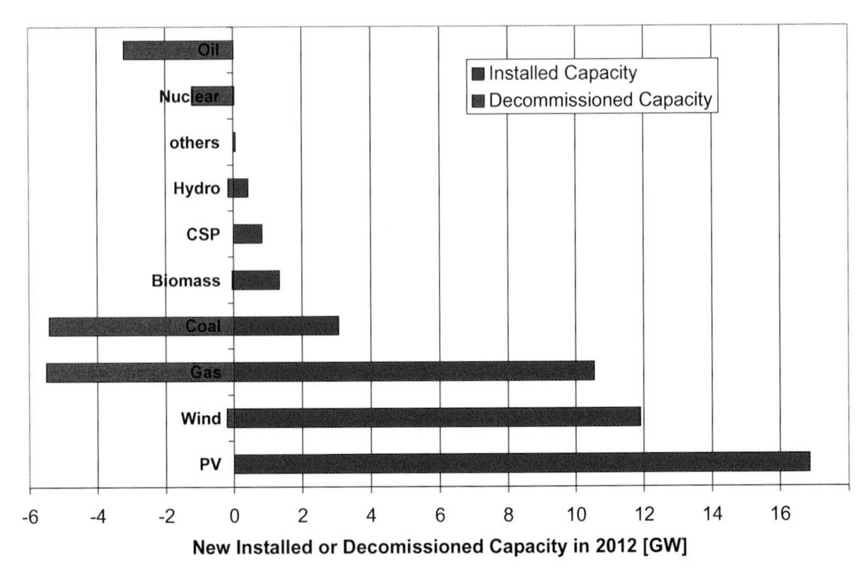

Figure 13.5 New connected or decommissioned electricity generation capacity in the EU in 2012.

In the following sub-sections, the market development in some of the EU Member States, as well as of Switzerland and Turkey, is described.

Belgium

Belgium showed another strong PV market performance in 2012, with new system installations of 882 MW bringing the cumulative installed capacity to 2694 MW. Photovoltaic power supplied 14% of Belgium's residential consumption, or 2.8% of the country's total electricity needs. However, most of this capacity was again installed in Flanders, where green certificates have been in place for 20 years. The certificate system was suspended at the end of July 2012 and replaced by a new regime for PV systems commissioned after 1 August 2012 (Belgisch Staatsblad, 2012, 2013): for systems smaller than 250 kWp, the tariff was reduced from 0.23 €/kWh to 0.21 €/kWh, and for larger systems from 0.15 €/kWh to 0.09 €/kWh for systems installed before the end of 2012. At the same time, the duration for which the certificates could be claimed was reduced from 20 to 10 years.

Since 1 January 2013, the right to receive green certificates has been determined by the duration of the amortisation period — for PV it was 15 years in 1H 2013 and the net metering scheme for systems below 10 kW continued. A technology-dependent banding factor, which is set twice a year by the Flemish Energy Agency, was introduced, so that one certificate is no longer for 1 MWh, but now depends on the type of installation (RES LEGAL Europe, 2013). Since 1 January 2013 the value of the certificates has been €93.

In Wallonia, the green certificate scheme called SOLWATT was replaced by a new scheme called QUALIWATT from 1 April 2013 (Government of Wallonia, 2013). The main change was that green certificates with a value of 0.065 €/kWh will only be granted until the time that the PV installation is fully reimbursed, and their maximum term will be 10 years. In addition, for systems below 10 kW, net metering continues and a progressive pricing system has been introduced. The value of green certificates in the Brussels region is €65 and for PV systems there is a multiplier of 2.2. In addition, the possibility of net metering exists for systems below 5 kWp so long as the generated electricity does not exceed the in-house electricity demand of the owner.

Bulgaria

A new Renewable Energy Source (RES) Act was approved in May 2011. This fixed the FiT levels and resulted in new installations of around 110 MW, increasing the total installed capacity to 134 MW by the end of 2011. At the end of 2012, 933 MW of PV systems had been cumulatively installed. In March 2012, the Bulgarian Parliament voted to revise the RES Act (Bulgarian State Gazette, 2012a). The most significant change was that the price at which electricity would be purchased was no longer fixed at the date of completion of installation, but at the date the usage permit was granted.

In September 2012, the Bulgarian State Energy and Water Regulatory Commission (SEWRC) published prices for the retroactive grid-usage fee, in accordance with the Energy Act amendments adopted by the Bulgarian Parliament in July 2012 (Bulgarian State Gazette, 2012b) for access to the transmission and distribution grid. For PV systems commissioned after 1 April 2010 and in 2011, the fee amounted to 20% of the FiT. For systems commissioned during the first half of 2012, the fee was 39% of the FiT, for those commissioned between 1 July and 31 August 2012 it was 5%, and after 1 September 2012, 1% of the respective FiT applied (SEWRC, 2012a).

Modified FiTs, mandating a 34–54% reduction, depending on system type, came into force on 1 June 2012. A further reduction ranging between 5 and 35% came into force in October 2012 (SEWRC, 2012b).

Denmark

Due to the introduction of a net-metering system and the country's high electricity prices of 0.295 €/kWh, 378 MW of PV systems were installed in Denmark in

2012. Following this rapid development, the regime was changed in November 2012 (Denmark, 2012). Under the new rules, full net metering was only possible within one hour from when PV electricity was produced, and the excess electricity exported to the grid was reimbursed at the rate of 1.30 DKK/kWh (0.174 €/kWh[16]) in 2013. To account for decreasing PV system prices, this rate will decrease to 1.17 DKK/kWh (0.157 €/kWh) in 2014 and to ~1.00 DKK (0.174 €/kWh) in 2015. After ten years, the rate will be further reduced to about 0.60 DKK/kWh (0.080 €/kWh).

France

In 2012, 1.08 GW of new PV systems were installed in France, which led to an increase of the cumulative installed capacity to over 4 GW, including about 300 MW in the French Overseas Departments. New PV installations in mainland France accounted for 35% of total new electricity production capacity commissioned in 2012. Of the total capacity, residential systems smaller than 3 kWp represented 16% or 0.64 GW, systems up to 250 kWp accounted for 40% or 1.6 GW, and systems larger than 250 kWp added 44% or 1.76 GW.

In 2013, France had three different support schemes for PV. For systems up to 100 kWp, there is the FiT (allocation of 200 MW for residential and 200 MW for commercial applications); for rooftop systems between 100 and 250 kWp there is a 'simplified' call for tender with a volume of 120 MW for 2013; and for systems larger than 250 kWp (large rooftop and ground-mounted systems) an additional call for tender with a volume of 400 MW was issued, which received more than 1.9 GW of project applications.

In 2012, four PV tenders were offered and the average electricity sale price proposed by the bidders fell from 229 €/MWh during the first round to 194 €/MWh in the fourth one. New FiTs were published in February 2013, foreseeing an adaptation every three months (Ministère de l'Écologie, 2013). Photovoltaic systems with defined European content were eligible for a bonus of 5 to 10%.

Germany

Photovoltaics installed in Germany in 2012 increased slightly compared with 2011, from 7.5 GW to 7.6 GW (Bundesnetzagentur, 2013). German market growth is directly correlated to the introduction of the Erneuerbare Energien Gesetz (Renewable Energy Sources Act) in 2000 (EEG, 2000). This introduced a guaranteed FiT for electricity generated from solar PV systems for 20 years and had a built-in

[16]Exchange rate: 1€ = 7.46 DKK.

annual decrease, which was adjusted over time to reflect the rapid growth of the market and the corresponding price reductions. As only estimates of the installed capacity existed before 2009, a plant registrar was introduced from 1 January 2009 onwards.

The German market performed strongly throughout 2012 with peaks of 1.2 GW installed in March, 1.8 GW in June and 1 GW in September. The total installed capacity at the end of the year was 32.7 GW. Since May 2012, the FiT has been adjusted on a monthly basis depending on the actual installations during the previous quarter. The fact that the tariff for residential PV systems (July 2013: 0.151 €/kWh) is now below the rate conventional electricity consumers are paying (0.287 €/kWh) makes the increase of self-consumption more attractive and is opening new possibilities for the introduction of local storage (Bundesnetzagentur, 2013; Strategies Unlimited, 2013). Since 1 May 2013, the Kreditanstalt für Wiederaufbau (KfW) has offered low interest loans with a single repayment bonus of up to 30% and a maximum of €600 per kW of storage for PV systems up to 30 kWp (KfW, 2013). The maximum repayment bonus is limited to €3000 per system.

Greece

In 2009, Greece introduced a generous FiT scheme but this had a slow start until the market accelerated in 2011 and 2012. In 2012, 687 MW of new PV systems were installed — more than 1.5 times the 439 MW that had been cumulatively installed up to the end of 2011. In April 2013, the total installed capacity surpassed 2 GW (over 1.9 GW in mainland Greece and over 115 MW in the Islands) (HEDNO S.A., 2013; LAGIE, 2013). On 10 May 2013 the Greek Ministry of Environment, Energy and Climate Change announced retroactive changes in the FiT for systems larger than 100 kWp and new tariffs for all systems from 1 June 2013 onwards.

Italy

Italy connected more than 3.5 GW during 2012, bringing cumulative installed capacity to 16.4 GW by the end of the year (Gestore Servizi Energetici, 2013a). The Quinto Conto Energia (Fifth Energy Bill) was approved by the Italian Council of Ministers on 5 July 2012 (Italy, 2012). This set new half-yearly reductions of the tariffs, and the annual expenditure ceiling for new installations was increased from €500 million to €700 million. In addition, a new requirement to register systems larger than 12 kWp was introduced. On 6 June 2013, Gestore Servizi Energetici announced that the €6.7 billion ceiling of the bill had been reached with 18.2 GW, out of which 17.1 GW were already operational, and that the Quinto Conto Energia would cease within 30 days, as foreseen (Gestore Servizi Energetici, 2013b).

Slovakia

After two years of rapid growth, the Slovakian market decreased by over 90% in 2012 and only 29 MW were newly installed. The total capacity of 517 MW is more than three times the original 160 MW capacity target for 2020, published in the National Renewable Energy Action Plan in 2010. From February 2011, support was limited to applications for systems smaller than 100 kW, and from 1 July 2013 onward, support was limited to systems of up to 30 kW placed on buildings.

Spain

Spain is still third in Europe for cumulative PV capacity, with 4.5 GW installed by the end of 2012. Most of this was installed in 2008, when the country was the biggest market, with close to 2.7 GW in 2008 (European Photovoltaic Industry Association, 2013). This was more than twice the expected capacity and was due to an exceptional race to install systems before the Spanish Government introduced a cap of 500 MW on the yearly installations in the autumn of 2008. Royal Decree 1758/2008 set considerably lower FiTs for new systems and limited the annual market to 500 MW, with the proviso that two-thirds should be rooftop-mounted and not free-field systems. These changes resulted in a steep drop in the number of new installations. In 2012, new system installations with a capacity of 194 MW increased total capacity to 4.5 GW. Photovoltaic-generated electricity contributed 7.8 TW, or 2.9%, of the Spanish power demand in 2012.

In January 2012, the Spanish government passed a further Royal Decree 1/12 (Government Gazette, 2012), which suspended the remuneration pre-assignment procedures for new renewable power capacity, affecting about 550 MW of planned solar PV installations. The justification given for this move was that Spain's energy system had by then amassed a €30 billion power-tariff deficit and it was argued that the special regime for renewable energy was the main reason. However, the Spanish government has prevented utilities from charging consumers for the true cost of electricity for over a decade. Instead of allowing utilities to increase rates every time electricity generation costs increased (due to rising coal or natural gas costs, inflation or changes in energy or environmental policy), the government allowed them to create schemes such as a deferral account whereby they could recover shortfalls in any individual year from revenues generated in subsequent years.

In January 2007, the European Commission opened an in-depth investigation to examine the potential support to large and medium-sized companies and electricity distributors in Spain, in the form of artificially low regulated industrial tariffs for electricity (EUR-Lex, 2007). In 2005, these regulated tariffs led to a deficit of

€3.8 billion in the Spanish electricity system, which increased to almost €9 billion in 2007, a time when payments under the special regime for renewable energy were still limited.

Despite the Royal Decree 1/12, the tariff deficit increased further in 2012 and reached €35.6 billion at the end of February 2013 (Comición National de Energía 2013). The question remains: is renewable electricity generation responsible for this? Despite the fact that the premium payments amount to €8.4 billion in 2012 and €9.1 billion for 2013, the answer is not clear. An often-neglected aspect is the fact that investment in renewable capacity has increased supply at the wholesale market, thus decreasing the system marginal price by the order of merit effect. It is argued that renewable energy 'pays for itself' because by bidding at the pool at zero prices these units substantially decrease the system marginal price. Therefore the cost of all electricity should be lower, if this price reduction is passed on to the customer. An analysis for the entire special regime concluded that for 2010 the decrease in the wholesale price of around 29 €/MWh covered 70% of the FiT costs (Ciarreta *et al.*, 2012b).

Switzerland

About 200 MW of PV systems were installed in Switzerland in 2012, almost doubling the total capacity to 411 MW. Prices for turnkey systems decreased by over 40% in 2012 (Photovoltaik Guide, 2013). In view of these decreasing prices, the FiT was reduced three times in 2012.

United Kingdom

The United Kingdom introduced a new FiT scheme in 2010, which led to the installation of approximately 55 MW that year and over 1 GW in 2012. This steep increase was caused by the announcement of a fast-track review of large-scale projects by the Department of Energy & Climate Change (DECC) in February 2011, which led to a rush to complete these projects during the first half of 2011 (DECC, 2011). A second rush occurred towards the end of the year to meet the deadline of 12 December 2011, when the DECC planned to decrease the residential tariff by about 50% as a result of another fast-track consultation. However, this decision was contested in court and the tariffs were only changed on 1 April 2012. The average reductions in April were 44–54% for systems smaller than 50 kWp and 0–32% for systems above 50 kWp. In November 2012, a further reduction of 3.5% for systems smaller than 50 kWp was imposed. No reduction was made for larger systems, as almost no such systems had been installed between May and July 2012. However, for larger systems it was possible to receive Renewables Obligation Certificates (ROCs) until April 2013; for PV the rate was two ROCs per MWh. Overall about 630 MW were installed, bringing the cumulative capacity to 1.8 GW at the end of 2012.

During 2012, the Energy and Climate Change Minister, Greg Barker, repeatedly declared his desire to see the UK solar market reach 22 GW by 2020. During the first six months of 2013, the UK added 802 MW of new solar PV installations, representing the strongest first-half year ever for the UK PV industry. This was comprised of 20 MW in the first quarter of 2013, followed by 282 MW in the second quarter of 2013 (Colville, 2013).

Austria

In 2012 Austria installed about 230 W of new PV systems and more than doubled its cumulative capacity to 417 MW. The Ökostrom-Eispeisetarifverordnung 2012 (eco-electricity decree) is the regulation that sets the prices for the purchase of electricity generated by green power plants. An investment subsidy with a budget of €36 million in 2013 is also in place. Regardless of the size of the systems, a maximum of 5 kWp is supported, with 300 €/kWp for add-on and ground-mounted systems and 400 €/kWp for building-integrated systems. Operators of PV systems larger than 5 kWp can choose to opt for the so-called net-parity tariff (Netzparitäts-Tarif) for a period of 13 years. Since 1 January 2013, this option has only been available for systems on buildings.

Portugal

Despite high solar radiation, solar photovoltaic system installation in Portugal has grown only very slowly, reaching a cumulative capacity of 229 MW at the end of 2012.

Ukraine

Ukraine again saw an impressive growth and almost doubled its capacity with over 180 MW new installed PV systems to 370 MW. In February 2013, the Ukrainian Parliament had the first reading of a bill to simplify the access of households to the feed-in scheme. Reduction in the feed-in tariffs by between 16% and 27%, depending on the type of installation, was finally adopted in February 2013 and went into force on 1 April 2013. Since then, tariffs have been adjusted on a monthly basis. In 2013 another doubling of the installed capacity is possible.

Turkey

In March 2010, the Energy Ministry of Turkey unveiled its 2010–2014 Strategic Energy Plan. One of the government's priorities was to increase the proportion of renewably generated energy to 30% of total energy generation by 2023. At the beginning of 2011, the Turkish Parliament passed renewable energy legislation that defined new guidelines for FiTs. The FiT is 0.133 $/kWh (0.10 €/kWh) for owners commissioning a PV system before the end of 2015. If components made in Turkey are used, the tariff will increase by up to $0.067 (€0.052), depending on the material mix. The first licensing round, with a volume of 600 MW, closed in June 2013 and was oversubscribed about 15 times, with close to 9 GW of projects submitted to the Turkish Energy Regulatory Authority. How much of this will be installed still has to be seen. At the end of 2012, around 15 MW of grid-connected and stand-alone systems were estimated to be installed cumulatively.

13.2.4 *Africa*

Despite its vast solar resources and the fact that there are areas in Africa where the solar potential is very considerable — with the same photovoltaic panel able to produce twice as much electricity in Africa as in Central Europe on average — only limited use of PV electricity generation is made in Africa. The main application is for small solar home systems, and the market statistics for these are extremely imprecise or non-existent. Therefore, all African countries are considered potential or emerging markets and some of them are mentioned below.

Algeria

In 2011, the Algerian Ministry of Energy and Mines published its Renewable Energy and Energy Efficiency Programme, which aimed to increase the share of renewable energy used for electricity generation to 40% of domestic demand by 2030. The plan foresaw 800 MW of installations by 2020 and a total of 1.8 GW by 2030. It is estimated that about 5 MW of small decentralised systems were installed at the end of 2012. For 2013, new installations of around 20 MW were planned.

Capo Verde

Capo Verde's Renewable Energy Plan (2010–2020) aims to increase the use of renewable energy to 50% of total energy consumption in 2020. The policy to achieve this is to use PPAs. Law n1/2011 established the regulations regarding independent energy production. In particular, it established the framework conditions for the set-up of independent power producers using renewable energy (15 years PPA), and for self-production at user level. It created the micro-generation regime, regulated rural electrification projects and established the tax exemption of all imported renewable energy (RE) equipment. At the end of 2012, 7.5 MW of centralised grid-connected PV systems were installed. In addition, there are a number of smaller off-grid and grid-connected systems. To realise the 2020 50% renewable energy target, about 340 MW of PV systems will be required.

Ghana

In 2011 the Parliament of Ghana passed a Renewable Energy Bill with the aims of increasing the proportion of renewable energy, particularly solar, wind, mini hydro and waste-to-energy schemes, in the national energy supply mix, and contributing to the mitigation of climate change (Parliament of Ghana, 2011). The bill set a goal of renewable energy constituting 10% of national energy generation by 2020. By the end of 2012, a few thousand solar home systems and a few of off-grid systems with an estimated 5 MW had been installed in the country. In 2012, Episolar, Canada, signed a PPA with Ghana's second largest utility, the Electricity Company of Ghana, for a 50 MW PV plant with an option to increase the overall project size to 150 MW (Episolar, 2012). In December 2012, Blue Energy, UK, announced that it had agreed with Ghana's Public Utilities Regulatory Commission on a 20-year PPA for the 155 MW PV plant in Nzema to be operational by 2015 (Blue Energy, 2012). These two projects are still in the development phase looking for strategic investors. In May 2013, the Volta River Authority inaugurated its first solar power plant at Navrongo with a capacity of 2 MW. The Volta River Authority plans to install a total of 14 MW by 2014.

Kenya

In 2008, Kenya introduced FiTs for electricity from renewable energy sources, but solar was only included in the revision dated 2010 (Ministry of Energy, Kenya, 2010). However, only a little over 560 kW of PV capacity was connected to the grid in 2011. The majority of the 14 MW of PV systems were off-grid installations. In 2011, Ubbink East Africa Ltd., a subsidiary of Ubbink B.V. (Doesburg, The Netherlands) opened a solar PV manufacturing facility with an annual output of 30,000 modules in Naivasha, Kenya. This plant produces modules for smaller PV systems such as solar home systems. The estimates for the PV market in Kenya are average sales of 20,000–30,000 home systems and 80,000 solar lanterns per year.

Morocco

The Kingdom of Morocco's solar plan was introduced in November 2009, with the aim of establishing 2000 MW of solar power by 2020. To implement this plan, the Moroccan Agency for Solar Energy was founded in 2010. Both solar electricity technologies, concentrated solar thermal power (CSP) and PV, were to compete openly. In 2007, the National Office of Electricity announced a smaller programme for grid-connected distributed solar PV electricity, targeting 150 MW of solar PV power. Various rural electrification programmes using PV systems have been ongoing for a long time. At the end of 2012, Morocco had about 20 MW of PV systems installed, mainly under the Global Rural Electrification Programme Framework, and about 1 to 2 MW grid-connected systems.

South Africa

South Africa has a fast-increasing electricity demand and vast solar resources. In 2008, the country enacted its National Energy Act, which called for a diversification of energy sources, including renewables, as well as fuel switching to improve energy efficiency (South Africa, 2008). In 2011, the Renewable Energy Independent Power Producer Procurement Programme (IPP) was set up with rolling bidding rounds. Two rounds took place in 2011 and 2012, and a third one closed in August 2013. The overall target is 3.725 GW, including 1.45 GW for solar PV. In the first two bidding rounds 1048 MW of solar PV projects were allocated to preferred bidders. The average bid price decreased between the first round (closure date: 4 November 2011) from 2.65 ZAR/kWh (0.252 €/kWh[17]) to 1.65 ZAR/kWh (0.157 €/kWh) in the second round (closure date: 5 March 2012). It is estimated that about 30–40 MW of PV systems were installed in South Africa by the end of 2012.

13.2.5 *The Americas*

Canada

In 2012, the Canadian market was about the same as in 2011, with 268 MW cumulative installed PV capacity increasing to about 830 MW. This development was driven by the introduction of an FiT in the Province of Ontario, enabled by Bill 150, Green Energy and Green Economy Act 2009. More than 77% of the total capacity is installed in Ontario. On the federal level, an accelerated capital cost allowance exists under the income tax regulations. On a province level, nine Canadian provinces have net metering rules, with solar photovoltaic electricity as one of the eligible technologies. Sales tax exemptions and renewable energy funds exist in two provinces and micro grid regulations and minimum purchase prices each exist in one province.

The Ontario FiT's were set in 2009, depending on system size and type, and were then reduced in various steps. In June 2013, the Minister of Energy announced an annual cap of 150 MW for the small FiT regime and 50 MW for the micro FiT regime for the following four years, with a transition in 2013, which meant that

[17] Exchange rate: 1€ = 10.5 ZAR.

systems larger 500 kW would no longer be eligible for the FiT. Following the WTO ruling in May on the local content regulations, the Minister said that Ontario would comply with it.

United States of America

With over 3.3 GW of newly installed PV capacity, the USA reached a cumulative PV capacity of 7.7 GW (7.2 GW grid-connected) at the end of 2012. Utility PV installations again more than doubled, compared with 2011 and became the largest segment with 1.7 GW in the year. The top ten states — California, Arizona, New Jersey, Nevada, North Carolina, Hawaii, Maryland, Texas, New York and Massachusetts — accounted for more than 88% of the market (SEIA, 2013). Market growth around 30% is estimated for 2013. Photovoltaic projects with PPAs with a total capacity of 10.5 GW are already under contract and over 3 GW of these projects are already financed and under construction (SEIA, 2013). If one adds those projects in an earlier development stage, the pipeline stands at almost 22 GW.

Many state and federal policies and programmes have been adopted to encourage the development of markets for PV and other renewable technologies. These consist of direct legislative mandates (such as renewable content requirements) and financial incentives (such as tax credits). One of the most comprehensive databases about the different support schemes in the USA is maintained by the Solar Centre of the State University of North Carolina. The Database of State Incentives for Renewable Energy (www.dsireusa.org) is a comprehensive source of information on state, local, utility, and selected federal incentives that promote renewable energy.

Argentina

In 2006, Argentina passed its Electric Energy Law, which established that 8% of electricity demand should be generated by renewable sources by 2016 (Argentina, 2006). It also introduced FiTs for wind, biomass, small-scale hydro, tidal, geothermal and solar for a period of 15 years. In July 2010, amongst other renewable energy sources, the government awarded PPAs to six solar PV projects totalling 20 MW. By the end of 2012, about 17 MW (7 MW grid-connected) of PV systems had been installed. According to the renewable energy country attractiveness indicator, the Argentinean government has set a 3.3 GW target for PV installations by 2020 (Ernst & Young, 2011).

Brazil

At the end of 2012, Brazil had about 20 MW cumulative installed capacity, mainly in rural areas. In April 2012, the board of the National Agency of Electric Energy approved new rules to reduce barriers to install small-distributed generation capacity. The rule applies to generators that use subsidised sources of energy (hydro, solar, biomass, cogeneration and wind). Over 2 GW of PV projects applied for approval. It is estimated that about 40 MW (6.6 MW grid-connected) had been installed by the end of 2012. In its mid-term market report, the IEA forecast a cumulative installed PV capacity of about 200 MW in 2013, and 1.2 GW by 2018 (IEA, 2013a).

Chile

In February 2012, the President of Chile announced a strategic energy plan to reach 20% of non-conventional renewable energy by 2020. In the first quarter of 2012, the first MW-size PV system was installed in the northern Atacam desert. By June 2013, Chile's Environmental Assessment Service had approved over 4 GW of projects, with an additional 2.2 GW still under review. At the end of 2012, it was estimated that about 10 MW of PV systems (5 MW grid-connected) had been installed. In June 2013, about 70 MW of PV systems were under construction and it is expected that about 50 MW of new systems will be operational at the end of 2013.

Dominican Republic

In 2007, the Dominican Republic passed a law promoting the use of renewable energy and setting a target of 25% renewable energy share in 2025 (Dominican Republic, 2007). At that time ~1–2 MW of solar PV systems had been installed in rural areas, which had increased to over 5 MW by 2011. In 2011, the first PPA for 54 MW was signed between Grupo de Empresas Dominicanas de Energía Renovable and Corporación Dominicana de Empresas Eléctricas Estatales (CDEEE). The first phase (200 kW) of the project became operational in July 2012, and the whole solar farm was connected to the grid early 2013. In 2012, CDEEE signed two more PPAs with a total capacity of 116 MW.

Mexico

In 2008, Mexico enacted a Law for Renewable Energy Use and Financing Energy Transition to promote the use of renewable energy (Mexico, 2008). In 2012, the country passed its Climate Change Law, which foresaw a decrease in greenhouse gas emissions of 30% below the business-as-usual case by 2020, and 50% by 2050 (Mexico, 2012). It further stipulated that renewable electricity should constitute 35% of total supply by 2024. By the end of 2012, about 52 MW of PV systems had been installed (IEA, 2013c).

Peru

In 2008, Peru passed its Legislative Decree 1002, which made the development of renewable energy resources a national priority. The decree stated that at least 5% of electricity should be supplied from renewable sources by 2013. In February 2010, the Organismo Supervisor de la Inversión en Energía y Minería held the first round of bidding and awarded four solar projects with a total capacity of 80 MW. A second bidding round was held in 2011, with a quota of 24 MW for PV. By the end of 2012, about 85 MW of PV systems had been installed. A doubling of capacity is regarded as possible in 2013.

13.3 The photovoltaic industry

The photovoltaic industry consists of a long value chain from raw materials to PV system installation and maintenance. So far, the main focus has been on solar cell and module manufacturers, but in addition there is the so-called upstream industry (e.g. materials, polysilicon production, wafer production, equipment manufacturing) and the downstream industry (e.g. inverters, balance-of-system [BOS] components, system development, project development, financing, installations and integration into existing or future electricity infrastructure, plant operators, operation and maintenance). In the near future, it will probably be necessary to add (super)-capacitor and battery manufacturers, as well as power electronics and IT providers for demand and supply management, including meteorological forecasting. The main focus in this section, however, will be on solar cell and module manufacturers and polysilicon manufacturers.

In 2012, world solar cell production grew by 10% to about 38.5 GW. The market for installed systems grew by almost 30% and values between 29 and 31 GW have been reported by various consultancies and institutions. This mainly represents the grid-connected photovoltaic market. To what extent the off-grid and consumer-product markets are included is unclear. The difference of roughly 7–9 GW has therefore to be explained as a combination of unaccounted off-grid installations (~1–200 MW off-grid rural, ~1–200 MW communication/signals, ~100 MW off-grid commercial), consumer products (~1–200 MW) and cells/modules in stock.

In addition, the fact that some companies report shipment figures, some report sales figures and others report production figures, adds to the uncertainty. An additional source of error is the fact that some companies produce fewer solar cells than solar modules, but the reporting does not always differentiate between the two, and there is a risk that cell production is double-counted, first at the cell manufacturer and second at the 'integrated' cell/module manufacturer. Difficult economic conditions contributed to decreased willingness to report confidential company data. Nevertheless, the figures show a significant growth in production, as well as an increasing installation market.

Despite the fact that a significant number of companies filed for insolvency, scaled back or even cancelled their expansion projects over the past year, the number of new entrants into the field, notably large semiconductor or energy-related companies, overcompensated. The announced production capacities — based on a survey of more than 300 companies worldwide — increased again in 2012, and is expected to continue to do so in 2013 and 2014, albeit at a slower pace. However, rapid changes in the sector and the difficult financing situation make forecasting for future capacity developments very speculative. It is believed that the current situation of overcapacity in solar cell manufacturing will continue until at least 2015, when the global PV market should exceed 50 GW of new installations annually.

13.3.1 *Technology mix*

As silicon prices fell drastically after the temporary silicon shortage, wafer-based silicon solar cells decreased in cost very rapidly and remain the main technology, with close to 90% market share in 2012. Commercial module efficiencies fall within a wide range between 12% and 21%, with monocrystalline modules between 14% and 21%, and polycrystalline modules between 12% and 18%. The massive manufacturing capacity increases for both technologies were followed by the necessary capacity expansion for polysilicon production.

In 2005, production of thin-film solar modules reached more than 100 MW per annum for the first time. Between 2005 and 2009, the compound annual growth rate of thin-film solar module production was greater than that of the overall industry, increasing the market share of thin-film products from 6% in 2005 to 10% in 2007 and 16–20% in 2009. Since then, the thin-film market share has decreased slowly as the ramp-up of new production lines did not follow that of wafer-based silicon.

The majority of thin-film companies are silicon-based and use either amorphous silicon or an amorphous/microcrystalline silicon structure. Next are companies using $Cu(In,Ga)(Se,S)_2$ as the absorber material, whereas only a few companies use CdTe or dye and other materials. These technologies are covered in Chapters 4 to 9.

Concentrated photovoltaics (CPV), discussed by Luque-Heredia and Luque in Chapter 11, is an emerging technology that is growing fast, although from a low starting point. About 60 companies are active in the field of CPV development, the majority focusing on high-concentration concepts. Over half of these companies are located either in the USA (primarily in California) or in Europe (primarily Spain).

Within CPV, there is a differentiation according to the concentration factor[18] and whether the system uses a dish (Dish CPV) or lenses (Lens CPV) to concentrate sunlight. The main parts of a CPV system are the cells, the optical elements and the tracking devices. The recent growth in CPV is based on significant improvements in all of these areas, as well as system integration. However, CPV is at the beginning of an industry learning curve, with considerable potential for technical and cost improvements. The most challenging task is to become cost-competitive with other PV technologies quickly enough in order to grow to reach factory sizes, which can count on economies of scale. Despite the small installed CPV capacity to date, various consultancy companies predict that the CPV market will grow to 500 MW by 2015, and 1 GW by 2020 (GlobalData, 2013; Strategies Unlimited, 2013).

The existing PV technology mix is a solid foundation for future growth of the sector as a whole. No single technology can satisfy all the different consumer needs, ranging from mobile and consumer applications, with the need for a few watts, to multi-MW utility-scale power plants. The variety of technologies is insurance against a roadblock for the implementation of solar photovoltaic electricity, if material limitations or technical obstacles restrict the further growth or development of a single technology pathway.

13.3.2 *Solar cell manufacturers*[19]

Worldwide, more than 350 companies produce solar cells. The solar cell industry over the last decade has been very dynamic, and the changes that have been happening since 2011 allow only for a snapshot picture of the current situation, which might already have changed a few weeks later. Despite the fact that a few dozen companies filed for insolvency, scaled back, idled or stopped production, the number of newcomers and their planned capacities still exceeds the retired capacity.

The following sections give short descriptions of the 20 largest companies, in terms of actual production/shipments in 2012. More information about additional solar cell companies and details can be found in various market studies. The capacity, production or shipment data are from the annual reports or financial statements of the respective companies or the cited references.

[18]High concentration > 300 Suns (HCPV), medium concentration $5 < x < 300$ Suns (MCPV), low concentration < 5 Suns (LCPV).

[19]Solar cell production capacities mean: in the case of wafer silicon-based solar cells, the cell; in the case of thin-films, the complete integrated module. Only those companies that actually produce the active circuit (solar cell) are counted. Companies that purchase these circuits and make cells are not counted.

Yingli Green Energy Holding Company Ltd. (*China*)

Yingli Green Energy (www.yinglisolar.com) went public on 8 June 2007. The main operating subsidiary, Baoding Tianwei Yingli New Energy Resources, is located in the Baoding National High-New Tech Industrial Development Zone. The company deals with the whole set, from solar wafers, cell manufacturing and module production. According to the company, production capacity reached 1.85 GW at the end of 2011. In its 2012 annual report, the company reported a capacity of ingot, wafers, cells and modules of 2.45 GW at the end of the year, and that in addition to their own cells they also purchased cells from other companies. For 2012, total reported shipments of solar modules were 2.3 GW. Solar cell production was estimated at 1950 MW for 2012.

In January 2009, Yingli acquired Cyber Power Group Ltd., a development-stage enterprise designed to produce polysilicon. Through its principal operating subsidiary, Fine Silicon, the company started trial production of solar-grade polysilicon in late 2009, and was still ramping up to full production capacity of 3000 metric tons per year at the end of 2011. However, recent financial results indicate that the company has written off its investment in Fine Silicon and according to other media reports, production has now closed down.

In January 2010, the Ministry of Science and Technology of China approved an application to establish a national-level key laboratory in the field of PV technology development, the State Key Laboratory of PV Technology, at Yingli Green Energy's manufacturing base in Baoding. In October 2013, BTU International signed a two-year agreement to support this laboratory, which has not yet been built.

First Solar LLC (*USA/Germany/Malaysia*)

First Solar LLC (www.firstsolar.com) is one of the few companies worldwide to produce CdTe thin-film modules. The company currently has three manufacturing sites, in Perrysburg (USA), Frankfurt/Oder (Germany) and Kulim (Malaysia), with a combined capacity of 2.376 GW at the end of 2011. A second Frankfurt/Oder plant, doubling the capacity there to 528 MW, became operational in May 2011 and expansion in Kulim increased production capacity there to 1.584 GW at the end of 2011. In April 2012, however, the company announced a major restructuring in response to changing market conditions (First Solar, 2012), closing all its manufacturing operations in Frankfurt/Oder at the end of 2012, and idling four production lines in Kulim. In addition, the company put its factory in Meza (AZ) on hold, and sold its factory in Dong Nam Industrial Park, Vietnam.

In 2012, the company produced 1.875 GW of product. For Q1 of 2013, the company reported production costs of 0.69 $/Wp (0.53 €/Wp), including under-utilisation and upgrading costs of 0.05 $/Wp (0.038 €/Wp).

JA Solar Holding Company Ltd. (*China*)

JingAo Solar (www.jasolar.com) was established in May 2005 by the Hebei Jinglong Industry and Commerce Group, the Australia Solar Energy Development Pty. Ltd. and Australia PV Science and Engineering Company. Commercial operation started in April 2006 and the company went public on 7 February 2007. According to company reports, production capacity was 2.5 GW for cells, 1.8 GW for modules and 1 GW for wafers at the end of 2012. For 2012, sales of 1.7 GW were reported.

Trina Solar Ltd. (*China*)

Trina Solar (www.trinasolar.com) was founded in 1997 and went public in December 2006. The company has integrated product lines, making ingots, wafers and modules. In December 2005, a 30 MW monocrystalline silicon wafer product line went into operation. According to the company, production capacity was 1.2 GW for ingots and wafers and 2.4 GW for cells and modules at the end of 2012, and shipments of 1.6 GW were reported for the year.

In January 2010, the company was selected by the Chinese Ministry of Science and Technology to establish a State Key Laboratory to develop PV technologies within the Changzhou Trina PV Industrial Park. This laboratory has been established as a national platform for driving PV technologies in China. Its mandate includes research into PV-related materials, cell and module technologies and system-level performance. It will also serve as a platform to bring together technical capabilities from the company's strategic partners, including customers and key PV component suppliers, as well as universities and research institutions.

Hanwha (*China/Germany/Malaysia/South Korea*)

Hanwha Group (www.hanwha.com) acquired a 49.99% share in Solarfun Power Holdings in 2010 and the name was changed to Hanwha SolarOne in January 2011. The company produces silicon ingots, wafers, solar cells and solar modules. Its first production line was completed at the end of 2004 and commercial production started in November 2005. It went public in December 2006, and reported the completion of its production capacity expansion to 360 MW in Q2 of 2008.

In August 2012, the company acquired Q-Cells (Germany/Malaysia), which had filed for insolvency in April 2012. Hanwha, with its two brands Hanwha Q Cells and Hanwha SolarOne, has combined production capacity of 2.3 GW of solar cells (1.3 GW China, 800 MW Malaysia, 200 MW Germany). In addition Hanwha SolarOne has 800 MW of ingot and wafer production capacity. Shipments of 830 MW for Hanwha Solar-One and an additional solar cell production of 570 MW for Hanwha Q Cells were reported for 2012 (Ikki, 2013b).

Suntech Power Company Ltd. (China)

Suntech Power (www.suntech-power.com) was founded in January 2001 by Dr. Zhengrong Shi and went public in December 2005. Suntech specialises in the design, development, manufacturing and sale of photovoltaic cells, modules and systems. In its preliminary financial statement, the company reported shipments of 1.8 GW for 2012. External reports give the cell production at 1.35 GW (PV News, 2013).

Due to financial problems, Wuxi Suntech Power, the Chinese subsidiary of Suntech Power Holdings, filed a petition for insolvency restructuring with Wuxi Municipal Intermediate People's Court in Jiangsu Province, China, which was accepted on 20 March 2013.

Motech Solar (Taiwan/China)

Motech Solar (www.motech.com.tw) is a wholly-owned subsidiary of Motech Industries Inc., located in Tainan Science Industrial Park. The company started mass production of polycrystalline solar cells at the end of 2000, with an annual production capacity of 3.5 MW. Production increased from 3.5 MW in 2001 to 1 GW in 2011. In 2009, Motech started the construction of a factory in Kunshan, China, which reached its nameplate capacity of 500 MW in 2011. Total production capacity at the end of 2012 was given as 1.6 GW. For 2012, total shipments of 1.28 GW were reported.

Canadian Solar Inc. (China)

Canadian Solar Inc. (CSI) (www.canadiansolar.com) was founded in Canada in 2001 and was listed on NASDAQ in November 2006. It has established six wholly-owned manufacturing subsidiaries in China, manufacturing ingots/wafers, solar cells and solar modules. According to the company, CSI had 216 MW of ingot and wafer capacity, 1.6 MW cell capacity and 2.4 GW module manufacturing capacity (2.1 GW in China, 330 MW in Ontario, Canada) at the end of 2012. For 2012, the company reported sales of 1.47 GW of modules, but no cell production figure was given. This must, however, be lower, because the company states in its financial reports that it buys cells from other manufacturers. External reports gave cell production as 1.24 GW (PV News, 2013).

Gintech Energy Corporation (Taiwan)

Gintech (www.gintech.com.tw) was established in August 2005 and went public in December 2006. Production at Factory Site A, Hsinchu Science Park, began in 2007 with an initial production capacity of 260 MW and increased to 1.17 GW at the end of 2011, with further expansion to 1.5 GW in 2012. In 2012, the company produced 1.1 GW (PV News, 2013).

JinkoSolar Holding Company Ltd. (China)

Jinko Solar (www.jinkosolar.com) was founded by HK Paker Technology Ltd. in 2006. Starting from upstream business, the company expanded operations across the solar value chain, including recoverable silicon materials, silicon ingots and wafers, solar cells and modules in 2009. The company went public in May 2010 and is currently listed on the New York Stock Exchange. According to the company, it had manufacturing capacities of 1.2 GW each for wafers, solar cells and solar modules at the end of 2012. It reported sales of about 1.2 GW (912 MW modules, 84 MW cells and 190 MW wafers) for 2012.

Neo Solar Power Corporation (Taiwan)

Neo Solar Power (www.neosolarpower.com) was founded in 2005 by PowerChip Semiconductor, Taiwan's largest DRAM (dynamic random-access memory) company, and it went public in October 2007. The company manufactures both mono- and multicrystalline silicon solar cells. Production capacity of silicon solar cells at the end of 2012 was 1.3 GW. In 2013, the company produced about 958 MW (PV News, 2013). In 2013, the company merged with DelSolar to become the largest Taiwanese cell producer.

SunPower Corporation (USA/Philippines/Malaysia)

SunPower (http://us.sunpowercorp.com) was founded in California in 1988 by Richard Swanson and Robert Lorenzini to commercialise proprietary, high-efficiency, silicon solar cell technology. The company went public in November 2005. SunPower designs and manufactures high-performance silicon solar cells, based on the interdigitated rear-contact design described in Section 3.4 of Chapter 3. The initial products, introduced in 1992, were high-concentration solar cells with an efficiency of 26%. SunPower also manufactures a 22%-efficient solar cell, called Pegasus, which is designed for non-concentrating applications.

SunPower conducts its main research and development (R&D) activity in Sunnyvale, California, and has cell-manufacturing plants in the Philippines and Malaysia. In 2011, the company had two cell-manufacturing plants outside of Manila, but decided to close down Fab. No 1 in early 2012. It then consolidated production at Fab. No 2 (with a nameplate capacity of 575 MW) in the Philippines and its newer facility, Fab. No 3 in Malaysia, which is operated as a joint venture with AU Optronics Corporation and had a capacity of 1.4 GW at the end of 2012 (PV Tech, 2012). The company has two solar module factories in the Philippines (600 MW) and, since 2011, also in Mexico (500 MW). In addition, modules are assembled for SunPower by third-party contract manufacturers in China, Mexico, Poland and California. Total cell production in 2012 was reported at 925 MW.

Hareon Solar Technology Company Ltd. (*China*)

Haeron Solar (www.hareonsolar.com) was established as the Jiangyin Hareon Technology in 2004 and changed its name to the Hareon Solar Technology in 2008. It has five manufacturing facilities in Jiangsu and Anhui provinces: Jiangyin Hareon Power, Altusvia Energy (Taicang), Hefei Hareon Solar Technology, Jiangyin Xinhui Solar Energy and Schott Solar Hareon. Solar cell production started in 2009, with an initial capacity of 70 MW. According to the company, production capacity was to be increased to over 2 GW for cells and 1 GW for modules in 2012. Production of 900 MW was reported for 2012 (PV News, 2013).

Kyocera Corporation (*Japan*)

In 1975, Kyocera (http://global.kyocera.com/prdct/solar/) began research on solar cells. The Shiga Yohkaichi Factory was established in 1980, and R&D and manufacture of solar cells and products started with mass production of multicrystalline silicon solar cells in 1982. In 1993, Kyocera became the first Japanese company to sell home PV generation systems.

Besides its solar-cell-manufacturing plants in Japan, Kyocera has had module-manufacturing plants in China in a joint venture with the Tianjin Yiqing Group (10% share) in Tianjin (since 2003), Tijuana, Mexico (since 2004) and Kadan, Czech Republic (since 2005).

The company also markets systems that both generate electricity through solar cells and exploit heat from the Sun for other purposes, such as heating water. Its Sakura factory in Chiba Prefecture is involved in everything from R&D and system planning to construction and servicing, and its factory in Shiga Prefecture is active in R&D, as well as the manufacture of solar cells, modules, equipment parts and devices which exploit heat. In 2012, Kyocera had a production of 800 MW and planned to increase its capacity to 1 GW in 2013 (Ikki, 2013ab).

Renewable Energy Corporation AS (*Norway/Singapore*)

The Renewable Energy Corporation's (REC) (www.recgroup.com) vision is to become the most cost-efficient solar energy company in the world, with a presence throughout the whole value chain. The company, located in Høvik, Norway, has five business activities, ranging from silicon feedstock to solar system installations, and is involved in all major aspects of the PV value chain through its various subsidiaries. In 2011, the company closed down REC ScanCell, which was located in Narvik, and had a production capacity of 180 MW. The next closure, announced in March 2012, was the wafer factory in Glomfljord, which had 300 MW capacity for multicrystalline wafers, while the 650 MW wafer plant at Herøya continued operations. From 2012, production of solar cells and modules continued only at REC Solar Singapore, which operates an integrated site for wafers, solar cells and module production, with a capacity of 750 MW. In 2012, production was reported at 750 MW.

ReneSola Ltd. (*China*)

Renesola (www.renesola.com) was established in 2005 and went public on the New York Stock Exchange in 2008. The company produces polysilicon, wafers, solar cells and solar modules. According to the company, it had a production capacity of 2 GW of wafers, 1.2 GW of modules and 800 MW of solar cells at the end of 2012. Shipments for 2012 were reported as 1.5 GW wafers and 710 MW modules.

Changzhou EGing Photovoltaic Technology Company Ltd. (*China*)

EGing PV (www.egingpv.com) was founded in 2003 and works along the complete photovoltaic industry value chain, from the production of monocrystalline furnaces, quartz crucibles, 5–8 inch monocrystalline silicon ingots, supporting equipment of squaring and wire sawing, monocrystalline silicon wafers, solar cells and solar modules. According to the company, it had a production capacity of 1 GW across the complete value chain of ingots, wafers, cells and modules at the end of 2011. Sales of 600 MW were reported for 2012 (ENF, 2013).

Astronergy Solar (*China*)

Astronergy Solar (www.astronergy.com) was established as a member of the Chint Group in October 2006. The first production line of 25 MW, for crystalline silicon cells and modules, was installed in May 2007 and an increase in production capacity to 100 MW was achieved in July 2008. Commercial production of micromorph® solar modules started in July 2009. Thin-film capacity was 30 MW in 2010 and increased to 75 MW in early 2011. Total production capacity for 2012 was reported as 800 MW, with 584 MW of production (ENF, 2013).

Solar Frontier (Japan)

Solar Frontier is a 100%-owned subsidiary of Showa Shell Sekiyu K.K. In 1986, Showa started to import small modules for traffic signals, and entered module production in Japan, cooperatively with Siemens (now Solar World). The company developed CIS solar cells and completed the construction of its first factory with 20 MW capacity in October 2006. Commercial production started in FY 2007. The second Myazaki factory (MP2) started manufacturing in 2009 with a production capacity of 60 MW. In July 2008, the company announced they would open a research centre 'to strengthen research on CIS solar-powered cell technology, and to start collaborative research on mass production technology of the solar modules with Ulvac, Inc.' The aim was to start a new plant in 2011 with a capacity of 900 MW. The ramp-up started in February 2011, and at the end of that year overall capacity was 980 MW. In 2011, the company changed its name to Solar Frontier, and production was reported as 450 MW (RTS Corporation, 2012). A moderate increase to above 500 MW was expected for 2012. Early in 2013, the company reported that its Kunitomi plant was operating at full capacity. Production at MP2 was halted at the end of 2012 to make adjustments for the production of new, differentiated products, and the plant resumed production in July 2013.

Jiangxi Risun Solar Energy Company Ltd. (China)

Risun Solar Technologies was established in 2008. The company manufactures mono- and multicrystalline solar cells and modules, with a stated production capacity of 700 MW for solar cells and 300 MW for modules. An expansion to 3 GW was planned without a specified date. Production of 500 MW was reported for 2012 (ENF, 2013).

13.3.3 *Polysilicon supply*

The rapid growth of the PV industry since 2000 led to the situation where, between 2004 and early 2008, the demand for polysilicon outstripped supply from the semiconductor industry. Prices for purified silicon started to rise sharply in 2007, and prices for polysilicon peaked at around 500 $/kg in 2008, resulting in higher prices for PV modules. This extreme price hike triggered a massive capacity expansion, not only by established companies, but attracting many new entrants as well.

These massive production expansions, as well as the difficult global economic situation, led to price decreases for polysilicon throughout 2009. Prices reduced to about 50–55 $/kg at the end of 2009, with a slight upwards tendency throughout 2010 and early 2011, before they dropped significantly again. The lowest level on the spot market was reached in December 2012 with prices below 16 $/kg (12.3 €/kg) before they started to recover slightly, trading between 16.5 and 17.5 $/kg (12.7–13.5 €/kg) for most of 2013.

China was and remains a major polysilicon producer. In January 2011, the Chinese Ministry of Industry and Information Technology tightened the rules for future polysilicon factories to combat the oversupply problem. New factories must be able to produce more than 3000 metric tons of polysilicon per year and also meet certain efficiency, environmental and financing standards. Their total energy consumption must be less than 200 kWh/kg of polysilicon, and China is aiming to create large companies with at least 50000 metric tons annual capacity by 2015. These two framing conditions, in addition to the enormous price pressure, are the reasons why a significant number of Chinese manufacturers closed down their polysilicon production in the first half of 2012. This is also why China imported 83000 metric tons of silicon in 2012, 32% more than 2011 (Bloomberg New Energy Finance, 2013g).

Projected global silicon production capacities in 2013 vary between 290,000 metric tons (Bloomberg New Energy Finance, 2013e) and 409,690 metric tons (Ikki, 2013a). It is estimated that about 27000 metric tons will be used by the electronics industry. The possible solar cell production will, in addition, depend on the material used per Wp; the current worldwide average is about 5.6 g/Wp.

13.3.4 *Polysilicon manufacturers*

Worldwide, more than 100 companies produce or have started up polysilicon production. This section gives a short description of the ten largest companies in terms of production in 2012. More information about additional polysilicon companies and details can be found in various market reports.

Wacker Polysilicon (Germany)

Wacker Polysilicon AG (www.wacker.com) is one of the world's leading manufacturers of hyper-pure polysilicon for the semiconductor and photovoltaic industry, chlorosilanes and fumed silica. In 2011, Wacker increased its capacity to over 40000 metric tons and reported sales of 32000 metric tons. Their 15000 metric ton factory in Nünchritz Saxony, started production in 2011. In 2010, the company decided to build a polysilicon plant in Tennessee with 15000 ton capacity. The groundbreaking of this new 18000 metric ton factory took place in April 2011, and construction should be finished by the end of 2013. In addition, the company also expanded its Burghausen capacity by 5000 metric tons in 2012 and, together with a further expansion of the Nünchritz factory by 5000 metric tons, it plans to have 70000 metric tons of production capacity in 2014. Total polysilicon sales are reported as 38000 metric tons in 2012.

GCL-Poly Energy Holdings Ltd. (China)

GCL-Poly (gcl-poly.com.hk) was founded in March 2006 and construction of their Xuzhou polysilicon plant (Jiangsu Zhongneng Polysilicon Technology Development) began in July 2006. Phase I has a designated annual production capacity of 1500 tons and the first shipments were made in October 2007. Full capacity was reached in March 2008. At the end of 2011, polysilicon production capacity had reached 65000 metric tons and 8 GW of wafers. For 2012, the company reported production of 37055 metric tons of polysilicon and 5.6 GW wafers. It has invested in the downstream business of solar: GCL Solar System Ltd. (SSL) is a wholly-owned subsidiary of GCL-Poly Energy Holdings Ltd. and provides solar system turnkey solutions for residential, governmental, commercial and solar farm projects, including design, equipment supply, installation and financial services. Another subsidiary is GCL Solar Power, which develops, operates and manages solar farms.

OCI Company (South Korea)

OCI Company Ltd. (www.oci.co.kr) (formerly DC Chemical) is a global chemical company with a product portfolio spanning the fields of inorganic chemicals, petro and coal chemicals, fine chemicals, and renewable energy materials. In 2006, the company started its polysilicon business and completed its 6500 metric ton P1 plant in December 2007. The 10500 metric ton P2 expansion was completed in July 2009, and P3 with another 10000 metric tons brought the total capacity to 27000 metric tons at the end of 2010. The debottlenecking of P3 in 2011 increased the capacity to 42000 tons at the end of the year. Further capacity expansions, P4 (20000 tons) and P5 (24000 tons), were put on hold due to the rapid price decline of polysilicon. Instead the company is pursuing a further debottlenecking of its existing plants to increase capacity by 10000 metric tons during 2013. For 2012, production of 33000 metric tons was estimated from the company's financial figures.

OCI has also invested in downstream business and holds 89.1% of OCI Solar Power, which develops, owns and operates solar power plants in North America. On 23 July 2012, the company signed a PPA with CSP Energy, Texas, for a 400 MW solar farm in San Antonio, TX.

Hemlock Semiconductor Corporation (*USA*)

Hemlock Semiconductor Corporation (www.hscpoly.com) is based in Hemlock, Michigan. The company is a joint venture of Dow Corning Corporation (63.25%) and two Japanese firms, Shin-Etsu Handotai Co. Ltd. (24.5%) and Mitsubishi Materials Corporation (12.25%). The company is the leading provider of polycrystalline silicon and other silicon-based products used in the semiconductor and solar industry. In 2007, the company had an annual production capacity of 10000 tons of polycrystalline silicon and production at the expanded Hemlock site (19000 tons) started in June 2008. A further expansion at the Hemlock site, as well as a new factory in Clarksville, Tennessee, was started in 2009. Total production capacity was 46000 metric tons in 2011 and expansion to 56000 metric tons was scheduled to be finalised in 2012. Production of 28000 metric tons was estimated for 2012.

Renewable Energy Corporation AS (*Norway*)

In 2005, Renewable Energy Corporation AS (REC) (www.recgroup.com) took over Komatsu's US subsidiary, Advanced Silicon Materials LLC (ASiMI), and announced the formation of its silicon division business area, REC Silicon Division, comprising the operations of ASiMI and REC Solar Grade Silicon LLC (SGS). Production capacity at the end of 2012 was around 22400 metric tons and according to the company, a total of 21405 metric tons of polysilicon, of which 18791 metric tons were electronic-grade silicon, was produced in 2012.

SunEdison Inc. (*USA*)

SunEdison, formerly MEMC Electronic Materials Inc. (http://sunedisonsilicon.com), has its headquarters in St. Peters, Missouri. It started operations in 1959 and it makes semiconductor-grade wafers, granular polysilicon, ultra-high purity silane, trichlorosilane, tetrafluorosilane and sodium aluminium tetrafluoride. In February 2011, the company and Samsung entered into a 50/50 equity joint venture to build a polysilicon plant in Korea with an initial capacity of 10000 metric tons in 2013. At the end of 2011, the company closed its 6000 metric ton Merano, Italy, factory and reduced its capacity in Portland, USA, during 2012. Production capacity at the end of 2012 was $\sim$10000 metric tons with estimated production of about 12000 metric tons.

SunEdison is developing solar power projects as well as acting as a solar energy provider. At the end of 2012, the company had 73 MW of photovoltaic power plants under construction and 2.6 GW of projects in the pipeline.

Tokuyama Corporation (Japan)

Tokuyama (www.tokuyama.co.jp) had an annual production capacity of 9200 tons in 2010. In February 2011, the company broke ground for a new 20000 ton facility in Malaysia. The first phase, with 6200 metric ton capacity, started trial production in March 2013, and the second phase, with 13800 metric ton capacity, should be fully operational in 2015. For 2012, production of 7800 tons was estimated.

M. Setek Company Ltd. (Japan)

M. Setek is a subsidiary of Taiwan's AU Optronics. The company is a manufacturer of semiconductor equipment and monocrystalline silicon wafers. It has two plants in Japan (Sendai, Kouchi) and two in China (Hebei Lang Fang Songgong Semiconductor Co. Ltd. in Beijing, and Hebei Ningjin Songgong Semiconductor Co. Ltd. in Ningjin). In April 2007, polysilicon production started at the Soma Factory in Fukushima Prefecture. For 2012, a production capacity of 7000 metric tons was reported (Ikki, 2013b), and production was estimated at 5600 metric tons.

Kumgang Korea Chemical Company (South Korea)

Kumgang Korea Chemical Company (KCC) (www.kccworld.co.kr/eng) was established by a merger in 2000 of Kumgang and the Korea Chemical Company In February 2008, KCC announced its investment in the polysilicon industry and began to manufacture high-purity polysilicon with its own technology at the pilot plant of the Daejuk factory in July of the same year. It started to mass-produce polysilicon in 2010, with an annual capacity of 6000 tons, and increased this to 12100 metric tons in 2012 (Ikk 2013a). Production for 2012 is given at around 5400 metric tons.

Daqo New Energy Company Ltd. (China)

Daqo New Energy (www.dqsolar.com) is a subsidiary of the Daqo Group and was founded by Mega Stand International Ltd. in January 2008. The company built a high-purity polysilicon factory with an annual output of 3300 tons in the first phase of construction in Wanzhou. The first polysilicon production line, with an annual output of 1500 tons, started operation in July 2008. Production capacity in 2009 was 3300 tons and reached more than 4300 metric tons at the end of 2011. In Q4

2012, expansion Phase 2 with 3000 metric tons came online, and according to the company it will reach a total production capacity of 9300 metric tons at the end of 2013. The company invested in the downstream business, ranging from wafers, cells, modules and projects. At the end of 2011, it had a manufacturing capacity of 125 MW wafers and 100 MW modules. Polysilicon shipments of 3585 metric tons were reported for 2012.

13.4 Outlook

The photovoltaic industry has changed dramatically in the last few years. China has become the major manufacturing country for solar cells and modules, followed by Taiwan, Germany and Japan. Amongst the 20 biggest photovoltaic manufacturers in 2012, only three had production facilities in Europe, namely First Solar (USA, Germany, Malaysia), Hanwha Q Cells (Germany and Malaysia) and REC (Norway and Singapore). However, REC closed down its production in Norway in early 2012 and First Solar closed its factory in Germany later in the year.

It is important to remember that the PV industry consists of more than just the cell and module production, and that the whole PV value chain is relevant. Besides the information in this chapter, the whole upstream industry (e.g. materials, equipment manufacturing), as well as the downstream industry (e.g. inverters, BOS components, system development, installations) has to be looked at as well.

The implementation of the 100,000 roofs programme in Germany in 1990, and the Japanese long-term strategy set in 1994, with a 2010 horizon, were the beginning of an extraordinary PV market growth. Before the start of the Japanese market implementation programme in 1997, annual growth rates of the PV markets were in the range of 10%, mainly driven by communication, industrial and stand-alone systems. Since 1990, PV production has increased by almost three orders of magnitude, from 46 MW to about 38.5 GW in 2012. Statistically documented cumulative installations worldwide accounted for almost 100 GW in 2012. This represents mostly the grid-connected photovoltaic market. To what extent the off-grid and consumer product markets are included is not clear, but it is believed that a substantial part of these markets are not included in these figures as it is very difficult to track them.

Even with the current economic difficulties, overall installations are increasing, due to overall rising energy prices and the pressure to stabilise the climate. In the long term, growth rates for photovoltaics will continue to be high, even if economic conditions vary and lead to short-term slowdowns.

This view is shared by an increasing number of financial institutions, which are turning towards renewables as a sustainable and stable long-term investment. Increasing demand for energy is pushing the prices for fossil energy resources higher and higher. In 2007, a number of analysts predicted that oil prices could well hit 100 $/bbl by the end of that year or early 2008 (Bloomberg, 2007). After the spike of oil prices in July 2008, at close to 150 $/bbl, prices decreased due to the worldwide financial crisis and hit a low of around 37$/bbl in December 2008. Since then, the oil price has rebounded and the IEA reported average prices for oil imports around 110 $/bbl since the second quarter of 2011, with peaks around 120 $/bbl between February and April 2012 and September 2013. Oil demand has increased from about 84 million bbl/day in Q1 2009 to around 91 million bbl/day in Q3 2013, and a moderate further increase to 93 million bbl/day by Q4 2014 is expected. Even though no significant changes are currently forecast by analysts for 2014, the fundamental trend, that increasing demand for oil will drive its price higher again, is intact and will be evident as soon as the global economy recovers.

Over the last 20 years, numerous studies of the potential growth of the PV industry and the implementation of PV electricity generation systems have been produced. In 1996, the Directorate-General for Energy of the European Commission (EC) published a study entitled *Photovoltaics in 2010* (European Commission, 1996). The medium scenario of this study was used to formulate the EC's *White Paper* target of 1997, which was to have a cumulative installed capacity of 3 GW in the EU by 2010 (European Commission, 1997). The most aggressive scenario in this report predicted a cumulative installed PV capacity of 27.3 GW worldwide and 8.7 GW in the EU for 2010. This scenario was called 'Extreme scenario' and it was assumed that a number of breakthroughs in technology and costs, as well as continuous market stimulation and elimination of market barriers, would be required to achieve it. The reality check reveals that even the most aggressive scenario is lower than what we expect from the current developments.

According to investment analysts and industry prognoses, solar energy systems will continue to grow at high rates in the coming years. The different photovoltaic industry associations, as well as Greenpeace, the European Renewable Energy Council (EREC) and the IEA, have developed new scenarios for the future growth of PV. Table 13.1 shows the various scenarios of the Greenpeace/EREC study, as well as the different 2013 IEA World Energy Outlook scenarios and the IEA PV Technology Roadmap of 2010. It is interesting to note that the 2015 capacity values of only two scenarios — the Greenpeace (revolution) and IEA 450 ppm Scenarios — were not reached at the end of 2013. With forecast new installations of between 93 and 106 GW in 2014 and 2015, even the Greenpeace revolution scenario is no longer fictional thinking (Bloomberg New Energy Finance, 2014).

The above-mentioned solar photovoltaic scenarios will only be possible if solar cell and module manufacturing are continuously improved and novel design concepts are realised, as with current technology the demand for some materials,

Table 13.1 Predicted cumulative solar electrical capacities until 2035. Data sources: Bloomberg New Energy Finance, 2014; Greenpeace, 2012; IEA, 2010, 2013d.

Year	2012 [GW]	2013 [GW]	2015 [GW]	2020 [GW]	2030 [GW]	2035 [GW]
Actual installations	100	138				
Bloomberg New Energy Finance Market Outlook			231–244			
Greenpeace* (reference scenario)			88	124	234	290
Greenpeace* (revolution scenario)			234	674	1764	2420
IEA Current Policy Scenario**			133	352	545	680
IEA New Policy Scenario			138	379	710	951
IEA 450 ppm Scenario**			144	422	950	1389
IEA PV Technology Roadmap***			76	210	872	1330

*2035 values are extrapolated, as only 2030 and 2040 values are given
**2015 and 2030 values are extrapolated, as only 2011, 2020 and 2035 values are given
***2015 and 2035 values are extrapolated, as only 2010, 2020, 2030 and 2040 values are given

like silver, would dramatically increase the costs for these resources within the next 30 years. Research to avoid such problems is underway and it can be expected that these bottle-necks will be avoided. With 100 GW cumulative installed photovoltaic electricity generation capacity installed worldwide by the end of 2012, photovoltaics still is a small contributor to the electricity supply, but its importance for our future energy mix is finally acknowledged.

References

Argentina (2006), Boletín Oficial de la República de Argentina, Número 31.064 (www.diputados-catamarca.gov.ar/ley/BO020107.pdf).

Belgisch Staatsblad (2012), 182e Jaargang, No 237, 20 Juli 2012, Derde Editie, N. 2012—2122 [C – 2012/35855] 13 JULI 2012.—Decreet houdende wijziging van het Energiedecreet van 8 mei 2009, wat betreft de milieuvriendelijke energieproductie (1), p. 40448ff.

Belgisch Staatsblad (2013), 183e Jaargang, No 11, 16 Januari 2013, [C – 2013/11001], 21 DECEMBER 2012.—Koninklijk besluit tot wijziging van het koninklijk besluit van 16 juli 2002 betreffende de instelling van mechanismen voor de bevordering van elektriciteit opgewekt uit hernieuwbare energiebronnen, p. 1574ff.

Bloomberg (2007), '$100 oil price may be months away, say CIBC, Goldman', 23 July 2007 (http://www.bloomberg.com/apps/news?pid=newsarchive&sid=ajxtV4oWcHk0).

Bloomberg New Energy Finance (2012), 'Analyst reaction—solar', 17 August 2012.
Bloomberg New Energy Finance (2013a), 'Press release', 14 January 2013 (http://about.bnef.com/press-releases/new-investment-in-clean-energy-fell-11-in-2012-2/).
Bloomberg New Energy Finance (2013b), 'Scaling up financing to expand the renewables portfolio', presentation given by Michael Liebreich at the IEA Renewable Energy Working Party, Paris, 9 April 2013.
Bloomberg New Energy Finance (2013c), 'PV market outlook Q2 2013', 13 May 2013.
Bloomberg New Energy Finance (2013d), 'Analyst reaction—solar', 31 May 2013.
Bloomberg New Energy Finance (2013e), 'China said to prepare import duties on polysilicon', 16 May 2013.
Bloomberg New Energy Finance (2014), 'Q2 2014 PV Market Outlook–A time for cautious expansion', 15 May 2014.
Blue Energy (2012), 'Press release', 8 December 2012.
Bridge to India (2013), *India Solar Compass*, 20 April 2013 (www.bridgetoindia.com).
Bulgarian State Gazette (2012a), Decree 147/2012 from 28 March 2012, 10 April 2012 (http://dv.parliament.bg).
Bulgarian State Gazette (2012b), 'Law amending the Law on Energy, ORDINANCE No 271 from 3 July 2012', 17 June 2012 (http://dv.parliament.bg).
Bundesgesetzblatt für die Republik Őeterreich (2012), Jahrgang 2012 Ausgegeben am 18 September 2012, Teil II, 307. Verordnung: Ökostrom-Einspeisetarifverordnung 2012 – ÖSET-VO 2012.
Bundesnetzagentur (German Federal Network Agency) (2013), Datenmeldungen für 2012 (www.bundesnetzagentur.de/DE/Sachgebiete/ElektrizitaetundGas/Unternehmen_Institutionen/ErneuerbareEnergien/Photovoltaik/DatenMeldgn_EEG-VergSaetze/DatenMeldgn_EEG-VergSaetze_node.html).
Ciarreta A., Espinosa M. P. and Pizarro-Irizar C. (2012a), 'Efecto de la energía renovable en el mercado diario de electricidad', *Escenario 2020, Cuadernos Económicos de ICE* **83**, 101–116.
Ciarreta A., Espinosa M. P. and Pizarro-Irizar C. (2012b), 'The effect of renewable energy in the Spanish electricity market', *Lecture Notes in Information Technology* **9**, 431–436.
CIS Countries Legislation Database (2009), 'Law of the Republic of Kazakhstan, from 4 July 2009 of No. 165-IV ZRK, "About support of usage of renewable energy sources"' (http://cis-legislation.com).
Colville F. (2013), 'UK enters global PV top six with record-breaking half year – Part Two', *Solar Business Focus*, 24 September 2013 (www.solarpowerportal.co.uk).
Comición National de Energía (2012), Annual Report 2012, 18 April 2013 (www.cne.es).
The Daily Star (2011), 'Target 500 MW solar project', 15 May 2011 (www.thedailystar.net).
DECC (2011), 'Huhne takes action on solar farm threat', 7 February 2011 (https://www.gov.uk).
Denmark (2012), Klima-, Energi- og Bygningsministeriet, Bekendtgørelse om nettoafregning for egenproducenter af elektricitet, Lovtidende A, BEK nr 1068 af 16/11/2012 (www.lovtidende.dk/Forms/L0700.aspx?s 31=10&s19=1068+&s20=2012).
Dominican Republic (2007), Ley No. 57-07 sobre Incentivo al Desarrollo de Fuentes Renovables de Energía y de sus Regímenes Especiales (Dominican Republic).
EEG (Erneuerbare-Energien-Gesetz) (2000), Bundesgestzblatt Jahrgang 2000 Teil I, Nr. 13, p. 305, 29 March 2000.

ENF (2013), 'China production database', 2013 (www.enfsolar.com).

Episolar (2012), 'Siginik Energy partners with the electricity company of Ghana to develop largest solar installation in West Africa', 4 June 2012 (www.marketwired.com/press-release/siginik-energy-partners-with-electricity-company-ghana-develop-largest-solar-installation-1664719.htm).

Ernst & Young (2011), 'Renewable energy country attractiveness indicator', November, Issue 31.

ESDM (Ministry of Energy and Mineral Resources) (2012), 'Development of new and renewable energy and energy conservation in Indonesia', presented at Global Workshop on Clean Energy Development, Washington DC, 3–8 December 2012.

EUR-Lex (2007), 'Tarifs réglementés de l'électricité en Espagne', *Official Journal of the European Communities*, **C 43/50**, 27 February 2007.

European Commission (1996), *Photovoltaics in 2010*, European Commission, Directorate-General for Energy, Office for Official Publications of the European Communities, ISBN 92-827-5347-6.

European Commission (1997), *Energy for the Future: Renewable Sources of Energy: White Paper for a Community Strategy and Action Plan*, COM (1997) 599 Final (26/11/97).

European Photovoltaic Industry Association (2013), *Global Market Outlook for Photovoltaics until 2017* (www.epia.org).

European Wind Association (2013), 'Wind in power—2012 European statistics', February 2013 (www.ewea.org).

Gestore Servizi Energetici (2013a), 'Rapporto Statistico sul solare fotovoltaico 2012', 8 May 2013.

Gestore Servizi Energetici (2013b), 'Conto Energia: raggiunti i 6,7 miliardi di euro', 6 June 2013 (www.gse.it).

Global Sphere (2013), 'Construction begins on 300 million USD solar panel factory in Hue', 17 January 2013 (www.global-sphere.com).

GlobalData (2013), 'Concentrated Photovoltaic (CPV), 2012 Global Market Size, Average Installation Price, Market Share, Regulations and Key Country Analysis to 2020', GlobalData Report Store, January 2013.

Government Gazette (2012), 'Spanish Royal Decree 1/12', 28 January 2012 (www.boe.es).

Government of Kazakhstan (2013), 'Government of Kazakhstan targets 1,040 MW of renewable energy by 2020', 6 February 2013 (www.kazakhembus.com).

Government of Wallonia (2013), 'Dossier électricité: traduction juridique de l'accord Solwatt et approbation de la tarification progressive et solidaire', 30 May 2013 (http://nollet.wallonie.be).

Greenpeace (2012), *The Energy [R]evolution 2012*, July 2012, ISBN 978-90-73361-92-8.

HEDNO S.A. (Hellenic Electricity Distribution Network Operator) (2013), 'Press release: issuance of power generation informative report for the non-interconnected islands, January 2013', 5 March 2013.

IEA (International Energy Agency) (2010), *PV Technology Roadmap*.

IEA (International Energy Agency) (2013a), *Medium-term Renewable Energy Market Report 2013—Market Trends and Projections to 2018*, IEA, Paris.

IEA (International Energy Agency) (2013b), 'Photovoltaic power systems programme, national survey report of PV power applications in Japan 2012', 31 May 2013.

IEA (International Energy Agency) (2013c), 'Photovoltaic power systems programme, annual report 2012', 30 April 2013.

IEA (International Energy Agency) (2013d), *World Energy Outlook 2013*, ISBN 978-92-64-2013-9.

IHS Global Demand Tracker (2013), 'Global PV installations to exceed 35 gigawatts in 2013; Asia & Americas more than compensate for ailing Europe', April 2013.

Ikki O. (2013a), *PV Activities in Japan*, Volume 19, No 4, April 2013.

Ikki O. (2013b), *Pv Activities in Japan*, Volume 19, No 7, April 2013.

Italy (2012), Supplemento n. 143 alla Gazzetta n. 159 del 10/07/2012, Ministero dello sviluppo economico, decreto ministeriale 5 luglio 2012; Attuazione dell'articolo 25 del decreto legislativo 3 marzo 2011, n. 28, recante incentivazione della produzione di energia elettrica da impianti solari fotovoltaici (cd. Quinto Conto Energia).

Kaz Silicon (2011), 'Production of photovoltaic modules based on Kazakhstan silicon' (www.kazsilicon.kz/en/node/16).

KfW (Kreditanstalt für Wiederaufbau) (2013), 'KfW-Programm Erneuerbare Energien "Speicher"', Programmnummer 275 (www.kfw.de).

Kruangam D. (2013), 'Current status of solar energy policy and market in Thailand', 9th Workshop on the Future Directions of Photovoltaics, Tokyo Institute of Technology, 7–8 March 2013.

LAGIE (Operator of Electricity Market) (2013), 'RES and CHP statistics, April 2013', 27 May 2013.

Mercom Capital Group (2013), 'Solar report', 18 March 2013.

Mexico (2008), DECRETO por el que se expide la Ley para el Aprovechamiento de nergías Renovables y el Financiamiento de la Transición Energética, Diario Oficial (Primera Sección) 88, 28 November 2008 (Mexico) (www.diputados.gob.mx/LeyesBiblio/ref/laerfte/LAERFTE_orig_28nov08.pdf).

Mexico (2012), Se Expide la Ley General de Chambio Climático, Diario Oficial (Segunda Sección) 1, 6 June 2012 (Mexico) (www.diputados.gob.mx/LeyesBiblio/ref/lgcc/LGCCorig_06jun12.pdf).

Ministère de l'Écologie, du Développement Durable et de l'Energie (2013), 6 February 2013 (www.developpement-durable.gouv.fr/Quels-sont-les-tarifs-d-achats).

Ministry of Economic Affairs (2013), 'Bureau of Energy, Press Release', 28 May 2013 (http://web3.moeaboe.gov.tw).

Ministry of Energy, Kenya (2010), 'Feed-in-tariffs policy for wind, biomass, small hydros, geothermal, biogas and solar', 1st revision, January 2010 (http://kerea.org).

MIIT (Chinese Ministry of Industry and Information Technology) (2012), 'Issuance of the solar photovoltaic industry "Second Five-Year Development Plan"', 24 February 2012 (www.miit.gov.cn).

NanoWin Thin Film Technology (2013), 'Zhambilskie Electricheskie Seti LLP', 9 May 2013 (www.nanowin.com.tw/news.php?id=60).

NEA (National Energy Administration) (2012), 'Press release', 8 August 2012 (www.nea.gov.cn).

NEA (National Energy Administration) (2013), 'China', 8 January 2013 (www.nea.gov.cn).

PEW Charitable Trusts (2013), 'Who's winning the clean energy race?' (www.pewenvironment.org).

PHOTON International (2012), 'PHOTON International market survey reveals Chinese companies dominate top 10', March 2012 (www.pv-tech.org/news/china_dominates_top_ten_global_solar_manufacturers).
Photovoltaik Guide (2013), 'Schweiz: Preise von Photovoltaikanlagen auf marktübliches Niveau gesunken', press release, 5 April 2013 (www.photovoltaik-guide.de/schweiz-preise-von-photovoltaikanlagen-auf-marktuebliches-niveau-gesunken-26098).
PV News (2013), published by Greentech Media, ISSN 0739-4829 (www.pv-magazine.com).
RES LEGAL Europe (2013), 'Legal sources on renewable energy', April 2013 (www.res-legal.eu/home).
RTS Corporation (2013), 'PV activities in Japan', Vol. 18 (www.rts-pv.com).
Parliament of Ghana (2011), 'Renewable Energy Bill' (www.parliament.gh).
Philippine Department of Energy (2008), 'An act promoting the development, utilization and commercialization of renewable energy resources and for other purposes', 16 December 2008 (www.doe.gov.ph).
The President of the Republic of Indonesia (2006), 'National Energy Policy, Presidential Regulation 5/2006', 25 January 2006 (http://faolex.fao.org).
SEDA (Sustainable Energy Development Authority) (2013), 'Current governing regulations of renewable energy for feed-in tariff', Seminar on RE Standards and FiT Implementation, 21 May 2013.
SEIA (Solar Energy Industry Association) (2013), 'US solar market insight, 2012 year in review, executive summary' (www.seia.org).
SERC (State Electricity Regulatory Commission, People's Republic of China) (2013), 15 January 2013 (www.serc.gov.cn).
SEWRC (Bulgarian State Energy and Water Regulation Commission) (2012a), 'Decision No. C-33', 14 September 2012 (www.dker.bg).
SEWRC (Bulgarian State Energy and Water Regulation Commission) (2012b), 'Decision No. C-28', 29 August 2012 (http://dker.bg).
Solarbuzz (2013), 'Marketbuzz 2013, press release', 11 March 2013.
Strategies Unlimited (2013), 'SPV market research product roadmap for 2013', April 2013.
Systèmes Solaires (2013), 'Photovoltaic energy barometer', 9 April 2013.
UNB Connect (2011), 'ADB assures fund for 500 MW solar system', 4 June 2011 (www.unbconnect.com).
World Bank (2010), 'Winds of change—East Asia's sustainable energy future', May 2010 (http://web.worldbank.org).
World Bank (2013), 'Bangladesh—lighting up rural communities', 15 April 2013 (www.worldbank.org).

APPENDIX I

FUNDAMENTAL CONSTANTS

Quantity	**Symbol**	**Value**
Speed of light in vacuum	c	2.998×10^{8} m s^{-1}
Planck constant	h	6.626×10^{-34} J s
Boltzmann constant	k	1.3807×10^{-23} J K^{-1}
Permittivity of vacuum	ϵ_{O}	8.854×10^{-12} F^{-1} m^{-1}
Unsigned charge on an electron	q	1.602×10^{-19} C
Stefan–Boltzmann constant	σ_{SB}	5.6705×10^{-8} W m^{-2} K^{-4}

APPENDIX II

USEFUL QUANTITIES AND CONVERSION FACTORS

$1\,\text{eV} = 1.6022 \times 10^{-19}\,\text{J} = 96.488\,\text{kJ mol}^{-1} = 8065.6\,\text{cm}^{-1} = 2.4180 \times 10^{14}\,\text{Hz}$
$1\,\text{cm}^{-1} = 1.9864 \times 10^{-23}\,\text{J} = 11.963\,\text{J mol}^{-1} = 0.12398\,\text{meV} = 2.9979 \times 10^{10}\,\text{Hz}$
$1239.8/(\lambda/\text{nm}) = E/\text{eV}$
kT at 298.15 K = 0.02569 eV
kT/q at 298.15 K = 0.02569 V
$(\ln 10)kT/q$ at 298.15 K = 0.05915 V
$h/2\pi = 6.582 \times 10^{-16}\,\text{eV s}$
$1\,\text{kWh} = 3.6 \times 10^{6}\,\text{J}$
1 thermochemical calorie = 4.184 J
1 horse power = 0.7457 kW
$1\,\text{atm} = 1.01325 \times 10^{5}\,\text{Pa}$
1 torr = 1/760 atm
$1\,\text{Å} = 10^{-8}\,\text{cm}$
1 mile = 1.6093 km
$1\,\text{hectare} = 10^{-2}\,\text{km}^2 = 2.471\,\text{acre} = 3.861 \times 10^{-3}\,\text{mile}^2$
$1\,\text{acre} = 4.047 \times 10^{-3}\,\text{km}^2 = 0.4047\,\text{hectare} = 1.5626 \times 10^{-3}\,\text{mile}^2$
$1\,\text{litre} = 1\,\text{dm}^3 = 0.26417$ US gallon = 0.21997 UK gallon
1 lb = 0.45359 kg
1 tonne (1 metric ton) = 10^3 kg
1 ton (UK) = 1 long ton (US) = 1.0160 tonne
1 ton (US) = 1 short ton (US) = 0.90718 tonne

APPENDIX III

LIST OF SYMBOLS

Symbol	Preferred Unit	Description
a		absorptance (absorptivity)
A		electron acceptor
A_e	eV	electron affinity
A_h	eV	hole affinity
c_i	mol dm^{-3}	molar concentration of species i
D		electron donor
$D(E)$	$cm^{-3}\ eV^{-1}$	density of electronic states of energy E
D_i	$cm^2\ s^{-1}$	diffusion coefficient of species i
$D_{e,h}$	$cm^2\ s^{-1}$	diffusion coefficient of electron or hole
E	eV $molecule^{-1}$ or kJ mol^{-1}	internal energy
E_c, E_v	eV	conduction or valence band-edge energy
E_g	eV $molecule^{-1}$ or kJ mol^{-1}	bandgap energy or minimum energy for electronic excitation
E_F	eV	Fermi level
E_{Fn}, E_{Fp}	eV	Fermi level in n or p material
E_F^e, E_F^h	eV	quasi-Fermi level of an electron or hole
g	$cm^{-3}\ s^{-1}$ or $m^{-3}\ s^{-1}$	volume rate of generation of excited states
G	eV $molecule^{-1}$ or kJ mol^{-1}	Gibbs free energy
H	eV $molecule^{-1}$ or kJ mol^{-1}	enthalpy
H	m	cell width
$H_{p,n}$	m	width of the p or n layer
H_{RP}	eV $molecule^{-1}$	electronic coupling matrix element between reactant (R) and product (P) states
i	mA cm^{-2}	current density
i_{ph}	mA cm^{-2}	photocurrent density
i_j	mA cm^{-2}	junction (dark) current density

Symbol	Preferred Unit	Description
i_o	mA cm^{-2}	reverse-bias saturation current density
I	A	current
I	W m^{-2}	irradiance
I_λ	W m^{-2} nm^{-1}	spectral irradiance per unit wavelength
I_S^o	W m^{-2}	solar constant: annual average normal solar irradiance just outside the Earth's atmosphere
j_{ph}	photons m^{-2} s^{-1}	flux of photons with $\lambda \leq \lambda_g$
j_g^o	photons m^{-2} s^{-1}	incident flux of solar photons with $\lambda \leq \lambda_g$
j_S^o	photons m^{-2} s^{-1}	incident solar photon flux of all wavelengths
j_λ	photons m^{-2} s^{-1}	spectral photon flux with respect to wavelength
$\mathbf{k}$		wave vector
$\boldsymbol{k}_{c,v}^{A}$	m^6 s^{-1}	band-to-band Auger recombination rate constant for the conduction or valence band
$\boldsymbol{k}_{c,v}^{TA}$	m^6 s^{-1}	trap-assisted Auger recombination rate constant for the conduction or valence band
$\boldsymbol{k}_{e,h}^{T}$	m^3 s^{-1}	Shockley–Hall–Reed trap recombination rate constant for an electron or hole
k_{nr}	s^{-1}	rate constant for nonradiative decay
k_r	s^{-1}	rate constant for radiative decay
$L_{e,h}$	m	minority carrier diffusion length of an electron or hole
$L_{e,h}^{\varepsilon}$	m	minority carrier drift length of an electron or hole
L	W m^{-2} sr^{-1}	radiance
L_λ	W m^{-2} sr^{-1} nm^{-1}	spectral radiance with respect to wavelength
$\boldsymbol{m}_e^*, \boldsymbol{m}_h^*$	kg	effective mass of an electron or hole
M	W m^{-2}	radiant emittance
M_S	W m^{-2}	radiant emittance of the Sun
n	cm^{-3} or m^{-3}	electron concentration in the conduction band of a semiconductor
n^o	cm^{-3} or m^{-3}	equilibrium electron concentration
n_r		refractive index
N_A, N_D	cm^{-3} or m^{-3}	density of acceptor or donor impurities
N_c, N_v	cm^{-3} or m^{-3}	effective density of states in conduction or valence band

Symbol	Preferred Unit	Description
p	atm or Pa	pressure
p	cm^{-3} or m^{-3}	hole concentration in the valence band of a semiconductor
p^{o}	cm^{-3} or m^{-3}	equilibrium hole concentration
P	$W\ cm^{-2}$ or $W\ m^{-2}$	power density
q	C	charge of an electron
Q	C	electrical charge
Q_p, Q_n	m	width of the p- or n-layer quasineutral region
r	$cm^{-3}\ s^{-1}$ or $m^{-3}\ s^{-1}$	volume rate of recombination of excited
r_λ	1	reflectivity at wavelength λ
R	Ω	resistance
$R_\square$	$\Omega\ square^{-1}$	sheet resistance
R_L	Ω	load resistance
S_e, S_h	$m\ s^{-1}$	surface recombination velocity for an electron or hole
T	K	absolute temperature
T		transmittance (transmissivity)
u_e, u_h	$cm^2\ s^{-1}\ V^{-1}$	electron or hole mobility
$u_\mathcal{E}$	$cm\ s^{-1}$	drift speed in electric field strength $\mathcal{E}$
$U_{Ox,Rd}$	V *vs.* ref.	electrode potential for the reaction $Ox + ne^- \rightleftarrows Rd$ *vs.* reference electrode
U^{o}	V *vs.* ref	standard electrode potential (based on activities)
$U^{o\prime}$	V *vs.* ref	formal electrode potential (based on concentrations)
V	m^3	volume
V	V	voltage
V_b	V	band-bending potential
V_b^{o}	V	equilibrium band-bending potential
V_{th}	V	thermal voltage $(\beta)\ kT/q$ (= 0.02569 V for $\beta = 1$ and 25° C)
α	cm^{-1} or m^{-1}	optical absorption coefficient
α_λ	cm^{-1} or m^{-1}	optical absorption coefficient at wavelength
β		diode quality factor
ε		emissivity
ϵ_o		permittivity of a vacuum
ϵ_{op}		optical dielectric constant ($\epsilon_{op} = n_r^2$)
ϵ_s		static dielectric constant
$\mathcal{E}$	$V\ m^{-1}$	electric field strength

Symbol	Preferred Unit	Description
η		power conversion efficiency
η_{mp}		power conversion efficiency at maximum power
κ		quantum mechanical transmission coefficient
λ	nm	wavelength
λ_g	nm	bandgap wavelength
μ_i	eV molecule^{-1}	chemical potential of species i
$\tilde{\mu}_i$	eV molecule^{-1}	electrochemical potential of species i
ν	Hz	frequency
$\bar{\nu}$	cm^{-1}	wavenumber
ρ	C cm^{-3}	space-charge density
ρ	Ω cm	electrical resistivity
ϕ	V	inner electric potential (Galvani potential)
ϕ		quantum yield
ϕ_b	V	barrier potential
$\Delta\phi_{scl}$	V	space-charge layer potential drop
Φ	eV	work function
χ	V	surface (dipole) electric potential
ψ	V	outer (Volta) electric potential
Ω	steradian (sr)	solid angle

APPENDIX IV

ACRONYMS AND ABBREVIATIONS

AM	Air Mass
BHJ	bulk heterojunction
BIPV	Building Integrated Photovoltaics
BOS	Balance Of System
BSF	Back-Surface Field
cb	(as a superscript or subscript) conduction band
CBE	Chemical Beam Epitaxy
CLEFT	Cleavage of Lateral Epitaxial Films for Transfer
CPV	Concentrated Photovoltaics
CSP	Concentrated Solar Power
CVD	Chemical Vapour Deposition
CZ	Czochralski (method of crystal growth)
DN(I)	Direct Normal (Irradiance)
Dk	(as a superscript or subscript) in the dark
DoS	Density of States
EFG	Edge-defined Film-fed Growth
EPC	Engineering, Procurement and Construction
EQE	External Quantum Efficiency
fb	(as a superscript or subscript) flat band
FZ	Floatzone (method of crystal growth)
G	Global (irradiance)
HCPV	High-Concentration Photovoltaics
IQE	Internal Quantum Efficiency
ITO	indium tin oxide
Lt	(as a superscript or subscript) in the light
LCL	lateral conduction layer
LGBC	Laser-Grooved Buried Contact
LOLP	Loss-Of-Load Probability
LPE	Liquid Phase Epitaxy
MBE	Molecular Beam Epitaxy
MIS	Metal|Insulator|Semiconductor (junction or device)
MOCVD	Metal Organic Chemical Vapour Deposition
mp	(as a superscript or subscript) maximum power
MS	Metal|Semiconductor (junction or device)
NOCT	Nominal Operating Cell Temperature
oc	(as a superscript or subscript) open circuit

Ox	oxidant
PCE	power conversion efficiency
PEDOT:PSS	poly(3,4-ethylenedioxythiophene):poly(styrenesulphonate)
PTC	Practical Test Conditions (for cell efficiency measurements): (plane-of-array irradiance 1000 W m^{-2} AM1.5G, ambient temperature 25° C, wind speed air 1 m s^{-1})
PPV	poly(*p*-phenylenevinylene)
PV	photovoltaic
QNR	quasi-neutral region
RAPS	Remote Area Power Supplies
Rd	Reductant
sc	(as a superscript or subscript) short circuit
scl	(as a superscript or subscript) space-charge layer
SHE	Standard Hydrogen Electrode
SHS	Solar Home System
SIS	Semiconductor|Insulator|Semiconductor junction
ss	(as a superscript or subscript) surface state
STC	Standard Test Conditions (for cell efficiency measurements): plane-of-array irradiance AM1.5G 1000 W m^{-2}, cell temperature $25 \pm 1/2$C
TCO	Transparent Conducting Oxide
TPV	thermophotovoltaic
vb	(as a subscript or superscript) valence band
Wp	peak watts

INDEX